The Whirlpool galaxy, also known as M51, as observed by the Hubble Space Telescope. The grand-design spiral galaxy has arms that are factories for star formation. A small yellowish galaxy, NGC 5195, appears to lie at the tip of one of the spiral arms. However, it is passing behind. The encounter between the two galaxies may be the reason why star formation is so active in the Whirlpool. The Whirlpool galaxy is located 31 million light years away in the constellation Canes Venatici.

FIFTH EDITION
DISCOVERING
Astronomy

Stephen J. Shawl
University of Kansas

Keith M. Ashman
University of Missouri-Kansas City

Beth Hufnagel
Anne Arundel Community College

With contributions from
R. Robert Robbins
University of Texas, Austin

William H. Jefferys
University of Texas, Austin

KENDALL/HUNT PUBLISHING COMPANY
4050 Westmark Drive Dubuque, Iowa 52002

Book Team

Chairman and Chief Executive Officer   Mark C. Falb
Senior Vice President, College Division   Thomas W. Gantz
Director of National Book Program   Paul B. Carty
Editorial Development Manager   Georgia Botsford
Developmental Editor   Lynne Rogers
Vice President, Production and Manufacturing   Alfred C. Grisanti
Assistant Vice President, Production Services   Christine E. O'Brien
Project Coordinator   Angela Puls
Designer   Jeni Chapman

Cover art courtesy photos.com.

Copyright © 1981, 1988, 1995, 2000, 2006 by Kendall/Hunt Publishing Company

ISBN 0-7575-1787-0

All rights reserved. No part of this publication may be reproduced,
stored in a retrieval system, or transmitted, in any form or by any
means, electronic, mechanical, photocopying, recording, or otherwise,
without the prior written permission of the copyright owner.

Printed in the United States of America
10 9 8 7 6 5 4 3 2 1

# Dedication

Steve Shawl dedicates this edition to Jeannette, who has put up with so many long nights and weekends and missed events through three editions of this book, and to Jim Hesser, for his personal and professional friendship.

Keith Ashman dedicates his part to his students over the years for helping him become a better teacher, and his long-time collaborator Steve Zepf for helping him become a better scientist.

Beth Hufnagel dedicates her work to Dave Adler for letting this book take priority over time for him, and for his pride in her accomplishments.

About the Authors

Stephen J. Shawl was educated at the University of California at Berkeley in the early 1960s and the University of Texas at Austin, where he received his Ph.D. in 1972. Research interests have involved the study of the fundamental properties of globular star clusters, the presence of gas and dust in these clusters, and cool variable stars whose light is sometimes polarized. He has been teaching all levels of undergraduate astronomy since that time. Because of that found interest in teaching, he has attended numerous Chautauqua short courses and other workshops for college teachers. As an extension of that interest, he has been the principal investigator for an astronomy workshop for junior high and middle school science teachers. He has been a member of the Education Advisory Board of the American Astronomical Society and has served as the Chairman of their Working Group on Astronomy Education. (Shawl@ku.edu)

Keith Ashman received a B.Sc. in astrophysics and a Ph.D. in theoretical cosmology from Queen Mary College, University of London. In 1989 he moved to the Space Telescope Science Institute in Baltimore, MD as part of the European Space Agency contingent working on the Hubble project. After a NATO fellowship held at the Canadian Institute for Theoretical Astrophysics in Toronto, he held positions at the University of Kansas and Baker University, KS before joining the physics faculty at the University of Missouri—Kansas City in 2002. His main research interests include the formation and evolution of globular clusters and globular cluster systems, galaxy formation and evolution, and merging galaxies. He is co-author of the graduate text "Globular Cluster Systems" and co-editor of "After the Science Wars." (Ashmank@umkc.edu)

Beth Hufnagel began her working career as an accountant after having earned an undergraduate degree in 1976 in accounting at Drexel University. At age thirty-two she decided to switch careers to astronomy. After undergraduate work in mathematics and physics at Columbia University, she received her Ph.D. in 1995 from the University of California, Santa Cruz in astronomy and astrophysics. She has been teaching adults since the late 1970's in subjects ranging from auditing to statistics to astronomy. Her astronomy research included a search for supernova progenitors, active galactic nuclei, star clusters, and the evolution of the Milky Way. Her current research field is astronomy education, and she has been exploring the astronomy understanding of undergraduates taking a course for non-majors since 1996. Products include the Astronomy Diagnostic Test, the first survey designed for undergraduates with measured reliability and validity, and modules for on-line astronomy courses. (Hufnagel@comcast.net)

R. Robert Robbins studied mathematics at Yale University and received his doctorate in astronomy from the University of California at Berkeley in 1966. He spent a year as a McDonald Observatory Fellow and a year at the University of Houston before joining the faculty at the University of Texas. His research interests have been in the fields of gaseous nebulae, atomic physics, and the transfer of radiation through gaseous media. In recent years his research interests turned towards archaeoastronomy and ethnoastronomy—in particular, studying the astronomy we can find in the structures and folklore of the indigenous cultures of Middle America and the American Southwest. He has received a variety of awards for excellence in teaching at the University of Texas and has served on a variety of national and international committees concerned with astronomy education. He has taught a variety of Chautauqua courses for college teachers, the most recent ones dealing with Mayan astronomy. He has now retired. [rrr@astro.as.utexas.edu]

William H. Jefferys was educated at Wesleyan University and Yale University, receiving the Ph.D. degree from Yale in 1965. He has taught at Wesleyan University and at the University of Texas and is currently the Harlan J. Smith Centennial Professor Emeritus in Astronomy at the University of Texas at Austin. His main research interests are in the fields of dynamical astronomy (the study of motions of planets, stars, and galaxies), astrometry (the measurement of the positions, orbits, and distances of celestial objects) and statistics. Since 1977 he has been the Astrometry Science Team Leader for the Hubble Space Telescope project. He has taught widely, both at the undergraduate and graduate levels, as well as teaching several National Science Foundation Chautauqua courses on teaching techniques to in-service teachers at both the high school and college levels. He believes strongly in the "hands on" approach to teaching that gives students a direct experience of what it is like to make, analyze, and understand scientific observations. [Bill@Clyde.as.utexas.edu]

Preface

The beauty of the night sky connects you to your universe. That beauty becomes even more inspiring when you have some understanding of what you are observing. When you see how humans approach an understanding of the universe, you begin to appreciate humanity and nature even more. The overall goal of this text, then, is to provide the reader with the opportunity to understand and appreciate the incredible place in which we exist.

The authors of this book feel that astronomy is not only an exciting and interesting subject, but also a living one in which everyone can directly participate. A considerable amount of astronomy can be experienced personally, and many things can be learned by simple and direct observations. We believe that the introductory course in astronomy, for which this book has been written, can be greatly enriched by the addition of observational activities that lead to a direct experience of the reader's universe. Such an approach, pioneered in astronomy by the first edition of *Discovering Astronomy*, reflects current discussions of science education at the national level. The idea of personal involvement is the guiding philosophy behind this book. Instructors wishing to include an observational component in their lecture course can obtain an Activities Manual to accompany the text by contacting the first author.

New to the Fifth Edition

Discovering Astronomy, since the time it first appeared in 1981, has been a pioneer in astronomy education. The fifth edition improves upon the previous editions while maintaining the philosophy of its unique approach. That approach is simple—engage the student in active involvement in his or her own learning process.

One way the book achieves that active involvement is through the inclusion of numerous Inquiries within the reading. These Inquiries purposely break up the reading so the student stops and thinks about what's been read before continuing. Answers to these inquiries at the end of each chapter allow students to gauge their understanding.

The other major way active learning is enhanced is through the supplemental *Discovering Astronomy Activities Manual* available separately from the first author; contact him if you are interested. These activities are not laboratory exercises in the sense that they are used in a laboratory course. Rather, they were specifically designed to provide students in both large and small lecture classes with hands-on experiences, giving them the opportunity to use various learning methods. Because *Discovering Astronomy* is a "main stream" text in and of itself (albeit with unique pedagogy), when it is used in conjunction with the *Activities Manual,* its unique approach is maximized.

The fifth edition has been revised in a number of ways. First, it has been brought up-to-date by including recent results from the Mars orbiters and landers and the Cassini mission to Saturn. Hubble Space Telescope results are discussed when appropriate. The discussion of extrasolar planets has been updated and expanded significantly.

A significant new chapter on the search for extraterrestrial life has been added. The cosmology chapter has been rewritten and reorganized to reflect modern perspectives.

We made significant efforts to simplify and update the text while maintaining the rigorous conceptual level users of *Discovering Astronomy* demand. Illustrations were revised as needed; some were dropped and others added.

While the overall organization is the same, some chapter sections were reorganized (and rewritten) as required by the astronomy community's changing understanding; in some cases, sections were moved to different chapters. For example, Chapter 15 on normal stars was reorganized to place the finding of stellar properties such as mass, radius, and luminosity before they are used in constructing the H-R diagram. Simplified and updated language and illustrations keep *Discovering Astronomy* user-friendly for students.

The Process of Science

Science can be defined as a process followed by humans involved in a search for an understanding of their surroundings. This understanding is usually expressed in terms of hypotheses whose predictions are verified or falsified by comparison with observation and experiment. The authors believe that an understanding of the scientific enterprise is important for people in a democratic society. For this reason, we do not present only the final results of scientific investigation, but we emphasize the process, which often contains false starts, side roads, and errors along the way to our current understanding. (Unfortunately, this precludes a short textbook!) These in turn also undergo modification and refinement by future generations of scientists. In placing an emphasis on process, students are better positioned to understand discussions on which they will be asked to make informed decisions. For these reasons, the process of science is a theme that is carried out throughout the book.

The Discovery Approach

We have written this book as a comprehensive introductory astronomy text to facilitate unassisted inquiry through active experiences for students. Our approach is a discovery-or inquiry-based approach. To translate this philosophy into a unique textbook, we have woven numerous questions, which we refer to as *Inquiries,* into the text at appropriate locations. Two kinds of activities may also be used—(1) those included in the text, which we call *Discoveries,* and (2) those presented in a separate Activities Manual available from the first author. *Discoveries* are highlighted at pertinent places in the text narrative. While we feel that learning is enhanced by using the *Inquiries* and *Discoveries,* individual instructors can choose to use them or not.

Discoveries

The Discovery activities are desired to emphasize that the nature of science is one of discovery. Short and simple, these activities include instructions and inquiry questions. An example is the discovery on Weightlessness in Chapter 5. They are noted at the appropriate place in the narrative and appear at the end of chapters.

Activities Manual

Some observational activities are available in a separate Activities Manual that can be ordered by your college bookstore by contacting the first author. These activities have the student make simple astronomical measurements using instruments such as a stellar protractor or quadrant. Instrument patterns are included in the Activities Manual. An observational activity may require a student to measure, for example, planetary positions, average the measurements, and discuss the uncertainties of the measurements. Other activities, such as Classification of Galaxies, do not require measurement but the use of photographs.

The activities and discoveries allow participation in astronomical exploration. They can be carried out by the thoughtful and careful student with no extra equipment or tutorial assistance.

Pedagogical Structure

We have implemented our inquiry-based approach in the pedagogical structure of this book. We have taken every opportunity to emphasize the scientific process and encourage observational activities to add to student understanding and enjoyment.

Inquiries

Each chapter contains in-chapter questions, which we call *Inquiries,* to emphasize the inquiry nature of the presentation. Placing these *Inquiries* within the text asks students to stop and ask of themselves, "I just read something important; what was it?" Sometimes, the *Inquiry* asks the student to use the knowledge already gained and to think about some idea immediately prior to its discussion in the text. In this way, the student is actively participating, not just reading passively. The answers to all Inquiries are provided at the end of each chapter.

Chapter Summaries

Discovering Astronomy pioneered a new style of Chapter Summary that emphasizes the nature of the scientific method. When appropriate, Chapter Summaries have three subsections: Observations, Theory, and Conclusions. Although many aspects of science belong in more than one category, this structure nevertheless serves to focus student attention onto these fundamental parts of the scientific process.

Summary Questions

Next are Summary Questions, which ask students straightforward questions about the concepts covered in the chapter. These questions can serve as a review.

Applying Your Knowledge

Knowledge should be applied to new and different situations. This section, at the end of each chapter, presents the reader with questions meant to require the application of concepts previously learned. Some, marked by a triangle (▶) require the use of simple mathematics.

Glossary and Scientific Appendices

The book contains an extensive Glossary of those terms likely to be new to students or with a narrow scientific meaning. These are highlighted with bold print in the text.

A set of appendices presents a review of scientific notation as well as other mathematical concepts that will be particularly helpful for students who need a refresher in high-school level mathematics. Additionally, tabulated data on planets, satellites, stars, and galaxies is available and may be used as the source material for additional activities.

Colorful and Instructive Illustrations

The art program includes full color photographs throughout to enable students to enjoy and appreciate the beauty of the universe with every page read. Computer generated color diagrams help to clarify complex concepts. In so far as possible within the limits of space, diagrams convey single concepts.

Organization

We have organized the book in the way astronomy developed: from observations to an understanding first of the solar system and only later to stars and galaxies. In this standard "inside out" approach, we have divided the book into six parts:

   I. Discovering the Science of Astronomy
  II. Discovering the Nature and Evolution of the Solar System
 III. Discovering the Techniques of Astronomy
 IV. Discovering the Nature and Evolution of Stars
  V. Discovering the Nature and Evolution of Galaxies and the Universe
 VI. Discovering if There is Life Elsewhere in the Universe

Physics Presented Only as Needed

Within our organizational structure, the observational basis of astronomy is further emphasized by generally having observation precede theory. Furthermore, we present physical ideas where and when they are needed rather than grouping them together in a single chapter. This approach allows students to see a reason why physics is being presented, and they immediately see applications for the ideas. We have purposely placed the solar system before the chapters on radiation and telescopes for two reasons: (1) to get the student into modern astronomy as soon as possible without their feeling bogged down by physics ("if I had wanted a physics course I would have taken one!"), and (2) because much of a modern survey of our knowledge of the solar system at the introductory level may be presented effectively without prior knowledge of the material on light, spectra, and telescopes. Instructors desiring to discuss electromagnetic radiation first may do so without difficulty.

Science versus Pseudoscience

We believe that the ideas of what distinguishes science from pseudoscience are so important that our Chapter 2 on Science and Pseudoscience comes immediately after the book's introduction to astronomy. In this way, time constraints at the semester's end will not cause it to be skipped. When completed, the student should no longer exclaim, "But, it's *only* a theory!"

Comparative Planetology

Part II, on the nature and evolution of the solar system, begins with an overview of observations important to understanding the solar system. We then examine both early and modern hypotheses concerning its origin. The next logical topic is a study of objects that are mostly unchanged since their formation and that provide astronomers with important information about the conditions at the formation time—comets, asteroids and meteoroids. Instructors preferring to discuss these objects after the planets may do so without loss of continuity.

The discussion of the planets was completely rewritten and modernized in the third edition and updated for this one. Chapter 8 on Earth and the Moon presents those concepts necessary to understand the other planets and moons in the solar system. It provides a basis for understanding the other bodies. Chapter 9 discusses the terrestrial planets, and Chapter 10 presents the Jovian planets and Pluto. The presentation within the chapters on the Earth-like and Jupiter-like planets differs from those in many books by approaching planets from the view of comparative planetology. Information about the planets is not isolated from comparable information on similar planets but is integrated together. "The Planets One by One," located after each of the terrestrial and Jovian planet chapters, summarize the information on each planet individually.

Logical Coverage of Stellar Evolution

The Sun is discussed in its role as a typical star at the point where the stellar evolution discussion begins (Part IV). Those instructors preferring to relate the Sun more closely to the solar system than to the stars can easily do so with no loss of continuity.

Following the discussion of the concepts required for understanding stellar evolution are chapters on star formation and evolution to the main sequence, post main-sequence evolution, and the terminal evolutionary events.

Interstellar Medium Covered as Needed

Rather than presenting the interstellar medium in isolation in a separate chapter as is frequently done in other books, we introduce it as integral to a number of processes. For example, some information on nebulae is given as an application of spectroscopy in Chapter 13. Additional discussion

occurs in Chapter 17 on star formation. Finally, the effects of interstellar material on observations of stars and its influence on our knowledge of the structure of the Milky Way comes in Chapter 20. This integrated approach is more consistent with our modern view of the interstellar medium as a key component of evolving galaxies.

Supplements

In order to enhance instructor's teaching resources, a comprehensive set of supplements accompanies Discovering Astronomy. For more information about these supplements, please contact Kendall/Hunt Publishing at 1-800-228-0810. These supplements include:

- **An Instructor's Manual** has been prepared by the authors to provide instructors with teaching hints, syllabi, and chapter overviews for effective classroom use of Discovering Astronomy. A number of short cooperative learning group activities that have been classroom tested are included.
- **A Test Bank,** prepared by the authors, provides instructors with over 2200 questions of various types. A Computerized Test Bank is also available for both Macintosh and PC platforms.

Acknowledgments

A large number of astronomers and astronomy teachers provided input to the fifth edition. Their help is warmly and gratefully acknowledged.

Reviewers

Narahari Achar, University of Memphis
Peter Anderson, Oakland Community College—
    Highland Lakes
Jay Ansher, Illinois State University
Morrie Barembaum, Santiago Canyon College
Nadine Barlow, Northern Arizona University
Greg Black, University of Virginia
Dan Britt, University of Central Florida
Debra Burris, Oklahoma City Community College
Juan Cabanela, St. Cloud State University
Michael Carini, Western Kentucky University
Karen Castle, Diablo Valley College
Ed Coppola, Community College of Southern
    Nevada, Charleston

Peter Detterline, Kutztown University of
    Pennsylvania
James Dickinson, Clackamas Community College
Alexander Dickison, Seminole Community College
William Dieterle, California University of Pasadena
Steven Federman, University of Toledo
Doug Franklin, Western Illinois University
Harold Geller, George Mason University
Perry Gerakines, University of Alabama,
    Birmingham
Igor Glozman, Highline Community College
Christopher Godfrey, Missouri Western State College
John Hamilton, University of Hawaii, Hilo
Russell Harkay, Keene State College
Paul Hinds, Pierce College District II
Ayorinde Idowu, Houston Community College
Richard Ignace, East Tennessee State University
Doug Ingram, Texas Christian University
F. Duane Ingram, Rock Valley College
Kenneth Janes, Boston University
Katherine Jore, University of Wisconsin,
    Stevens Point
Robert Joseph, University of Hawaii at Manoa
Christopher Keating, University of South Dakota,
    Vermillion
Andrew Kerr, University of Findlay
David Kriegler, University of Nebraska Omaha
Kristine Larsen, Central Connecticut State University
Paul Lee, Middle Tennessee State University
Marilyn Listvan, Normandale Community College
Bernie McNamara, New Mexico State University
Jose Mena-Werth, University of Nebraska Kearney
Basil Miller, Henderson State University
Dinah Moche, Queensborough Community College
Stephen Mojzsis, University of Colorado
Windsor Morgan, Dickinson College
Thomas Morin, Plymouth State University
David Morris, Eastern Arizona College
Steve Mutz, Scottsdale Community College
Richard Olenick, University of Dallas
Fritz Osell, Northern Oklahoma College Enid
Terry Oswalt, Florida Institute of Technology
Judith Parker, Muhlenberg College
Roger Philips, Lane Community College
Eric Preston, Indiana State University
Herbert Ringel, Manhattan Community College
Randall Scalise, Southern Methodist University
Anahita Sidhwa, Brookhaven College
G. Roger Stanley, San Antonio College
Martin Stringfellow, Mississippi Gulf Coast
    Community College, Jackson
Brent Studer, Kirkwood Community College

Chris Taylor, California State University, Sacramento
Hugh Tornabene, Bowie State University
Alex Umantsev, Fayetteville State University
Nate Van Wey, Kent State University Stark
Richard Wainscoat, University of Hawaii, Honolulu
Ed Wehling, Anoka-Ramsey Community College
Richard Wheeler, SUNY College at Cortland
Dan Wilkins, University of Nebraska, Omaha

We are pleased to acknowledge Boris Starosta, whose skill provided the vast majority of the illustrations. His continuous questioning about what we wanted the students to learn from a given illustration helped to make the entire set of illustrations the invaluable contribution that it is.

A number of scientists kindly provided the authors individually with scientific information, materials, or comments for the fifth edition. We are indebted to them for their willingness to help:

David Adler
Lori Allen
Donald Brownlee
Humberto Campins
Robin Canup
Caryl Gronwall
David H. Hathaway
Ivan Hubeny
Hideyuki Izumiura
Jennifer Johnson
Kathryn Johnston
Paul Kreiss
Larry Lebofsky
James LoPresto
Laurence Marschall
Chris Mihos
Axel Melinger
Joe Orman
Michael Seeds
Carol Veil
Ray Villard
Wayne Waldron

Steve Shawl
(shawl@ku.edu)

Keith Ashman
(ashmank@umkc.edu)

Beth Hufnagel
(hufnagel@comcast.net)

Brief Contents

Contents

PART TWO Discovering the Nature and Evolution of Planetary Systems 113

PART THREE Discovering the Techniques of Astronomy 283

Chapter 13

Spectra: The Key to Understanding the Universe 329

PART FIVE   Discovering the Nature and Evolution of Galaxies and the Universe   527

PART SIX   Discovering If There Is Life Elsewhere in the Universe   641

Helpful Study Aids

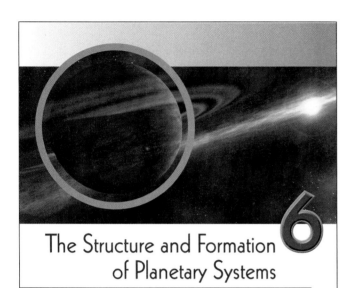

The Structure and Formation of Planetary Systems

6

First of all there came Chaos, . . .
From Chaos was born Erebos, the dark, and black night,
And from Night again Aither and Hemera, the day,
were begotten . . .

Hesiod, Theogony, c. 800 B.C.E.

115

Chapter Overview

A brief paragraph gives an overview of concepts discussed in the chapter. It serves as an organizing device to help you preview chapter coverage.

Some years ago, an astronomer was on a train in Europe. In the dining car, he was seated opposite an elderly gentleman with a long, white beard, dressed in a long robe, who looked like one of the patriarchs straight out of the Old Testament.

The elderly gentleman evidently wanted to talk, and so the astronomer just listened, not saying what his profession was. It turned out that the patriarch was the world leader of the Hollow Earth Society, whose adherents assert that Earth is a hollow ball and that we are living on the *inside* of it (**Figure 2-1**). The Hollow Earth movement has many adherents. Our astronomer had happened to meet up with the biggest Hollow Earther of them all!

The patriarch explained to the astronomer why we mistakenly think we are on the outside of the Earth even though we live on the inside; how the Sun, Moon, planets, and stars only *look like* they go around the Earth; and how an optical illusion causes satellites to take pictures of the Earth that make it appear round. Finally, he explained the *direct* evidence that the Earth is hollow. There were only two pieces of evidence of interest to him. The first was a survey made of the Great Lakes around the turn of the twentieth century that showed that the surface of the Earth was concave instead of convex. And the second? **Figure 2-2** shows the side view of a shoe. Do you notice how the sole of the shoe curves upward instead of downward? That's because you have been walking around on the *inside* surface of the hollow Earth all these years!

This story may sound made up, but it really happened. The elderly gentleman was a typical example of a pseudoscientist. Although his particular hokum may seem pretty hard to swallow, he had quite a few convinced followers. As psychologist Robert Thouless has pointed out, there is no idea so absurd that you can't find someone who believes it. For this reason, educated people should know what is and what is not science.

Astronomy is a science in every sense of the word. In the first part of this chapter you will learn what science is, using astronomy as an example. In the second part, we further define science by showing what is not science. Finally, we examine various characteristics of what often pretends to be science but isn't.

Opening Quote

Each chapter opens with a quotation from popular fiction or a scientific source that helps to set the stage for topics covered in the chapter. These quotes often give a perspective from outside the immediate field of astronomy.

2.1 An Expedition to Earth

To give you a feeling for what science is, and for the difficulties involved in studying objects with extremely long lifetimes, we will discuss the problems that would confront an alien astronaut who is visiting Earth for only a short period of time, and whose responsibility is to search for life on Earth and try to understand the life cycle of its inhabitants. Suppose we give this visitor just 15 seconds to take as many pictures as possible (being advanced, the alien's civilization has ultra-high-

Chapter opening photo: A scene from *Mars Attacks*.

speed cameras). We choose 15 seconds because it bears approximately the same proportion to the lifetime of a human being as the 100 years we have been observing the universe photographically does to the age of the universe:

$$\frac{15 \text{ seconds}}{\text{Human lifetime}} = \frac{100 \text{ years}}{\text{Universe's lifetime}}$$

When the astronaut returns home with the photographs, the scientists there must try to understand Earth and all its life forms by studying the photos (**Figure 2-3**). It is almost like studying a still picture, because 15 seconds is not long

Inquiry 5-4

Assuming that the angle at C is 7° and that the distance of Alexandria from Syene is 5,000 stadia (the units used by the ancient Greeks; stadia is the plural of stadium), what is the diameter of the Earth in stadia? Although the exact length of a stadium is unknown, it is thought to be about 0.16 of a kilometer (one-tenth of a mile). Using this value, compute the radius of the Earth.

Inquiry 5-5

What does Eratosthenes' experiment assume about the shape of the Earth?

Inquiry 5-6

If the Earth were flat, what would be the value of the angle at C?

HIPPARCHUS

Perhaps the greatest of all ancient astronomers was **Hipparchus** (c. 150 B.C.E.). Many of the conclusions he drew were so sophisticated that it takes some knowledge of astronomy to appreciate how great his contributions were. He built an observatory, constructed the best astronomical instruments up to that time, and established a program of careful and systematic observations that resulted in the compilation of a great star catalog, with 850 entries, using a celestial coordinate system similar to our modern one for cataloging the sky. It was Hipparchus who originated a system, which is still in use today in modified form, for estimating the brightness of stars. In addition, he used the older Babylonian observations and deduced Earth's precession (Section 4.6), which is so slow that it takes almost 26,000 years for it to complete one cycle. Finally, he greatly developed trigonometry, which was, and still is, a useful tool for astronomy.

OTHER DEDUCTIONS OF THE GREEK ASTRONOMERS: THE DISTANCES OF THE PLANETS

The Greeks estimated the relative distances of the planets from Earth using principles still in use today for determining distances to astronomical objects. They reasoned that the more distant a planet was, the more slowly it would appear to move across the sky. The effect is similar to what happens when we compare the apparent motion of a high-flying airplane with that of one that is fly-

ing low. The distant airplane appears to move slowly across the sky, whereas the low-flying one is seen for only a short time and then is gone. In the same way, the Greeks could put most of the naked-eye planets in order of their distance from Earth by assuming that increasing distances corresponded to slower motions. The argument fails with Mercury and Venus, however, because it places Mercury closer to Earth.

Inquiry 5-7

What assumption results in Mercury being placed closer to Earth than Venus?

We have another, independent determination of relative planetary distances from their brightnesses. We use an analogy: when you are driving at night on a two-lane road and wish to pass the car in front of yours, you pass only if the headlights of an oncoming car are faint. When you do this, you are making an implicit assumption: all car headlights have about the same intrinsic brightness, with their apparent brightness depending on the distance. Similarly, if we assume that all planets have the same intrinsic brightness, then their apparent brightness as seen from Earth would depend on their distances from us. Of course, all the planets do not have the same intrinsic brightness, because their differences in size, distance from the Sun, surfaces, and atmospheric properties affect the amount of light they reflect in our direction. However, even allowing for these uncertainties, it is still possible to use this principle to rank the planets approximately in order of distance from the Earth.

THE APPARENT MOTIONS OF THE PLANETS IN RELATION TO THE STARS

Three additional observations of planetary motion were important in determining the details of the models the Greeks developed. We now summarize these observations, which were discussed in detail in the previous chapter:

1. Because the planets are considerably closer to us than the fixed stars, they appear to move against the starry background. Observations of Mars, Jupiter, and Saturn showed them to move generally eastward on the celestial sphere. In other words, stars move *with the*

Chapter 5 • The Historical Quest to Model the Solar System   87

Inquiries

Throughout the text, Inquiries (questions) are included in appropriate places to ask you to use the information presented in the text. By actively participating and not just passively reading, you gain understanding of the concepts. Answers to all Inquiries are presented at the end of each chapter.

Although the alien would find the surfaces of all these planets to be rocky in appearance, the Earth is distinctive in that its surface is largely covered by water, whereas the other planets show little evidence of water. Again, she would wonder about the reason for this.

At this point, the alien astronaut might notice an anomaly—the planet Pluto. Like the dwarf planets, Pluto is small. Unlike them, it is located a large distance from the Sun beyond the giant planets, and has a low density. Moreover, its orbit is more elliptical and out of the plane of the orbits of the other eight planets. Might it be that this planet had an origin different from the rest of the planets?

Summarizing this picture, the astronaut might draw up a table similar to **Table 6-1**. In it are listed the characteristics of the dwarf, or **terrestrial**, planets (after the Latin word for Earth) and the giant, or **jovian**, planets (after the Latin name for Jupiter). These characteristics must be explained by any successful theory for the origin of the solar system.

Appendix C contains reference tables that summarize some of the fundamental data about each of the planets. The significance of some of the data was suggested in the previous paragraphs and will be discussed further as you learn more about the solar system. To help you better appreciate the relative sizes of the planets, they are illustrated to scale in **Figure 6-2**. Similarly, the relative sizes of the planetary orbits are displayed in **Figure 6-3**. If Jupiter were drawn to scale in Figure 6-3, it would

be only about 1/10,000 of an inch in diameter, and the Sun would be about 1/1,000 of an inch in diameter. Therefore, to put Figure 6-3 on the same scale as Figure 6-2, we would have to draw it about 10,000 times bigger!

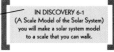

IN DISCOVERY 6-1
(A Scale Model of the Solar System)
you will make a solar system model
to a scale that you can walk.

Table 6-1   Characteristics of the Two Major Groups of Planets

TERRESTRIAL	JOVIAN
Mercury, Venus, Earth, Mars	Jupiter, Saturn, Uranus, Neptune
Near the Sun	Far from the Sun
Small diameter	Large diameter
Low mass	High mass
High density	Low density
Primarily composed of heavier elements	Primarily composed of hydrogen and helium (composition similar to the Sun)
Rocky surface	No surface
Thin atmosphere	Thick atmosphere
High temperature	Low temperature

Figure 6-2   Relative sizes of the planets compared with the Sun.

118   Part 2 • Discovering the Nature and Evolution of Planetary Systems

References to Discovery Activities

Although Discoveries appear at the end of the chapter so as not to disrupt the flow of reading, they are referenced at the appropriate places in a green shaded notation. [It is not necessary to do these activities to understand the concepts.]

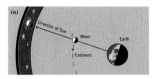 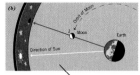

Figure 4-24 *(a)* At new Moon, the Sun and Moon both are seen in the same part of the sky. *(b)* After the Moon has gone exactly once around the Earth, it is seen in the same location in relative to the stars. But the Sun has moved to a different position in the sky, so the Moon is not yet new. (Not to scale.)

Figure 4-25 The geometry of a solar eclipse. (Not to scale)

Instructive Illustrations

Computer-generated color diagrams help to clarify complex concepts and encourage an appreciation of the beauty of the universe.

the background stars during the month. **Figure 4-24a**, a view from outside the Earth–Moon system, shows the situation at new Moon; the Sun and Moon are seen from Earth against the same background stars. In **Figure 4-24b**, one *sidereal* period later, the Moon has returned to the same place among the stars, but the Sun too has moved eastward. Thus, the Moon requires about two more days to reach the position of the Sun and establish the new-Moon geometry once again.

ECLIPSES

Eclipses have awed and frightened observers since the dawn of time. What takes place during eclipses of the Sun and Moon? Every body in the Solar System carries a shadow along with it as it moves around the Sun. When one body enters the shadow of another, an eclipse occurs. **Figure 4-25** shows the geometry of the Sun, Moon, and Earth at the time of a **total solar eclipse**, as seen by an observer in space. The shadow of the Moon sweeps across the surface of the Earth at a high rate of speed. Observers temporarily located within the dark spot will see the bright disk of the Sun completely covered by the Moon (see Figure 3-9).

A typical shadow consists of a dark inner part, called the **umbra**, and a lighter outer part called the **penumbra** (shown in Figures 4-25 and 4-9 for both lunar and solar eclipses). When the Moon's umbra passes over an observer, he or she sees a total solar eclipse. However, because the umbra is, at most, only about 150 miles wide, only those observers along the umbral path will see a total eclipse. Even then, the longest period of totality at any one location is slightly over seven minutes. People not within the umbra but still inside the larger penumbra will see part of the Sun's face blocked, a **partial solar eclipse**. (You can see a partial solar eclipse prior to totality in Figure 3-9.)

The distance of the Moon from the Earth varies somewhat during the year, since it has an elliptical orbit. Whenever the Moon is relatively far from the Earth, the apparent size of the lunar disk will be smaller than average and may not quite cover the Sun. At such a time, an observer on Earth will see an **annular eclipse**, with a bright ring, or annulus, of sunlight around the dark disk of the Moon. **Figure 4-26** shows the appearance of the Sun for an annular and a total eclipse.

Eclipses do not occur every month because the 5° tilt between the Moon's orbit and the ecliptic means that the new Moon is generally above or below the Sun. The end result is that solar eclipses

Chapter Summary

The Chapter Summary not only reviews the topics covered but presents the information in a way that reflects the scientific method. The material is organized in subsections of Observations, Theory, and Conclusions to give you practice in thinking in these scientific ways.

Chapter Summary

Observations

- The **Hertzsprung-Russell diagram**, the most important graph in astronomy, plots **luminosity** and **spectral type** (which is determined by the star's temperature). The majority of stars lie on the **main sequence**, with higher-luminosity groupings called **giants** and **supergiants**, and lower-luminosity **white dwarfs**.

- The brightest stars in the sky appear bright because they have high luminosities, not because they are close to the Sun.

- For stars on the main sequence there is a **mass-luminosity relation** showing that stellar luminosity increases strongly with stellar mass.

- There are three types of binary star systems: **visual binaries**, **spectroscopic binaries**, and **eclipsing binaries**. Such systems are most important in determining stellar masses and sometimes their diameters.

Theory

- A star's average surface temperature can be calculated from its spectral type.

- Luminosity is proportional to the fourth power of the surface temperature and to the star's surface area (which is proportional to the square of the radius).

- The **inverse square law** of light states that the apparent brightness varies inversely with the distance squared.

Conclusions

- Most stars in our galaxy are main-sequence stars.

- The **spectroscopic parallax** method of distance determination requires the star's spectral type and **luminosity class**, both of which are readily obtained from the spectrum. The star's luminosity is then inferred from the H-R diagram. When inferring the luminosity, we implicitly assume the star whose distance is to be found is similar to the other stars in the H-R diagram. The determination of luminosity from the H-R diagram is uncertain because we do not know if the star in question actually has the average luminosity of its luminosity class, or if it is more or less luminous. The distances determined using these luminosities, therefore, contain uncertainties.

- Stellar masses are inferred from observations of the period and semi-major axes of stars in binary systems. If a star's luminosity is known, and if the star is a main-sequence star, its mass can be inferred from the mass-luminosity relation.

- Stellar diameters may sometimes be obtained from analysis of binary stars. The techniques of interferometry and occultations also give astronomers information on stellar radii.

Summary Questions

1. Sketch and label the Hertzsprung-Russell diagram for all types of stars and identify each of the various groups of stars on it. Explain the terminology main sequence, red dwarf, white dwarf, giant, and supergiant.

2. Why is the Hertzsprung-Russell diagram for the nearest stars different from that for the brightest stars in the sky?

3. How and why are a star's radius, temperature, and luminosity interrelated?

4. How do the various groups of stars on the Hertzsprung-Russell diagram differ in terms of their radii?

5. Where on the main sequence do stars of different masses lie? Draw a schematic H-R diagram and indicate such locations.

6. What is the range of orders of magnitude for stellar mass, luminosity, radius, surface temperature, and density?

7. What is the method of spectroscopic parallax? What are its advantages and disadvantages?

8. What are the three main types of binary star systems? Describe them.

9. How can astronomers determine the masses and radii of stars directly?

DISCOVERY 5-1

Weightlessness

When you have completed this Discovery, you should be able to do the following:

- Describe what is meant by weightlessness.

An astronaut floating in a space shuttle has mass but is not held down to the floor of the shuttle. This is not caused by a lack of gravity in space.

- **Discovery Inquiry 5-1a** Suppose you were standing on a scale in a stopped elevator. What would happen to the scale reading when the elevator suddenly accelerates upward? What would happen to the scale reading when the elevator rapidly descends from rest?

Although the elevator has accelerated, the force of gravity on the elevator occupants and therefore their weight has not changed. The scale reading, however, has changed because of the acceleration. The force of gravity has been partially offset by the changing force from the elevator's floor. This effect is commonly called *weightlessness*, even though the pull of gravity and your weight never changed.

You can easily demonstrate weightlessness for yourself. Take a paper cup or an aluminum pop can and punch two holes on opposite sides near the bottom. With your fingers over the holes, fill the container with water.

- **Discovery Inquiry 5-1b** *Before removing your fingers from the holes,* describe what you expect to happen. Then, while taking necessary precautions, remove your fingers from the holes and observe what happens.

Again place your fingers over the holes and fill with water.

- **Discovery Inquiry 5-1c** *Before dropping the container* (into a garbage can, the bath tub, or *safely* outside!), describe what you expect to observe. After thinking about the answer, drop the container from as high as you can reach, and describe what you observed.

In this experiment, you should have observed no water flowing from the container while it was dropping. Although gravity was still there, the water and the container both accelerate downward at the same rate. Similarly, because the astronaut and the space shuttle both accelerate toward the center of the Earth *at the same rate,* the astronaut is not pulled to the floor (or ceiling!) of the shuttle. Since we usually judge weight by how hard we push on the floor or a scale, it seems as if the astronaut is weightless.

Chapter Summary

Observations

- Ancient Babylonians and Egyptians could predict seasons, eclipses, and make calendars.
- **Aristarchus** first suggested that the Earth circles the Sun, and found the relative sizes of the Earth, Moon, and Sun, as well as the relative distances of the Moon and Sun. **Eratosthenes** first found the size of the Earth. **Hipparchus** made a great star catalog, began the magnitude system used to specify the brightness of stars, and discovered precession. **Ptolemy** advanced the hypothesis of planetary epicycles and helped preserve Greek knowledge in the *Almagest.*

Discovery Activities

The nature of science is one of discovery. Short and simple Discovery Activities appear at the end of many chapters to reflect this emphasis by getting you actively involved with the material.

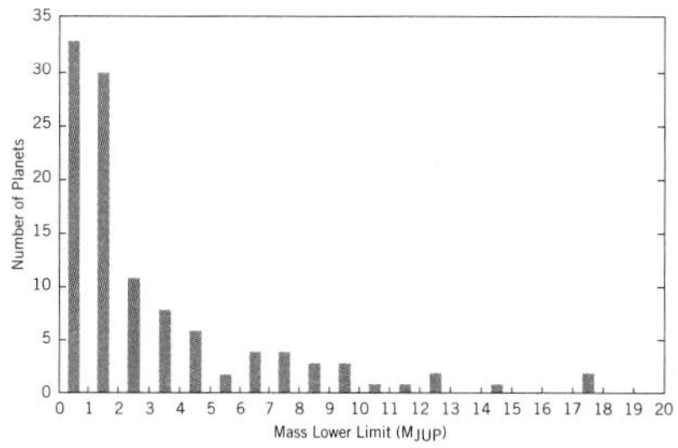

Figure 6-20 The distribution of masses for the 111 planets known as of July 2004.

DISCOVERY 6-1

A Scale Model of the Solar System

When you have completed this Discovery, you should be able to do the following:

- Describe the size of the solar system on a human scale.

A model of the solar system, which can help you understand better the sizes and distances in our planetary system, can be built on a human scale. For this model, let one astronomical unit be represented by 25 steps. After completing the questions below, walk your model. Stop at each planet and look back to where you started.

- **Discovery Inquiry 6-1a** Make a table of the number of steps you need to take to walk from the Sun to each planet.
- **Discovery Inquiry 6-1b** Determine the length of your step by measuring the length of 10 steps and dividing by 10.
- **Discovery Inquiry 6-1c** How large in yards or meters is your model solar system?
- **Discovery Inquiry 6-1d** On the scale of your model, how many miles or kilometers are there to the nearest star, which is approximately 200,000 AU away?
- **Discovery Inquiry 6-1e** On the scale of your model, determine the sizes in inches or centimeters of the Sun and the various planets.

Discovery Inquiries

To reinforce the key concepts of these Discoveries, questions (Inquiries) are included to guide your thinking.

Answers to Inquiries

6-1. They would be composed of the lowest mass elements, hydrogen and helium.

6-2. As the star's light passes through the planet's atmosphere, the light passes through successively denser layers of gas, which causes the light to gradually fade until the planet itself blocks the star's light.

6-3. The low gravity makes it easier for atoms to escape. The high temperature increases the average velocities of the atoms, so that a greater fraction of them are moving fast enough to escape Mercury's gravity.

6-4. The distance between the Earth and Sun is approximately 10,000 times the Earth's diameter. In Figure 6-2, the Earth's diameter is about 0.4 cm, which means the Earth's distance, on this scale, is 4,000 cm or 40 m. Pluto, which has an average distance 40 times farther away, would be about 1,600 m away.

6-5. The hypothesis would not be consistent with the fact that the planes of the planetary orbits are nearly the same, because it would be expected that the planets would be captured into random orbits with their masses also distributed randomly. To capture a planet into a nearly circular orbit requires special, rare circumstances. The hypothesis is consistent with having a large range of planetary masses, but not their split into terrestrial and jovian groups. The hypothesis is not consistent with most of the planets and moons revolving in the same direction.

6-6. The hypothesis would not be consistent with the fact that the Earth, Moon, and meteorites have the same age because the various bodies could be captured at different times. However, if all these objects had actually been formed at the same time, they would have the same age.

6-7. The hypothesis would not be consistent with the near-circularity of the planetary orbits, because capture would more likely create highly eccentric orbits. It would not be

consistent with the fact that the moons mostly orbit in one direction (the same as the planets), and it would not place all the giant planets together in the solar system as we see them today. It is consistent with the variable masses and the variable compositions that we see.

6-8. The simple nebular hypothesis would explain the basic traffic pattern of motions. It would not explain irregularities like the spin of Venus or the odd inclination of Uranus's spin axis, nor the distribution of angular momentum in the solar system.

6-9. Perhaps a process similar to the one proposed for the formation of planets themselves could be used to explain the formation of moons around a planet.

6-10. The nebular hypothesis as presented would not be able to explain the chemical compositions of the planets.

6-11. The collision hypothesis would lead us to expect that planetary systems are rare in the Galaxy, whereas the nebular hypothesis implies that they are common, a natural companion of star formation. Direct observational evidence strongly favors some type of nebular hypothesis.

6-12. This hypothesis has the same basic problems as the nebular hypothesis. Also, it is awkward to fit the formation of the moons of the planets into this picture.

6-13. The modern theory explains the overall counterclockwise rotation and spin of most of the planets and their moons. It does not explain the anomalous spins of Venus and Uranus. And all hypotheses still have some trouble with Pluto.

6-14. Clockwise rotation is not explained.

6-15. The low-mass group, because the gravity of these planets is lower. Because the low-mass elements present in the disk (hydrogen and helium) would be the ones most readily blown away, the inner planets would tend to be composed of the heavier elements that remain.

Answers to Inquiries

As a check on how you are doing as you progress through the chapter, the answers to the periodic Inquiries are included at the end of the chapter.

Summary Questions

Straightforward questions at the end of the chapter help you to review the material you have read and to check your understanding of it. Answering these questions is good practice for taking tests.

Summary Questions

1. What are the names of the planets in order of their distance from the Sun?
2. What are the two major groups of planets? List the characteristics that distinguish members of each group.
3. What are the major patterns in the orbits and motions of the planets that need to be explained by any successful theory of the origin of the solar system? What exceptions are there to these patterns?
4. What observational evidence can you use to criticize the early nebular and collisional hypotheses of the origin of the solar system?
5. What are some possible reasons for the dramatic differences in composition between the terrestrial and jovian planets?
6. Why were dust grains important to the formation of the solar system?
7. Where is most of the angular momentum in the solar system located? Why was the distribution of angular momentum a problem for earlier hypotheses of solar system formation? Describe a process thought to solve the angular momentum problem.
8. What techniques have astronomers used to detect planets orbiting other stars?
9. Describe the differences in the formation of terrestrial and jovian planets.

Applying Your Knowledge

The best way to test your comprehension of new material is to be able to apply this knowledge to new and different situations. These end-of-chapter questions are designed to require the application of concepts previously learned. Those questions requiring the use of simple mathematics are marked by a triangle ▶.

Applying Your Knowledge

1. Hypothesis: The solar system formed when a passing star pulled material from the Sun that then formed the planets. Present evidence both for and against this hypothesis, and state your final conclusion.
2. List all observations about the solar system that are exceptions to the generally observed features.
3. Explain why the following two definitions of volatile materials are the same. (a) Volatile materials are those that condense at low temperatures. (b) Volatile materials are those that turn to gas at low temperatures.
4. Use the condensation sequence of Figures 6-12 and 6-13 to explain in your own words the reasons different planets differ in chemical composition.
5. Refer to Figure 6-5b, which shows the inclinations of the planets to their orbital planes. For

each planet, discuss the seasonal variations you expect throughout the planet's year.
▶ 6. What fraction of planetary orbital angular momentum is held by Jupiter? Saturn? (We are neglecting rotational angular momentum, which is less than that from the orbital motion.)
▶ 7. Compute Pluto's density if its mass is 0.0022 the mass of the Earth and its diameter is 2,302 km. Express the density relative to the density of the Earth. What conclusions about the nature of Pluto might you draw from the observed density?
8. What would be the angular separation between the Sun and Jupiter if they were observed from the distance of Alpha and Proxima Centauri, the nearest stars, at a distance of 4.3 light-years? Compute the distance a dime would have to be for its angular size to match the angle you computed.

PART ONE

Discovering the Science of Astronomy

We begin where astronomy began—with people observing the sky. Astronomy is first and foremost an observational science; only after observations of phenomena have been made does interpretation based on theory come into play.

The first chapter is a grand tour of the universe. In addition, we take a first look at the range of sizes and distances of objects in the universe.

Astronomy is, in every sense of the word, a *science*. In Chapter 2 we discuss what modern science is and how it allows us to understand the universe. To emphasize what science is, we provide a counterpoint by discussing what is meant by pseudoscience. We examine those qualities that distinguish scientific findings from those of pseudoscience.

We begin learning about astronomical observations in Chapter 3 with a discussion of angles and their measurement on the sky. The relationship between an object's apparent size, physical size, and distance is discussed in detail. Because an understanding

of random and systematic uncertainties is crucial to understanding the scientific enterprise, we discuss the source and meaning of measurement uncertainties.

Basic observations of the sky made by ancient peoples, and observations you can make, are discussed in Chapter 4. These observations include daily and yearly motions of the Sun, motions and phases of the Moon, and movements of the planets. In addition, we discuss the reasons for the seasons.

To complete Part One, we look in Chapter 5 at how the observations presented in Chapter 4 were used by various civilizations to produce models leading to an understanding of the universe. This discussion leads naturally to the advances provided by the work of Copernicus, Brahe, Kepler, and Galileo, and finally to Newton's invention of the concept of gravity and his laws of motion. At this point we begin our comparison of observation with theory by looking at how Newton's theoretical ideas have been verified by a variety of observations.

Beyond the Blue Horizon: A Grand Tour of the Universe

My suspicion is that the universe is not only queerer than we supposed, but queerer than we can suppose.

J. B. S. Haldane, Possible Worlds, *1927*

The universe is filled with a bewildering variety of objects. We begin our study of the cosmos with a tour to give you a sense of the size and scale of the universe as astronomers now understand it. We will briefly survey the universe beginning at Earth and expanding our horizons outward to ever-increasing sizes and distances. As we progress from figure to figure, our scale will increase until we have covered the presently known universe. This chapter is meant to intrigue you—to raise questions—but not to explain. Explanation and understanding will come from the rest of the book.

As we foreshadow the contents of the book, we will also introduce you to some of the special difficulties that astronomers encounter as they try to understand the universe, and some of the unique problems they face as scientists who are not able to manipulate or experiment with the objects they study.

1.1 What Is Astronomy?

The word **astronomy** comes from the Greek *astro* meaning star and *nemein* (pronounced neh-main) meaning law, order, and arrangement. Thus, traditionally, astronomy was the study of the arrangement of the heavenly bodies in the sky. Early astronomers spent their time observing the relative locations of heavenly bodies and mapping them. Changes in those positions, for some bodies, would be observed. Now and then, new objects would appear and then disappear. It was those changes that provided input for some of the greatest philosophical, religious, and scientific debates in the history of humanity. To many people, astronomy is the observational study of the universe.

The view of many early astronomers was that the stars and planets were fundamentally different than the Earth. But they were wrong! Eventually, advances in our understanding of nature provided us with knowledge and techniques to probe more deeply and to learn about celestial bodies as physical objects. With that change, the science of astrophysics was born. **Astrophysics** is the application of the laws of the physical world to the universe beyond the Earth. It can be thought of as the theoretical rather than observational study of the universe.

Chapter opening photo: The range of distances observed by astronomers is dramatically illustrated in this photograph by Dr. William Liller, which shows Halley's comet on the right (within the Solar System), many stars outside the Solar System but in our galaxy (the Milky Way), the globular star cluster Omega Centauri on the left (within the Milky Way about 17,000 light-years away), and a distant galaxy, Centaurus A in the upper right corner (at 16 million light-years away).

Today, the terms *astronomy* and *astrophysics* are used pretty much interchangeably because we understand that observations need a theoretical explanation and that theoretical ideas must be constrained by observations that tell us how nature actually behaves.

1.2 Our View of the Cosmos

You may have grown up with pictures like **Figure 1-1**, taken by the *Apollo* astronauts. That familiarity may make it difficult for you to appre-

Figure 1-1 Lunar module *Eagle* returning to rendezvous with the *Apollo 11* command module.

ciate what a milestone in human history such photographs represent. We refer not so much to the effort and expense of landing a person on the Moon—although that was considerable—but, rather, to the changes in our perspective that have come about in the last few decades.

Most of what we know about the universe was learned in the twentieth century; indeed, most of our modern astronomical knowledge has been acquired since World War II. As recently as 1920 the accepted size of the entire universe was no more than 30,000 light-years across. A **light-year** (abbreviated **ly**) is the distance that light travels in one year while traveling at the **speed of light** (approximately 300,000 km/sec); in more familiar units, it is about 6 million million (or 6 trillion) miles, or about 10 trillion kilometers (abbreviated **km**). Although this may sound like an enormous distance, by 1930 astronomers had found that the universe was in fact at least a *million* times larger than that. More incredibly, before 1917 the best information we had showed that the Earth's star, the Sun, was at the center of the universe. Our current ideas are about as far from this self-centered notion as possible. They suggest that we are probably just one among billions of planetary systems, located in the peripheral regions of an immense spiral-shaped stellar system containing several hundred billion stars. We call this huge stellar system in which we live the Milky Way—it is a galaxy, just one of billions of similar galaxies in the universe.

These spectacular changes in our understanding of the universe have triggered profound changes in the way we view ourselves. Some argue that the revolution in our perspective is even greater than the much-touted intellectual shake-up that accompanied the Copernican revolution in the sixteenth century. With this new knowledge, we are trying to formulate a cosmic perspective in which we not only consider the possibility that we may not be alone in the universe, but we recognize that our Earth is limited in size and resources and requires care and nurturing if it is to survive as a place for us to live.

ASTRONOMY USES LARGE AND SMALL NUMBERS

Because astronomy considers both the largest and smallest objects in the universe, astronomy uses both the largest and smallest of numbers. **Table 1-1** shows the equality between ordinary

Table 1-1 Large Numbers

1	10^0	one	
10	10^1	ten	
100	10^2	hundred	
1,000	10^3	thousand	
1,000,000	10^6	million	thousand thousand
1,000,000,000	10^9	billion	thousand million
1,000,000,000,000	10^{12}	trillion	million million

numbers, their powers of ten notation, and the word used for the number. The final column emphasizes that large numbers can also be expressed as multiples of smaller numbers.

You can visualize some large numbers using a piece of graph paper 10 centimeters by 10 centimeters (abbreviated **cm**) with millimeter divisions. There are 100 millimeter squares along each edge, and 10,000 squares in the 10 cm by 10 cm area. Extending this area 10 cm into the third dimension would produce a cube having a million small cubes. If you now increase each side of this $10 \times 10 \times 10$ cm cube 10 times so it is slightly more than one meter on a side, the cube will enclose a billion small cubes! A typical small theater might be 10 times longer on each side and would contain a trillion of the small cubes.

Another method of visualizing large numbers is to count at the rate of one per second; you will reach 1,000 in about 17 minutes, a million in 12 days, and a billion in nearly 32 years!

Distances within the Solar System are generally measured by using the average Earth–Sun distance as a standard measuring stick. In familiar units, this distance is about 93,000,000 miles (150,000,000 km[1]), but astronomers generally refer to it simply as one **astronomical unit** (abbreviated **AU**). The average distance of Saturn from the Sun is almost 10 times farther, or 10 AU, which is much easier to say and to visualize than is 1,000,000,000 miles. You can get a better feeling for the size of the AU when you realize that the astronomical unit is about 100 (10^2) times larger than the diameter of the Sun, or 10,000 (10^4) times larger than the diameter of the Earth.

Astronomy also deals with the smallest objects in the universe. For example, an atomic nucleus is 10^{15} times smaller than a person.

---

[1]See Appendix A4 for a discussion of conversion from one set of units to another.

Using scientific notation it is possible to write extremely large or extremely small numbers with ease (Table 1-1). Scientific notation also greatly simplifies the arithmetic of multiplying and dividing large and small numbers together. If you are not familiar with scientific notation, you should read Appendix A1 now, because scientific notation will be used frequently throughout this book.

Another way of speaking about the relative sizes of objects or their distances is in terms of **orders of magnitude**. An order of magnitude is the same as a power of ten. That is, rather than speaking of Saturn as being 10 times farther from the Sun than is the Earth, we would say it is one order of magnitude farther. Similarly, the astronomical unit is two orders of magnitude larger than the Sun's diameter, which itself is two orders of magnitude larger than Earth's diameter. From the previous discussion, then, the atomic nucleus is 15 orders of magnitude smaller than a person.

In many cases, having an approximate comparison is sufficient, and one would round to the closest order of magnitude. For example, for our purposes it is not necessary to state that an astronomical unit is really 107 times the diameter of the Sun; saying two orders of magnitude is sufficient. That is, for many purposes it is necessary to know only the approximate size of a number. This is true of most of the numbers that appear in this book.

We have used examples of numbers that differed from each other by factors of ten. We compared the relative sizes of such numbers simply by comparing the number of zeros. In the general case, the number of times one needs to multiply a smaller number by 10 to make it approximately the same as a larger one is the number of orders of magnitude between the two numbers. For example, 5,000 is three orders of magnitude larger than 5, since you need to multiply 5 by 10 three times to make 5,000.

We can now extend this concept to enable us to make rough comparisons between any two numbers. We consider two numbers to be of the same order of magnitude if they are *approximately* equal to each other. There are many things about the universe that we know only roughly (and, of course, some things we don't yet know at all), and it would be misleading to describe an approximate piece of knowledge by using a number that looks precise.

Inquiry 1-1

Which one of the following pairs of numbers are of the same order of magnitude? (a) 12,000 and 96,000 (b) 500 and 300 (c) 3,000,000 and 4,000

A hydrogen atom is about 10 orders of magnitude smaller than a human being, and its nucleus is several orders of magnitude smaller than that. The nucleus of the atom is approximately one ten-trillionth of a centimeter in extent, or 22 orders of magnitude smaller than the Earth's diameter. The diameter of the observed universe (the distance to which we have seen objects) is approximately 10 billion ly, which is about 19 orders of magnitude larger than the Earth. Thus, astronomy deals with sizes that range over more than 41 orders of magnitude. The Solar System is approximately in the middle of this range of sizes. The final figure of the chapter, Figure 1-20, provides a summary that allows you to determine the number of orders of magnitude difference in size between any two of the objects discussed.

A detailed example will demonstrate a general procedure that will always work. The question is: How many orders of magnitude larger than a person's height is the Earth's radius? A typical person is about 2 meters tall (abbreviated **m**); an exact number is not important. Earth has a radius, obtained from Appendix A2, of 6,378 km. Writing the radius in powers of ten notation gives 6.378×10^3 km. Because there are $1,000 = 10^3$ meters in a kilometer, the Earth's diameter in meters must be a larger number. Thus, we have $(6.378 \times 10^3 \text{ km})(10^3 \text{ m/km}) = 6.378 \times 10^6$ m. For our purpose of obtaining an approximation, this number can be roughly written as $10 \times 10^6 \text{ m} = 10^7$ m by rounding 6.378 up to 10. The solution is a five-step process shown in **Table 1-2**.

Inquiry 1-2

How many orders of magnitude larger than the orbit of the planet Pluto (diameter averages about 40 AU) is the Milky Way galaxy (diameter is 100,000 ly)? Be sure all your numbers are expressed in the same units before comparing the numbers.

Table 1-2

STEP	PROCEDURE	EXAMPLE
1	Express the larger number in powers of ten notation and round the coefficient (the number in front of the power of 10) up to 10 if 5 or greater or down to 1 if less than 5.	6.378×10^3 km $\sim 10 \times 10^3$ km $= 1 \times 10^4$ km
2	Convert the numbers to the same units, if necessary.	$(1 \times 10^4 \text{ km})(10^3 \text{ m/km})$ $= 1 \times 10^7$ m
3	Express the smaller number in powers of ten notation.	2×10^0 m $\sim 1 \times 10^0$ m
4	Divide the larger by the smaller number. Divide the numbers and subtract the exponents: (Notice that the units divide out.)	$\dfrac{1 \times 10^7 \text{ m}}{1 \times 10^0 \text{ m}} = \dfrac{1}{1} \times \dfrac{10^7}{10^0}$ $= 1 \times 10^{7-0} = 1 \times 10^7 = 10^7$
5	The number of orders of magnitude is the exponent of the 10.	7
Answer	Earth is 7 orders of magnitude larger than a person!	

1.3 The Solar System

Our tour beyond the blue horizon of the Earth begins with the Sun, which because of its great amount of material rules over the Solar System. Next come the planets closest to the Sun, which are at the same time both similar to and different from the Earth. As our distance from the Sun increases, we will encounter the Solar System's giant planets, which are all more similar to Jupiter than the Earth. As we travel yet farther within the Solar System, we reach the outermost objects—the planet Pluto and the comets. Finally, we go to the distant stars to see planets around them.

THE STAR OF THE SOLAR SYSTEM

At the surface of the Sun (**Figure 1-2**), we find violent activity that churns the gases of the solar atmosphere. Sometimes particles ejected from the Sun even reach out far enough to disturb communications on Earth. The diameter of our star is about 1,400,000 km, approximately 10 times that of Jupiter. This means that the Sun's diameter is one order of magnitude larger than Jupiter's and two orders of magnitude larger than the Earth's. If the Earth were represented by a basketball, the Sun would be a sphere the height of a 10-story building!

While the diameter of the Sun is only one order of magnitude larger than Jupiter, the Sun contains about 1,000 times more material than Jupiter. This enormous amount of material generates extremely high pressures and temperatures in the Sun's central regions, high enough to start nuclear reactions

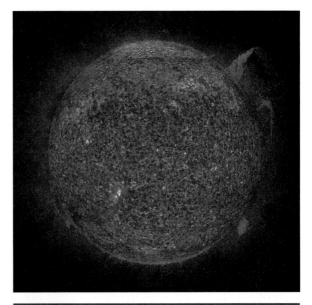

Figure 1-2 The Sun photographed in ultraviolet light. Violent surface activity in the form of solar prominences is seen sticking out from the Sun.

there. The generated energy streams out through the Sun and radiates into space. The tiny fraction intercepted by the Earth's surface generates and sustains life on our planet.

The power of the Sun is difficult to imagine by Earthly standards. It produces more energy in a second than the world has used in the last 10,000 years. Expressed differently, the Sun produces energy at the rate of 10 billion nuclear bombs going off every second. If we figured energy costs at approximately seven cents per kilowatt-hour, the Sun would be radiating 7 million million million dollars worth of energy into space every *second*.

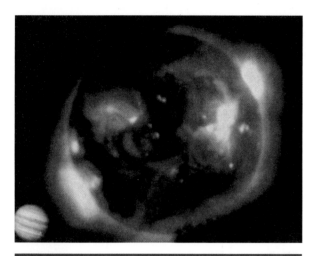

Figure 1-3 The Sun, as photographed by *Skylab* in X-rays. The inset shows Jupiter to the same scale.

Figure 1-4 The Earth and the Moon photographed by the *Galileo* spacecraft on its way to Jupiter in 1992.

All the bombs exploded during World War II would power the Sun for only a millionth of a second. Yet the Sun is a fairly ordinary star. In Chapter 19, we will study stars that generate a million times more energy (that is, an amount six orders of magnitude greater). However, there are also large numbers of stars almost a million times less powerful.

Like most other celestial bodies, the Sun radiates energy in the form of gamma rays, X-rays, ultraviolet radiation, infrared radiation, microwaves, and radio energy, as well as visible light. After World War II, astronomers began in earnest to develop instruments that would measure other radiation in addition to visible light. Such instruments have provided the data for the rapid growth in astronomical knowledge in the last few decades. **Figure 1-3** shows the Sun as it appears to a detector sensitive to X-rays, a view dramatically different from the view in ultraviolet light shown in Figure 1-2. From such observations, astronomers conclude that the glowing white regions are hotter than the darker parts by three orders of magnitude. What makes the Sun work? Why is it so hot inside? Will it remain the same forever, or does it change? The discussion of such questions begins in Chapter 16 and continues for a few chapters afterward.

THE EARTH-LIKE PLANETS

Earth is an island in the universe, just like every other planet. But it is *our* island. We see that our planet, whose diameter is about 8,000 miles (13,000 km, twice its radius of 6,378 km,

rounded), is so large compared to the size of a human that it seems flat to us. It has continents, active volcanoes, lots of water, and an atmosphere. A magnetic compass shows that it exhibits magnetism. Further examination shows that the continental bodies and mountain ranges change over time.

Our nearest neighbor, the Moon (**Figure 1-4**), at an average distance of about 240,000 miles (380,000 km), is the only other celestial body humans have visited in person. The Moon's distance is about 30 times the diameter of the Earth. Although this sounds close, think of the Earth as a basketball and the Moon as a softball; they would have to be on the opposite corners of a room to be at the correct scale. A small body having roughly a quarter the diameter and containing 1 percent of the material of the Earth, the Moon nevertheless has a strong effect on its parent body. Among other things, its gravitational attraction causes the tides on Earth. A few questions dealt with in Chapter 8 include: Do the Earth and Moon have identical chemical compositions? Are processes that occur on Earth present on the Moon? Did the two bodies form together or separately?

Mercury, Venus, and Mars show some similarities to Earth. People have speculated that the red planet Mars (**Figure 1-5**) was a possible location for life since spacecraft have found ancient volcanoes and huge canyons spanning the planet, as well as a mineral that can form only in standing water. Questions discussed in Chapter 9 include: What processes occur on Mars to produce such

Figure 1-5 Mars. A mosaic of some 100 *Viking Orbiter* photographs.

Figure 1-6 Jupiter, photographed from 17,500,000 miles by one of the *Voyager* spacecraft. The inset shows Earth in its true size relative to Jupiter.

large volcanoes? Were the canyons formed from flowing water? What happened to Mars's water? Chapter 24 reviews the history of our fascination with Mars as another location of life and asks questions like: Can life still exist on Mars?

THE GIANT PLANETS

Figure 1-6 shows the giant planet Jupiter photographed in 1979 as the *Voyager 2* spacecraft flew by. Unlike rocky Earth, Jupiter is a huge gas ball; it may not even have a solid core. We will see that solid ground like the surface of the Earth is a rarity in the universe because most of the observable matter in the cosmos is gaseous.

Jupiter contains more matter than all the other planets put together. Its diameter is approximately 10 times larger than Earth's. Jupiter's diameter is therefore one order of magnitude larger than that of the Earth, which means Jupiter's volume is $10^3 = 1,000$ or three orders of magnitude larger. The inset of Figure 1-6 shows the Earth to scale.

Jupiter's largest moons (one of which is seen in Figure 1-6) are perhaps even more fascinating than the giant planet itself. As we will study in Chapter 10, each was found to have its own distinct character. Why is each one unique? Why is the surface of one heavily cratered while the surface of another is smooth without impact craters? Why does the position of a moon in its orbit about Jupiter influence the planet's observed radiation?

Figure 1-7 Saturn and its fabulous ring system as viewed by *Cassini* from a distance of 48 million kilometers.

In Chapter 24, the possibility of a salty ocean harboring life underneath the surface of Jupiter's moon Europa will be explored.

Smaller than Jupiter, but still a giant by earthly standards, Saturn (**Figure 1-7**) is famous for its ring system. Several rings are visible from Earth, but 1982 close-up views from the *Voyager* probes showed an incredibly complex system of tens of thousands of *ringlets* engaged in a complicated

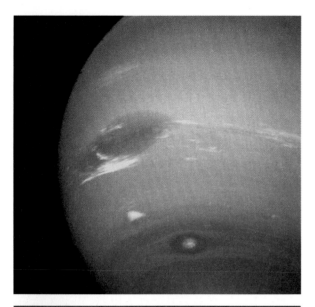

Figure 1-9 Halley's comet.

Figure 1-8 Neptune, as seen by *Voyager 2*.

gravitational dance. Research has shown that the other giant planets (Jupiter, Uranus, and Neptune) also have rings, although none are as prominent as those of Saturn. Saturn also boasts Titan, a moon with an atmosphere so thick we could only explore the surface by sending the *Huygens* probe into it in 2005. To what extent are Jupiter and Saturn similar? What are the rings, and why do the giant gas planets have them? Why does the ring system of each planet differ from that of the others? Why do they have such complex systems of moons and why are they so different? These questions are studied in Chapter 10.

While passing Saturn, *Voyager 2* received a gravitational boost to propel the space probe on to Uranus and then to Neptune (**Figure 1-8**); no other probes have been sent since then. In addition to features resembling some of those on Jupiter and Saturn, Neptune showed a complete ring system and additional moons. Its moon Triton was especially exciting, as you will see in Chapter 10. There, too, you will read about how Neptune differs from Uranus, what surface processes occur on Triton, and in what ways the orbits and motions of some of Neptune's moons are strange and unexpected.

Pictures are transmitted to Earth from probes such as *Voyager* as strings of numbers, then are reconstructed into two-dimensional images by computer. The color may or may not be true color because computers can be instructed to manipulate colors to enhance contrast or make certain features more visible. Many of the spectacular images that adorn astronomy books these days are actually enhanced or artificially colored by computer. The colors in the photographs in this chapter are all true colors except as follows: Figures 1-6 and 1-8 are close to real but have been intensified by computer to show features in greater contrast; Figure 1-10 is computer colored.

THE OUTER REACHES OF THE SOLAR SYSTEM

Pluto, the smallest of the traditional planets, has such an unusual orbit and composition that some astronomers have questioned its stature as a planet. However it is classified, it remains a fascinating object to study, not the least because it has such a large moon (Chiron) that it can be considered a double-planet system. In Chapter 10 you will learn more about what makes a body that is so small and distant (6 billion km) that it appears as little more than a point of light in telescopes so interesting.

Figure 1-9 is a photograph of Halley's comet taken during its 1986 visit to the inner Solar System. Because the previous visit by this comet was 76 years earlier, 1986 was the first opportunity to analyze it with instruments in space, and several spacecraft from many nations closely studied the comet. Comets merit this kind of intense scrutiny because they consist of dust and frozen gases that may well be remnants of the material from which the Solar System formed. In 2004 a new object in a cometary orbit beyond Pluto was discovered and called Sedna. It is absolutely enormous for a comet, being close to the size of Pluto. It may be an example of a newly discovered population of

Figure 1-10 The red dwarf star AW Microscopii. This young 12-million year old star is surrounded by a disk of dust that reflects its starlight.

more distant minor planets that could number in the tens of thousands. Where do comets come from? What are they made of? Did they have any influence on the development of life on Earth? Comets, along with asteroids, meteorites, and this newly discovered class of objects that are yet to be fully understood are studied in Chapter 7.

PLANETS AROUND OTHER STARS

Astronomers have expected that the processes that make stars will also make planets. That expectation started to receive some observational confirmation when the Infrared Astronomical Satellite (IRAS) observed the star Beta Pictoris to have a disk of material surrounding it that was interpreted as a possible planetary system in formation. More recently, the Hubble Space Telescope has observed a disc around the red dwarf star AW Microscopii, showing that they may be common (**Figure 1-10**). Seeing this disc around a star only 32 ly away is not surprising because red dwarfs are the most common type of star in our neighborhood of the Galaxy. In addition, definitive observations have discovered more than 100 planets to be orbiting stars other than the Sun. Having many samples of planetary systems will provide an improved understanding of how our Solar System formed. Chapter 6 includes a discussion of other planetary systems.

1.4 Stars, Galaxies, and Beyond

Stars are the fundamental building blocks for larger structures in the universe, and they are the subject of Chapters 14 and 15. The nearest star beyond the Sun has a distance of about 270,000 AU, or about 25 million times the diameter of the Sun. Clearly, the distances between the stars are great compared to the sizes of the stars themselves. With so much space between them, collisions or even close encounters between stars are rare.

The distance to the nearest star is about 270,000 AU, so expressing distances in astronomical units results in inconveniently large numbers. However, the light-year is ideal, because it is between four and five orders of magnitude larger than the astronomical unit. It is easier to talk about 4.3 light-years than 270,000 AUs! This distance is also about the average distance between stars in our galaxy. One of the more important questions we can address is how astronomers determine distances to stars. We begin looking at stellar distances in Chapter 15.

The stars are so far away from us that they appear only as points of light. Even high telescopic magnification does not reveal a surface. (The apparent sizes of star images in pictures, produced by image spreading within the photographic emulsion or electronic detector, are determined by a star's apparent brightness, not its actual size.) But there are some special observing techniques described in Chapter 15 that do allow us to record surface features on a few stars that are not only nearby but also very large. The supergiant Betelgeuse (pronounced *beetle juice*), in the constellation of Orion, is such a star; indeed, it is one of the largest and brightest stars in our galaxy. Its diameter is about three orders of magnitude greater than the diameter of the Sun. If the Sun were to be replaced by a supergiant star like Betelgeuse, the star would stretch out to nearly the orbit of Saturn. Such a star generates about 25,000 times more energy than the Sun. What does the surface of a star look like? What makes some stars large and others small? These questions and others will be dealt with in Chapters 16, 17, and 18.

INTERSTELLAR MATERIAL AND STAR FORMATION

The space between stars is not empty but is filled with observable gas and dust called the **interstellar medium**. Astronomers have concluded that

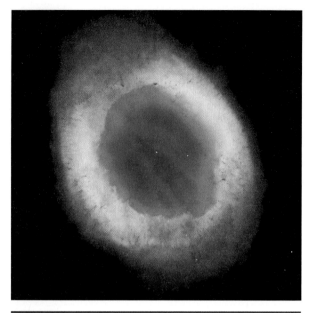

Figure 1-12 The Ring Nebula, a cloud of gas ejected by a dying star in the center.

Figure 1-11 The Orion region is an extensive region of gas and dust excited to glow by the ultraviolet light of young stars recently formed inside the gas. This region is about 15 light years away.

stars are born from large globs of interstellar gas that collapse into stars under the influence of gravity. The dust and luminous gases pictured in **Figure 1-11** form a much-studied region of star formation not too far from the Sun, called the **Orion nebula**. A newly formed cluster of hot stars inside the cloud excites the gases to radiate from a region about 15 ly across. Although striking when viewed through even a small telescope, the Orion nebula is approximately a million times (six orders of magnitude) less dense than the best vacuum we can create on Earth. Even more interesting, the *Hubble Space Telescope* observations show objects in Orion that may be forming new planetary systems.

STELLAR EVOLUTION

Stars evolve. They typically remain stable and fairly constant in energy output for at least millions if not billions of years. But even stars gradually feel the effects of age. Although the Sun has been nearly the same for about five billion years and will be for another five billion years, it is slowly changing, giving off more energy and slowly expanding. Other stars evolve at different

rates, depending on the amount of matter they begin life with.

Betelgeuse is a star that many astronomers feel may be nearing the end of its life. The changing chemical composition inside the star, resulting from nuclear reactions there, eventually causes significant changes in the outer (observable) parts of the star. Betelgeuse is swollen into its supergiant configuration by forces we will examine in Chapter 16, when we begin to discuss stellar evolution in more detail.

Two other objects in advanced stages of evolution are pictured in **Figures 1-12** and **1-13**. The objects in these figures are a few light-years across. The object in Figure 1-12, called the **Ring nebula**, is an example of a **planetary nebula**. The star in its center is in the process of sloughing off its outer atmosphere. Perhaps a tenth of a solar mass (an amount of mass equal to 0.1 the mass of the Sun) of gases move outward at a speed of about 30 km per second; this gas then mixes back in with the gases of interstellar space. The central star eventually becomes a **white dwarf**, a dead star having the amount of material in the Sun compressed into a space with the diameter of the Earth.

Stars much more massive than the Sun undergo a catastrophic mass ejection, tossing material outward at speeds of tens of thousands of kilometers per second. The result of one such **supernova** explosion is the **Crab nebula** (Figure 1-13), the remains of an exploding star recorded by Chinese

Figure 1-13 The Crab Nebula, the remnant of a spectacular stellar explosion witnessed by Chinese astronomers in 1054 C.E.

Figure 1-14 The globular cluster M13, a giant aggregate of hundreds of thousands of stars formed early in the history of our galaxy.

astronomers in the year 1054 C.E.[2] Near the center of the Crab Nebula is an object that has more material than the Sun squeezed into a ball only about 10 to 15 km across; it is called a **neutron star**, also a type of dead star or stellar remnant. Other supernovae may result in the production of an ultra-dense object from which no light can escape—a **black hole**. The closest star to us that could go supernova at any time is Betelgeuse. When it explodes, it will appear as bright as the full moon because it is relatively close to the Sun—only about 500 ly away.

CLUSTERS OF STARS

Both observations and theory indicate that if the massive interstellar clouds fragment into condensations, they will form a cluster of stars. In fact, the best modern studies have shown that *all* stars probably form in stellar groups. Isolated stars like the Sun are in the minority; they presumably result from the breakup or disintegration of a star cluster.

The large spherical star cluster known as M13 (**Figure 1-14**) is called a **globular cluster**. It is an impressive aggregate of some 100,000 stars bound together by gravity. Such clusters tend to be found in a spherical halo around the center of our galaxy.

Globular clusters are among the oldest objects in the galaxy, having formed between 10 and 14 billion years ago.

GALAXIES

Planets, stars, star clusters, and interstellar gas and dust are all combined together into larger units called **galaxies**. The galaxy pictured in **Figure 1-15** is popularly called the **Andromeda galaxy** (named for the constellation in which it appears). As the thirty-first object in the 1781 catalog of the French astronomer Charles Messier, it is also named Messier 31, or simply **M31**. The Andromeda Galaxy is thought to be similar to our own galaxy, the **Milky Way**. At a distance of at least 2.2 million light-years, the Andromeda Galaxy is the most distant object you can see with the naked eye. The Milky Way and M31 are **spiral galaxies**, thought to be approximately 100,000 ly in their longest dimension. Each galaxy contains a few hundred billion stars, with some very large galaxies containing a trillion stars. The photograph shows that the Andromeda Galaxy has two smaller satellite galaxies gravitationally associated with it. We need to readjust our sense of scale here: the smallest of these companion galaxies is approximately 100 times the size of the globular cluster in Figure 1-14. Like the Andromeda Galaxy, our own Milky Way also has several known companion galaxies orbiting around it. Chapter 20, on the

---

[2]C.E. is the "Common Era," corresponding to A.D. B.C.E. is "Before the Common Era," corresponding to B.C.

Figure 1-15 The Andromeda Galaxy, also known as M31, is a spiral galaxy somewhat similar to our own Milky Way galaxy. It is also relatively close, "only" two million light-years away, and under dark skies can be seen by the naked eye as a faint patch of light. It is orbited by two elliptical galaxies.

Figure 1-16 The Large Magellanic Cloud, a small irregular galaxy orbiting near our Milky Way galaxy.

Figure 1-17 M82 is a galaxy that has clearly been disturbed in some way.

Milky Way, addresses questions such as these: Where is the solar system located in the Milky Way? Why do we think our home galaxy is a spiral? How did the Milky Way form?

Galaxies come in forms other than spirals. Two examples of **elliptical galaxies**, which are shaped like giant ellipses, are the companions around M31, shown in Figure 1-15. Others defy easy classification and are known as **irregular galaxies**. The irregular galaxy pictured in **Figure 1-16** is a relatively nearby object called the **Large Magellanic Cloud**. Easily visible to the naked eye from Earth's Southern Hemisphere, it is one of the satellite galaxies attracted by gravity to our own Milky Way. Explaining the diverse shapes and forms of galaxies is another challenge to modern astronomy. The various colors shown by the galaxies reveal different types of stars in their different parts. Using these observations along with a little detective work, astronomers can document the history of star formation within a galaxy. One important question we can ask is: Why are some galaxies spiral shaped while others are elliptical or irregularly shaped? We discuss galaxy formation in Chapter 21.

The galaxies we have seen are just a few representatives of the vast population of galaxies whose images can be captured with modern telescopes. Using the world's largest telescopes and modern observing techniques, astronomers estimate that they could detect billions, perhaps trillions, of galaxies.

Some galaxies emit little visible light but huge amounts of invisible light and so require sophisticated modern instruments to be studied. Observations show them to be in violent states of agitation. Astronomers began to detect large numbers of such objects after World War II, when the first generation of radio telescopes was constructed. **Figure 1-17** shows such a **radio galaxy**, known as M82. It contains large amounts of gas and dust surrounding a nucleus undergoing a burst of star formation. Other radio galaxies may be two objects in collision. The evidence is strong for radio galaxies to contain supermassive black holes at their centers.

Quasars are enigmatic objects, all appearing to be moving away from us at large fractions of the speed of light. They are thought to be among the

Figure 1-18 The Coma cluster of galaxies.

Inquiry 1-3

Why do we observe galaxies colliding with each other, but do not observe stars colliding with stars?

most distant objects known to astronomers, emitting one to two orders of magnitude more energy than the entire Milky Way from a location the size of the Solar System. The most important unanswered question is: What produces all the energy emitted by these quasars and radio galaxies? What produces jets of gas observed in many of them? Why are they moving at large fractions of the speed of light? These and other questions will be considered in Chapter 22.

CLUSTERS OF GALAXIES

As enormous as they are, galaxies are not the largest known entities. In fact, galaxies are generally bound by gravitational forces into **clusters of galaxies** some millions of light-years in diameter. **Figure 1-18** shows a galaxy cluster in the constellation of Coma Berenices. One of the more complex and vexing unanswered questions about galaxy clusters is whether galaxies formed before clustering occurred, or whether matter clustered and then fragmented into galaxies.

Although research suggests that all the galaxies in this photograph formed at approximately the same time, we find an amazing diversity of shapes. Because galaxies in clusters, unlike stars in galaxies, are close together relative to their size, we find examples of galaxies in the act of colliding or otherwise interacting with each other. Such interactions may determine the form of many galaxies.

The Milky Way galaxy is one of nearly 40 galaxies forming a small cluster of galaxies called the **Local Group**. The Andromeda Galaxy pictured in Figure 1-15 and the Large Magellanic Cloud (Figure 1-16) are members of the Local Group.

In the last 30 years, astronomers have come to appreciate that the clusters of galaxies themselves are collected into even larger clusters of clusters of galaxies, or **superclusters**. These superclusters, 19 orders of magnitude larger than the Earth, are the largest objects so far detected in the universe and are too large to show in a single photograph. At a distance of 10 billion ly, a cluster of galaxies is barely resolved into individual objects on a photograph. Our Local Group of galaxies is affiliated with other relatively nearby galaxy clusters into what we call the **Local Supercluster**. Astronomers have also come to realize that there are large regions of space called **voids** where fewer galaxies exist.

Most luminous matter is clumped into stars, which, as we saw, are often organized into star clusters. Star clusters, in turn, are parts of galaxies, which themselves group into clusters and superclusters. One of the goals of modern astronomy is to understand why the universe exhibits this hierarchical structure. How this structure relates to the formation and evolution of galaxies is included in Chapter 21.

Inquiry 1-4

What characteristics other than size also scale with the hierarchy of the objects in the universe, such as asteroids, planets, stars, and galaxies?

THE UNIVERSE AS A WHOLE

Our tour of the universe has looked at the variety of types of objects contained in the universe. How about the universe as a whole? Exciting advances in the study of the structure and evolution of the

Figure 1-19 The Hubble Ultra Deep Field, which shows 10,000 galaxies, shows the faintest visible light objects ever observed.

universe, the subject of **cosmology**, have been made in the last 10 years. For example, the *Hubble Space Telescope* observed the faintest galaxies ever seen when it observed what is now called the Hubble Deep Field. Even fainter objects have now been seen by the Hubble Ultra Deep Field (**Figure 1-19**). Most recent are the findings of the Wilkinson Microwave Anisotopy Probe (called WMAP), which provides information on the earliest times and structures in the universe. These topics are discussed in Chapter 23 when we ask some of the ultimate questions about what the universe looked liked just after the Big Bang and its long-term fate.

1.5 Where Does Astronomy Go from Here?

The meager amount of light from distant objects that falls on astronomers' telescopes limits our ability to observe the universe. To fully appreciate what the received light tells us, astronomers seek to furnish an explanation for the celestial events and objects based on astrophysics. This means that astronomers use modern physics to explain how the various forms of radiation received from the cosmos were created. Although much of our knowledge of physics comes from Earth-based experiment and observation, all tests done thus far indicate that the laws of physics are valid throughout the universe and throughout all time. Thus, an object such as a star can be understood as a collection of an enormous number of atoms bound together by gravity. The radiation given off by the star is actually the total of the radiation emitted by all the individual atoms in the star, so we must descend to the atomic scale to understand the atom's process of giving off radiation.

Smaller than the atomic scale is the nuclear one, where nuclear reactions produce the energy emitted by stars. To understand the first few minutes after the birth of the universe, astronomers also delve into *sub*nuclear physics, the realm of the smallest things in the universe. Because the material of an atom is concentrated into its nucleus, and most of an atom is relatively empty space, it appears that the hierarchical structure we previously observed continues on down from the largest to the smallest of scales we are able to detect.

All the relative sizes of the objects in the universe that we have been discussing can be put onto a chart that may help you visualize them. Using orders of magnitude, **Figure 1-20** shows the various dimensions in the universe and how much larger they are than the Earth. Note that each step up the ladder means that the object named is 10 times larger or farther than the previous step; by looking at the chart, you see that the universe is 10,000,000,000,000,000,000 times (19 orders of magnitude) larger than the Earth.

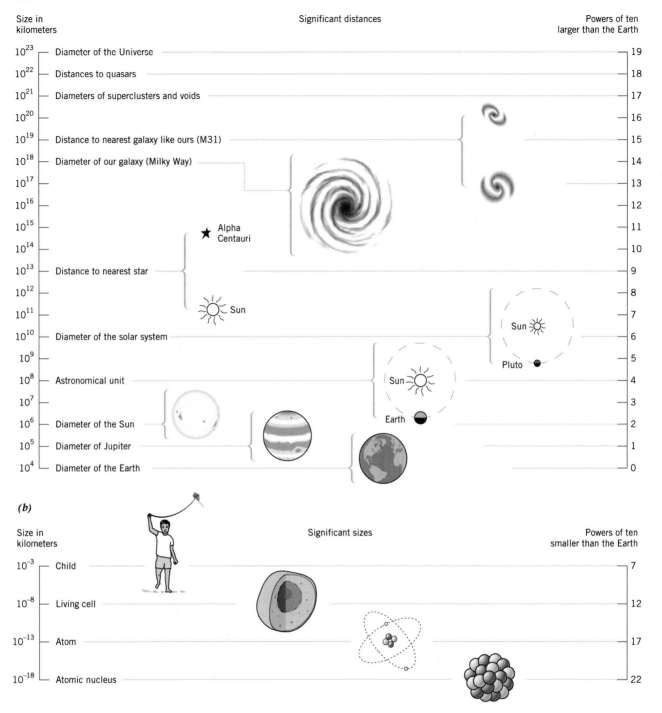

(a)

Size in kilometers	Significant distances	Powers of ten larger than the Earth
10^{23} — Diameter of the Universe		19
10^{22} — Distances to quasars		18
10^{21} — Diameters of superclusters and voids		17
10^{20} —		16
10^{19} — Distance to nearest galaxy like ours (M31)		15
10^{18} — Diameter of our galaxy (Milky Way)		14
10^{17} —		13
10^{16} —		12
10^{15} —	Alpha Centauri	11
10^{14} —		10
10^{13} — Distance to nearest star		9
10^{12} —		8
10^{11} —	Sun	7
10^{10} — Diameter of the solar system	Sun	6
10^{9} —	Pluto	5
10^{8} — Astronomical unit	Sun	4
10^{7} —		3
10^{6} — Diameter of the Sun	Earth	2
10^{5} — Diameter of Jupiter		1
10^{4} — Diameter of the Earth		0

(b)

Size in kilometers	Significant sizes	Powers of ten smaller than the Earth
10^{-3} — Child		7
10^{-8} — Living cell		12
10^{-13} — Atom		17
10^{-18} — Atomic nucleus		22

Figure 1-20 *(a)* The scale of astronomical sizes. *(b)* The scale of people and microscopic things. The number of orders of magnitude difference between two objects in the figure can be found by taking the difference between the powers of ten on the right side of the figure.

Chapter Summary

- The universe contains a variety of objects, including planets, stars, star clusters, galaxies, clusters of galaxies, and clusters of clusters of galaxies (*superclusters*). These objects form a hierarchical structure.

- An *astronomical unit* (**AU**) is the average distance from the Earth to the Sun. An astronomical unit is two orders of magnitude larger than the Sun, which is two orders of magnitude larger than the Earth. A *light-year* (**ly**) is the distance light travels in one year.

- An *order of magnitude* is a term used to describe the relative sizes of numbers. Two numbers differ by one order of magnitude for each factor of 10 by which one number is larger than the other. The concept is especially useful in astronomy, where we deal with a vast range of sizes and distances.

- Our Solar System is gravitationally anchored by its only star, the Sun, with many objects such as asteroids, comets, and planets orbiting around it. Moons orbit around most of the planets. In turn, the Sun orbits around the center of our galaxy, the Milky Way.

- Stars in our part of the Galaxy are about four light-years apart, extremely distant compared to their size. Clouds of gas and dust fill the spaces in between the stars.

- Stars are born, evolve, and eventually die. In doing so, they pass through various stages, beginning with gas and dust, evolving through middle age like the Sun, and eventually dying. The final form of a dead star may be a **white dwarf**, **neutron star**, or **black hole**, depending on the amount of matter the star contains when it dies. Most stars are born in clusters of tens to millions of other stars.

- Our Solar System is located on the outer part of a spiral-shaped galaxy made up of billions of other stars embedded in gas and dust. Many of those other stars also have planetary systems. Our galaxy, the Milky Way, is similar to the other billions of galaxies in the universe.

- Galaxies also are born, evolve, and eventually will die. The stages they pass through are not yet well understood, but probably include interacting and merging with other galaxies as well as the aging of the stars within them.

- Stars, star clusters, and interstellar gas and dust gravitationally clump together into galaxies. Galaxies, in turn, cluster, and these clusters and the voids between them fill up the universe.

- Cosmology is the study of the structure and evolution of the universe.

- Most of the observations by which we perceive the universe are of radiation. To understand the universe we must first understand the source of this radiation, which are atoms and their nuclei.

Summary Questions

1. State the hierarchy of objects in the universe, from the smallest to the largest.
2. Define the astronomical unit (AU) and the light-year (ly) and state their approximate sizes in kilometers. Calculate the size of the light-year in astronomical units.
3. What factors make stars appear to be different from each other?
4. Are all objects called stars really stars?
5. Briefly describe the appearance of several varieties of galaxies.

Applying Your Knowledge

Questions marked with ▶ require computation.

1. Which photograph in this chapter intrigues or interests you most, and why?

2. From simply looking at the photographs in this chapter, can you say which object is the largest? How do you know?

3. Go to the Astronomy Picture of the Day (APOD) Web site *http://antwrp.gsvc.nasa.gov/apod/astropix.html* and examine the picture and read about the image. Where does this object fit into the hierarchy of the universe? Is it something mentioned in this chapter, or is it something completely different?

4. Why do astronomers use astronomical units rather than miles or kilometers?

▶ 5. How many orders of magnitude larger than an atom is a typical person?

▶ 6. How many orders of magnitude are there between each of the following pairs: (a) the diameters of the Earth and Sun, (b) the diameter of the Earth and an astronomical unit, (c) the length of an astronomical unit and the distance to the nearest star beyond the Sun, (d) the distance to the nearest star beyond the Sun and the diameter of the Milky Way galaxy, (e) the diameter of the Milky Way and the distance to the nearest galaxy like ours, and (f) the size of a nucleus and the diameter of the universe?

7. Use information in Appendix C to determine the number of orders of magnitude between the diameters of Uranus and Neptune, the Moon and Jupiter, Earth and Jupiter, and Pluto and Saturn.

8. Write the following numbers in powers of ten notation. (a) 100,000,000 (b) 0.000,000,000,001 (c) 25,000,000,000,000

9. Use Appendix C at the back of the book as your source of data for placing the planets into groups having diameters of the same order of magnitude. Repeat for the planetary masses.

10. Charles Messier cataloged objects in the sky that might be mistaken for comets. Go to *http://www.seds.org/messier/Messier.html* and select either nebulae, star clusters, or galaxies. Look at the images in that category, and explain how their appearance determines which type of object it is.

11. New discoveries are constantly changing our view of the universe. Visit the home page of one of NASA's Great Observatories, the Hubble Space Telescope (*http://www.stsci.edu/spitzer*), the Spitzer Space Telescope (*http://www.spitzer.caltech.edu/*) or the Chandra X-Ray Observatory (*http://chandra.harvard.edu/*) and explore the latest discovery there. Does it update or change something that you read about in this chapter?

Answers to Inquiries

1-1. (b) Because in the other cases you must multiply them by 10 at least once to make them of the same order of magnitude.

1-2. First, each number must be expressed in the same units. From Appendix A we find the conversion factors from AU to kilometers and from light-years to kilometers. Thus, the Milky Way is 10^5 ly $\times$ 9.5×10^{12} km/ly = 9.5×10^{17} km across; Pluto is 40 AU $\times$ 1.5×10^8 km/AU = 6×10^9 km. Thus, following the procedure in Table 2-1, the Milky Way is eight orders of magnitude larger. This particular problem can be solved using the scales of Figure 1-20; the difference of the number of powers of ten is 8.

1-3. The stars are far apart from each other compared to their size even in clusters; galaxies in clusters are fairly close to each other compared to their diameters.

1-4. The mass of each of the named objects also scales with their size. The stellar remnants such as white dwarfs, neutron stars, and black holes are exceptions, however. Another characteristic that usually scales with size is the brightness. Quasars are an exception to this rule.

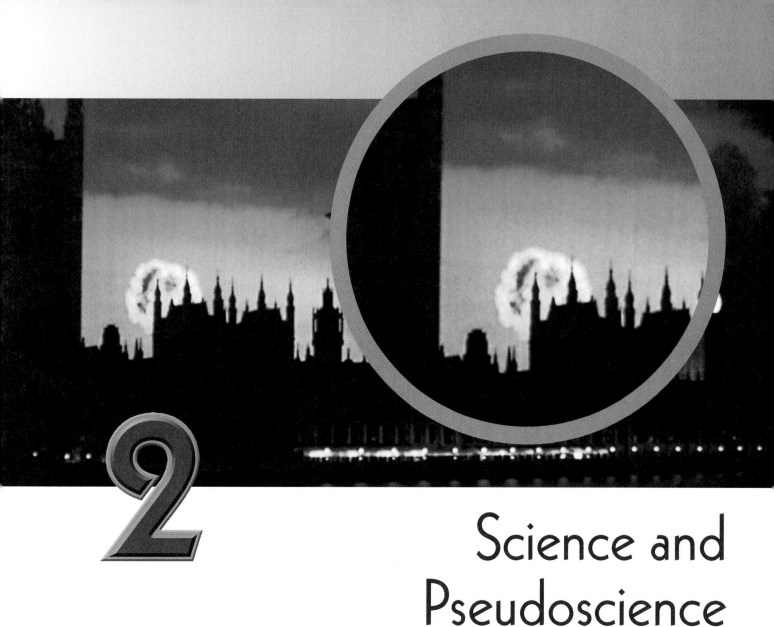

Science and Pseudoscience

Science is the great antidote to the poison of . . . superstition.

Adam Smith, The Wealth of Nations, 1776

Science is facts; just as houses are made of stones, so is science made of facts; but a pile of stones is not a house, and a collection of facts is not necessarily science.

Henri Poincaré

S ome years ago, an astronomer was on a train in Europe. In the dining car, he was seated opposite an elderly gentleman with a long, white beard, dressed in a long robe, who looked like one of the patriarchs straight out of the Old Testament.

The elderly gentleman evidently wanted to talk, and so the astronomer just listened, not saying what his profession was. It turned out that the patriarch was the world leader of the Hollow Earth Society, whose adherents assert that Earth is a hollow ball and that we are living on the *inside* of it **(Figure 2-1)**. The Hollow Earth movement has many adherents. Our astronomer had happened to meet up with the biggest Hollow Earther of them all!

The patriarch explained to the astronomer why we mistakenly think we are on the outside of the Earth even though we live on the inside; how the Sun, Moon, planets, and stars only *look like* they go around the Earth; and how an optical illusion causes satellites to take pictures of the Earth that make it appear round. Finally, he explained the *direct* evidence that the Earth is hollow. There were only two pieces of evidence of interest to him. The first was a survey made of the Great Lakes around the turn of the twentieth century that showed that the surface of the Earth was concave instead of convex. And the second? **Figure 2-2** shows the side view of a shoe. Do you notice how the sole of the shoe curves upward instead of downward? That's because you have been walking around on the *inside* surface of the hollow Earth all these years!

This story may sound made up, but it really happened. The elderly gentleman was a typical example of a pseudoscientist. Although his particular hokum may seem pretty hard to swallow, he had quite a few convinced followers. As psychologist Robert Thouless has pointed out, there is no idea so absurd that you can't find someone who believes it. For this reason, educated people should know what is and what is not science.

Astronomy is a science in every sense of the word. In the first part of this chapter you will learn what science is, using astronomy as an example. In the second part, we further define science by showing what is not science. Finally, we examine various characteristics of what often pretends to be science but isn't.

2.1 An Expedition to Earth

To give you a feeling for what science is, and for the difficulties involved in studying objects with extremely long lifetimes, we will discuss the problems that would confront an alien astronaut who is visiting Earth for only a short period of time, and whose responsibility is to search for life on Earth and try to understand the life cycle of its inhabitants. Suppose we give this visitor just 15 seconds to take as many pictures as possible (being advanced, the alien's civilization has ultra-high-

speed cameras). We choose 15 seconds because it bears approximately the same proportion to the lifetime of a human being as the 100 years we have been observing the universe photographically does to the age of the universe:

$$\frac{15 \text{ seconds}}{\text{Human lifetime}} = \frac{100 \text{ years}}{\text{Universe's lifetime}}.$$

When the astronaut returns home with the photographs, the scientists there must try to understand Earth and all its life forms by studying the photos (**Figure 2-3**). It is almost like studying a still picture, because 15 seconds is not long

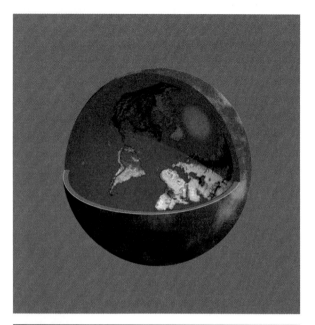

Figure 2-1   The Hollow Earth hypothesis. According to this notion, we are living on the inside of a hollow ball.

Figure 2-2   The sole of your shoe curves upward. According to its proponents, this is "evidence" for the Hollow Earth theory.

Figure 2-3   An interstellar visitor examines data about Earth.

enough for any serious or important changes to show up in any one individual.

How might the alien scientists determine the dominant form of life on Earth? If size is their main criterion, they might choose whales or trees for study. If they count sheer numbers, insects might win out. A sophisticated criterion might be to determine the amount of land space controlled by one species, in which case the automobile might be selected as the dominant species, at least in many urbanized areas. Indeed, thinking of ways by which scientists might determine that human beings are the dominant forms of life on our planet is itself a challenging exercise.

Inquiry 2-1

By what criteria might the extraterrestrials decide that humans are living organisms and automobiles are not?

Suppose the alien scientists decide that human beings are worthy of further study. At this point the problem has just begun, because close observation would reveal a considerable diversity of characteristics among humans (**Figure 2-4**). The scientists would first have to set up a system for classifying humans on the basis of observable

Figure 2-4   A visiting alien would observe that the inhabitants of Earth show a large variety of characteristics.

characteristics. (Later on, we will see how astronomers have done the same kind of thing with stars and galaxies.) It would be immediately obvious that humans come in a variety of sizes (as do stars). Although the alien scientists would note a fairly continuous distribution of sizes, suppose they decide to classify all humans into the categories small, medium, and large. Suppose they also establish the color categories black, brown, yellow, white, and red. By careful observation and much thought, they might also detect two sexes, call them male and female. But they might initially establish other categories that could prove less fruitful in their analysis of human beings and cause them to waste a great deal of time; categories such as hair length, permanent versus removable teeth, color of clothes, and so on. Each irrelevant category would spawn unproductive hypotheses concerning the life cycle of humans, and much thought would have to go into distinguishing important from unimportant characteristics.

Inquiry 2-2

By what criteria might the alien scientists be able to establish that hair length is not relevant, but that sex is, given the fact that in some societies there is a degree of correlation between the two?

Let us consider just the characteristics of size, color, and sex mentioned in the preceding paragraph. The scientists can now begin to construct hypotheses for the life cycle of humans, asking questions such as: Do small, brown, female humans grow into large, red, male humans? Or, instead, do large, black, male humans grow into medium-sized, yellow, male humans? Or is there perhaps no change at all—the small stay small, and the large stay large forever? Even in this extremely simple analogy with only three straightforward characteristics to compare (**Table 2-1**), there are $3 \times 5 \times 2 = 30$ different possible combinations, and $30 \times 30 = 900$ possible combinations of starting and ending points for humans. Obviously, the universe will present us with many more possible combinations and a much longer time frame!

Table 2-1 Characteristics of People

SIZE	COLOR	SEX
Small	Black	Male
Medium	Brown	Female
Large	Red	
	White	
	Yellow	

Inquiry 2-3

Because the alien astronaut cannot take photographs of every human being, but of only a small number, what fundamental assumption must be made to interpret the data meaningfully?

You can gain some direct personal experience of the dilemma of studying objects with long lifetimes by going outside at night and contemplating the stars for a while. You will undoubtedly be struck by the fact that nothing happens, other than the daily rotation of the sky, which is, of course, just the Earth spinning. Bring back the architect of the Roman Coliseum and stargaze with him. He will assure you that all the star patterns still look about the same as they did in his time! The naked-eye sky appears eternal and unchanging. For thousands of years, one of humanity's most enduring metaphysical concepts was the idea of the unchanging universe. Only in the twentieth century did we break through the time barrier and appreciate the life cycle of astronomical objects.

To draw one final analogy, astronomers are in the position of being shown just three or four out-of-sequence frames of a movie that runs for about a year and being asked to reproduce the whole plot. Amazingly, to some extent we have been able to do just that.

2.2 Astronomy as an Observational Science

In this section we will begin to see the ways in which astronomy is a science and how it begins to make advances in understanding the universe. The universe viewed by all physical sciences must be consistent; for example, the 4.5-billion-year age of the Earth determined by geology must be either integrated into biology and astronomy models or disproved by them. We will also look at how the science of astronomy is fundamentally different from sciences such as physics and chemistry.

THE PROCESS OF DOING ASTRONOMY

We will see throughout this book that many physical laws and theories are concerned with how one quantity varies with another. Examples include how the force of gravity between two objects varies with the distance between them and how the time for a planet to make one orbit around the Sun depends on its average distance from the Sun. Unlike a physicist or chemist, however, an astronomer cannot perform *controlled experiments* to test theoretical models. When a physicist wants to know how the electrical properties of a conductor change with its temperature, he or she will perform experiments in which every variable other than temperature (pressure, humidity, magnetic field) is kept at the same value. In this way, the physicist can be sure that the measured changes in the electrical properties are really a result of a change in temperature.

An astronomer cannot set up controlled experiments because the objects cannot be studied in the laboratory. What an astronomer can and does do is collect the light and other forms of radiation that come from celestial objects and then use all available information, as well as ingenuity and creativity, to try to interpret these signals from afar (**Figure 2-5**). An astronomer must become an expert in studying objects from a distance. Indeed, some astronomers feel that once an object has been visited—as the Moon has—it is no longer the subject of astronomical study but should instead be turned over to the geologists and chemists.

You might think that without being able to visit a star or another galaxy, astronomers might come up with lots of wild ideas about how astronomical objects behave. The main reason that they do not is that ideas will not be accepted unless they are consistent with the laws from other branches of science (particularly physics). Before humans visited the Moon, we were not certain what the Moon was made of. However, the idea that the Moon was made of cheese would not be one an astronomer would make because the density of cheese is only about one third the density of the Moon. And the density of the Moon can be determined without

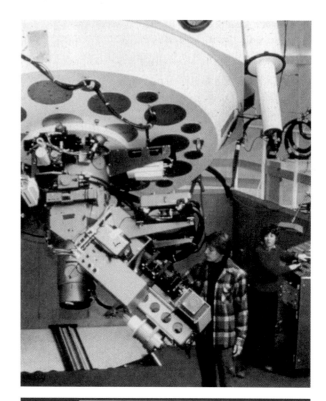

Figure 2-5 The 2.1-meter telescope at Kitt Peak National Observatory is checked out by astronomers Drs. Catherine Pilachowski and Carol A. Christian in preparation for another night's observing.

visiting it. This consistency of science is one of its defining features, and one we return to in our discussion of what science is not.

A scientist can make observations, which, in turn, suggest speculations, hypotheses, and perhaps eventually theories. A **hypothesis** is a conjecture that is used as a suggestion for describing the results of observations and experiments. A hypothesis can be wildly speculative, but to be useful, a hypothesis must make predictions about nature that can either be confirmed or refuted by observation (**Figure 2-6a**). For example, I can hypothesize that students completing an introductory astronomy course will know more about the phases of the Moon than students who have not taken such a course. From this hypothesis, I can make predictions that can be confirmed or refuted by giving a test to both groups.

In astronomy, predictions coming from a hypothesis usually result in the astronomer returning to the telescope to determine whether these predictions are supported by further observations. Reasoning from a hypothesis to a set of particular

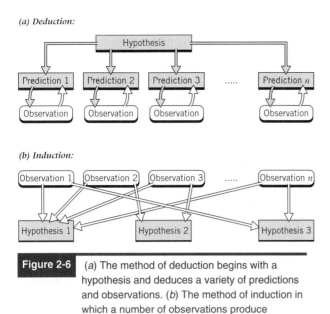

(a) Deduction:

(b) Induction:

Figure 2-6 (*a*) The method of deduction begins with a hypothesis and deduces a variety of predictions and observations. (*b*) The method of induction in which a number of observations produce hypotheses that allow explanation of past behavior and the prediction of future results.

Inquiry 2-4

The following are examples of inductive or deductive reasoning, or situations in which induction or deduction can be used. For each case, which type of reasoning is used? Explain your answer. *(a)* A detective who finds a bloodstained weapon next to a victim, blood of the same type on the clothes of a suspect, scratches on the suspect's face, and skin under the fingernails of the victim would use what process in concluding who committed the crime? *(b)* Economist A believes the health of the U.S. economy is dependent only on the money supply allowed by the Federal Reserve. What reasoning process is used in attempting to validate the idea? *(c)* A social scientist believes that women make better scientists than men. What reasoning process is used to try to validate the idea? *(d)* A person suggests that UFOs prove the existence of extraterrestrial intelligence. What reasoning process might be used to study this question?

predictions is called **deductive reasoning**, or simply **deduction**. The new observations may suggest refinements to the hypothesis, which then provides new predictions, and so on.

Another approach, pioneered by Francis Bacon and later by Isaac Newton, begins with a series of observations. These observations are then understood in terms of one or more possible explanations. One of the central problems of science is choosing which hypothesis is best. As diagramed in **Figure 2-6***b*, hypothesis 1 would be considered the best, because it encompasses the most observations. The fact that observations 2 and 3 do not fit with hypothesis 3 serves to argue against it. The reasoning involved in this process is **inductive reasoning**, or simply **induction**.

Often the evidence in support of a particular hypothesis is indirect, and often the evidence will support more than one hypothesis (Figure 2-6*b*). Induction is the process of determining which one of the available hypotheses is most likely to be correct, based on the fact that it does a more satisfactory job of accounting for the available evidence than the competing hypotheses do. As an example, if you go into the kitchen and see the cookie jar broken on the floor, and the cat mewing piteously, you might conclude that the cat pushed it off the shelf. If, however, there has just been an earthquake, your conclusion might be quite different.

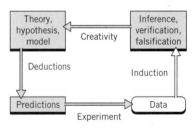

Figure 2-7 Aspects of science fit together in a never-ending cycle of hypothesis, prediction, data gathering, and verification.

A hypothesis may reach the status of a **theory** once the evidence for its validity becomes strong. Strong evidence is both *repeatable* and *verifiable*. A good theory comes from observations and experiments that can be repeated by all scientists in a particular discipline, as well as tested for accuracy. It is probably not fixed for all time; new observations may require modification of the theory. Such modification may be made without throwing out the entire theory. The interplay between theory and observation continues indefinitely; in good science, neither can exist without the other. **Figure 2-7** diagrams the process. From a hypothesis, one uses deduction to make predictions. Predictions bring about experiments, which produce data. The data, along with inductive reasoning, either verify or invalidate the hypothesis. Finally, using creativity, a revised hypothesis is made from which the process continues in an endless loop.

As an example, consider Einstein's theories of relativity. These theories make a wide variety of testable predictions that scientists have been checking continuously for nearly 100 years. Every tested prediction of Einstein's theories have been verified. Therefore, we say that Einstein's theories are valid scientific theories.

When scientists invalidate a hypothesis, they are showing it to be false. When that occurs, we say that the hypothesis has been **falsified**. That is not the same as saying that someone *falsified* the data, by which is meant faking it, cheating, misrepresenting, or deceiving. Science advances through the falsification of hypotheses, because hypotheses cannot be proven true but only proven false.

But not all theories that agree with the observational data are of equal scientific worth. Some hypotheses are so vague that they would agree with any observations that might conceivably be made. Here is an example: "The position of the planets affects events in the lives of people." This hypothesis is so general that, like the typical newspaper horoscope, it can fit anyone.

Other conjectures that agree with observations may be unscientific because they are untestable. Consider, for example, a conjecture that claims that lightning is caused by the god Zeus throwing bolts at the Earth, and that Zeus does this whenever he is angry. There is no evidence that could possibly disprove this idea, even in principle. Supernatural causes are not observable; thus, they cannot be tested scientifically. A supernaturally based hypothesis cannot ever be proven wrong (or falsified), no matter what it claims. Science deals with the natural, observed universe, not the supernatural one. Experience has shown that speculations on the supernatural do not lead to a better scientific understanding of nature. This is not to say that supernaturally based conjectures are wrong or not of value, only that they are not in the realm of science or scientifically useful. Such conjectures give us no help in how to proceed further and acquire more understanding.

The scientific use of the word *theory* is not the same as the everyday, nonscientific use, which is more similar to the word *hypothesis*—a conjecture. For example, a detective may have a gut feeling (a theory) about who perpetrated some crime without having any hard supporting evidence in the case. A scientific theory, by contrast, not only requires detailed supporting evidence, but demands that the theory be able to predict future events.

If two competing theories encompass all observations equally well, which is to be preferred? One school of thought in the philosophy of science tells us that the theory using the fewest arbitrary assumptions is to be preferred. This idea is often called the **Law of Parsimony**, or **Occam's razor**, after the fourteenth-century English philosopher William of Occam. Such a theory seems more likely than one having numerous ad hoc assumptions.

A scientific theory can never be proven to be true. What science can do is to show that observations and experiments are consistent or inconsistent with the theory. Consistency does not prove the theory to be true, but to be reasonable. However, a lack of consistency can falsify a theory, thereby showing it to be invalid.

The Astronomer's Challenge

Astronomers have an absolutely incredible display of phenomena to study, many of which are incapable of being reproduced in any way in Earthly laboratories. A tantalizing challenge indeed! We have already pointed out that diffuse gas clouds such as the Orion Nebula (Figure 1-11) consist of gases in such a rarefied (low-density) state that they give off important types of radiation not readily seen on Earth. On the other end of the scale are black holes, objects of such high density that their enormous gravitational fields prevent light itself from leaving the object.

The range in properties of celestial objects furnishes an almost endless diversity of objects for study. We could not reproduce the energy output of even the most feeble star in a lab on Earth, and the most powerful stars are beacons emitting more energy than a million Suns.

The scale of the cosmos is so vast that many things that are unlikely or impossible on Earth become possible. Indeed, in such a vast volume, any behavior that is not specifically *prohibited* by physical law may well be taking place somewhere, and consequently, astronomers can find exotic phenomena simply by searching. The detection of black holes, antimatter, and neutrinos are three examples of this. These objects were predicted by theory, which made people interested enough to go out and look for them—sometimes with the expectation of *disproving* the theory!

The Astronomical Time Machine

One final tool astronomers have comes, strangely enough, from the very large distance and time scales that cause them so much difficulty. Because light travels at the finite speed of 300,000 kilometers per second (186,000 miles per second), it takes time for the radiation from celestial objects to reach us. Examination of objects that are a great distance away from us also gives us a certain perspective in time, because the farther an object is from us, the longer it has taken the light from that object to reach us (**Figure 2-8**). Thus, we see the Sun now as it was eight minutes ago, the next nearest star as it was over four years ago, and relatively nearby galaxies as they were a few million years ago. The most distant galaxies are seen as they were around 12 to 13 billion years ago, an epoch that is not far removed from the earliest moments of the universe itself. It is as if we have an astronomical time machine that is capable of transporting our view backward in time to extremely remote epochs and finding out what the universe was like then. For this reason, the analogy with the alien astronaut is not quite perfect, because she has no opportunity to compare different epochs in the situation we put her in.

Inquiry 2-5

Making conclusions based on the properties of objects at different distances requires making an important assumption. What might that be?

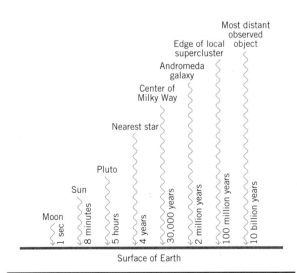

Figure 2-8 The length of time that has passed since the light we observe today left the objects shown.

This ability to look back in time is vital in helping us study galaxies whose time scales for significant change are much longer than the lifetime of an astronomer. To study the life cycles of such long-lived objects, astronomers try to use their perspective in time in the following way. If similar galaxies can be observed at different epochs (by observing them at different distances) and still be recognized as being the same type of object, then with patience it may be possible to construct a life history of that particular type of galaxy, because we may be seeing examples of it at different ages and life stages. Note that this view through time becomes most useful if we push back to the most distant epochs. But it means looking at the most distant objects, which must, of necessity, also be extremely faint and difficult to observe. So our perspective in astronomical time must be purchased with a great expenditure of human time and effort.

Inquiry 2-6

What would you conclude about galaxies if you were told there was evidence that the most distant galaxies show explosions taking place in their cores, whereas the nearby galaxies do not?

2.3 Science as a Process

The most important aspect of what we call science is the process used to come to a set of conclusions. It is through that process that all the facts and conclusions of science become known. This section's purpose is the illumination of that process.

The Explosion of Knowledge in the Twentieth Century

Over the centuries, ideas about the nature of the universe have changed dramatically. The universe was once thought to be small and relatively uncomplicated, extending over distances not much larger than a tribe could walk in a year. Now we know that it would take us a lifetime to travel even the short distance to the most distant known body in the solar system. However, the rate at which astronomical knowledge has been acquired in recent years has increased dramatically, and most of what we now know about the universe has been learned in the past half-century. Such an increase

Figure 2-9 A complete year of The *Astrophysical Journal* in 1900 and in 1991.

to illustrate the rapid changes that have taken place in our awareness of the scale of our cosmos.

Inquiry 2-7

Express the estimated size of the universe in kilometers in scientific notation.

If astronomers were so wrong about the scale of the universe only one or two lifetimes ago, what confidence can anyone have that the description of the structure and history of the universe as it is given in this book is any more correct? Scientists tend to look at the history of science and conclude that the methods and practice of science have provided descriptions of the world that agree increasingly well with observation. Ultimately, you will have to assess the evidence for yourself. Our goal is to provide you with enough understanding of basic astronomical ideas and methods so you will be able to appreciate not only the picture of the universe described here but also new discoveries that will be made in the future. In a field expanding as rapidly as astronomy, mere memorization of facts and theories current at one moment is not sufficient, because some of these facts and many theories will soon become obsolete. More useful is for you to understand the methods by which astronomers obtain observational data and the underlying physical principles we use to interpret the observations.

It is also necessary to see how observations and theory are combined into a **model** that attempts to describe accurately some aspect of reality. We like to think that the universe described by current astronomical theory is essentially correct in its broad outlines—for example, that the size and age of the universe are approximately known, and that most of the major types of objects in the universe are at least roughly understood. At the same time, there are some notable exceptions that will be pointed out, and there are doubtless many surprises in store that will significantly alter our understanding of aspects of the universe previously thought to be well understood. We must always keep in mind that the history of science shows that all scientific theories are ultimately refined, and then often discarded in favor of better ones.

in knowledge is a characteristic of an active, vital science. One possible measure of the growth in astronomical knowledge is the amount of research results published in one year. **Figure 2-9** shows one year of *The Astrophysical Journal* in 1900 and in 1991. The actual growth is even more dramatic than shown because many new journals have come into existence throughout the world in that time span.

One way to illustrate this explosion of knowledge is to compare the size of the universe as it was estimated in the early 1900s with today's ideas. In 1920, the universe was generally accepted to be only about 10,000 light-years across, with the Sun located roughly in the center. In many ways, this number is respectably large—it would take an astronaut traveling at 40,000 kilometers per hour (25,000 miles per hour), which is the speed required to escape from the Earth more than 25,000 years to cover a single light-year, and 200 million years to cross a distance of 10,000 light-years. Yet it turned out to be a gross underestimate of the actual size of the universe. We now accept the universe to be not 10,000 ly across, but more than 10 billion ly (six orders of magnitude larger), containing literally billions of galaxies, each, in turn, composed of billions of stars. These large numbers are intended not to intimidate but merely

The late astronomer Carl Sagan argued that we are living in the most intellectually exciting era in the history of the Earth, either past or future, because we are in the first stages of leaving the Earth and exploring the universe beyond our home. In a real sense, we are the first people on Earth with the potential for developing a true cosmic consciousness—a real understanding of our place in the vast universe. If not abused, perhaps our heightened awareness will also increase our likelihood of surviving the growing perils of this exciting era in which we live.

FROM IDEA TO TEXTBOOK—HOW SCIENCE PROCEEDS

How does science advance from a rough idea to the point where information appears in a textbook such as this? **Figure 2-10** shows the system of filters built into the scientific enterprise that, we hope, prevents the distribution of poor information. Filters are never perfect, however, and some results pass further through the system than they should.

Because people do science, science contains all the foibles held by those doing the research. The top of Figure 2-10 contains everything that a person contributes to his or her research. In addition to education, we all have human traits that have a bearing on everything we do. Because much of modern science requires at least a minimal amount of funds, ideas that are nonsensical might be caught in the process of writing proposals for funds.

Astronomy, like every field of human endeavor, has its own culture. The culture changes, so that in one decade astronomers may dismiss a line of inquiry as uninteresting—yet eventually, its importance may be recognized. A recent example of this change is the search for extraterrestrial life, which is discussed in the final chapter of this book. That subject is now taken to be one of valid scientific inquiry.

When the research is completed, a paper is written for publication. The work is submitted to a scientific journal, which then sends the paper out for review by peer scientists chosen by the editor of the journal. Not all papers are accepted for publication. The rate at which papers are rejected depends on the particular scientific field and the prestige of the journal to which the paper was submitted. Once published, the work becomes part of the **primary literature** of the field.

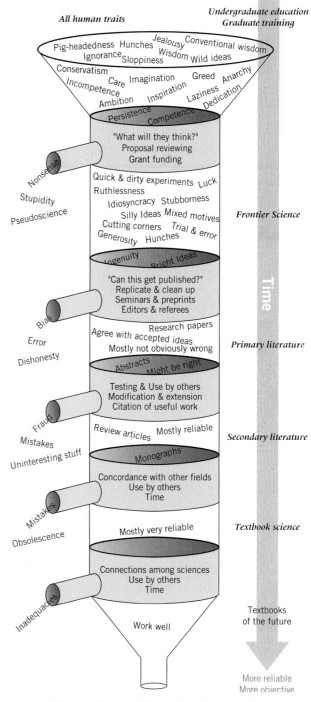

Figure 2-10 The knowledge filter, in which deficiencies in science are (ideally) removed as an idea proceeds from formation through publication in textbooks.

Inquiry 2-8

Newton was reluctant to publish his Universal Law of Gravity (to be studied in Chapter 5) that contradicted the classical Greek ideas held for 2,000 years. Which part of the knowledge filter would have made publication difficult?

The work is now part of the public domain and may be tested by other researchers. If the research can be neither replicated nor validated, or if previously uncaught errors are found, the work will be rejected by the scientific community through the publication of other papers by other scientists. If the work is validated, other researchers may begin to use the results. Furthermore, the results may begin to appear in articles reviewing the status of the field in general. These articles form a field's **secondary literature**. After getting this far through the system of filters, some of the research may then begin to appear in textbooks, either for the training of future scientists or for the liberal arts education of the majority of our population.

TWO EXAMPLES OF SCIENCE AS A PROCESS

The "Face" on Mars

One of the images returned by the *Viking* spacecraft during its 1976 study of Mars included a northern hemisphere area called the *Cydonia region*. The terrain is relatively flat with few craters. It contains a number of buttes and mesas protruding from the surface. One of these features, shown in **Figure 2-11a**, is about a mile across and has the appearance of a human face. For that reason, the popular press has published the suggestion that the feature was the work of aliens, and it has received a lot of attention.

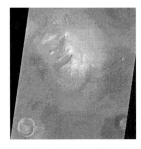

Figure 2-11 The "face" on Mars as observed at low resolution by (a) the *Viking* spacecraft and by (b) the *Mars Global Surveyor*.

Scientists noted that only one image was taken, under only one set of lighting conditions. Illumination angles have been known for years to be an extremely important consideration in photo interpretation. Finally, they noted that extraordinary claims, such as the one that the feature was produced by intelligent beings, require extraordinary evidence. Such evidence was not provided by the *Viking* spacecraft.

The *Mars Global Surveyor* passed over the Cydonia region in the spring of 1998 and obtained new images under different lighting conditions and showing details 10 times better than before. **Figure 2-11b** shows the results of the final processed images. The appearance is considerably changed. Real science requires that data be repeatable and verifiable and that additional data over time will continue to be collected. We will let you be the judge of the alien hypothesis. Again, remember that extraordinary claims require extraordinary evidence.

The Martian Meteorite and Life

In the summer of 1996, NASA scientists announced that analysis of a meteorite previously inferred to have come from Mars contained fossilized life forms. This claim, if true, would be one of the most important discoveries in human history.

We will not go into the arguments on either side of the question here. They are extensive and detailed. The important point is the way in which the methods of science have been applied. The original announcement was followed by publication of the results in a mainstream scientific journal. The techniques used were described in detail; alternate conclusions were examined and discussed.

The act of publication provides the rest of the scientific community with the ability to examine fully what was done. Experts in a wide range of scientific fields and having a wide range of techniques available to them were then able to examine and critique the work in an open manner. It is this mode of operation that marks science and makes it different from pseudoscience. As of this writing (2005), the (cautious) consensus of the scientific community is that the interpretation of ancient life is incorrect and that the data can be explained as resulting from nonbiological events. Of course, the consensus can be wrong, and it can change as the weight of evidence changes.

2.4 Is It Science or Pseudoscience?

How is it possible to tell good science from pseudoscience? While the answer is not always easy, there are a number of characteristics of pseudoscience which, if present in a piece of work, can help you decide.

THE "GAME" OF SCIENCE

Science may be thought of as an intellectual game, in the sense that it has certain rules that need to be followed. These rules have evolved over time as scientists have learned by trial and error how to develop a procedure that generates new and useful knowledge. For example, if a scientist carries out an experiment and claims the results overthrow a law of physics, one of the rules for acceptance of this finding is that other scientists can reproduce the results. Those who play by rules different from the rules of science may be engaged in worthwhile activities, perhaps even activities that are more valuable than science, but they are *not* doing science. The term for those who claim to be doing science when, in fact, they are not playing by the rules of science is *pseudoscientist,* and what pseudoscientists produce is called **pseudoscience**. The prefix *pseudo* literally means "that which deceptively resembles or appears to be something else."

Like all scientists, astronomers are contacted frequently by pseudoscientists who want to explain their latest ideas. Astronomy seems to be one of their favorite subjects. The list of pseudoscientific ideas relevant to astronomy includes astrology, flat and hollow Earthism, geocentrism, theories about ancient astronauts, belief in UFOs, intelligent design, and creation science. Although some of these ideas have small followings (such as flat Earthism), others (like astrology) appear to influence large numbers of people. But the differences between flat Earthism and astrology are differences of degree, not of kind. The problem for most people is in being able to determine whether someone claiming to be a scientist is talking sense or nonsense.

Despite the fact that most people are not trained as scientists, it is still possible to distinguish between genuine science and pseudoscience. Here are some characteristics of good science:

1. Science is a never-ending quest for new knowledge.
2. The results of science must be both repeatable and verifiable.
3. The rules of science dictate that a model or hypothesis should make predictions, and that these predictions are then tested by new observations. The inability of astrology to provide meaningful predictions is illustrated in **Figure 2-12**, which shows the lack of agreement in the predictions made by four astrologers from the same information on human characteristics.

A classic stumbling block for astrology and its practitioners has been their inability to predict or explain why the personalities and lives of twins (especially identical twins) can be so different. A related approach is to examine the variation in personality and lifestyle shown by people who were born on the same day of the same year. To give just a couple of examples, Microsoft Founder Bill

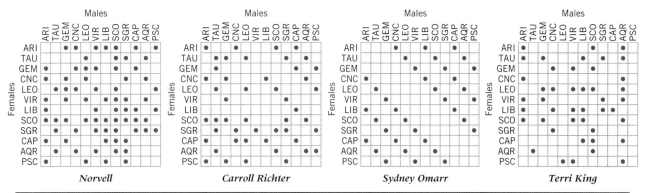

Figure 2-12 Sun-sign incompatibilities in marriage as published by four astrologers. If astrology were a scientific model with verifiable and repeatable results, the results of the four astrologers should show a high degree of correlation. However, the total lack of agreement shows astrology's inability to make verifiable predictions.

Gates and actress Julia Roberts have the same horoscope, as do President Clinton and Orville Wright.

When people begin to find evidence that contradicts established ideas, they may be skeptical at first, particularly if the new evidence conflicts with well-entrenched ideas. Eventually, however, if the evidence is strong enough, and if enough of it is found, the old ideas inevitably give way to the new. It is this tentative but evolving nature that characterizes science—the certainty that at any given moment in history we do not know everything, and the belief that even well-established ideas may be wrong and therefore should be subjected to constant and rigorous verification by experiment and observation. You will see numerous instances of this throughout the book. In contrast, pseudoscientists typically do not use the deductive approach and make specific predictions that could endanger their ideas. The predictions they make tend to be *ex post facto;* that is, made after something is known. Only then can a pseudoscientist "explain" it.

Science is always an unfinished business. There always remain new and undreamed-of discoveries, and there are always opportunities to overturn old and widely believed ideas. At any moment a certain number of observations are unexplained. This does not necessarily mean that current theories are wrong, because there can be many reasons why something is not known (e.g., the difficulty of making detailed calculations, or a lack of sufficient data). And, of course, there is always the possibility of inadequate theories. Therefore, although scientists do work within a framework, or **paradigm**, of established theory, they generally strive to maintain a healthily skeptical attitude toward accepted ideas.

A common view among nonscientists (shared by some scientists as well) is that science is engaged in finding the truth. Modern science has abandoned this as arrogant and unrealistic. Historically, this resulted in labeling accepted theories as *laws,* as in Newton's universal law of gravity. The goals of science are really much more limited. Fundamentally, science aims to construct models that accurately predict the future behavior of physical systems and give a consistent account of the past behavior of such systems. The confusion over "truth" stems largely from faulty thinking that views the models of physical systems as identical to the physical systems they describe. A map of your hometown is a useful tool for finding your way around, but it is not the same as your hometown—it is a model of it.

Many things loosely described as scientific facts are not really facts at all. For example, you might have the impression that this book stated the "fact" that the universe is nearly 14 billion years old. But such usage of the word *fact* is a habit of speech that is seen on close examination to be imprecise. In reality, the age astronomers assign to the universe is an **inference** made from the large amount of observational data that we have, as interpreted with our best understanding of physics and in accordance with those theories that have best stood the test of time and evidence. At present, all the credible evidence points to an age around 14 billion years. The creationist's claim that the universe is only 10,000 years old is, at present, not scientific, *regardless of whether it is true or false.*

What, then, do we mean by a **fact**? A fact can be thought of as something that is accepted as real or true by the community using it. In this book, we will use the professional astronomy community's consensus as to what is a fact. Sometimes facts result from definition: the Earth is 1 AU from the Sun is an undisputed fact. However, in astronomy, a fact is more likely to result from the process of inference as discussed above. For example, the Earth is 93 million miles from the Sun, which is a fact that results from a series of logical steps and inferences. Astronomical facts are always open to change, since new observations or insights in how to interpret existing observations will always be a *fact of life* for astronomers!

IS THE HYPOTHESIS AT RISK?

The first test that can help us tell whether we are dealing with pseudoscience or genuine science is to ask: Is the hypothesis at risk? By this we mean, are there practical experiments or observations that might, at least in principle, show that the hypothesis is wrong? If the hypothesis is not at risk of being disproved, then it is not scientific. But by the same token, every time a hypothesis that is at risk is found to be in agreement with new evidence, we gain new confidence in it.

Consider the earlier example of Zeus and the lightning bolts: if there is lightning, Zeus is angry; if not, he is not angry. The theory that lightning bolts are due to an angry Zeus is never at risk, regardless of the circumstances. By contrast, the early-twentieth-century theory of the observable universe that put the Sun at the center, and which seemed to be so well confirmed by the evidence, fell to pieces when the distribution of globular clusters was studied in 1917. Even the present-day picture of the Galaxy we will give you in this book is at risk; our current theories may fall victim at any time to new evidence or better ideas. Pseudosciences, however, do not change with time. They have a body of ideas that rarely changes significantly.

To give an example from an existing pseudoscience, consider the claim of some creationists that the universe is only 10,000 or so years old. When astronomers point out that we appear to be receiving light from objects that are millions or billions of light-years away (hence, the light has been in transit that long), some creationists reply that all those photons from all the different galaxies were created supernaturally, in motion and at the proper place in space, to arrive at Earth in such a way as to give us that impression. The evidence of Earth's longevity (and of biological evolution) contained in the fossil record is explained by some creationists as evidence of Noah's Flood. Such responses can never be at risk and are unscientific and unhelpful in acquiring a greater understanding of the world in which we live.

Don't Confuse Me with the Facts

To pseudoscientists, evidence is interesting only if it tends to support their ideas; indeed, they tend to ignore or even deny the validity of any evidence that does not support their ideas. Consider the Hollow Earther, for example. You could have easily given a dozen or more persuasive pieces of evidence against the Hollow Earth theory, yet the only evidence the patriarch was interested in were an obsolete survey and the shape of your shoes. You can "prove" anything if you ignore all the evidence that disagrees with it.

Inquiry 2-9

What are three arguments that would persuade others that the Hollow Earth theory is wrong?

Inquiry 2-10

What arguments might a Hollow Earther use to explain away each of the arguments you thought of in Inquiry 2-9?

Simple Answers to Complex Problems

Scientists have great faith that nature is fundamentally simple. Yet the phenomena that nature presents us with can, at first, appear to be frustratingly complex. Although everything in the physical world may ultimately be explained by the interactions of a few elementary particles, experience teaches us that the path to knowledge is often a difficult one, and that long, arduous training is needed just to learn to use the basic tools of a scientific field. The growth of the body of scientific knowledge has become so large that scientists are forced to become experts in narrow disciplines within their fields. For example, among astronomers there are specialists in double stars, stellar atmospheres, celestial mechanics, planetary science, interstellar matter, galaxies, and so on. Within each of these specialties there are even smaller specializations. Specialties even exist in the techniques used to observe. No one can become an expert on more than a small part of science, and even then constant study is required just to stay current with the new discoveries made in that little corner of knowledge.

In contrast, pseudoscientific theories tend to be easily understood. Pseudoscientists are usually untrained in the fields in which they claim expertise, and their theories are unconventional in terms of the accepted scientific ideas of the day. In fact, the conflict with conventional science is sometimes presented as proof of their truth. Perhaps most importantly, pseudoscientists will rarely admit the possibility that their idea or theory is wrong. This is also true of most major religions, where dogma and infallibility are a major component of faith. Such an approach may work well for religions, but religion is not science, and neither is pseudoscience. It is true that individual scientists can be arrogant and that they sometimes appear dogmatic, but those who are engaged in the process of science cannot avoid the idea that even their most important contributions could be overthrown at any time. That is how science moves forward.

Pseudoscientists frequently try to exploit the fact that science has not yet explained everything satisfactorily. Scientists can and often do disagree with one another! Scientists are patient and accept that there will always be many things that an evolving understanding will not have explained yet. Pseudoscientists uncritically give all discordant facts equal weight. They will argue that any discrepancy unexplained by current science is evidence that the whole of modern science is wrong (and that their particular brand of pseudoscience is therefore right). It is illogical to argue that any evidence against some standard scientific view is evidence in favor of a pseudoscientific view; this ignores the fact that there may be many other much more satisfactory explanations that have not yet emerged through the process of science.

PLAYING THE UNDERDOG

It serves a psychological purpose for pseudoscientists to portray themselves publicly as heroic Davids fighting the Goliath of the scientific establishment. They often claim that scientists are prejudiced against their ideas and won't give them a fair hearing. They frequently accuse scientists of being closed-minded because scientists don't unquestioningly accept their claims. The reason scientists often reject pseudoscience is that they understand the issues very well and are in the habit of thinking critically about evidence. The astronomy community (and also the general community of scientists) is conservative in the sense that new ideas are accepted slowly and only when confirmed by independent observations or experiments.

CONSPIRACY THEORIES

A favorite myth among pseudoscientists is that there is a conspiracy of scientists cooperating to lock out new ideas. Perhaps the best example of this is the oft-repeated claim that we have been visited by Little Green Men in Flying Saucers, but that the federal government is keeping them all locked up on a military base so we won't find out! Typically, no rationale is given as to why the government might do such a thing, not to mention how we might be able to lock up someone so technologically superior that they are capable of interstellar travel while we are not. What reward could possibly induce scientists to hide knowledge like this? Any scientist who was the first person to prove that he or she had talked to an extraterrestrial would, for a while at least, become the most famous person in the world. The idea of a conspiracy to hide knowledge neglects the fundamental competitiveness of scientists; they are paid to try and shoot down each other's ideas. If you are a scientist, the quickest road to fame and fortune is to rock the established boat, but you have to have the

necessary evidence to convince a large body of aggressive and skeptical competitors.

PLAYING ON FEAR AND EMOTION

It is unfortunately true that many people fear and distrust science and scientists, and pseudoscientists typically try to exploit these fears by playing on people's emotions and inducing them to agree with their ideas. Emotional appeals that obscure areas of rational debate provide the reason why scientists usually try to avoid pseudoscientists as much as possible. Yet scientists ought to speak out against pseudoscience. Probably the best way to do this is indirectly, by devoting more time to helping the public understand the power, fascination, and beauty of science. Although most people are uncomfortable with uncertainty and yearn to understand themselves and the world around them, science and scientists frequently seem to be incomprehensible, forbidding, and unapproachable. Yet people's desire to understand is so strong that they may turn to the simplistic ideas put forth by various pseudoscientists appearing to have all the answers. People will even ignore the fact that the pseudoscientist profits mightily from their faith in him or her.

DO THEY DO RESEARCH?

Pseudoscientists rarely perform experiments or make observations, as true scientists do. At the 1982 trial of the Arkansas legislation that mandated the teaching of creationism if evolution is taught, leading creationists admitted under oath that they did not do research and did not consider it important. On those rare occasions that they do make observations, pseudoscientists do it primarily for the purpose of finding more support for their ideas, and not with the expectation, much less the hope, that the idea might be shown to be wrong. Often the purpose of the experiments is to garner evidence that can be used for public propaganda. True scientists do not do experiments to affect public opinion; on the contrary, they are usually all too oblivious to public opinion.

Pseudoscientists do write books attempting to show that famous scientists themselves were skeptical about their ideas, and presenting this as proof that the science is flawed. This is in opposition to the scientific philosophy that the observations and experiments are what is important, not the opinion of the scientist who presents the data to the scientific community. It is surprisingly easy to distort or misinterpret lengthy scientific works by taking a brief phrase or sentence out of context.

Inquiry 2-11

Can you find a statement in this chapter that, if quoted out of context, would support a conclusion opposite to that intended by the authors?

Inquiry 2-12

There is only one way to be sure that a quotation has not been made out of context. What is it?

FOR WHOM DO THEY WRITE?

Scientists are professionally interested in advancing scientific knowledge. Although some scientists occasionally write books like this one that are intended for the general reader, scientists generally do not gain professional recognition for their popular writing. Their reputations are made or unmade on the basis of the research that they publish in professional journals especially dedicated to this purpose. Before an article is printed in such a journal, it is subjected to extensive critical reviews from other scientists working in that field. Frequently, the editor of the journal sends the paper to a reviewer who is known to oppose the ideas being put forth, just to increase the hurdles it must overcome.

The articles that scientists submit to such journals are usually highly technical and difficult even for scientists not working in that field to understand, much less a lay person. Indeed, in astronomy, someone working in a narrow field such as interstellar matter, for example, might well have difficulty fully comprehending an article on cosmology. For lay people, the problem is usually the extensive use of mathematics.

Pseudoscientists seldom publish their ideas in the professional journals. Indeed, they seldom even submit their articles to such journals, arguing that prejudiced scientists would reject their ideas anyway. Therefore, pseudoscientists normally write books and articles for popular consumption. The publisher is often the author, who is unable to find a reputable publisher of scientific materials to publish his or her ideas. They are very good propagandists, and their books are easy to read and understand. Superficially, their arguments may appear to be convincing, and it may be difficult to sort out truth from fiction. Bear in mind that authors of pseudoscience publications profit enormously from public belief. A skeptical and critical attitude, combined with reading both sides of an issue, is the best way to separate science from pseudoscience.

2.5 Do New Ideas Displace the Old Ones?

Scientists have good reason to be extremely wary of theories that claim to solve all the problems of science, because truly revolutionary ideas in science are few and far between, and even they build on what was known before. Such ideas are revolutionary in enabling us to understand new and unsuspected phenomena, or to understand familiar ones better, but not in contradicting what is already well established.

Newton's theories, for example, completely changed the way people looked at nature, yet Newton himself acknowledged the debt he owed to those who went before him. "If I have seen farther," he wrote, "it is by standing on the shoulders of giants." Newton's theories did not contradict those of Copernicus, Galileo, and Kepler, but perfected them, albeit in a way that they could not have anticipated. After Newton, the planets still moved around the Sun, in orbits that were nearly circles, just as Copernicus and Kepler had said. But observations of bodies such as the moons of Jupiter, discovered by Galileo, could now be better understood.

In our own century the two great revolutionary ideas in physics have been Einstein's theory of relativity and the theory of quantum mechanics. Yet even though they fundamentally changed the way we look at nature, neither altered by much our understanding of the many areas of knowledge where the earlier concepts of Newton were applicable. To this day astronomers use Newton's equations, with slight modifications at most, to calculate where the planets are. To give another example, although the Sun-centered model of the universe was ultimately overturned, the actual *data* were still valid and could be understood in the light of previously known physics once the presence of large amounts of obscuring dust in the

Galaxy was recognized. In the same way, it is probable that our present-day ideas about the Galaxy need revision, but any such revisions are almost certain to fit in with most of our facts and much of our theory.

CONCLUDING THOUGHTS

We readily admit that science is sometimes wrong, and that some scientists can be both pig-headed and arrogant. This book gives a number of examples from the history of astronomy; many others could be cited. Frequently, a mistake by science results when an important part of the scientific process is omitted. For example, it is critically important in science for investigators to publish their studies for the entire world to see. This exposes any piece of research to the largest possible critical audience. Secret or classified research done by a small group is much more liable to wander into error, partly because it lacks that exhaustive scrutiny.

An example of the problems that can result from secrecy in research is the 1989 controversy over cold fusion, which could replace oil and other sources of energy. As you will learn in later chapters when we study the life cycles of stars, there are reasons why scientists conclude that fusion can occur only at extraordinarily high temperatures.

Were it possible to produce controlled fusion at room temperatures, our energy problems would vanish. The idea of cold fusion is therefore an idea of practical importance, even if fanciful. In March 1989, the chemists B. Stanley Pons and Martin Fleischmann reported in a news conference that they had successfully produced fusion at room temperature. Skeptical scientists the world over attempted to duplicate their work based on the sketchy reports provided. Unfortunately, Pons and Fleischmann did not publish the details of their techniques and their work could not be verified. Researchers were not allowed into their laboratory because they had placed a cloak of secrecy over their work. Pons and Fleischmann had been well-known and respected chemists. However, they failed to play the game of science according to its rules; they tried to go around the knowledge filter shown in Figure 2-10. Although the controversy died down, it is not over. From this example we see that even the best of minds will make mistakes, and the best protection against mistakes is to have as many minds as possible thinking about something.

At the end of this chapter we suggest a number of books and articles on both sides of several pseudoscientific issues related to astronomy. We hope that you will read some of them.

Chapter Summary

- A *hypothesis* is a reasonable supposition made in describing the results of experiments and observations. It attempts to make predictions of future behavior. Once the evidence for its validity is strong, it becomes a *theory;* such validity means that many scientists have repeated and verified (confirmed) the experiments/observations. No theory can be proven to be true. However, data can prove a theory to be false.

- A fact is something that is accepted as real or true by the community using it, comes from definition, or results from a valid process of inference.

- Science progresses through the interaction of observation and experiment with theory. Theory makes predictions; observation and experiment place constraints of reality on the-

ory and thus provide a means of determining the best description of nature.

- In *deductive reasoning* scientists start with a hypothesis from which specific predictions are made.

- *Inductive reasoning* selects as the most likely hypothesis the one that is in best agreement with all of the available data.

- An *inference* is a conclusion reached by incorporating many lines of reasoning, including observation, experiment, and theory.

- A *model* is a description of a phenomenon, based on observation, experiment, and theory. It is not necessarily the truth or reality, but a description that allows prediction of future behavior.

- The characteristics of good science include the continuous search for new knowledge, repeatability and verifiability, and the ability to make predictions and test them against observations.
- **Pseudoscience** can generally be distinguished from true science by determining whether the ideas espoused can ever be at risk, whether they are open to modification and evolution into new ideas, whether they claim to supply simple answers to complex questions, whether practitioners play the role of an underdog, and whether they do research and publish it where trained scientists can examine the purported results.

Summary Questions

1. Distinguish between a scientific theory and a scientific hypothesis.
2. Distinguish between a fact and an inference.
3. Describe what scientists mean by a model and its relationship to *truth.*
4. Explain the difference between an observational and an experimental science. List and explain the particular difficulties that face astronomy and describe several tools that astronomy has over earthbound sciences.
5. Describe why the great distances in astronomy are both a disadvantage and an advantage to the study of astronomy.
6. For the hypothetical expedition to Earth of this chapter, formulate and evaluate hypotheses that might be made by the alien scientists about life on Earth. Specify the kinds of observations that might be made when attempting to validate the hypotheses.
7. Explain how the position of the alien scientists is similar to that of present-day astronomers (and also how it is different).
8. Compare and contrast the characteristics of pseudoscience presented in the chapter with those of science.
9. Explain how theory and observation interact in the development of scientific ideas.
10. Describe and explain the knowledge filter.

Applying Your Knowledge

1. State and analyze some of the conclusions an alien scientist might make about people on Earth through an analysis of their eating habits.
2. About which of the following sources of information might you tend to be more skeptical on questions of science? *(a)* Time or *Newsweek* magazine; *(b)* The National Enquirer; *(c)* Scientific American; *(d)* The New York Times.
3. Which of the following publishing houses might you expect to be the least reliable on questions of science? *(a)* Random House Inc. *(b)* Bantam Press *(c)* Creation-Life Publishers *(d)* Oxford University Press.
4. Which of the following true statements are facts, and which are inferred results? *(a)* The average density of material in the Earth's core is 11 times that of water. *(b)* Jupiter's average diameter is 11 times that of Earth. *(c)* All stars like the Sun get their energy from conversion of hydrogen to helium.
5. Critically analyze the following statement: "The first generation of human beings ever to have grown up educated about the true nature of the universe is still alive."
6. Discuss why science is not degraded when talking about it in terms of the *game of science,* as done in the chapter.

Answers to Inquiries

1. One could note that humans are self-propelled, while cars need a human before they can move; or that human characteristics show a smooth continuum of properties such as size (suggesting growth), while cars are restricted to a small and fixed range of properties.

2. They would have to examine carefully examples of extreme cases such as women without hair and men with long hair to see that it was a superficial property. If one of the photographs showed the inside of a barbershop, it would probably supply sufficient information.

3. It must be assumed that the data gathered fairly represent the entire group and are not biased.

4. *(a)* induction (begin with data and then develop a hypothesis) *(b)* deduction *(c)* deduction *(d)* deduction (Inquiries b, c, and d all start with a hypothesis and only later will predictions be made)

5. We assume that intrinsic properties do not depend on distance.

6. The more distant galaxies are seen as they were a long time ago (when they and the universe were younger). One conclusion might be that when galaxies are younger, their cores are more likely to explode than when they are older.

7. Ten billion ly is 10^{10} ly. Converted to kilometers, this is 10^{10} ly $\times 9.5 \times 10^{12}$ km / 1 ly, which rounds up to 10^{23} km. Written out, this is 100,000,000,000,000,000,000,000 km.

8. Newton's new ideas certainly were against conventional wisdom. Jealousy by other scientists might have hindered publication.

9. Some arguments might be: *(a)* if stars are on the inside of a sphere, they cannot be very far away; *(b)* if you look up, you should see the other side of Earth; *(c)* we see the Sun rise and set every 24 hours; *(d)* we can travel all the way around the Earth and measure its size.

10. Responses might include: *(a)* you cannot prove that stars are far away; *(b)* the great thickness of the atmosphere scatters light and makes it impossible to see the other land on the other side; *(c)* the rising and setting is caused by a body that revolves about the Sun every 24 hours, thus producing night.

11. One example might be "Earth is a hollow ball and we are living on the *inside* of it." The complete statement is in the second paragraph of the chapter-opening vignette.

12. Go to the original source and see the context in which it was made.

Further Readings

A good listing of sources relevant to various pseudoscience topics that involve astronomy, titled *Astronomical Pseudo-Science: A Skeptic's Resource List,* is available on the Web at **www.astrosociety.org/education/resources/pseudobib.html**. In the following listing, specific pseudoscience topics, writings by both proponents (Pro) and contradictory writers (Con) are included.

General

The Fringes of Reason, Ted Schultz (ed.). Harmony Books (1989). Short descriptions of many unusual beliefs, and resources to find out more about them.

The Art of Deception, by Nicholas Capaldi. Prometheus Books, 1971. Pseudoscientists frequently make use of dishonest rhetorical tricks to convince their audience. This is an excellent book to alert you to these tactics.

Betrayers of the Truth: Fraud and Deceit in the Halls of Science, by William Broad and Nicholas Wade. Simon and Schuster, 1982. Scientists are humans just like everyone else, and sometimes they are led to do things that are dishonest and contrary to the spirit of science. This book recounts a number of examples and helps to put the scientific enterprise in a more realistic perspective.

Science—Good, Bad and Bogus, by Martin Gardner. Prometheus Books, 1981. The one to read if you can only read one.

Myths of the Space Age, by Daniel Cohen. Dodd Mead, 1965. Another excellent summary of the arguments against many astronomically related pseudosciences.

Flim-Flam, by James Randi. Lippincott and Crowell, 1980. Has sections on UFOs and ancient astronauts.

Paranormal Borderlands of Science, Kendrick Frazier (ed.). Prometheus Books, 1981.

Science and the Paranormal, by G. Abell and B. Singer (eds.). Scribners, 1981.

Science and Unreason, by D. and M. Radner. Wadsworth, 1982.

Debunked! ESP, Telekinesis, and Other Pseudoscience, by Georges Charpak and Henri Broch, translated from the French by Bart K. Holland, Johns Hopkins Press, 2004. An examination by two physicists, on a Nobel Prize winner, of pseudoscientific claims.

Ancient Astronauts

Pro: *Chariots of the Gods?* by Erich von Daniken. Bantam, 1970. This is the book that started it all.

Con: *The Space Gods Revealed,* by Ronald Story. Harper and Row, 1976. Answers the claims made by von Daniken.

Con: *Ancient Astronauts, Cosmic Collisions, and Other Popular Theories about Man's Past,* by William Stiebing. Prometheus Books, 1984.

Velikovskyism

Pro: *Worlds in Collision,* by Immanuel Velikovsky. Doubleday, 1950. The classic example of pseudoscience.

Con: *Beyond Velikovsky: The History of a Public Controversy,* by Henry H. Bauer. University of Illinois Press, 1984. The definitive work on Velikovsky. Contains several excellent chapters on pseudoscience in general.

Con: *Scientists Confront Velikovsky,* by Donald Goldsmith (ed.). Cornell University Press, 1977. A collection of articles, mostly presented at a debate held before the American Association for the Advancement of Science in 1976, between Velikovsky and his critics. The article by Asimov is especially recommended.

Astrology

Pro: Virtually any bookstore has dozens to hundreds of books on astrology, often in the "occult" section, so specific recommendations seem unnecessary.

Con: *The Gemini Syndrome: Star Wars of the Oldest Kind,* by R. B. Culver and P. A. Ianna. Pachart, 1979. An excellent introduction to why astronomers discount astrology.

Con: *Arachne Rising,* by James Vogt. Granada, London, 1977. An apparent piece of pseudoscience arguing for a 13th constellation of the zodiac that turns out to be a devastating parody of pseudoscience.

Con: "The Scientific Case against Astrology," by Ivan Kelly. *Mercury* (January/February 1981), p. 13, and (November/December 1980), p. 135.

Creationism

History: *The Creationists,* by Ronald L. Numbers. Alfred A. Knopf, 1992. History of the modern creationist movement by a distinguished historian.

Pro: *Scientific Creationism,* by Henry Morris. Creation-Life Publishers, 1974. The definitive statement of what young-Earth Creationists believe.

Pro: *What Is Creation Science?* by Henry Morris and Gary Parker. Creation-Life Publishers, 1982. A more popularized account of Creationism.

Con: "Astronomy and Creationism," by David Morrison. The claims of Creationists are totally at variance with what modern astronomy has discovered. *Mercury* (September/October 1982), p. 144.

Con: "The Science-Textbook Controversies," by Dorothy Nelkin. The "equal time" demands by creationists are investigated. *Scientific American* (April 1976), p. 33.

Con: *Science on Trial: The Case for Evolution,* by Douglas J. Futuyma. Pantheon Books, 1983. Exposes the fallacies in creationist arguments. Has an excellent discussion of evolution.

Con: *Scientists Confront Creationism,* by Laurie R. Godfrey (ed.). Norton, 1983. A collection of articles by scientists in different disciplines that demonstrate the flaws in a number of creationist arguments. Extensive bibliographies.

Con: *The Meaning of Creation: Genesis and Modern Science,* by Conrad Hyers. John Knox Press, 1984. Written from the perspective of a theologian, this discusses the thesis that creationism is not only bad science but also flawed theologically.

Con: *Christianity and the Age of the Earth,* by Davis A. Young. Zondervan, 1982. Written by an Evangelical Christian who is also a geologist, this exposes the flaws in creationist estimates of the age of the Earth.

Con: *Science and Creationism,* by Ashley Montague (ed.). Oxford University Press, 1984. A valuable collection of essays on the evolution/creation controversy.

Con: *Science and Earth History: The Evolution/ Creation Controversy,* by Arthur N. Strahler. Promethean Press, 1987. A thorough and authoritative discussion of all creationist arguments.

No discussion of the evolution/creation controversy would be complete without mentioning Darwin's *The Origin of Species.* There is a continuing series of books by Stephen Jay Gould. Particularly good ones include *The Panda's Thumb* and *The Mismeasure of Man.* Like Darwin, Gould is often quoted out of context by creationists.

UFOs

Again, countless books exist at most bookstores to represent the "pro." The most cogent would probably be anything by J. Allen Hynek.

Con: *The Scientific Study of Unidentified Flying Objects,* by Daniel Gillmore (ed.). Bantam, 1969. The classic Condon Report issued at the termination of the military's extensive review of UFO evidence.

Con: *UFOs Explained* (Random House, 1974) and *UFOs: The Public Deceived* (Prometheus Books, 1983), both by Philip J. Klass. Klass has for years been the most informed and most aggressive debunker of "flying saucer" theories.

Con: *The World of Flying Saucers,* by Donald Menzel and Lyle Boyd. Doubleday, 1963. A perspective from the point of view of a professional astronomer.

Con: *The UFO Verdict: Examining the Evidence,* by Robert Sheaffer. Prometheus Books, 1981.

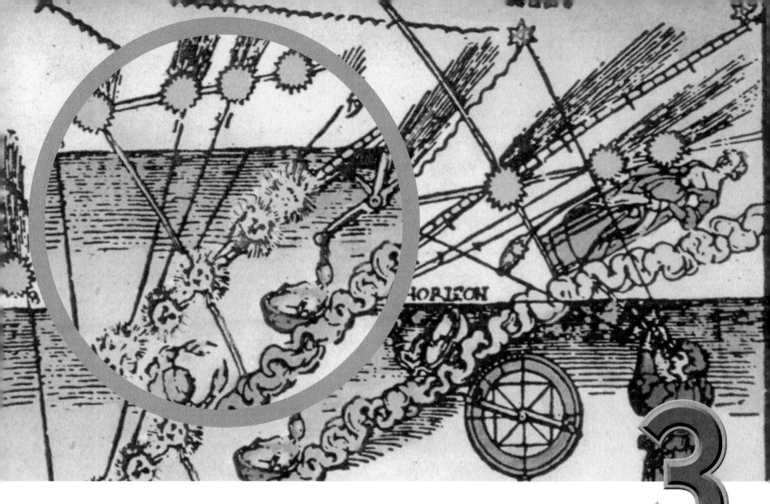

Astronomical Observations: Angles and Uncertainties

The comet was so horrible and frightful . . . that some . . . [people] died of fear and others fell sick. It appeared as a star of excessive length and the color of blood; at its summit was seen the figure of a bent arm holding a great sword in its hand, as if about to strike. On both sides . . . were seen a great number of axes, knives and spaces colored with blood, among which were a great number of hideous human faces with beards and bristling hair.

Ambrose Pare, Physician, 1528

The chapter opening quotation is a dramatic example of what we would describe today as unscientific thinking. The rapid progress made in the physical sciences over the last few hundred years stems largely from the evolution of a method of investigation in which the systematic and objective measurement of the phenomena of nature is the ultimate arbiter of the validity of our ideas. On *one* hand, speculations in the absence of careful observation can suggest an infinite number of possibilities, with no indication of which one, in fact, has been chosen by nature. On the other hand, data gathered in a random manner without the guiding inspiration of theory may well remain puzzling and may indeed be useless. It is only from the interplay between theory and careful measurement that new knowledge is attained.

Beginning with this chapter, this book contains a variety of activities called *Discoveries* at the end of some chapters. Working with these *Discoveries,* you can become actively involved in learning astronomy and have the opportunity of making simple but meaningful measurements. These *Discoveries* also give a deeper understanding of both the subject matter and the methods of astronomy. A guiding philosophy of this book is to make active learning experiences available wherever it is possible to do so. *Discoveries* are self-contained in the text and allow you the opportunity to do something with available data.

3.1 Angles and Angular Measurement

Astronomers describe the location of objects in the sky in terms of the angles between two objects or the angle between the object and the horizon, which is where the Earth blocks the sky. Early astronomers measured angles using simple instruments, such as the astrolabe and quadrant shown in the chapter opening illustration and **Figure 3-1**.

The basic unit of angular measure is the **degree**. There are 360 degrees (written 360°) in a full circle. A quarter of circle, 90°, is a right angle. **Figure 3-2** shows several different angles. Astronomers must frequently deal with angles considerably smaller than one degree, particularly when telescopic observations are considered. Traditionally, one degree has been subdivided into 60 **minutes of arc** (written

Figure 3-1   An old quadrant, used to measure altitudes of stars and planets.

Chapter opening illustration: An ancient astronomer measuring the changing positions of a comet.

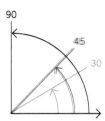

Figure 3-2 | The sizes of various angles.

60′), and each minute of arc has been further subdivided into 60 **seconds of arc** (written 60″).[1] Do not confuse minutes and seconds of *arc* (measures of angles) with minutes and seconds of *time,* which are also based on the sexagesimal system.

The angular unit to be used depends on the context. Large angles are best given in units of degrees. For example, degrees can be expressed as degrees and decimal fractions of a degree, and that is the practice we will follow in this book. Instead of writing 3°12′, we will write 3.2°, because 12′ is equal to 12/60, or 2/10 of a degree. The use of degrees and decimal fractions of a degree is also more compatible with hand-held calculators.[2] However, a small angle is best given in seconds of arc; writing 1″ rather than 2.7×10^{-4} degrees is generally more meaningful to the reader.

3.2 Angles on the Celestial Sphere

Models of the universe invented by the earliest civilizations visualized the sky and the bright spots visible on it as something like a large bowl surrounding the Earth, but at a considerable distance from it, with the bright lights of the heavenly bodies attached to the bowl. This "bowl" is known as the **celestial sphere** (**Figure 3-3**). It is not obvious from naked-eye observations of the sky how far away the stars are, but we can certainly map out

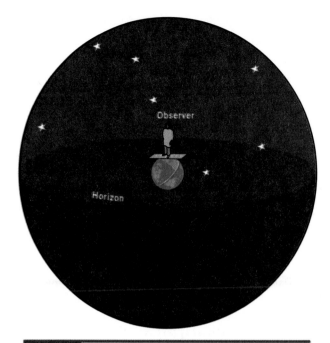

Figure 3-3 | The celestial sphere, in which the observer appears to be at the center of the universe.

Figure 3-4 | The northern sky and horizon, showing the Big Dipper, Cassiopeia, and the North Star, or Polaris.

the relative positions of various stars and constellations by specifying the angles separating them as they appear on the celestial sphere.

For example, most people in the Northern Hemisphere have at some time noticed the grouping of stars called the Big Dipper (in Great Britain, the Plough). **Figure 3–4** shows the Big Dipper's stars, with their Arabic names[3] indicated, as they would appear above the northern horizon just after

---

[1]The traditional method of designating angles in degrees, minutes, and seconds of arc can be traced back to the Mesopotamians, who used the sexagesimal notation based on units of 60.

[2]There is, of course, nothing sacred about the decimal, or base ten, number system, since it probably came about simply from a habit of counting on the fingers. There is a great deal of linguistic evidence for this idea; for example, many primitive tongues use the same words for fingers of the hand and the numbers one through five. The Mayans counted both on fingers and toes and had a number system based on units of 20. Modern computers are most easily designed to work on the binary system, where numbers are expressed in units of 2.

[3]The naming of stars is explained in Chapter 4.

dark in mid-October. Polaris, which is called the North Star or pole star, lies close to the **north celestial pole**, which is the point on the celestial sphere where the Earth's rotation axis intersects it. The stars Merak and Alkaid are separated by an angle of 24°. This means that the angle between the two lines of sight from the eye to the two stars is 24°, as shown in **Figure 3-5**.

The stars Merak and Dubhe are 5° apart. These two stars are frequently referred to as the Pointer Stars, because a line drawn through them and extended another 30° will pass near the North Star (Figure 3-4).

Objects such as the Sun and Moon, which show a visible disk as opposed to a point of light, can be characterized by their **angular size**, also called **apparent size**. The angular size of an object is the angle between two lines of sight from the eye, one to each side of the object. The angular size of the Moon is 0.5°, as shown in **Figure 3-6**.

You can measure angles very roughly using the fist method; that is, using your hand and fingers held at arm's length as a rough gauge of angles. At arm's length a thumbnail has an angular size of about 1° for most people. The first two knuckles of the fist are about 2° apart, whereas the fist itself spans about 10°. If the hand is spread out as far as it will go, the distance from the tip of the thumb to the end of the little finger is approximately 20° (**Figure 3-7**).

Inquiry 3-1

Assume that a child's hand and arm, being smaller than an adult's, are still in the same proportion to body size. Are the angles given above still correct for the child? Explain.

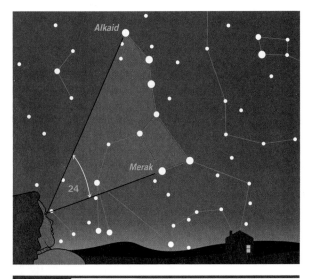

Figure 3-5 Angle between Merak and Alkaid, showing a 24° angular separation.

Figure 3-6 The angular size (apparent size) of the Moon is ½ degree.

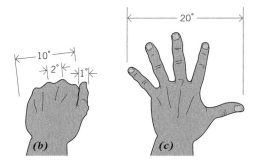

Figure 3-7 Using the human hand to measure angles roughly. *(a)* Extend the hand at arm's length and sight along it. *(b)* The thumb measures about a degree, two knuckles about 2°, and the fist about 10°. *(c)* The hand measures about 20°.

3.3 The Relationship between True and Apparent Size

Figure 3-8a shows an observer measuring the angular size of a person located a short distance away. **Figure 3-8b** shows a similar situation with the same person located at a greater distance. This figure illustrates that the observed angle decreases as the distance increases. In fact, when the apparent size of the object is expressed in degrees and the physical size[4] and distance of the object are expressed in the same units (e.g., meters), then the relationship can be expressed in this formula:

$$\text{Angular size (degrees)} = 57.3° \times \frac{\text{Physical size}}{\text{Distance}}.$$

This important relationship, called the **angular size formula**, is derived in Appendix A5. The formula gives us the angular size of the object expressed in degrees if the true size and distance of the object are expressed in the same units (e.g., meters). This relationship is *not* valid for large apparent sizes, but for angles less than 20° or so it will give satisfactory results.

Astronomers are mostly concerned with small angles, usually a fraction of degree. Small angles occur when the object size is small compared to the distance. For this reason, a better unit is the minute or the second of arc. If we express the angle in seconds of arc, the formula looks like this:

$$\text{Angular size (seconds of arc)} = 2.1 \times 10^5 \times \frac{\text{Physical size}}{\text{Distance}}.$$

As an example of the use of the angular size formula, let us calculate the angular size of the Moon, given that its physical size is 3,475 km in

---

[4]By the term *physical size* we mean the actual physical size measured in, for example, millimeters, centimeters, or kilometers.

diameter and its average distance from Earth is 385,000 km.

$$\begin{aligned}
\text{Angular size of Moon} &= 57.3° \times \frac{3,475 \text{ km}}{385,000 \text{ km}} \\
&= 0.52° \\
&= 31.2 \text{ minutes of arc} \\
&= 1,872 \text{ seconds of arc.}
\end{aligned}$$

Inquiry 3-2

If the Sun's physical size is 1.39 million km in diameter and its distance from Earth is 149.6 million km, what is its angular size (angular diameter) in degrees? How does it compare to the angular diameter of the Moon?

SOLAR ECLIPSES

A total **eclipse** of the Sun will take place if the Moon happens to pass directly between the Earth and Sun at a time when the Moon's angular size is equal to or greater than that of the Sun. At such a time, the Moon blocks the bright disk of the Sun (**Figure 3-9**), and the faint, tenuous outer atmosphere of the Sun, called its **corona**, can be seen. The corona is only one-millionth as bright as the Sun's disk and therefore is usually lost in the glare of the Sun. It is best seen at total eclipse time, and for this reason astronomers often go on long and arduous eclipse expeditions to out-of-the-way parts of the globe to study the corona.

Viewed from space (**Figures 3-10 and 3-11**), the Moon casts a small moving shadow spot onto the surface of the Earth. Only if you are within this shadow do you see the total eclipse. Even then, you do not see it for long, because the shadow spot sweeps rapidly past you. Because the angular sizes of the Sun and Moon are so close, the longest period when the eclipse is total at any one location on Earth is slightly over seven minutes. A more detailed discussion of eclipses occurs in Chapter 4.

(a)　　　　*(b)*

Figure 3-8　The farther the object, the smaller the observed angular size.

Figure 3-9   Multiple exposures during a total eclipse of the Sun in July 1991 as observed on Mauna Kea, Hawaii. The solar corona shows itself at mid-totality.

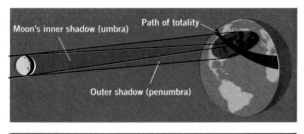

Figure 3-10   The shadow of the Moon on the Earth during a total eclipse of the Sun (not to scale).

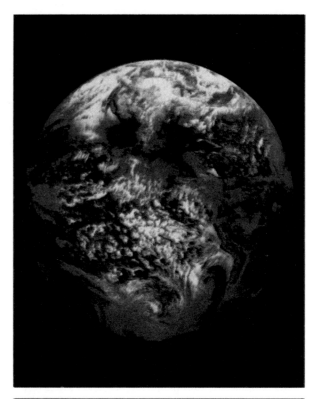

Figure 3-11   The Moon's shadow passing over the Earth during the solar eclipse of July 11, 1991. This photograph is a composite of 4 exposures taken by the GEOS-7 satellite.

3.4 Measurement Uncertainty in Science

Dealing with uncertainties in measurements is one of the most important and critical things scientists do. People untrained in the scientific method often tend to imagine that the results of scientific measurements are necessarily exact and without uncertainty. Nothing could be further from the truth. Any measurement has associated with it a certain amount of uncertainty, no matter how sophisticated or expensive the measuring device. Although it may be true that more elaborate instruments have smaller uncertainties than simple instruments, uncertainties are still present and still need to be considered in detail, especially when the measurement process is being pushed to its limit to extract as much information as possible from the data.

The measurement process is fundamental to the scientific method as practiced today. Without data, there is no way to approach a correct understanding of nature, simply because without observational and experimental data one cannot distinguish among the many hypotheses that exist.

Historically, however, observation was not always so important. The ancient Greeks assumed that they could simply ponder how the universe must be if it were to be a logical place (to their minds). However, they seriously underestimated the universe and the number of possibilities it presents to us.

Is what we observe and measure reality? The ancient Greeks made a strong distinction between appearances (or *phenomena*) and reality (what they called *noumena*). They feared that our senses might merely reveal the superficial appearance of the world while missing underlying fundamentals. Hence, the school of thought led by the mathematician Pythagoras adopted a purely logical approach without observation, because it would not be subject to such errors. It was not apparent at the time that this approach was subject to other, more serious errors. In the process of evolving the most productive procedures for scientific inquiry, science ended up dropping the distinction between appearance and reality as unproductive. Today, we assume that what we measure is real and we move on from there.

Measurements always contain uncertainties. Scientists generally refer to uncertainties as *errors*. In this sense, *error* does not mean there is something wrong—a mistake—but simply that the measurement or result has some degree of uncertainty to it. Throughout this book, we will use the term uncertainty rather than error. If a news article or report talks about scientific errors, you need to be aware that they may be referring to uncertainties rather than mistakes.

PRECISION AND ACCURACY

When discussing uncertainties in science, scientists make a distinction between **precision** and **accuracy**. By *precise* we mean internal consistency within a set of measurements. In other words, each measurement is close to all the others. By *accurate* we mean that the average value of a set of measurements is close to the correct value. **Figure 3-12** illustrates the meaning of these words using holes in a target as an example. A tight grouping, away from the center, is precise but not accurate. A wide group centered on the bull's-eye is accurate but not precise.

Scientists have to worry about two types of uncertainty: **random uncertainty** and **systematic uncertainty**. We expand on these ideas in the next two subsections.

RANDOM UNCERTAINTIES

We illustrate these different uncertainty types through a specific measurement activity—the measurement with a ruler of the length of the room in which you sleep.

It is probably evident to you that when you take several readings with a ruler, you rarely if ever get exactly the same answer. You lay the ruler down

Figure 3-12 Accuracy and precision.

differently each time, you read it differently, you line it up differently. When you take any measurement repeatedly, there is no way you can do everything in exactly the same way you did it before. These uncertainties are essentially random; that is, some measurements will be larger than the true value and some will be smaller.

But if we get a different number every time we measure something, what should we do to obtain the most accurate value possible? A simple method is to **average** (or take the mean of) all the different measurements. The mathematical process of taking the average tends to make the values that are too large cancel out the values that are too small.

Inquiry 3-3

Suppose you measured the length of your bedroom five times and got the following values: *(a)* 4.4 meters *(b)* 4.3 meters, *(c)* 4.5 meters, *(d)* 4.6 meters and *(e)* 4.2 meters. What would be the best value you could give for the length of your room?

Appendix A6 describes a method by which the random uncertainty of a set of measurements can be found.

SYSTEMATIC UNCERTAINTY

It would be nice if we also had a simple method of analyzing systematic uncertainties, but unfortunately, such is not the case. A *systematic* uncertainty is one that is always in the same direction, as opposed to the random uncertainty's tendency to be sometimes larger and sometimes smaller than the true value. To illustrate a systematic uncertainty in the context of the example just given, suppose you measured the length of your room with a meterstick but thought you were using a yardstick (because a meterstick is 39.37 inches in length; they are not much different). You might say the room is four yards across, when in reality it would be four meters, almost a foot longer than you said. Every measurement you made with the meterstick (thinking it was a yardstick) would be systematically off. Another example of a systematic uncertainty would be if the first centimeter were missing from a meterstick because of a manufacturing error.

Inquiry 3-4

Show that your measurement would be systematically off by calculating how many inches across the room is if you measure four yards when in fact you had measured four meters.

Inquiry 3-5

What kind of uncertainties, random or systematic, are shown in each part of Figure 3-12? In the diagram showing precision without accuracy, are the random and systematic uncertainties of equal importance?

Inquiry 3-6

What are the random and systematic uncertainties for a set of measurements of boiling water, whose temperature should be 100° C? The measurements are: 102.4C, 102.6C, 102.2C, 102.7C and 102.2C. An explanation of how to find random uncertainties is given in Appendix A6.

Inquiry 3-7

Determine the importance of the uncertainty by comparing the size of the uncertainty to the average value. Make this comparison by dividing the uncertainty by the average value and then multiplying by 100 to get the percentage uncertainty.

The example in Inquiry 3-4 illustrates one reason why systematic uncertainties are dangerous and tricky; they frequently result from problems you do not know exist. Furthermore, there is no standard process by which you can discover with certainty that these problems are present. Indeed, a systematic uncertainty may sometimes be as subtle as an incorrect assumption, or a failure to consider all the possibilities when a problem is reasoned out. Hidden systematic uncertainties have been responsible for many of the great disasters of the history of science.

Sometimes a systematic uncertainty is clearly present and a correction can be made. For example, suppose you were to make a temperature measurement of a water ice solution, whose temperature should be 0° C. If your thermometer gave you an average reading of, say, 10° C, you then know that all your temperature readings with this thermometer will be systematically 10 degrees too high. You could correct for the systematic uncertainty by subtracting 10 degrees from each measurement. Without that correction, the measurement would be inaccurate; with the correction for the systematic uncertainty, it becomes more accurate, even though it will still have random uncertainties.

Figure 3-13 summarizes our discussion of random and systematic uncertainties. The figure shows two sets of measurements, labeled A and B. The solid line is a graph of the angular size formula, which represents how the angular size of an object changes with distance. We notice that while dataset A scatters around the theoretical angular size formula, a smooth curve drawn through those points would conform well to the theoretical curve. Thus, dataset A is precise and accurate. For dataset B, however, all the observations are well above the angular size formula curve, which indicates the presence of a systematic uncertainty; the data are not accurate. However, since the points would fit the shape of the angular size curve well (the dashed line), they are precise.

> Since understanding measurements is crucial to understanding the methods of science, we encourage you to do Discovery 3-1, Measuring Your Room, at this time.

3.5 Systematic Uncertainties in Astronomy: The Kapteyn Universe as an Example

In the years before World War I, the principal method that astronomers used for mapping the region of space in which we live was to count the number of stars of various brightnesses that could be seen in different directions in space. To make the principle behind this method clear, assume for

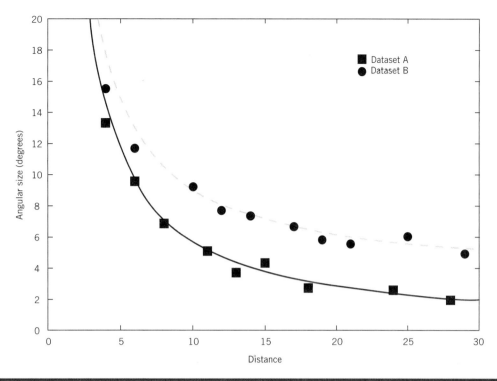

Figure 3-13 Uncertainties in data and a comparison with theory. Dataset A is both accurate and precise, while Dataset B is precise but inaccurate. The solid red line is the correct theoretical relationship (the angular-size formula).

a moment that all stars have the same intrinsic brightness. Under this assumption, variations in the *apparent* brightness of stars would be due entirely to their differing distances from Earth, with the nearer stars appearing brighter and the more distant ones fainter (**Figure 3-14** illustrates an analogy with approaching cars). In counting fainter and fainter stars, we would be observing stars at greater and greater distances. Once Galileo in 1609 used a small telescope to demonstrate that the visible Milky Way was, in fact, myriads of distant, point-like stars, star counts could be used to study the apparent distribution of stars in space.

Star counts were first carried out in a systematic and detailed way by the English astronomer William Herschel in the eighteenth century. Herschel found that beyond a certain brightness the number of stars began to dwindle rapidly, and he interpreted the decline as evidence that there were no stars at greater distances. To put it another way, it was presumed that he had counted out to the edge of our stellar system.

The culmination of this line of research was contained in a paper by the Dutch astronomer Jacobus Kapteyn in 1922. He analyzed decades of laborious research and concluded that the system of stars in which we live was shaped somewhat

Figure 3-14 When driving at night, we automatically assume that fainter headlights come from more distant cars rather than from fainter headlights on nearby cars. Similarly, if all stars were equally luminous, their brightnesses would depend on their distance from us.

like a large, fat pancake, as shown in **Figure 3-15**. The long dimension was about 10,000 ly and the short dimension only about one-fifth of that. The Sun was located almost in the center of this system, a preferential position that made some astronomers extremely uneasy with this picture. Most astronomers had to admit, however, that Kapteyn had done careful work, and they found it quite convincing. For example, Kapteyn did not assume that all stars had the same energy output. He used the knowledge that was developing at that time about the differences that exist between stars of different types.

Kapteyn's figures were taken to be not just the dimensions of our own stellar system but those of the entire known universe, because it was not known at that time that there are other systems like our own. So in 1922 the entire universe was considered to be about 10,000 ly in diameter, with the Sun at the center.

The history of astronomy has generally not been kind to chauvinistic concepts such as this one, and sure enough, within five years, developments in astronomical research made astronomers realize that our galaxy was actually 10 times larger in

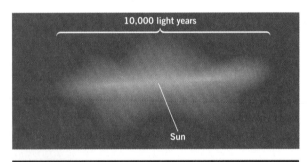

Figure 3-15 The Kapteyn Universe, from about 1920, pictured the Sun near the center.

size, and the universe 100,000 times larger in extent, than had been thought. Measured by its volume, the universe was actually a factor of 10^{15} times larger than previously thought. Today, the most distant object known is thought to be more than 12 *billion* light-years away. Furthermore, the Sun turned out to be far from the center.

How could astronomers have been so wrong about the size of the universe? Why did so many elaborate and carefully made star counts give such incredibly misleading results? **Figure 3-16**, a mosaic photograph of the Milky Way, shows what you would observe on a very clear night in a dark location away from lights. (*Note:* This flat photomosaic covers a large section of the sky and would wrap almost completely around a sphere representing the sky.)

Looking at the Milky Way, we see a band of faint light encircling the sky. This light comes from an enormous number of faint stars at great distances, as was first shown by Galileo. But if you examine Figure 3-16 closely, you will see numerous dark regions, called rifts, crossing the Milky Way. These rifts cannot be places where there are simply no stars, because such regions would have to be enormously long, skinny tunnels that cross through our stellar system, all pointing toward the Sun, as illustrated in **Figure 3-17**. If the tunnels were not pointing toward the Sun (direction C), no rift would be seen. Furthermore, even if such tunnels were to exist, they could not persist for long, because the stars are constantly in motion in relation to each other. The explanation of these

Figure 3-16 Mosaic photograph of the Milky Way.

rifts is that between the stars are clouds of dust particles—small, solid particles of matter floating in space between the stars—that block light from the more distant stars in certain directions just as heavy clouds in Earth's atmosphere block the light of the Sun.

Astronomers were aware of this dust in the galaxy, but they had not discovered a way to tell how much dust there was. The dust will make stars seen through it appear fainter than they really are, but how much fainter? A factor of two? A factor of 10 million? Kapteyn was aware of the problems that could be caused if absorbing matter were in space, so he did some research to test for the possibility of dust. Along with many other astronomers at that time, he concluded (incorrectly, it turned out) that there were indeed some regions of the sky where heavy dust obscuration was encountered, but

that most lines of sight outward from the Earth passed through a space that was almost totally transparent.

The point of the Kapteyn story is to provide a real example to demonstrate that systematic uncertainties—in this case, the presence of significant amounts of unknown interstellar dust that hid so many stars from our optical telescopes—can lead one to wrong conclusions. **Figure 3-18** diagrams a galaxy we think is similar to our galaxy, with the position of the Kapteyn "universe" indicated in it. Because of a serious systematic uncertainty, for more than a century astronomers were under the impression that what they could easily see in the solar neighborhood was actually the entire universe.

Astronomers today feel more confident about the picture they have developed describing the size and evolution of our physical universe, but the possibility of serious systematic or conceptual errors always remains. To judge for yourself the confidence you will want to place in our current models, you need an understanding not only of the models themselves, but also of the *process* by which they were constructed. Astronomy developed so rapidly in the latter half of the twentieth century that any specific set of facts may well be obsolete in a decade. Our goal in this book, therefore, is not to encourage you to learn dry facts but instead to help you understand how the interaction between theory and observation leads, in a gradual way, to models of the physical universe. Of course, you must learn *some* facts in order to have a context in which to discuss theory and the changes brought about by new observations.

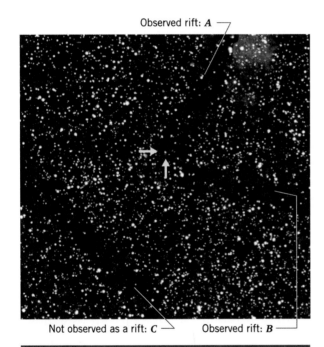

Observed rift: *A*

Not observed as a rift: *C*   Observed rift: *B*

Figure 3-17 For the observed rifts to be produced by regions having no stars, the places without stars would have to be long, skinny tunnels (directions A and B) pointed toward the Sun (marked by arrows). For a tunnel in the direction C not pointing at the Sun, a large number of stars would still be seen.

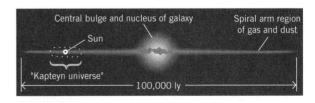

Figure 3-18 The modern picture of the galaxy compared to the universe as proposed by Kapteyn.

DISCOVERY 3-1

Measuring Your Room

After completing this Discovery, you will be able to

- Take a set of measurements, find their average, and estimate their random uncertainty.

Obtain a ruler; you may use either a meterstick, yardstick, or foot ruler. Make five repeated measurements of the length of your room. Tabulate your data. Find the average and estimate the random uncertainty by finding the range in your values (the largest minus the smallest) and dividing by two. Write your results as an average plus or minus your uncertainty.

- **Discovery Inquiry 3-1** What possible sources of systematic uncertainties might enter into your measurements?

Chapter Summary

To emphasize the interrelationships between observation and theory, and the importance of uncertainties, beginning with this chapter we organize summaries into three parts as applicable: (1) observations, (2) theory, and (3) conclusions.

Observations

- The **apparent size** of an object is expressed as an angle, which is determined by the object's physical size and distance from the observer. The apparent size is also called the **angular size**.

- Stars appear to be located on the **celestial sphere**. The **north celestial pole** is a point on the celestial sphere where the Earth's rotation axis and the celestial sphere intersect. Any two objects on the celestial sphere are separated from one another by angles, which are measured in **degrees**, **minutes of arc**, and **seconds of arc**.

- Total **solar eclipses** in which the Moon exactly covers the disk of the Sun, occur mainly because at the time the angular size of the Sun and Moon are equal. Solar eclipses can also occur when the Moon's angular size is slightly larger and covers more of the Sun.

Theory

- The **angular size formula** is a theoretical relationship between the apparent size of an object, the object's physical size and the distance from the observer.

- All scientific measurements have one or both types of uncertainty: **random** uncertainty and **systematic** uncertainty. Random uncertainties are handled by taking the **average** of the measurements. Systematic uncertainties are difficult to detect, and scientists must take great care to avoid them. Scientists make a careful distinction between the concepts of **precision** and **accuracy**.

Summary Questions

1. What do scientists mean by random and systematic uncertainty? Give an example of each.

2. Explain how systematic uncertainties affected Kapteyn's estimates of the size of the universe.

Applying Your Knowledge

1. Discuss the uncertainties involved in comparing your car's odometer readings against the one-mile markers on a highway. Assume the comparison continues for 20 miles. (Please do not do this while you are driving!)

2. Suppose you made a number of precise but inaccurate angular measurements on the sky with either your fist or a protractor. What might you do to improve the accuracy?

3. Think about and discuss the possible uncertainties present the last time you made a measurement of any type. (Such a measurement might include weighing yourself, measuring an ingredient when cooking, filling your car with fuel, etc.)

▶ 4. Suppose the dark rifts in the Milky Way were long, skinny tunnels without stars. Assume a typical tunnel has a 10 ly diameter, and that a typical star moves with a speed of 10 km/sec. How long should it take for stars to fill the tunnel? (Remember, the distance something moves is given by its speed times the length of time it travels.)

▶ 5. What would be the apparent size of a quarter if it were at a distance of 3,000 miles (i.e., across the length of the United States)? (Such an angular size could easily be seen by a modern radio telescope.)

▶ 6. What would be the width of the "canals" on Mars if they were to have an angular size of 0.25 seconds of arc when Mars is at its closest approach to the Earth (roughly 50 million miles)?

▶ 7. What would be the angular size of the Earth for an observer on the Moon?

▶ 8. If the Sun were at the distance of the nearest star—4.3 ly—what would be its angular size? Convert your answer (which the formula gives as a decimal fraction of a degree) into a decimal fraction of a second of arc.

▶ 9. Suppose a student makes a set of measurements of the angular size of a 100-meter-tall building from a distance of 1 km (1,000 m). The measured angles are 10.5°, 11.0°, 9.8°, 10.1°, and 9.5°. Determine both the random and systematic uncertainties. *Hint:* To compute the systematic uncertainty, you need to use the angular size formula to determine what the angular size of a 100 m high object observed from 1,000 m should be.

Answers to Inquiries

3-1. Yes. Since angular size measurements are ratios, the proportion is the same, and so is the result. For example, if a 6-foot-tall adult had a 3-foot arm length and a 6-inch hand size, then a 2-foot-tall child would have a 1-foot arm length and a 2-inch hand size.

3-2. $0.53°$; $0.53 \times 60 = 31.8'$; $31.8 \times 60 = 1908''$

3-3. The average value of 4.4 m is found by summing the five readings and dividing by the number of measurements $(4.4 + 4.3 + 4.5 + 4.6 + 4.2)/5 = 4.4$.

3-4. 4 yards $\times$ 36 in./yard = 144 in.
4 m $\times$ 39.37 in./m = 157 in

3-5. *(a)* random and systematic; *(b)* random; *(c)* random. You always have random uncertainties. In this example, random uncertainties are less than systematic uncertainties.

3-6. The average value, found by adding the numbers and dividing by 5, is $102.4°C$. To estimate the random uncertainty, you determine range of measurements: the maximum value measured is 100.7 while the minimum is 102, for a range of $0.5°C$. The Snedecor rough check (Appendix A6) estimates the random uncertainty as half the range, or $0.5/2 = 0.25°C$. Thus, the measured value is written as 102.4 ± 0.3. There appears to be a systematic uncertainty in the measurement because the average value (102.4) is $2.4°$ away from the value it should have, and the random uncertainty is small compared to that difference.

3-7. The random uncertainty, compared with the average value, is $0.3/102.4 \sim 0.003$, or 0.3%. This is a small uncertainty and thus the random uncertainties are small and not too significant. The systematic uncertainty is about 2%. However, that can be corrected for by subtracting the $102.4 - 100 = 2.4$ from each measurement.

Basic Observations and Interpretations of the Sky

The objects which astronomy discloses afford subjects of sublime contemplation, and tend to elevate the soul above vicious passions and grovelling pursuits.

Thomas Dick, Geography of the Heavens, 1843

The night sky is incredible when viewed from a dark location. Until recent times, most people lived in such locations, and they followed the heavens throughout the year and throughout their lives. In this chapter, we will examine some of the observations made by ancient peoples (observations you can make if you desire) and the conclusions they drew about the heavens. With the chapter appendix, you will have the opportunity to learn how to use seasonal star maps to help you learn the locations of constellations and bright stars. We take an inductive approach in which we look at fundamental observations in this chapter. Then, in the next chapter, we build a model that best explains the observations while also making testable predictions.

4.1 Early Observations of the Sky

Three to four thousand years ago, the ancient Chinese, Egyptians, and Mesopotamians were systematically studying the skies and keeping records of their observations. In later centuries, similar activities were carried out on the Indian subcontinent, and also by the great cultures of the New World that flourished in what is now Mexico and most of Central America and in the South American regions we now call Bolivia and Peru. Thanks to their systematic observations, carefully made over a long time, ancient priests were able to detect many regularities in the motions of the Moon, planets, Sun, and other stars. For example, from sunrise to sunrise is one day; the interval of time for the Moon to return to appear to be fully lit is one month, and so on. The priests also became quite skilled in predicting certain astronomical events, such as eclipses. Even today, some of the ancient eclipse observations are useful in helping astronomers determine the rate at which the Earth's rotation is slowing down (Chapter 8).

The famous Aztec calendar stone shown in **Figure 4-1** is not actually a calendar. The stone is a complex visual representation of a cyclical cosmological model showing the Aztec belief that the Earth and the cosmos undergo alternating periods of creation and destruction. What it shows is that careful sky observations played an important role in this civilization.

The monthly repetition of the phases of the Moon presents the most obvious periodic event in

Figure 4-1 An Aztec calendar stone. The stone is 4 meters in diameter and weighs 24 tons.

the sky after the daily rising and setting of the Sun. Caused by the changing position of the Moon while it orbits the Earth, the lunar phases provide a clear basis upon which to base a calendar and to make predictions. Lunar calendars are used to this day by many of the world's major religious groups. The ancient Babylonian calendar, based on the phases of the Moon, began each month at sundown on the day the crescent Moon was first visible in the west. Because the time from one new Moon to the next is about 29.53 days, the length of the Babylonian months generally alternated between 29 and 30 days. The year of 365.24 days contains an average of 12.37 lunar months, so their years had sometimes 12 and sometimes 13 months. The ancient Hebrews brought the

Chapter opening photo: Star trails around the north celestial pole as seen from Arches National Monument.

56 Part 1 • Discovering the Science of Astronomy

Figure 4-2 | The Sun rising over the Heelstone at Stonehenge on the first day of summer.

Figure 4-3 | Many believe that *Medicine Wheel*, a circular rock formation in *Wyoming's Bighorn Mountains*, is a solar calendar built by the Plains Indians to track the sun throughout the year.

Babylonian calendar back with them after their captivity, and it forms the basis of the present-day calendar of Judaism. Christians use a modified version of this calendar in setting the date of Easter, which was originally associated with Passover. The Islamic calendar is also a lunar calendar, and the holy month of Ramadan is linked to phases of the Moon, which is why it occurs on different dates of our Western calendar each year. The Buddhist New Year, which is celebrated all over the Far East, is also set by the Moon.

Our calendar, by contrast, is based on the apparent motion of the Sun around the sky. One year is the length of time for one complete cycle of the Sun, which is caused by the Earth's annual motion about the Sun. The ancient Egyptian calendar was a solar calendar with 12 months of 30 days each, with a few extra days added at the end of the year. It was important for the Egyptian priests to keep an accurate calendar in order to predict the annual flooding of the Nile, which brought fertility to the land. The Egyptian calendar was a precursor of the conceptually similar Julian calendar, which was ordained by Julius Caesar, and of its successor, the modified Gregorian calendar, which we use today.

At about the same time the pyramids were being built in Egypt, amazing circular constructions were being built in England and in the French province of Brittany. At Stonehenge in southern England, many of the lines of sight from one stone to another point to the places on the horizon where the Sun and Moon rise and set. The principal alignment, shown in **Figure 4-2**, runs from the center of the monument to the northeast at which is a stone known as the Heelstone. On the first day of summer each year, the Sun rises directly over the Heelstone when observed from

this central vantage point. The people who built Stonehenge clearly had considerable astronomical knowledge.

One of the most intriguing constructions in North America is the Big Horn Medicine Wheel in Wyoming (**Figure 4-3**), which appears to have numerous celestial alignments based on early Native American observations of the sky. Not only does it seem to indicate the point on the horizon where the Sun rises on the first day of summer, it also shows alignments to the horizon points where the bright stars Aldebaran, Sirius, and Rigel rise just before the Sun at one-month intervals during the summer. Like that of Stonehenge, the astronomical interpretation of the Big Horn Medicine Wheel is debated.

Other North American Indian groups made astronomical observations as well. A thousand years ago the Anasazi built elaborate cities of stones and mud mortar that incorporated significant solar alignments in their construction. Some well-known Anasazi sites are Chaco Canyon, Mesa Verde, Hovenweep, and Chimney Rock, all located in the Four Corners area of Utah, Colorado, New Mexico, and Arizona. We also find astronomical alignments in the Cahokia mounds just east of St. Louis, and in the Council Circles in Central Kansas.

Studies by one of this book's original authors, R. Robert Robbins, of the art on the burial bowls of the Mimbres Indians, located in southwestern New Mexico, revealed that these Indians had a sophisticated appreciation of the Moon's motion, as well as of eclipses. One of the bowls studied

Figure 4-4 A burial bowl of the Mimbres Indians, showing an image of the supernova explosion in 1054 C.E. that produced what we now call the Crab Nebula.

Figure 4-5 This instrument, preserved at Purple Mountain Observatory, Nanjing, China, operates on the same principle as the modern equatorial telescope mount.

(**Figure 4-4**) shows a representation of the supernova explosion of 1054 C.E. that produced the object modern astronomers call the Crab Nebula (Figure 1-13). Astronomers had found many examples of rock art suggesting that the Indians of the Southwest observed this cataclysmic event. But because dating rock art without harming it was impossible, these findings had remained speculative. In the case of the Mimbres bowl, it was buried and untouched until its discovery, thus protecting it from vandals and also making it possible to use tree-ring dating on the timbers of the room where the bowl was found. Thus, the archaeological dating confirms the astronomical content.

By far the most advanced of the New World astronomers, however, were the Maya of Southern Mexico, Central America, and, later, the Yucatan. Tragically, Spanish invaders intent on wiping out the Maya religion destroyed nearly all of their records. The full extent of their astronomical sophistication may never be known, but it is clear that at the peak of their civilization, the Maya were second to none in their understanding of the cosmos. Their cosmology was particularly concerned with cycles of time. They tracked celestial events on a number of different calendars, including a 365-day solar calendar, a 260-day sacred calendar used to schedule rituals (whose origin is not known), and a sequential count of days elapsed "since the creation of the universe" that kept time in multiples of 360. This *Long Count* was extended by Mayan astronomers in their calcula-

tions forward and backward in time for millions of years.

Current research continually increases our awareness of the scope and sophistication of early Asian astronomy. Chinese observations can be traced without difficulty to 1300 B.C.E. The Chinese, Japanese, and Koreans recorded eclipses, comets, and *guest stars,* stars that flared into brilliance and then slowly faded away. The Chinese, as the Mimbres did, recorded a guest star in 1054 C.E. at the same location as the Crab Nebula (Figure 1-13).

The Chinese invented a system for locating the stars in the sky that is similar to the latitude and longitude we use for locating places on the Earth. Introduced into Europe in the Middle Ages, it gained wide acceptance and is the system used by modern astronomers. The Chinese also invented the principle of the clock-driven equatorial telescope mounting, hundreds of years before it was used in the West. This type of mounting makes it possible to follow stars automatically as they move across the sky (**Figure 4-5**).

In observing the sky, most civilizations saw distinctive patterns of stars in the sky that reminded them of their cultural heritage or their everyday life. Such groupings of stars are called **constellations**. The constellations we refer to today are often based on Greek mythology, but different civilizations had their own constellations. Today, the sky is divided into 88 constellations that define areas of the sky; each has a well-defined border.

4.2 Early Greek Observations: The Round Earth

Greek civilization was the first one to organize its observations and then attempt to find a cohesive framework in which to place them. The Greeks made great progress in measuring the positions of stars and determining the relative sizes and distances of the various objects in our solar system. Greek astronomers were the first to formulate and apply the scientific method successfully to observational sciences such as astronomy. Greek astronomy is discussed in more detail in Chapter 5; here we look at the Greek's earliest attempts to make sense of their observations.

As an example of the way that conceptual models interact with observations to produce further understanding of observed events, let us go back about 2,500 years and consider how Babylonian astronomers saw the shape of the Earth. Elementary observations seemed to establish that the Earth was flat. The Babylonians pictured a finite Earth; they assumed that if one traveled far enough, one would eventually reach the point where Earth and sky meet (**Figure 4-6**). This assumption is not surprising, because the idea of infinity (in this case, an Earth that extended endlessly through space) is a difficult one to grasp and was not invented until about 200 B.C.E.

The model of a flat Earth surrounded by the heavens (which comes to the Western world from Babylonia via the generally accepted interpretation of the biblical account in the book of Genesis) poses some problems. For one thing, if all things

fall down, as we observe that they do, why doesn't the Earth? To someone who lacks an understanding of gravity, this is a serious question. To avoid a "fall," the Earth would have to extend infinitely downward, but the Babylonians were not yet ready for the concept of infinity. Although the Babylonians seem to have ignored the problem, the Greeks did not. Anaximander thought that the Earth was cylindrical in shape (with Greece on the "upper" face), and that it was placed in the exact center of the universe (**Figure 4-7**). Because it was in the center of the universe, it had no particular direction to go and naturally remained in the same place. Although Anaximander's concept solved the problem of an infinitely extended Earth, it conflicted with several familiar facts.

Inquiry 4-1

What phenomena would you expect to see in the sky if you were observing from a flat Earth? A round one?

The Greeks were, of course, seafarers. Sailors had long known that when a ship sailed away from the shore, not only did it diminish in apparent size but, in addition, the bottom of the ship disappeared before the sails did. This is easily explained if the Earth is curved. Furthermore, the Greeks knew that if one traveled in a northerly or southerly direction, some stars disappeared from view, whereas other stars became visible. That this is impossible on a flat Earth is easy to see (**Figure 4-8a**), but it can be explained simply if the

Figure 4-6 The Babylonian universe was enclosed in a domelike shell called the Firmament.

Figure 4-7 Anaximander's universe was infinite in extent, with the Earth at the center.

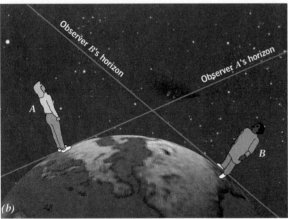

Figure 4-8 *(a)* On a flat Earth, all observers see the same stars. *(b)* On a round Earth, observers have different horizons, and they see only those stars above their own horizon.

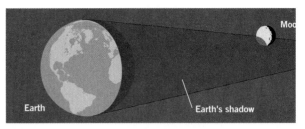

(a)

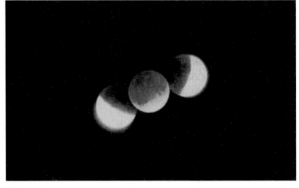

(b)

Figure 4-9 *(a)* Eclipse of the Moon (not to scale). The shadow of the Earth on the Moon is always a segment of a circle, proving that the Earth is round. *(b)* A lunar eclipse showing the Earth's curved shadow in the Moon's face. The eclipsed part of the Moon is illuminated by light refracted through the Earth's atmosphere.

Earth is round (**Figure 4-8***b*). However, it was not until the sixteenth century that the Portuguese explorer Magellan circumnavigated the globe between 1519 and 1522 and directly showed that the Earth had no end.

The Greeks were also well aware that eclipses of the Moon occur when the Earth passes between the Sun and Moon, with its shadow falling on the Moon (**Figure 4-9***a*). Furthermore, they realized that the line between light and darkness on the Moon's face during the eclipse was observed always to be curved and always to have the same amount of curvature (**Figure 4-9***b*). Because the only object that always casts a circular shadow, regardless of its orientation, is a sphere, it followed that the Earth was spherical. The idea of a spherical Earth is attributed to the mathematician Pythagoras (c. 570–500 B.C.E.), and the arguments employed here came down to us through the writings of the philosopher Aristotle some 200 years later.

Inquiry 4-2

Suppose the Earth were shaped like a coin that still spins once a day and that, when seen from the Moon, is observed edge-on. What shape would the Earth's shadow be on the Moon if a lunar eclipse took place just after sunset?

By assuming that the Earth was spherical, the Greeks also resolved the question of which way was down: it was toward the center of the Earth. Their model satisfied them completely, because in Greek philosophy, the sphere was the most perfect geometrical form.

The questioning approach that Greek thinkers took toward nature, together with their considerable strength in geometry and mathematics, made it possible for them to progress rapidly in astronomy. The Greeks determined the sizes of the Earth, Moon, and Sun (although the method used for the Sun was not accurate). They understood eclipses and determined the proper order of the planets with respect to their distance from the Earth by studying their apparent speeds as they crossed the sky. We will discuss further aspects of Greek astronomy in the next chapter.

Inquiry 4-3

Which of the reasons just given for thinking the hypothesis that the Earth is round are primarily supported by observations? Which ones are primarily supported by theory?

4.3 The Observed Motions of the Sky

To understand our observations further, we require a better means of describing the sky. We must bring together into one unified picture of the heavens the different apparent motions as seen by observers at different locations on Earth.

THE CELESTIAL SPHERE: A MODEL OF THE SKY

When looking at the sky, the stars give the appearance of being positioned on a bowl or sphere with the observer at the center. We call this apparent sphere the **celestial sphere** (**Figure 4-10**). The

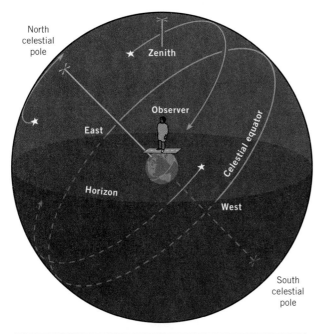

Figure 4-10 The celestial sphere, as seen from the "outside" for an observer at intermediate latitudes. The figure is not to scale.

stars are imagined as fixed to the surface of the celestial sphere, whose radius is so large that the Earth, at the center, is insignificant in comparison. Thus, the stars move in their fixed patterns with the celestial sphere. The concept of the celestial sphere is an example of a useful model that helps us describe and visualize the observed motions of objects but is not physical reality—modern astronomers don't think the Earth is at the center of the Universe!

The daily apparent east-to-west motion of bodies on the celestial sphere is just a reflection of the Earth's west-to-east daily rotation on its axis. The stars appear to rotate about a point on the celestial sphere called the **north celestial pole** (the **NCP**), as shown in the chapter opening image. It is simply the point where the extension of the Earth's rotation axis intersects the celestial sphere. By coincidence, during our millennium there happens to be a moderately bright star, **Polaris**, relatively close to the NCP. For this reason, Polaris is also known as the pole star. There is a south celestial pole, but there is no southern pole star. In a similar way, the extension of the Earth's equator to the celestial sphere defines the **celestial equator**.

The **horizon** is a circle on the celestial sphere that divides the parts of the sky an observer can see from those that are hidden by the Earth (Figures 4-10 and 4-11a). Stars rise when they become visible above the horizon in the east and set when they go below in the west. There is a group of stars that never sets; which stars depends on the observer's latitude. **Circumpolar stars** appear to rotate about the celestial poles in a circle, always keeping the same angular distance from the pole. In the chapter opening photograph, they are the stars between the horizon and the center of the star trails.

A star's **altitude** is its angle above the horizon. The altitude of the pole star depends on the observer's location on Earth. For example, **Figure 4-11a** shows an observer at the North Pole (at a latitude of 90°). In this case, Polaris is directly overhead, at the observer's **zenith**. The observer's horizon is perpendicular to the direction to the zenith and is represented as a plane that is tangent to the Earth where the observer is located. For this observer the altitude of Polaris therefore is 90°. The celestial equator, then, is along the horizon. At the North Pole (or South Pole), the stars always stay at the same altitude above the horizon, moving in circles around the pole directly above the

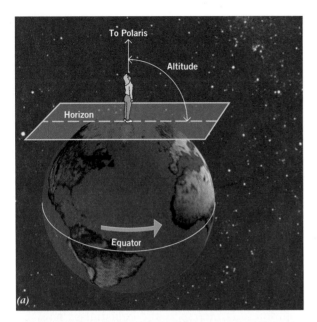

(a)

Figure 4-11 *(a)* Polaris, as observed from the North Pole, would appear overhead. *(b)* As seen from the pole, stars appear to move on horizontal paths parallel to the horizon.

observer. Stars near the horizon will appear to move parallel to the horizon (**Figure 4-11*b***). At the poles, therefore, all visible stars are circumpolar—they never set. Half the stars are below the horizon, and because they can never rise, they are always invisible to an observer at a pole.

Inquiry 4-4

For an observer at the equator, where on the celestial sphere will the north celestial pole be located?

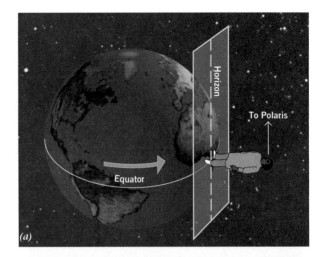

(a)

(b) East

Figure 4-12 *(a)* Polaris, as observed from the equator, would lie just on the horizon. The altitude would be 0°. *(b)* At the equator, stars in the east and west rise and set vertically.

Inquiry 4-5

Do stars move parallel to, perpendicular to, or at some random angle to the celestial equator?

In **Figure 4-12*a***, the observer is at the equator, where the latitude is 0°. In this case the observer, sighting on a line parallel to the horizon, sees the distant Polaris just on the northern horizon. The picture is not to scale, and in reality the observer's height is tiny compared with the diameter of the Earth. For this reason, Polaris does not appear above the horizon but, instead, at an altitude of 0°. The celestial equator is now perpendicular to the horizon. Therefore, because stars are attached to the celestial sphere and move with it, a star's

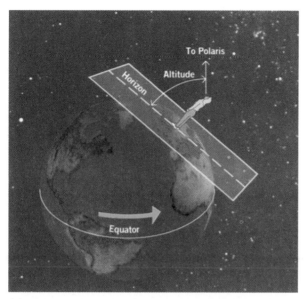

(a)

(b)

Figure 4-13 *(a)* Viewed from an intermediate latitude, Polaris would appear at some angle above the north point on the horizon. *(b)* At intermediate latitudes looking west, the stars set at an angle with respect to the horizon. The angle depends upon the observer's latitude.

motion with the celestial sphere is parallel to the celestial equator (see Inquiry 4-5). When a star rises or sets, it does so on a path that is perpendicular to the horizon, as shown in **Figure 4-12b**. For an observer at the equator, every star in the sky always rises and sets and is above the horizon half the time and below the horizon the other half. For an equatorial observer, every star in the sky is visible at *some* time of the year, and there are no circumpolar stars.

In **Figure 4-13a**, the observer is at an intermediate latitude, for example, in the USA. At such latitudes, the pole star is above the horizon, at an altitude greater than 0° and less than 90°, as in the chapter opening photograph. In the general case, the altitude of the pole equals the latitude of the observer. Some stars are circumpolar and never set, while others rise and set. Over the course of a year, an observer can see more than half the celestial sphere, but there is a significant portion of the sky that never rises and is always hidden from view. Stars rise and set at an angle that depends on the observer's latitude on Earth. **Figure 4-13b** shows the setting stars for an observer at an intermediate northern latitude.

4.4 The Motion of the Sun

Observations of the Sun show that it *appears* to move on the celestial sphere in relation to the background stars. Such motions determine our methods of timekeeping. In the following sections, we describe these motions.

THE DAILY MOTION

The **daily motion** of the Sun is its apparent motion across the sky resulting from the Earth's rotation on its axis. It is the cause of day and night. A **solar day** is the length of time between two successive passages of the Sun through the observer's **meridian**, which is a circle running north–south that divides the celestial sphere into eastern and western halves (**Figure 4-14**). By observing the daily motion of the Sun during the course of the day, it is possible to obtain a good understanding of the apparent daily motions of other stars as well.

The daily motion of the Sun is easy to study by using a shadow stick, or **gnomon**—a time-honored device that has been used for thousands of years. A gnomon is simply a vertical stick that casts a shadow on the ground. By interpreting the motion of the shadow, one can learn about the apparent motion of the Sun. For example, when the shadow has its shortest length of the day, the shadow is aligned north–south, the Sun is at its highest point in the sky, and it is crossing the meridian. This is the definition of noon.

THE MOTION OF THE SUN IN RELATION TO THE STARS

It is not easy to observe the changing position of the Sun in relation to the stars, because the stars are not visible when the Sun is up. But it is not

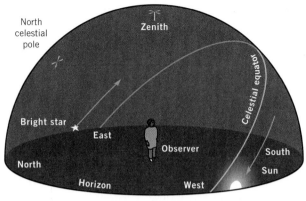

(a) Star rises opposite the sun

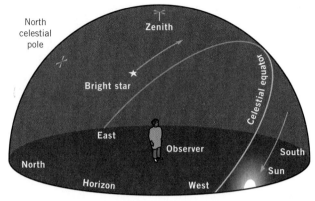

(b) Star rising one month later

Figure 4-14 (a) The Sun setting, as a bright star rises on a particular day of the year. (b) When the Sun sets one month later, the same bright star shown in part a is observed to be already well above the horizon.

difficult to *infer* the Sun's motion in relation to the stars from the observation that different stars are seen in the sky at different times of the year.

We begin with **Figure 4-14a**, which shows the location of the Sun on the celestial sphere at sunset. Just as the Sun is setting in the west, a bright star is rising in the east. One month later, as shown in **Figure 4-14b**, we observe the bright star to be already well above the eastern horizon when the Sun is setting. In other words, the star rose earlier than it did the previous month. The Sun appears to have moved in an *eastward* direction, across the celestial sphere toward the bright star. In other words, the Sun appears to have moved eastward in relation to the background stars. This movement can also be described as counterclockwise.

The reason for this apparent eastward motion of the Sun is that the Earth moves in an eastward direction around the Sun; the Sun, when viewed from the moving Earth, appears to move eastward against the background of fixed stars. **Figure 4-15** illustrates this. When the Earth is at point *A* in its orbit, the Sun will be seen in the direction of star *A′* when viewed from the Earth. The stars toward the right side of the figure will be the ones visible at night from the Earth's dark side, the side not facing the Sun. One month later, the Earth has moved eastward and is at point *B*; the Sun has shifted its position in relation to the stars and is seen in the direction of star *B′* to the east of *A′*. By the time the Earth moves over to point *C*, the stars on the left side of the figure will be nighttime stars, and the ones toward the right side will have become daytime stars, not visible due to the glare from the Sun.

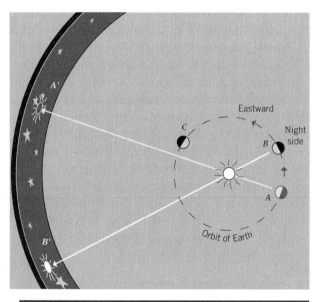

Figure 4-15 The apparent eastward (counterclockwise) motion of the Sun, from point *A′* to point *B′*, against the background of the fixed stars caused by the Earth's eastward (counterclockwise) motion from point A to point B. The view is from above the Earth's North Pole.

Because the Earth completes its 360° orbit around the Sun in 365.24 days, the Sun's apparent motion in relation to the background stars is nearly one degree per day. This apparent motion is about twice its 0.5° angular diameter. The appearance of the night sky changes little from one night to the next because this angular motion of the Sun on the celestial sphere is small. However, observations made a few weeks apart will show obvious changes in the sky. Do not confuse the one degree per day apparent motion of the Sun in relation to the background stars that is caused by the Earth's orbital motion about the Sun with the Sun's 360° daily motion (that is its rising and setting) that is caused by Earth's rotation.

The movement of the Sun on the celestial sphere traces a repeatable annual path among the stars. This apparent path of the Sun in the sky with respect to the fixed stars is called the **ecliptic**. The entire ecliptic is a circle extending all the way around the celestial sphere.

Inquiry 4-6

We define the ecliptic in terms of what the observer on Earth sees. If the observer were located at the Sun and looking at the motion of the Earth, how would the ecliptic then be defined?

The relationship between the ecliptic and the celestial equator appears in **Figure 4-16**. The ecliptic results from the Earth's *orbital motion* around the Sun. The celestial equator is established by the Earth's *rotation* on its spin axis. Because the Earth's spin axis is tilted 23.5° from the perpendicular to its orbit, the ecliptic and the celestial equator are inclined at an angle of 23.5° with respect to each other, intersecting at two points 180° apart.

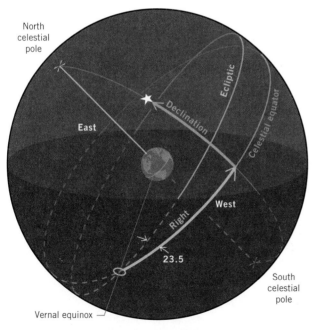

Figure 4-16 The celestial sphere, showing the ecliptic tilted 23.5° in relation to the celestial equator, and the vernal equinox. The definitions of the right ascension and declination of a star, which are defined in section 4–6, are illustrated.

Solar and Sidereal Days

If the bright star Sirius rises at 8:00 P.M. one evening, it will rise about 7:56 the next evening, 7:52 the next, and so on—about four minutes earlier each night. Thus, after two weeks it will rise about one hour earlier (4 minutes/day × 14 days = 56 minutes); after four weeks (one month), two hours earlier, six months twelve hours earlier, and after one year it will be again rising at 8:00 P.M. To understand the reason fully requires looking carefully at the locations of stars seen by an observer that lives on a rotating, revolving Earth. **Figure 4-17** shows an observer with the Sun directly above, on the meridian; suppose, too, that a distant star is also on the meridian (but behind the Sun). After a certain interval of time has passed, the same distant star will again be on the meridian. This time interval for a complete rotation of the Earth in relation to a background star is called the **sidereal day** (from the Latin *sider* meaning star). The Sun, however, will not yet be at the meridian because during the rotation the Earth moved in its orbit to position number 2. Although the Earth moved in its orbit, the Sun appeared to move 1° eastward in relation to the background stars. Therefore, for the Sun to be on the meridian again requires the Earth to rotate through an additional 1°. To complete the 1° rotation will require almost four minutes. The end result is that the time between successive overhead passages of the Sun is nearly four minutes longer than the sidereal day, which measures the Earth's true rotation period. The cumulative effect of the four minutes per day

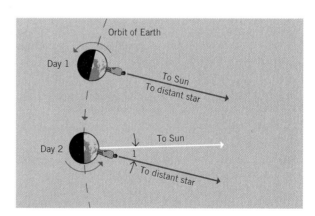

Figure 4-17 At noon on day 1 the Sun and a distant star are on the observer's meridian. On day two, one rotation relative to the star later, for the Sun. To be on the meridian, the Earth must rotate through about 1°, which requires nearly four minutes. (Note that the angles in the figure are exaggerated.)

is that you see different stars and constellations at different times of the year.

Inquiry 4-7

If a star is on the meridian at midnight on June 1, at approximately what time will it be on the meridian July 1?

4.5 The Reasons for the Seasons

In addition to its apparent easterly motion among the stars during the year, the Sun has another apparent motion—one that is responsible for the seasons. Even casual observers in North America find the Sun to be higher in the sky in summer than in winter. These changes in position, which result from an apparent motion of the Sun north and south of the celestial equator, are due to the 23.5° tilt of the Earth's axis of rotation in relation to the plane of the Earth's orbit around the Sun. It is these changes that cause seasons on Earth.

Figure 4-18 shows how this happens. On the left side of Figure 4-18*a*, the Earth is shown where it would be about December 21, at the beginning of winter in the Northern Hemisphere and summer in the Southern Hemisphere. At this time, the North Pole of the Earth is tilted *away* from the Sun. (The NCP is always pointing at Polaris, so the spin axis of the Earth stays fixed. The Earth's tilt in relation to the Sun is changing.) If you imagine the Earth's equator projecting out into space, you should be able to convince yourself that the

Sun will be south of the celestial equator, and low in the sky at noon. As time progresses, because the Earth's position in its orbit changes, the Sun's position on the celestial sphere moves northward. About March 21 the Sun's location on the ecliptic crosses the celestial equator. At this time day and night are equal, and the Sun is said to be at the **equinox**. This is the first day of spring in the Northern Hemisphere; it is called the **vernal** or **spring equinox**. As the Earth continues in its orbit, the Sun's position moves further north of the celestial equator. The Sun is highest in the sky about June 21, when Earth reaches the point in its orbit shown on the right side of Figure 4-18*a*. This point, known as the **summer solstice** (for the Northern Hemisphere), occurs when the Sun is at its maximum angle north of the celestial equator (23.5°). Hereafter, the Sun's angle north of the celestial equator begins to decrease until the Sun again reaches the celestial equator, at the time of the **autumnal equinox** in September. After another three months, during which the Sun is moving further south on the celestial sphere, the Sun reaches its lowest point again at the **winter solstice**.

A further effect is illustrated in Figure 4-18*b*. In the Northern Hemisphere summer, when the Sun is north of the celestial equator, the Sun is higher in the sky, and the Sun's rays are more perpendicular than in the winter, which results in a higher concentration of solar energy per unit area on the Earth's surface. In addition, the Sun's arc across the sky is longer, which means that the Sun is above the horizon a greater fraction of each 24-hour day. Daytime is longer for those in the

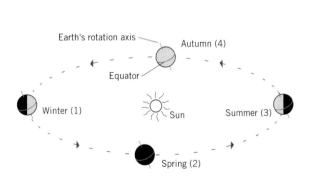

(a) Seasons for northern hemisphere

(b)

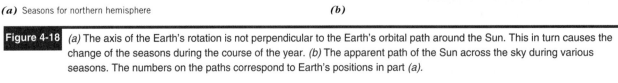

Figure 4-18 (a) The axis of the Earth's rotation is not perpendicular to the Earth's orbital path around the Sun. This in turn causes the change of the seasons during the course of the year. (b) The apparent path of the Sun across the sky during various seasons. The numbers on the paths correspond to Earth's positions in part (a).

Northern Hemisphere, which means there is more time for heating of the Earth's surface. The combination of these two effects produces the heat of summer in the Northern Hemisphere. The Southern Hemisphere, of course, experiences winter at this time.

It is apparent from Figure 4-18*b* that the Sun rises north of east in the northern summer and sets north of west, whereas in the northern winter it rises south of east and sets south of west.

You may have noticed that this discussion of the seasons said nothing about the distance from the Earth to the Sun. That is because distance is not involved in Earth's seasonal changes. As an illustration of this point we note that Earth is farther from the Sun during summer in the northern hemisphere than in the winter. Furthermore, for an observer near 40° north latitude (for example San Francisco, Denver, New York), the difference in the distance from the observer to the Sun caused by the tilt of the rotation axis being toward or away from the Sun amounts to only 2,000 miles out of the Sun's 93 million mile distance. This means that the difference in distance we experience over the course of the year due to the Earth's rotational axis tilt is about 50,000 times smaller than the actual distance between the Earth and the Sun—an insignificant amount.

Inquiry 4-8

Where in the sky is the Sun at the Northern Hemisphere's winter solstice? How would you describe the length of the day (for the Northern Hemisphere) near December 21?

Inquiry 4-9

Do you expect seasonal variations to be large or small at the equator? Explain your answer. How about at the North Pole?

4.6 The Location of Stars on the Celestial Sphere

Earlier discussion explained that the apparent location of stars changes with time—daily as the Earth rotates on its axis and throughout the year as it revolves around the Sun. Not only that, but the altitude of a given star is different for observers at different latitudes. If you were on the telephone describing to a friend who lived across the country the location of a celestial event in relation to your horizon and zenith that you had just witnessed, your description would not help your friend find the event. For this reason, astronomers had to find a way to describe the positions of celestial objects that is independent of the position of the observer on Earth.

A coordinate system that more or less satisfies our needs is the **equatorial coordinate system**. It may be described with Figure 4-16, which is a modification of Figure 4-10 that we used to define the celestial sphere. Just as two coordinates, latitude and longitude, are required to uniquely specify a location on the Earth's surface (Austin, Texas, for example, has a longitude of 97° west and a latitude of 30° north), two coordinates are required to uniquely locate an object on the celestial sphere. The angular distance of a star north or south of the celestial equator is called **declination**; it is analogous to latitude on Earth. The equator is zero degrees, the NCP +90°, and the SCP –90°. The longitude of a place on Earth is the angle east or west of the meridian through the Royal Observatory at Greenwich, England. The choice of Greenwich as the zero or reference point is arbitrary and was based on political considerations in the 1800s. On the celestial sphere, the location where the celestial equator and the ecliptic cross when the Sun moves from the southern into the northern hemisphere, the vernal equinox, is the reference point. This east–west coordinate is called the **right ascension** and is shown in Figure 4-16. Rather than degrees, right ascension is measured in a total of 24 hours, with the vernal equinox at zero hours.

The model of the celestial sphere has the stars fixed in place, and this is consistent with the observations the ancients could make. However, images of the same stars taken years apart show that the stars do move slowly in relation to each other.

PRECESSION

The pole star, Polaris, is close to but not at the north celestial pole. It is less than 1° away at present. However, observations show that the NCP moves in relation to the stars. In fact, the NCP is moving closer to Polaris and, at its closest approach in about the year 2100, it will be within 1/2° of Polaris. In ancient times, however, the NCP

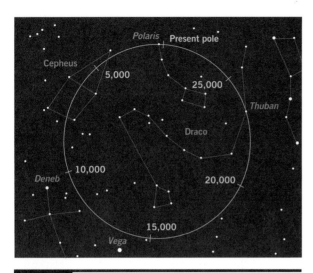

Figure 4-19 The path of Earth's rotation axis on the celestial sphere during its 26,000-year precession. Dates indicate the year the pole will be in a given location.

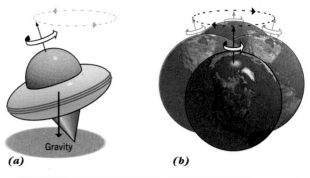

Figure 4-20 (a) The precession of a spinning top is due to the force of gravity acting on the top. (b) The precession of the Earth is due to the force of the Moon and Sun acting on the Earth's equatorial bulge.

was nowhere near Polaris. Referring to **Figure 4-19**, in about 3000 B.C.E., the bright star closest to the pole was Thuban. Between 2,500 and 5,000 years from now, the stars in the constellation of Cepheus will be closer to the pole, and in about 12,000 years Vega, one of the brightest stars in the sky, will be closest to the pole. About 26,000 years from now, Polaris will again be the northern pole star.

How can this happen? You may be familiar with the behavior of a spinning top (**Figure 4-20a**). As it spins, its rotation axis revolves in a circle as a result of an interaction between the spin of the top and the downward pull of gravity. The movement of the axis, which is often described as a *wobble,* is called **precession**. The same kind of thing happens to the Earth as it spins in space. The Moon and, to a lesser extent, the Sun, exerts a gravitational force on the equatorial bulge of the spinning Earth; that force causes the axis of the Earth's rotation to change direction slowly. As a result, the north celestial pole moves slowly around the celestial sphere, taking about 26,000 years to trace a circle among the stars (**Figure 4-20b**).

The Greek astronomer Hipparchus discovered this motion in about 100 B.C.E. using Babylonian data. Even in his time, naked-eye observations had accumulated over a long enough time for this very slow motion of the NCP to be detected. This is a dramatic example of the value of carefully recorded, continual observations. Casual observations would not have detected precession.

Precession has a practical effect for astronomers: it changes the right ascension and declination of stars in the sky. If astronomers pointing their telescopes to a star did not *precess* the coordinates to the date of observation, the telescope would point to the wrong location.

4.7 The Motion and Phases of the Moon

The Moon is extremely bright compared to the stars, and even the most casual observer is able to observe its changes easily. We will now learn the reasons for its changing appearance.

THE MOTION OF THE MOON

The Moon's rapid motion in relation to the fixed stars has given it a unique role in the history of astronomy. From prehistoric times, it has been used as the basis for calendars, and it was the Moon that gave Sir Isaac Newton the crucial information he needed to discover the law of gravity (Chapter 5). Even today, the Moon remains an object of fascination for astronomers and other scientists who study not only its motion, but also the precious rocks brought back from its surface.

The Moon's period of revolution around the Earth in relation to the background stars—its *sidereal* period—is 27.3 days. Its movement among the background stars is counterclockwise (eastward) and amounts to 360° every 27.3 days, or about 13° per day. Because this counter-clockwise angular motion is about 26 times the Moon's 0.5° angular size, the motion is easy to see. Do not confuse the 13° per day apparent motion of the Moon in relation to the background stars that is caused by the Moon's orbital motion about the Earth with the Moon's daily rising and setting motion that is caused by Earth's rotation.

Inquiry 4-10

How many lunar diameters per hour does the Moon move in its eastward motion with respect to the stars?

The Moon's movement is counterclockwise in relation to the stars because as the Moon's orbital motion carries it eastward in its orbit, an observer on Earth will see it farther to the east with each succeeding hour (**Figure 4-21**).

The Moon's apparent path on the celestial sphere is close to the ecliptic but tilted to it by an average of 5°. This 5° tilt also prevents eclipses from occurring monthly, since an eclipse requires nearly perfect alignment.

We always see the same face of the Moon. Although an elementary observation, it leads us to an important conclusion: it proves that the Moon rotates. In particular, it shows that the Moon turns exactly once on its axis each time it revolves around the Earth, just as two people turn while dancing face to face. Because the Moon takes a sidereal month to revolve around the Earth, it takes that long for it to rotate 360° as well.

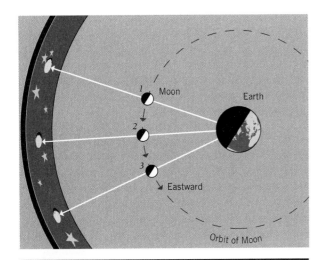

Figure 4-21 The apparent counterclockwise motion of the Moon in relation to the background stars, caused by the counterclockwise motion of the Moon around the Earth.

THE PHASES OF THE MOON

The half of the Moon facing the Sun is always fully lit, just like the daytime half of the Earth. However, from our viewpoint here on Earth, the Moon is constantly changing its appearance. Sometimes we see the unlit surface, sometimes the fully lighted surface, or (most often) something in between. The Moon shows **phases**—from **new** to **waxing crescent**, **first quarter**, **waxing gibbous**, and **full**, and then reversing to **waning gibbous**, **last quarter**, **waning crescent**, and new to begin the cycle again (**Figure 4-22**). **Figure 4-23** illustrates this statement. The figure shows the Moon as it orbits the Earth. The Moon's appearance to an observer on Earth is shown by the Moon's projection onto a plane. The phase changes as the angle between the Sun, Earth, and Moon changes. For example, when the Moon is on the opposite side of the Earth from the Sun, we see its entire illuminated surface and call it *full*. New Moon occurs when the Moon is in approximately the same direction as the Sun, with its unlit surface facing us, and quarter Moon is when it is 90° from the Sun.

Everyone on Earth sees the Moon in the same phase in one day although, since you only see the Moon when the part of the Earth you're on faces it, you see the Moon for about twelve hours a day. During a full Moon, the 12 hours occur at night. The crescent Moon is visible during the day and for a while after sunset. The new Moon is above the horizon only during daylight hours, but it cannot be seen. Thus, you can imagine the following situation: the Sun is setting on the western horizon and the first quarter Moon will be on the meridian. Furthermore, if the Sun were on the meridian (which it is at noon), the first quarter Moon would be rising on the eastern horizon.

Inquiry 4-11

If the Sun sets at 6 P.M., at what approximate time would the full Moon rise?

The time interval from new Moon to the next new Moon is 29.5 days and is called the **synodic period**. It differs from the Moon's 27.3-day sidereal period. The difference between the Moon's sidereal and synodic periods is caused by the Earth's orbital motion about the Sun, which causes the Sun's apparent eastward motion in relation to

4.0 days: Waxing Crescent   5.0 days: Waxing Crescent   7.1 days: First Quarter

9.1 days: Waxing Gibbous   11.1 days: Waxing Gibbous   15.0 days: Full

17.0 days: Waning Gibbous   19.0 days: Waning Gibbous   19.5 days: Waning Gibbous

21.4 days: Last Quarter   23.5 days: Waning Crescent   26.6 days: Waning Crescent

Figure 4-22 The Moon's phases throughout the month. The age of the Moon in days after new Moon (where age is 0), and the lunar phase is given to the right.

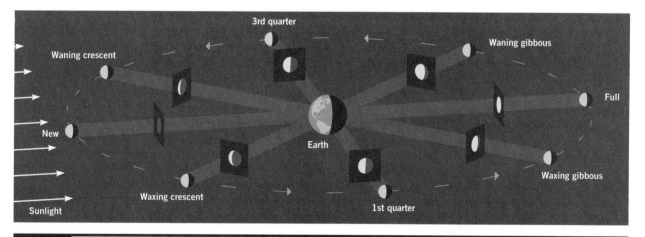

Figure 4-23 The lunar phases. The Moon's illumination as it orbits the Earth is shown. The images projected on to the eight planes surrounding the Earth indicate what fraction of the illuminated disk an observer on Earth will see.

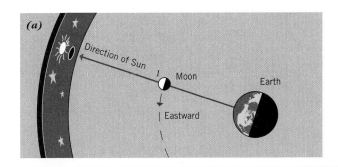
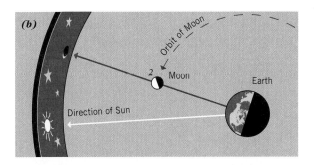

Figure 4-24 (a) At new Moon, the Sun and Moon both are seen in the same part of the sky. (b) After the Moon has gone exactly once around the Earth, it is seen in the same location in relative to the stars. But the Sun has moved to a different position in the sky, so the Moon is not yet new. (Not to scale.)

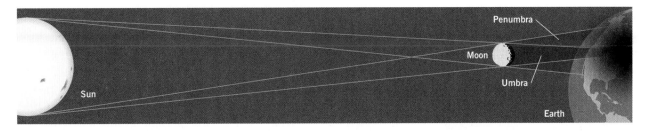

Figure 4-25 The geometry of a solar eclipse. (Not to scale)

the background stars during the month. **Figure 4-24a**, a view from outside the Earth–Moon system, shows the situation at new Moon; the Sun and Moon are seen from Earth against the same background stars. In **Figure 4-24b**, one *sidereal* period later, the Moon has returned to the same place among the stars, but the Sun too has moved eastward. Thus, the Moon requires about two more days to reach the position of the Sun and establish the new-Moon geometry once again.

ECLIPSES

Eclipses have awed and frightened observers since the dawn of time. What takes place during eclipses of the Sun and Moon? Every body in the Solar System carries a shadow along with it as it moves around the Sun. When one body enters the shadow of another, an eclipse occurs. **Figure 4-25** shows the geometry of the Sun, Moon, and Earth at the time of a **total solar eclipse**, as seen by an observer in space. The shadow of the Moon sweeps across the surface of the Earth at a high rate of speed. Observers temporarily located within the dark spot will see the bright disk of the Sun completely covered by the Moon (see Figure 3-9).

A typical shadow consists of a dark inner part, called the **umbra**, and a lighter outer part called the **penumbra** (shown in Figures 4-25 and 4-9 for both lunar and solar eclipses). When the Moon's umbra passes over an observer, he or she sees a total solar eclipse. However, because the umbra is, at most, only about 150 miles wide, only those observers along the umbral path will see a total eclipse. Even then, the longest period of totality at any one location on Earth is slightly over seven minutes. People not within the umbra but still inside the larger penumbra will see part of the Sun's face blocked, a **partial solar eclipse**. (You can see a partial solar eclipse prior to totality in Figure 3-9.)

The distance of the Moon from the Earth varies somewhat during the year, since it has an elliptical orbit. Whenever the Moon is relatively far from the Earth, the apparent size of the lunar disk will be smaller than average and may not quite cover the Sun. At such a time, an observer on Earth will see an **annular eclipse**, with a bright ring, or annulus, of sunlight around the dark disk of the Moon. **Figure 4-26** shows the appearance of the Sun for an annular and a total eclipse.

Eclipses do not occur every month because the 5° tilt between the Moon's orbit and the ecliptic means that the new Moon is generally above or below the Sun. The end result is that solar eclipses

(a)

(b)

Figure 4-26 *(a)* In an annular eclipse, the Moon's apparent disk is not quite large enough to cover the apparent disk of the Sun. The result is a bright ring around the dark Moon. *(b)* In a total solar eclipse, the Moon completely covers the bright disk of the Sun. Then the sky becomes dark enough for us to see the faint outer part of the solar atmosphere, known as the corona.

of all types (partial, total, and annular) occur somewhere on Earth a minimum of two times and a maximum of five times each calendar year. There are from zero to three annular or total solar eclipses each calendar year. Should the opportunity arise for you to see a total eclipse of the Sun, take it! It will be one of life's most memorable events.

You should never stare directly at the Sun, even during an eclipse. Permanent eye damage can occur rapidly and without your immediate knowledge. If you are fortunate *enough to view an eclipse of the Sunm, either project the solar image onto a piece of paper, or visit your local library to obtain information on properly protecting yourself from damaging radiation.*

Inquiry 4-12

What must the Moon's phase be at the time of a solar eclipse?

A total eclipse of the Sun is a spectacular sight, but you may have to travel to see one. On March 29, 2006, a 4-minute-long total eclipse will be visible in Northern Africa, Europe, and Asia. On August 1, 2008, northern Canada, Europe, and Asia will have a short 2.5-minute eclipse, while Asia and the Pacific Ocean will have a nearly 7-minute eclipse on July 22, 2009. A trip to Chile, Argentina, and Easter Island on July 11, 2010, will allow you to see a 5-minute total solar eclipse.

There will be no total solar eclipses visible from North America until August 21, 2017. On the average, if you stand in one place and wait for a total eclipse to come to you, your waiting time will be about 450 years.

When the Moon moves into the shadow of the Earth, a **lunar eclipse** occurs (Figure 4-9*a, b*). At this time, we can see the curved edge of the Earth's shadow creeping across the illuminated face of the Moon until it is darkened. At the time of a total lunar eclipse, the only light reaching the surface of the Moon is light that has been bent around the Earth by its atmosphere, which behaves like a weak lens. Because of scattering of blue light by particles in the Earth's atmosphere, the light that reaches the Moon is redder than normal sunlight (this is also what makes the setting Sun red and the daytime sky blue). As a consequence, the Moon's surface has a reddish cast.

Because any observer on the side of the Earth facing the Moon can see it, a lunar eclipse can be seen by anyone anywhere on that side of the Earth. The conditions for lunar eclipses allow for two to five lunar eclipses of all types per year.

Inquiry 4-13

What must the Moon's phase be at the time of a total lunar eclipse?

4.8 The Motions of the Planets

The paths of the planets in relation to the background stars are more difficult to trace than those of the Sun and Moon, because they generally appear to move more slowly on the celestial sphere. In addition, their phases cannot be seen without telescopic assistance. Still, you can learn much by following their motions with the naked eye, just as all astronomers did until the start of the seventeenth century when Galileo turned his telescope to the heavens.

Because the planets appear to move mostly counterclockwise (eastward) in relation to the background stars (for observers in the Northern Hemisphere) and vary dramatically in brightness, they readily attracted the attention of sky watchers. In fact, the word **planet** in Greek means *wanderer*. **Figure 4-27** shows the positions of Mercury, Venus, and Earth in relation to the constellations at the beginning of January and February in some year for an observer in space. In January, while the Sun appeared in Sagittarius, Mercury appeared in Scorpius and Venus between Virgo and Libra.

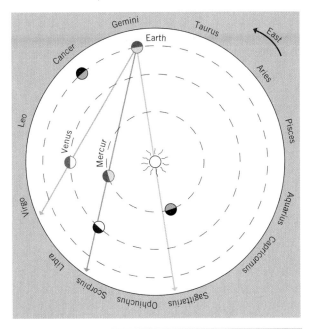

Figure 4-27 The solar system as viewed from above on January 1 (light symbols) and February 1 (dark symbols) of some year. Lines from the Earth to the Sun, Mercury, and Venus show their positions in relation to the constellations.

Inquiry 4-14

Using Figure 4-27, in what constellations were the Sun, Mercury, and Venus during February?

If this apparent eastward movement among the stars were all there was to their motion, it is likely that the ancients would not have been so interested in the planets. However, on occasion they appear to slow down, cease their apparent eastward motion, and begin to move westward for a while. Such apparent backward motion in relation to the background stars is called **retrograde motion**. After some weeks or months, the planets begin to move eastward again. They appear to execute a little loop in the sky; **Figure 4-28** shows retrograde loops for Mars, which in this specific case lasted a few months. Explaining this looping motion was historically a great challenge, as we will see in Chapter 5.

Two planets, Mercury and Venus, have a rather special motion of their own. They always remain

Figure 4-28 Retrograde motion of Mars. The dates give an indication of the time interval over which retrograde motion occurs.

relatively close to the Sun, first moving away from it, then pausing, then moving back toward it. Venus never gets more than 48° from the Sun, whereas Mercury never gets more distant than 28°.

Because of this behavior, Venus and Mercury can be seen in both the morning and evening skies, but they are never seen at midnight (except in polar latitudes, where the Sun can be observed at midnight). For example, if Venus is east of the Sun, it will be an evening object and set after the Sun. If Venus is west of the Sun, then it will be a morning object, rising in the east before the Sun.

The planets do not move all over the sky but are observed only in a rather narrow band near the ecliptic. The ancients noticed this behavior and attributed special significance to this region, which was called the **zodiac**.

Throughout their orbital cycles, planets assume various positions, or **configurations**, in relation to the Earth-Sun line as seen from the Earth. When a planet farther from the Sun than the Earth—Mars, for example—appears to be opposite the Sun and is closest to the Earth we say the planet is in **opposition** (**Figure 4-29**). When in line with but on the exact opposite side of the Sun, and thus farthest from the Earth, the planet is said to be in **conjunction** with the Sun. When the planet is 90° (a quarter of a circle) from the Sun as seen from the Earth, the planet's configuration is eastern or western **quadrature**. The two planets closer to the Sun

than the Earth—Mercury and Venus—are said to be at **inferior conjunction** when they are in between the Earth and the Sun, and at **superior conjunction** when they are on the other side of the Sun and aligned with it. When the angle between the Earth–Sun line and the Earth–planet line—an angle called **elongation**—is at its maximum, the planet is said to be at **maximum (eastern or western) elongation**. Although planets can show phases, not all planets can show all phases. Planets that do not come in between the Earth and the Sun cannot appear new from Earth; Mars can never appear new. Venus's phase is full at superior conjunction and new at inferior conjunction.

Inquiry 4-15

What phase will Venus have at maximum elongation? Use a drawing to determine your answer. What phase will Jupiter have at quadrature?

The length of time from one opposition to the next is a planet's synodic period. The synodic period differs from the planet's actual period around the Sun in relation to the distant stars, which is called the sidereal period (just as the Moon's period around the Earth in relation to the background stars is its sidereal period). The synodic period, rather than the sidereal period, is what astronomers actually observe. Then, from the observed synodic period, the planet's sidereal period can be calculated. The relationship is presented in Appendix A11.

Inquiry 4-16

Why is it easy to observe a planet's synodic period, but difficult to observe its sidereal period?

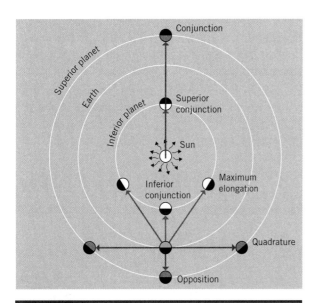

Figure 4-29 Planetary configurations are locations in relation to the Sun-Earth-planet line. When the angle is 90° for a superior planet, we have quadrature; when the angle is its maximum for an inferior planet, we have maximum elongation.

4.9 How Stars Get Their Names

Giving names to stars makes it easier to refer to specific objects in the sky. Stars are named in various ways. Most of the brighter stars have common names dating back to antiquity, such as Castor and Pollux, named for the twins of Greek mythology. A number of stars have Arabic names, such as Betelgeuse, Altair, and Algol, because during the

Table 4-1 The Greek (lowercase) Alphabet

α	Alpha	ι	Iota	ρ	Rho
β	Beta	κ	Kappa	σ	Sigma
γ	Gamma	λ	Lambda	τ	Tau
δ	Delta	μ	Mu	υ	Upsilon
ε	Epsilon	ν	Nu	φ	Phi
ζ	Zeta	ξ	Xi	χ	Chi
η	Eta	o	OmIcron	ψ	Psi
θ	Theta	π	Pi	ω	Omega

Middle Ages much of the scientific knowledge of the ancient world that was lost to the rest of the world survived only because the rising Moslem culture of the Middle East preserved it for posterity. Before and during the Renaissance, astronomical knowledge came to Europe from the Arab world, naturally with many Arabic names.

In 1603, the German lawyer Johannes Bayer published a star map in which a new method for naming the stars was introduced. Bayer named the stars after the constellation in which they appeared, distinguishing one star from another with letters of the Greek alphabet (**Table 4-1**). In general, the brightest star in a constellation was called alpha (α), the next brightest beta (β), and so on. Thus Sirius, the brightest star in the constellation of Canis Major, the Great Dog, is called α Canis Majoris, using the Latin genitive ending on the name of the constellation. However, Bayer was not perfectly consistent in his schemes. For example, he named the stars in the Big Dipper (part of Ursa Major, the Great Bear) in order from the bowl to the handle, and *not* in order of brightness.

After all 24 letters of the Greek alphabet are used, stars are numbered 1, 2, etc. For example, the star after ω Orionis would be called 1 Orionis. Stars whose brightness varies with time are designated with Latin letters. For instance, a famous variable star in the constellation of Lyra is RR Lyrae. Finally, most stars are so faint that even this notation becomes too cumbersome, and they are known only by their number in a catalog. One such star that has emerged from obscurity is known as HDE 226868, that is, the 226,868th star in the catalog known as the Henry Draper Extension. This is an object that consists of two stars, one of which is a black hole.

Thus, stars may have many names. For example, the variable star Mira is known as o Ceti,

HD 14386, BD −3° 353, GCRV 1301, BS 681, Boss 2796, IRC 100030, IRAS 02168−0312, SAO 129825, and so on, after various catalogs in which it has been listed. Finally, it is always possible to refer to an object by its right ascension and declination. The myriad of names a given object can have often leads to confusion even among professionals! Modern computer databases have been immensely helpful in keeping names straight.

Many objects, especially faint ones and those discovered by other than optical means, are referred to by the right ascension and declination. Object names such as PSR 0540−69 and 01061013 are not uncommon.

Star names are sold by some organizations for fundraising. They may even send you a star chart showing "your" star. However, these names will not be valid and will not be recognized by astronomers or anyone else. The naming of astronomical bodies is done by the International Astronomical Union, which is the international organization of professional astronomers. If you purchase a star name, know that you are doing it for fun and not for immortality. You might save your money and print your own certificate on your computer!

4.10 The Brightness of Stars

Amateur and professional astronomers still describe how bright stars appear from Earth with a system used more than 2,000 years ago by the Greek astronomer Hipparchus. He grouped all the brightest stars together, calling them stars of *first magnitude*. The stars in the next brightest group were called *second magnitude,* and so on, down to the faintest stars that he could detect with the naked eye, which he assigned to the class of *sixth magnitude.* Because human eyes are not all the same, magnitudes determined with the eye are highly subjective. Modern astronomers have refined this **magnitude system** greatly and now measure brightness to tenths, hundredths, and even thousandths of a magnitude with precision electronic equipment. In the modern system, a star of magnitude 1.0 is approximately 2.5 times brighter[1] than one of magnitude 2.0; a star of magnitude 2.0 is 2.5 times brighter than one of magnitude 3.0; and so

---

[1]While we will use 2.5, the number is 2.511886, which approximates the fifth root of 100.

Table 4-2 The Brightness of Objects In the Magnitude System

Object	Magnitude
Sun	−26.7
Full Moon	−12.7
Venus (at brightest)	−4
Jupiter, Mars (at brightest)	−2
Sirius (brightest star in sky)	−1.5
α Centauri (nearest star)	−0.1
Andromeda galaxy (most distant naked-eye object)	3.5
Naked-eye limit (average)	6.5
Binocular limit	10
6-inch telescope limit	13
Hubble Space Telescope limit	28–29

on. A star of magnitude 1.0 is 100 times brighter than one of magnitude 6.0, because $2.5 \times 2.5 \times 2.5 \times 2.5 \times 2.5 \approx 100$. The magnitude system has been extended to objects fainter than the naked eye can see (magnitude greater than 6), whereas the brightest objects (Sun, Moon, some planets, and a few stars) are so bright that their magnitudes are counted as negative (less than magnitude 0). For example, the brightest star, Sirius, has magnitude −1.5, whereas the Sun's magnitude is −26.7. **Table 4-2** shows the brightness of a variety of objects expressed in the magnitude system.

The word *magnitude* used to describe the brightness of stars is not the same as the factor of 10 *order of magnitude* discussed in Chapter 1.

> Readers who would like to make observations of the night sky will find the chapter appendix valuable. You might want to use a pair of binoculars.

Appendix: Observing Hints and the Use of Star Charts

After reading this chapter appendix and observing the sky, you will be able to do the following:

- Use a map as a guide to find the principal constellations and stars in the northern part of the sky at any time of year.
- Use a map of the full sky to locate the principal constellations and stars for your time and season.

BRIGHTNESS OF STARS

The brighter stars are indicated on star maps with larger dots, and the fainter stars with smaller dots. (*Note:* This is a convention and does not mean that the brighter stars are physically bigger. Brighter stars do appear larger on photographs, but this is because the increased brightness exposes a larger area of the photographic emulsion.)

On the star maps in Appendix H is a key that gives the range of magnitudes corresponding to each symbol used to plot the various stars. The star maps show all the stars down to the third magnitude and some fainter stars, but the faintest (fifth and sixth magnitude) are not shown because most observers would not be able to see them. Don't be discouraged if you live in a place with lots of light—it's easier to learn the sky if you can see only the brightest stars! A few other objects of interest, such as bright nebulae and star clusters, are also indicated and can be observed with binoculars or a small telescope.

OBSERVING STARS IN THE NORTHERN PART OF THE SKY

Because the appearance of the sky changes throughout the year, we begin by showing you how to use the star maps to observe the northern sky at one particular time—in September. Afterward, we look at other times of the year and other parts of the sky.

THE NORTHERN SKY IN SEPTEMBER

The appearance of the northern part of the sky on September 1 at 9 P.M. standard time (10 P.M. daylight saving time) is shown in Figure 3-4. The horizon is drawn for observers at 40° north latitude, but the appearance of the sky would be substantially the same for observers anywhere in the United States, except Alaska and Hawaii.

At this time of year, the two most obvious groupings of stars are the Big Dipper (part of the constellation of Ursa Major, the Great Bear) in the northwest and Cassiopeia (named since ancient times for a queen of Ethiopia) in the northeast. Although most of the stars in the Great Bear are not very prominent, you should have no difficulty locating the Big Dipper. At this time of night and year, the handle of the Dipper points northwest.

The brighter stars of Cassiopeia form a group shaped somewhat like an M or a W, depending on the season. They are supposed to resemble the queen's throne, but it is easier to remember the M or W pattern. At this time of year, the W is lying on its left side.

Midway between Cassiopeia and the bowl of the Big Dipper lays the star **Polaris**. This star is located near the north celestial pole in the sky. Notice that two of the stars in the bowl of the Big Dipper (called the Pointer Stars) can be used to locate Polaris, as shown by the arrow in Figure 3-4.

There are stars below Cassiopeia and the Big Dipper, but they are fainter and more likely to be obscured by haze, dust, and pollution when looking at the sky close to the horizon. Most city dwellers will not be able to see objects closer than 20° to the horizon, except those that are unusually bright. Objects near the horizon will also tend to twinkle a great deal, varying rapidly in both brightness and color. This is why observing the sky through an atmosphere is so difficult.

THE NORTHERN SKY AT OTHER TIMES OF THE YEAR

If the appearance of the sky were always the same, it would be easy to become familiar with it. But the rotation of the Earth and its orbital motion around the Sun complicate the picture and cause the appearance of the sky to vary. The changes are regular, however, and you can use **Figure 4-30** to help you find the northern stars at any time of year. Facing north, hold Figure 4-30 in front of you so that the name of the month is at the bottom. This will show the orientation of the various stars at 9 P.M. (standard time) on that date. For example, if you put September at the bottom, the configuration will be similar to Figure 3-4.

Inquiry 4-17

What are the locations of the polar star groupings at 9 P.M. on February 1?

Inquiry 4-18

At what time of year is the Big Dipper highest in the sky at 9 P.M.?

Inquiry 4-19

At what time of year would it be hardest to see Cassiopeia because of horizon haze at 9 P.M. standard time, or 10 P.M. daylight saving time?

The Little Dipper, shown in Figure 4-30, is actually part of the constellation of Ursa Minor, the Little Bear. When you look at the stars in this constellation, it will be immediately apparent that they are much fainter than the stars of the Big Dipper and are therefore much harder to see. If either city lights or the Moon make the sky bright, you will probably be able to see only the brightest stars of the Little Dipper—Polaris, at the end of the handle, and the two stars at the end of the bowl. If all the stars in the Little Dipper are visible, then the sky is exceptionally dark and clear, and you have very good observing conditions. On such an evening, you will also be able to see the stars of Draco, the Dragon, snaking between the bowls of the two dippers; and Cepheus, a king of ancient Ethiopia, between the Little Dipper and Cassiopeia. These constellations are shown in Figure 4-30.

OBSERVING THE REST OF THE SKY: USING THE STAR MAPS

The maps in Appendix H will be your guide to the other parts of the sky that can be observed at any time of year. You may view the following paragraphs as a guide in preparing to go outside, or you can read them using a red flashlight while outside under the stars.

To see how they can be used, use the map index at the start of Appendix H to select the map that represents the sky on September 5 at 9 P.M. standard time (10 P.M. daylight saving time). Hold the map so that the label NORTH can be read (i.e., turn the page upside down). As shown in **Figure 4-31**, you will see a representation of the stars in the northern part of the sky in the lowest of the four segments into which the map is divided. Compare this with Figure 4-30. Notice that groupings such as the Big Dipper and Cassiopeia are located similarly on the two maps.

Now hold the map so that the label WEST can be read at the bottom. This time, the lower portion of the map shows the appearance of the sky over the western horizon. You will see that the star Arcturus, the brightest star in the constellation of Boötes, the Herdsman, dominates the western sky. In a similar way, the stars of the southern and eastern skies can be found by holding the map with the appropriate direction at the bottom. To observe the stars overhead, hold the map up over your head and look at the central part of it; be sure to match directions on the map with the directions on your horizon.

Directions:
Hold map so that month of observation is at the bottom.

December

November

January

Alkaid

Mizar

Alioth

Phecda   *Megrez*

October

February

"Big Dipper"

Merak

Dubhe

Draco

Ursa Major

Ursa Minor

September

March

Polaris

Lynx

Cepheus

Camelopardalis

August

April

Cassiopeia

Magnitude
◇ ● ● · ·
0 1 2 3 4 5

Capella

July

May

Perseus

June

Figure 4-30 Perpetual map of the northern sky. When held with the current month at the bottom, it shows the northern sky at 9 P.M. standard time.

Like all maps, these try to represent the spherical surface of the sky on a flat piece of paper, and this cannot be done perfectly. You can see how it works by imagining that you have cut out the map along its edges and then taped the edges together. The resulting construction would resemble an inverted bowl, like the sky.

There are 12 maps in Appendix H, each of which represents the sky for a specific combination of date and time throughout the year. Use the map index to choose the map that most closely corresponds to the time and date of your observations. If you are observing at a time not given on the maps, you can easily calculate which map would be the right one using the fact that a difference in time of two hours corresponds to a difference in date of one month. For example, the ninth map lists September 5 at 9 P.M. and August 5 at 11 P.M. as the times for its use. It could also be used on July 5 at 1 A.M. or June 5 at 3 A.M.

Figure 4-31 Using the full-sky star maps.

Inquiry 4-20

In which direction (north, south, east, west, or overhead) would you look to find the Great Square of Pegasus at 9 P.M. on September 5? At 8 P.M. on November 20?

Inquiry 4-21

At 8 P.M. on March 20, what bright star is closest to the zenith (the point overhead)?

Inquiry 4-22

Where in the sky would the constellation of Orion be found on January 1 at 8 P.M.?

OBSERVING TIPS

You will find it much easier to find your way among the stars if you follow these eight tips; they are the product of much experience.

1. *Observe under a dark sky.* The darker and clearer the sky, the more you will be able to see. It is best to leave the city entirely, but this may not always be possible. At the very least, you should avoid obvious city lights and find a spot that is as dark as possible. Furthermore, you should try to avoid times of the month when the Moon is nearly full (consult your local newspaper or a calendar for information about when the Moon will be full), as a nearly full Moon makes observations of faint stars difficult, if not impossible. If you are able to see all the stars in the Little Dipper, you know you have a good night to observe. Be aware of what other people around you are doing, and leave an area if you feel unsafe; consider observing with a partner.

2. *Adapt to the dark.* The pupil of your eye changes its size to regulate the amount of light it admits. In bright sunshine, the pupil contracts to a small diameter, whereas at night, it opens up wide to admit as much light as possible. It takes time for the eye to become fully adapted to the dark, and you will not be able to see the fainter stars at first. They will begin to appear after about five minutes; full adaption to the dark takes about 15 minutes. Many more stars can be seen when your eyes are fully adapted. (Smoking and drinking alcoholic beverages will both impair your ability to become fully dark-adapted.)

3. *Use a red flashlight.* A bright flash of light can destroy your dark adaption for many minutes, yet you will need to use a flashlight to read your star map and make notes from time to time. A compromise is to cover a flashlight with red cellophane, both to cut down its brightness and because red light is less likely to harm your eyes' dark adaption. When using your flashlight, try to keep it faint and use it as little as possible.

4. *Use averted vision.* Different parts of the retina of the eye have different sensitivities. There is a blind spot where the optic nerve joins the retina, and there is also a region of maximum sensitivity that you can use by looking slightly to one side of the object you are observing rather than directly at it. You can best learn this trick by experimentation: pick a faint object and see how its brightness appears to vary as you look at spots in its immediate vicinity.

5. *Be alert for clouds and haze.* Certain types of clouds, especially a thin layer of cirrus, can be difficult to detect when they appear during an observing session. Even though the brighter stars are easily visible, the fainter ones may disappear completely. Thus, you must be on the alert for changing sky conditions.

6. *Take your time.* The most common mistake among beginning observers is hastiness. If you are observing an object or a region of the sky,

don't just glance at it and expect to take in all that there is. Not even an experienced astronomer can do that. Instead, concentrate on it and, if what you are observing is large, break it up into smaller regions and observe each smaller region carefully.

7. *Move from the known to the unknown.* It would be difficult to learn all the constellations at once. Start with easily recognized groupings near ones you already know, using a star map as a guide. Look for simple geometrical figures: triangles, squares, rectangles, and so forth. Gauge distances by consulting your map and comparing them with the sizes of familiar groupings.

8. *Make reliable records.* Write down or draw everything you observe. Do *not* entrust anything to memory. Without detailed and accurate written records, you will not be able to observe the various changes in the sky that are discussed in this book. Finally, the process of making a written record will encourage you to observe more carefully.

Chapter Summary

Observations

- A variety of ancient civilizations observed celestial phenomena and built structures that allowed them to make astronomical predictions.

- Most ancient civilizations used the motions of the Moon and/or Sun around the celestial sphere as the basis of their various calendar systems.

- Early observations consistent with the Earth being spherical included the disappearance of ships over the **horizon** and the curved shape of the Earth's shadow on the Moon during a **lunar eclipse**.

- The Sun, Moon, planets, and stars appear to rise in the east and set in the west daily. Night after night the Sun appears to move farther eastward in relation to the background stars. The Moon shows a similar but larger apparent eastward movement. While planets generally have an eastward movement among the stars, they sometimes reverse direction, showing a **retrograde** motion. **Precession** is the 26,000-year apparent motion of the celestial poles on the celestial sphere caused by the wobbling of the Earth's spin axis.

- Observations of the sky can be described in terms of a **celestial sphere** on which celestial objects are located. The **celestial poles** are the projection of the Earth's rotation axis onto the celestial sphere; the **celestial equator** is the projection of the Earth's equator onto the celestial sphere. The path of the Sun on the celestial sphere is called the **ecliptic**. The location of objects on the celestial sphere is given by **declination** and **right ascension**, which are analogous to latitude and longitude on Earth. The detailed appearance of the sky depends on the observer's latitude.

- Stars that are not circumpolar rise four minutes earlier each day, thus changing the stars observable throughout the year. A **solar day** is about four minutes longer than a **sidereal day**.

- Seasons are produced by the Sun's changing altitude throughout the year. Such changes result from the Earth's rotation axis having a 23.5° tilt with respect to the orbit's perpendicular.

- The Moon exhibits monthly **phases** produced by the changing angle between the Sun, Earth, and Moon.

- **Eclipses** occur when the Sun, Moon, and Earth are closely aligned. **Solar eclipses** occur at **new Moon**, while **lunar eclipses** are at **full Moon**.

- The observed time interval between successive planetary **configurations** such as **opposition** to opposition or **conjunction** to conjunction is a planet's **synodic period**.

Theory

- The sphere, according to Greek philosophy, is a perfect geometrical form.

Conclusions

- From a variety of observations (given above), the Greeks concluded that the Earth is round.

- A planet's **sidereal period** is not directly observable but must be derived from the observed synodic period.

Summary Questions

1. What observations did ancient people use to indicate that the Earth was round? How did these observations interact with the culture's theoretical and philosophical underpinnings?

2. What are the principal motions of the Earth? Explain the effects that each of these motions has on the apparent motions of the stars and other objects in the sky.

3. What effect does an observer's latitude have on what he or she will see in the sky?

4. Explain why the solar day is longer than the sidereal day.

5. What is precession? What is its effect on the positions of stars?

6. How can you determine your latitude from the observed altitude of Polaris?

7. What do we mean by the terms *celestial sphere, celestial equator, ecliptic, meridian, declination,* and *right ascension?*

8. What two factors result in seasonal variations on Earth? How do they operate to create differences between summer and winter in the Northern Hemisphere?

9. How do the different phases of the Moon come about? Your answer should include drawings.

10. What causes solar eclipses? Lunar eclipses? At what phase of the Moon does each occur?

11. What are the observed apparent motions of the planets with respect to the background stars? Describe them.

Applying Your Knowledge

1. Is the model of the Earth suggested by Anaximander any better in explaining observations than the Babylonian model? In particular, compare the observations of the sky that could be predicted using the Babylonian model with those that could be predicted using the Greek model.

2. What effects would precession have on the seasons?

3. What would the Earth's phase be for an observer on the Moon if the lunar phase for an observer on Earth is waxing crescent?

4. Where in the sky would you look at midnight to see Jupiter when it is at quadrature?

5. If the star Vega rises tonight at 3 A.M., at what time should you be looking for it to rise three months from now? Explain.

6. If the Sun rises at 6 A.M., at what approximate time does the waxing gibbous Moon set?

7. Use tracing paper to trace stars from the maps in Appendix H; then try making your own con-stellations. Are the traditional constellations any better or more useful than yours?

8. What is the maximum possible lunar altitude for an observer at the North Pole? (Hint: A drawing will be helpful.)

9. Use a drawing to help you explain which conditions produce the most favorable total solar eclipse: Earth nearest or farthest from the Sun, and Moon nearest or farthest from the Earth.

10. At what latitude is an observer for whom the stars within 25° of the pole are circumpolar?

11. Refer to Figure 4-13b and determine the latitude at which the photograph was taken.

12. How many degrees into the southern celestial hemisphere can an observer at a latitude of 40° north see?

13. For what southern latitudes will the Sun never be observed at the zenith?

4-1. *Flat Earth:* The same constellations would be observed from all locations; the height of the Sun above the horizon at noon would be constant throughout the year; it would always be either night or day everywhere on Earth; the shadow on the Moon during lunar eclipses would be a straight line. *Round Earth:* There would be a change in constellations as the observer moves north or south; the height of the Sun above the horizon at noon *might* vary throughout the year; half of the Earth would be in night while the other half has daylight; the shadow on the Moon during lunar eclipses would always be part of a circle.

4-2. It would produce a straight-line shadow on the Moon.

4-3. *Observational:* Disappearance of boats, different stars seen at different points on Earth, and the shape of the Earth's shadow on the Moon. *Theoretical:* Solves the problem of "which way is down" and fits in with the idea that the sphere is the most perfect shape.

4-4. On the horizon; it's 90° away from the equator.

4-5. Stars must move parallel to the celestial equator because an observer's motion caused by Earth's rotation is parallel to Earth's equator.

4-6. It would be the apparent path of the *Earth* around the celestial sphere. Thus, it would be the plane defined by the Earth's orbit.

4-7. 10:00 P.M. Stars rise 4 minutes earlier per day; at 30 days in June, in one month we would have 4 minutes per day times 30 days equals 120 minutes or two hours earlier.

4-8. The Sun is 23.5° south of the celestial equator. The Sun is moving low across the southern sky during the day. Most of its daily motion is below the horizon, giving the shortest number of daylight hours of any day of the year.

4-9. Since at the equator the altitude of the Sun at noon varies by only 23.5° on either side of the zenith throughout the year, it is always nearly overhead at noon. For this reason, seasonal variations are small. At a pole, however, the variations are extreme because for six months of the year the Sun never gets above the horizon.

4-10. The Moon moves eastward at about 13° per 24 hours, or about 0.5° per hour. This equals a little more than the Moon's diameter every hour.

4-11. 6 P.M.

4-12. New Moon.

4-13. Full Moon.

4-14. The Sun is in Capricornus (Cap); Mercury is between Capricornus (Cap) and Sagittarius (Sgr); Venus is between Ophiuchus (Oph) and Scorpius (Sco).

4-15. Venus would show a quarter phase; Jupiter, gibbous.

4-16. The sidereal period is difficult to observe because the positions of both the planet and Earth change, making it difficult to know when the planet has completed one full cycle measured against the stars.

4-17. Cassiopeia and the bowl of the Dipper will both be at about 30° altitude, on either side of the pole star.

4-18. May.

4-19. April, May.

4-20. East, overhead.

4-21. Castor and Pollux.

4-22. In the southeast, about 50° up from the horizon.

The Historical Quest to Model the Solar System

The progress of science is generally regarded as a kind of clean, rational advance along a straight line; in fact, it has followed a zig-zag course, at times almost more bewildering than the evolution of political thought. The history of cosmic theories, in particular, may without exaggeration be called a history of collective obsessions and controlled schizophrenias; and the manner in which some of the most important individual discoveries were arrived at reminds one more of a sleepwalker's performance than an electronic brain.

Arthur Koestler, The Sleepwalkers, *1959*
The Granger Collection, New York

n this chapter we follow the development of our understanding of our own corner of the universe, the Solar System. We will be interested in how the ideas that developed in one era were a natural outgrowth of the state of astronomical observations and degree of sophistication of the times. In previous chapters we briefly discussed some historical aspects of astronomy; for example, we considered the idea of a spherical Earth, and how this concept slowly came to be accepted by 100 B.C.E. by Greek thinkers through the gradual accumulation of evidence from observations of both the Earth and the heavens. Yet this concept was established relatively rapidly when compared to the slow acceptance of the idea of a Sun-centered Solar System.

5.1 Greek Astronomy

Although we saw in the last chapter that a variety of cultures were intimately involved with phenomena of the sky, it was Greek culture that made progress toward the models we have today. We now look more closely at the astronomical contributions of ancient Greek culture.

WHAT THE GREEKS INHERITED

The Babylonians and Egyptians bequeathed to the Greeks an extensive body of astronomical knowledge. However, the astronomy of both ancient Egypt and Babylonia was the province of a priestly aristocracy; as a consequence, practical and political considerations often took precedence over theoretical inquiries. Egyptian astronomers, for example, had discovered that when the bright star Sirius could just be seen rising in the east before the Sun, the flooding of the Nile was imminent. Such knowledge gave tremendous power to the priesthood and inevitably involved its members closely with the state. Similarly, in Babylonia, astronomer-priests had acquired a considerable amount of information concerning the motions of the Moon, Sun, and planets and had found certain regular cycles that enabled them to predict some eclipses, a power that was frequently used for political purposes.

The ancient Egyptians and Babylonians knew the length of the year and the different types of calendars, both solar and lunar. The Egyptians had learned the rudiments of simple mathematics, algebra, and geometry. Sundials had been invented, and systems of timekeeping were in exis-

Figure 5-1 A clay tablet from ancient Mesopotamia containing astronomical observations of Jupiter in the top part, and a description of the method of calculation in the bottom part. The British Museum

tence. The Babylonians had made systematic observations of the positions of heavenly objects and had practical methods for predicting the positions of the Moon, Sun, and planets. **Figure 5-1**, for example, shows a table of data for Jupiter. The bottom part of the figure describes the method of calculation. The Babylonian value for the length of the synodic period of the Moon was not surpassed in accuracy until the end of the nineteenth century. Both cultures had attempted to construct a cosmology that placed Earth and humanity in their proper position in relation to the universe and the gods, but they never attempted to construct a truly consistent theoretical framework for their cosmology in the way the Greeks did.

Chapter opening illustration: Galileo showing the heavens to the public in Venice with one of his telescopes. The Granger Collection, New York.

ARISTOTLE

The Greeks enjoyed philosophy, by which they meant a broad attempt to understand all things. **Aristotle** (384–322 B.C.E.) viewed the universe in the abstract and thought such thinking was sufficient to provide an understanding of it.

Thus, Aristotle claimed that an object sought to be at rest in its natural place, and all natural motions expressed the striving of an object to rest as close as possible to its natural place. Given that the thinking of the time said there were four elements making up the universe—earth, water, air, and fire—objects made of "earthy material" fell to Earth because their natural place was close to Earth.

In addition, he taught that motions required continuous contact between the mover and the moving object to perpetuate the motion. As discussed later in this chapter, he argued that planets would move in circular orbits. Such ideas came from philosophy and were not tested; they were the mainstay of scientific knowledge until the seventeenth century.

ARISTARCHUS OF SAMOS

When Greek philosophical thinking was coupled with their highly developed mathematical skills, the Greeks advanced so rapidly in astronomy from 600 B.C.E. to 100 B.C.E. that their era was without question one of the shining examples of scientific discovery prior to the year 1500 C.E. Progress in understanding our solar system came especially quickly with the emergence of the Alexandrian school of Greek astronomers, particularly the work of **Aristarchus of Samos** (c. 300 B.C.E.), who combined careful observations and sharp reasoning to draw inferences about the relative sizes and distances of the Earth, Moon, and Sun.

Aristarchus demonstrated that the Sun was many times farther from the Earth than the Moon by using a reasoning process illustrated in **Figure 5-2**. You can see that if the Sun were quite close to the Earth–Moon system, then the time interval from first quarter to third quarter would be *longer* than the time interval from third quarter to first quarter. Aristarchus observed that these two time intervals were in fact nearly equal, implying that the Sun was many times farther away than the Moon. Hence the rays of light from the distant Sun would be arriving in nearly parallel lines for all parts of the Moon's orbit.

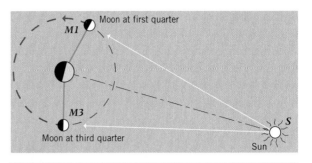

Figure 5-2 Aristarchus's method to show that the Sun is farther from the Earth than the Moon. If the Sun were very near the Earth-Moon system, the times between quarter phases would differ measurably. However, if the Sun were much more distant than the Moon, the lines *S-M1* and *S-M3* would be nearly parallel, and the time interval between first and third quarters would be nearly equal to the time interval between third and first quarters.

Inquiry 5-1

What did Aristarchus assume about the Moon's orbital shape and motion?

Aristarchus was also able to estimate the relative sizes of the Earth and Moon by timing the duration of lunar eclipses. **Figure 5-3** shows the principle of this determination. The length of time it takes the entire Moon to enter the Earth's shadow depends directly on (that is, is proportional to) the diameter of the Moon. The length of time it takes for one part of the Moon—say, its leading edge—to completely cross the Earth's shadow is proportional to the diameter of the Earth. By comparing these two times, we can estimate the ratio between the diameters of the Earth and Moon, as Aristarchus did.[1]

Inquiry 5-2

Suppose it takes an hour for the entire Moon to enter the Earth's shadow and four hours for the edge of the Moon to cross the shadow. How many times larger than the Moon's diameter is the Earth's diameter?

---

[1]Actually, the diameter of the Earth's shadow at the distance of the Moon is slightly less than the diameter of the Earth; however, this leads to only a small error.

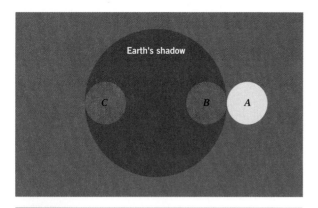

Figure 5-3 The principle of Aristarchus's measurement of the Moon's diameter relative to that of Earth. The time it takes the Moon to enter the Earth's shadow (between points *A* and *B*) is proportional to the Moon's diameter. The time the Moon takes to cross the Earth's shadow, between *A* and *C*, is proportional to the Earth's shadow diameter, which is nearly that of the Earth.

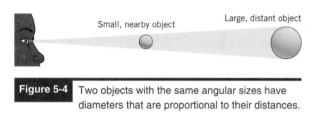

Figure 5-4 Two objects with the same angular sizes have diameters that are proportional to their distances.

Finally, Aristarchus knew that during a total eclipse of the Sun, the Moon was just able to cover the Sun, and therefore, the angular sizes of the Moon and Sun in the sky were about the same. He could then reason that the actual sizes of the Moon and Sun were proportional to their distances, and in this way he could estimate the diameter of the Sun (**Figure 5-4**). Unfortunately, although his estimate of the distance to the Moon was quite good, his estimate of the distance to the Sun was about 10 times too small. For this reason, his estimate of the Sun's diameter was also 10 times too small. Nevertheless, he was able to show that the distance to the Sun was considerably greater than that to the Moon, and that the Sun's diameter was much greater than the Earth's.

Inquiry 5-3

Aristarchus was the first to propose that the Earth goes around the Sun, rather than vice versa. Suggest one factor that may have led him to this conclusion.

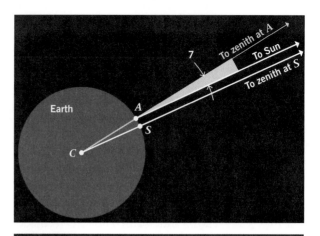

Figure 5-5 Eratosthenes' measurement of the diameter of the Earth. The difference in altitude of the Sun is proportional to the distance between the two points, which is known. This allows the length of one degree on the Earth's surface to be determined.

ERATOSTHENES

Another classic experiment of antiquity was the determination of the Earth's circumference by **Eratosthenes** (c. 200 B.C.E.). The conclusions of Aristarchus concerning the relative sizes and distances of the Earth, Moon, and Sun were all in terms of the then-unknown size of the Earth. Their actual sizes could not be known until the Earth's size was known.

Eratosthenes had heard that at Syene, near the modern Aswan in Egypt, there was a deep well, and that on a certain day of the year the Sun stood directly overhead so that its reflection could be seen in the water in the bottom of the well. Eratosthenes was also able to observe that on that same day of the year in Alexandria, where he lived, the Sun was not directly overhead but was 7° south of the zenith. He determined this angle with a gnomon, which is a pointed stick perpendicular to the ground. **Figure 5-5** shows the geometry of Eratosthenes' experiment. Point *A* represents Alexandria and point *S* is Syene; point *C* is at the center of the Earth. The Sun is so far away that the lines *A*-Sun and *S*-Sun are nearly parallel. The angle at *A* is nearly equal to the angle at *C* at the center of the Earth, and we may write the following proportion:

$$\frac{\text{Circumference of Earth}}{\text{Distance from } A \text{ to } S} = \frac{360°}{\text{Angle at } C}.$$

Inquiry 5-4

Assuming that the angle at *C* is 7° and that the distance of Alexandria from Syene is 5,000 stadia (the units used by the ancient Greeks; stadia is the plural of stadium), what is the diameter of the Earth in stadia? Although the exact length of a stadium is unknown, it is thought to be about 0.16 of a kilometer (one-tenth of a mile). Using this value, compute the radius of the Earth.

Inquiry 5-5

What does Eratosthenes' experiment assume about the shape of the Earth?

Inquiry 5-6

If the Earth were flat, what would be the value of the angle at *C*?

HIPPARCHUS

Perhaps the greatest of all ancient astronomers was **Hipparchus** (c. 150 B.C.E.). Many of the conclusions he drew were so sophisticated that it takes some knowledge of astronomy to appreciate how great his contributions were. He built an observatory, constructed the best astronomical instruments up to that time, and established a program of careful and systematic observations that resulted in the compilation of a great star catalog, with 850 entries, using a celestial coordinate system similar to our modern one for cataloging the sky. It was Hipparchus who originated a system, which is still in use today in modified form, for estimating the brightness of stars. In addition, he used the older Babylonian observations and deduced Earth's precession (Section 4.6), which is so slow that it takes almost 26,000 years for it to complete one cycle. Finally, he greatly developed trigonometry, which was, and still is, a useful tool for astronomy.

OTHER DEDUCTIONS OF THE GREEK ASTRONOMERS: THE DISTANCES OF THE PLANETS

The Greeks estimated the relative distances of the planets from Earth using principles still in use today for determining distances to astronomical objects. They reasoned that the more distant a planet was, the more slowly it would appear to move across the sky. The effect is similar to what happens when we compare the apparent motion of a high-flying airplane with that of one that is fly-ing low. The distant airplane appears to move slowly across the sky, whereas the low-flying one is seen for only a short time and then is gone. In the same way, the Greeks could put most of the naked-eye planets in order of their distance from Earth by assuming that increasing distances corresponded to slower motions. The argument fails with Mercury and Venus, however, because it places Mercury closer to Earth.

Inquiry 5-7

What assumption results in Mercury being placed closer to Earth than Venus?

We have another, independent determination of relative planetary distances from their brightnesses. We use an analogy: when you are driving at night on a two-lane road and wish to pass the car in front of yours, you pass only if the headlights of an oncoming car are faint. When you do this, you are making an implicit assumption: all car headlights have about the same intrinsic brightness, with their apparent brightness depending on the distance. Similarly, if we assume that all planets have the same intrinsic brightness, then their apparent brightness as seen from Earth would depend on their distances from us. Of course, all the planets do not have the same intrinsic brightness, because their differences in size, distance from the Sun, surfaces, and atmospheric properties affect the amount of light they reflect in our direction. However, even allowing for these uncertainties, it is still possible to use this principle to rank the planets approximately in order of distance from the Earth.

THE APPARENT MOTIONS OF THE PLANETS IN RELATION TO THE STARS

Three additional observations of planetary motion were important in determining the details of the models the Greeks developed. We now summarize these observations, which were discussed in detail in the previous chapter:

1. Because the planets are considerably closer to us than the fixed stars, they appear to move against the starry background. Observations of Mars, Jupiter, and Saturn showed them to move generally eastward on the celestial sphere. In other words, stars move *with* the

celestial sphere whereas the Sun, Moon, and planets move *on* the celestial sphere.

2. Occasionally, however, as discussed in Chapter 4, a planet's motion becomes, retrograde, in which it changes from eastward to westward for up to several months before it slows down and again reverses its direction, resuming its normal easterly motion (see Figure 4-28).

3. Venus and Mercury are never more than 48° and 28°, respectively, from the Sun.

THE GEOCENTRIC MODEL OF THE SOLAR SYSTEM

Where should the center of the system be? There were really only two obvious candidates—the Earth and the Sun. This question was considered carefully by Greek philosophers, and the fact that ultimately they reached an incorrect conclusion provides an interesting example of why the scientific method is not the simple turn-the-crank-and-the-answers-fall-out process that some sources describe it to be. If the Sun is at the center of the solar system, then Earth moves around it in space. Such a hypothesis provides a prediction. As shown in **Figure 5-6**, some of the stars ought to shift their apparent positions in the sky as the Earth moves from one side of its orbit to the other. Such **parallax** effects, as they are called, were looked for by many Greek observers, including Hipparchus, but were never found. The Greeks therefore concluded that the Earth was stationary in space.

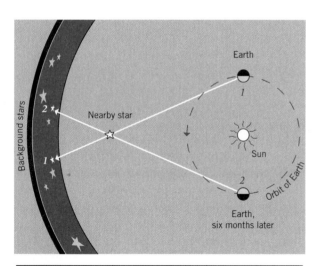

Figure 5-6 The prediction of stellar parallax in a heliocentric universe. The apparent position of a star with respect to the background stars would appear to change if the Earth went around the Sun. (Not to scale.)

Aristarchus, however, apparently supported the hypothesis that the Earth orbits the Sun, if surviving works of Archimedes and Plutarch are correct. Unfortunately, the work in which he put forth his hypothesis is lost, and apparently no other Greek astronomers held to this opinion. The model of a **geocentric** (Earth-centered) system easily won out over the **heliocentric** (Sun-centered) one.

Inquiry 5-8

What hypothesis might explain why stellar parallaxes were not observed, even though the Earth does in fact orbit the Sun?

THE HYPOTHESIS OF CIRCULAR MOTION

A second feature of the Greek models of the solar system provides another excellent example of the way in which scientific models can go astray—inflexible assumptions. Ever since the time of Pythagoras (c. 570–500 B.C.E.) and Aristotle, the circle and the sphere had been considered to be the most perfect geometrical figures. Such perfection was expected in the natural world. (Even today, symmetry and simplicity in scientific thought are important concepts.) Thus, the great astronomer Hipparchus assumed the widely accepted idea that celestial objects, being perfect, could move only in circular orbits. Furthermore, this circular motion had to be uniform: an object moved the same angular distance in its orbit each day. Because of the observations of retrograde motion, however, Hipparchus cleverly extended the idea to include motion that was a combination of circular motions. In this way it was eventually possible to model the retrograde motion of the planets.

The idea of combining circles geometrically is due to Apollonius of Perga (c. 265–190 B.C.E.), but Hipparchus was the first to apply the idea to actual celestial objects when he proposed a hypothesis of the motion of the Sun and Moon. However, he did not have enough data to apply it to the planets; this final step was carried out by the astronomer Ptolemy (whom we discuss shortly).

The explanation of retrograde motion by means of combinations of circular motion is illustrated in **Figure 5-7a**. The planet moves around a small circle called an **epicycle**, and the center of the epicycle moves around on a larger circle called the **deferent**. Because one can adjust the relative sizes of the epicycle and deferent, and the speeds with which motion takes place on each, it is not difficult for the model to produce complicated apparent motions. In particular, if the planet moves around the epicycle faster than the epicycle moves around the deferent, retrograde motion will be observed at certain points along the orbit, as shown in **Figure 5-7***b*.

Using the epicycle and deferent, it became possible to explain the special motions of Mercury and Venus by adding only one more feature to the model. If the center of the epicycle of Venus, for example, is attached firmly to the line joining the Earth and Sun, then Venus must always remain near the Sun, as shown in **Figure 5-8**. In a similar fashion, Mercury's epicycle is attached to the Earth-Sun line. Now, as the Sun moves along its deferent at a rate of about 1° per day, it carries Mercury and Venus along with it.

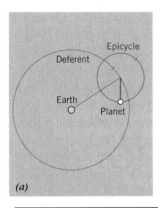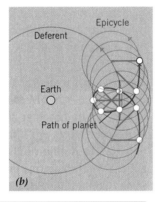

Figure 5-7   A geocentric explanation of retrograde motion. *(a)* The planet moves on an epicycle, which moves about the Earth on a deferent. *(b)* The motion resulting from a planet moving about an epicycle as it moves around the deferent.

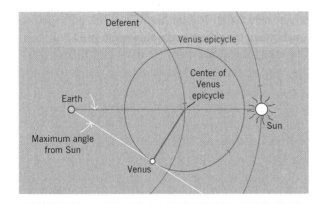

Figure 5-8   An explanation of the motion of Venus, using epicycles. According to this model, Venus is always closer to Earth than the Sun is.

Inquiry   5-9

What phases of Venus are possible for the model shown in Figure 5-8? (Hint: Make an enlarged drawing.)

PTOLEMY

The last of the great ancient astronomers was **Claudius Ptolemy** (c. 150 C.E.). In fact, the geocentric system that was previously discussed was passed down bearing his name—the **Ptolemaic system**. Ptolemy completed the explanation of planetary motion in terms of combinations of circles and added a number of complex refinements to Hipparchus' system to improve the model's agreement with observations.

Ptolemy's aim was to produce a model that correctly predicted observations, which remains a key goal of science today. In that sense, he succeeded admirably. The Greeks believed that a true knowledge of reality was confined to the gods alone, and the best that humans could do was to produce descriptions of the observed world that would correctly predict the results of observation and experiment. Leaving aside the gods, these ideas are close to the spirit of modern science, in which even everyday concepts such as force and mass, which modern physicists think they understand quite well, have meaning only in reference to models and the measuring processes that are appropriate to them.

Models are powerful tools not only to interpret observations, but to suggest possible new observations. In fact, an incorrect model can often lead one to ask the wrong questions, and thus can delay understanding. For example, Kapteyn's model for the universe (described in Section 3.5) was wrong and delayed our understanding of Earth's position in the Galaxy. In astronomy, models necessarily become more and more uncertain as the objects they represent become more distant, because we have less and less data on which to base the model. Despite this, models play a central role in our quest to understand the universe.

Inquiry   5-10

What are some models, in the sense just described, that are used in everyday life? (You may want to think about such fields as politics, economics, education, philosophy, or psychology.)

5.2 Astronomy during the Middle Ages

No significant advances were made for the next 13 centuries of astronomical history. Greek civilization declined and the Greeks became subservient to the Romans. (This had already taken place by the time of Ptolemy.) If the Greeks were scientists, the Romans were engineers, doing little basic science. Progress in science came to a virtual standstill in the West.

A different story emerged in the Middle East. The astronomical knowledge that originated in Greece reached the Arab world, and scientists and scholars there continued the pursuit of knowledge. There was a flowering of literature, art, and science throughout the Middle East. One of the most important events was the preservation of much of the scientific knowledge of the ancient world, at least in translation. Ptolemy's great work, for example, survived as a book called the *Almagest*, Arabic for "the greatest." In this form the Arabs eventually transmitted Ptolemy's writings back to the West.

A considerable amount of original work was done in the Islamic world. Much of it was directed toward practical matters, such as navigation. Beautiful and intricate instruments such as the astrolabe (**Figure 5-9**) were perfected. Much theo-

retical work was accomplished, particularly in mathematics, aided by the Arabic system of numbers (that had come from India), as well as by perfection of the methods of algebra (the word *algebra* is Arabic). Imagine doing your income taxes using Roman numerals!

At the same time in China, an indigenous tradition of astronomy made important progress during long periods of political stability. Astronomy was always closely connected with the state, and changes in dynasty were frequently accompanied by calendar reforms, providing secure employment for court astronomers. Early on, the Chinese developed the view that space was infinite in extent and that the stars floated independently in it. They were assiduous and systematic observers of heavenly events; often our only records of novae and comets are Chinese. They invented advanced instrumentation for observing the sky and measuring time, including elaborate water-driven clocks. In many ways their science was more advanced than what the Greeks had accomplished. Joseph Needham, the eminent scholar of Chinese science and technology, has shown that the influence of Chinese science on the West was much greater than has generally been recognized.

In Europe, by contrast, intellectual decadence was widespread. The principles of Greek science, received from the Arabs, were adopted as dogma by the increasingly powerful Roman Catholic Church. It was argued that there was only a finite amount of knowledge, and that the Greeks had discovered it all! Discovery was no longer necessary; humans had only to memorize what the Greeks had found to be true. The Ptolemaic picture became frozen into a rigid image of the universe with no room for independent thought and inquiry that would challenge this view of the universe.

5.3 The Heliocentric Hypothesis

We will now look at the contribution of three scientists whose work challenged the status quo and helped move Western thought to a Sun-centered view.

NICOLAUS COPERNICUS

During the Middle Ages, tinkering with the Ptolemaic model had continued, but in an uncreative way—for example, by adding more epicy-

Figure 5-9 A Persian astrolabe. The small leaves mark the positions of bright stars. By suspending the astrolabe by the ring and sighting along the movable sight, we can measure the altitude of a star.

cles. In fact, this model had become so complicated that King Alfonso of Castile is said to have remarked, when having the contemporary version of the Ptolemaic hypothesis explained to him, that if *he* had been around when the world was created he could have taught the Creator a thing or two.

The human mind cannot be contained forever. By the early sixteenth century the Renaissance was shaking the antique castles of Western intellectualism. **Nicolaus Copernicus** (1473–1543), a Polish prelate with a strongly mathematical bent as well as a new vision of the heavens (**Figure 5-10**), found himself in this changing atmosphere. As a student in Italy, Copernicus read of Aristarchus's heliocentric hypothesis. Wherever he got the idea of placing the Sun at the center of the universe (**Figure 5-11**), there is no doubt that he felt this was the proper place, for as he himself says, "For who would place this lamp of a very beautiful temple in another or better place than this wherefrom

it can illuminate everything at the same time?" An important benefit of the heliocentric model is that it allowed the sidereal orbital periods and relative distances of the planets to be determined from the observed synodic periods of the planets (as discussed in the previous chapter).

Moreover, in a heliocentric model, the apparent retrograde motion of a planet like Mars is a simple consequence of the relative motions of the Earth and Mars. **Figure 5-12** shows how this comes about. At some time Mars will appear in direction *1*. As each planet orbits the Sun, Mars's line-of-sight

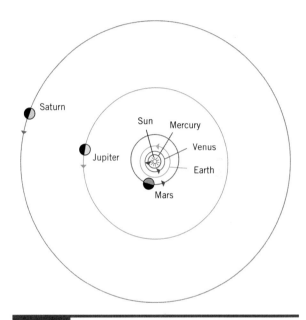

Figure 5-11 The Copernican model of the Solar System.

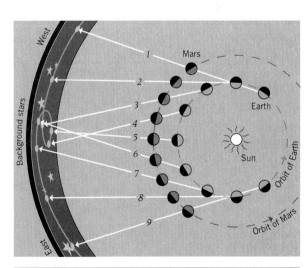

Figure 5-12 Retrograde motion is a natural consequence of the heliocentric hypothesis and occurs as Earth approaches and then passes the planet. The loop appears because the orbital planes of the Earth and the planet are not identical.

Figure 5-10 Nicolaus Copernicus (1473–1543) proposed the daring idea that the Sun, not the Earth, was at the center of the universe.

position in relation to the background stars will move toward the east, as indicated by lines *1* through *4*. However, between times *4* and *6* Mars will appear to move backward, in a retrograde direction, toward the west. At point *6* Mars will have resumed its normal easterly track. As you can see, Mars shows retrograde motion whenever the Earth laps it in its orbit. The greater orbital speed of the Earth results in the apparent backward motion of Mars, which is actually a parallax effect. One sees a similar effect when a fast automobile passes a slower one on the highway. Viewed from the faster car, the slower one appears to move backward when seen against the distant landscape.

Inquiry 5-11

What would be the relative positions of the Earth, Sun, and Mars for the retrograde motion to occur?

Despite its simple explanation of retrograde motion and elegant positioning of the Sun in the center of the solar system, many objections were raised against the Copernican hypothesis. For example, some contended that if the Earth moved it would leave the Moon behind, an argument that was hard to answer because the concept of gravitation had not yet been developed. More seriously, Copernicus's hypothesis in this simple form could not explain the observations as accurately as Ptolemy's hypothesis. It was necessary to introduce a large number of small epicycles to explain numerous small deviations from circular motion. Moreover, Copernicus was opposed to Ptolemy's complicated refinements, feeling that they detracted from the perfection of uniform circular motion and were unworthy of celestial objects. This conservatism on Copernicus's part meant that he actually needed *more* epicycles than Ptolemy to get equivalent accuracy, and this marred the aesthetic simplicity of his basic heliocentric hypothesis, which used circular orbits.

Inquiry 5-12

Employing the principle of Occam's razor (the simplest model is preferred), from the point of view of astronomers of the time, which model of the solar system would have been preferable, Ptolemy's or Copernicus's?

Unquestionably, Copernicus was aware of the difficulties with his model. For example, the model predicted that parallax should be observed for nearby stars, which it was not. However, he realized the lack of parallax meant the stars were far away and too small to be observed with the instruments then available. In addition, he was a rather timid man, and it is possibly for these reasons that publication of his hypotheses was delayed until he was literally on his deathbed. His work appeared, in the year 1543, titled *De Revolutionibus Orbium Cœlestium,* which means *On the Revolutions of the Heavenly Spheres.* It contained a preface stating that the book expounded a mathematical model and was not to be construed as a representation of reality. It is certain that Copernicus didn't write the preface, and it is doubtful that he would have sanctioned its inclusion had he known of it.

Copernicus hypothesized a heliocentric universe; he did not prove it. Validation of his ideas was to require input from additional scientists over the following decades and centuries.

TYCHO BRAHE

In 1546, a Danish nobleman was born who was to become the first great observational astronomer of the modern era—**Tycho Brahe** (pronounced Tee-ko Bra-he) (**Figure 5-13**). Tycho became interested in astronomy as a teenager when he observed a predicted eclipse of the Sun. Later, in 1572, he observed a supernova, an exploding star, that was so bright it could be observed in daytime. He favored the heliocentric hypothesis, but the lack of supporting observations caused him to rethink his position. Through careful observations, Tycho showed that the daily rotation of the Earth caused no parallactic shifts in the supernova's position—in fact, the star did not move at all during the many months it was visible. He therefore concluded that it must be well beyond the Moon, and probably as distant as the stars. This caused instant problems for the older ideas of the universe, which had assumed that the heavens were unchangeable. Five years later, Tycho proved that the comet of 1577 was also more distant than the Moon. Because comets had been believed to be "exhalations of the Earth" this posed additional problems for the older ideas.

The fortunate combination of Tycho's noble birth and his astronomical talents soon earned him the finest observatory up until that time in the Western world. It was financed from the coffers of

Figure 5-13 Tycho Brahe (1546–1601) was the greatest naked-eye observer in the history of astronomy. The Granger Collection, N.Y.

Figure 5-14 Tycho's great mural quadrant, his principal instrument for measuring star positions. The Granger Collection, N.Y.

King Frederick II of Denmark and was located on an island off the Danish coast. Since optical instruments had not yet been invented, the instruments were sighting devices similar to the quadrant, protractor and cross-staff, but larger in size and constructed with the greatest possible precision. Tycho's principal instrument was the great mural quadrant (**Figure 5-14**). By sighting along the movable pointers in a manner not unlike sighting a gun, he could aim the pointers at two objects and read the angle between them with great accuracy (to about 0.5 minute of arc).

Tycho distinguished himself by the great care he brought to his observations. This included repeated observations and averaging them together to minimize random uncertainties. So painstaking was his work that his observations had unprecedented accuracy, not to be improved upon until invention of the telescope. Another valuable aspect of his observations was that he observed continuously and systematically over many years, as Hipparchus had done. As a result, he amassed a large body of data of consistently high accuracy that was suitable for further investigation. It is something of a mystery why an aristocrat such as Tycho could become as obsessed with astronomy as he was, and it is miraculous that his obsession led him to produce such excellent astronomical data. Another fortunate historical accident is that this body of work eventually fell into the hands of Johannes Kepler (1571–1630), who, as it turned out, was uniquely equipped to put it to good use.

JOHANNES KEPLER

Although Tycho was a great observer, he realized that he was not a strong mathematician, and he longed for a collaborator who could properly interpret his excellent observations. His accumulated observations on the positions of Mars posed the greatest difficulties for theoretical interpretation; Mars deviated from its predicted position more than any other planet. Having moved to Prague toward the end of his life, Tycho acquired a young assistant who, by his brilliance in mathematics, transcended his lower-class origins. **Johannes Kepler** (**Figure 5-15**) had sufficient genius to overthrow the sterile hypotheses of the geocentric universe and perfect circular motion and to substitute a new and truer description of the Solar System.

In his book *The Sleepwalkers,* Arthur Koestler describes the contrast between Tycho and Kepler. Tycho was an aristocrat by birth and arrogant by nature, accustomed to power and privilege. He ran his observatory like a court, arriving for the night in full formal dress and imperiously ordering his assistants about. Always sure of himself, in his college days he had lost his nose in a duel over a point of mathematics, and ever afterward he wore a false nose made of silver or bronze. He had a dwarf servant named Jepp, who followed his master about like a pet and received scraps of food at the supper table. Kepler, by contrast, had no advantages of birth and reached a position of eminence by virtue of brilliance and sheer tenacity. It appears that he had little in common with Tycho, other than irascibility and an interest in astronomy. During much of his life, he had to earn his living

by casting horoscopes, and in fact he cast his own horoscope every day. He was a compulsive individual who kept meticulous notes on everything he thought and did. A hypochondriac as well, Kepler recorded an hour-by-hour chronicle of his physical maladies. But because his diary also included his scientific efforts, we have been left a detailed account of the paths through which he wandered in making his momentous discoveries.

Kepler took Tycho's observations of Mars (literally took them, because Tycho's heirs had other plans for the data!) and set out to find a geometric curve that would represent its motion accurately. The calculations were extremely tedious, and he made many mistakes. Without calculators or even logarithms, every calculation had to be done by long multiplication or division. Yet, something in Kepler's character kept him working persistently at the problem. Time after time he rejected solutions that had taken him months, even years, to work out, because they failed to agree with the observations as accurately as he knew they should. At last, after a total of eight years, he boldly rejected the hallowed idea that planetary motion must take place on circular paths, thus ending two millennia of tradition. He describes in his diaries the fear and trembling he suffered in his mind when he took this step.

Figure 5-15 Johannes Kepler (1571–1630), seen here with Käiser Rudolf II, refined Copernicus's heliocentric hypothesis with the laws of elliptical motion.

5.4 Kepler's Laws of Planetary Motion

Kepler's laws of planetary motion are not *laws* in the sense that we use the word today, because there was at that time no conception of the physical forces that caused these motions. Although he came close to the idea of gravitational force, Kepler didn't quite succeed—he tried to construct a hypothesis in which the force was repulsive rather than attractive. His laws would be better described as *empirical* descriptions of planetary motions, meaning they were derived solely on the basis of observation with no theoretical underpinning. However, for the first time they gave a description that was as accurate as the best available data allowed.

Kepler found that Mars moved around the Sun on a mathematical curve called an **ellipse**. **Figure 5-16** shows a simple way to draw this curve. The two points through which the pins are pushed are called the **foci** of the ellipse (foci is the plural of **focus**). Ellipses are characterized by the length of their longest dimension (called the **major axis**—*B* in the figure) and their degree of noncircularity. If the two foci coincide, we have a circle. The further apart the foci, the greater is the **eccentricity** of the ellipse.

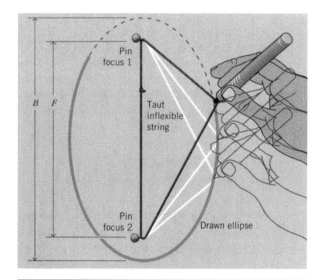

Figure 5-16 Drawing an ellipse with two pins and a loop of string. The pins mark the position of the two foci of the ellipse. The eccentricity of an ellipse is defined to be $e = F/B$, where F is the distance between the two foci and B is the longest axis (the major axis) of the ellipse. It characterizes how far the ellipse deviates from being a perfect circle.

Kepler's first law can be stated in this way:

 The orbits of the planets are ellipses, with the Sun at one focus.

Figure 5-17 illustrates Kepler's first law. The flattening of the ellipse, the eccentricity, is much exaggerated—no planet has an orbit this elliptical. In fact, if one were to observe these orbits from outside the solar system, it would be hard using the naked eye to distinguish most of them from circles. This is the principal reason why it took so long to discover the true shapes of the planetary orbits. As shown, the second focus is empty. Eccentric ellipses are also found; comets are examples of objects that move in highly eccentric ellipses.

The distance of a planet from the Sun varies as the planet moves. For example, Earth is almost 5 million kilometers closer to the Sun in January than in July. Kepler found that as Mars's distance from the Sun varied, its orbital speed also varied, being greatest when the planet was closest to the Sun and least when it was farthest away. He found even more: there is a definite relationship between a planet's distance from the Sun and its speed in its orbit.

Figure 5-18 indicates four positions on an elliptical orbit. During the interval of time between points *A* and *B,* a line from the planet to the Sun sweeps out the long, skinny area shaded on the left. Between the *equal* time interval from *C* to *D,* the line from the planet to the Sun sweeps out the fat area on the right. Kepler found that this area is equal in size to the area swept out between times

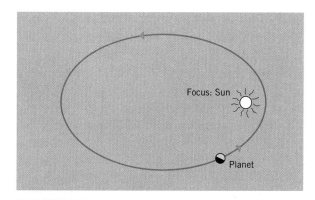

Figure 5-17 Kepler's first law: Planets move in ellipses with the Sun at one focus.

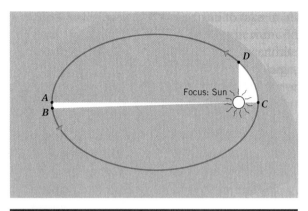

Figure 5-18 Kepler's second law: If the planet takes the same amount of time to go from *A* to *B* as it does to go from *C* to *D*, then the two shaded areas are equal.

A and *B*. In other words, because the time intervals from *A* to *B* and *C* to *D* are equal, and because the distance from *C* to *D* is greater than that from *A* to *B*, the planet must be moving faster from *C* to *D* than from *A* to *B*. For example, the motion of the Earth is more rapid in January than in July.

Kepler's second law of planetary motion can be formulated as follows:

The line joining a planet and the Sun sweeps out equal areas in equal amounts of time.

Kepler's third law was discovered much later than the first two. It appears, almost as an afterthought, in a rambling, mystical work of Kepler's titled *Mysterium Cosmographicum,* or *Cosmic Mystery.* The emphasis of this work lay in mystical speculation on the cosmos rather than on the mathematical relationship expressed in the third law.

Kepler's third law is a relationship between the average distance of a planet from the Sun (which equals half the length of the major axis, or **semi-major axis**) and the planet's *sidereal period,* the length of time it takes for the planet to orbit once around the Sun. The third law is most easily expressed mathematically. It states that if *P* is the orbital period of a planet *measured in years* and *A* its average distance from the Sun *measured in astronomical units,* then

$$P^2 = A^3.$$

This equation clearly works for planet Earth. Earth's orbital period is 1 year and its average dis-

tance from the Sun is 1 AU, so $P = 1$ and $A = 1$, and $1^2 = 1^3$.

We can express the equation in equivalent forms that are easier to use with your calculator:

$$P = A \sqrt{A}, \text{ and } A = \sqrt[3]{P^2}.$$

For example, if a planet's semi-major axis is 9 AU, one finds that its orbital period is

$$P = 9 \sqrt{9} = 9 \times 3 = 27 \text{ years.}$$

However, a planet that has an orbital period of 8 years would have an average distance from the Sun of

$$A = \sqrt[3]{P^2} = \sqrt[3]{8^2} = \sqrt[3]{64} = 4 \text{ AU.}$$

because $4 \times 4 \times 4 = 64$. Problems using Kepler's third law are more easily solved with a pocket calculator.

Inquiry 5-13

The average distance of Mars from the Sun is 1.52 AU. To the nearest tenth of a year, what is its orbital period?

Inquiry 5-14

Uranus's orbital period is 84 years. To the nearest tenth of an AU, what is its average distance from the Sun? (*Hint:* $20 \times 20 \times 20 = 8,000$.)

Unfortunately, the sidereal period is not a directly observable quantity. What is observable, however, is the planet's synodic period, which is the time interval between successive occurrences of a given planetary configuration (described in Section 4.8). For example, the synodic period is the time interval between, say, two successive oppositions or conjunctions.

An analogy can be drawn using two runners on a track in the place of planets. The *sidereal* period is the time from the starting line back to the starting line. The *synodic* period is the time for one runner to lap the other one—that is, to move ahead of and then pass the other runner again. Because both runners are moving, one runner cannot measure the sidereal period of the other. However, each runner could easily determine the time it takes to lap the other runner—the synodic period. If you then know your own sidereal

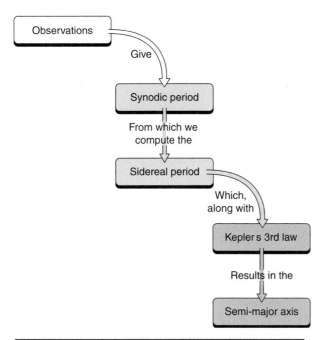

Figure 5-19 The procedure for finding a planet's semi-major axis given an observation of the synodic period.

period, you can compute the other runner's sidereal period.

In summary, we observe an object's synodic period, compute the sidereal period, plug it into Kepler's third law, and determine the planet's semi-major axis (**Figure 5-19**).

Kepler's three laws, as described here, are exact only in the idealized case of two isolated objects. In the real solar system, the large number of objects that interact with one another causes deviations from the ideal. Nevertheless, the laws are accurate enough to describe the motions of the planets and agree with the best naked-eye observations, which happened to be Tycho's. Had Tycho's observations been somewhat less accurate, or had Kepler been less exacting in his criteria for agreement between hypothesis and observation, Kepler's three laws might not have been discovered until much later in history.

Kepler's three laws overthrew the earlier ideas; no longer did we believe that the planets moved with uniform speeds in circular orbits. These laws are general in that they apply to any two objects circling each other. They may be applied to natural or artificial satellites orbiting planets, stars orbiting stars, stars orbiting galaxies, or even galaxies orbiting galaxies. Thus, we will return to Kepler's laws at many places throughout our study of astronomy.

5.5 The Search for Underlying Laws

Kepler's works are purely empirical, meaning they were derived solely on the basis of observation with no theoretical underpinning. Nevertheless, Galileo's experimental work on the motion of objects and Newton's beautiful theoretical development of his laws of motion show the validity of Kepler's work.

GALILEO GALILEI

Galileo Galilei (1564–1642), shown in the chapter opening illustration, was a contemporary of Kepler's, was an early convert to the heliocentric hypothesis, and argued vigorously for it. In his later years this brought him into conflict with the Inquisition, and he was forced to recant his beliefs and live under a form of house arrest—a pitiful, partially blind old man, fearful of the instruments of torture that apparently had been shown to him.

Kepler's laws, successful as they were in predicting the positions of the planets in the sky, were no proof that the *Earth* moves. Skeptics could reasonably claim that the three laws were merely mathematical tricks that happened to give accurate predictions and not in any way the truth about the physical nature of the universe.

Galileo clung to the hypothesis of circular motion for heavenly objects while conducting experiments in dynamics that eventually overthrew it. Remember, the ideas of Aristotle had been the accepted ones for centuries. And, they were what the Catholic Church, as the arbiter of authority, said was truth.

Galileo, in a revolutionary move, decided to let *observation,* not human reason, be the ultimate arbiter of physical reality. He performed experiments to decide whether a particular hypothesis was true. Galileo's insistence on making and using observations to find scientific "truth" led the way to the modern approach to science with the use of inductive logic. For example, he investigated the motions of objects under various conditions by rolling balls down inclined planes, thus showing that a state of motion was as natural as a state of rest, and that to change either state required an outside influence. In another famous case, it had been claimed by believers in Aristotelian physics that an object that was twice as heavy as a second one would fall twice as fast. Perhaps by actual demonstration, or maybe only with thought experiments, Galileo showed that such objects accelerated at nearly the same rate and hypothesized that only the resistance of the air kept the rates from being identical. The truth of Galileo's position was demonstrated dramatically when the Apollo 15 astronauts dropped a feather and a hammer together in the vacuum of the Moon. The two objects reached the lunar surface simultaneously. You, too, can perform Galileo's experiment by simultaneously dropping, say, a paper clip and a heavy shoe.

In performing his experiments, Galileo formulated the concept of **inertia**, which is an object's tendency to resist a change in its motion. The amount of inertia is measured by what we call **mass**, which is, loosely, a measure of the amount of matter in an object.

Galileo's greatest fame comes from his discoveries in astronomy, even though his contributions to physics were probably equally important. Upon learning of the invention of the telescope, he immediately constructed one of his own and turned it to the heavens. His most important astronomical discoveries were the following:

1. He found that the diffuse band of light across the sky, known as the **Milky Way**, actually consists of myriad stars too faint to be seen by the naked eye. This discovery was the first step toward a modern view of the nature of our galaxy.

2. He examined the Moon in detail, and discussed and named many of its surface features—its mountains, the dark areas called *maria* ("seas"), craters, and the like.

3. While observing with a telescope, he discovered that the planets had a visible disk, whereas the stars remained infinitesimal points even at the highest magnification.

4. He discovered the four brightest satellites of Jupiter by following the variations in their positions from night to night and showing that they always remained close to the moving planet.

5. He studied sunspots and, by following them across the visible disk of the Sun, showed that the Sun spins. Together with his observations of the Moon, this discovery struck a blow to the older hypotheses of astronomy, which had maintained that the heavenly objects were perfect, without blemishes. (These observations

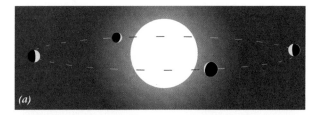

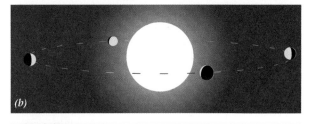

Figure 5-20 *(a)* Venus's phases under the Ptolemaic model as viewed from Earth. Note that only crescent phases are seen. See Figure 5-8 for another view.
(b) Venus's phases in a heliocentric model, as viewed from Earth. Because Venus can go behind the Sun, a full set of phases—including gibbous phases—is possible.

caused or contributed greatly to his later blindness. Let that be a warning to the reader!)

6. He found that the planets could show phases, as the Moon does. This was particularly important in the case of Venus, which exhibits all phases. In the Ptolemaic model, Venus could only show crescent phases, because it would *always* be closer to Earth than is the Sun (**Figure 5-20a**). Only if Venus were sometimes farther from the Earth than the Sun is could it be more than half-illuminated and exhibit a gibbous phase (**Figure 5-20b**). This discovery showed that, in at least one respect, the Ptolemaic picture of the universe must be incorrect. An observation such as this is an example of a **critical observation**, which is one that, by itself, is sufficient to favor one model over another. In this case, however, preconceived ideas won out over observation.

Inquiry 5-15

What might be the reason for Galileo's third observation?

Inquiry 5-16

One of the arguments against Copernicus's hypothesis was that the Earth could not move because it would leave the Moon behind. How does Galileo's fourth observation help disprove that contention?

Figure 5-21 Sir Isaac Newton (1642–1727) laid a firm theoretical basis in mathematics and physics for planetary motions, which was to reign unchallenged for over 200 years.

ISAAC NEWTON

As important as the contributions of Galileo and Kepler were in altering long-held ideas about astronomy and physics, they did not create a single comprehensive system that unified the physics of the heavens with the physics of the Earth. Such unification awaited the genius of **Isaac Newton** (1642–1727), who was born in the year of Galileo's death. Newton provided us with a system of physical laws that is so fundamental that even today they are the basis for much scientific activity (**Figure 5-21**). It would be difficult here to give an adequate idea of the power of Newton's intellect and the importance of his accomplishments; we can only scratch the surface. As a young man, Newton invented the tools of the branch of mathematics known as calculus, did fundamental work on optics and color vision, and developed the reflecting telescope, as well as proposing his famous laws of motion and gravity. He made such an impression on his contemporaries that the poet Alexander Pope was moved to write: "Nature and Nature's laws lay hid in night:/God said *Let Newton be!* and all was light."

To begin to understand Newton's contributions to the study of moving objects we must define some terms. **Velocity** describes the change in position of an object divided by the time interval over which the change occurs. Velocity describes not only the speed of an object but its direction. For example, while a car traveling 100 km per hour north moves at the same *speed* as a car traveling 100 km per hour east, their *velocities* are different because their directions of motion are different.[2] Motion is rarely constant in either amount or direction. Any change in an object's velocity, in either its speed or direction, is called an **acceleration**. Acceleration is the term used even if you are slowing down. When traveling at a constant speed on a Ferris wheel, you are accelerating because the direction is continuously changing.

Newton is most renowned for his three laws of motion and his law of gravity. His famous three laws of motion, first published in his *Philosophiae Naturalis Principia Mathematica (The Mathematical Principles of Natural Philosophy),* are generalized statements concerning the motions and interactions of objects. The first law concerns inertia.

> In the absence of an outside influence, an object at rest will stay at rest, while an object in motion will continue to move at a constant speed in a straight line.

Examples of the first law can be seen if you consider a truck with a box in the back that is not tied down. If the box is located toward the front of the truck bed and the truck accelerates quickly, the box will move toward the rear because it "wants" to stay at rest in relation to the street as the truck moves forward. On the other hand, if the box is toward the rear of the truck bed and the brakes are applied rapidly, the box will move toward the front, since it continues moving in the original direction. The more mass the box has, the more inertia it has.

If the statement of Newton's first law, the **law of inertia**, seems obvious to you, a little history might illustrate how revolutionary it was. Since the time of Aristotle, it had been assumed that an object required some continual action on it to

---

[2]Velocity is an example of a *vector,* a quantity having both size and direction. Speed, on the other hand, is called a *scalar* because it has only a size associated with it.

remain in motion, unless that motion were a part of the natural motion of heavy or light substances. The philosophers of the Middle Ages had refined Aristotle's view by asserting that an object was imparted a certain quantity of "impetus" by the hand or bow string that launched it, and that it would continue to move until it had used up its impetus. Impetus, however, was an example of what Newton referred to as an *occult quality,* for it could not be perceived in any fashion. It was merely supposed to exist in order to preserve Aristotle's ideas. Thus, all motion seemed to require the invocation of an unknown substance to explain it. Newton's simple statement swept away all that chaos and mental clutter, replacing it with brilliant simplicity.

Inquiry 5-17

According to Aristotle, the planets continue to move because that is their nature. What reason would Newton have given?

What happens if there is an outside influence? In answering this question, Newton called any outside influence acting on an object and causing it to accelerate a **force**. Thus the second law of motion:

 The total force on an object is equal to the product of the object's mass times its acceleration.

This expresses the famous formula $F = Ma,$ which is the basis for studying the motions of particles of all sizes, from submicroscopic atoms to stars in their orbits in the Galaxy. This law turned out to be enormously powerful. Namely, forces produce accelerations, and the size of the acceleration depends on the mass. Furthermore, an object can accelerate *only* if there is a force acting on it; remove the force and the object no longer accelerates, but continues moving at the same speed and direction. Finally, because acceleration has size and direction, so does force.

If an object is accelerated, it means that either its speed or its direction of motion is changing, or both. If an object is accelerated in the direction of motion, speed increases. Acceleration in the direction opposite to the direction of motion decreases the speed. A sideways acceleration changes its direction of motion. Any acceleration (forward, backward, or sideways) requires an external force that points in the direction of the acceleration.

Further, the greater an object's mass, the larger the force required to produce a given acceleration.

Inquiry 5-18

As an example of Newton's second law, consider a game of kick the can played with two cans—one empty and one filled with concrete. Which can has the greater mass? If someone came along and kicked the two cans with exactly the same force, which can would have the greater acceleration? Explain this in terms of what you have learned.

In the third law, Newton shows that forces always come in pairs.

 For every force that an object exerts on a second object, there is an equal and opposite force exerted by the second object on the first.

As an example of pairs of forces, when someone touches you, you also touch that person; one touch cannot occur without the other. If you and a friend are in rowboats on a lake and one of you pushes against the other with an oar, *both* boats will move, but in opposite directions. The simple act of sitting in a chair illustrates this law: you exert a downward force on the chair, and the chair exerts an equal and opposite force on you. While a falling brick exerts a force on the air molecules it encounters, pushing them ahead of it, the air molecules exert an equal and opposite force on the brick, thus slightly slowing its motion. A final example is a rocket. If one force is the rocket engine pushing gases backward, the opposite force is that of the gases pushing on the rocket engine in the forward direction. That is why rockets work in the vacuum of space; rockets *do not* operate because of the exhaust gases pushing back on the Earth.

Inquiry 5-19

In the kick-the-can game in Inquiry 5-18, what does this third law predict about the effect on the foot that kicks the two cans?

Newton's Law of Gravity

Newton's three laws, especially the second, enable us to calculate the acceleration of an object and hence its motion, but we must calculate the forces first. It is evident that heavenly objects such as the

Sun, Moon, and planets cannot exert forces on each other directly, since they do not touch each other. Newton proposed that they exert an *attractive* force on each other at a distance, that is, across empty space. He called this force *gravitation*. His **universal law of gravitation** expresses the gravitational force of attraction between *any* two objects. That is, gravity is a characteristic of any object that has mass.

Let's apply this to two objects, say the Moon and the Earth. The gravitational attraction between the Moon and the Earth must be the same as the attraction between the Earth and the Moon because of Newton's third law. To make this happen, the force must be proportional to the product of the masses. To generalize this for any two objects, one object is called M_1 and the second object M_2; it does not matter which object you call M_1 and which M_2. Gravitational force decreases as the distance between the centers of the objects increases, but the exact relationship was found using reasoning based on geometry. Newton argued that the force had to vary inversely with the square of the distance, d, between the two attracting objects. Combining the variation with mass and distance, Newton suggested that gravitational force could be calculated from this formula:

$$F = \frac{GM_1M_2}{d^2}$$

where F is the force exerted by an object of mass M_1 on an object of mass M_2, d is the distance between the centers of the two objects, and G is a numerical constant called the **universal gravitational constant**. The value of the universal constant depends on the units used to express the other quantities in the equation. From this formula, it is possible to calculate how the force between the objects varies with the distance between them, and therefore what the acceleration is at each moment.

Inquiry 5-20

If the distance between mass M_1 and mass M_2 is made four times *greater*, by what factor is the gravitational force between them changed? Is it increased or decreased?

Inquiry 5-21

Which force is greater: Earth's gravitational pull on the Moon or the Moon's gravitational pull on the Earth?

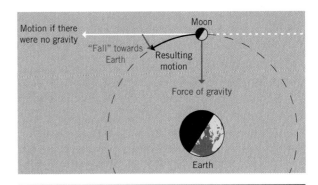

Figure 5-22 The motion of the Moon around the Earth according to Newton. The Moon "falls" toward the Earth just enough to keep it on a curved elliptical path around the planet.

Although the idea that objects of different masses fall at the same rate was experimentally shown by Galileo to be true, this does not make sense to most people. However, it is consistent with Newton's universal law of gravitation and his second law, as shown in Appendix A7.

A measure of Newton's genius is that he proposed that his laws of motion and gravity applied not only to objects on Earth but also to all objects in the universe. One of the first things he did was to compare the force of gravity exerted by the Earth on an object, allegedly a falling apple, with the gravitational force that would be required to keep the Moon in its orbit around the Earth. As shown in **Figure 5-22**, if there were no gravitational force exerted by the Earth on the Moon, it would travel in a straight line past the Earth in accordance with Newton's first law. To make the Moon travel around the Earth requires a force toward the Earth. The Earth's gravitational force on the Moon causes it to deviate from a straight line and follow a curved path around the Earth. For each kilometer the Moon moves in its orbit, the Moon must fall 0.0014 m (0.14 cm) toward the Earth in order to stay in its elliptical orbit. In other words, the Earth's force of gravity causes the Moon to accelerate just enough to maintain its distance from the Earth. For this reason, the Moon's orbital motion can be described as resulting from the Moon's falling toward the Earth's center! From such considerations Newton found that the actual force required to keep the Moon in its orbit around the Earth agreed well with the value he computed theoretically. Thus, observations showed his description to be a valid scientific theory. It is one that can be applied throughout the universe anywhere and at any time.

WEIGHT

Weight is the force one object exerts on another object. In particular, your weight is the gravitational force exerted by the Earth's mass on your body. Your weight depends on the mass of your body, the mass of the Earth, and your distance from the Earth's center, which is the Earth's radius. For this reason, your weight would be different should you travel to a different planet having a different mass and size than the Earth; your mass, however, would be the same. Interested readers can find weight expressed mathematically in Appendix A8.

> To understand why an astronaut floats in the space shuttle even though there is still gravity in space, you should do Discovery 5-1 at this time.

MOMENTUM

You probably have some intuitive feel for the word **momentum**. A train moving at 30 km per hour has more momentum than a bicycle moving at the same speed; it would have more impact and effect if it ran into something! For an object moving in a straight line, **linear momentum** is defined as the object's mass times its velocity.

Newton's first law can also be expressed as the principle of the **conservation of linear momentum**. This says that for a system of objects without any outside force, the sum of the linear momenta of all objects in a system is always the same. For example, if a billiard ball slows down by hitting another ball, the bill hit must speed up by the same amount.

Most objects move in curved paths, and the concept of **angular momentum** comes into play. The amount of angular momentum possessed by a planet in a *circular* orbit around the Sun is defined as follows:

Angular momentum = Mass × Speed × Distance from planet to Sun.

Expressed symbolically, if M is the mass of an object and v its speed when at a distance d, the angular momentum is given by

Angular momentum = Mvd.

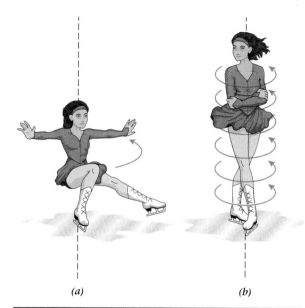

(a)   *(b)*

Figure 5-23 An ice skater's spin rate increases due to conservation of angular momentum when the arms are pulled in.

Like linear momentum, angular momentum is conserved for isolated systems. If the mass of an orbiting planet in an elliptical orbit remains constant, the only way for angular momentum to remain constant is for the speed of the planet to increase as the distance decreases, and vice versa. This is exactly what Kepler's second law says. The concept of angular momentum conservation is important in understanding not only the motions of planets but also such diverse subjects as the formation of the solar system, the motions of stars, and the shapes of galaxies. As applied to the orbits of the planets, the law states that the angular momentum of an orbiting planet remains constant.

A spinning object also has angular momentum. In this case, d represents the distance from the rotation axis to the surface. As a spinning gas cloud collapses to form a star, conservation of angular momentum requires that it increase in speed, just as an ice skater increases rotation speed as the arms are brought in (**Figure 5-23**).

For simplicity we have defined angular momentum in terms of circular orbits, but the concept can be extended to elliptical orbits.

Inquiry 5-22

Suppose a planet's farthest distance from the Sun is twice as great as its nearest distance. How many times greater is its orbital speed when the planet is nearest the Sun than when it is farthest away?

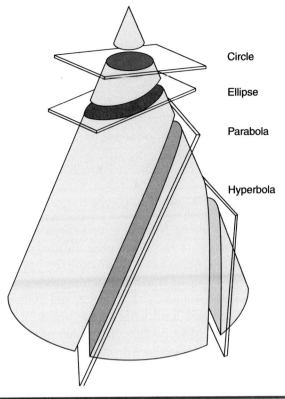

Circle

Ellipse

Parabola

Hyperbola

Figure 5-24 Defining the circle ellipse, parabola, and hyperbola in terms of sections of a cone.

NEWTON'S GENERALIZATION OF ORBITAL MOTION

While Kepler showed that planets moved in elliptical rather than circular orbits, Newton was able to show that moving objects could be in other types of orbits, too. The other possible orbital shapes are the **parabola** and the **hyperbola**. (Objects in such orbits are said to be in parabolic and hyperbolic orbits.) The various orbital shapes are best visualized geometrically and then understood by looking at the energy the orbiting object has.

Circles, ellipses, parabolas, and hyperbolas are all shapes known as conic sections because they are most readily visualized as formed from a cone (**Figure 5-24**). Conic sections result where a plane intersects the cone. For example, the curve formed by the intersection between a plane perpendicular to the cone's axis and the cone will be a circle, as shown. If the plane is not perpendicular to the axis the intersection will be an ellipse. If the plane is parallel to the cone's axis, the intersecting shape is a hyperbola. Finally, if the plane is parallel to the cone's sides, the shape will be a parabola.

Thus, objects moving in elliptical and circular orbits are in closed, repeating orbits; planets in elliptical orbits around the Sun are gravitationally bound to the Sun, which governs its motion. Objects in parabolic or hyperbolic orbits, however, are not bound but are in open orbits; the motions are not repeated and the orbiting object will continue moving away from the Sun forever. Examples of such objects are the *Voyager* spacecrafts that are escaping from the solar system and the nonperiodic comets (Chapter 7).

The type of orbit a body has is determined by the amount of energy the orbiting body has. Orbital motion includes two types of energy. The first is **potential energy**. A rock sitting on a cliff has gravitational potential energy because it has the potential for motion if it were to fall off. Orbiting bodies have potential energy because of the gravitational attraction of the attracting body.

The second type of energy is **kinetic energy**, which is energy of motion. All moving bodies have kinetic energy. The more massive the body, the greater the kinetic energy, and the faster it is moving, the greater the kinetic energy. Thus, even a small asteroid moving at, say, 20 km/sec contains significant kinetic energy such that, if it collided with Earth, it would make a big dent in the surface.

The total energy, which is the sum of the kinetic and potential energies, is constant. This concept is called the **conservation of energy**. Thus, since the potential energy decreases as an orbiting body approaches the Sun, its kinetic energy (that is, its speed) must increase.

Objects in elliptical orbits have more potential energy than kinetic energy and will therefore stay bound. If there is more kinetic energy than potential energy, the orbit is open and the orbiting body will escape.

We can summarize this discussion by considering the motion of a rocket. A rocket with the right amount of fuel would go into a bound orbit around the Earth (a circular or elliptical orbit). With just the right amount of additional energy, the rocket would travel in an open, parabolic orbit with exactly the right speed to escape from Earth. Additional energy beyond that amount and the rocket would escape Earth's gravity entirely in an open hyperbolic orbit.

The ideas presented here will be revisited in numerous contexts, from the orbits of comets, to molecules in planetary atmospheres, to the escape of stars and galaxies in collisions.

NEWTON'S FORM OF KEPLER'S THIRD LAW

Newton was able to show, using the calculus that he invented to solve the problem of orbital motion, that Kepler's empirical laws of planetary motion were direct consequences of his own more fundamental laws of motion. For example, he was able to show mathematically that a planet governed by a force of gravity that varies inversely with the square of the distance from the Sun would follow an elliptical orbit. This was a major accomplishment! In addition, using his laws he found Kepler's third law was even more general than originally proposed. In Newton's formulation, it becomes

$$(M_1 + M_2)P^2 = A^3$$

where again P is measured in years and A in astronomical units, and the additional factor $(M_1 + M_2)$ is the sum of the masses of the two objects in orbit about each other, expressed in terms of the mass of the Sun. When applied to the planets, $(M_1 + M_2)$ is almost precisely equal to 1, because the mass of the Sun is so much greater than that of any of the planets. (This is why Kepler did not discover it.) However, the form in which Newton wrote Kepler's third law turns out to be much more useful than Kepler's form. For example, if you observe the orbital period and separation of a planet's satellite, you can compute the mass of the planet; or, by observing the size of a double star's orbit and its orbital period, you can calculate the total mass of the stars in the binary system. The revised ideas even apply to galaxies in orbit around each other. In fact, much of the observational data we have on the masses of objects in the universe come from applying Newton's form of Kepler's third law.

As an example of its use, if two stars identical to the Sun (so the mass of each in solar units is 1) orbit about each other at a distance of 0.1 AU, we can use Newton's form of Kepler's third law as follows:

$$(1 + 1)P^2 = (0.1)^3$$
$$P^2 = 1/2 \times 10^{-3}$$
$$P = 2.2 \times 10^{-3} \text{ years} = 8.2 \text{ days.}$$

However, if the binary system had a star identical to the Sun and one six times as massive, the period would be 4.4 days.

In this example, the masses of the two objects are the same or not much different. If one object is many times more massive than the other (e.g., a star orbiting a galaxy), the sum of the two masses is basically the same as the more massive object.

Inquiry 5-23

Suppose a double star is observed to have a period of 10 years and an average distance between the two stars of 8 AU. What is the sum of the masses of the two stars in the system?

SUCCESSES OF NEWTON'S LAWS

The publication of Newton's laws of motion set the imagination of the world afire. Educated people of all backgrounds and disciplines applied them to all kinds of situations to see whether they were indeed as universal as Newton had claimed. Newton himself had difficulty explaining the motion of the Moon with accuracy comparable to observations, because the gravitational force of the Sun, as well as that of the Earth, was important and altered the orbit away from purely elliptical motion. Eventually, however, the mathematical problems of computing the Moon's motion were solved satisfactorily. In a similar way, as telescopes and observational techniques improved, the gravitational effects of one planet on another explained observed deviations from purely elliptical motion.

An early triumph of Newton's theory was the discovery by Edmund Halley, a contemporary of Newton, that several comets that had been observed in the past were in fact one and the same object. Halley predicted that it would reappear in 1758. Although Halley died in 1742, its reappearance right on time electrified Europe and led to a great spate of comet hunting. The comet, of course, was named for Halley. It returned again, right on schedule during late 1985; its next appointment near the Sun will be in the year 2061. Due to gravitational perturbations by the planets, the exact date cannot yet be determined.

In 1781, another discovery was made that at first led to one of Newtonian theory's greatest challenges and later to one of its greatest triumphs. Sir William Herschel, during a routine survey of the sky, discovered a new object that, unlike the stars, showed a clearly visible disk in his telescope. It was soon apparent that the object was moving with respect to the stars, and when the theory of gravity was used to establish its orbit, it was found to be a new planet, farther from the Sun than any previously known. Eventually the new planet was named Uranus, after the Greek god of the heavens.

The discovery of a new planet was not in itself a challenge to Newtonian physics, and at first seemed to be a confirmation of it. But after many years of observations, it became evident that something was wrong. No matter what was done, the orbit of Uranus could not be brought into agreement with the predictions of the law of gravity. In late 1845 and early 1846, two young mathematicians simultaneously proposed that there was another, as yet undiscovered planet that was affecting Uranus's orbit and causing it to deviate from the predictions of Newton's laws. The English astronomer John Couch Adams sent his calculations to the Astronomer Royal, who failed to appreciate the importance of the young man's work. The Frenchman U.J.J. Leverrier had better luck. His predictions were sent to the Berlin Observatory, where the new planet was found almost immediately, close to the predicted position. Both Adams and Leverrier now share the honor of Neptune's discovery.[4]

In conclusion, the point is that Newton's theory makes testable predictions that observations over hundreds of years verify.

5.6 Evidence in Favor of the Heliocentric Hypothesis

You have now learned about the history of the ideas leading up to the development of the heliocentric model of our solar system. In addition, you have learned the concepts that govern all motions in the universe. The question now is, what observational evidence shows the heliocentric model is valid? All that is required to validate it is a clear indication that the Earth moves. Astronomers now sought observational evidence for the Earth's motion.

It took so long to demonstrate the Earth's motion because it is a difficult problem. Although one might argue that the observation of different

---

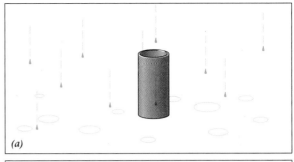

Figure 5-25 *(a)* A stationary bucket in the rain. Raindrops fall straight to the bottom. *(b)* Moving a bucket in the rain. It must be tilted to keep the drops from hitting the side.

constellations at different times of the year or the presence of seasons might do the trick, they do not. There are other explanations that could not be readily explained away as these could be.

But it was not until the year 1729 that any definitive evidence was discovered. In that year James Bradley, the British Astronomer Royal, who was trying to observe stellar parallax (see Figure 5-6), discovered instead a phenomenon called the **aberration of light** (the name does not imply that there is anything wrong with light). The aberration of light is an apparent shift in a star's position as a result of the motion of the Earth and the finite velocity of light. It is readily understood by means of an analogy. Imagine that you are in a rainstorm, with the raindrops coming straight down. If, on one hand, you want to catch the rain in a bucket, you hold the bucket pointing upward, as shown in **Figure 5-25a**, and the raindrops will, of course, fall straight down to the bottom of the bucket. On the other hand, if you run with the bucket, you will have to tilt the bucket in the direction of your motion to keep the drops from hitting the bucket's side (**Figure 5-25b**).

In the same way, a moving telescope on the moving Earth must be tilted very slightly in the direction

---

[4]Galileo recorded an object in his notes in the year 1613 that turns out to have been Neptune, according to Charles Kowal of Palomar Observatory. Of course, Galileo did not recognize that he had seen a planet, so the honor of Neptune's discovery rightfully belongs to Adams and Leverrier.

of its motion if light entering the top of the telescope tube is to pass all the way through and arrive at the eyepiece. The effect is not large: during the course of the year, a star's observed position deviates a maximum of 20 seconds of arc on either side of its true position. However, eighteenth-century telescopes were sufficiently precise to enable this effect to be detected. Because the deviation depends directly on the speed of the Earth in relation to the speed of light, a measurement of the deviation gives a numerical value for the speed of the Earth as it moves through space in its orbit around the Sun. The value is about one ten-thousandth of the speed of light, or 30 km/sec (19 miles per second). Thus, the first actual verification of the moving Earth came in 1729 with the discovery of the aberration of starlight.

The discovery of parallax, predicted by the Greeks and searched for in vain by Tycho, was delayed for yet another century. In the year 1838 the great astronomer Friedrich Wilhelm Bessel finally succeeded in measuring the parallax of the nearby faint star 61 Cygni. Why did parallax take so long to find? The answer is that the stars are so far away that the amount of their parallactic shift is extremely small. In the case of 61 Cygni, for example, the total shift is only about 0.6″ (0.00017 degrees), less than a thirtieth of the effect of aberration. This is approximately the angular size of a U.S. penny viewed from six kilometers away—a very small angle!

Figure 5-26a illustrates the effect of parallax for a star that is located in the plane of the Earth's orbit (the easiest case to visualize). When the Earth is at point A, the nearby star will be seen at position A′ with respect to the more distant background stars. Six months later, when the Earth is at point B, the nearby star appears to be located at point B′. The **parallactic shift** in the position of the star is the angle A-star-B. Clearly, the more distant the star is, the smaller the parallactic shift will be.

If we were to take photographs of the star field when the Earth is at points A and B in its orbit, they might look like **Figure 5-26b**. The nearby star would appear to shift its position in relation to the more distant stars, as shown. Assuming that the background stars are much more distant than the nearby stars, the angle A-star-B of Figure 5-26a will be practically equal to the angular shift in position shown in Figure 5-26b.

Using parallax to find the distance to stars will be discussed in Chapter 15.

Inquiry 5-24

If the parallax to the nearby star in Figure 5-26 was observed from a planet farther from the Sun than the Earth, would the parallactic shift be larger or smaller than when observed from the Earth? If observed from Mars, which has a distance of 1.5 AU, what would be the parallactic shift to the same nearby star?

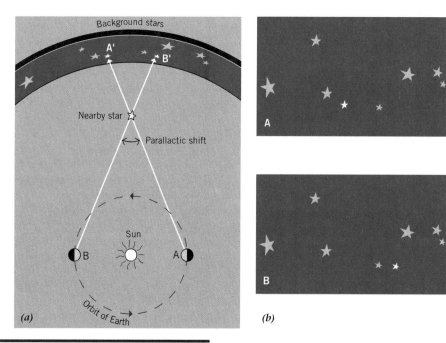

Figure 5-26 (a) The effect of parallax on a nearby star (not to scale). The apparent direction toward the star shifts slightly as a result of the Earth's motion around the Sun. (b) Two drawings of the same star field, six months apart. Note the shifted star.

5.7 Observational Evidence of the Earth's Rotation

Although every schoolchild can tell you that the Earth spins, few college graduates can explain *how* we know this simplest of all astronomical facts. As evidence, most people would cite the observation of the Sun and stars rising in the east and setting in the west. One could argue, as the Greeks did, that what is observed is simply the motion of a star-studded crystalline sphere around a stationary Earth once every 24 hours, with the Sun moving on its own sphere at its own slightly slower rate!

FOUCAULT PENDULUM

A famous demonstration of the rotation of the Earth is the pendulum experiment first performed by the nineteenth-century French physicist Jean-Bernard-Léon Foucault. In this experiment, a massive pendulum is hung from a long wire and set in motion. Pegs placed in a circle around the pendulum are successively knocked down over time due to the apparently changing plane in which the pendulum swings (**Figure 5-27**). To construct the sim-

plest possible explanation of what is observed, imagine that we suspend a pendulum on a long cable from a support that has been placed above the North Pole (**Figure 5-28**). We make the coupling between the cable and the support as friction-free as possible. If we set the pendulum to swinging back and forth, it will swing along a line that we can mark by driving pegs into the ice. Because of Newton's first law of motion, the plane defined by the swinging cable will be fixed in space in relation to the distant stars. If the Earth were not turning, the line between the pegs in the ice along which the pendulum would swing would not change. However, because the Earth turns beneath the pendulum, the line will *appear* to spin (as viewed from the turning Earth) in a direction that is opposite to the Earth's rotation.

At other latitudes, such as the latitude of Paris where the experiment was originally performed, the rotation will be slower but still evident. Only at the equator does the pendulum swing not change.

Inquiry 5-25

Why doesn't the plane of oscillation of the pendulum swing for an observer at the equator? (*Hint:* Consider how the pendulum would spin in the Southern Hemisphere.)

Although the motion of Foucault's pendulum is a demonstration of Earth's rotation on its axis, the concepts involved in the explanation depend on theories and models from other areas, in particular Newton's first law. Without it, the Foucault pendulum experiment would have made no sense.

Figure 5-27 The Foucault pendulum in the Reuben H. Fleet Science Center in San Diego. Notice the pegs knocked over by the pendulum as the Earth rotates beneath the swinging pendulum.

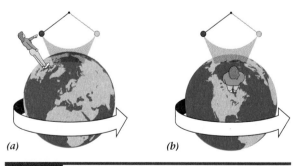

(a)   *(b)*

Figure 5-28 The Foucault pendulum. *(a)* Setup of the experiment in which the pendulum swings toward and away from the observer. *(b)* Six hours later, the pendulum still vibrates in the same plane relative to the stars, but because the observer has rotated, the pendulum now *appears* to move back and forth in front of the observer.

Coriolis Effects: Additional Evidence for a Spinning Earth

Because each and every part of the Earth spins full circle in 24 hours, different parts of its surface are actually moving through space at different speeds. A point on the Earth's equator moves about 40,000 km in 24 hours, or about 1,700 km per hour. A piece of ground 1 m from the North Pole moves only 6 m in 24 hours, for a slow speed of 6 m per day!

These speed differences have an observable effect. Suppose a cannonball is fired north from the Earth's equator at 1,700 km/hour (**Figure 5-29**). Not only is the cannonball moving northward at 1,700 km/hour, but it also has an eastward speed of 1,700 km/hour (from the speed of the ground on which the cannon sits). From Newton's first law, the path in space is a straight line. However, *in relation to the moving Earth's surface,* as the projectile moves northward, it will be passing over ground that moves east *less rapidly* than the projectile. As a consequence of the ground's slower motion, *in relation to a northward facing observer on the ground* the projectile will appear to be deflected toward the right (east) of the intended target.

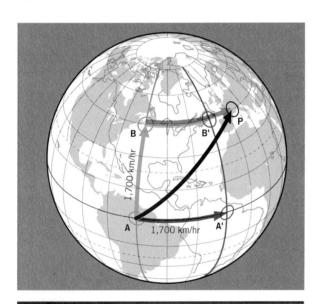

Figure 5-29 A cannonball fired northward from the equator (from A toward B) at 1,700 km/hour is also moving eastward at 1,700 km/hour (represented by the red arrow AA'), because this is the eastward speed of the cannon itself. The land over which the projectile travels moves more slowly than the equator. As a result, the projectile will arrive at P rather than B'. From the point of view of a north-facing gunner at A, the cannonball has been deflected to the right of the targeted location, landing at P rather than B'. Note that the length of BP is the same as AA'.

To summarize: Although from point *A* it was fired straight north toward point *B*, the projectile goes toward point *P* because of Earth's rotation. During the motion, point *B* moves to *B'*, so that the projectile falls to the east of *B*. (You can check this out by using a ruler to measure the length of the arrow *AA'* and seeing that it is the same as *BP!*) The observation demonstrates Earth's rotation.

This effect was largely a curiosity until World War I, when the Germans built a new generation of long-range cannons and found that they could not hit anything that wasn't exactly eastward or westward if they aimed directly at it; they had to aim to the left.[5] If you understand this discussion, you should be able to convince yourself that a cannonball fired southward (toward the equator) from some point in the Northern Hemisphere will be deflected to its right (west). Effects such as these that come about because we live on a rotating sphere are referred to as **Coriolis effects**. Their existence is evidence for a rotating Earth.

Coriolis effects control the general circulation of winds on the Earth and other planets, with the exception of Venus, which spins very slowly. Most of the time, these effects are too gentle to notice, but during hurricane season we can have dramatic illustrations of their reality.

Inquiry 5-26

How will projectiles fired north and south in the Southern Hemisphere be deflected?

Inquiry 5-27

Can you explain the motion of the Foucault pendulum as a Coriolis effect?

Precession and the Rotating Earth

In the previous chapter you learned about the precession of the Earth's rotation axis and its analogy to a spinning top. From the astronomical observation of precession, and from the understanding that a top precesses only if it spins, one can readily infer that Earth must spin. This understanding, however, comes only from a detailed understanding of Newton's accepted theories of motion.

---

[5]If the target was exactly east or west of the gun's position, the gun would be aimed directly at it.

Weightlessness

When you have completed this Discovery, you should be able to do the following:

- Describe what is meant by weightlessness.

An astronaut floating in a space shuttle has mass but is not held down to the floor of the shuttle. This is not caused by a lack of gravity in space.

- **Discovery Inquiry 5-1a** Suppose you were standing on a scale in a stopped elevator. What would happen to the scale reading when the elevator suddenly accelerates upward? What would happen to the scale reading when the elevator rapidly descends from rest?

Although the elevator has accelerated, the force of gravity on the elevator occupants and therefore their weight has not changed. The scale reading, however, has changed because of the acceleration. The force of gravity has been partially offset by the changing force from the elevator's floor. This effect is commonly called *weightlessness,* even though the pull of gravity and your weight never changed.

You can easily demonstrate weightlessness for yourself. Take a paper cup or an aluminum pop can and punch two holes on opposite sides near the bottom. With your fingers over the holes, fill the container with water.

- **Discovery Inquiry 5-1b** *Before removing your fingers from the holes,* describe what you expect to happen. Then, while taking necessary precautions, remove your fingers from the holes and observe what happens.

Again place your fingers over the holes and fill with water.

- **Discovery Inquiry 5-1c** *Before dropping the container* (into a garbage can, the bath tub, or *safely* outside!), describe what you expect to observe. After thinking about the answer, drop the container from as high as you can reach, and describe what you observed.

In this experiment, you should have observed no water flowing from the container while it was dropping. Although gravity was still there, the water and the container both accelerate downward at the same rate. Similarly, because the astronaut and the space shuttle both accelerate toward the center of the Earth *at the same rate,* the astronaut is not pulled to the floor (or ceiling!) of the shuttle. Since we usually judge weight by how hard we push on the floor or a scale, it seems as if the astronaut is weightless.

Chapter Summary

Observations

- Ancient Babylonians and Egyptians could predict seasons, eclipses, and make calendars.
- **Aristarchus** first suggested that the Earth circles the Sun, and found the relative sizes of the Earth, Moon, and Sun, as well as the relative distances of the Moon and Sun. **Eratosthenes** first found the size of the Earth. **Hipparchus** made a great star catalog, began the magnitude system used to specify the brightness of stars, and discovered precession. **Ptolemy** advanced the hypothesis of planetary epicycles and helped preserve Greek knowledge in the *Almagest.*

- The Greeks observed that the planets moved with individual speeds against the starry background; sometimes changed directions and moved with a *retrograde* motion; and, in the case of Mercury and Venus, were never far from the Sun.

- **Tycho Brahe** built large instruments and made observations over extended periods of time. **Johannes Kepler** found three empirical laws of planetary motion: (1) planets orbit about the Sun in elliptical orbits with the Sun at one focus; (2) planets move more rapidly when close to the Sun than when farther away; (3) if the period of a planet in its orbit is P years and the semi-major axis is A astronomical units, then $P^2 = A^3$.

- **Galileo** made the first telescopic observations and discovered that the Milky Way consists of a large number of stars. He observed lunar surface features, four Jovian satellites, sunspots, and phases of planets. He showed experimentally that objects of different masses fall to the ground with identical accelerations.

- The **aberration of starlight** is a small apparent shift in the direction to an object caused by the motion of an observer. The observation of this effect illustrates the motion of the Earth.

- **Stellar parallax** is the apparent change in the direction to an object caused by a change in position of an observer.

- The **Coriolis effect** is the apparent rightward drift of a projectile fired from the equator toward the North Pole, or from the North Pole toward the equator. There is a corresponding leftward drift in the southern hemisphere. It is caused by the fact that equatorial regions move more rapidly than polar regions.

Theory

- Greek science and philosophy assumed that planets exhibit uniform motion in circular orbits with the Earth at the center. They derived the relative distance of the planets from the observed speed.

- **Nicolaus Copernicus** proposed the heliocentric hypothesis.

- **Mass** is a quantity that measures the amount of **inertia** an object contains. An object's mass is independent of its location. **Velocity** is a change in location or direction of motion divided by the time over which the change occurs. **Acceleration** is a change in an object's velocity divided by the time over which the change occurs.

- **Isaac Newton's three laws** describe how objects move both in the absence (inertia) and presence ($F = ma$ and paired forces) of forces. They are general laws applicable to a wide variety of situations. From them, Kepler's laws can be derived. These concepts show that orbiting objects can have **elliptical**, **parabolic**, and **hyperbolic** orbits. Objects in elliptical orbits are bound to the more massive object, while objects in hyperbolic orbits are not bound and escape.

- **The universal law of gravitation**, found by Newton, provided a means to connect the motion of the Moon with the motion of falling objects on Earth.

- From Newton's laws, a modification of Kepler's third law results that allows astronomers to determine the masses of orbiting objects:

$$(M_1 + M_2)P^2 = A^3.$$

- The gravitational force exerted on an object by the Earth defines (and therefore equals) its **weight**. The weight depends on the object's location.

- The amounts of **linear momentum** and **angular momentum** contained in an isolated system are both conserved.

Conclusions

- From observations made with the **Foucault pendulum**, astronomers infer the rotation of the Earth.

- From observations of the **aberration of starlight**, astronomers infer the revolution of the Earth about the Sun.

- Newton's laws provide a model that successfully explains past events and predicts future ones. This framework explains empirical rules such as Kepler's laws of planetary motion.

Summary Questions

1. What is the significance of the observations of Aristarchus, Eratosthenes, and Hipparchus? What are the principles used to make their measurements of the sizes of the Earth and Moon?

2. What are two methods by which the order of the planets from the Sun could be determined?

3. What planetary observations must a reasonable model of the solar system incorporate? In your answer use diagrams to show how the Ptolemaic and heliocentric hypotheses explain the observations.

4. What is meant by stellar parallax? Explain its cause. How can parallax be used to distinguish between heliocentric and geocentric hypotheses? Why did the Greeks not observe stellar parallax?

5. What role did the concept of uniform circular motion play in the history of astronomical thought? Give examples from several eras.

6. What were specific astronomical contributions made by Copernicus and Brahe? What was the importance of these contributions?

7. State and explain Kepler's three laws. Explain Kepler's second law in terms of the conservation of angular momentum.

8. What are the principal astronomical discoveries of Galileo? Explain how these discoveries may have accelerated the acceptance of the heliocentric hypothesis.

9. State Newton's laws of motion and the law of gravity. Explain how, in principle, they can be used to explain the orbiting of one object about another.

10. What are some specific astronomical examples of how Newton's laws have been shown to be valid descriptions of nature?

11. How were the heliocentric hypothesis and the Earth's rotation finally confirmed observationally? How did Newton's laws play a role in demonstrating the Earth's rotation and revolution?

Applying Your Knowledge

1. Make up a table that lists, chronologically, the contributions to astronomy made by Aristarchus, Eratosthenes, Hipparchus, Ptolemy, and Pythagoras. Include dates.

2. If you had been a traditional scholar in the mid-1500s, what arguments would you have presented *against* the Copernican system?

3. Of the following people, who in your opinion made the most important contribution to astronomy: Copernicus, Brahe, Kepler, Galileo, or Newton? Explain.

4. Use the definition of acceleration to list the accelerators in a standard passenger car.

5. Use Newton's second law to explain why a planet in an elliptical orbit moves more rapidly when near the Sun than when farther away.

6. Explain why your weight would be different on Mars than it is on Earth.

7. Why can a rocket escape from the Moon's surface with a smaller speed than is needed to escape from the Earth's surface?

▶ 8. What is the distance from the Sun to a planet whose period is 129.14 days?

▶ 9. Find the ratio of the gravitational attraction between (*a*) a man of mass 100 kg and a woman of mass 50 kg who is 10 meters away from him, and (*b*) the attraction between the woman and the Earth. (*Hint:* See Appendix A8.)

▶ 10. If the Sun is 2×10^9 AU from the center of the Milky Way, and if it has a period of 200 million years, what is the mass of the Milky Way galaxy?

▶ 11. If identical galaxies at a distance of 650 million ly and having a mass of 10^{10} solar masses each were observed to have an angular separation of 10 seconds of arc, how long would their orbital period be? Assume the galaxies are seen at their maximum separation. (*Hint:* This problem contains a number of steps along the way.)

▶ 12. How much would a 100 lb. person weigh on the Moon? (*Hint:* Write an expression for the person's weight on Earth and an equivalent one for the Moon. Then divide one equation by the other; you will find that some quantities drop out of the problem.)

▶ 13. How fast would a person of mass 50 kg have to run to have the same linear momentum as a 5,000-kg bus traveling 100 km/hour?

Answers to Inquiries

5-1. A circular orbit and uniform motion.

5-2. The Earth is four times larger.

5-3. Because he knew the Moon was smaller than the Earth and was in orbit about the Earth, and because he now knew the Sun was larger than the Moon, it was logical that the Earth would go around the Sun.

5-4. 5,000 stadia $\times$ 360°/7° = 260,000 stadia. This is 26,000 miles (41,600 km); divide by π to get the diameter.

5-5. It is perfectly spherical.

5-6. Zero degrees.

5-7. The planets are moving in their orbits at about the same speed, so their apparent speeds are due only to their respective distances. Although the assumption is incorrect, the result ends up being correct for the more distant planets, all of which are moving more slowly because they are far away from the Sun.

5-8. Stars are very far away.

5-9. Only crescent phases would occur.

5-10. There are countless examples. Some are the germ model of disease, economic models used to set government fiscal policy, and educational models used to design instruction.

5-11. Nearly on a straight line, with the Earth in between.

5-12. Each has some points of simplicity. Ptolemy's has fewer epicycles, but Copernicus's has a neater explanation of retrograde motion, which in addition, predicts retrograde motion only when it is actually observed.

5-13. From $P^2 = A^3$, we have $P = \sqrt{A^3} = \sqrt{1.52^3} =$ 1.87 years.

5-14. From $P^2 = A^3$, $A = \sqrt[3]{P^2} = \sqrt[3]{84^2} = 19.2$ AU.

5-15. The planets are closer than stars.

5-16. While Jupiter was clearly moving (they thought around Earth), it was also dragging its moons along.

5-17. The planets continue to move because they have mass and hence inertia.

5-18. From $F = Ma$, the can with more mass (the one filled with concrete) would have the smaller acceleration if the forces were equal.

5-19. The reaction force on the foot will certainly be felt more in the case of the concrete can.

5-20. The force is 16 times smaller, because it depends on the square of the distance.

5-21. The forces are the same (Newton's third law)!

5-22. The orbital speed is also twice as great.

5-23. $(M_1 + M_2) = 8^3/10^2 = 512/100 = 5.1$ solar masses.

5-24. The parallactic shift (the parallax angle) would be larger. From Mars, the distance from the Sun is 1.5 AU, compared with 1.0 AU for the Earth; thus, the distance is 50% greater, which means the parallactic shift will be 50% greater. If the nearby star had a parallactic shift of 1 second of arc from Earth, it would be 1.5 seconds of arc from Mars.

5-25. In the Southern Hemisphere, the pendulum would appear to spin in the opposite direction from its rotation in the Northern Hemisphere. Therefore, for the direction of oscillation to change as the pendulum is carried from one hemisphere to the other, there must be no rotation at the equator.

5-26. In the Southern Hemisphere, projectiles will always deflect to the left. When fired directly east or west, there is no deflection.

5-27. Yes. Imagining the ball of the pendulum to be a projectile deflecting to the right in the northern hemisphere on each swing gives it the rotation we observe.

PART TWO

Discovering the Nature and Evolution of Planetary Systems

Because of fantastic images of distant worlds returned by a flotilla of spacecraft and published by television and print media, most people are more at home with the planets than with stars and galaxies. For this reason, we begin our study of modern astronomy with the solar system. Spacecraft missions to all the planets but Pluto, as well as missions to comets and asteroids, have provided an unprecedented opportunity for "up close and personal" exploration of the solar system.

We begin in Chapter 6 by looking at patterns observed within the solar system. These patterns lead to questions about how the system as a whole and the bodies within it formed. An important part of our study is a look at various models of the solar system's formation, and the evidence for and against each model. The chapter ends with a look at the evidence for planetary systems around other stars.

To help our understanding of the solar system's formation, we observe objects whose characteristics preserve both the materials and the physical conditions that were present at the time of their formation. For this reason, in Chapter 7 we examine comets, asteroids, and meteorites. Although these objects may constitute only a minor part of the mass of the solar system, most have undergone

fewer modifications than the larger planetary bodies and therefore provide crucial information about the conditions when the solar system formed.

Processes that occur within, on, and around planetary bodies are studied in Chapter 8 on the Earth and Moon. These objects are taken as the standard against which the properties of other bodies can be measured. The processes studied include the formation and modification of surface features by volcanic, tectonic, and impact forces; the formation, composition, and properties of the atmosphere; the response of the interior to seismic waves; and planetary magnetism.

The Earthlike planets are investigated in Chapter 9 by comparing and contrasting each with the other, and each with the Earth and Moon. Although the emphasis is on comparison and the processes that give each body the properties that we observe, we also will examine the unique characteristics of each body.

The final chapter of Part Two, Chapter 10, is similar to Chapter 9 but covers the jovian planets and Pluto. We also discuss the characteristics of the moons and the planetary ring systems. Through a comparison of one body with the other, we are better able to make sense of the reams of information sent to Earth by various spacecraft.

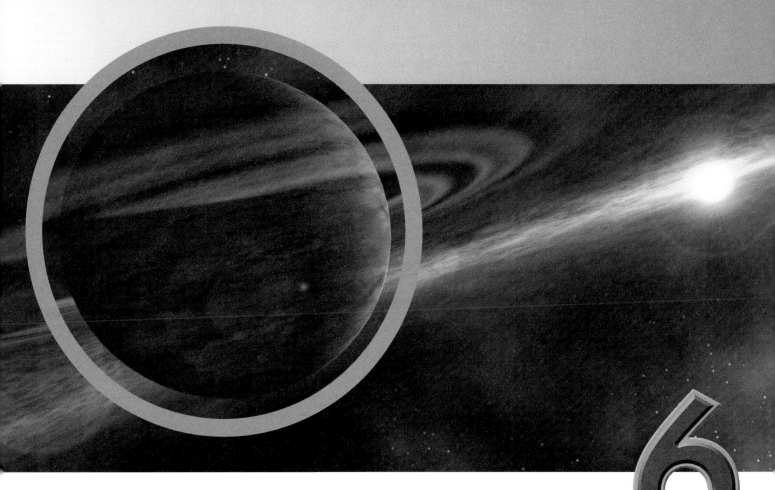

The Structure and Formation of Planetary Systems

6

First of all there came Chaos, . . .
From Chaos was born Erebos, the dark, and black night,
And from Night again Aither and Hemera, the day,
were begotten . . .

Hesiod, Theogony, c. 800 B.C.E.

W e begin our survey of the universe close to home with our Sun and its system of planets. In this chapter, we will look at the overall structure of the solar system (we defer the details to Chapters 8 to 10). We will find some striking patterns that can lead us to an understanding about how the solar system formed. We will discuss some of the hypotheses of the solar system formation and see how well each fits the observable data. Once we have done this, we will compare our own solar system to known planetary systems elsewhere in the universe.

6.1 An Overview of the Solar System

Let us imagine that the alien astronomer-astronaut introduced in Chapter 2 has been sent to our solar system to investigate the possibility of life in the vicinity of the Sun. The first thing she would see after the Sun would be the brightest planet, the giant Jupiter, about 5 AU from the Sun. After determining Jupiter's period about the Sun, she would then be able to use Kepler's third law to determine the Sun's mass. Translating into metric units, she would find that the mass of the Sun is about 2×10^{33} grams. (A mass of 1 kilogram weighs 2.2 pounds on Earth.) Once she observes our stellar system for a while, she might produce an illustration of it similar to that in **Figure 6-1**.

From a more careful observation, which would show that the Sun is affected slightly by Jupiter's gravity, the alien would discover that the mass of Jupiter is about 1/1,000 that of the Sun. However, it would be clear that the Sun is the dominant mass in the solar system and that all other bodies in the system must march to its gravitational tune.

Soon other planets, fainter and less massive than Jupiter, become obvious: first Saturn with its rings, about 9.5 AU from the Sun; then Uranus, at a distance of almost 20 AU; and finally Neptune, 30 AU distant. Coming even closer, our visitor would notice that these four bodies rotate at different rates and that each of them has moons and rings. By measuring the moons' periods and distances from their planets, the alien could also accurately determine the mass of each of the planets. From the planets' radii, she could determine their volumes,

Figure 6-1 A schematic view of the solar system from afar.

Chapter opening photo: An artist's conception of the appearance of a Jupiter-sized planet in the planet forming disc of the star CoKu Tau 4.

and finally their **densities**, defined as mass divided by the volume containing the mass:[1]

$$\text{Density} = \frac{\text{Mass}}{\text{Volume}}.$$

For example, a cubic centimeter of water contains a mass of 1 gram. The same volume of gold contains 19.3 grams; gold is therefore much denser. She would find that these planets have a density about the same as that of water. The density of Saturn, in fact, is less than that of water, and it would float in a bathtub, if one large enough could be found!

Inquiry 6-1

Considering the low density of this group of planets, would you expect them to be composed primarily of low-mass elements such as hydrogen and helium or high-mass elements such as silicon, iron, and nickel?

The astronaut would also notice that these planets have thick atmospheres and heavy cloud cover. Thanks to the thickness of the clouds, she could not determine whether there was any solid surface at all on these planets. We know that this type of planet is common, and such planets may very well be in her home system as well. Better look elsewhere for a landing site, she would conclude.

By this time, the alien would have detected another group of planets, which she might call *dwarf* planets due to their small size compared to the first group of *giant* planets. She would notice that, with the exception of Pluto, this group of planets was much closer to the Sun than the giant planets. In order of their distance from the Sun, she would find Mercury, Venus, Earth, and Mars.[2] Because Earth and Mars have moons, she would be able to determine their masses easily; she would find that the largest of the bodies, Earth, has a mass only 1/300 that of Jupiter and a density roughly five times that of water on Earth. Consequently, she would conclude that the dwarf planets were composed primarily of heavier elements, such as oxygen, silicon, iron, magnesium, and calcium.

---

[1]If the body is assumed to be a sphere of radius *r*, the volume is given by $4\pi r^3/3$.

[2]There are a variety of mnemonics for remembering the planets in order of their average distance from the Sun. One, compliments of Earl Phillip, is **M**y **V**ery **E**ducated **M**other **J**ust **S**howed **U**s **N**ine **P**lanets. Another, compliments of Jay Rambo, is **M**y **V**ery **E**asy **M**ethod: **J**ust **S**et **U**p **N**ine **P**lanets.

Compared to the giant planets, some of the dwarf planets have thin atmospheres that could be detected in several ways. Surface detail might be blurred a bit, or clouds might be seen, or, when the planet passed in front of a distant star, the star would fade out gradually as its light passed through the planet's atmosphere.

Mercury, the innermost planet, has only an insignificant amount of atmosphere. In attempting to understand this, the astronaut would notice two facts. First, Mercury's mass is the smallest of the planets examined so far. Second, Mercury is so close to the Sun that its atmosphere would be heated to high temperature, giving any molecules of gas in its atmosphere a velocity high enough to escape the planet's gravity.

Inquiry 6-2

Why would the light from a distant star fade gradually as one of the dwarf planets having an atmosphere passed in front of the star? Make a drawing to explain your answer.

Inquiry 6-3

How would Mercury's low gravity and high temperature each tend to cause Mercury's atmosphere to disperse into outer space after a time?

The next planet, Venus, has a much thicker and more extensive atmosphere—so thick that its surface could not be seen directly. Nevertheless, the alien would be able to show that the surface of the planet was not far below the visible cloud layer by bouncing radar signals off the planet, just as air traffic controllers locate the positions of airplanes through radar. A little probing would show that the principal component of Venus's atmosphere is 96 percent carbon dioxide (CO_2) and 3 percent nitrogen. The alien would see that it is bone dry, with only about 0.1 percent water vapor. By contrast, the next planet, Earth, which is almost the same size and mass as Venus, has a very different atmosphere—about 78 percent nitrogen (N_2) and 20 percent oxygen (O_2), with a readily detectable amount of water vapor. Our visitor would make a mental note to try to understand this difference, because the next planet, Mars, has a very thin atmosphere that is largely carbon dioxide.

Although the alien would find the surfaces of all these planets to be rocky in appearance, the Earth is distinctive in that its surface is largely covered by water, whereas the other planets show little evidence of water. Again, she would wonder about the reason for this.

At this point, the alien astronaut might notice an anomaly—the planet Pluto. Like the dwarf planets, Pluto is small. Unlike them, it is located a large distance from the Sun beyond the giant planets, and has a low density. Moreover, its orbit is more elliptical and out of the plane of the orbits of the other eight planets. Might it be that this planet had an origin different from the rest of the planets?

Summarizing this picture, the astronaut might draw up a table similar to **Table 6-1**. In it are listed the characteristics of the dwarf, or **terrestrial**, planets (after the Latin word for Earth) and the giant, or **jovian**, planets (after the Latin name for Jupiter). These characteristics must be explained by any successful theory for the origin of the solar system.

Appendix C contains reference tables that summarize some of the fundamental data about each of the planets. The significance of some of the data was suggested in the previous paragraphs and will be discussed further as you learn more about the solar system. To help you better appreciate the relative sizes of the planets, they are illustrated to scale in **Figure 6-2**. Similarly, the relative sizes of the planetary orbits are displayed in **Figure 6-3**. If Jupiter were drawn to scale in Figure 6-3, it would

be only about 1/10,000 of an inch in diameter, and the Sun would be about 1/1,000 of an inch in diameter. Therefore, to put Figure 6-3 on the same scale as Figure 6-2, we would have to draw it about 10,000 times bigger!

> ## IN DISCOVERY 6-1
> ## (A Scale Model of the Solar System)
> you will make a solar system model
> to a scale that you can walk.

Table 6-1 Characteristics of the Two Major Groups of Planets

TERRESTRIAL	JOVIAN
Mercury, Venus, Earth, Mars	Jupiter, Saturn, Uranus, Neptune
Near the Sun	Far from the Sun
Small diameter	Large diameter
Low mass	High mass
High density	Low density
Primarily composed of heavier elements	Primarily composed of hydrogen and helium (composition similar to the Sun)
Rocky surface	No surface
Thin atmosphere	Thick atmosphere
High temperature	Low temperature

Figure 6-2 Relative sizes of the planets compared with the Sun.

Inquiry 6-4

On the scale of Figure 6-2, about how far would Earth
be from the Sun? How far would Pluto be from the Sun?

To complete our overview of the solar system
we note that it also contains **minor planets** (also
known as **asteroids**), **comets**, and other debris
throughout the solar system making up the **interplanetary medium**. These components give additional insight into the origin of the solar system and
are important enough to warrant their own chapter
(Chapter 7). However, for the time being we will
ignore them as we try to visualize the Big Picture.

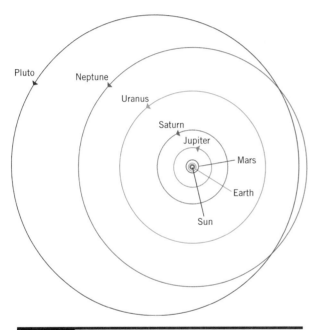

Figure 6-3 The solar system, showing the scale of the
planetary orbits, and the direction of motion as
observed from above Earth's North Pole.

6.2 Observations: Orbital Patterns

Our space explorer would probably find some of
the similarities in the motions of the planets more
striking than the various differences between them.
For example, she would find that the orbits of the
planets all lie nearly in the same plane. (This is
why they are always seen in the sky near the ecliptic, which defines the fundamental plane of the
solar system as seen from Earth.) The planets
whose orbital planes are farthest from the ecliptic
are Mercury and Pluto, as shown in **Figure 6-4**.

The orbits of the planets are nearly circular.
Even in the case of Mercury and Pluto, which
have the most highly elliptical orbits, the deviation from a precise circle amounts to only about a
3 percent difference in the lengths of the long and
short axes of the ellipse. For these two planets, the
Sun is well offset from the center of the orbits. In
the case of Pluto, although its *average* distance is
about 40 AU, it is sometimes even closer to the
Sun than Neptune.[3] (From Figure 6-3, it might
appear that Neptune and Pluto are in danger of
colliding; however, the tilt of Pluto's orbit prevents this from happening. Studies have shown
that Neptune and Pluto never get closer to each
other than 18 AU.)

Our astronaut would find that, viewed from the
north pole of the solar system, all the planets orbit
the Sun in the same counterclockwise direction
(Figure 6-3). She would realize that this must have
come about from the way the planetary system
formed.

---

[3]Pluto was closer to the Sun than Neptune during 1979–1999.

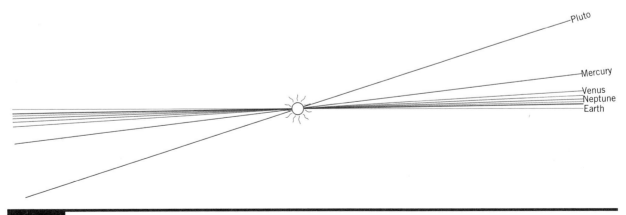

Figure 6-4 The solar system viewed from the side, showing the planes of the planets' orbits.

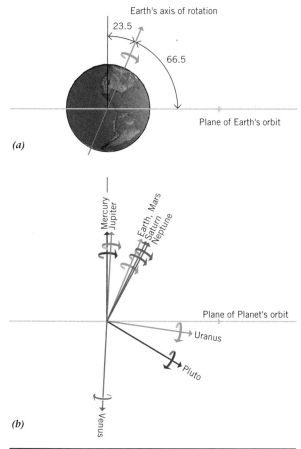

(a)

(b)

Figure 6-5 *(a)* The tilt between the Earth's axis of rotation and its orbital plane. *(b)* Similar to *(a)* for each of the planets. The arrows point in the directions of the planets' north poles. Venus's arrow points down because of the planet's retrograde rotation. The direction of rotation of each planet is indicated.

Except for Venus, Uranus, and Pluto, six planets and the Sun rotate on their axes in a counterclockwise direction. In addition, the rotational (spin) axes of most of the planets are roughly perpendicular (90°) to their orbital planes. For example, Earth's axis is at an angle of 66.5° (90°–23.5°) to its orbital plane (**Figure 6-5a**). The inclination for the other planets relative to their orbital planes is shown in **Figure 6-5b**. Venus's arrow points downward because it rotates in a clockwise direction—opposite to its orbital motion. The axis of Mars is at approximately the same angle as Earth's, and the axes of most of the other planets are relatively close to perpendicular. The two striking exceptions to the perpendicular rule are the planets Uranus and Pluto, whose spin axes are almost in their orbital planes; **Figure 6-6** illustrates the situation for Uranus and its moons. A successful theory of

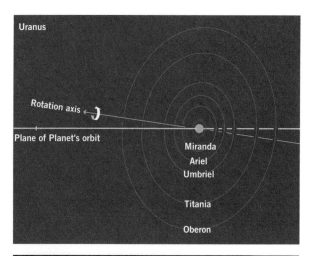

Figure 6-6 The anomalous orientation of Uranus and its moon system.

the formation of the planets would be expected to explain all three anomalies.

Planetary moons can be divided into two groups: those that are relatively large and close to a planet, and those that are smaller and farther away. The large ones close to the planet orbit counterclockwise, in circular orbit, and in orbital planes that roughly coincide with the planet's equator. A few of the outermost moons of Jupiter, Saturn, and Neptune orbit clockwise (retrograde) in elliptical orbits and in orbital planes that are highly inclined to the planet's equator. This is also true for the outermost moons of Uranus and Pluto's one moon. whose equatorial planes deviate from the ecliptic plane.

6.3 The Distribution of Angular Momentum

Our visitor would increase her understanding of our solar system if she measured the amount of angular momentum (defined in Chapter 5) possessed by each of the bodies in it. A body can possess angular momentum in one of two ways—either because of its orbital motion around another body (*orbital* angular momentum, as discussed in Section 5.5), or because of its rotation about an axis (which is called *spin* angular momentum). The total angular momentum in the solar system is the sum of all the orbital and spin angular momenta of all the planets, their moons, and the Sun.

One of the most striking properties of the solar system is the observation that 98 percent of the total angular momentum in the system comes from the orbital angular momentum of the planets, even though the Sun contains 99.8 percent of the mass. Because the Sun has the lion's share of mass, we might expect it to carry a similar proportion of the angular momentum. But for the Sun to have even half the angular momentum in the solar system, it would have to rotate 25 times faster than it actually does, or about once a day. Explaining why the Sun contains such a small percentage of the solar system's angular momentum (2 percent) has been one of the more difficult tasks facing hypotheses of the solar system's origin; from now on, we will refer to this difficulty as the *angular momentum problem*. We will return to it later in the chapter.

6.4 Hypotheses of the Origin of the Solar System

How might our alien astronaut put together a hypothesis to explain these broad aspects of the solar system that we have discussed? She will try to explain how a dominant mass, the Sun, might have acquired a system of planets and moons exhibiting all the patterns just described. In addition, a hypothesis that also explained the anomalies we mentioned would be even better.

A SIMPLE HYPOTHESIS THAT DOES NOT WORK

The simplest hypothesis to explain the existence of the Sun's planetary system is the idea that the Sun, in its motion through the Galaxy, has simply acquired its planets and their moons by encountering these bodies from time to time and capturing them by its gravitational attraction. How well does this hypothesis fare in explaining the observed patterns in the solar system? Read and answer the following inquiries to answer this question.

Inquiry 6-5

Would this hypothesis be consistent with the fact that the inclinations of the planes of planetary orbits are nearly the same? With the near circularity of most of the planetary orbits? With the fact that some planets have low mass and some high mass? With the fact that most of the planets and their moons orbit in the same direction?

Inquiry 6-6

It has been established that the Earth, Moon, and meteorites have about the same age. Is this observation consistent with the first hypothesis? Justify your answer.

Inquiry 6-7

What are several objections to this hypothesis? What facts can you think of that are *consistent* with the hypothesis?

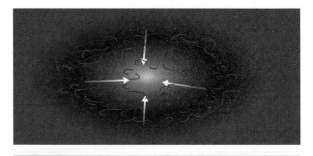

Figure 6-7 The collapse of a gas cloud according to the nebular hypothesis.

In looking carefully at the hypothesis and asking the right questions, we have been able to invalidate it easily. We will now consider some other possibilities.

EVOLUTIONARY HYPOTHESES

In an original flash of insight, the French philosopher René Descartes (pronounced *day-CART*) suggested in 1644 that space might have been filled with swirling gases and that the collapse of spinning whirlpools of material into denser condensations was responsible for forming celestial bodies. This is an evolutionary hypothesis, which is one in which formation occurs gradually, over a long period of time. What is especially surprising is that Descartes made these speculations *before* Newton formulated his fundamental laws of motion and theory of gravity.

The German philosopher Immanuel Kant suggested a more complete evolutionary hypothesis in the eighteenth century; the French mathematician Pierre-Simon Laplace developed it further. Kant's **nebular hypothesis**, as it was called, proposed that a vast cloud of interstellar gas collapsed to form the Sun and planets, as shown in **Figure 6-7**. Once contraction started, the gravitational force of the cloud acting on itself would accelerate its collapse into a relatively small volume.

For a cloud to start out with no rotation at all is virtually impossible. The rotation might be small, but chances are it will never be exactly zero. As it collapses, a cloud's rate of spin must increase due to the conservation of angular momentum discussed in

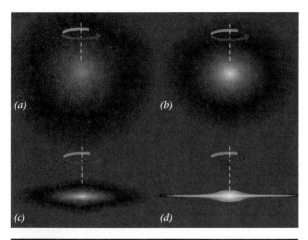

Figure 6-8 The forces acting on a spinning gas cloud cause it to flatten into a disk in the sequence *a–d*.

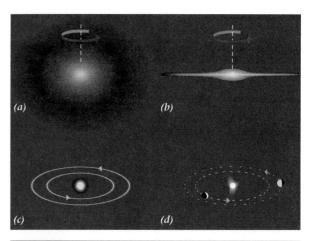

Figure 6-9 The original nebular hypothesis of Kant and Laplace. *(a)* Initially. *(b)* Formation of a ring. *(c)* Several rings have formed. *(d)* Condensation into planets.

the previous chapter. A familiar example of this principle is an ice skater in a spin.

Kant and Laplace pointed out that clouds of gas and dust tend to assume a disklike shape. The particles in a large cloud of gas and dust collide often. The resulting friction causes the particles that lost energy to move toward the equatorial plane. The parts of the cloud at the poles will fall toward the center, but the parts of the cloud at the equator have sufficient angular momentum to retain their original distance from the spin axis. The end result, then, is that the cloud flattens into a disk-like shape (**Figure 6-8**). In fact, any rotating body will flatten (which is why Jupiter and Saturn are noticeably not spherical, and planetary rings and spiral galaxies are disks). At this point, Laplace and Kant surmised that the collapsing mass of gas would leave behind a ring of material and that this ring would eventually gather itself into a planet. As the cloud collapsed further, more rings would be left behind. In this way, a planetary system might be formed, with the remaining gas in the middle forming the Sun (**Figure 6-9**).

Inquiry 6-8

Which of the patterns observed in the solar system are consistent with the nebular hypothesis? Which are inconsistent?

Inquiry 6-9

How might you explain the existence of planetary moons using the nebular hypothesis?

Inquiry 6-10

Would the nebular hypothesis be able to explain the differences in composition between the jovian and terrestrial planets?

The nebular hypothesis, as *originally* presented, has three fatal flaws. First, it is difficult to think of a way for the gaseous material left behind in the rings to coalesce into planets. It is much more likely that the internal pressure of the friction-heated gas would cause it to disperse. Second, the high spin rate required to throw off rings of material would result in a Sun whose spin is so great that it would contain most of the angular momentum of the solar system, instead of the small amount it does have. The final flaw is that the original nebular hypothesis does not explain the differences in compositions between the giant and the terrestrial planets. These flaws mean that other hypotheses must be considered.

CATASTROPHIC HYPOTHESES

A catastrophic hypothesis for the origin of the solar system is one that involves a sudden, unusual event. Collision hypotheses are an example. In 1745, the French naturalist Georges Buffon (pronounced *boo-fon*) suggested that the debris left over from a collision between a comet and the Sun was responsible for forming the planets. A renewed interest in collisional hypotheses occurred by the late nineteenth century when it became

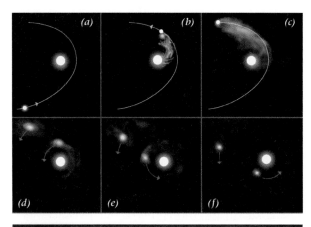

Figure 6-10 The Chamberlin-Moulton hypothesis. A passing star pulls matter from the Sun. The matter condenses into planets all revolving in the same direction.

apparent that nebular hypotheses were unable to explain the angular momentum distribution throughout the solar system.

The most popular form of a collision hypothesis is attributed to the geologist Thomas Chamberlin and astronomer Forest Moulton, who worked on the problem at the beginning of the twentieth century. Their hypothesis is illustrated in **Figure 6-10**. Chamberlin and Moulton surmised that when the Sun was relatively young, another star in the Galaxy passed close to it and pulled away an enormous streamer of gas (Figure 6-10). The streamer would tend to follow the star, so that when planets formed out of it, they would end up going in the same direction around the Sun. A variation of this hypothesis proposed that at one time the Sun was part of a binary system (i.e., it had a companion star) and that a third passing star disrupted the Sun's companion, thereby producing the material that coalesced into the planets. The distances between stars are enormous compared with their physical dimensions. As a consequence, collisions of the sort envisioned in the Chamberlin-Moulton hypothesis must be extremely rare.

Inquiry 6-11

What implications does the rarity of stellar collisions have for the number of planetary systems in the Galaxy? Compare this number to the number that would be expected if the nebular hypothesis were correct.

Inquiry 6-12

Which of the patterns observed in the solar system are consistent with the Chamberlin-Moulton hypothesis? Which ones are inconsistent?

Even though collision hypotheses were formulated to try to explain the angular momentum distribution in the solar system, they too fell short in this area. Furthermore, they offered no good mechanism to make the material that had been pulled away from the Sun condense into planets. Finally, they offered no explanation for the differences in composition between the jovian and terrestrial planets. Thus, they had basically the same problems as the original nebular hypothesis; neither provided a picture of solar system formation that was consistent with observations. We therefore need to look for other explanations.

ADDITIONAL HYPOTHESES

Other hypotheses exist that do not fall directly into either category of evolutionary or catastrophic hypotheses. One such model supposes the planets formed by accretion[4] directly from interstellar material swept up by the Sun. To keep the material from falling into the Sun, however, requires another nearby star. Furthermore, it is unclear how to explain the various observed patterns within the solar system with this model. A recently suggested model has the Sun encountering a protostar from which material was pulled to form the planets. Although such a model appears to account for new data on the abundances of the different elements, the scientific community does not accept it at this time.

6.5 Modern Ideas

Our modern ideas of the solar system's formation are evolutionary and branch out from the nebular hypothesis. Our current understanding, however, is still far from complete. We can approach a better understanding of solar system formation by including the roles played by energy and dust. From there we can move on to an understanding of

---

[4]*Accretion* is growth by gradual addition of material. The verb form is *to accrete.*

the chemical differences among the solar system's members, and the distribution of angular momentum within the solar system.

ENERGY AND SOLAR SYSTEM FORMATION

A cloud of gas and dust has potential energy, which is a body's ability to make something happen based on its position. For example, a rock held above a nail has the potential to push the nail into a board when the rock is dropped onto it. In this case, the source of the potential energy is gravity, and we speak of gravitational potential energy. Once the rock is dropped and it gets closer to the nail, its potential energy decreases while its energy of motion, its kinetic energy, increases. The increase in kinetic energy is exactly matched by the decrease in potential energy because of the law of conservation of energy, which says that in an isolated system the total amount of energy does not change.

The potential energy contained in a cloud of gas and dust exists because the cloud's gravity will make it collapse. When a cloud collapses, its potential energy decreases. The corresponding increase in kinetic energy that must result appears partly as an increase in the speeds of the cloud's gas particles and partly as an increase in its emitted radiation. The increased motion of the particles is measured as an increase in the cloud's temperature.[5]

---

[5]The temperature of a gas is actually defined in terms of the average kinetic energy of the particles in it. For a gas composed of one type of atom, the greater the particle speed, the greater the temperature.

Thus, a collapsing cloud will both heat up and radiate energy. The increase in temperature will tend to slow the collapse, while the loss of energy through radiation tends to cause the cloud to collapse further. Clearly, since stars do form, the loss of radiation wins out in the competition and collapse continues.

THE ROLE OF DUST GRAINS IN PLANET FORMATION

According to modern hypotheses, dust played a key role in the formation of our planetary system. Dust—fine particles of solid matter on the order of 10^{-5} to 10^{-6} cm in diameter—is spread throughout interstellar space and is concentrated in regions such as the Orion Nebula pictured in Figure 1-11. Newborn stars, which are formed in clouds of gas and dust, are usually invisible to optical telescopes because dust is opaque to visible radiation. But infrared radiation, which is radiation you can feel as heat coming from the heating element of an electric stove, can penetrate the dust and allow us to observe the stellar embryos inside. This is similar to how a night-vision scope allows us to detect the presence of a person at night by detecting the heat emitted by the skin. In the Orion region, infrared telescopes have detected entire clusters of collapsing stars not yet visible to the naked eye. **Figure 6-11** shows an area of recent and current star formation in the Monoceros region of the sky. This region is also permeated by both gas and dust.

Figure 6-11 A region of star formation in Monoceros. The Rosette Nebula surrounds a cluster of bright stars in the center (known as NGC 2244).

Dust was important in planet formation for two main reasons. First, dust particles have a strong tendency to stick together when they collide. Sticking made it possible for the dust particles to accrete into larger and larger aggregates as the material forming the solar system (called the **solar nebula**) collapsed into a **protoplanetary disk**. Second, the infrared radiation emitted by the warm dust particles allows a collapsing cloud to lose energy. That energy loss allows collapse to continue. In other words, without infrared-radiating dust, planet formation (and even star formation) could not occur.

CHEMICAL COMPOSITION

The chemical composition of planets is a result of the chemical composition of the gas and dust from which they formed. What was the dust in the embryonic solar system made of? The lowest mass elements, hydrogen and helium, were not a major constituent, for these two elements were too **volatile**, which means they readily vaporized even at low temperature. For this reason, hydrogen would appear only in the form of frozen compounds of materials such as water (H_2O), ammonia (NH_3), and methane (CH_4). The principal constituents of the dust, then, were the heavier elements such as carbon (C), oxygen (O), silicon (Si), and the like.

In addition to the dust that was already present in the solar nebula before it collapsed, more dust would be produced during the cooling of the nebula. Initially, the collapsing solar nebula heated up as atoms were squeezed into a smaller space and collided with each other. In such collisions, material loses kinetic energy and thus must emit light— mostly in the form of infrared radiation. This process of heating by cloud collapse took place most rapidly in the center of the collapsing disk. Here, material collapsed most rapidly into the **protosun**, the large aggregate of material that ultimately became the Sun. Eventually, however, the collapsing gas cloud began to cool again, especially in its outer regions as dust radiated infrared energy away. At that time, material that was originally in the form of gas began to condense into new dust grains. (This is exactly the same process by which a cooling cloud in the Earth's atmosphere forms ice crystals.) In this way, new dust grains formed, and older dust grains (if they survived the heating phases) grew in size through deposits of new dust on their surfaces.

Unlike a cloud forming ice crystals, however, an interstellar gas cloud contains a wide variety of chemical elements such as those shown in **Table 6-2** for the Sun. The different elements in a gas cloud will condense to solids at different temperatures (**Figure 6-12**). Experiments show that all the elements are in gaseous form at temperatures above about 2,000 Kelvin (written 2,000 K).[6] (A temperature of 2,000 K is about 3,100° F.) As shown in Figure 6-12, oxide flakes begin to form at about 1,600 K, whereas iron flakes form at a temperature of about 1,450 K and silicate particles at about 1,300 K. Because different regions of the collapsing solar nebula would be at different temperatures, different substances would condense out in different parts of the cloud and at different times.

This picture neatly explains the differences in the densities and compositions of the various planets. The nebular temperatures would decrease with distance from the newly formed Sun (Figure 6-12). In the innermost, hottest parts of the cloud, we would expect only elements that condense at high temperatures, such as iron, nickel, and silicates, to condense out and produce solid materials. Hydrogen-bearing materials, which cannot

Table 6-2 Abundances of Elements in the Sun

The elements are listed in order of increasing atomic weight.

ELEMENT	RELATIVE MASS OF ATOMS (PERCENT OF TOTAL)
H	74.5
He	23.5
C	0.41
N	0.099
O	0.96
Ne	0.17
Na	0.0035
Mg	0.075
Al	0.0065
Si	0.091
S	0.040
Ca	0.0065
Fe	0.16
Ni	0.0081

[6]Astronomers measure temperatures in units of Kelvin (designated by K). A Kelvin temperature is simply centigrade temperature plus 273; on this scale there are no negative temperatures, so it is often called *absolute temperature*. See Appendix A3 for the relationships between temperature scales.

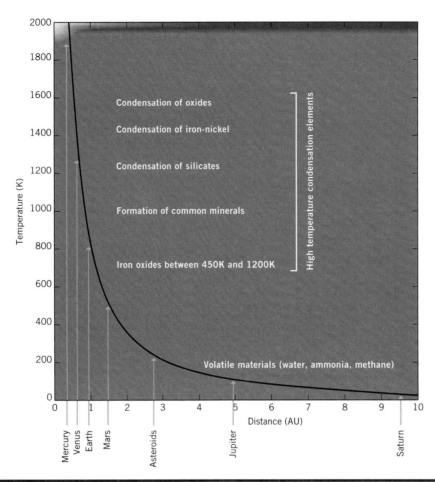

Figure 6-12 The temperatures at which various materials in the solar system condense. The temperature variation with distance from the center of the solar nebula allows astronomers to understand the observed compositions of the planets and moons.

condense in the hot inner solar system, would be absent. The previous discussion is in agreement with studies of the terrestrial planets that show their principal constituents to be those elements that condense at high temperatures.

In the cooler outer portions of the solar nebula, the first elements to condense as the temperature decreases would be those that condense at high temperatures. Later, when the temperature drops further, the low-condensation elements will condense. These materials would include ices of water (H_2O) and, even farther out, ammonia (NH_3) and methane (CH_4). As the condensed particles come together, the higher-density particles will move toward the center of the solar nebula and the lower-density ices will move toward the outside. We find that such ices do, in fact, make up the compositions of the moons of the outer planets and comets. The condensation processes we have just discussed are summarized in **Figure 6-13**.

BRINGING IT ALL TOGETHER

We can argue, based on physical and orbital similarities, that all the terrestrial planets (and the cores of the jovian planets) probably formed by the accumulation of dust grains rather than by the gravitational collapse of gases. In fact, the dust and gas in the solar nebula tend to separate from each other. The motions of gas particles produce a pressure that keeps the gaseous material somewhat extended in a thick rotating disk. The dust grains, which are more massive than the gas particles, are pulled toward both the central regions and the equatorial plane, thus forming a much thinner rotating disk. This packing allows the dust grains to collide more rapidly and to stick together. Studies that model the growth processes suggest that only a thousand years are required to produce 10 km diameter objects that are loosely bound. These bodies are called **planetesimals**. They form

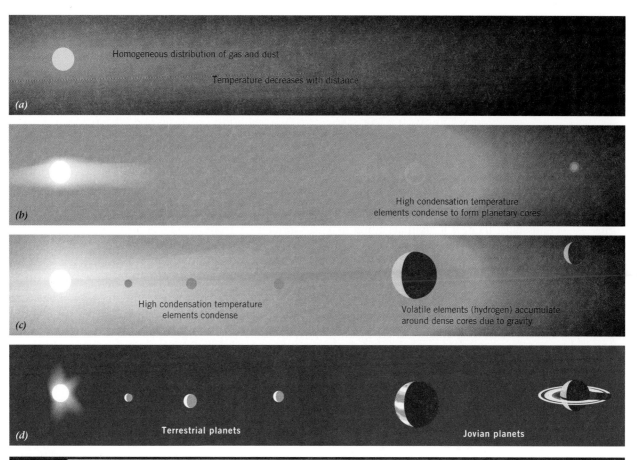

(a) Homogeneous distribution of gas and dust

Temperature decreases with distance

(b) High condensation temperature elements condense to form planetary cores

(c) High condensation temperature elements condense

Volatile elements (hydrogen) accumulate around dense cores due to gravity

(d) Terrestrial planets

Jovian planets

Figure 6-13 One possible scenario explaining why the inner planets contain elements that condense at high temperature while the outer planets contain primarily volatile elements. The temperature decreases with distance from the Sun. As time progresses, from part *a* to part *d*, the temperature decreases further.

rapidly compared with the overall time of some 50 million years for planets to form after the formation of the solar nebula.

The planetesimals then collided and merged with each other until only one planet—or a planet with moons—existed at a given distance from the Sun. Models show that a consequence of a planet's buildup out of many planetesimals is that it would acquire a rotation in the same direction as the disk. In effect, the direction of rotation of a body was forced to follow the direction of revolution of the disk from which it was formed.

In a similar way, a planet may acquire a disk around itself, consisting of those planetesimals that did not strike the planet's surface. These planetesimals could then coalesce into one or more moons. Because this disk will orbit the planet in the same (counterclockwise) direction as the planet spins, so will any resulting moon (**Figure 6-14**).

Figure 6-14 Acquisition of a protomoon disk by an accreting planet. The disk's direction of rotation is shown.

The description of terrestrial planet formation has been the accepted model for some years. However, as is the case for any science, new information could cause the ideas to change significantly. For example, recent chemical information indicates that the formation times may be 10 to 100 million years, which is much longer than when a gravitational runaway occurs. Other ideas also indicate that terrestrial planet formation may involve a long phase of accretion near the end caused by a significant number of impacts by large bodies—a time period often referred as the great bombardment. In fact, as will be discussed in Chapter 8, such an impact is thought to have formed the Moon.

If terrestrial planet formation was thought to be pretty well understood, that is not the case for jovian planet formation. The most widely accepted hypothesis involves the formation of a rocky core by accretion in the same way a terrestrial planet forms. A second stage, then, is a rapid accretion of gas from the protoplanetary disc. Rapid accretion increases the protoplanet's mass, which increases the accretion rate, and a runaway accretion occurs until the surrounding gas is depleted. Unfortunately, our understanding is incomplete and we do not know if the disc will last long enough in the protoplanet's strong gravity for the accretion to occur. An alternative process is one in which the planet forms directly from a massive and cool region of instability in the gaseous disc. Such formation could occur rapidly. However, this process will only occur if the conditions are just right, which makes it less likely to occur very often; it may only work for the most massive of such planets.

The description we have given is not fact, but theory. Remember, however, that theory in science is backed up by observation and experiment. The ideas presented in this section are consistent with our knowledge and are likely to be correct in general, if not in detail.

Inquiry 6-13

Which of the patterns observed in the solar system are consistent with the modern hypotheses of solar system formation? Which ones are inconsistent?

Inquiry 6-14

How can this theory, as it stands, account for the moons that travel around the planets in *clockwise* orbits?

REFINEMENTS OF THE THEORY

The basic theory outlined in the foregoing section does not explain all the observed features of the solar system. For example, those ideas do not explain the moons that revolve in a direction oppo-

site to their planet's rotation. Almost all these moons are, as it happens, quite far from their planet, and it has been suggested that the planet captured them somewhat later than the swarms of particles that formed the inner moons. Calculations show that far from a planet, moons revolving in the same direction as the planet's rotation will quickly escape; thus only distant moons revolving in the direction opposite the planet's rotation will remain.

Theory shows that moons formed close to a planet will tend to have orbits that lie near the equatorial plane of the planet, whereas those formed far from the planet will tend to be near the fundamental plane of the solar system. Our observations more or less confirm this theoretical expectation, although there are exceptions.

Some of the differences in composition between the two major groups of planets could also be explained once we accounted for the fact that the central condensation in the disk is becoming a star at the same time the planets are forming. By studying stars in the process of formation, astronomers have found that, just before becoming a stable star, protostars appear to go through a stage in which they eject a considerable amount of mass, which is returned to the interstellar medium. If the Sun went through such a stage, it must have started out considerably more massive than it is now. When its extra mass was ejected, gases in the disk that had not been incorporated into planets would have been blown away, along with some or all of the gases in the atmospheres of the protoplanets.

Inquiry 6-15

Which group of planets would be more likely to lose their atmospheres at this time—the low-mass or the high-mass group? What about the group that is nearest the Sun compared with the group that is farthest away? What effect would this have on the chemical composition of the planet?

Because the early chronology of the solar system, particularly the speed of planet formation in relation to the Sun's formation, is far from definitely established, the following alternative sequence of events is also a possibility. As the Sun settled into its final equilibrium state as a star, the force of solar radiation started pushing the lighter gases out of the inner part of the solar system. If this occurred before a significant amount of condensation into protoplanets occurred, then the inner protoplanets would have formed entirely out of the ices and dust particles of the heavier elements. Furthermore, the protoplanets would have had negligible atmospheres. Because hydrogen and helium constituted all but a small percentage of the early solar nebula, most of the gas would have been swept out of the inner solar system in this picture, with the inner protoplanets forming out of the tiny residual of heavy elements left behind. **Figure 6-15** summarizes this process, which is different from that summarized in Figure 6-13. Note that the end results of these different models appear to be the same, so it is difficult to choose between them.

No matter which chronology is correct, the terrestrial planets are thought either to have lost all their original **primitive atmospheres** at an early era or else to have formed without an atmosphere.

(Of course, some of the terrestrial planets have atmospheres now, and the reason for this will be discussed in Chapter 9.) However, the strong gravity of the jovian planets would have caused them to retain most of their primitive atmospheres, which would explain why today we think we see them much as they were when they accumulated their gas from the solar nebula. A comparison of the present compositions of the giant planets with those of the terrestrial planets confirms these theoretical predictions. For example, studies of Jupiter show that its abundances are much like those of the Sun; that is, mostly hydrogen with a little helium and only a tiny fraction of everything else. Earth, on the other hand, has little hydrogen and helium (most of its hydrogen is bound up in water, H_2O) and is composed primarily of heavier elements such as silicon, oxygen, and so forth.

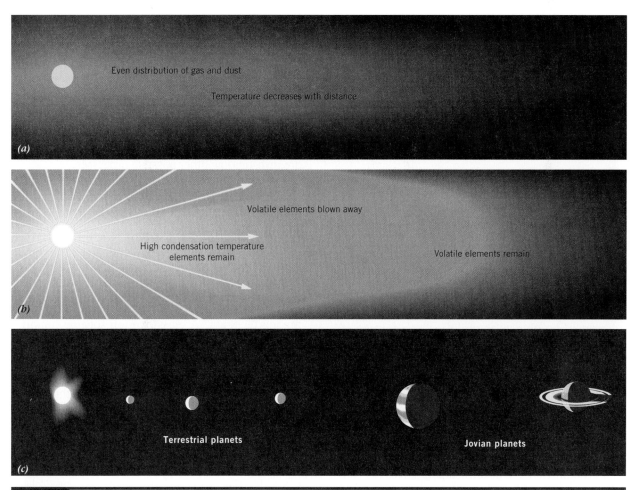

(a) Even distribution of gas and dust

Temperature decreases with distance

(b) Volatile elements blown away

High condensation temperature elements remain

Volatile elements remain

(c) Terrestrial planets

Jovian planets

Figure 6-15 Another possible scenario explaining why the inner planets contain mostly elements that condense at high temperatures while the outer planets contain primarily volatile elements. Compare with the scheme in Figure 6-13.

LEFTOVER ODDS AND ENDS

Several patterns in the solar system disturb the orderly picture already sketched. The most vexing, yet intriguing, are the following:

1. Venus's rotation is opposite in direction to that of most of the other planets. Its period of rotation is long—243 days. One suggested explanation for the anomalous spin is that it once had a normal rotation but was slowed down by the tidal effects of the Sun and later by those of the Earth. Collision with another body, however, is the most generally accepted idea today.

2. Uranus's spin axis lies nearly in the ecliptic instead of roughly perpendicular to it, as with all the other planets (Figure 6-5b). The anomalous rotation of Uranus and its moon system is more difficult to explain. Possibly many of the blobs of material that formed Uranus collided with the protoplanet at unusual angles, but such speculation is difficult to prove. Similarly, the idea of a single catastrophic collision while the system was forming might explain why the planet, rings, and moon system are all tilted; computer models indicate that such a collision will work.

3. The outer solar system, which is from Pluto outward, is of interest to astronomers more than ever. Pluto's orbital plane is inclined by a rather large angle (17°) to the ecliptic (Figure 6-4). In addition, the orbit has a somewhat large eccentricity (Figure 6-3). Furthermore, Pluto's small mass (only some 0.2 percent that of the Earth) and diameter implies that its density must be low, near that of the giant planets. Pluto appears to be the closest and largest known member of a group of objects known as trans-Neptunian objects (TNOs). The objects, which are discussed in more detail in Chapter 7, are cometlike objects thought to have formed in the vicinity of the giant planets and then migrated further out. An understanding of the orbits and chemistry of these objects is crucial to understanding the formation of the solar system.

THE ANGULAR MOMENTUM PROBLEM

As it stands, the modern theory of the origin of the solar system we have described *still* leaves the Sun with too much angular momentum. That is, it predicts that the Sun should be rotating much faster than it actually does. If this theory is basically correct, then some mechanism must have either given the excess angular momentum to the planets or removed it from the system entirely.

Several hypotheses have been advanced to resolve the angular momentum problem. One suggests that the gas in the solar nebula may have contained a magnetic field,[7] as is often the case for the gas in interstellar space. As the solar nebula collapsed and increased in temperature, some of the gas atoms were broken up into positively charged particles (called ions) and electrons, which have a negative charge. Such a gas composed of charged particles is known as a **plasma**. If a plasma contains a magnetic field, the gas and magnetic field are coupled and are forced to move with each other. As the cloud collapsed, the strength of the magnetic field would have become stronger.

If this were the case in the solar nebula, then the rotating protosun would have dragged its magnetic field around with it, and the rotating magnetic field would have dragged the plasma in the disk around with it. This would have transferred angular momentum from the Sun to the disk, thus slowing the Sun's rotation substantially. Evidence in favor of this **magnetic braking** includes the observations that old stars rotate much slower than young stars, assuming that sufficient time has passed for the magnetic braking to have slowed the stars' rotation. While not proven to occur, magnetic braking is the most widely accepted mechanism, although astronomers are investigating other possible processes.

WHICH HYPOTHESIS OF SOLAR SYSTEM FORMATION IS PREFERRED?

The models just discussed can be classified according to their answers to two questions: Were the planets and the Sun formed at the same time? Did the solar system form from material from the Sun, or directly from the gas and dust present among the stars? A classification scheme for finding the preferred hypothesis is shown in **Table 6-3**. For example, because in the original nebular hypothesis planets formed directly out of interstellar material at the same time as the Sun, this model goes into the upper left box. In the case of the collision hypothesis, planets formed from material

---

[7]Although the concept of the magnetic field will be presented more completely in Chapter 8, we can describe it for now in terms of the ability of a magnet to attract magnetic materials from some distance away. The stronger the magnetic field, the stronger the attraction.

Table 6-3   Classification of Solar System Formation Models

	PLANETS FORMED DIRECTLY FROM INTERSTELLAR MATERIAL	PLANETS FORMED FROM SOLAR MATERIAL
The Sun and the planets formed at the same time	Original nebular hypothesis	
	Modified nebular hypothesis	
The Sun and the planets did not form at the same time	Accretion of interstellar material	Collision theories in which tidal effects of passing star pull off material from which planets form
	Encounter with two nebulae	
	Encounter with a protostar whose material formed the planets	Sun is part of a binary system; planets formed from a second star

Source: T. Encrenaz, J.-P. Bibring, M. Blanc, The Solar System (Springer-Verlag), page 45, Table 3.2.

pulled from the Sun; that is, material that had been previously inside the Sun. Furthermore, in this model, the planets were produced well after the Sun's formation. Therefore, this model belongs in the lower right box. The models involving accretion of planets directly from interstellar material, from an encounter with two nebular gas clouds, and from an encounter with a protostar, all involve stellar material that formed the Sun and planets at different times. These three models therefore go into the lower left box.

Now let's see what the observations tell us, because they place severe constraints on our theoretical models. The answer to the question, "Did the solar system form from solar material?" may be answered by looking at the abundance of heavy hydrogen (**deuterium**) relative to the abundance of hydrogen. Deuterium is readily broken down into hydrogen by the temperatures found in stars. However, because the planets have measurable amounts of deuterium, the material from which the planets formed could not have been previously inside a star. Therefore, all hypotheses on the right side of Table 6-3 may be discarded. The answer to the question, "Were the planets formed at the same time as the Sun?" can be examined using radioactive dating techniques (discussed in the next chapter). From such studies, astronomers have determined that the time interval between the protosun's separation from the interstellar medium and planetary formation is less than 10^8 years. This result shows that the Sun and planetary system did form together. The bottom boxes of Table 6-3 are therefore eliminated, strengthening the idea that some type of evolutionary hypothesis formed the Sun and our planetary system.

MODEL SOLAR SYSTEMS

Is our planetary system the only one that could have resulted from the conditions present when our solar system formed? The discussion of the next section provides the observational answer: no. Prior to having the observational result, astronomers could take a theoretical approach by taking the ideas discussed in this chapter—orbital patterns, relative sizes and masses and their variation with distance from the Sun, the conservation of angular momentum, the elemental condensation sequence—and computing model solar systems. **Figure 6-16**, which pictures the results of several such calculations, shows the distances of the resulting planets from the Sun in astronomical units. The number by each planet is the mass in percentage of the Earth's mass, which is defined as 100. Our solar system is shown for comparison. Because the accumulation of material into a planetary body is a random process, these models show that a variety of solar systems could result, even when the starting conditions are the same.

6.6 Other Planetary Systems

As we have seen, the more recent hypotheses of the origin of the solar system view the formation of a planetary system as a more or less natural consequence of the formation of a star out of interstellar gas and dust. They suggest, on the one hand, that many of the single stars in the Galaxy—if not most of them—ought to have planets. On the other hand, hypotheses such as the collision hypothesis would predict that the formation of planets is an extremely rare event, perhaps unique

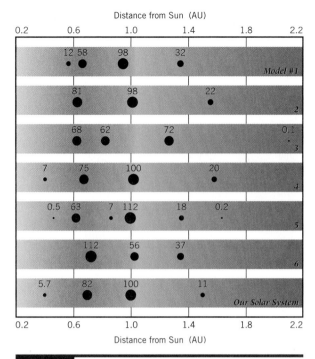

Figure 6-16 Model possible solar systems. The number above each dot is the mass of the planet expressed as a percentage of the Earth's mass, which is 100.

in the history of the Galaxy. How are we to select among these reasonable hypotheses? Observations must be used to select the best hypothesis, knowing that future additional observations will require modification of that hypothesis.

The answer to the question, "Do other stars have planets?" is now yes. The definitive, long-sought announcement came in 1996—nearly 400 years after the Italian monk, Giordano Bruno, was burned at the stake by the Roman Catholic Inquisition for, among other things, suggesting that the stars were other suns having planets circling them. More than 135 **extrasolar planets** (planets beyond our solar system) are known (**Table 6-4**). Now that astronomers know how to find them, more will continue to be found as techniques and technologies mature.

Why did it take so long to find these extrasolar planets? One reason is that a planet emits little light. Jupiter, for example, is 10^9 times fainter than the Sun. Finding such an object in the glow of a star that is a fraction of a second of arc away is a difficult problem. In fact, none of the currently

Table 6-4 Some Extrasolar Planetary Systems

All the stars are somewhat similar to the Sun. Star names with letters after them are the planets themselves; the other cases list the star to which a planet belongs.

Star	Distance (LY)	Orbital Period (Days)	Minimum Mass (Jupiters)	Semi-Major Axis (AU)	Number of Planets
Gliese 876b	15	61	1.89	0.21	2
Gliese 876c	15	30.1	0.56	0.13	—
55 Cancri b	44	14.65	0.84	0.12	3
55 Cancri c	44	44.28	0.21	0.24	—
55 Cancri d	44	5360	4	5.9	—
47 Ursa Majoris	46	1089	2.54	2.09	2
Tau Bootes	49	3.31	4.13	0.05	1
51 Pegasus	50	4.23	0.46	0.05	1
Rho Corona Borealis	57	39.8	1	0.22	1
Upsilon Andromeda b	57	4.62	0.69	0.06	3
Upsilon Andromeda c	57	241.5	1.89	0.83	—
Upsilon Andromeda d	57	1284	3.75	2.53	—
70 Virginis	59	116.7	7.4	0.48	1
14 Herculis	60	~2000	4	;3	1
HD 195019	65	18.2	3.57	0.14	1
HD 210277	68	434.3	1.3	1.12	1
16 Cygnus B	72	798.9	1.69	1.67	1
HD 114762	90	83.9	11	0.35	1
HD 75289	94	3.5	0.44	0.05	1
HD 168443b	108	58.1	7.73	0.3	2
HD 168443c	108	1770	17.2	2.87	—
HD 217107	120	7.13	1.3	0.07	1

known planets has actually been seen; their presence is inferred from observations, which we discuss next.

METHODS OF SEARCH AND DISCOVERY

Four different methods of searching for extrasolar planets are possible. The first three involve observing the star itself for effects of the planet, and the fourth one simply images the planet directly.

The first method of searching for extrasolar planets looks for a change in the apparent motion of a star over a period of many years. This technique will be described in detail in Chapter 14, but we will give a brief description of it here. All single stars without a planetary system follow a relatively smooth orbital path around the center of the Galaxy, since the stars are too far apart to drastically affect each other's orbits. However, if a star has a companion, the center of mass of the system will follow that smooth path and we would observe the star's path to be oscillating (**Figure 6-17**). Stellar companions have been inferred in this way, notably the white dwarf companion to Sirius, but to date no planets have been discovered using this method.

The second method looks for the effect of a planet on the color of light of emitted by a star. You probably have the experience of driving down a highway with the car windows open and being passed by a car or truck, and hearing a change in the pitch of the vehicle as it approaches and then passes. Specifically, you probably heard the pitch of the sound become higher as the vehicle approached and then decrease as it passed you. You may have heard the same thing happen while waiting for a train to pass. This change in pitch results from something called the *Doppler effect*, to be described in detail later. This change in the tone results from the motion of the vehicle in relation to you, the observer. The faster the relative motion, the greater the change in pitch; in fact, the

change in the pitch is directly proportional to the relative speed toward or away from you.

The Doppler effect also happens with light. Light emitted from a star that is moving toward us will have its color become very slightly bluer than it would be when stationary. And, a star moving away from us will have its color become very slightly redder. Thus, by measuring the tiny change in color, the astronomer can determine two things: that one object is moving in relation to the other, and the relative speed.

As already described, if a planet is orbiting a star, both the planet and the star actually orbit around their common center of mass. Thus, although a planet will not be seen, its *effect* on the star will produce a motion around the center of mass. That motion is observable through the Doppler effect. In fact, the amount of the Doppler effect will change periodically as the star revolves around the planet, and it is through that periodic change that the presence of planets is discovered and the orbital properties determined. **Figure 6-18** is a graph of the relative speed of the star 51 Pegasi over time. The smooth and periodic nature of this graph is indicative of orbital motion around a center of mass. From this graph, the mass of the unseen planet can be determined, as well as the distance of the planet from its parent star.

The planet's characteristics are found by a detailed comparison of the observations with model calculations having different input parameters. For example, if a researcher picks a planet's mass, orbital eccentricity, and distance from the

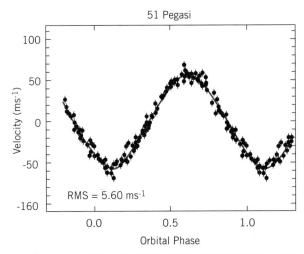

Figure 6-18 The variation of the observed velocity of a star around its center of mass with time for the star 51 Pegasi.

Figure 6-17 The expected motion of a star (heavy line) that has a planet orbiting around it. The thin line represents the motion of the supposed center of mass.

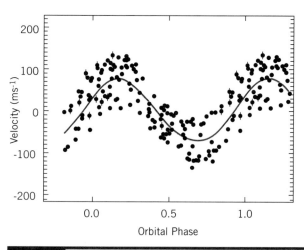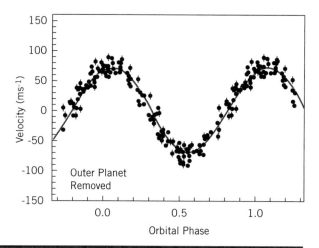

Figure 6-19 The extrasolar planetary system 55 Cancri. *(a)* shows a model based on one planet orbiting the star (solid line), the points are the difference between the model and the observations. *(b)* Same as *(a)* but for two orbiting planets. Note the decrease in the scatter around the model when two planets are assumed. In fact, a third planet has been found in this planetary system.

parent star, she can compute what the graph of velocity against time should look like for that combination of parameters. Additional models for different input parameters are also computed and compared with the observations. Once a model is computed that agrees with the observations, the researcher then knows that planet's characteristics (assuming it is a unique answer).

One very interesting case is that of the star 55 Cancri in the constellation Cancer. The comparison of the observations with the best model is shown in **Figure 6-19a**. Although the general agreement is good, a large scatter from the theoretical curve exists among all the observations. This scatter is indicative of the presence of a second body that is perturbing the motion of the star! If models are computed to account for the presence of a second planet, the result is **Figure 6-19b**, where we see that the agreement with observations is considerably improved. In fact, the scatter is still larger than expected and a third planet has been inferred for this system.

Yet another technique for discovering extrasolar planets takes advantage of the slight change in light from a star when an orbiting planet eclipses some of the stellar light. For example, if an observer on another planet were observing the Sun when Jupiter crossed between the Sun and the observer, a small but measurable decrease in brightness would occur. The observation is a difficult one to make. Thus far, three planets have been discovered in this way. This method will be used to search for Earthlike planets with the NASA spacecraft called *Kepler,* which has a launch planned for 2007. If Earth-sized planets are as common as we think,

Kepler should discover a significant number of them during its four-year mission.

An eclipsing planet can also cause a slight increase in brightness as a result of the star's light being bent by the gravity of the planet. This effect, called **microlensing**, is a consequence of Einstein's theory of general relativity. The planet is acting like a gravitational lens and focusing the light, which causes a temporary enhancement in the apparent brightness. As strange as this sounds, mass can cause light's path to bend, and the concept will be used in other contexts in this book later.

The last technique is to image the planet, but to date no extrasolar planets have yet been imaged. Imaging them is extremely difficult because you are trying to observe a faint object very close to a much brighter one when both of them are far away. Difficult problems rarely stop astronomers, however! Thus, both Earth-based and space-based instruments and techniques are now being developed and built that should eventually allow some extrasolar planets to be imaged within 10 years.

The search for Earthlike extrasolar planets and the desire to assess their life-support possibilities are the goals of NASA's *Terrestrial Planet Finder* project. The concept involves two separate missions to be launched sometime between about 2014 and 2020. One mission will have a coronagraph, which is a 4- to 6-m-diameter telescope with the capability of blocking the light of a star to make the surroundings visible. The second mission would consist of five 3 m to 4 m mirrors at fixed distances from one another (forming an interferometer, as discussed in Chapter 12) to look for heat emitted by planets orbiting stars.

Now that more than 100 extrasolar planetary systems have been found—and in a short period of time!—we can conclude that planetary systems are common in the Galaxy. The fact that most hot stars seem to rotate rapidly (in a day or less) whereas cooler stars appear to rotate slowly (taking weeks or months) indicates that the cooler stars have transferred their angular momentum to a protoplanetary disk through the process of magnetic braking. Thus, we predict that such cool stars would be surrounded by a planetary system. In fact, more than 95 percent of all stars appear to be in the slowly rotating group, suggesting that most of the 100 billion (10^{11}) stars in our galaxy could have planets. But if the difference in rotational speeds has another explanation—for example, if the angular momentum is instead transferred to the interstellar gases—then slow rotation may not imply a planetary system.

PROPERTIES OF EXTRASOLAR PLANETS AND THEIR STARS

The study of extrasolar planets has become a growth industry in astronomy! We now have sufficient data to begin to compare and contrast them with our solar system.

Searches for extrasolar planets have mostly been for planets around solar-type stars. However, planets have also been discovered orbiting cooler, larger stars known as K giants (to be further discussed in Chapter 14) and a pulsar. An important result is that the greater the abundance of iron in a star, the greater the probability that the star will have planets. Since, as we will study further in later chapters, iron is manufactured in massive stars and then incorporated into later generations of stars, this means that planet formation was unlikely to occur in the earliest times in the history of the universe. Once iron was produced, and especially when stars having more iron than the Sun formed, the possibility of planet formation increased.

The first extrasolar planet to have its atmosphere observed by astronomers orbits the star known as HD 209458, which is 150 light-years away. The atmosphere was discovered when the planet, whose radius is 1.6 times that of Jupiter, passed in front of its parent star. The first observations found the planet's atmosphere to have less sodium than expected for this type of planet—perhaps indicating blocking by high-altitude clouds. Subsequent results have detected both oxygen and carbon. Not only that, but this planet's separation from its star is only 0.05 AU (1/8 the size of Mercury's orbit), making it one of the closest planets to its parent star. And, with its mass near that of Jupiter, it is known as a *hot Jupiter,* which is a Jupiter-type extrasolar planet orbiting very close to its star. The result of the planet's proximity is that it is observed to be evaporating hydrogen into space in an almost cometlike tail! For this reason, the planet is being called *Osiris,* after the Egyptian god whose brother killed him and cut him into pieces to prevent his return to life.

The rapid loss of mass shown by Osiris might amount to as much as 10 percent over its lifetime. The discovery of the evaporation of Osiris may have found the first of a new class of extrasolar planets that has been termed the *chthonian planets,* in reference to the Greek gods of the hot underworld. Such planets would be the solid remnant core of a hot Jupiter whose orbit took it too close to its star.

How do hot Jupiters form? Accepted ideas say that gaseous planets of that size cannot form so close to hot stars. Thus, astronomers hypothesize that the planets formed at much greater distances and then have their orbits decay through gravitational encounters with other bodies or a protoplanetary disc to move the planet closer.

Extrasolar planets are also found to have orbits that are, more often than not, considerably more elliptical than those in our solar system. The reason is not yet fully understood, although ideas include gravitational effects within the protoplanetary disc or of one planet on another.

The first observation of magnetism for an extrasolar planet has been reported. The star HD 179949 has been found to have a hot spot that revolves around the star every 3.093 days—the identical period of its planet's revolution. Such a hot spot had previously been predicted to result when a planet's magnetism interacts with a star's magnetism. Thus, the observation of the hot spot leads one to infer the presence of extrasolar magnetism.

Figure 6-20 shows the distribution of the masses of 111 extrasolar planets. Most are Jupiter-sized. The smallest planet discovered thus far is 0.05 times Jupiter's mass, or about 14 Earth masses. As technology improves, we expect that many more lower-mass planets will be discovered.

Until recently, astronomers had only one example of a planetary system to study—our own. Generalization based on only one system is notoriously unreliable. Now, however, we have more than 100 systems to test our hypotheses. An exciting golden age of understanding planetary system formation is about to dawn.

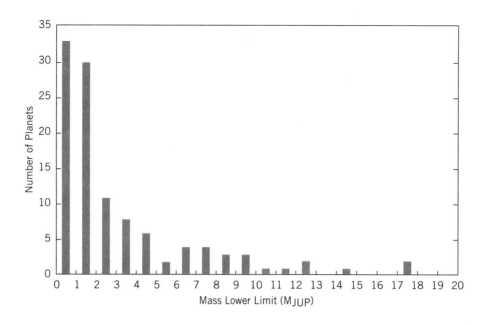

Figure 6-20 The distribution of masses for the 111 planets known as of July 2004.

A Scale Model of the Solar System

When you have completed this Discovery, you should be able to do the following:

- Describe the size of the solar system on a human scale.

A model of the solar system, which can help you understand better the sizes and distances in our planetary system, can be built on a human scale. For this model, let one astronomical unit be represented by 25 steps. After completing the questions below, walk your model. Stop at each planet and look back to where you started.

- **Discovery Inquiry 6-1a** Make a table of the number of steps you need to take to walk from the Sun to each planet.
- **Discovery Inquiry 6-1b** Determine the length of your step by measuring the length of 10 steps and dividing by 10.
- **Discovery Inquiry 6-1c** How large in yards or meters is your model solar system?
- **Discovery Inquiry 6-1d** On the scale of your model, how many miles or kilometers are there to the nearest star, which is approximately 200,000 AU away?
- **Discovery Inquiry 6-1e** On the scale of your model, determine the sizes in inches or centimeters of the Sun and the various planets.

Chapter Summary

Observations

- *Terrestrial planets* are near the Sun, are relatively small in mass and diameter, have high density, are composed of heavy elements, and have a high-temperature, low-density atmosphere. *Jovian planets* are far from the Sun, are relatively large in mass and diameter, have low density, are composed of low-mass elements, and have a low-temperature, high-density atmosphere.

- Planets move in orbits that are nearly circular and lie in the ecliptic plane, which also is in the Sun's equatorial plane.

- With the exception of Venus, Uranus, and Pluto, planets rotate about their axes in a counterclockwise direction. With the exception of Uranus and Pluto, planetary rotation axes are within 30° of the perpendicular to the planet's orbital plane.

- The largest planetary moons revolve around their planet in the same direction the planet rotates. Moons close to a planet are near the planet's equatorial plane.

- The Sun contains 99.8 percent of the mass of the solar system but only 2 percent of the angular momentum.

- More than 100 extrasolar planets were observed just as soon as our technology could do so, implying that planets are common.

- An extrasolar planetary atmosphere has been observed. This planet's star is also observed to be evaporating hydrogen away from the atmosphere.

Theory

- The *nebular hypothesis* states that a rotating cloud of gas and dust collapsed to form the Sun and left rings of material behind that formed the planets, which are located in a flat plane. The *collision hypothesis,* also known as the catastrophic hypothesis, states that tidal forces from a passing star pulled material from the forming Sun. This material then condensed into the planets.

- When dust particles in a collapsing cloud collide, they stick together, thus providing the first step in the eventual growth of planetesimals. These planetesimals can form terrestrial planets or the more massive cores of jovian planets. Jovian planets, then, accrete gas from the protoplanetary cloud to become giant planets. Dust particles provided the heavy elements present in the terrestrial planets and jovian planet cores.

- Moons forming near a planet should be in the equatorial plane.

- The Sun may have transferred its angular momentum to the protoplanetary disk by means of an interaction between the particles it constantly ejects and its magnetic field (magnetic braking).

- From our understanding of star and planet formation, we expect our galaxy to be filled with planets orbiting other stars.

Conclusions

- The Sun and planets formed at nearly the same time.

- From observations of deuterium, we infer that the planets did not form from previously processed stellar material.

- Chemical differences between the terrestrial and jovian planets can be understood in terms of the different temperature-dependent condensation rates of the elements while the solar nebula was cooling.

- The Sun went through a phase in which it lost much of its original material. This rapid loss caused gas and dust remaining within the inner solar system also to be removed from it. This loss of mass may also have contributed to the loss of volatile elements by the terrestrial planets.

- The currently accepted model of solar system formation is a nebular-type theory that incorporates a wide range of observations.

- Our galaxy must contain large numbers of planets.

- The observations of extrasolar planets require changes to our simple hypotheses of planetary system formation.

Summary Questions

1. What are the names of the planets in order of their distance from the Sun?

2. What are the two major groups of planets? List the characteristics that distinguish members of each group.

3. What are the major patterns in the orbits and motions of the planets that need to be explained by any successful theory of the origin of the solar system? What exceptions are there to these patterns?

4. What observational evidence can you use to criticize the early nebular and collisional hypotheses of the origin of the solar system?

5. What are some possible reasons for the dramatic differences in composition between the terrestrial and jovian planets?

6. Why were dust grains important to the formation of the solar system?

7. Where is most of the angular momentum in the solar system located? Why was the distribution of angular momentum a problem for earlier hypotheses of solar system formation? Describe a process thought to solve the angular momentum problem.

8. What techniques have astronomers used to detect planets orbiting other stars?

9. Describe the differences in the formation of terrestrial and jovian planets.

Applying Your Knowledge

1. Hypothesis: The solar system formed when a passing star pulled material from the Sun that then formed the planets. Present evidence both for and against this hypothesis, and state your final conclusion.

2. List all observations about the solar system that are exceptions to the generally observed features.

3. Explain why the following two definitions of volatile materials are the same. (a) Volatile materials are those that condense at low temperatures. (b) Volatile materials are those that turn to gas at low temperatures.

4. Use the condensation sequence of Figures 6-12 and 6-13 to explain in your own words the reasons different planets differ in chemical composition.

5. Refer to Figure 6-5b, which shows the inclinations of the planets to their orbital planes. For each planet, discuss the seasonal variations you expect throughout the planet's year.

▶ 6. What fraction of planetary orbital angular momentum is held by Jupiter? Saturn? (We are neglecting rotational angular momentum, which is less than that from the orbital motion.)

▶ 7. Compute Pluto's density if its mass is 0.0022 the mass of the Earth and its diameter is 2,302 km. Express the density relative to the density of the Earth. What conclusions about the nature of Pluto might you draw from the observed density?

8. What would be the angular separation between the Sun and Jupiter if they were observed from the distance of Alpha and Proxima Centauri, the nearest stars, at a distance of 4.3 light-years? Compute the distance a dime would have to be for its angular size to match the angle you computed.

Answers to Inquiries

6-1. They would be composed of the lowest mass elements, hydrogen and helium.

6-2. As the star's light passes through the planet's atmosphere, the light passes through successively denser layers of gas, which causes the light to gradually fade until the planet itself blocks the star's light.

6-3. The low gravity makes it easier for atoms to escape. The high temperature increases the average velocities of the atoms, so that a greater fraction of them are moving fast enough to escape Mercury's gravity.

6-4. The distance between the Earth and Sun is approximately 10,000 times the Earth's diameter. In Figure 6-2, the Earth's diameter is about 0.4 cm, which means the Earth's distance, on this scale, is 4,000 cm or 40 m. Pluto, which has an average distance 40 times farther away, would be about 1,600 m away.

6-5. The hypothesis would not be consistent with the fact that the planes of the planetary orbits are nearly the same, because it would be expected that the planets would be captured into random orbits with their masses also distributed randomly. To capture a planet into a nearly circular orbit requires special, rare circumstances. The hypothesis is consistent with having a large range of planetary masses, but not their split into terrestrial and jovian groups. The hypothesis is not consistent with most of the planets and moons revolving in the same direction.

6-6. The hypothesis would not be consistent with the fact that the Earth, Moon, and meteorites have the same age because the various bodies could be captured at different times. However, if all these objects had actually been formed at the same time, they would have the same age.

6-7. The hypothesis would not be consistent with the near-circularity of the planetary orbits, because capture would more likely create highly eccentric orbits. It would not be consistent with the fact that the moons mostly orbit in one direction (the same as the planets), and it would not place all the giant planets together in the solar system as we see them today. It is consistent with the variable masses and the variable compositions that we see.

6-8. The simple nebular hypothesis would explain the basic traffic pattern of motions. It would not explain irregularities like the spin of Venus or the odd inclination of Uranus's spin axis, nor the distribution of angular momentum in the solar system.

6-9. Perhaps a process similar to the one proposed for the formation of planets themselves could be used to explain the formation of moons around a planet.

6-10. The nebular hypothesis as presented would not be able to explain the chemical compositions of the planets.

6-11. The collision hypothesis would lead us to expect that planetary systems are rare in the Galaxy, whereas the nebular hypothesis implies that they are common, a natural companion of star formation. Direct observational evidence strongly favors some type of nebular hypothesis.

6-12. This hypothesis has the same basic problems as the nebular hypothesis. Also, it is awkward to fit the formation of the moons of the planets into this picture.

6-13. The modern theory explains the overall counterclockwise rotation and spin of most of the planets and their moons. It does not explain the anomalous spins of Venus and Uranus. And all hypotheses still have some trouble with Pluto.

6-14. Clockwise rotation is not explained.

6-15. The low-mass group, because the gravity of these planets is lower. Because the low-mass elements present in the disk (hydrogen and helium) would be the ones most readily blown away, the inner planets would tend to be composed of the heavier elements that remain.

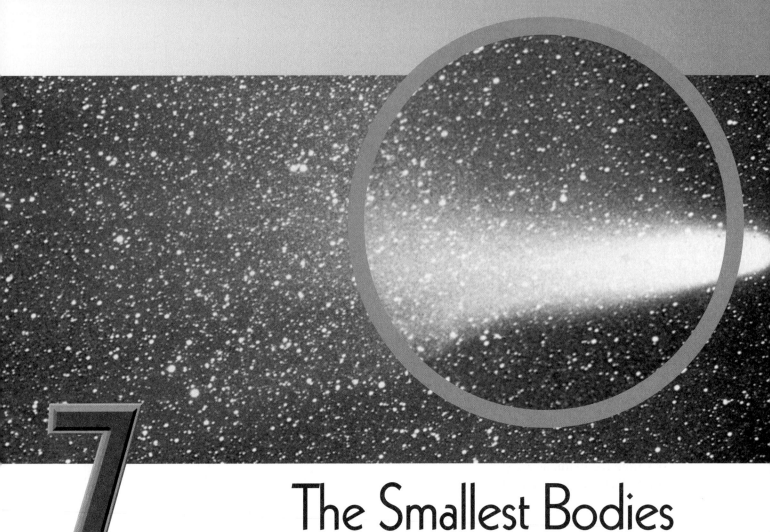

7

The Smallest Bodies
of the Solar System: Comets,
Minor Planets, and Meteorites

The search for the "primordial" composition [of the Solar System] may be likened to the search for the Holy Grail. The analogy is apt insofar as fervor is concerned, though medieval bloodshed has been replaced by mere acrimony.

Edward Anders, *Annual Review of Astronomy and Astrophysics, 1971*

C omets, asteroids, and meteorites can shed a great deal of light on the origin of the Solar System. Although the surfaces of the planets have undergone many changes since the Solar System was formed, the smaller objects we will discuss in this chapter have changed much less and can tell us about the early Solar System. During this discussion, we will also fill in some of the details of the theoretical scenario proposed in Chapter 6 for the origin of the Solar System.

7.1 Comets

Comets are frequent and sometimes spectacular visitors to our part of the Solar System. In former times they caused consternation and even panic among those who viewed them because comets were often interpreted as signs and omens of the future. One such example is the comet of 1066 C.E., which the Saxons saw as an evil omen before their defeat by the Normans at the Battle of Hastings. A portion of the Bayeux Tapestry, which commemorates this event, shows the presence of the comet in the sky (**Figure 7-1**).

Aristotle contended that comets were exhalations from the Earth, and it was Tycho Brahe who showed that they were too distant from the Earth for this to be true. After the discovery of Newton's laws of motion, Edmund Halley (who later became England's Astronomer Royal) calculated the orbit of the comet of 1682; he found that its period was approximately 76 years and that it traveled in an orbit that took it about 35 AU from the Sun (**Figure 7-2**). Halley suggested that the comets of 1531 and 1607 were actually the same object and predicted that it would return in 1758. Although he did not live to see his prediction come true, the comet's reappearance in 1758 caused great excitement and sealed the triumph of Newton's theory by showing that all objects in the Solar System are subject to the same physical laws.

The comet was named in honor of Halley; it turned out to be the same comet that was seen in 1066, and sightings of it have been traced in Chinese records as far back as 240 B.C.E. The Earth actually passed through the extended tail of Halley's comet during its spectacular 1910 appearance. Although the comet returned on schedule in 1985–1986 (**Figure 7-3**), that was its least impressive appearance in 2,000 years, because Earth and the comet didn't come close to each other, and the comet was low in the sky for Northern Hemisphere viewers. However, during this visit, several spacecraft were sent to rendezvous with it, and we were able to learn more about comets in just a few months than in all previous history.

Figure 7-1 The Bayeux Tapestry. The legend in the tapestry reads: "They marvel at the star."

Chapter opening photo: Halley's comet during its 1985/1986 approach to the Sun.

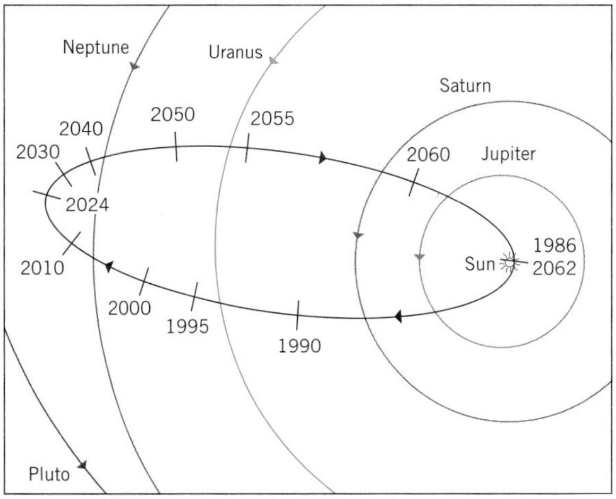

Figure 7-2 The orbit of Halley's comet, approaching and receding from the Sun.

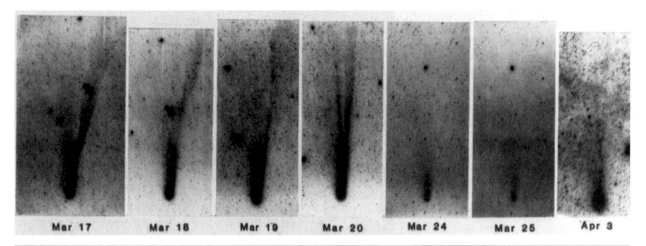

Figure 7-3 Halley's comet, as it appeared on various dates during its 1985–1986 appearance.

Mar 17   Mar 18   Mar 19   Mar 20   Mar 24   Mar 25   Apr 3

Prior to the year 1995, an average of about a dozen new comets were discovered each year, often by amateurs. Because it is not possible to predict the first appearance of a new comet, a certain amount of luck is required, along with diligence, patience, and enough familiarity with the sky to recognize an intruder. The situation changed in 1995 with the launch of *SOHO (Solar and Heliospheric Observatory)*. Although designed primarily to study the Sun, the satellite is able to block out the Sun's radiation to look for comets near the Sun. SOHO has become the all-time record holder for comet discoveries, with nearly 1,000 discoveries as of mid 2005. Many of these discoveries have been by amateur astronomers obtaining images over the World Wide Web and searching them for comets. A large fraction of these comets are sungrazers—comets that closely approach or even collide with the Sun (**Figure 7-4**).

THE STRUCTURE OF COMETS

Despite their impressive appearance when near the Sun, comets are actually rather small objects. They have been described as ". . . the nearest thing to nothing that anything can be and still be called something." When far from the Sun they do not show a visible disk in the telescope as does a planet, and it is estimated that the cometary body itself, called the **nucleus**, averages only about 10 km in diameter. Furthermore, they are not very massive; comets have been observed to pass close to the moons of Jupiter without noticeably altering the moons' orbits. The average mass of a comet is approximately 10^{14} kg, only one ten-billionth that of Earth.

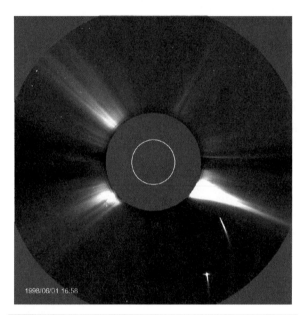

1998/06/01 16:58

Figure 7-4 Two of the many sun-grazing comets discovered by the *SOHO* spacecraft.

As a comet nears the Sun, it appears to get larger and to develop a fuzzy appearance. The fuzziness comes about because solar energy heats volatile ices located within the nucleus, causing the ices to change directly to a gas. The process is the same as when dry ice changes from solid to gas. The central nucleus of the comet appears as a small, bright point surrounded by a nebulous region known as the **coma** (**Figure 7-5**). A coma, when near the Sun, is typically 10^5 km in diameter. As the comet approaches even closer to the Sun, a long **tail** forms, which may extend more than 100 million km from the comet's nucleus.

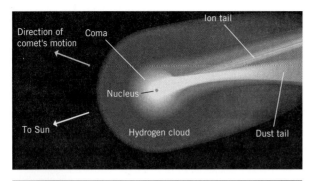

Figure 7-5 The structure of a comet near the Sun.

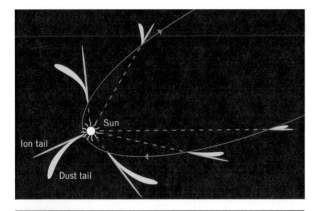

Figure 7-6 As a comet goes around the Sun, the tail always points away from the Sun.

The tail can appear highly complex, and there may be more than one tail present (Figure 7-5). The tail (or tails) tends to point *away* from the Sun; the tail thus precedes the comet when it leaves the vicinity of the Sun (**Figure 7-6**). This fact was known to Chinese astronomers at least 900 years before Europeans learned it in the sixteenth century. Finally, observations from satellites have found that comets have a hydrogen halo extending some 1 to 10 million kilometers around them. The hydrogen halo is thought to come from the break-up of H_2O molecules.[1]

What causes these spectacular changes in the appearance of a comet as it rounds the Sun? How can such a small object provide such a fantastic display? To answer these questions, let us look at a model for comets that appears to explain things reasonably well.

---

[1]A molecule is the smallest particle into which a compound can be divided without changing its physical and chemical properties. It is a group of atoms held together by chemical forces.

THE DIRTY SNOWBALL MODEL

Astronomers now know that the nucleus of a comet consists of frozen gases along with solid materials. The frozen gases are mostly water ice, along with some carbon monoxide, and carbon dioxide. The comet expert Fred Whipple has aptly described this body of ice, dust, and rocks orbiting the Sun as a sort of **dirty snowball** or *dirty iceberg*. When the nucleus approaches the Sun, solar energy vaporizes some of the ice and dust, releasing gas and dust from the nucleus into surrounding space. This produces the extended coma. Eventually, all that will be left will be a loosely connected swarm of rocks.

Evidence supporting the dirty snowball model came from Comet Shoemaker-Levy 1993e (**Figure 7-7**) when it spit into numerous pieces after a close approach to Jupiter (0.007 AU) in May 1992. Prior to their colliding with Jupiter in July 1994, the pieces were studied to learn more about cometary structure.

As the comet approaches even closer to the Sun, two effects that are important in producing the comet's tail increase in significance. First, pressure from the Sun's radiation pushes dust particles in the comet in a direction away from the Sun. Second, a **solar wind**, which consists of charged particles that are constantly ejected from the Sun, exerts a force on the material released from the coma. Because both the **radiation pressure** and the solar wind are directed outward, the tail will tend to point away from the Sun.

The two effects that produce tails tend to act in different directions and to affect the gas and dust particles somewhat differently, which is why two tails often are seen. The difference between them can be detected by a careful study of the light from each tail. For example, because the smooth-appearing tail has light similar to sunlight, this tail must be composed of small dust particles that simply reflect the sunlight. It is therefore called the **dust tail**.

The other tail shows a complex and variable structure composed of gases excited into giving off radiation. Within this **gas tail**, we find emissions characteristic of charged molecules such as CH^+, OH^+, N_2^+, and CO_2^+, to name a few. Presumably, the charged molecules are present because the charged particles of the solar wind knock electrons off some of the gas molecules vaporized from the comet's coma. The interaction of the charged particles with

Figure 7-7 The nucleus of Comet Shoemaker-Levy 1993e was observed to have broken into numerous pieces, probably from a close passage to Jupiter in 1992.

the magnetic field of the Solar System produces the complex structure that we observe in comet tails. The smooth dust tail is not charged and is not affected by magnetism.

Inquiry 7-1

Observations show that comets tend to become fainter and fainter with successive returns near the Sun. Can you explain this in terms of the dirty snowball model?

Until recently, the small, faint nucleus could not be analyzed directly, and it was necessary to infer the actual composition of the nucleus's frozen gases from the elements observed in the gas tail—hydrogen, oxygen, carbon, and nitrogen. The molecules we see are *daughter* products formed when the Sun's ultraviolet light and the solar wind break down more complex substances, a fact that was confirmed by the International Comet Explorer (ICE) satellite in a rendezvous with Comet Giacobini-Zinner in late 1985. It is generally thought that the parent gases are frozen methane (CH_4), ammonia (NH_3), and water (H_2O). Other satellite studies of Halley's comet in the spring of 1986 greatly enlarged our knowledge of the nucleus.

When a comet is close to the Sun, atomic emissions from metals such as sodium, calcium, silicon, and iron are also observed. These are presumably the result of vaporization of dust particles by the intense solar heat, followed by excitation of the vapor by the Sun's radiation; these emissions are additional evidence of the presence of solid material in the cometary body.

A final piece of evidence in favor of the dirty snowball model is the fact that the motion of some comets is slightly erratic. An example is Comet Encke, which has been well studied because of its short three-year period. It has been observed to slow down or speed up unpredictably, especially when near the Sun. Fred Whipple's model explains this phenomenon: after many passages near the Sun, much of the comet's surface material has evaporated, leaving behind a hard, black crust on the surface. Solar heating of pockets of gas near the surface might force the gases through a thin part of the crust as stronger parts hold firm (**Figure 7-8**). This released gas acts like a jet that accelerates or decelerates the comet, depending on which direction the rotating nucleus happens to be pointed at the time. It is these jets that make cometary behavior difficult to predict reliably.

ORBITS OF COMETS

Cometary orbits tend to be highly elongated ellipses (see Figure 7-2), oriented randomly with respect to the ecliptic plane. In general, it is just as likely that a comet will be found traveling around its orbit in a clockwise direction as in a counterclockwise direction. This is different from the situation for planetary orbits, which are nearly circular, close to the ecliptic plane, and always traversed in a counterclockwise direction.

The orbits of comets fall into two basic types. Comets in *elliptical* orbits, such as Halley's comet, are clearly members of the Sun's family, orbiting the Sun indefinitely. In contrast to the closed circuit that a body in an elliptical orbit follows, a body in a *hyperbolic* orbit follows an open path that does not close back on itself (**Figure 7-9**). These comets in hyperbolic orbits have high velocities that allow them to escape from the Sun's gravitational hold and become interstellar outcasts.

Inquiry 7-2

Apply Kepler's law of areas to a comet following a highly elliptical orbit. When is the comet moving most rapidly? Where in its orbit does the comet spend most of its time?

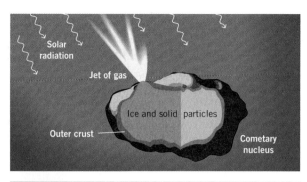

Figure 7-8 Whipple's explanation of the erratic motion of Encke's comet. "Jets" of gas squirting out through fissures in the nuclear crust alter the motion of the comet in an unpredictable way.

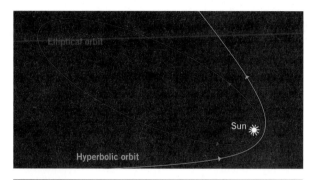

Figure 7-9 Cometary orbits can be either elliptical or hyperbolic. Comets in hyperbolic orbits escape from the Solar System and are never seen again.

Are comets with hyperbolic orbits interlopers from interstellar space? It appears that this is not the case. Rather, there is evidence that comets with hyperbolic orbits were once in elliptical orbits but came too close to a planet such as Jupiter, which accelerated them to high speeds and altered the shape of their orbits from elliptical to hyperbolic.

Inquiry 7-3

Comets are observed to come toward the Sun from all directions in space. We also know that the Sun is traveling rapidly toward a point in the constellation of Hercules. If we hypothesize that the Sun were picking up comets from interstellar space during its travels toward Hercules, from what direction would the comets tend to come? How do the observations support the idea that hyperbolic comets were originally part of the Sun's family?

Inquiry 7-4

What might you conclude about the distribution in space of cometary nuclei, given that the long-period comets are observed to come equally from all directions?

THE OORT AND KUIPER COMET CLOUDS

Because comets with long periods come toward the Sun from all directions in space, their distribution around the Sun must be roughly spherical. The Dutch astronomer Jan Oort hypothesized that there exists a large spherical cloud of comets surrounding the Solar System, between about 50,000 AU and 100,000 AU from the Sun. Although well beyond the orbit of Pluto, the **Oort cloud** is still a part of the Solar System. Recent estimates are that the Oort cloud contains some 6 trillion comets and has a total mass 40 times that of the Earth.

All useful theories about the origin of comets presume that they were formed early in the history of the Solar System. For example, they could have condensed out of the solar nebula at an early stage, before it began to flatten. If this idea is correct, then the cloud of gas and dust from which the Solar System was formed must have been large. However, the density of material in the outer reaches of the Solar System would have been so low that it is difficult to imagine how nuclei some 10 km in diameter would form. For this reason, today we have reasons to accept the idea that the comets formed in a higher-density region, in the general area between the orbits of Saturn and Neptune, at 10–20 AU from the Sun. Such a model is consistent with the icy composition of the comets. We can then easily imagine that gravitational perturbations from the giant planets (especially Jupiter) could toss them out of the inner Solar System, thus forming the Oort cloud.

The entire discussion of the Oort cloud is based on inference from observation; it has yet to be actually observed—until, perhaps, early 2004 when the discovery of the most distant object in the Solar System was reported. Sedna, as it has been temporarily named, is observed to be at 120 AU from the Sun (three times Pluto's distance). Its orbit is highly eccentric, which means its distance from the Sun varies from as far as 900 AU to as close as 76 AU. Such a large range is more characteristic of a cometary orbit than anything else. From an estimate of Sedna's reflectivity, which is thought to be relatively high, a size of about 75 percent that of Pluto is found. Although highly reflective, its surface does not appear to be icy as might be expected; furthermore, it is nearly as red as Mars, which has planetary scientists baffled. Thus, Sedna may be considered to be a member of the inner part of the Oort Cloud—the first such object to be found, or a part of another comet system called the Kuiper Belt that is discussed shortly. Keep in mind, though, that these results are so new that modifications are bound to occur quickly, and you may want to check the Internet for the most up-to-date information.

Inquiry 7-5

Given that Sedna's greatest distance from the Sun is 900 AU and its nearest distance is 76 AU, determine its orbital period around the Sun. (*Hint:* The semi-major axis of the orbit is the same as the average distance.)

Most of the objects in the Oort cloud would have nearly circular orbits, so that they would never get close enough to the Sun to be seen. From time to time, however, a passing star might give cometary nuclei in the Oort Cloud a small gravitational nudge that changes the orbits so that they begin their long trips toward the inner Solar System. With such a large number of comets in the Oort cloud, a large number of cometary nuclei can be perturbed toward the Sun. Such an event triggering showers of comets might occur when the star Gliese 710 passes through the Oort cloud in about 1.4 million years. Gravitational tidal forces with the disk of the Milky Way galaxy and interactions with gas clouds might also send comets on their way for a rendezvous with the Sun. Thus, the Oort cloud is a reservoir of comets that every now and then, have their orbits disturbed so that the comet nucleus can enter the inner Solar System.

An additional complexity appears from analysis of the orbits of short-period comets—those having periods less than about 200 years. These comets are more confined to the ecliptic plane than the longer-period ones. Mathematical models of orbits show that short-period comets could not have come from the spherical Oort cloud. For this reason, astronomers hypothesized the existence of a nearer, disk-shaped belt of comets some 30–50 AU from the Sun, just outside Pluto's orbit. This belt, called the **Kuiper belt** (pronounced KY-per) after the astronomer who supposedly first proposed it, is estimated to have about the mass of the Earth, much less than the 40 Earth masses hypothesized for the Oort cloud. Unlike the Oort cloud, objects belonging to the Kuiper belt have been observed. Such objects have often been referred to as *Kuiper belt objects,* but the preferred term nowadays is

trans-Neptunian object, and we will use this term from here on.

Trans-Neptunian objects (TNOs) have been observed since 1992. As of spring 2005, the number of cataloged objects nears 900. They are chemically different from asteroids and have orbital properties different from comets. Being formed far from the Sun, trans-Neptunian objects might be expected to be composed of water ice and various frozen gases, and thus be similar to cometary nuclei. This expectation is borne out by observations showing that the first observed TNO, known as 1992 QB_1, contains dark, carbon-rich material similar to at least some comets. Another object, Chiron, with a diameter of 200 to 300 km, was originally thought to be a distant asteroid. However, it has been observed to have a coma, meaning it must be a large comet and a member of the Kuiper belt. Some trans-Neptunian objects of significant sizes have been found: 2000 WR 106, known as Varuna, is about 40 percent the size of Pluto; 2002 LM60 (tentatively known as Quaoar) is about 50 percent the size of Pluto, and 2004 DW is estimated at 70 percent. (The year of discovery is the first part of the object's name.)

Nearly all the observed TNOs are found beyond the orbit of Neptune, suggesting that it might play a role in defining the belt (**Figure 7-10**). All the objects have orbits that are never far from the ecliptic. Planetary scientists estimate from the observations that the Kuiper belt contains at least 70,000 objects larger than 100 km in diameter. The mass of the Kuiper belt is then inferred to be a hundred times greater than the asteroid belt between Mars and Jupiter (discussed in the next section).

A number of trans-Neptunian objects are found to be at the same distance from the Sun as Pluto. These objects, known as **Plutinos**, are at a distance that corresponds with the 3:2 resonance with Neptune, which means that they revolve around the Sun twice for every three orbits of Neptune. The existence of Plutinos provides astronomers with new ways of explaining many of the unanswered questions concerning the formation of the outer Solar System. The Kuiper belt, then, might be considered a storehouse of short-period comets. Some astronomers have asked whether this storehouse might, in fact, have been the birthplace for Pluto and its moon, and the large Neptunian moon Triton. These objects, which are discussed further in Chapter 10, have similarities to one another but

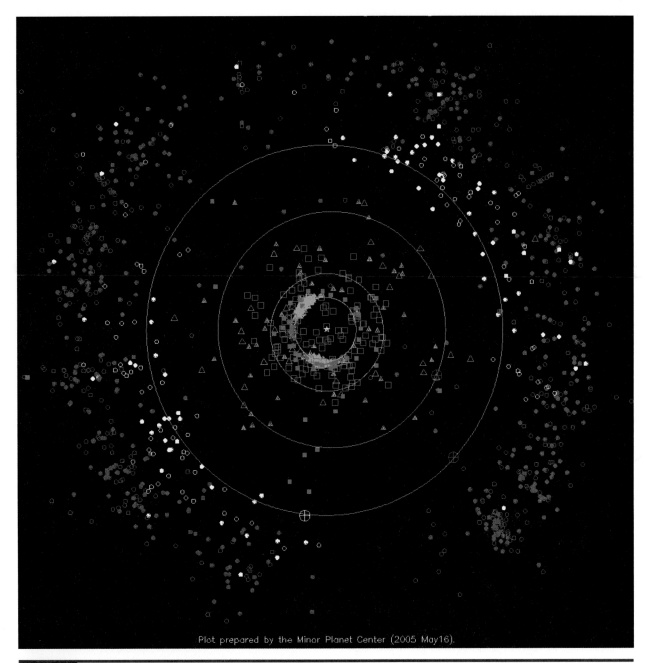

Plot prepared by the Minor Planet Center (2005 May16).

Figure 7-10 Currently known objects thought to belong to the Kuiper belt. The orbits of the Jovian planets are indicated, as is the May 2005 position of Pluto.

are unlike their other neighbors. Thus, Pluto is probably the largest of the currently known trans-Neptunian objects.

Small differences in chemical composition may exist between short period comets from the Kuiper belt and the long-period comets from the Oort Cloud. The Kuiper belt is thought to have formed at a greater distance from the Sun than where the Oort Cloud objects originally formed (before being ejected to their more distant current loca-

tion). Therefore, the temperature was less at that greater distance. The conclusion, then, is that the objects that belong to the Kuiper Belt (the TNOs) may be the most primitive icy objects remaining in the Solar System, which makes their study by astronomers extremely valuable.

Finally, we note that the discussion of comets is consistent with the discussion of Solar System formation by means of a nebular-type theory of formation as discussed in Chapter 6.

Inquiry 7-6

After many passages near the Sun, a comet will eventually wear out, becoming depleted of gas and unable to produce a coma or a tail. If the period of a typical comet in its orbit around the Sun is 1,000 years, and it returns 100 times before using up its material, what is the lifetime of a typical comet? What fraction of the age of the Solar System (about 4.5 billion years) is this comet's lifetime?

Inquiry 7-7

Is it likely that the comets we see today have been coming near the Sun since the Solar System was created? Explain.

Inquiry 7-8

The short lifetimes of comets, and the fact that we continue to observe them today, presents a dilemma. How might Oort's theory be able to resolve this dilemma?

SPACECRAFT STUDIES OF COMETS

During 1985 and 1986, several spacecraft missions were sent to rendezvous with comets. The *International Comet Explorer (ICE)* was originally designed to study the Sun, and it had been in orbit for several years doing just that. However, it turned out to be possible to change the orbit of the satellite and send it to the neighborhood of Comet Giacobini-Zinner, which was passing close to the Earth. There was enough spare fuel on board to send the satellite on a trip around the Moon, and from the Moon's gravity it acquired additional acceleration. In fact, it made several passes around the Earth-Moon system before achieving enough of a *slingshot effect* to travel toward the comet. This virtuoso feat of celestial mechanics allowed the satellite to study in detail the interactions between the comet and the solar wind.

After Halley's comet rounded the Sun in the spring of 1986, a veritable flotilla of spacecraft from several different nations (Russia, Japan, and a European consortium) converged on it. Each mission was designed to study the comet from different distances and acquire different kinds of information. Two Russian *Vega* spacecraft returned the first close-up images of Halley, passing within 5,523 miles (almost 8,900 km) of the nucleus. Next, the Japanese spacecraft *Suisei* (the word means *comet*) viewed the comet in ultraviolet light. Even though *Suisei* came no closer to the comet than 94,000 miles (150,000 km), it was still rocked by collisions with two large particles moving through space with the comet.

The most spectacular (and risky) mission was the close pass by the European spacecraft *Giotto,* only five days after the *Vega 2* rendezvous. Named for the Renaissance artist who depicted an earlier passage of Halley's comet in a painting, *Giotto* passed within 376 miles of the nucleus. *Giotto* was struck by one collision that aimed its antenna in the wrong direction for a while, but the spacecraft's gyroscopes restored control after 34 minutes of anxiety. No further TV pictures were returned after the collision, but some 2,000 images of the comet were collected before *Giotto's* cameras were literally sandblasted away by the dust and debris moving through space with the comet. *Giotto* was designed to resist the destructive forces of the comet as well as its designers could anticipate, but there is still a certain element of luck involved when you are aiming a precision instrument into a storm where sandlike particles are moving at thousands of miles per hour!

Giotto confirmed that the nucleus of Halley's comet was about 8 by 16 km (5 by 10 miles) in size and shaped somewhat like a peanut (**Figure 7-11**).

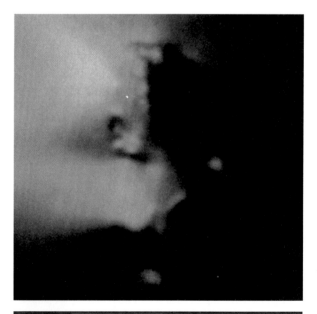

Figure 7-11 Jets of material are seen streaming away from the black, elongated nucleus of Halley's comet in this photograph taken from a distance of 15,950 miles by the European spacecraft *Giotto.*

It showed jets of evaporating material concentrated at the two ends of the nucleus. Perhaps most surprising was the color of the nucleus. It turned out to be jet black, one of the darkest objects in the Solar System, reflecting only 5 percent of the light incident on it. At least for Halley's comet, the traditional picture of a dirty snowball should be modified to something like "a lump of coal with frozen ices inside." Much to everyone's surprise, observations found that the nucleus was hot, with a temperature of 330 K (or 135° F). The surface of the nucleus was definitely not covered with ice or the temperature would have been lower. The dark surface absorbs sunlight efficiently, and the evaporating gases pour out from the interior in jets and streams.

These spacecraft determined that Halley's comet has a low density, only some 0.1 to 0.25 grams/cm³. Such a low density implies that the nucleus is porous, actually some 90 percent empty space. Another surprise discovery was the presence of magnetism around the nucleus. The charged molecules in the coma causes this magnetism.

Inquiry 7-9

How might studies of the composition of comets help us learn about the primordial composition of the Solar System?

On the all-important subject of chemical composition, the instruments on the *Giotto* and two *Vega* spacecraft revealed the presence of water, carbon dioxide, and carbon monoxide gases. The gaseous jets, in fact, contained about 80 percent water vapor. Such a large amount of water, when broken down by energy from the Sun, can explain the extensive hydrogen halo surrounding comets. But observers also noted signatures of heavier compounds containing silicon, carbon, oxygen, sulfur, and iron, to name a few. The relative amount of carbon to some other elements is the same as that found in the Sun, confirming the idea that comets formed from raw interstellar material. The dirty snowball model has also been confirmed.

The Stardust mission encountered Comet Wild 2 (pronounced *vilt two*) in January 2004. While the primary goal was to collect cometary dust and volatile materials during a close encounter and return it to scientists on Earth, *Stardust* also collected interstellar dust. During the approach,

particles were trapped on sticky panels on the spacecraft. These will be dropped off at a reentry capsule to be parachuted to Earth in 2006. The image of the nucleus (**Figure 7-12**) provides the most detailed view of a comet nucleus we have ever seen.

Stardust found Comet Wild 2 to have spires of material rising as high as 100 m. This finding provides evidence that this comet is a single porous mass and not just a loosely bound jumbled mass of material. Another important finding is the presence of abundant amounts of organic materials that have been hypothesized to be the source of the chemicals that eventually formed life on Earth.

FUTURE SPACECRAFT STUDIES OF COMETS

Astronomers who yearned for more spacecraft studies of comets thought their dreams were coming true early in the twenty-first century. However, the planned *Comet Nucleus Tour* (CONTOUR) failed, as the spacecraft broke into at least two pieces when leaving Earth orbit. The *Space Technology 4 Champollion* rendezvous and landing mission to Comet Tempel 1 that was planned for 2003 was cancelled in 1999 for budgetary reasons.

The European Space Agency joined the fun with the Rosetta mission in 2004. The spacecraft will fly by some asteroids, and arrive at comet Churyumov-Gerasimenko for a rendezvous in November 2014. An instrument package will land to take measurements on the comet's surface.

Figure 7-12 The nucleus of Comet Wild 2, which is 5 km (3 miles) across, as viewed by *Stardust*.

Deep Impact is a NASA flight to Comet Tempel 1 launched in January 2005. When the comet was reached on July 4, 2005, an impactor traveling 10.2 km/sec (6 miles/sec) struck the surface with the energy of nearly 5 tons of TNT, equivalent to the total monthly energy used by a typical American household. The size of the crater formed depends on the details of the comet's structure, but it may be around 200 m across and 10–50 m deep. The scientific goals are to observe crater formation, how large a crater is formed, measure the chemical composition of the interior of the crater and the ejected material, and determine how future outgassing changes as a result of the impact. The impactor will be accompanied by a flyby spacecraft that will photograph the impact and collect data from the impactor for later transfer back to scientists on Earth.

7.2 Minor Planets

Ever since astronomers have had good estimates of the distances of the planets from the Sun, it has been suggested that there is an unusually large gap between Mars and Jupiter and that there ought to be a planet there. In 1801, the Italian astronomer Giuseppe Piazzi discovered a new object in an area of the sky he had been studying. Greatly interested, he followed its motion on successive nights until it neared the Sun and was lost. Enough observations had been made for the German mathematician Karl Friedrich Gauss to determine its orbit, and the next year it was found again where Gauss predicted it would be. Piazzi named the object Ceres, in honor of the patron goddess of Sicily. The orbit Gauss calculated for Ceres lay between Mars and Jupiter and many thought the object to be the long-sought planet. Yet, it was an unusually small planet.

In 1802, a second such planet, Pallas, was discovered, and, in 1804, a third one, Juno. In time, many others were found, and it was slowly realized that the region between Mars and Jupiter was literally swarming with small bodies, all much smaller than the major planets. For this reason, they have come to be known as **minor planets**, although the terms **planetoid** and especially **asteroid** are also used. More than 11,000 asteroids have names. Modern surveys have found thousands of these objects, and orbits have been determined for nearly 2000 of them. The region of space they occupy has come to be known as the **asteroid belt**.

Figure 7-13 The asteroid 433 Eros, as imaged by the *NEAR* spacecraft.

CHARACTERISTICS OF THE MINOR PLANETS

Ceres is the largest of the minor planets, with a diameter of about 950 km (about the size of Texas); next largest is Pallas at 560 km (about the size of Kansas). The sizes must range all the way down to particles as small as dust grains. Because of the large number of asteroids, collisions must take place frequently, slowly grinding them down to smaller and smaller sizes. When spacecraft to the outer planets have passed through the asteroid belt, collisions with only a small number of tiny particles occur. It appears that some force is at work sweeping most of the smaller particles out of the asteroid belt. Matching pairs of dust bands above and below the asteroid belt, thought to be thrown off by colliding asteroids, have been observed.

Most asteroids are too small for their shapes to be observed directly from Earth. However, in 2000 the *Near Earth Asteroid Rendezvous (NEAR)* spacecraft took careful observations while for the first time orbiting an asteroid, 433 Eros. It then landed on the asteroid. Eros is an irregularly shaped object 33 × 13 × 13 kilometers in size—a fractured rock covered with a layer of rocks and dust shown in true colors in **Figure 7-13**. Asteroid Ida (**Figure 7-14**), observed by the *Galileo* spacecraft from a distance of 10,000 km, shows the

Figure 7-14 The asteroid Ida, as imaged by the Galileo spacecraft.

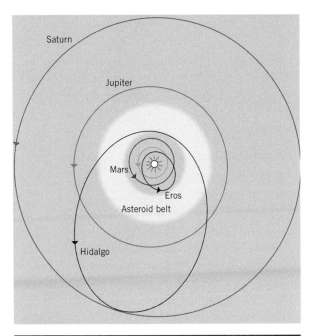

Figure 7-15 The orbits of some minor planets. Most minor planets stay between Mars and Jupiter, but there are some exceptions.

56-km-long asteroid not only to be pitted with craters but also to have a 1.5 km diameter moon orbiting it only 100 km away.

Minor planets, especially the smaller ones, are rather irregular in shape, because most of them are not sufficiently massive for their own gravity to force them into spherical shape, as in the case of the planets. They come in two types, either large chunks of solid rock or rubble piles of loosely held, gravitationally stuck-together chunks of debris.

Inquiry 7-10

The brightness of a typical minor planet is observed to vary significantly over periods of an hour or two. Assuming that minor planets rotate on an axis, what are two possible explanations of these observations?

> **You should do discovery 7-1, Asteroid Brightness Variations, at this time.**

ORBITS OF THE MINOR PLANETS

Although the majority of the minor planets have nearly circular orbits situated between Mars and Jupiter and fairly near the ecliptic plane, there are some notable exceptions. One is Hidalgo, whose orbit is shown in **Figure 7-15**. Its orbit is elongated and inclined by more than 40° to the ecliptic plane.

Inquiry 7-11

Can you suggest a possible reason why Hidalgo's orbit is so different from those of most minor planets? (*Hint: What kind of object does Hidalgo's orbit remind you of?*)

An important and unusual group of asteroids are the few dozen known Apollo asteroids—asteroids whose orbits come close to the Earth's. These objects could have been sent into their present orbits by gravitational perturbations from the giant planet Jupiter. There are probably some 1,000 such Apollo asteroids larger than one kilometer in diameter. Some objects with near-Earth orbits could in fact be burned-out comets. The risk of collision with one of these objects is not negligible, as we will discuss later.

One of the most important asteroids historically has been 433 Eros, which in 1901 and 1931 came within 26 million km of the Earth. On the second occasion it was possible for astronomers to measure with increased accuracy its distance in kilometers. This, in turn, made it possible for the length of the astronomical unit, in kilometers, to be determined accurately for the first time, because the distance to the minor planet in astronomical units was well known from Kepler's laws.

Inquiry 7-12

Suppose a beam from a radar dish on Earth is sent to Eros and returns 173.3 seconds later. What would be the distance in kilometers from Earth to Eros?

Inquiry 7-13

If astronomers calculated that Eros's distance from Earth at this time was 0.17 AU, how many kilometers would this indicate that there are in an astronomical unit?

Today, even more accurate methods of measuring the astronomical unit are employed, but the principle is similar. With radar, it has become possible to measure the distance from Earth to other planets with high precision (considerably better than 1 km). Again, Kepler's laws give the same distance in astronomical units, so the number of kilometers in an astronomical unit can be worked out.

THE TROJANS AND THE KIRKWOOD GAPS

An unusual group of minor planets is the **Trojan group**, located in Jupiter's orbit. The possibility of such a group was already known in the eighteenth century, when the mathematician Joseph-Louis Lagrange showed that if a small body were placed in a circular orbit around the Sun in such a way that it, the Sun, and Jupiter formed an equilateral triangle, then it would stay in that position, moving at the same rate around the Sun as Jupiter (**Figure 7-16**). Such gravitationally stable locations, where the overall gravitational force is zero, are called **Lagrangian points**. The concept will appear again in our discussion of the moons of the Jovian planets. The Lagrangian points for the Earth-Moon system have been suggested as possible locations for space colonies.

The Trojan asteroids are an example of the importance of Jupiter's gravitational effects. Another example is found within the asteroid belt itself; there are certain distances from the Sun where few, if any, minor planets are found. These regions are known as **Kirkwood gaps**, after the American astronomer who discovered them. The influence of Jupiter on the formation of these gaps is clear, because they occur at distances from the Sun at which a minor planet (if it were there) would have a period around the Sun that is in a simple ratio to Jupiter's period. For example, as shown in **Figure 7-17**, which shows the variation of the number of asteroids with distance, there is a prominent gap at the place where the period of a minor planet would be just one-half Jupiter's period, and another where its period would be just one-third Jupiter's period. An asteroid in these locations would find itself close to Jupiter at regularly spaced intervals in time and would suffer a cumulative perturbing force that would sooner or later send it into an entirely different orbit. We say that gaps occur at **resonances** of, for example, 2:1 and 3:1. As we will see, the concept of the orbital resonance occurs in other situations, too, such as Saturn's rings, various moon systems, and even the planetary system. The astronomer Jack Wisdom has shown that many of the bodies forced out of the Kirkwood gaps by Jupiter would eventually collide with Mars or Earth.

WHY ARE THE MINOR PLANETS SO SMALL?

The bare facts about minor planets raise more questions than they answer. In particular, why are there so many of them, and why are they so small? The total mass of all the minor planets put together

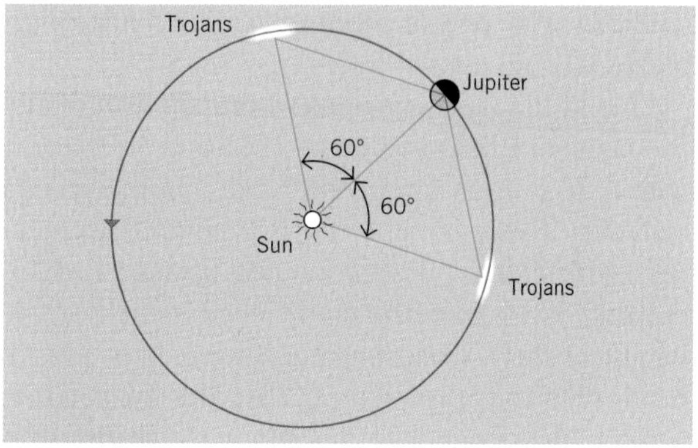

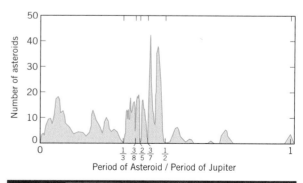

Figure 7-16 The Trojan asteroids, which are located at Lagrangian points in Jupiter's orbit.

Figure 7-17 The Kirkwood gaps.

is considerably less than that of any of the major planets—where did the rest of the mass go, if it was ever present?

There is no question that the presence of Jupiter has been of great importance in answering these questions. Calculations show that if there were a large number of small bodies between Mars and Jupiter early in the history of the Solar System, most of the ones fairly near Jupiter would have been so strongly affected by it that they would have either been captured by Jupiter or else thrown entirely out of the Solar System. This would have greatly reduced the amount of mass in the region.

We might also speculate that the mere presence of a body as large as Jupiter would have prevented the material between Mars and Jupiter from gathering itself together into a major planet, although it is not clear how this would have happened. Alternatively, several modest-sized bodies may have formed there, despite the presence of Jupiter, and later collided with each other, fragmenting into many smaller bodies. Indeed, calculations show that numerous collisions take place in the asteroid belt at present, so such a scenario is certainly possible.

Inquiry 7-14

Groups of minor planets, called *Hirayama families* after the Japanese astronomer who discovered them, have orbits that are remarkably similar to each other, with nearly the same size, shape, and orientation with respect to the plane of the Solar System. How might this observation tend to support the theory that the present-day asteroids were formed as a result of the collisional fragmentation of larger bodies?

CHEMICAL COMPOSITION

Chemical information about asteroids can be obtained through detailed studies of their ability to reflect light of different colors. There are four main compositional groups: silicon rich (S type), nickel-iron rich (M type), carbon rich (C type), and the D types, which may be rich in organic compounds. For example, the asteroids Gaspra and Ida (Figure 7-14) are both relatively light in color, meaning they are silicon-rich (S-type) objects. An important finding is that the locations of these chemical groupings within the asteroid belt depend on distance from the Sun. The silicon-rich asteroids are reddish, reflect about 15 percent of the incident light, and are found toward the inner edge of the asteroid belt. The carbon-rich asteroids are dark like a piece of black coal, reflecting only some 3 percent of the incident light. They are rare in the inner belt, but common in the outer regions, near the orbit of Jupiter. The nickel-iron asteroids, which are only about 5 percent of the total asteroid population, contain little silicate material and are thought to be the fragmented cores of larger bodies. Such correlations provide important tools to be used in our attempts to understand the formation of these bodies and of the Solar System itself.

When the *NEAR* spacecraft flew to and then landed on the asteroid 433 Eros in 2001 (Figure 7-13), it investigated Eros's chemical composition and found the relative amounts of certain elements to be the same as the Sun. This argues that Eros did not undergo heating and cooling after it formed 4.5 billion years ago, which means it is made of primitive Solar System material. Eros's density of 2.7 g/cm^3 (the density of water is 1 g/cm^3) is about that of Earth's crust.

The *NEAR* mission flew past the asteroid Mathilde (**Figure 7-18**) while on its way to Eros. Like Eros, Mathilde's appearance is that of an object that has undergone violent collisions. Mathilde is dark, reflecting only some 3 percent of the light incident on it. Thus, it is similar to the carbon-rich (C-type) asteroids described before. *NEAR* flew close enough to Mathilde to measure small spacecraft accelerations caused by the asteroid. Its density was then determined to be only 1.3 g/cm^3, which is a low value for a carbon-rich moon. For this reason, planetary scientists infer that Mathilde is either

Figure 7-18 The asteroid Mathilde, as observed by the *NEAR* spacecraft from a distance of 2400 km, shows details as small as 1250 feet. The asteroid's dimensions are about 59 by 47 km. The dark crater on the front surface is estimated to be 10 km deep!

Table 7-1 Summary of Observed Properties of Comets and Asteroids

PROPERTY	COMETS	ASTEROIDS
Orbit shape	elliptical	circular
Orbital tilt to ecliptic	random	flat
Physical structure	ice-rock conglomerate	rock
Chemical composition	primarily water ice	rocky (iron, nickel, silicon, carbon)
Formation location	>=30AU	asteroid belt
Current location	Oort belt (~100,000 AU) Kuiper belt (~50 AU)	asteroid belt
Source of	meteor showers	meteorites

porous or an example of a loosely bound pile-of-rubble asteroid mentioned earlier. Although many asteroids have companions (Figure 7-14 shows Ida and its companion, Dactyl), none was found for either Mathilde or Eros.

In conclusion, we see that the observed properties of asteroids are consistent with the nebular-type hypotheses discussed in Chapter 6 on Solar System formation.

FUTURE SPACECRAFT STUDIES OF ASTEROIDS

One mission to study asteroids is currently on its way: the Japanese *Hayabusa (Falcon)* spacecraft. Asteroid Itokawa will be reached in October 2005, remain close to it for 5 months at which time it will collect samples, and then return the samples to Earth in summer 2007.

Two large asteroids, Ceres and Vesta, will be studied by NASA's *Dawn* mission, which is scheduled for launch in 2006. It will arrive at Vesta in the summer of 2010, where it will remain for a year before leaving for an arrival at Ceres in the summer of 2014. These two asteroids appear to have evolved in dramatically different ways, in which Ceres remained cool with no melting while Vesta was hot with a melted interior. Having one spacecraft studying two such different objects should provide answers to many questions on the Solar System's origin.

Table 7-1 summarizes the observed properties of comets and asteroids.

7.3 Meteors, Meteor Showers, and Meteorites

Every day in its travels around the Sun, the Earth sweeps up tons of interplanetary debris. Most of this material is in the form of small particles the size of a grain of sand or less, although some of it is substantial in size. Such particles are called **meteoroids**. When a meteoroid traveling at typical speeds of tens of kilometers per second hits the atmosphere, friction with the air heats the meteoroid to a high temperature. The high temperature causes the surrounding atmosphere to glow, creating a bright streak of light called a **meteor**. Meteors are popularly but erroneously known as *shooting stars*. The meteoroids usually vaporize completely; those that are massive enough to survive the journey to Earth's surface are known as **meteorites**.

Where do these particles originate? Photographs taken with special automatic meteor cameras indicate that many of them came from the asteroid belt and were perturbed by Jupiter's gravity into an orbit that crosses Earth's.

METEORITE CRATERS

Massive bodies striking Earth at the speeds of tens of kilometers per second are likely to leave visible evidence in the form of craters. A mass moving at a speed of 10 to 30 km/sec or more possesses a formidable amount of kinetic energy. Kinetic energy is energy caused by a body's motion and depends on its mass and speed.[2] The kinetic energy of a meteoroid will be released explosively on impact. For example, the Barringer crater in northern Arizona (**Figure 7-19**), nearly one mile in diameter, was caused by the impact of a body perhaps the size of a medium-sized apartment house that may have weighed more than 50,000 *tons*. On striking the ground, the enormous kinetic energy of the meteoroid must have vaporized portions of both it and the Earth's surface, causing a tremen-

---

[2]The kinetic energy is given mathematically as $MV^2/2$, where M is the body's mass, and V is the speed.

Figure 7-19 Barringer crater (Meteor Crater), in Arizona, is the best known of Earth's impact craters. It is quite young (50,000 years), which explains its excellent state of preservation.

dous explosion and leaving the crater we see today. The kinetic energy of this event at impact would have been equivalent to some 100 nuclear bombs of the size dropped on Hiroshima.

The Barringer crater is by no means the largest one known. Since it is now possible to photograph the surface of the Earth from high-altitude aircraft and satellites, many large craters have been found. Most of these were previously unknown; their great size and the eroded condition due to their age made them difficult to detect from the ground. A large one of these, a couple of hundred kilometers across, was discovered in 1992 off the east coast of Mexico.

The extensive cratering of the Earth is also evidence for the planetesimals we discussed in the previous chapter. After the planets formed, there was a period of intense bombardment from planetesimals in noncircular orbits until the Solar System was swept free of debris. The Moon and other planets and satellites also show the effects of such an epoch of bombardment.

Meteoroids large enough to produce craters more than a hundred kilometers across are extremely rare, and even those as large as your fist are uncommon. The vast majority of meteoroids range downward in size from a pea to a grain of sand to dust particles. The motion of extremely small particles is rapidly slowed by friction with the atmosphere, and they drift slowly downward to settle on the Earth. These small bodies do not burn up because they are able to radiate away their heat

rapidly. They are called **micrometeorites**, and it is difficult visually to distinguish them from ordinary atmospheric dust. Because some micrometeoritic material is iron, and therefore magnetic, it may be gathered using a magnet. The most effective way to study micrometeoroids is with satellites, which can collect them above the Earth's atmosphere on large, specially designed surfaces.

DID AN ASTEROID OR COMET IMPACT SEND THE DINOSAURS INTO EXTINCTION?

A far-reaching hypothesis was made by geologist Walter Alvarez and his Nobel-prize-winning physicist father Luis, who suggested that the impact of a large body on Earth was responsible for a massive extinction of life-forms, including the dinosaurs, about 65 million years ago. It turns out that layers of clay deposited in the Earth's crust at the end of the Cretaceous period contain an anomalously high amount of certain elements (particularly iridium) that are abundant in meteorites but not generally so on Earth (**Figure 7-20**). This enrichment in meteoritic material has been found in geological strata from the same era but at many different locations on Earth. Furthermore, researchers have discovered heavy deposits of soot in those same layers, indicating the occurrence of a global firestorm at this time. The scenario is not unlike the sequence of events that had been proposed to produce a nuclear winter: One or more impacts produce explosions that set off forest fires

Figure 7-20 The boundary layer clay between the Cretaceous and Tertiary periods.

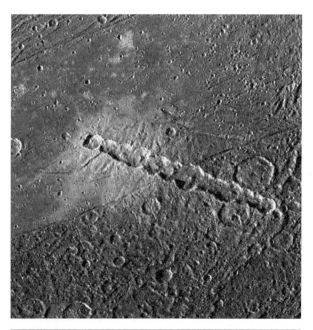

Figure 7-21 A chain of 13 craters on Ganymede probably resulted when a comet, which was pulled to pieces by Jupiter's immense gravity, collided with the satellite.

over extensive regions of the Earth. Enormous clouds of smoke darken the sky, blocking the Sun and bringing on months of deep winter and disruption of the food chain, which, in turn, could result in massive extinctions of plant and animal life.

Geological strata show evidence of similar worldwide catastrophes at other epochs in the past, and some researchers claim that they have detected a periodicity in these events, coming at intervals of 26 to 34 million years. This alleged periodicity has been strongly challenged by others, who have shown how it could have arisen from faulty data-analysis techniques. However, this hypothesis has triggered speculation as to what type of effect might cause periodic extinctions. A periodic event would seem to argue against asteroid impacts. Further, the total energy required for this scenario to work is greater than the typical large asteroid hitting the Earth would carry.

But suppose, for example, that the Sun was gravitationally bound to a stellar companion. The motion of a faint distant companion star through the Oort cloud could perturb nearby comets that would then hurtle off in many different directions. A shower of comets could well be released in the direction of the Sun, resulting in multiple hits on Earth—a comet storm. A search for a red dwarf companion star, a dim star that is the most common type of star, found nothing. However, we can imagine reasons why such a companion star might have escaped detection. For example, it could be a star that has lived out its lifetime, used up its sources of fuel, and is now burned out and dark. (In later chapters on stellar evolution we will see how such a thing can happen.) The Sun and its

companion would form a binary system, moving about a common center of gravity over a period of many years. Detecting this motion would be difficult because it would take a long time for one orbit to be completed. Thus, at present no observations exist to verify an astronomical source for a hypothesized periodic event.

An indication of the possible effects of a comet storm, but on a much smaller scale, is the crater chain observed by the *Galileo* spacecraft on Jupiter's moon Ganymede (**Figure 7-21**). As with the case of Comet Shoemaker-Levy (Figure 7-7), a comet undoubtedly fragmented as it approached too close to Jupiter. Each of the resulting impact craters is about 12 km in diameter.

A spectacular impact that took place in Tunguska, Siberia, on June 30, 1908, might have been due to a cometary remnant. Trees were blown down for more than a thousand square kilometers, animals were killed, and a man 80 km away was supposedly thrown from a chair and knocked unconscious. No meteoritic fragments have been found, leading to the suspicion that the object may have been a fragile one that broke up on its way through the atmosphere. It is hypothesized that the 60-m-diameter body weighed some 100,000 tons and exploded about 8 km from the Earth's surface, producing a blast wave equivalent to a 10-megaton nuclear bomb.

Did an extraterrestrial impact cause the extinction of the dinosaurs? That question is still debated. Wouldn't such a large impact have left behind a substantial crater? A candidate for this impact site has been identified just off the Yucatan Peninsula of Mexico.

The possibility of future devastating impacts on Earth has prompted a search for undetected asteroids whose orbits cross Earth's orbit. The pioneer in the field is planetary scientist Tom Gehrels of the University of Arizona, whose Spacewatch camera scans the sky on clear, moonless nights searching for such objects. In addition to new asteroids, many new comets have also been discovered. Should an object on a collision course with Earth be discovered well in advance of collision, we might have an opportunity to change its orbit.

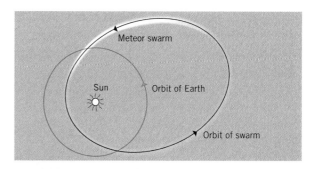

Figure 7-22 The path of a meteor stream intersects the orbit of the Earth. Each year, when Earth passes through the region, a meteor shower is seen.

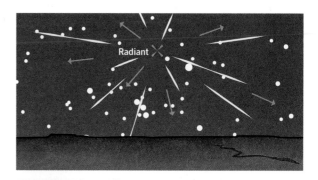

Figure 7-23 The radiant of a meteor shower.

Inquiry 7-15

Why would a search for faint Earth-crossing asteroids be made at new moon rather than any random lunar phase?

METEOR SHOWERS

In 1845, an unusual event occurred: Comet Biela, with a period around the Sun of about seven years, was observed to split into two fragments. On Comet Biela's next return two comets were observed; on the following return, no comet was seen. Instead, there was a shower of meteors. Nor is this the only example of a known comet being associated with a meteor shower. Spectacular displays in 1833 and 1866 were associated with Comet Tempel, and many other comets, including Halley, have been associated with meteor showers.

How does this happen? Recall that a comet's nucleus is considered to be a rather loosely packed mass of ices and particles, which can lose matter when the ices are vaporized near the Sun. Repeated visits to the Sun's neighborhood will cause considerable shedding of matter that will, of course, continue to orbit the Sun on paths that are similar but not identical to that of the comet. Each particle will have a slightly different speed, and eventually the particles will spread themselves all along the comet's orbit (**Figure 7-22**).

What will happen when Earth passes near this stream of meteoroids? If it gets close enough, we should see quite a few meteors coming from approximately the same region of the sky. Because the orbit of the meteor stream is fixed in space,

one would expect a shower every year on the date that Earth passes near the orbit.

Inquiry 7-16

The particles in a meteor stream should not be distributed uniformly around the orbit but, instead, should be thicker in some parts of the orbit than others due to random ejections by jets. These clumps of particles would travel around the orbit in a time roughly equal to the period of the comet. How could this fact help to explain the unusually heavy meteor showers in 1833 and 1866?

Individual particles in a stream travel along nearly parallel tracks in space, and as a result meteor trails seen during a meteor shower are not scattered at random all over the sky but appear to emanate from a point in the sky called the **radiant** (**Figure 7-23**). This is due to an effect of perspective, which can make parallel lines *appear* to radiate outward from a point, much as parallel railroad tracks appear to converge at a distant point. Showers are designated by the position of their radiants. Thus, the Leonids appear to radiate from the constellation Leo, the Perseids from Perseus, and so on.

Table 7-2 Meteor Showers

METEOR SHOWER AND RADIANT LOCATION	APPROXIMATE DATES	ASSOCIATED COMET (IF KNOWN)
Quantrantids (Boötes)	January 1–4	—
Lyrids (Lyra)	April 19–23	1861 I
May Aquarids (Aquarius)	May 2–6	Halley
July Aquarids (Aquarius)	July 25–31	—
Perseids (Perseus)	August 10–14 (some visible in early August)	1862 III (Swift-Tuttle)
Draconids (Draco)	October 9–19	Giacobini-Zinner
Leonids (Leo)	November 14–19	Tempel[a]
Geminids (Gemini)	December 8–14	—

[a]In 1966, the Leonids again staged a good show after missing 1899 and 1933; they put on a good show in 1998 as well.

Most meteor showers are modest events, presenting on average a couple of objects per minute, but some can be spectacular. The 1866 Leonid shower is reputed to have been one of the finest displays of modern times, with more than a quarter of a million meteors estimated to have been visible from one observing station, and the 1966 Leonids were almost as abundant. **Table 7-2** lists some prominent meteor showers, giving the location of their radiants and the date in the year that the shower can be expected to be at its maximum.

Meteor showers are best observed on cloudless evenings away from bright lights and when the Moon is not up to brighten the night sky. Observing after midnight is best. Just as a car moving into a snowstorm gathers more snow on the front window than the rear one, so, too, the meteor display is best when Earth is moving into the particles. Such movement occurs after midnight, because at that time the observer is on the side of the Earth that is moving into the meteor stream.

Occasionally, a meteor is observed that is so bright it can be seen even by day. One such **fireball** was sighted by numerous observers in the western part of the United States in July 1972 (**Figure 7-24**). This was an unusual body that, unlike most that the Earth encounters, did not land, but instead skipped back out into space after grazing our atmosphere, much as a stone can be skipped across water. Fireballs sometimes have a sound associated with them. Such a fireball is

Figure 7-24 A fireball visible in daylight over Jackson Lake at the Grand Tetons.

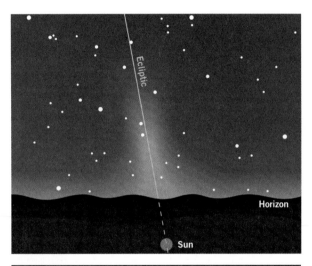

Figure 7-25 The zodiacal light is due to light reflected from dust particles in the ecliptic plane.

called a **bolide** and occurs when the meteoroid explodes within the atmosphere.

7.4 Interplanetary Dust

If you are located in an area blessed by dark, clear night skies, you might be able to detect the presence of dust in the Solar System. Approximately 90 minutes before sunrise (or after sunset), look for a faint glow of light extending away from the Sun along the ecliptic (**Figure 7-25**). This zodiacal light is due to dust, concentrated in the plane of the Solar System, that reflects the light of the Sun toward our eyes. Reflection of light from dust also produces a faint patch of light in a direction exactly opposite to the Sun, called the gegenschein (from the German *gegen* meaning *opposite*).

Small dust particles cannot remain in orbit around the Sun forever, because the pressure of solar radiation impinging on them will greatly alter their orbits over a period of time. The smallest of particles will be blown out of the Solar System by this radiation pressure. Larger particles, which orbit around the Sun just like the planets, will experience a deceleration from solar radiation and spiral toward the Sun. Eventually they will be vaporized by the intense heat. Because this happens quickly by astronomical standards, somewhere in the Solar System there must be a source that is constantly replenishing the dust that has fallen toward the Sun.

Inquiry 7-17

What might be the possible sources of such dust?

7.5 Meteorites and the Early Solar System

Until recently, meteorites were the only objects from space that we could actually touch. They provided the only direct information on the early physical conditions and early chemistry of the Solar System. What can we learn about the early Solar System by studying meteorites? To what extent do the meteorites, as random samples of asteroidal material, qualify as primeval material? Such questions are the subject of this section.

CHEMICAL COMPOSITION

There are three basic types of meteorites: **irons**, **stones**, and **stony-irons**. The irons are composed of about 90 percent iron alloyed with about 10 percent nickel and traces of other elements. Their composition is similar to what most geologists reason prevails in the Earth's core. The stones are primarily silicate materials similar to the Earth's rocks, whereas the stony-irons are a mixture of the two types.

If a body is observed to pass through the atmosphere to Earth and is successfully tracked and recovered, it is called a meteorite **fall**. If a meteorite is found on the ground but is not known to be freshly fallen, the discovery is referred to as a meteorite **find**. Contrary to popular belief, meteorites are not hot when they hit the ground. In fact, after so many years in cold space, the interior is extremely cold, and meteorite falls have been seen with frost on the surface.

Inquiry 7-18

More than two-thirds of meteorite *finds* are irons, but more than 90 percent of *falls* are stones. Why do you think this difference exists?

Inquiry 7-19

Which group is probably most representative of the actual meteorite population in its proportion of stones to irons: the finds or the falls?

Once a meteorite has been found and examined in the laboratory, its composition can be determined with great accuracy. Allowing for the correct proportions of each type of meteoroid, it is found that the proportions of the elements in meteorites are not unlike those of the Earth's crust.

The similarity of this composition to that of the Earth, and the fact that most meteorites originate in the asteroid belt, leads us to suppose that the meteorites we find were once part of larger bodies in the asteroid belt. These bodies were large enough that some separation of the elements must have taken place, just as in the Earth the lighter elements formed the crust and the heavier elements the core. If these large bodies broke up (say, by collisions), this would produce some bodies that were primarily iron and others that were primarily stony material.

However, if all the mass in the asteroid belt were summed, even the most generous estimates put it at about one-tenth of an Earth mass. If this is so, there may never have been a body of any substantial mass in the region between Mars and Jupiter, due perhaps to the influence of Jupiter. The stony-irons, in particular, may have come from parent bodies so small that no significant heating and segregation of material took place. However, estimating the total amount of mass that may have been swept out of the asteroid belt over time is difficult, so this argument is not conclusive.

One particular type of meteorite is thought to have remained much as it was when the Solar System was formed—the **carbonaceous chondrites**, a relatively rare type of stony meteorite containing an unusually large proportion of hydrocarbons and as much as 20 percent water. Hydrocarbons and water are both highly volatile materials and tend to vaporize even at low temperatures, so any object that has retained its hydrocarbons must have avoided the heating most of the asteroidal material appears to have undergone. The elemental abundances of carbonaceous chondrites are thought to represent that of the early Solar System better than any other bodies we have studied.

The presence of organic compounds in carbonaceous chondrites is fascinating. Not only have hydrocarbons been found, but also complex substances such as amino acids, the fundamental building blocks of proteins. Although care must be taken not to identify an amino acid that is actually a contaminant (e.g., a fingerprint of the investigator), it seems clear that some of these compounds are actually of extraterrestrial origin. That the precursors of life can be formed under the conditions prevailing in outer space certainly gives hope to those who would like to believe that life is common in the universe.

Chemical analysis of meteorites has led to a surprising conclusion: Some meteorites that have been picked up in Antarctica definitely came from the Moon (50 in 2004), while others appear to be from Mars (**Figure 7-26**)! Meteorite ALHA 81005 was found to have a chemical composition extremely similar to the rocks collected on the Moon by the Apollo astronauts (Chapter 8) and substantially different from that of other meteorites. Presumably, it was a piece of the Moon that was knocked off when a large meteoroid hit it. The conclusion that 30 meteorites now in hand come from Mars was bolstered by chemical analysis of Martian rocks performed by the *Sojourner* spacecraft. In addition, the relative abundances of magnesium and nitrogen gases trapped inside a few meteorites are similar to those found in the Martian atmosphere.

Internal Structure

The internal structure of a meteorite can be studied if it is cut into sections. If an iron meteorite is sliced, polished, and etched with acid, it will reveal a characteristic pattern of lines, called *Widmanstätten patterns* (**Figure 7-27**). These are actually the boundaries of iron crystals that may have been formed when the material slowly cooled from a state of high temperature and pressure, such as would be expected in the interior of a large body. Research suggests that, under some circumstances, bodies as small as 20 km in diameter could form these figures, but their existence still suggests the preexistence of a parent body.

Inquiry 7-20

Why couldn't the Widmanstätten patterns have been formed as a result of the heating of the meteoroid during its passage through our atmosphere followed by subsequent cooling?

The vast majority of stony meteorites, when sliced open, show a beautiful and complex structure (**Figure 7-28**). These structures, called **chondrules**, appear as separate pieces of material that were included in the meteorite at the time it formed. Chondrules exhibit a wide range of chemical composition. Some appear to have formed from rapid cooling of small liquid droplets, while others are highly irregular, indicating slower cooling. The most

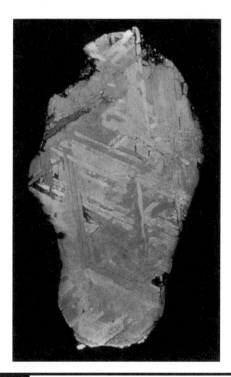

Figure 7-27 Widmanstätten patterns reveal that iron meteorites once existed in a molten state and cooled slowly.

Figure 7-26 Meteorites found in Antarctica. Chemical analyses indicate that *(a)* came from the Moon, and *(b)* from Mars.

vexing question in the study of stony meteorites continues to be that of the formation of the chondrules. Because they are probably representatives of the oldest materials in the Solar System, they hold important clues about the history of the bodies in which they are found, and of the Solar System itself. They are not giving up that information easily.

Many meteorites show evidence of strong stressing, such as might be expected as a result of collisions occurring at some 20 to 30 km/sec. An example is the presence of small black diamonds in some meteorites. We don't know whether these structures were formed in collisions within the asteroid belt or on impact with the Earth. Such small diamonds have been found in the layers at the end of the Cretaceous period and provide further evidence for an extraterrestrial origin to mass extinctions at that time.

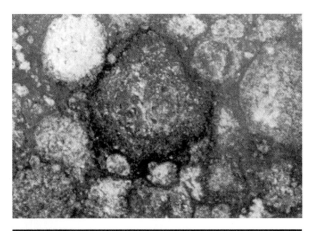

Figure 7-28 The inside of a chondritric meteorite. The round inclusions are the chondrules.

THE ORIGIN OF METEORITES

Observational hints to the origin of the meteorites include the presence of irons, stones, and stony-iron compositions, the implication that they were originally part of the asteroid belt, the low total mass of all material in the asteroid belt, the properties of the carbonaceous chondrites, the cooling implied by the Widmanstätten patterns, the presence or absence of chondrules, and the stressing that implies the presence of collisions. A schematic drawing of the formation of meteorites is shown in **Figure 7-29**, which suggests a possible scenario for a large asteroid, perhaps 650 km in diameter. Internal heating melts the interior and causes a separation of heavy from light elements *(a)*. Collisions with other bodies occur, causing the surface to shatter *(b)*. Further collisions break off large chunks; additional fragmentation produces larger numbers of small chunks of material *(c–e)*.

The carbonaceous chondrites are thought to be fragments from the asteroid's surface, because the high abundance of volatiles in carbonaceous chondrites implies they were not subjected to the high temperatures or the high pressures in the asteroid's interior. The broken fragments will have different and complex chemical compositions. Some will show the effects of physical stressing from the collisions, while some will have cooled at different rates from others. Small asteroids, whose interior temperatures never became high enough to produce melting, may have gone through similar shattering processes, but meteoroids from them will have different chemical properties.

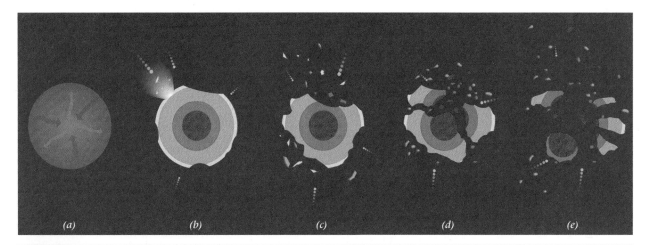

    (a)   *(b)*   *(c)*   *(d)*   *(e)*

Figure 7-29 The formation of various types of meteorites from the fragmentation of an asteroid. Different regions of the asteroid have different chemical properties and produce a variety of types of meteorites depending on the details of the internal heating process and the degree of break-up of the asteroid.

Although no asteroid collisions have been observed, they no doubt occur (see Figures 7-14 and 7-18). Meteoroids identical to those hypothesized to be produced in such collisions have been observed. From such circumstantial evidence, we can infer that the general picture of meteorite formation is, therefore, correct.

METEORITE DATING

Radioactive elements are ones that spontaneously emit radiation because of instabilities in their atomic nuclei. The study of radioactive elements and their decay products in meteorites gives us valuable information about when they were formed, and hence about the age of the Solar System. For example, uranium-238 (^{238}U)—uranium whose nucleus contains a total of 238 protons and neutrons—decays in a sequence of steps into lead-206 (^{206}Pb). (In this case, the uranium is often referred to as the *parent* nucleus, while the lead is referred to as the *daughter* nucleus.) These events can be characterized by what is called the **half-life** of the element, and is precise and accurate because it is based on extensive and reliable laboratory experiments. For example, the half-life of ^{238}U is about 4.5 billion years; this is the length of time it takes for half the uranium originally present to decay, no matter what the size of the sample. The greater the age of an object initially containing ^{238}U, the greater the amount that has decayed to lead and the greater the ratio between the amount of ^{206}Pb and the amount of ^{238}U. By measuring this ratio, we can estimate the age of the object (**Figure 7-30**).

By refining the method, we can overcome the uncertainties caused by the presence of primeval ^{206}Pb. For example, another product produced during the decay of ^{238}U is helium-4 (^4He). If, during the formation of a minor planet, the material was heated to a high temperature, originally present light gases such as ^4He would be driven off, leaving the object free of this primordial gas. After cooling, the ^{238}U would continue to decay, and any ^4He produced as a result would become trapped in the solid material. Analyzing the object today, we can be sure that all the ^4He we see is due to radioactive decay. Therefore, by measuring the ratio of ^4He to ^{238}U, we can estimate the age of the object more accurately. By such means, the ages of meteorites have been determined to be about 4.5 billion years.

There are many different radioactive decay processes going on in nature, with half-lives ranging from short to long. **Table 7-3** lists the parent atom of some radioactive elements, the daughter it decays into, and the measured half-life. Astronomers typically try to construct their chronologies using a variety of radioactive decay cycles of many different chemical elements. By using a variety of decay cycles, uncertainties in the ratio of initial abundances are decreased. The agreement that we find between these various processes gives us confidence that our estimates of primeval abundances are not leading us systematically astray. Among the best results currently available, dating using the decay of ^{87}Rb to ^{87}Sr gives an average age for meteorites of 4.498 ± 0.015 billion years.

Inquiry 7-21

If the original material of which an object was formed already contained some ^{206}Pb, an incorrect age would be inferred for the object. Would the estimated age be too young or too old in this case? Explain your answer.

Inquiry 7-22

How useful would you expect ^{14}C, which has a half-life of 5,568 years, to be in dating astronomical objects?

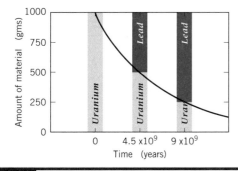

Figure 7-30 Radioactive decay of uranium into lead has a half-life of 4.5 billion years. After each half-life 50% of the remaining uranium transforms into lead.

Table 7-3   Half-lives of Some Common Isotopes

ISOTOPE (PARENT → DAUGHTER)	HALF-LIFE (YEARS)
Potassium 40 (^{40}K) → Argon 40 (^{40}Ar)	1.3×10^9
Rubidium 87 (^{87}Rb) → Strontium 87 (^{87}Sr)	47×10^9
Uranium 235 (^{235}U) → Lead 207 (^{207}Pb)	0.7×10^9
Uranium 238 (^{238}U) → Lead 206 (^{206}Pb)	4.5×10^9

Yet another piece of evidence helps us understand the evolution of meteorites. Like all objects in space, meteorites are constantly being bombarded by **cosmic rays**, subatomic particles (generally protons, electrons, and helium nuclei) that move through interstellar space at speeds approaching the speed of light. When a cosmic ray strikes a body in the asteroid belt, atomic nuclei in the body can fragment into much lighter nuclei. Among the elements produced are lithium, beryllium, and boron—all rare elements. After being exposed to cosmic rays for many millions of years, the outer crust (to a depth of about 1 meter) of a large asteroid will develop an excess of these elements. Bodies smaller than a meter in size will show an excess throughout. From an analysis of the abundances of these elements, the time since the rock's surface was first exposed to cosmic rays can be estimated. In this way it is possible to determine how long it has been since the breakup of the parent body from which the asteroid came. Cosmic-ray ages for meteorites range from tens of thousands to millions of years. Such young ages of their surfaces provide further circumstantial evidence for their formation from the collisional breakup of asteroids.

Inquiry 7-23

In some meteorites, the ages determined by the cosmic ray exposure method are different on different sides of the object. What might be the cause of this?

DID A SUPERNOVA EXPLOSION TRIGGER THE FORMATION OF OUR SOLAR SYSTEM?

In the early 1980s, studies of the carbonaceous chondrite meteorite known as the *Allende meteorite* (named after the Mexican village Pueblito de Allende, where it was found) revealed the presence of some unusual and unexpected chemical isotopes—atoms that resulted from the radioactive decay of unstable elements not normally seen in the Solar System. For example, embedded in the meteorite were small chunks of material found to be highly enriched in magnesium-26 (^{26}Mg), a byproduct of the radioactive decay of aluminum-26 (^{26}Al).

On one hand, the observation of ^{26}Mg in meteorites is significant because current research into the formation of chemical elements suggests that the most likely way to produce the unusual isotope ^{26}Al, which decays into ^{26}Mg, is in a supernova explosion, a catastrophically explosive event that occurs toward the end of the lives of massive stars (Chapter 19; see Figures 19-7 through 19-9). During such a cataclysm, extreme temperatures are generated that can create all sorts of exotic chemical elements in brief episodes of nuclear fusion inside the star. The explosion then ejects this material (^{26}Al) from the star and scatters it around interstellar space and into gas and dust clouds, where it then decays into the ^{26}Mg. At the same time, the blast wave from the explosion propagates outward from the star through the nearby clouds that now contain the ^{26}Mg. Such a blast wave, if it moves faster than the speed of sound, is called a *shock wave*. A shock wave can actually squeeze interstellar gas clouds and cause some of them to collapse into stars (Chapter 17). The evidence seems to indicate that this mechanism started the formation of our Solar System.

Age-dating studies suggest that the solar nebula collapsed rapidly (in several hundred thousand years), and if a supernova triggered the collapse, the primeval gases would indeed be able to capture enough undecayed ^{26}Al to account for the ^{26}Mg found in the Allende meteorite.

On the other hand, observations by the Chandra X-ray space telescope have provided data suggesting that the supernova hypothesis may not be needed. Chandra found the observed ^{26}Al could have been produced during a phase of the Sun's formation when it was thought to have produced a large number of X-ray flares. Such flares would have the capability of producing a number of exotic isotopes including ^{26}Al.

Which model is correct? Although further evidence is needed, we will see later on that star formation triggered by stellar explosions is probably a fairly common event in our galaxy and in other galaxies as well. Thus, for now, we will stick with the supernova model.

DISCOVERY 7-1

Asteroid Brightness Variations

After completing this discovery you should be able to do the following:

- Describe how the brightness of an asteroid changes as it rotates.

To study how an asteroid's brightness changes as it rotates and tumbles, find an object such as a shoe box, a book, or an audio- or videocassette box.

- **Discovery Inquiry 7-1a** If one side of your object has an area three times that of another side, what will be the ratio of the brightness when one side faces you compared with when the other side faces you?

Hold your object as shown in **Discovery Figure 7-1-1**, with the long axis facing you. Assume the Sun is behind you, so the side of the object facing you receives full illumination. Rotate it about the long axis as shown (the axis coming towards you, labeled *A*), and imagine how the brightness would vary. (If possible, use a bright light to illuminate the object and observe the changes in brightness at it rotates.) Draw a graph of the variation of brightness with time as you rotate the object. Repeat for the other two axes shown in the figure (still with the long axis facing you). Think about the relative sizes of the faces as you draw your three graphs. Try to imagine the complications if your object were rotating about all three axes at once!

Curves like those you drew are similar to those that astronomers determine from their brightness observations of asteroids. The name of the game for the astronomer is to determine the shape of the tumbling asteroid from observations of the brightness changes with time!

Figure 7-1-1 This figure shows an object that is able to rotate about three perpendicular axes. It is illuminated from behind the observer.

Chapter Summary

Observations

- **Comets** are observed to have a nucleus, a **coma**, one or more **tails**, and a **hydrogen halo**. Comet tails are of two types, one consisting of dust particles and the other having charged particles.
- **Long-period comets** follow elliptical orbits that are at random orientations to the ecliptic plane. **Short-period comets** have orbits that are closer to the ecliptic. Some comets may have **hyperbolic orbits**.
- A significant number of trans-Neptunian objects have been observed.
- Halley's comet was observed to be dark and to contain a large amount of water ice.
- The **asteroids minor planets** are small, rocky bodies found mostly, but not exclusively, between the orbits of Mars and Jupiter. Their orbits are generally nearly circular and located in the ecliptic plane.
- The **Trojan asteroids** are two special groups of asteroids located in Jupiter's orbit at distances from Jupiter equal to its distance from the Sun. The Trojan asteroids thus form equilateral triangles with the Sun and Jupiter.
- The **asteroid belt** is not uniformly filled with asteroids but contains **Kirkwood gaps**, regions devoid of asteroids. These gaps in the asteroid belt occur at locations from the Sun that correspond to simple fractions of Jupiter's period about the Sun.
- There are four chemical groups into which asteroids generally fall: silicon rich, nickel-iron rich, carbon rich, and those that may be rich in organic compounds.
- Meteorites have composition types that are **iron**, **stony**, or **stony-iron**. The cooling of certain of these materials forms chondrules and Widmanstätten patterns inside the rocks.
- Meteor showers, which result when Earth crosses the orbit of a comet, appear to come from a particular direction in space called the **radiant**.
- Interplanetary dust is observed as the zodiacal light and the gegenschein when sunlight is scattered by small dust particles located in the ecliptic plane.

Theory

- Sunlight acting on dust particles pushes them away from a comet's coma. Particles from the solar wind collide with particles in a comet's tail and cause them to become charged. The motions of these charged particles then change in response to magnetism present throughout the Solar System.
- Asteroids can sometimes be used to determine the length of the astronomical unit by obtaining their distance in both kilometers and astronomical units.
- The **half-life** of **radioactive decay** is the time required for half the material to decay from one element into another. From the known half-life of radioactive decay and from the inferred amount of material originally present, ages of meteorites can be found.

Conclusions

- If a comet has a gravitational interaction with a massive planet, the comet's orbit may change from elliptical to hyperbolic.
- Long-period comets come from the Oort cloud, which is located 50,000 to 100,000 AU from the Sun. Short-period comets come from the Kuiper belt outside the orbit of Pluto.
- The dirty snowball model of a comet provides a valid model of a cometary nucleus, halo, and tails.
- Asteroids may be small because Jupiter's strong influence never allowed larger bodies to form, or because numerous collisions fragmented larger bodies.
- Comet and asteroid formation are both consistent with the nebular-type models for the formation of the Solar System.
- The composition types of meteorites are best explained as resulting from the fragmentation of asteroids.
- From observations of excess amounts of iridium in certain rocks, scientists hypothesize that species extinctions may have occurred because of the impact of a large meteoroid with the Earth.

- **Cosmic rays** are energetic subatomic particles that fill the galaxy. Collisions between cosmic rays and asteroids produce chemical changes on the asteroid's surface. From an analysis of the elements present, astronomers can infer the length of time the surface has been exposed to cosmic rays and, thus, how long ago the meteoroid fragmented from a larger body.

- Certain meteorites have chemical compositions that indicate they came from the Moon and Mars.

Summary Questions

1. What is the icy conglomerate model of a comet? What evidence is there for it? How can it explain the various phenomena observed in comets?

2. Why does the tail of a comet always point *away* from the Sun? Why may a comet have more than one tail?

3. What evidence do astronomers have in favor of Oort's hypothesis of a comet cloud beyond the orbit of Pluto?

4. What do we mean by the term *minor planet?* What are they? Where are they found? How can they be used to determine the length of the astronomical unit?

5. What do we mean by the term *Trojan asteroid?* What is the relationship between the Trojans and the planet Jupiter? What are the Kirkwood gaps, and what was Jupiter's role in forming them?

6. What role did gravitational interactions play in the development of asteroid and comet orbits?

In the production of meteroids and the zodiacal light?

7. What is a meteor? Distinguish a meteor from a meteorite. What are the two main sources of meteors?

8. What are the three main types of meteorites? Why is the proportion of stones to irons seen in finds on Earth different from the true proportion in space?

9. How do Widmanstätten figures and chondrules provide information on the conditions under which some meteorites were formed?

10. Why do meteor showers repeat themselves year after year in the same part of the sky? Why do meteors in a shower emanate from a single location in space?

11. How do various radioactive decay processes and cosmic ray exposure times give us clues to the formation and history of the Solar System?

Applying Your Knowledge

1. Make a drawing to demonstrate that the Earth encounters more meteoroids after midnight than before midnight.

2. Would you expect there to be many asteroids having moons orbiting them? Explain your reasoning.

3. What arguments might you present to people who believe that comets bring disaster and evil, to convince them their ideas are wrong?

4. From what you know of comets, would you expect their motions to be immediately apparent to the casual observer? Why or why not? How about for a meteor?

5. Meteorites are observed to have an amount of xenon-129 (^{129}Xe) far in excess of that normally found on Earth. ^{129}Xe comes from the radioactive decay of iodine-129 (^{129}I), which is formed in supernova explosions. Use this information to argue that the formation of the Solar System was initiated by the nearby explosion of a supernova.

6. Suppose you came across a rock you thought might be a meteorite. How might you determine whether it is a meteorite, given that you had no special equipment? How about if you had all the special equipment you desired?

7. Make a list of easily obtainable household items you might use to make a cometary nucleus.

8. What is the distance of the Trojan asteroids from Jupiter?

▶ 9. If the period of Halley's comet is 76 years, approximately what is its maximum distance from the Sun? Relate this distance to that of the planets.

▶10. What would be the age of a rock in which you measured the ratio of potassium-40 to argon-40 to be 1 to 7? What assumption are you making? (*Hint:* A drawing will be helpful.)

▶11. What would be the age in the previous question if the rock originally contained equal amounts of potassium-40 and argon-40?

▶12. One of the Kirkwood gaps appears at a location where the asteroid's period is exactly one-half Jupiter's orbital period about the Sun. At what distance from the Sun (in AU) is this gap located?

Answers to Inquiries

7-1. As a comet returns again and again to the neighborhood of the Sun, it loses more and more of its volatile materials. Eventually, they are completely "boiled away" and the object might be difficult to distinguish from a minor planet.

7-2. When the distance from the Sun is small, the comet moves rapidly. Therefore, it spends little time close to the Sun and most of its time far from it.

7-3. One would expect that we would encounter more comets coming from the direction of Hercules, just as when we run in the rain the front of our body gets wetter than the back. The hyperbolic comets arise from interactions with Jupiter and are thus clearly part of the Sun's family and moving with it toward Hercules.

7-4. They are distributed uniformly around the Solar System.

7-5. The average distance is $(900 + 76)/2 = 488$ AU. From Kepler's third law, we have $P^2 = 488^3 = 1.16 \times 10^8$, so that $P = 1.1 \times 10^4$ years.

7-6. 100,000 years. Comparison with the age of the Solar System: $10^5/(4.5 \times 10^9) \cong 2 \times 10^{-5}$.

7-7. The comets we see today probably started to come near the Sun only recently. If they had always had orbits that brought them near the Sun regularly, they would long since have lost their supply of volatile materials and faded.

7-8. Oort's theory provides a way for new comets to enter our part of the Solar System, replacing those that have worn out.

7-9. If comets are indeed composed of near-primeval material, they could clue us in to the composition of the early solar nebula.

7-10. Irregular shapes and irregular distribution of light and dark regions over the surface.

7-11. Perhaps Hidalgo is a worn-out comet.

7-12. The round-trip distance is 173.3 seconds times 3×10^5 km/sec $= 5.2 \times 10^7$ km. The distance of the asteroid is one-half this amount, or 2.6×10^7 km.

7-13. 2.6×10^7 km / 0.17 AU = 153 million km/AU.

7-14. If an asteroid is broken up in a collision, one would expect the pieces to continue to follow orbits similar to the original body.

7-15. The night sky is much darker at new moon than at full moon, making it easier to find faint objects.

7-16. Presumably, in 1833 and 1866, the Earth passed through a denser region of a stream of particles orbiting the Sun with a period of roughly 33 years. It has been suggested that perturbations of the particles by Jupiter moved the entire orbit in such a way as to substantially miss the Earth in 1899, because there was no spectacular display that year (nor in 1933, but there was in 1966).

7-17. Fragments of asteroids and dust from comets are the two most likely (listed in order of probable importance).

7-18. Meteorite stones are so similar to Earth rocks (especially after undergoing weathering and erosion) that after a time they are difficult to distinguish and so are not noticed. Irons, by contrast, are dense, distinctive, and easily recognized.

7-19. The falls, because we should recover stony and iron meteorites in proportion to the number that actually fell.

7-20. The heating is too brief and only affects the skin of the meteorite.

7-21. Too old. If the original ^{206}Pb is not accounted for, the scientist will conclude that the lead came from radioactive decay over a longer time interval.

7-22. Of no use at all; its half-life is too short.

7-23. Different sides must have been exposed to cosmic rays for different lengths of time. Such an observation is a further indication that meteoroids were produced when colliding bodies broke pieces from larger bodies. Only when fragmentation occurs will a surface that was originally inside a larger body be exposed to cosmic rays.

The Earth and Moon: Processes and Facts

Cold-hearted orb that rules the night
Removes the colors from our sight
Red is grey and yellow white
But we decide which is right
And which is an illusion

Moody Blues, Days of Future Passed, *1967*

E arth is a planet, circling the Sun in its own time like all the other planets. Our presence on it makes the Earth of special interest to us. It makes sense to begin our study of planets by learning some of the basic facts about our home body, as well as the variety of processes that have made the Earth the way it is.

The Earth's moon has been an object of study and wonder throughout the ages. During most of history, observing was done without using telescopes (referred to as **naked-eye** observations); it was not until the seventeenth century that Galileo and other pioneers turned their crude telescopes to the Moon. In the nearly four centuries since then, earthbound observations made important progress in understanding the Moon—progress that was indispensable in the design of the enormously complex and highly successful *Apollo* missions to the Moon. An understanding of the Moon, and the physical processes that have occurred to make it the way it is, will allow us to understand better what recent spacecraft observations of planets and their moons have been teaching us.

Our approach to the study of planets and moons is a comparative one. The emphasis is on comparing and contrasting the various bodies with the goal of understanding how the many pieces of the planetary puzzle fit together. For example, the Earth and Moon exhibit both chemical similarities and differences that we must understand to figure out how each body formed and evolved. We will not be comprehensive but will consider only those ideas and processes that are necessary for understanding the other bodies of the Solar System.

8.1 Earth as an Astronomical Body

Earth is an astronomical body. We want to know the same types of information about it as we know about other planets. We therefore first look at how scientists learned the most fundamental properties of our home planet.

DETERMINATION OF BASIC PROPERTIES

The diameter of the Earth has been known ever since about 200 B.C.E., when Eratosthenes determined its circumference (Chapter 5). Modern data show the Earth to be slightly pear-shaped. The diameter measured through the equator is greater than that measured through the poles because rotating bodies are flatter at the poles than nonrotating ones. The average diameter is given in **Table 8-1** along with other basic properties.

The Earth's mass can be determined using a variety of techniques, but they all should give the same answer since they all ultimately depend on

Table 8-1   Properties of the Earth

PROPERTY	VALUE
Orbital properties	
Average distance from Sun	149,600,000 km (1 AU)
Minimum distance	0.983 AU
Maximum distance	1.017 AU
Orbital period	365d6h8m24s (1.000 years)
Orbital inclination	0° 0′ 0″
Physical properties	
Diameter (average)	12,756 km (1.000 $D_{\oplus}$[a])
Mass	5.974×10^{24} kg (1.000 $M_{\oplus}$)
Density (average)	5.518 g/cm^3
Surface gravity	9.80 m/s^2 kg (1.000 Earth gravity)
Escape velocity	11.2 km/s
Rotation period	23h56m4.091s
Tilt of rotation axis from orbit perpendicular	23° 27′
Reflectivity[b]	37%
Surface temperature	240 to 320 K (–27 to +117° F)
Moons	1

[a]*The symbol $\oplus$ is the symbol for the Earth. When used as a subscript, such as $D_{\oplus}$, and $M_{\oplus}$, it refers to the diameter, and mass of the Earth, respectively.*
[b]*The **reflectivity** is the fraction of the incident sunlight reflected from a body.*

Chapter opening photo: The quarter Earth and Moon, as observed by the Galileo spacecraft on its way to Jupiter.

Newton's laws. For example, Kepler's third law may be applied using the period and distance of the Moon. For the Earth and Moon, whose masses are $M_{\oplus}$ and M_{m}, respectively, we have

$$(M_{\oplus} + M_{m})P^2 = A^3.$$

Assuming the mass of the Moon is far less than that of the Earth, we can ignore the Moon's much smaller mass and solve for the Earth's mass:[1]

$$M_{\oplus} = \frac{A^3}{P^2}.$$

Remembering, too, that the distance must be expressed in AU and the period in years, with an Earth–Moon distance of 380,000 km (2.53×10^{-3} AU), and a period of 27.3 days (7.48×10^{-2} year), we find

$$M_{\oplus} = (2.53 \times 10^{-3} \text{ AU})^3/(7.48 \times 10^{-2} \text{ year})^2$$
$$= (2.9 \times 10^{-6} \text{ solar masses}).$$

Multiplying by the mass of the Sun, 2×10^{30} kg/solar mass gives

$$M_{\oplus} = 5.8 \times 10^{24} \text{ kg}.$$

This simple example gives a result close to the more precise one in Table 8-1.

The density of the Earth is found by dividing its mass by its volume. The average density of Earth is calculated to be 5.5 g/cm³, or 5.5 times that of water. As we saw in Chapter 6, this value is substantially larger than that of a typical Jovian planet, whose density is near that of water. A body's density gives an idea of its overall, average properties, and we can use it to make useful comparisons between different objects.

As we saw in Chapter 5, a person's weight is determined by the force of gravity between the Earth and the person. That force of gravity is determined by the person's mass, the mass of the Earth, and the radius of the Earth. For example, if you drop a shoe the force of gravity accelerates it toward the center of the Earth. If you were to drop the same shoe on the Moon, it would accelerate more slowly toward the Moon's center. This acceleration due to gravity is described by Newton's second law ($F = ma$) and his universal law of gravitation. When acceleration is caused by gravity, physicists use the symbol g in Newton's second law instead of a. If the gravitating mass is the Earth ($M_{\oplus}$) and the distance of the accelerating body is the radius of the Earth ($R_{\oplus}$),

the acceleration[1] caused by gravity is $g = GM_{\oplus}/R_{\oplus}^2$. Similar equations are used for each planet. Thus, the greater a planet's mass, the faster it will accelerate a falling body. Put another way, the greater a planet's value of g, the more a person would weigh on that body. The acceleration of gravity on a planet is one important characterization of the planet. The point here is that one can experimentally determine the acceleration of gravity on a body and, from that, determine the mass of the gravitating body.

One experimental method of finding the value of g is by dropping an object and observing how its rate of fall changes with time. Using such an experimentally determined value, the known value of the gravitational constant G, and the known value of a body's radius, its mass is readily found using the given equation.

The value determined in this way and the value determined from the use of Kepler's third law are not identical because of observational and experimental uncertainty and other complicating factors.

We are held to Earth's surface by gravity. To escape Earth's gravity requires a speed in excess of 11.2 km/sec. The speed required to escape is called the **escape velocity** and is proportional to $\sqrt{\dfrac{M}{R}}$, where M and R are the mass and radius of the planet.[1] The formula shows that high-mass planets have a greater escape velocity than low-mass planets of the same size. Similarly, given two planets of the same mass, the smaller one will have the greater escape velocity because its gravity at its surface will be greater. The escape velocity for each of the planets is given in Appendix C.

8.2 Earth's Interior

The interior structure of the Earth cannot be studied directly. Although drilling is taking place, the deepest drill hole, which is located in Russia some 250 km north of the Arctic Circle, reached the 12.4 km mark in 1998, some 20 years after beginning. (The goal is 15 km; beyond that, they predict that the temperature is too high to drill.) We can, however, learn about the interior of the Earth from studies of earthquakes.

---

[1]Calculations do not always have to be precise as long as they are accurate. Here, the Moon's mass is small, relative to that of the Earth, and does not significantly affect the calculation. Therefore, we can neglect it.

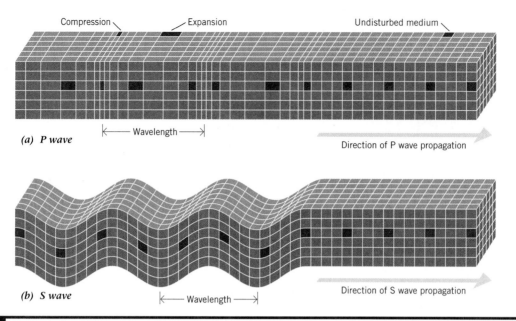

(a) P wave

Compression   Expansion   Undisturbed medium

←— Wavelength —→

Direction of P wave propagation

(b) S wave

←— Wavelength —→

Direction of S wave propagation

Figure 8-1 Two types of seismic waves are illustrated here. (a) A P-wave contains regions of high and low compression, which periodically pass through each point. The direction of motion of the vibration is in the direction the wave travels (here, to the right). Although a given square continually becomes a rectangle, it always returns to the original square. (b) An S-wave vibrates perpendicularly to the direction in which the wave travels. A given region distorts to a parallelogram, and then back to a square again.

SEISMIC STUDIES

An earthquake produces waves of energy that travel in all directions, both over the surface and through the interior, to be detected and analyzed by seismograph stations the world over. Earthquake energy compresses and deforms the rocks through which it passes. The rebounding of the compressed rock causes it to transmit the wave throughout the interior. A **compression wave**, which consists of alternating pulses of compression and expansion (**Figure 8-1***a*), is characterized by a wave's motion in the same direction as its vibration. Because the compression causes a change in the pressure on the rock, a compression wave is also known as a **pressure wave**, or **P-wave**. An example of such a wave in everyday life is a sound wave.

A second type of wave generated by a **seismic event** deforms rock without compressing it. Because such a stress is called *shear stress,* the type of wave producing it is often called a **shear wave** or **S-wave**. In contrast to the P-wave, the shear wave's direction of motion is *perpendicular* to its vibration (**Figure 8-1***b*).

The properties of S-waves differ from those of P-waves in important ways. P-waves travel with higher speeds than S-waves and thus arrive at a

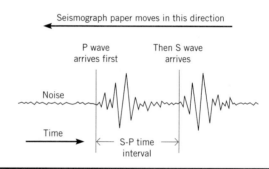

Seismograph paper moves in this direction

P wave arrives first   Then S wave arrives

Noise

Time

←— S-P time interval —→

Figure 8-2 A seismogram of an earthquake. Note the arrival of the P-waves before the S-waves.

given seismic station first. For this reason, P-waves are also known as **primary waves**, and S-waves as **secondary waves**. Because of the speed difference, the greater the distance from a seismic event to a recording station, the greater the delay between the P- and S-waves. By observing the delay between the times of arrival of the two waves (**Figure 8-2**), seismologists can determine the distance to the point where the earthquake occurred.

Although P-waves easily pass through liquids, S-waves do not. For this reason, a particular seismograph station may only detect P-waves. Knowing this, a comparison of data taken from

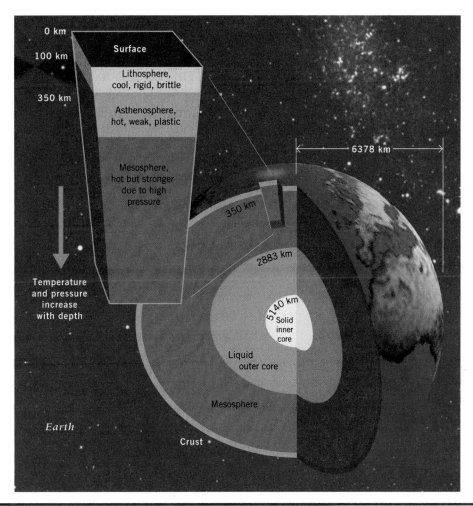

0 km

100 km

350 km

Surface

Lithosphere,
cool, rigid, brittle

Asthenosphere,
hot, weak, plastic

Mesosphere,
hot but stronger
due to high
pressure

Temperature
and pressure
increase
with depth

350 km

2883 km

6378 km

5140 km
Solid
inner
core

Liquid
outer core

Mesosphere

Crust

Earth

Figure 8-3 The interior structure of the Earth. The numbers are the depths below the surface.

seismograph stations all over the world has allowed geologists to determine the internal structure of the Earth (**Figure 8-3**).[2] The Earth is found to consist of three general regions: the **core**, the **mantle**, and the **crust**. The core contains 16 percent of the Earth's volume and is further subdivided into a solid **inner core** and a molten **outer core**. The subdivision occurs because, although the interior temperature is high (thought to be in the vicinity of 3,000 to 5,000° C, which is 5,400 to 9,000° F), the extreme pressures in the core cause the material in its inner region to solidify.

The mantle is a thick region of dense rock surrounding the core. The lower part of the mantle contains highly compressed rock that has great strength even though the temperature is high. Farther up in the mantle, the pressure decreases, causing the strength of the rock to become weaker. The region

of weaker rock is called the **asthenosphere** (*weak sphere*). This region is important because the rock here easily deforms, like soft plastic. Rock on top of the softened asthenosphere will be capable of floating on it. The important consequences of this will be discussed later in the chapter.

The outer 100 km of the Earth is cooler and consists of strong, rigid rock. This *rock sphere* is the **lithosphere**. Although the rock composition of the lithosphere and the asthenosphere are the same, the difference in rock strength distinguishes them.

The crust is the uppermost part of the lithosphere and contains rock of lower density than the material underneath. It is not uniform but is thicker in the continental regions (**continental crust**) than under the oceanic regions (**oceanic crust**). Furthermore, the continental crust is slightly less dense (2.8 g/cm³) than the oceanic crust (3.0 g/cm³). These observed differences are important for understanding the evolution of the Earth's crust, as we discuss later in this chapter.

---

[2]The real situation is more complex because waves bend and reflect inside the Earth.

Inquiry 8-1

The average density of the Earth is 5.5 g/cm^3; the density of a typical rock you might pick up is about 3 g/cm^3. What can you conclude from this information about the density of the Earth's interior?

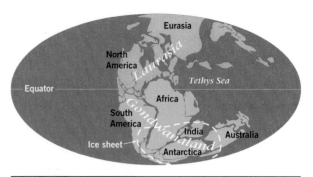

Figure 8-4 The supercontinent, Pangaea, 200 million years ago.

The chemical makeup of the Earth is not uniform throughout. Gravity causes denser materials to sink toward the center, with lighter elements rising toward the top. For this reason, the core consists of iron and nickel, with the crust containing lighter silicon, magnesium, and aluminum. Bodies whose chemical structure is determined by the effects of gravity have undergone the process of **differentiation**. For example, the inner core has a density of about 13.3 g/cm^3; the outer core, 11 g/cm^3; and the mantle, about 4 g/cm^3. Thus, the Earth is highly differentiated. Are the other planets?

PLATE TECTONICS

Even a cursory look at the Earth's surface raises an interesting question: Why do the eastern part of South America and the western part of Africa appear to fit together? Furthermore, detailed studies of the flora and fauna of these distant regions show some surprising similarities. For these reasons, some scientists suggested many years ago that these continents had at one time been part of a larger continent that had broken up and moved apart. The suggestion was ridiculed, however, because no forces were known that could overcome friction sufficiently to move the continents.

The modern theory of continental motions was suggested by Alfred Wegener in 1912. The theory is that some 200 million years ago, the continents existed in the form of a supercontinent we call **Pangaea** (pronounced *pan-jeé-ah,* meaning "all lands"), as shown in **Figure 8-4**. Furthermore, the theory suggests that the crust is not a solid sphere but is similar to a loosely fitted jigsaw puzzle whose pieces can be easily moved. These pieces of lithosphere, known as **continental** and **oceanic plates**, are diagramed in **Figure 8-5**. Evidence in its favor mounted, until in the 1960s it was accepted as an important theory.

Two pieces of evidence helped its acceptance. First, data showed that the crust moved. Figure 8-5 shows the velocities of the plates in centimeters per year.[3] The study of the motion of the continental plates is what we mean by **plate tectonics**. Second, scientists discovered the plastic nature of the underlying asthenosphere, which provided a medium on which continents could float. Because the continental plates contain material of lower density than the underlying mantle, the continents float, just as lower-density ice floats on top of higher-density water. Floating reduces the pressure, which then reduces friction, just as there is less friction when you gently press your hands together and rub them rather than when you press your hands together hard and then rub. (Try it!)

But to move the giant continents requires a force, according to Newton's second law. Although the details are still not fully understood, we know that the rising of hot, low-density rock, along with the subsequent falling of cool, high-density material produces a slow circulatory motion of the rock within the asthenosphere, similar to the movement of boiling water (**Figure 8-6**). This motion, called **convection**, plays a major role in plate tectonics. The combination of continental buoyancy with the resulting lowered friction and convective motions within the asthenosphere provides the necessary ingredients for continental motion.

Plates interact with one another in various ways. For example, two plates can move apart, leaving a break (called a *spreading center*) in the lithosphere. Asthenospheric material, which has been heated from below, can flow through the break to form new crust, as shown at the spreading center in **Figure 8-7**.

---

[3]This figure provides a good example of the importance of *velocity* compared with *speed* because direction is clearly an important consideration here.

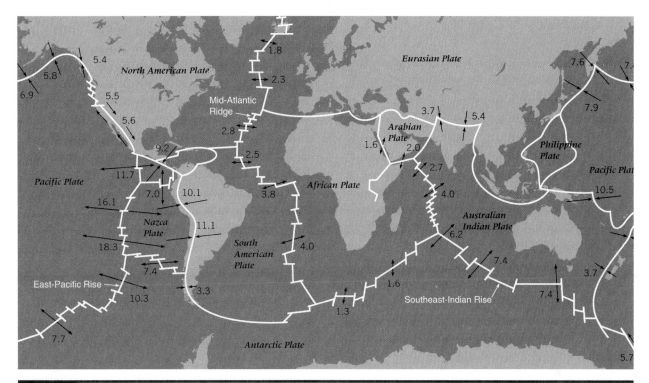

Figure 8-5 The motions of the continental plates. The numbers indicate the speed of one plate relative to that of the adjoining plate in centimeters per year. The arrows indicate the direction of relative plate motion.

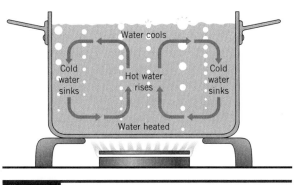

Figure 8-6 Convection in a pot of boiling water.

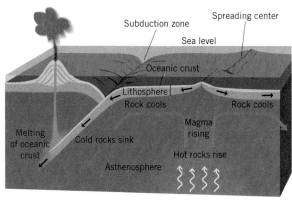

Figure 8-7 Schematic diagram showing the separation of two plates at a spreading center and the movement of magma in producing new crust. A subduction zone, where one plate moves under another, is also shown.

Another form of interaction occurs when two plates slide by each other in opposite directions, as along the San Andreas Fault in California. Here, the North American Plate on the east side (which contains San Francisco) moves south against the Pacific Plate (on which Los Angeles sits), which moves north. The plates grind against one another; sometimes, rock edges catch against each other. The continued motion increases stress, which eventually releases, causing an earthquake.

The other possible type of plate interaction occurs when plates converge on each other. In one instance, the cool lithosphere of one plate can sink underneath another plate, in a process called **subduction**. At the **subduction zone** where this occurs, deep oceanic trenches result (Figure 8-7). In another instance, a piece of low-density continental crust may collide with another piece along a subduction zone and thrust it upward, making spectacular mountain ranges. The Himalaya, Alps, and Appalachians resulted from such collisions.

Do other planets show evidence of plate tectonics?

THE SOURCE OF HEAT IN THE INTERIOR

One of the byproducts of the radioactive decay discussed in Chapter 7 is heat. Radioactive elements in the Earth's core have been decaying ever since the Earth's formation, and because this energy dissipates slowly, it provides the bulk of the heat in the Earth's interior. This results in temperatures high enough to melt rock. Do other planets have molten interiors?

8.3 Earth's Surface

A planet's surface provides scientists with their first clues about its nature. For this reason, we will now study a variety of surface features and processes that we can apply as we view the surfaces of other planetary bodies. Our study will include the types of rocks present on Earth, the chemical nature of the surface, processes involving volcanoes and mountains, impact craters, and the effects of tidal forces.

ROCK TYPES, PROCESSES, AND AGES

Geologists classify rocks into three major families that are determined by the formation process. **Igneous rock** (from the Latin for *fire*) is formed from the cooling and solidification of molten material. Rocks formed from volcanoes are a prime example. **Sedimentary rock** forms when loose materials held in water, ice, or air settle onto a surface, stick, and then build up. The final group, **metamorphic rock** (from the Greek *meta* meaning change, and *morphe* meaning form) consists of material whose original form has been modified by high temperature, high pressure, or a combination of both. Although 75 percent of Earth's surface rocks are sedimentary, 95 percent of the crustal material underneath the surface is igneous or metamorphic rock derived from igneous materials.

Inquiry 8-2
Why does sedimentary rock occur near the surface rather than deep within the crust?

Rocks undergo a continuous cycling and can undergo change from one type to another. For example, an igneous rock on the surface might have been formed from a volcanic eruption. This rock is subject to erosion from wind, water, ice, and impact with other rocks. The eroded particles form a sediment that can be carried by wind and water, eventually settling, becoming cemented onto the surface, and producing a sedimentary rock. Movement of Earth's crust can carry the sedimentary rock to areas where the temperature and pressure are high enough to change the rock into a metamorphic one. If the rock is brought into yet hotter regions, it may melt and form a new igneous rock. Thus, rocks participate in a **rock cycle** involving a slow and continuous interplay of external and internal processes.

Whether a particular material is a solid or a liquid (or a gas) depends critically on its temperature and pressure. An example of the effects of varying temperature and pressure is shown in **Figure 8-8**. Material on the left side (low temperature) is a solid, while material on the right (high temperature) is a liquid. For a given temperature, an increase in pressure can change a material from a liquid to a solid. There is an intermediate region in which solid and melted rock coexist, like ice and

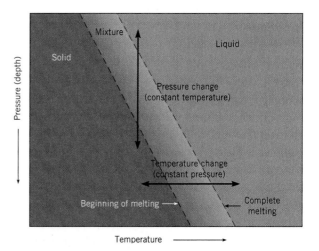

Figure 8-8 The change between solid and liquid depends on pressure and temperature. A change in pressure, holding temperature constant (a vertical change in the figure), can change a rock from solid to liquid.

water coexist in a crushed ice drink. The concepts illustrated in Figure 8-8 are factors that determine the detailed internal structure of a planetary body.

Inquiry 8-3

Explain why the inner core of the Earth is solid, while the outer core is liquid.

Rocks on Earth are subject to numerous types of change. For this reason, we should not expect to find any that are as old as the Earth itself. The same radioactive-dating techniques discussed in the context of meteorites in Chapter 7 can be applied to terrestrial rocks. We find that the oldest age determined using radioactive decay is 4.1 billion years for individual mineral grains in a sedimentary rock from Australia. Ages of 3.9 billion years have been determined for continental crust composed of igneous rock.

CHEMICAL COMPOSITION

The chemical elements most abundant in the Earth's continental crust are shown in **Table 8-2**. These elements combine into some 3,000 minerals that comprise the Earth.

Because silicon and oxygen are the most abundant elements in the crust, minerals made from them are the most abundant on Earth. *Silicates* are minerals having silicon, oxygen, and one or more of the other abundant elements. Another major mineral group is the *oxides,* which include quartz (SiO_2) and limonite ($Fe_2O_3.H_2O$), also found on Mars. Other elemental combinations produce other mineral groups, including the *carbonates* that we return to later in this chapter.

VOLCANISM

The 40 or more volcanic eruptions and the 15 earthquakes of magnitude 7.0 or more in the year 2003 serve as continual reminders that the Earth's surface is highly dynamic. However, volcanic activity need not be dramatic; sometimes lava simply runs out from a series of parallel fissures in the crust, as shown in **Figure 8-9**. Such eruptions provide the only opportunity to study liquid rock, or **magma**, in its natural state. From the study of magma we are able to learn about conditions and processes inside the Earth.

A step toward understanding volcanism comes from examining the locations of volcanoes around the Earth. Rather than being random, they form a well-defined *ring of fire* around the Pacific basin (**Figure 8-10**). This distribution of volcanoes is significant because it corresponds closely to the places where tectonic plate subduction occurs. As a plate sinks, rock is heated by friction and by the hotter asthenosphere. The exact temperature at which the rock melts depends on the pressure and the amount of water present in the rock. The more water, the lower the temperature necessary for melting. Liquid rock, under high pressure, then works its way to the surface.

Table 8-2 Elemental Abundances in the Continental Crust by Percentage of the Mass

ELEMENT	%
Oxygen (O)	45
Silicon (Si)	27
Aluminum (Al)	8
Iron (Fe)	6
Calcium (Ca)	5
Magnesium (Mg)	3
Sodium (Na)	2
Potassium (K)	2
Titanium (Ti)	0.9
Hydrogen (H)	0.1
Manganese (Mn)	0.1
Phosphorus (P)	0.1
All Others	0.8

Figure 8-9 A fissure eruption, in which lava flows through a series of parallel fissures, on Mauna Loa, Hawaii, in 1984.

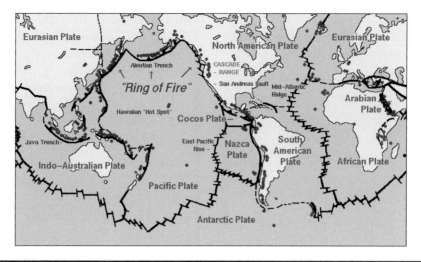

Figure 8-10 The Ring of Fire outlined by volcanoes around the Earth.

The Hawaiian Islands are volcanic but are located far from plate boundaries. They are thought to have formed as the moving Pacific Plate passed over a mid-oceanic hot spot that melted the thin oceanic crust and produced successive volcanoes on the moving plate. The process continues today; a new island, Loihi (pronounced low-EE-hee), that is still under water, is forming.

MOUNTAINS

Some of the processes that produce mountains on Earth have been mentioned in earlier sections. These include repeated volcanism, and the convergence of continental plates that can fold continental crust and thrust it upwards. Examples of such fold-and-thrust mountain ranges include the Appalachians and the Alps mentioned earlier, and the Canadian Rockies.

What goes up must come down, mountains included! The downward pull of gravity causes the destruction of mountains. The process of degradation may begin by landslides occurring near regions of active volcanism. Once a boulder begins its downward plunge, it may loosen other material that also will move downward, in a process that helps slowly destroy mountains. Regions containing water can be eroded in a variety of ways. Flowing water can directly carry away loose material and cause other materials to weaken. Tall mountains, which have continuous freeze-thaw periods, are eroded by pressure from freezing water. Slowly moving glaciers dislodge rocks and grind the mountains through which they move. Vegetation can force open cracks and chemically degrade rocks.

IMPACT CRATERS

Any extraterrestrial bodies coming into the Earth's vicinity will be attracted by its gravity. We saw in the previous chapter that the Solar System is full of meteoroids, ranging in size from dust grains to boulders kilometers across. Figure 7-19 shows the result of a relatively recent impact in the desert of Arizona some 50,000 years ago. Although this crater is an impressive example, it is certainly not the only one. Since satellite imagery of the Earth began, detailed reconnaissance of the Earth's surface has found many ancient craters (**Figure 8-11**). Undoubtedly, the number of impacts was high in earlier years. The history of these impacts has been erased, however, because the craters have been subjected to the same processes that destroy and build mountains.

OCEAN TIDES

Some 75 percent of the Earth's surface is covered with water. The large mass of water readily responds to the gravitational pulls of the Sun and Moon, which cause tides. In fact, the *solid* surface also responds to the gravitational tugging of the Moon and Sun.

The Moon and Earth are represented in **Figure 8-12**. The Moon's gravitational attraction on the Earth's water closest to the Moon is stronger than the Moon's attraction on the Earth's center, which is 6,500 km more distant. (This follows from Newton's law of gravitation, discussed in Chapter 5; the effect is exaggerated in Figure 8-12.) Hence, there is a difference in the gravitational force acting on the near side and the force acting on the Earth's center. The result is a net force toward the Moon that we call a

Figure 8-11 An impact crater in Chad, Africa.

Figure 8-12 The production of tidal bulges on the Earth by the Moon (not to scale). In (a), the straight arrows represent the size of the Moon's gravitational force at each point. In (b), the location A on the Earth is under the bulge and has a high tide. The curved arrow shows the direction of Earth's rotation, such that six hours later (c), the location of point A has rotated relative to the Moon and has a low tide.

differential gravitational force, or **tidal force**. The result of the tidal force is that, because water flows so easily, it "piles up" at the point nearly under the Moon (point *A* in Figure 8-12*b*).

In a similar manner, the force of the Moon's gravity on the center of the Earth is greater than that on the water located on the side of the Earth away from the Moon. Therefore, the solid Earth experiences a stronger acceleration than the water located away from the Moon, and is pulled away

from the water. The end result is a bulge of water on the far side of the Earth nearly equal to that on the side facing the Moon.

The Earth is rotating underneath this bulge of water, which always remains pointed nearly toward the Moon. At some point in time, a given location (point *A* in Figure 8-12*b*) will be nearly underneath the Moon and will experience high tide; six hours later, the Earth's rotation will have moved as in Figure 8-12*c*, and it will experience a low tide.

Inquiry 8-4

How many high tides will location *A* in Figure 8-12 experience in one day? How many low tides?

Inquiry 8-5

Will the time of high tide be the same each day, or will it change? Explain your reasons. If it changes, by how much?

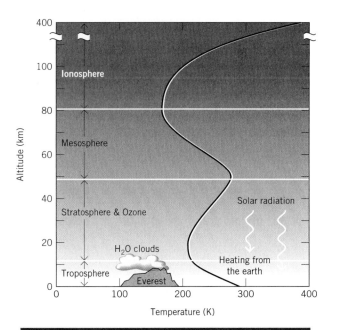

Figure 8-13 The temperature structure of the Earth's atmosphere. Note the location of ozone in the region, where the temperature increases with increasing altitude.

The Sun also produces tides. Although the Sun's gravitational attraction on the Earth is more than 100 times that of the Moon, its greater distance means that the *difference* in the Sun's pull on the two sides of the Earth is less than the *difference* in the Moon's pull. Therefore, using Newton's second law, astronomers can readily show that the Moon is twice as important as the Sun in producing ocean tides.

Consequences to the Earth–Moon system that result from the lunar[4] tides are discussed in Section 8.11.

WATER

The Earth's surface is covered with water. Why? Where did all the water come from? Planetary scientists are looking at and testing various hypotheses. The first is that the water molecules were captured directly from the solar nebula. However, that hypothesis can be ruled out because the observed abundances of deuterium relative to hydrogen on Earth are vastly different than that thought to be present in the solar nebula. A second hypothesis is that the Earth was built up from planetesimals containing significant amounts of water. A third is that water came from early accretion of asteroids and comets. The fourth one is bombardment by asteroids and comets later in Earth's history. No definitive answer is available at this time, as results are varied and sometimes contradictory. More than likely, multiple mechanisms contributed to our watery planet.

8.4 Earth's Atmosphere

The Earth's atmospheric volume is composed of 78 percent nitrogen, 21 percent oxygen, 1 percent argon, 0.03 percent carbon dioxide (CO_2), and a

---

[4]The Latin name for the Moon is *Luna,* the source for the adjective *lunar.*

variable amount (less than 2%) of water vapor (H_2O). The nitrogen and oxygen are present as the molecules N_2 and O_2, not as atoms. Ozone, which is frequently in the news, is made from three oxygen atoms (O_3). Ozone is present as a trace element, providing only a small amount (0.00004%) of the volume of the atmosphere. In addition to these gases, the atmosphere also contains varying amounts of solid particles, which come from wind-blown dust, fires, volcanic eruptions, and industrial pollutants.

We measure the amount of atmosphere in terms of its pressure on us: at sea level, a column of atmosphere having a cross-section of one square inch weighs 14.7 pounds. For convenience, we refer to this pressure as 1 atmosphere.

Scientists have divided the atmosphere into layers that are determined by the variation of temperature with height, as shown in **Figure 8-13**. Within the lowest layer, the temperature decreases with height to about 10 km. This layer (the troposphere) is where weather occurs. Within the next layer (the stratosphere), the temperature increases with increasing height, due to absorption of solar ultraviolet energy by the ozone located within it. Destruction of ozone by the injection of chlorofluorocarbons (CFCs) allows more ultraviolet radiation to reach the surface, where its absorption by living cells, and resulting damage, can occur.

Because there is no ozone above the stratosphere, the temperature again decreases with increasing height. Above about 100 km, the temperature again increases with height because of ultraviolet radiation from the Sun. This ultraviolet energy is capable of ejecting electrons from gas atoms and leaving charged particles behind. Because such charged particles are called *ions* (the processes involved in producing them are discussed in Chapter 13), this "sphere of ions" is called the **ionosphere**. Prior to communications satellites, the only way we could send radio signals beyond the horizon was to reflect them off the ionosphere.

THE GREENHOUSE EFFECT

Although the atmosphere absorbs ultraviolet radiation from the Sun, visual radiation (that which you can see) does pass through and strike the ground, releasing heat. This heat energy radiates into the atmosphere, giving it a warm temperature at low altitudes. You have probably seen an effect of re-radiation from an asphalt highway in the summertime: the blacktop absorbs visual radiation and then becomes hot. Atmospheric ripples and mirages appear as the re-radiation from the road heats the air above it. However, this emerging heat radiation is absorbed by carbon dioxide and water vapor in the atmosphere, which acts like a blanket by slowing the escape of the heat. Because this phenomenon is related to what occurs in a greenhouse,[5] it is called the **greenhouse effect**. A schematic diagram of the process is shown in **Figure 8-14**.

The greenhouse effect occurs naturally. A problem comes about only if the naturally occurring balance between energy input and energy radiated is upset by a dramatic increase in carbon dioxide emissions into the atmosphere.

ATMOSPHERIC CIRCULATION

The atmosphere is far from static. Weather satellites show us the movement of cloud systems, low- and high-pressure systems, and hurricanes. Atmospheric circulation is driven by heat from the Sun that is re-radiated into the atmosphere by the Earth's surface. If this were the only source of energy input, the atmospheric circulation patterns

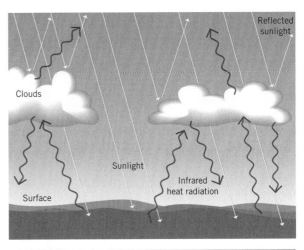

Figure 8-14 The greenhouse effect. Sunlight coming through the atmosphere heats the ground, which then emits infrared (heat) radiation. The heat radiation is partially trapped primarily by carbon dioxide and water vapor.

would be simple. The complication is that 75 percent of the Earth's surface is covered with water, a tremendous storehouse of solar energy. Furthermore, the continents are warmer than the water, thus adding to uneven heating of the atmosphere. We now examine the effect of these sources of heat on the atmosphere.

When heat is injected into the atmosphere from water or continents, the atmospheric temperature increases. The heated gas expands and the atmospheric pressure changes. These changes, which also cause the density of air to decrease, force heated air to rise. It is analogous to the low-density continental crust floating on top of the higher-density asthenosphere. Atmospheric convection, therefore, is a process that moves energy through the atmosphere. Because the uneven reservoir of heat on the Earth's surface unevenly heats the atmosphere, it contains numerous regions of low and high pressure. For example, air at the equator increases in temperature and rises. Due to the rising air, the atmospheric pressure at the equator lessens. Atmospheric gases from more northerly and southerly latitudes will move toward the low-pressure area at the equator and thus produce surface winds. The gas that rose over the equator will move toward the poles, cool, sink back to the surface, and begin to circulate back to the equatorial regions.

The Earth's rotation complicates the picture dramatically. In Chapter 5 we introduced the con-

---

[5]The greenhouse effect in a real greenhouse is somewhat simpler than in the atmosphere. In both cases, the infrared radiation is partially trapped and cannot escape easily, but in a real greenhouse, the glass cover also prevents the heated air from rising or flowing away.

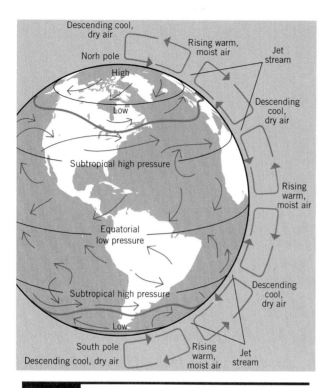

Figure 8-15 The complex general circulation patterns of the Earth's atmosphere. The oval shapes with the arrows on the right indicate the cell structure of the atmosphere, in which heated air rises and cool air falls. Jet streams are located where adjacent cells meet.

cept of the *Coriolis effect,* in which a northward traveling rocket appears to veer to the east because of the Earth's rotation. In a similar way, the Coriolis effect causes northward moving surface winds to veer toward the east. For this reason, the Coriolis effect controls the general circulation of winds on the Earth and other planets. Most of the time, these effects are part of the typical weather patterns, but during hurricane, monsoon, or tornado seasons we can have dramatic illustrations of their reality.

The end result is that the atmosphere is broken up into cells, as shown in **Figure 8-15**. Low pressure occurs where the cells rise while high pressure occurs where they fall. The upward and downward movements within the cells at latitudes near 30° and 60° produce high-speed, eastward-moving winds (the jet stream we often hear about on weather reports). We expect that rapidly rotating planets would show an even more complex circulation pattern, as we will see when we study the Jupiter-like planets in Chapter 10.

ORIGIN AND MAINTENANCE OF THE ATMOSPHERE

The chemical makeup of the Earth's atmosphere is not what we might expect. Most of the universe is made up of hydrogen and helium with all the other elements comprising only some 2 percent of the total. We might therefore anticipate an atmosphere of hydrogen and helium, with trace amounts of everything else, but this is not the case.

Astronomers think that the original, primitive atmosphere of the Earth was lost and later replaced by the atmosphere we have today. Until recently, because of the vast amount of hydrogen in the universe, it was thought that the original atmosphere was composed primarily of hydrogen and hydrogen-bearing compounds, such as methane (CH_4), ammonia (NH_3), and water (H_2O). There should also have been an amount of neon approximately equal to the fraction found in the Sun. From the observed lack of neon—which could not have come about as a result of chemical reactions because neon is chemically inert—astronomers conclude that the Earth's original atmosphere must have been lost.

What happened to the compounds that formed the primitive atmosphere? Gas atoms and molecules are constantly in motion. The greater the temperature, the greater their speed. Furthermore, less massive particles move more rapidly than heavy particles having the same temperature. Therefore, if the temperature of a gas is high enough, the particles may reach speeds in excess of the Earth's escape velocity of 11.2 km/sec. At the temperature of the Earth, both hydrogen and helium attain velocities high enough to escape. And because methane and ammonia are readily broken into their constituent molecules, their hydrogen atoms can also escape, leaving a planet without an atmosphere.

In the planet formation process, pockets of gas can be trapped in the planet's interior. As the interior temperature increases due to the decay of radioactive elements in rocks, the trapped gases can slowly escape from the interior and form a new atmosphere by a process called **outgassing**. Such gases can also escape through volcanic eruptions. Even today, volcanoes spew large amounts of water, carbon dioxide, sulfur dioxide, hydrogen sulfide, chlorine, and nitrogen into the atmosphere. In fact, a more modern hypothesis on the composition of the primitive atmosphere suggests it was

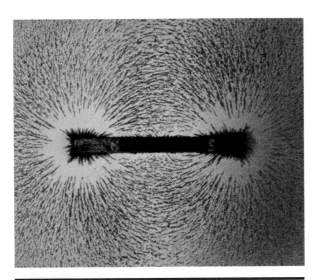

Figure 8-16 Magnetic force fields, as shown by the alignment of iron filings around a magnet. Grant Heilman Photography, Inc.

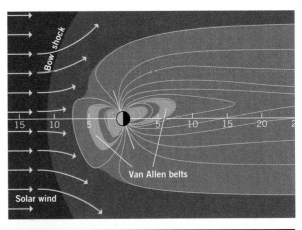

Figure 8-17 The Earth's magnetic field as observed from space. The magnetic field lines are distorted by charged particles from the Sun, located in the solar wind. Note the compression of the field on the side toward the Sun, and the stretching on the side away from it. The Van Allen belts are two regions surrounding the Earth in which charged particles are trapped by the magnetic field.

these molecules, and not methane and ammonia, that composed the original atmosphere.

There is little carbon dioxide in the atmosphere because it is tied up in carbonate rocks. You may have noticed that oxygen has not been mentioned. Little was present in the atmosphere until substantial amounts of plant life evolved some 2 to 2.5 billion years ago. Plants frequently absorb nitrogen from the ground (this is why fertilizers contain nitrogen); this nitrogen gets released into the atmosphere after they die.

8.5 Earth's Magnetism

A magnetic compass points toward the north *magnetic* pole (which differs from the north *geographic* pole), showing that the Earth has magnetism associated with it. The effects of magnetism are readily seen when small needles of iron are placed over a magnet (**Figure 8-16**). The pattern formed by the iron filings traces the magnetic force field, and we often speak of the lines formed by the pattern in terms of **magnetic lines of force** or **magnetic field lines**. The closer to the magnet the stronger the magnetic force and the closer together the magnetic lines of force.

What produces magnetism? Scientists have known for over 100 years that an electric current, which is produced by electrons moving in a wire, will produce a magnetic field. Whenever electrons move, a magnetic field is produced.

What produces the Earth's magnetic field? Although the detailed answer is still not known, we think the circulation of molten iron and other metals in the liquid outer core causes it. Metallic elements such as iron have an outer sphere of loosely bound electrons; it is these electrons that allow metals to conduct electricity and heat so well. The convection of the molten outer core, along with the rapid rotation of the Earth (and, by assumption, its core), provide a large mass of moving electrons, which we think produces the magnetic field. This process is known as the **dynamo effect**. This process explains a number of observations, but it does not explain everything. For example, geological measurements show that the north and south magnetic poles vary in strength and even swap positions irregularly. The dynamo effect does not easily account for these observed switches in the magnetic field's direction and strength.

The Earth's magnetic field, illustrated in **Figure 8-17**, is similar to that of the simple bar magnet in Figure 8-16. The magnetic lines of force, which surround the Earth in a sphere of magnetism we call the **magnetosphere**, extend out into space. The Earth is bathed in rapidly moving (supersonic) charged particles because, as we will see in Chapter 16, the Sun is constantly ejecting electrons and protons into space in all directions. The place where the speeding particles and Earth's

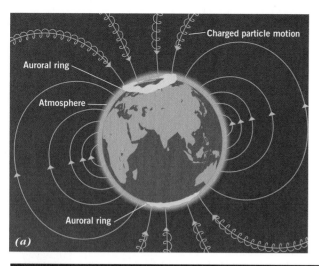
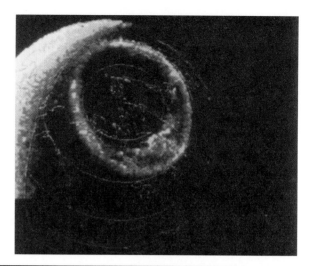

Figure 8-18 The aurora borealis. (a) Charged particles spiral around magnetic field lines, and the magnetic field intersects the atmosphere in a ring. (b) The resulting auroral ring as observed by a satellite about three Earth radii from the North Pole.

magnetic field collide forms a boundary region called a *bow shock*. Once these charged particles encounter a strong magnetic field, the particles' path changes and they slow down. The *bow shock* is a specific example of a more general concept, the **shock wave**, which results when something is moving faster than the speed of sound. The concept of the shock wave will appear in various contexts throughout the rest of this book.

The charged particles from the Sun modify the Earth's magnetic environment. The solar particles push on the field lines, compressing them in the direction from which the particles came. In addition, those particles that pass near the Earth pull the field lines along with them, thereby stretching them out in the direction away from the Sun. The compression and stretching of the magnetic field distorts the magnetosphere. Spacecraft passing through the magnetosphere can detect the presence of the field and any distortion in it.

The path followed by charged particles in a magnetic field is not a straight line but a spiral around the magnetic field lines (**Figure 8-18a**). The particles move toward locations where the field is strongest, such as the magnetic poles. Whenever large concentrations of particles moving along the field lines strike the upper atmosphere, the gas atoms in the atmosphere can be made to glow, forming the **aurora borealis** in the Northern Hemisphere and the **aurora australis** in the Southern Hemisphere (**Figure 8-18b**). Because the magnetic field is three-dimensional, it intersects

the atmosphere in a circle (Figure 8-18*a*), and an auroral ring is produced surrounding the north magnetic pole (Figure 8-18*b*).

Finally, the Earth is surrounded by two donut-shaped regions of charged particles that are trapped by the magnetic field. These belts of particles were discovered when the first U.S. satellite was launched in 1958. Named the **Van Allen Belts** after the scientist who discovered and analyzed them, the two belts are centered on regions about 1.5 and 3.5 Earth radii away and are included in Figure 8-17.

8.6 The Moon: Large-Scale Characteristics

In the rest of this chapter we explore the closest extraterrestrial body—the Moon. The study of the Moon is ancient, going from naked-eye studies by the Greeks, to the first telescopic observations by Galileo, to the early unmanned and then later manned *Apollo* project in the 1970s and then to newer spacecraft studies in the 1990s.

Inquiry 8-6

Before going any further, imagine that you are an official of the space program planning a mission to the Moon. What would you want to know about the Moon?

Table 8-3 Properties of the Moon

PROPERTY	VALUE
Orbital properties	
Average distance from Earth	384,401 km (60.4 $R_\oplus$)
Minimum distance	363,297 km
Maximum distance	405,505 km
Orbital period	27d7h43m12s
Lunar month (from new moon to new moon)	29d12h44m3s
Orbital inclination	5° 8′ 43″
Physical properties	
Diameter	3,476 km (0.273 $D_\oplus$)
Mass	7.35×10^{22} kg (0.0123 $M_\oplus$)
Density (average)	3.34 g/cm^3
Surface gravity	0.165 Earth gravity
Escape velocity	2.4 km/s
Rotation period	27d7h43m12s
Tilt of rotation axis from perpendicular to its orbit	6° 41′
Reflectivity	7%
Surface temperature	100 K night to 400 K day (−279 to +260° F)

From a study of eclipses, Aristarchus, in the third century B.C.E., determined the size of the Moon relative to the Earth and found it to be some four times smaller (Chapter 5). Knowing the size and observing the angular diameter to be about 1/2° allows us to find the distance, which is about 384,000 km. The Moon's mass is most easily determined by examining the motion of an artificial satellite in lunar orbit and applying Newton's version of Kepler's third law; the Moon's mass is 1/81 that of Earth. From this we find a density of 3.3 g/cm^3 (i.e., 3.3 times the density of water). The currently accepted values, along with other data for the Moon, are given in **Table 8-3**.

A close look at the table shows that the Moon's period of rotation is exactly equal to its period of revolution about the Earth. The Moon is said to be in **synchronous rotation** with the Earth. That is why earthlings always see the same side of the Moon! Similar effects will be seen with other planets and moons in later chapters.

8.7 The Moon's Atmosphere

If the Moon had a substantial atmosphere, starlight passing through it would gradually become fainter and fainter until, at last, the solid surface of the Moon would extinguish it completely. However, as the Moon moves in its orbit around the Earth, from time to time passing in front of stars, we observe the starlight to disappear from view extremely rapidly. From such an event, which is called a **lunar occultation**, astronomers have long since concluded that the Moon has virtually no atmosphere. This is consistent with the Moon's low escape velocity and high daytime temperature. Escape of the Moon's atmosphere, if it ever had one, must have been extremely rapid.

Without an insulating layer of atmosphere, the Moon's surface undergoes extreme variations in temperature, from 400 K (which is equivalent to 260° F, just above the temperature of boiling water) during the day down to only 100 K (−280° F) at night.

Spacecraft visits to the lunar vicinity showed it to have a tenuous but essentially insignificant atmosphere that varies with time. Part of the atmosphere comes from outgassing from deep within the interior, and it may be released during the Moon's weak seismic events, which are discussed later.

Other parts of the atmosphere come from the solar wind, which contains hydrogen and helium along with some heavier elements. Any of these materials that strike lunar areas receiving sunlight will eventually escape. However, as *Lunar Prospector* showed, regions that remain cold by being in permanent shadow are able to hold hydrogen onto the surface indefinitely. Whenever slight warming occurs, the hydrogen can be released and a very weak hydrogen-containing atmosphere can result.

8.8 The Lunar Surface

A primary goal of the manned lunar missions was to return the astronauts safely to Earth. One result of this priority was that the astronauts were sent to safer (and hence possibly less scientifically interesting) areas of the Moon, where they would be able to land with the least danger. Inevitably, the limited amount of sampling that was done would leave many unanswered questions. Nevertheless, due to the *Apollo* missions, many important new scientific results have been obtained that could not have been discovered in any other way. Recent unmanned spacecraft have also provided new and exciting results.

Figure 8-19 FIGURE 8-19. The full Moon with the principal features indicated.

> All readers are strongly encouraged to do Discovery 8-1 at the end of the chapter *before* continuing to read.

GENERAL SURFACE CHARACTERISTICS

The visible features on the Moon's surface seem permanent, indicating that the surface is solid. Is it like a cement slab, a frozen pond, a sand dune, a dust pile, or something else entirely? Galileo, on first observing the dark areas of the Moon's surface (**Figures 8-19** and **8-20**), thought they looked like water and named them **maria** (pronounced mar′-ay-ah; singular form **mare**, pronounced mar′-ay), the Latin word for *sea*. However, it has long been clear that the maria contain no water and that the dark color of the maria has to have another cause.

A number of things can affect the reflectivity of a surface. For example, a smooth surface will directly reflect a large proportion of the light falling on it, whereas a rough surface will reflect the light in various directions, thereby decreasing the surface's overall ability to reflect. Thus, surface texture is one possible cause of variations in the Moon's reflectivity. A second possible cause is that the maria may contain a different type of material than that found in other, lighter-colored areas of the Moon. Just as some rocks and soils on Earth are light in color and others are dark, the same could be true of the Moon.

We can infer something about the nature of the lunar surface from observations of how the Moon's temperature changes when sunlight is blocked during a lunar eclipse. The lunar surface cools slowly, telling us that it must be composed of a good insulator for it to be able to hold the heat of the lunar day so effectively.

One model that was consistent with the observed temperature variation during an eclipse was that a substantial layer of porous dust covered the Moon's surface. There would be no surprise in this because the Moon is constantly bombarded by meteoroids of all sizes that would slowly pulverize its surface rocks. Even the extreme temperature changes that the Moon undergoes every month would tend, in time, to weaken and even crumble the rocks by causing them to expand and contract.

The possibility that there might be a thick layer of dust on the Moon raised fears that the first astronauts might simply disappear into the bottom of a dust bowl. For that reason the planners of the

Figure 8-20 The Moon at first quarter. Note how much more visible the craters are than in the previous figure.

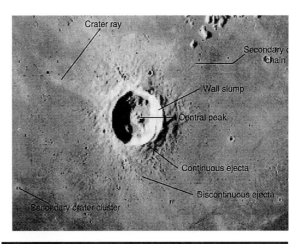

Figure 8-21 The impact crater Euler, with a well-defined ejecta blanket and secondary craters. Euler is 27 km in diameter.

first manned mission sent an unmanned vehicle first! Fortunately, it turned out that the dust had a sufficient amount of cohesion to prevent such a disaster from taking place, but this could not have been known for certain before the first lunar lander in 1966.

The American program to visit the Moon started with a number of unmanned missions. First was the *Lunar Ranger* project, in which several probes relayed close-up television pictures to Earth prior to crash-landing on the lunar surface. The *Lunar Orbiter* program provided pictures of the lunar far side, along with high-quality images of the near side. The final unmanned landing was the *Surveyor* program, which landed spacecraft to carry out the first experiments on lunar material density and properties.

We will now examine the variety of surface features present on the Moon, paying particular attention to the processes that formed them. We want to find out whether the processes that shaped the Moon are similar to or different from those that shaped the Earth.

CRATERS

Ever since Galileo first discovered lunar craters, scientists have speculated about their origin. Two possible hypotheses immediately presented them-

selves: they may be due to meteoroid impacts, similar to those that formed the Barringer crater in Arizona (Chapter 7); or they might be volcanic.

There is no question that many, if not most, of the lunar craters have their origin in meteoroid impacts. Evidence for impact is plentiful. In **Figure 8-21** the relatively young crater Euler is surrounded by material that has been ejected from the crater; such an **ejecta blanket** is characteristic of impact. Furthermore, its circular shape is in contrast to the various volcanically produced craters, which are often somewhat elongated. The figure also shows a chain of craters along a line from Euler; such **crater chains** are produced when clots of excavated material from one impact strike the surface, producing **secondary craters**.

Further evidence of the impact nature of craters comes from the light-colored **rays** emanating from the crater *Tycho* (Figure 8-19), suggesting that material was thrown outward by tremendous forces during impact. The distribution of rays around a crater can be understood in terms of the angle at which the impacting body hit the surface. A perpendicular impact produces an even, symmetric distribution of rays, while an oblique impact produces asymmetric rays.

Impact can readily explain the presence of central mountain peaks in a significant number of the largest craters (Figure 8-21). These peaks result from rebounding of the compressed surface after the impact. The crater in this figure has terraced walls, caused by the gravitational pull of the Moon that makes the walls slump downward.

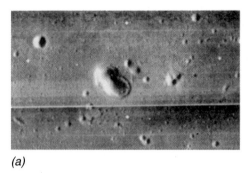

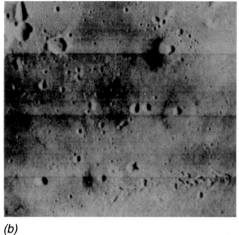

(a)   (b)

Figure 8-22 Lunar volcanic craters. (a) This crater is elongated with dimensions of 3 km by 5 km, lacks exterior walls, and is similar to basaltic volcanoes on Earth. (b) A volcanic dome with a crater on its summit. The domes are roughly 10 km across.

However, not all lunar craters are the results of impact. The craters in **Figure 8-22a** are not impact craters because they lack walls and are not circular. In addition, the Moon shows numerous **volcanic domes** similar to those on Earth (**Figure 8-22b**). Such domes result when lava's resistance to flow, its **viscosity**, is high enough to prevent it from readily flowing across the surface. Although the Moon has numerous domes, it does not have any tall volcanoes similar to Mt. Etna or the large Hawaiian volcanoes.

Craters do erode with time. The continuous bombardment by micrometeorites, meteorites, and cosmic rays over billions of years gives old craters a softened appearance.

Although much research has been done into how craters, large and small, form, details are highly complex and still not well understood. When a large meteoroid, perhaps moving at 20 km/s, strikes the surface, rapid compression of the surface occurs. The impact produces pressure waves moving at speeds greater than the speed of sound; these shock waves travel throughout the impacted body. The impact melts or vaporizes the impacting meteorite and produces molten lunar rock that is squeezed out to the sides at high speed. Some of this molten material moves at speeds greater than the Moon's escape velocity and is ejected into space. Shortly afterward, new outward-moving material is deflected upward by the forming walls. Velocities are now below the escape velocity, so the material falls back to the lunar surface. As the energy of the impacting body is expended, the speed of the ejected material goes

to zero, and the cratering event is over. The compressed point of impact can rebound and form a central peak. Crater walls may collapse and modify the crater. These events are summarized in **Figure 8-23**. Material originally on the surface becomes buried, and material originally under the surface ends up on the surface. For a typical large crater, this process takes perhaps a few minutes. The size of the crater produced is typically about 10 times the size of the impacting body.

MARIA AND HIGHLANDS

The maria can be explained as being the result of a gigantic impact that produced a large basin, subsequently filled in by flowing lava. The sizes of these maria indicate impacting bodies having diameters from about 30 to 100 km. Their relative flatness comes from fluid flow. For example, in Mare Imbrium (the Sea of Rains), which has a diameter of nearly 800 km, extensive lava flow fronts are visible, coming from the lower left of **Figure 8-24**. Because no volcanic cones are present to account for the flows, the material must have come from long fissures similar to those on Earth, shown in Figure 8-9. Not all lunar volcanoes show flow fronts. This indicates a variation in the lava's viscosity from one place to another. An analysis of rocks returned by the *Apollo* astronauts shows some rocks, when melted, flow easily compared with lava flows on Earth, and have the viscosity of engine oil. Finally, an analysis of the sizes of crystals in the lunar maria shows that the material cooled relatively rapidly, over a period of a few years.

Meteoroid impact

Meteoroid is vaporized

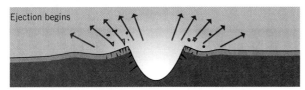

Ejection begins

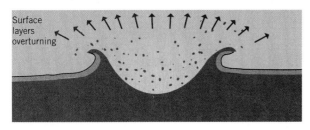

Surface layers overturning

Ejecta blanket

Figure 8-23 How a large crater is formed. After initial compression, the surface reacts by throwing material into the surrounding area. Note that material originally on the surface ends up buried, while material originally underground is on top. Only large craters form central mountain peaks.

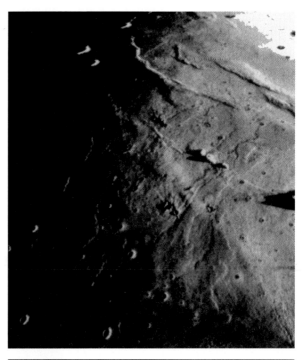

Figure 8-24 Lava flows in Mare Imbrium. The lava flows shown here traveled distances up to 600 km. The average flow fronts have heights of 30 m. Since there are no obvious craters from which the lava erupted, it must have come from fissure eruptions, similar to that in Figure 8-9.

Figure 8-25 Mare Orientale.

The most dramatic feature on the lunar surface is Mare Orientale (**Figure 8-25**), located right on the dividing line between the near and far sides. This is an example of a **multiringed basin**. It consists of four rings, two of which are easily seen. The inner ring is called the Rook Mountains and has a 620-km diameter, while the outer one, called the Cordillera Mountains, is 900 km in diameter. The central region has been flooded with lava (as we discuss shortly).

There are two main hypotheses for the formation of the Orientale Basin. The more widely accepted one is that a gigantic impact produced tidal-like waves that froze in place. The other

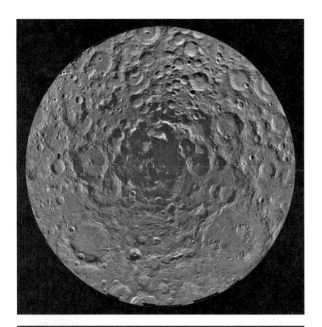

Figure 8-26 The South Pole Aitken basin.

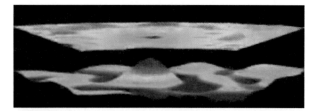

Figure 8-27 The gravity field around Mare Serenitatis. The top part shows a topographic map while the bottom shows how the strength of gravity varies over the mare.

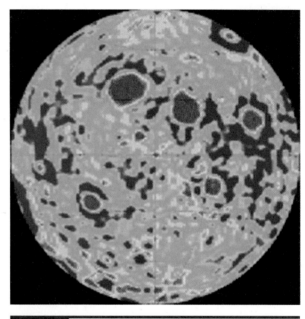

Figure 8-28 The gravity field of the Moon's near side. Red signifies areas of stronger gravity.

hypothesis looks at the entire feature as a single crater, in which the rings are terraces that resulted from collapse of the walls.

The largest impact basin on the Moon is the South Pole's Aitken basin. The impacting body that produced its 2,600 km size excavated the surface to a depth of some 8 km below the surface. The adjacent region rises to some 8 km above the surface (**Figure 8-26**). Thus, the range in elevation is 16 km—approaching Earth's 20 km. However, Earth's elevation range comes from tectonic activity that produces high mountains and deep trenches; on the Moon, the elevation range comes from meteoroid impacts.

Shortly after the first spacecraft missions to the Moon, scientists discovered concentrations of mass—that is, regions of enhanced gravitational attraction for the spacecraft—at various locations of the lunar surface. These mass concentrations, given the name *mascon,* are generally located in the centers of the circular maria. The 1998 Lunar Prospector provided more refined data. The top of **Figure 8-27** shows the flatness of the lunar topography over Mare Serenitatis (Sea of Serenity), one of the Moon's largest maria (see Figure 8-19). The bottom part of the figure shows the variation of gravity over the same area; the peak is where the gravity is greatest and is, therefore, where a mascon exists. **Figure 8-28** is a gravity map of the entire near side of the Moon. This map shows that the mascons, which are the red areas, have a gravi-

tational attraction a few percent above the average. Lunar scientists now think the mascons are plugs of higher-density mantle rock that rose toward the surface after the impacts occurred.

The low-elevation lunar maria are distinct from the higher elevation areas that are known as the lunar **highlands**. The highlands are higher than the maria because the highlands contain material of lower density than the material underneath. The highlands are saturated with craters.

The near and far sides of the lunar surface are compared in **Figure 8-29**. The most obvious differences are that the far side is more heavily cratered and lacks the extensive maria of the near side. However, it does contain some large-impact basins, such as Hertzsprung, that are not so obvious because a thicker crust on the lunar far side prevented lava in the interior from reaching the surface. Hertzsprung appears to have an inner ring.

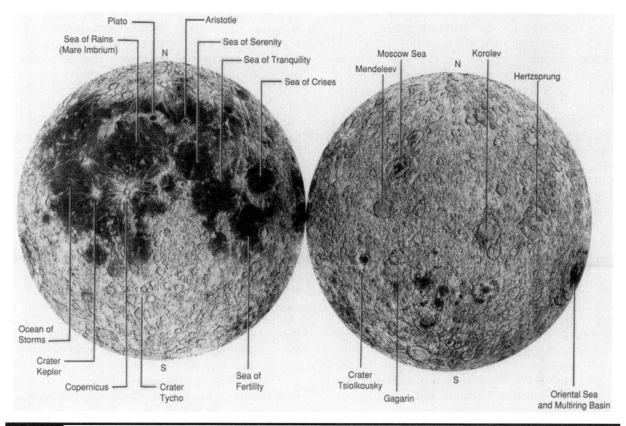

Plato — — Aristotle

Sea of Rains (Mare Imbrium) — N — Sea of Serenity

— Sea of Tranquility

— Sea of Crises

Moscow Sea — Korolev

Mendeleev — N — Hertzsprung

Ocean of Storms

Crater Kepler

Copernicus — Crater Tycho — S

Sea of Fertility

Crater Tsiolkousky — S

Gagarin

Oriental Sea and Multiring Basin

Figure 8-29 The far side of the Moon lacks the maria of the near side.

SURFACE MOVEMENT

Since the earliest days of lunar observing, nearly straight lines were observed at numerous locations. They were called **rilles**, which means *tiny water courses*. One of the rilles easily observable on ground-based photographs is the Straight Wall (**Figure 8-30**). More detailed study shows it to be the result of ground movement—a fault 115 km long, in which the ground on the one side has dropped 400 m relative to that on the other. Thus, although the original meaning of *rille* doesn't describe this particular feature, they are still called rilles. This process, where the original name is retained even though it is no longer apt, is common in sciences like astronomy where new interpretations of old observations occur all the time. Unfortunately, it is confusing at first!

The result of more complex crustal movement can be inferred from ridges visible in some maria (**Figure 8-31**). Note the alignment of the ridges, which is suggestive of tectonic motions within the crust, along with crustal folding and buckling. However, the Moon shows no large-scale plates or plate motion like that on Earth.

LAVA CHANNELS AND TUBES

In addition to the straight rilles, the lunar surface shows a large number of rilles that appear to wander like a meandering stream across the lunar surface. These are called **sinuous rilles** to distinguish them from the straight ones. The majority of them have one end at a rimless crater located higher than the other end. Sinuous rilles are interpreted as volcanic lava channels that start at the rimless crater and then flow downhill.

The outer surface of lava flowing in a channel will cool rapidly, forming a skin above the hot lava flowing underneath. This skin thus forms a roof, enclosing the molten rock inside a **lava tube**. The roof insulates the liquid, slowing its cooling and allowing it to flow farther from the source. Once the source of lava runs out, the liquid will continue to flow until the tube is drained. Thereafter, the tube may collapse. Such collapsed tubes are observed over parts of some sinuous rilles and provide additional evidence for lunar volcanism. **Figure 8-32** shows the largest sinuous rille on the Moon, called Schroeter's Valley. It is more than 150 km long, 4 to 6 km wide, and 500 m deep, and

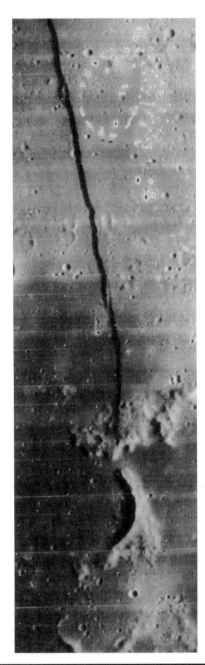

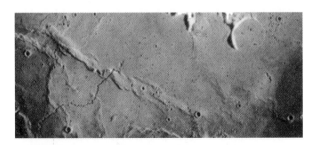

contains some collapsed lava tubes. It begins at a crater and eventually ends at a mare.

THE COMPOSITION AND STRUCTURE OF THE LUNAR SURFACE MATERIAL

As the astronaut's footprints in **Figure 8-33** illustrate, the lunar surface is sufficiently cohesive to provide solid support despite its dusty nature. The top layer of the lunar surface is called the **regolith** (rocky layer); it is composed primarily of rocky material ejected from lunar impact craters. The powdery soft nature of some areas is a result of erosion by the constant bombardment of small particles from space. Of course, there are occasional larger impacts, too, but, as on the Earth, most of the mass falling on the Moon today is in the form of particles the size of a grain of sand or smaller. The extreme heating and cooling of lunar rocks is also a source of erosion. Driven by temperature changes, rocks will expand and contract until finally they weaken and crumble.

The Moon's surface chemical composition can be studied both from orbiting spacecraft and from the surface. The 1994 *Clementine* and the 1998 *Lunar Prospector* missions provided maps showing the existence of different *compositional provinces* in the highlands and the overall distribution of some specific elements. For example, **Figure 8-34** shows the distribution of iron and

Figure 8-30 The Straight Wall resulted from crustal movement.

Figure 8-32 This sinous rille, called Schroeter's Valley, is an excellent example of the connection between a lava source at a crater, and the channels and tubes that transport it to an outlet at a mare. The valley is over 150 km long, 4–6 km wide, and 500 m deep.

Figure 8-31 Mare ridges in the Mare Serenitatis. The region shown is 115 km by 155 km. The ridges are a result of the folding of solidified lava.

potassium. These results come from a detection of gamma rays, which are radiations produced when cosmic rays (Chapter 7) interact with the surface elements. Because the details of the gamma rays emitted by different elements vary, scientists are able to determine the large-scale chemical composition of the lunar surface.

Chemical analysis of some of the 800 pounds of lunar rocks returned to Earth by the various *Apollo* missions shows that they fall into three main cate-

gories: **basalts**, **breccias**, and a type that has been called **KREEP**. Maria materials are primarily basalts; that is, igneous rocks that have resulted from the cooling of molten material. Breccias, by contrast, are formed by the fusing together of rock fragments that may well be older than the breccias themselves. Such fusing occurs because of impacts by external bodies, which increase the pressure and temperature of the region that was impacted. KREEP is a type of basalt that is unusually high in potassium (chemical symbol K), rare earth elements (REE), and phosphorus (P), relative to terrestrial rocks. These elements are unable to fit into the crystal structure of other materials and are left behind to float on the magma surface. On the basis of the rocks returned to Earth, lunar geologists thought the entire surface would be covered with KREEP. However, from the global perspective provided by *Clementine* and the *Lunar Prospector,* we now know that the samples returned by the astronauts are atypical of the entire lunar surface. An important lesson has thus been learned: one must be careful in drawing global conclusions from local data. **Figure 8-35** shows lunar basalt and a breccia.

The lunar highlands also consist of igneous rocks. Understanding why the entire lunar surface is igneous is a major part of understanding the Moon's history, which will come later in the chapter.

Rocks on Earth almost always include water molecules chemically bonded into the interior

Figure 8-33 The footprints show the surface material to be strong, yet compressible. The retroreflector left on the Moon by the first astronauts is in the foreground.

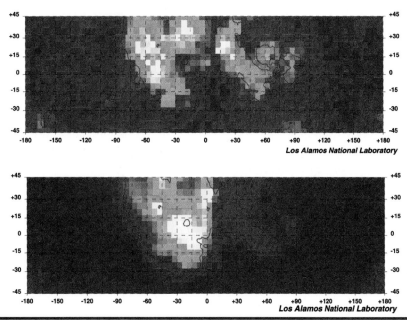

Figure 8-34 The distribution of iron (top) and potassium (bottom) over the lunar surface, as determined by the Lunar Prospector mission in 1998. The axes are lunar latitude and longitude.

(a)   (b)

Figure 8-35 Lunar rocks. (a) A basalt from one of the lunar maria. (b) A breccia, showing fragments cemented together.

structure. The Moon rocks picked up by the *Apollo* astronauts, however, are distinguished by their complete lack of water. Any water that was ever present in any of them must have escaped rapidly during the Moon's early stages, since these rocks formed then.

The rocks that were examined by the *Apollo* astronauts were from equatorial and low-latitude regions that receive sunlight. However, any water located in areas that never receive sunlight, for example in craters around the polar areas, would be locked there indefinitely. **Figure 8-36** is a mosaic of the Moon's south polar regions and shows some of these never-illuminated regions. When the *Clementine* spacecraft found indications of pockets of water ice in never-illuminated polar areas, scientists became excited. Thus, one of the main objectives of the 1998 *Lunar Prospector* mission was the search for water, which might have been deposited on the Moon by water-rich comets and asteroids. While not finding water *directly, Lunar Prospector* did find an excess amount of hydrogen in the dark polar regions. From these observations of hydrogen, scientists infer the presence of perhaps 10 billion gallons of water buried in shallow craters (Lake Erie is about one billion gallons). Because continuous bombardment has "gardened" the lunar material to a depth of two meters over the past two billion years, water could exist down to that depth. Our knowledge of the actual amount of water, if any, is, of course, highly uncertain at this time. The European *SMART-1* spacecraft, launched in 2003, should provide new results on questions of ice in the lunar polar

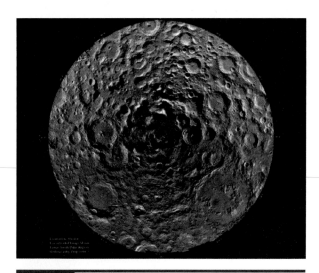

Figure 8-36 A mosaic of the Moon's south polar regions, as observed by Clementine.

regions by 2006. A detailed comparison of elemental abundances between lunar and Earth rocks provides fundamental information required to understand their origins. When compared with the Earth's rocks, lunar rocks are deficient in volatile elements (those with low melting points). This important finding suggests two possible explanations. First, the Earth and Moon could have formed from materials having slightly different chemical compositions. Second, the materials out of which the Moon was made could have formed at somewhat higher temperatures than was the case for the Earth; a higher temperature would have driven the more volatile elements away from the forming Moon. A higher temperature would have been present if the Moon formed in a hotter location than where the Earth formed, or if it formed earlier than the Earth. At this time, observations do not indicate which idea is preferred.

A final comparison with the Earth's materials is a detailed examination of the abundance of the various types of oxygen on the Earth and Moon. Although all oxygen atoms have eight protons in the nucleus, some have different numbers of neutrons. Atoms of a given element having different numbers of neutrons are called **isotopes**. A finding of great importance to understanding the evolution of the Moon is that the relative abundances of the various oxygen isotopes are identical on both bodies, although they differ from the abundances in meteorites.

The European Space Agency's (ESA) *SMART-1* orbiter launched in 2003 is investigating lunar chemistry and searching for ice at the lunar South Pole. This signals the beginning of an exciting new era in Solar System exploration, as this was ESA's first moon shot. Other goals are to test new rocket drives and miniature instruments.

Inquiry 8-7

What might you conclude about the origin of the Earth and the Moon, given the equality of the oxygen isotopes? What would you conclude if the abundances had been different?

THE AGE OF THE LUNAR SURFACE

Relative ages of lunar surface features can be determined using the geological concept of superposition, which says that younger features are on top of older ones. For instance, a crater on top of a mare will have formed after the lava flows in the mare stopped.

Absolute ages can be determined using the same dating techniques that were discussed for meteorites. Lunar rocks became available for radioactive-dating analysis with the *Apollo* manned flights to the lunar surface, which began in 1969. The results show impact basins, which were to be filled in by later lava flows, formed in the narrow time frame of 3.8–3.9 billion years ago, at a time when the impact rate was high for a short period of time that is referred to as the *lunar cataclysm*. The rocks returned from the lunar mare lava flows turned out to be about 3.0 to 3.8 billion years old (slightly younger than the oldest mea-

sured Earth rocks). What does this age range mean? The time determined by radioactive dating techniques tells us the time since the rock last solidified. In other words, the radioactive clock is reset to zero when melting occurs. Thus, these ages mean that the maria regions were molten (and then solidified) throughout an 800-million-year period. Later *Apollo* landings were more adventurous and also sampled the highland regions. Radioactive ages of highland rocks were found to be about 4.1 billion years—hundreds of millions of years older than the maria. Although the highland region is cratered and does not appear strongly volcanic, the highland material turned out to be composed of solidified magma, just like the maria. The conclusion is inescapable and important: There was an early epoch when the entire surface was molten. The basaltic leftovers that resulted when the surface cooled were then chopped up by heavy cratering to form the breccia material of the highlands.

How can the above observations be interpreted? We can now put together a rough chronology of the Moon's history. The Moon apparently formed about 4.5 billion years ago, about the same time as the Earth. Whether it was hot or cold at that time is unknown. Within a few hundred million years, however, the surface must have melted, fusing together the breccias in the lunar highlands; this was the time of the lunar cataclysm. During this time, radioactive decay in the interior regions was producing heat, which after about a billion years was sufficient to melt the interior. Therefore, the melting of the highlands could have come about either from internal radioactivity whose heating melted rock that then flowed to the surface, or from a short period of intense meteoroid bombardment. (Recall that the Solar System was filled with planetesimals while the solar nebula condensed.) After some cooling, during which the crust formed, the bombardment that formed most of the highland craters took place, about 4.1 billion years ago. Some of the impacts at that time would have involved asteroids, and the primitive maria basins would have been formed. Impacts of that large size would have produced deep cracks in the crust. Then, between 3.1 and 3.9 billion years ago, radioactively melted material found its way to the surface through these cracks, filling in the basins and producing the smooth maria plains. For the last 3 billion years, the Moon has remained cool, quiescent, and geologically dead. This history is summarized in the timeline of **Figure 8-37**.

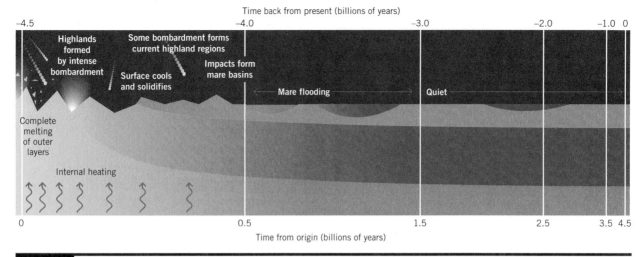

Time back from present (billions of years)

Highlands formed by intense bombardment

Some bombardment forms current highland regions

Surface cools and solidifies

Impacts form mare basins

Mare flooding

Quiet

Complete melting of outer layers

Internal heating

Time from origin (billions of years)

Figure 8-37 An evolutionary timeline of lunar history.

8.9 The Moon's Interior

Much uncertainty remains in our picture of the Moon, partly because the astronauts were not able to find any truly primeval material in the few regions they sampled. All the rocks they brought back to Earth had undergone some changes during the course of lunar history. Our understanding of the Moon's interior is even less clear.

The region of the Moon just below the surface can be investigated by analyzing impact craters and their surrounding areas, because such impacts invert the layers from what they were before impact. Data from the *Clementine* mission (**Figure 8-38**), such as that for the 95-km-diameter Copernicus, shows an example of the type of data available for such a study.

Some information about the interior can be obtained simply from the Moon's density. The average density of rocks gathered by the astronauts was 2.96 g/cm³, only slightly less than the overall average of 3.34 g/cm³. From this low density we conclude that the Moon is lacking in iron relative to the Earth. We can further infer that the lunar interior is not nearly as strongly differentiated as the Earth's interior. In fact, the Moon's relative iron abundance is lower than that of the Sun.

Further information about the interior has been learned from data transmitted back by seismometers left by the astronauts. Although *moonquakes* have been detected, their frequency and strength is far less than that of seismic events on Earth. Artificial moonquakes were caused by crashing the third stages of *Apollo* spacecraft into the Moon.

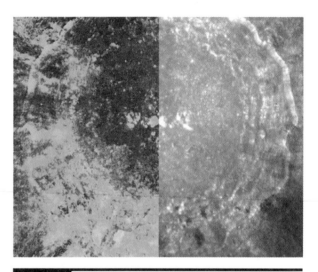

Figure 8-38 A mosaic of Copernicus, which shows mirror-images of the same crater side. The left-hand image was taken through a single filter, while the right hand half is a composite through three filters that provide important geologic information. For example, bright blue tones represent fresh material, mottled red-orange indicates soils, vivid red indicates deposits resulting from melting caused by impact, and green-yellow tones indicate a higher abundance of iron-bearing compounds.

One important characteristic is that they last much longer than quakes on Earth. The seismic waves bounce around the interior for a long time (up to an hour compared with seconds in the Earth). Lunar seismic events fall into two classes. For one type, the occurrence is correlated with the motion of the Moon around the Earth; roughly half these quakes occur when the Earth and Moon are at their closest, while the other half occur when they are farthest apart. Clearly, the Moon's interior is responding to tidal forces from the Earth, just as the Earth's oceans respond to lunar tidal forces. The second class is unrelated to the Earth, and its events originate deeper in the interior. Some 100 sources of deep moonquakes have been identified, and each produces a seismic event on a rough average of once a month. Some may be triggered by meteoroid impact. A deeper understanding of the lunar seismic events and the lunar interior will result from the Japanese *Lunar-A* spacecraft scheduled for launch no earlier than 2006.

As a result of lunar seismology studies, we now know that the Moon has an outer crust about 60 km deep, a lithosphere some 1,000 km thick, and a core. The size of the core is uncertain, and depends on the chemical composition. For example, if the core is pure iron, we think the size is 300 to 400 km in radius, while if it consists of iron sulfide, the size would be 500 to 600 km. The lithosphere is thick because the small size of the Moon allowed it to cool relatively rapidly. Such a thick lithosphere cannot break up into plates. Additional observations show that the crust is thinner on the side facing the Earth than on the lunar far side. The thinness of the crust is consistent with the observed asymmetry of maria on the near side and their lack on the far side. The seismic results show that there *may* be a small region of the core that is at least partially molten; the presence of a molten core has not been unequivocally established (**Figure 8-39**).

Improved knowledge of the lunar interior will come from the *Lunar-A* spacecraft, which will place heat-flow probes into surface penetrators on both the lunar near and far sides and penetrate the surface to between 1 and 2 meters.

> ## Inquiry 8-8
>
> How might the observed thickness of the crust on the lunar far side explain the lack of lava-filled basins there?

Magnetic field measurements show that the Moon does not have an overall magnetic field as the Earth does. If magnetic fields are formed by the dynamo effect discussed for the Earth, the lack of an overall lunar magnetic field is consistent with the lack of a molten interior indicated by the seismic data, and with the Moon's slow rotation.

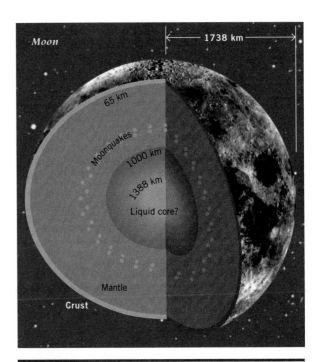

Figure 8-39 The interior structure of the Moon. Distances are from the surface.

However, there are locations on the Moon's surface where magnetic fields exist that are as strong as 1 percent of the Earth's field. These localized *mini-magnetospheres*, discovered by *Lunar Prospector*, are about 100 km across and, while weak, are nevertheless strong enough to divert the solar wind particles. Surprisingly, such regions of enhanced magnetism are found diametrically opposite to the young, large impact basins on the Moon. While hypotheses for this exist, no firm understanding is yet available. On one hand, this surface field might be left over from an earlier time when the Moon did have an overall field—a time when the Moon had a liquid core and rotated more rapidly. On the other hand, perhaps it results from a time in the past when the Moon was closer to the Earth and interacted with its magnetic field, thus magnetizing its own surface.

8.10 The Origin of the Moon

Drawing firm conclusions about the Moon's origin is difficult, because thus far no primeval lunar material has been found. Nevertheless, the question of lunar origin is an important one, and several basic classes of hypotheses have been proposed. Remember, a hypothesis becomes a viable theory when it is consistent with the wealth of data we have.

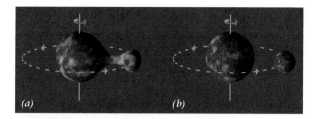

Figure 8-40 A fission hypothesis of the Moon's formation. (a) Material is torn out of the Earth and (b) forms into the Moon.

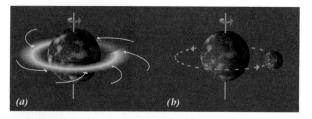

Figure 8-41 Stages in the formation of the Moon by accretion. (a) Earth gathers a ring of material that (b) collects to form the Moon.

One group of hypotheses could be called the **fission hypotheses**, because they propose that the material that was to form the Moon was somehow thrown off the rapidly spinning young Earth (**Figure 8-40**). The astronomer George Darwin, the son of Charles Darwin, proposed the first fission hypothesis in 1879. Evidence in favor of these hypotheses included the fact that the density of the Moon—about 3.3 g/cm^3—is close to the density of the outer layers of the Earth. An additional argument was the observation that the Moon is about the same diameter as the Pacific Ocean, and might have been ripped from there.

There are, however, four good reasons to suspect the fission hypotheses. One is that material thrown from the Earth would probably not have remained in orbit around the Earth but would have fallen back to its surface. The second is that the fission hypothesis would have formed the Moon in the Earth's equatorial plane, not as close to the ecliptic plane as it is today. Although recent calculations show that a transition from one orbital plane to another may be possible, the result is not yet firm. Third, due to continental drift, the present shape and distribution of the Earth's continents is vastly different from what it was at the time of lunar formation. Continental drift clearly indicates that the basin that is now the Pacific Ocean was not created by something being torn away from the Earth.

Inquiry 8-9

There is a fourth reason why the fission hypothesis is weak. How does the anomalous composition of material on the Moon help discredit the fission hypothesis?

A second group of hypotheses explaining the Moon's origin suggests that the Moon was formed somewhere else in the Solar System and was later captured by the Earth. However, because capture requires the incoming body to lose kinetic energy, the assistance of a third planet-sized body is required to carry away the lost energy. Although not impossible, such **capture hypotheses** are highly improbable.

Inquiry 8-10

Would the existence of KREEP material on the Moon tend to support or discredit the capture hypothesis? What about the abundances of the oxygen isotopes?

The third major class of hypotheses is a group of **accretion hypotheses**. According to these, the Moon was formed at about the same time as the Earth out of material that collected in orbit around the Earth. **Figure 8-41** illustrates this process.

The straightforward accretion hypothesis runs into the same difficulty that the fission hypothesis did—explanation of the compositional differences—because if the Moon and Earth were formed at the same time out of the same material, they should have the same composition. However, it is possible that the Earth formed first. Slightly later the Moon may have formed from small chunks of material that came from elsewhere in the Solar System to be captured into orbits around the Earth, where they accreted into the Moon. Such material must have been at a higher temperature than the Earth to explain the Moon's lack of volatile elements. Unlike the improbable capture required by the capture hypothesis, the collection of debris into a ring around the Earth is a likely occurrence. Nevertheless, scientists were discouraged and puzzled for many years when the detailed knowledge of lunar chemical composition determined from rocks returned by the *Apollo* expeditions did not clear up the question of the Moon's origin.

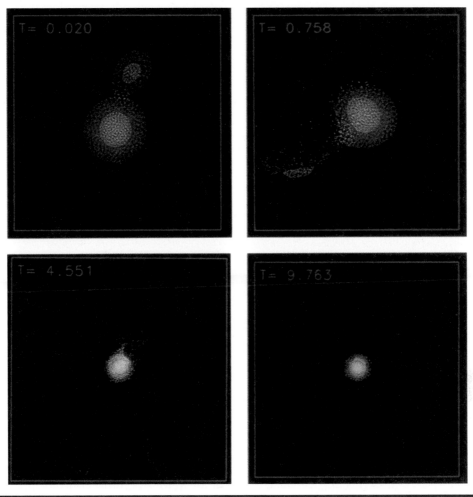

Figure 8-42 A computer simulation of the giant impact model. An object one-tenth Earth's mass strikes at 11 km/sec. The time covered by the simulation is 9.763 hours.

In recent years, a new line of thought has been gaining an increasing amount of acceptance among astronomers; originally proposed by astronomer William Hartmann of the Planetary Science Institute, it has come to be called the **giant impact theory**. It proposes that the Earth suffered a major impact during its earliest stages (more than 4 billion years ago) when the molten planet was still cooling and developing a surface crust. Recall that the early Solar System was probably packed with planetesimals, so that collisions would be more likely at that time. The theory envisions a collision with a body perhaps the size of the planet Mars.

Computer simulations of such a collision (**Figure 8-42**) indicate that the surfaces of the two planets would have been vaporized in the collision, and that material would squirt outward in a high-pressure jet with temperatures of thousands of degrees. The simulations further show that some of this material might be able to reform in Earth orbit,

eventually coalescing into the Moon. Thus, the Moon would be composed partly of Earth material and partly of material from the colliding body.

This theory has been highly successful in explaining the details of the Moon's chemical composition. For example, a serious problem with all previous origin hypotheses has been the distinct deficiency in lunar iron (when compared to the Earth's chemical composition). This "anemia" is easily explained by the giant impact theory, because the lunar material would have come primarily from the silicon-rich/iron-poor crustal portions of the two colliding objects, whereas the Earth's iron supply is primarily locked up deep in its central regions. Another anomaly of lunar composition that has been hard to explain is a shortage of water, sodium, and other volatile materials (relative to the Earth), but these are precisely the types of substances that would boil away in the vaporization process.

Table 8-4   A Comparison of Lunar Formation Theories

FACTOR[a]	FISSION	CAPTURE	ACCRETION	GIANT IMPACT
Lunar mass	D	B	B	B
Earth–Moon angular momentum	F	C	F	B
Lack of volatiles	B	C	C	B
Lack of iron	A	F	D	B[b]
Equality of oxygen isotopes	A	B	A	B
Lava covering entire surface	A	D	C	A
Physical plausibility	F	D2	C	Incomplete

After John Wood, in *Origin of the Moon,* edited by W. K. Hartmann, R. J. Phillips, and G. J. Taylor (Houston: Lunar & Planetary Institute). John Wood supplied the updated grades in the final column.

[a]*The factor in the first column represents the observation that each theory must successfully account for to be considered valid.*
[b]*This grade depends on the assumption that the Earth's iron core had already formed. Eventually the Earth's rotation will slow to the point that it always keeps one face toward the Moon, just as the Moon has been slowed by Earth tides so that it keeps one face toward us.*

This theory may even explain some long-standing puzzles about the Earth's composition. For example, it has been difficult to explain why the Earth has so much gold and platinum in its surface layers. Typically, such heavy elements should have sunk toward the Earth's core. Now it seems possible that these elements came from the material in the body that collided with the Earth.

Table 8-4 summarizes the various classes of lunar formation theories by assigning a grade to each of the factors that workable theories must explain. A grade of A means the factor is well explained by the theory, while a grade of F means the theory fails at explaining the observation. Lunar mass is a factor because the Moon is larger relative to its parent planet than any other moon in the Solar System (with the exception of Pluto's moon). Angular momentum is a factor because the Earth–Moon system appears to have more angular momentum than expected from a comparison with other planetary systems. The importance of the depletion of the volatile elements and iron and the equality of the oxygen isotopes were discussed earlier. A viable theory must account for the overall melting of the lunar surface. Lastly, formation mechanisms must be physically plausible.

Our knowledge of lunar origins should move forward significantly with the *SELENE* project, which stands for SELenological and Engineering Explorer. With a launch planned for 2006 or later, SELENE is billed as the largest lunar mission after *Apollo.* Its mission is to develop technologies for future lunar exploration while obtaining data dealing with lunar origin and evolution.

8.11 Tides and the Future of the Earth–Moon System

As the Earth turns beneath the tidal bulges, it attempts to drag them along with it. This produces a substantial amount of friction, which in turn tends to slow down the rotation of the Earth. As the Earth's rotation slows, the day grows longer and longer. The effect is small—the day only lengthens by about 0.0016 of a second each century—but over the course of time this effect accumulates and becomes important.

The first persons to notice the slowing of the Earth were astronomers trying to compare records of ancient solar eclipses with their predictions of those eclipses. They found substantial disagreements but saw that their predictions could easily be explained if they assumed that the Earth rotated more rapidly in the past than it does today. These results have since been confirmed by counting growth rings in ancient corals several hundred million years old. Corals add one growth ring each day, and the width of the ring depends on such things as water temperature, which, in turn, varies with the season. Ancient corals have been found with more than 400 growth rings per year. This means that there must have been more than 400 days in a year at one time, implying that the Earth was rotating about 10 percent faster on its axis than it does today. In fact, at one time, the Earth may have rotated once every five hours.

If the Earth is rotating on its axis more and more slowly with time, it must be losing angular momentum. But the angular momentum of the

Earth–Moon *system* is conserved. The Moon picks up the Earth's lost angular momentum, with the result that its orbit is slowly increasing in size. The reason the Moon picks up the angular momentum lost by the Earth is that just as the Moon exerts a force on the Earth's tidal bulge, the tidal bulge exerts an *equal and opposite* force on the Moon, which tends to accelerate the Moon forward in its orbit. This makes the Moon move slightly further away from the Earth.

As a result of the slow increase in the size of the Moon's orbit—only a few centimeters per year— there eventually will come a time when the angular size of the Moon will be too small to completely cover the Sun's face. At this time, Earth will lose the spectacular phenomenon of the total solar eclipse. This is a little sad, for there are few things in nature more beautiful and awe-inspiring than a total eclipse of the Sun.

8.12 Understanding the Universe Using the Moon

One of the items left by the astronauts on the Moon has made possible an important series of experiments that have far-reaching results. A **retroreflector** (visible in Figure 8-33) is a device similar to a bicycle reflector that has the ability to reflect the light that falls on it directly back to the source from which it came. If that light comes from a **laser**, this is called **laser-ranging**.

Several observatories around the world have been beaming short pulses of laser light at several retroreflectors on the Moon and then detecting the returning pulse of light (see **Figure 8-43**). The time it takes for the round trip is measured with great precision, and, from the known constant speed of light, the distance between the observing station and the Moon can be calculated. The round-trip time of about 2.5 seconds can be measured to a fraction of a *billionth* of a second, so the uncertainty in measurement of the distance to the Moon is small—only a few centimeters.

What have we learned from this program? For one thing, we now know the orbit of the Moon with far greater accuracy than was possible before. This in itself is of great value, but it may also make possible the measurement of some extremely small natural effects that are of great importance. For example, physicists have always assumed that the gravitational constant, *G,* does not change with

Figure 8-43 A telescope sending out a laser beam to be reflected off the moon.

time, but if it were slowly decreasing as the universe aged, there could be profound effects on the way the universe evolved. The lunar ranging program has shown that any variations in G are far smaller than some hypotheses predict. Already the program has led to an important confirmation of Einstein's theory of general relativity by showing that a certain motion predicted by some theories of gravity, but not by Einstein's theory, is not present in the motion of the Moon.

Finally, it is now possible to use the Moon as a standard reference point for mapping the Earth. By measuring the distance to the Moon from two different points on the Earth's surface, scientists can measure the distances between the two observing stations to a high degree of accuracy. We then can measure precisely the motions of the Earth's crust by measuring changes over time in the locations of observing stations. There are some important applications of such knowledge: for example, earthquake prediction (from measurements taken along both sides of fault lines) and, most exciting, measurement of tectonic plate motion.

Lunar Surface Features, as Seen from Earth

After completing this activity, you will be able to do the following:

- Describe different types of lunar surface features.
- Explain why less detail is seen at full Moon than at other phases.
- Determine the sizes of the largest and smallest dark maria and the largest and smallest craters observable from Earth.
- Offer a hypothesis for the formation of lunar rays.

Readers having a magnifying glass may want to use it to discover details in the prints. Figures 8-19 and 8-20 are photographs of the full Moon and the Moon's first quarter, respectively. Study the photographs carefully, using a magnifier if available, and then do the following Discovery Inquiries.

- **Discovery Inquiry 8-1a** List the different types of features you can see on these photographs—for example, craters, mountains, etc. Don't worry about giving them astronomically correct names; just call them something descriptive of what you see. You may find some kinds of features that don't even have names.

- **Discovery Inquiry 8-1b** Did you find any features in common between the two photographs? How can this be when they show the Moon at different positions in its orbit around the Earth? Draw diagrams of the relative positions of the Earth, Moon, and Sun (a) when the Moon is full and (b) when the Moon is seen at a quarter phase (half illuminated) from the Earth.

- **Discovery Inquiry 8-1c** Examine Figure 8-20 along the **terminator** (the line between the light and dark areas of the Moon). Why do you see more detail in the quarter Moon? Why are craters more easily seen? Use the diagrams you drew in the previous inquiry.

- **Discovery Inquiry 8-1d** Using the photograph in Figure 8-19, how many maria can you count on the side of the Moon facing the Earth? Consider two maria to be separate if each occupies a roughly circular area on the Moon, even if the two regions are connected. (For example, the Mare Serenitatis [Sea of Serenity] and the Mare Tranquilitatis [Sea of Tranquility] count as two distinct maria in Figure 8-19.) If the Moon is 3476 km in diameter, what are the sizes of the largest and smallest of the maria? (*Hint:* The Moon's diameter is about 3,476 km, which corresponds to about 3.6 inches in the picture. Therefore, 1 inch = 695 km and 1 mm = 38 km.)

- **Discovery Inquiry 8-1e** Note the rayed crater Tycho marked in Figure 8-19. The rays extend from it in all directions to great distances from the crater. What could have done this, and why does the material of the rays appear so much lighter than its surroundings?

- **Discovery Inquiry 8-1f** What are the diameters of the largest and smallest craters you can see in Figure 8-19 and 8-20? Would you expect there to be smaller craters on the Moon than you can see in these diagrams?

- **Discovery Inquiry 8-1g** Do the maria appear to be at the same elevation as the surrounding light-colored material? If not, do they appear to be higher or lower?

Chapter Summary

Table 8-5   The Earth and Moon: A Comparison

	EARTH	MOON
Orbital data		
Orbit shape	nearly circular	nearly circular
Orbit inclination to ecliptic (°)	0	5.1
Rotation period	24 hours	27.32 days
Axial tilt (°)	23.5	6.7
Orbital period	365d6h8m24s	27d7h43m12s
Physical data		
Diameter (km; Earth units)	12,756	3,476; 0.273
Mass (kg; Earth units)	5.974×10^{22}	7.35×10^{24}; 0.0123
Average density (g/cm^3)	5.518	3.34
Surface gravity (cm/sec^2; Earth units)	9.8	0.165
Escape speed (km/sec)	11.2	2.4
Temperature (surface)	260 to 310 K	100 to 400 K
Surface features		
Craters	few	many
Impact basins	perhaps one	many
Continental-sized areas	yes	no
Mountains	yes	yes
Ancient lava flows	?	yes
Active volcanism	yes	no
Plate tectonics	yes	no
Surface composition	silicates	silicates; iron deficient; deficient in volatile elements
Other characteristics		
Atmosphere	nitrogen, oxygen	practically none
Magnetic field (global)	yes	no
Differentiated core	yes	no?

- **Table 8-5** presents a general comparison of the properties of the Earth and Moon. Most of the information in this chapter involves observation of well-known phenomena and some inferences from them.

Observations

- The observed average density of the Earth (5.5 g/cm^3), when combined with the density of surface rocks (3.3 g/cm^3) leads to the conclusion that the Earth's core contains dense materials such as iron. The surface chemical composition is primarily silicon and oxygen, with various amounts of other elements grouped together into a variety of minerals. The surface of the Earth is constantly changing due to erosion by water, wind, glaciers, and continental drift.

- The three types of rocks are **igneous**, **sedimentar**, and **metamorphic**. Igneous rocks result from cooling of molten material; sedimentary result from the deposition of material by wind and water; metamorphic rocks result when the effects of temperature and pressure change the characteristics of a pre-existing rock.

- The difference in the strength of the Moon's gravity on different sides of the Earth produces ocean tides. These tides are causing the Earth's rotation to slow and the day to lengthen. Similarly, the Earth produces tides on the Moon, whose distance is slowly increasing due to conservation of angular momentum.

- The Earth's atmosphere contains nitrogen and oxygen with traces of argon, carbon dioxide, and water vapor. The carbon dioxide and water vapor cause the greenhouse effect to be present. The atmosphere comes from outgassing during volcanic emissions. Atmospheric circulation is governed by a combination of heat from the oceans and continents, and movement caused by the Coriolis effect.

- The Earth has a global magnetic field that is thought to be formed by rotation and convection in the molten outer core in a process known as the **dynamo effect**. For this reason, the Earth is surrounded by a **magnetosphere**. Charged particles from the Sun that strike the magnetosphere have their paths changed so that they spiral around the magnetic field into the polar regions, where they produce **aurorae**.

- **Occultations** of stars by the Moon show it to lack a substantial atmosphere. Spacecraft have found a highly tenuous one.

- Surface features on the Moon include **craters**, **rays**, **maria**, **rilles**, **domes**, and **highlands**. Rays result from material ejected during crater formation from impact. Some rilles result from surface movement, while others, the **sinuous rilles**, result from lava flowing in channels and tubes. Domes result from volcanism.

- The lunar surface is entirely covered with **basalts** in the form of rocks and a powdery **regolith**. In addition, the rocks contain **breccias** and **KREEP**. Lunar rocks are deficient in water molecules and other volatile elements, and also deficient in iron relative to the Earth. However, the relative amounts of various oxygen **isotopes** are the same on the Moon as on the Earth.

- Locations that never receive sunlight may contain significant amounts of water.

Theory

- **Fission hypotheses** for the Moon's formation are physically implausible; they predict that the Moon would be in the Earth's equatorial plane, and that the Moon should be chemically identical to the Earth. **Capture hypotheses** do not require the Earth and Moon to be chemically identical. Although not impossible, capture is unlikely. **Accretion hypotheses** predict strong similarities between the Earth and Moon. The **giant impact theory** of lunar formation involves the impact of a Mars-sized body on the forming Earth; material ejected into orbit later coalesced into the Moon. The theory predicts that the Moon would have formed from material that is poor in both volatile elements and iron.

Conclusions

- From seismic studies, we infer that the Earth has a solid inner core that is surrounded by a liquid outer core. The lower-density mantle exhibits convection in which hot, low-density material rises and cools, and higher-density material descends.

- From a variety of data scientists conclude that Earth's continents were once together in a supercontinent we call **Pangaea**. Thereafter, the land masses separated due to **continental drift**, in which lower-density continental plates float on the higher-density mantle underneath. In this model, mountains form when plates collide.

- Most lunar craters result from meteoroid impact, although some are volcanic in origin. Maria result from lava flows.

- Younger features are on top of older ones. Absolute ages can be obtained using the methods of **radioactive dating**, which determine when material last solidified. From such techniques we deduce that the lunar highlands were produced 4.1 billion years ago, and the maria 3.1 to 3.9 billion years ago.

- The Moon's interior contains a core that may be partially melted. The lack of a liquid core along with slow rotation accounts for the Moon's lack of an overall magnetic field. However, the entire lunar surface was molten at a time early in its history.

Summary Questions

1. What is the Earth's distance from the Sun? The Earth's diameter? Its average density?

2. What is the Moon's distance from the Earth? The Moon's approximate diameter relative to the Earth? Its average density?

3. What are the meaning and importance of the concepts of the acceleration of gravity and the escape velocity?

4. How do seismic data give us information about the interiors of the Earth and Moon? What were the results of the lunar seismology experiments?

5. What are the various layers of the Earth's interior and their important properties?

6. What is continental drift, and what produces it? How is volcanism related to it?

7. What chemical elements make up the Earth's atmosphere? What is the percentage of each?

8. What is the greenhouse effect, and how is it produced?

9. Where did the Earth's current atmosphere come from? Why do scientists think it is not the same as the original atmosphere?

10. What is the Earth's magnetosphere? Describe its appearance. What effect does the magnetosphere have on charged particles in its vicinity, and what is its role in producing an aurora?

11. How did astronomers know the Moon lacked an atmosphere before NASA sent spacecraft there?

12. What are the characteristics of the various types of lunar surface features? Describe how these features formed and present evidence in favor of the formation mechanism. How does the appearance of the lunar near side differ from that of the far side?

13. What basic types of rock are found on the Moon? What are the implications of these rock types for the origin and history of the Moon?

14. What is the probable scenario for the early history of the Moon's surface as determined from the evidence provided by radioactive dating?

15. What are the four main ideas of the Moon's origin? Describe them, and give the evidence for and against each hypothesis. Which one is most likely to be valid? Explain why.

16. Why does the Moon produce two tidal bulges on the Earth? What are the long-term effects of tides on the Earth–Moon system?

Applying Your Knowledge

1. Compare and contrast the processes that modify the surfaces of the Earth and the Moon. Consider separately processes that occur internally as well as those involving externally produced events.

2. Apply the idea that theories cannot be proven true, but only falsified, to the hypotheses of lunar formation discussed in the chapter. What is your conclusion?

3. Because the Earth's atmosphere consists of 78 percent nitrogen (N_2), why does the atmosphere not have substantial amounts of ammonia (NH_3), as does Jupiter?

4. Consider each rock type. If specimens of each type were found on a given planetary body, what would you be able to conclude about conditions and processes on and within that body?

5. How might the Earth have differed if it had been 0.7 AU from the Sun rather than where it is now?

6. The Earth's tidal force acting on the Moon has slowed the Moon's rotation to the point where it is synchronous (the same face always faces the Earth). Given that the Moon also exerts a tidal force on the Earth, why is the Earth not in synchronous rotation?

7. Describe a baseball game being played on the Moon. Include discussions of the distance the ball might be hit, the strides of the base runners, the ease or difficulty of fielding the ball, the likelihood of a rain check or calling the game due to darkness, and how the fans might boo the umpire.

▶ 8. Suppose you measured g to be 10 m/sec^2. What, then, is the mass of the Earth?

▶ 9. Compute the escape velocities of the Earth and Moon. How much larger is the Earth's escape velocity? (Use data for the Moon from Table 8-3.)

▶ 10. Because the uncertainty in the time for a laser beam to reach the Moon and back is about a billionth of a second, compute the uncertainty in the distance by finding how far light travels in a billionth of a second.

▶ 11. Using Figure 8-19, determine the diameter of Mare Serenitatis, the Sea of Serenity.

▶ 12. What would be the period of a satellite in orbit about the Earth if the satellite were 385,000 km from it?

▶ 13. Assuming the South American and African plates have always moved with the same speed they are moving today, compute how long ago the continental land masses were joined. (Make a reasonable estimate of their current separation.)

▶ 14. Explain how you might go about computing the height of the central peak shown in Figure 8-21. What information would you need to have?

Answers to Inquiries

8-1. The interior density must be much greater than 5.5 g/cm^3.

8-2. Sediments, which are carried by water, wind, and ice, are located near the surface and will be deposited there, not below the surface. Furthermore, the churning regions below the surface would destroy the soft sediments.

8-3. The pressures are higher in the central regions than they are farther out. According to Figure 8-8, the rock would therefore be solid. However, away from the center, even if the temperature were about as high, the lower pressure would cause it to be a liquid. (We are ignoring any effects of differences in chemical composition.)

8-4. Two high tides and two low tides, because there is a bulge on either side of the Earth and a point on the Earth's surface will pass beneath each one during any 24-hour period.

8-5. It will change due to the Moon's daily eastward movement in its orbit relative to the Sun. As discussed in Chapter 4, the Moon moves eastward some 13° per day;

this corresponds to a delay in the rise time of about 53 minutes each night. The tides will be about an hour later every day.

8-6. Some questions you might think of include: What is the composition of the lunar rocks? Their ages? Are they volcanic in origin? Is there evidence of water either now or in the past? Does the Moon have a magnetic field like the Earth? Does it have seismic activity?

8-7. They must have formed from the same primordial material. Had they been different, the conclusion would be the opposite.

8-8. A thick crust, like that on the lunar far side, is harder to crack than a thin crust, like that on the near side. Once cracked, lava is then free to flow and fill the large basins on the side facing the Earth.

8-9. If the Moon were made of terrestrial material, it would have the same composition as the Earth.

8-10. While the presence of KREEP could support the capture hypothesis, the equality of the various oxygen isotopes would not.

The Earth-like Planets 9

This planet [the Earth] is the cradle of the human mind, but one cannot spend all one's life in a cradle.

Konstantin Tsiolkovsky, a Russian rocket and space pioneer

A time would come when Men should be able to stretch out their Eyes . . . they should see the Planets like our Earth.

Christopher Wren, Inauguration Speech, Gresham College, 1657

The four planets closest to the Sun exhibit a variety of similarities, including their small size, small mass, and high density compared to the outer planets. They also have a similar chemical composition. Due to these likenesses, astronomers group Mercury, Venus, Earth, and Mars together as the **terrestrial,** or Earthlike, planets. In this chapter, we study the terrestrial planets from the viewpoint of comparative planetology, contrasting them with each other.[1] We use Earth and the Moon as the standards of comparison, simply because we know more about them and the processes that determined the way they are.

Although our discussion emphasizes comparisons among the terrestrial planets, you should also keep in mind that each planet is a separate and unique world. We want you to know each of these worlds individually and understand why they are the way they are, because doing so will give you a new appreciation and a new view of the cosmos.

9.1 Introduction to the Terrestrial Planets

Our study of the terrestrial planets relies on the understanding we have gained of the processes occurring on Earth and the Moon and the observable consequences of these processes. We begin with our knowledge of these bodies gleaned from ground-based observations and the early days of planetary exploration by spacecraft.

---

[1]For detailed information on each of the planets, refer to Table 9-1 and Appendix C.

Chapter opening photo: The rover *Opportunity* studies Martian rocks in the crater Endurance.

EARLY GROUND-BASED STUDIES

The Earthlike planets, Mercury, Venus, and Mars, are shown side by side in **Figure 9-1**. Each presents astronomers with its own impediments that make Earth-based study difficult. **Mercury**, for example, is small and never strays far from the Sun. Telescopic observations of it found only indistinct and hazy surface markings. By following their apparent motion, astronomers concluded (incorrectly, as we will see) that **Mercury** rotated on its axis once every 88 days. Because this rotation period was the same as its orbital period around the Sun, astronomers inferred (again incorrectly) that Mercury always kept the same face toward the Sun, just as the Moon keeps the same face toward Earth. The reason in both cases is the same—strong tidal forces. Figure 9-1*a* is about as good an Earth-based photograph of Mercury as we can get.

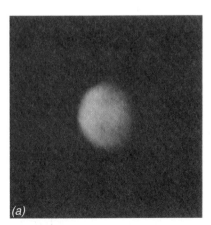 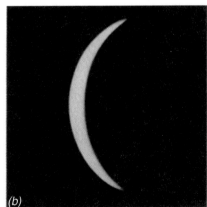 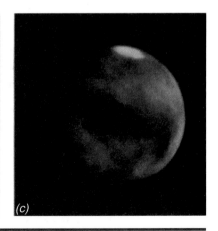

Figure 9-1 Earth-based telescopic views of *(a)* Mercury, *(b)* Venus, and *(c)* Mars.

Although **Venus** had been observed by Galileo to have phases, no further details could be seen. Like Mercury, Venus never appeared far from the Sun. However, some indistinct markings observed by numerous observers provided information about an apparent 225-day rotation period. Here, too, the rotation was thought (incorrectly) to be the same as the period of orbital revolution. The visual appearance of Venus is indistinct to us because, as we now know, a perpetual thick layer of clouds floating in its CO_2 atmosphere covers the planet (Figure 9-1b). For this reason, the observed rotation period referred to the observed clouds, not the planetary surface itself. Venus's varying phases, which are similar to those of the Moon, are shown in **Figure 9-2**.

At one time **Venus** was thought to be a benign place for life to exist, but Earth-based studies in the 1960s began to paint a rather unpleasant picture of it, which included a thick carbon dioxide atmosphere with a surface pressure almost 100 times that of Earth, a surface hot enough to melt lead, and howling winds. Studies revealed a minute amount of water vapor, suggesting a dry, desert-like terrain—a Death Valley carried to the extreme. We now know that the clouds on Venus consist of high concentrations of corrosive compounds such as sulfuric and hydrofluoric acid. The cloud systems move around Venus in four days, indicating that typical wind speeds in the upper atmosphere are greater than 300 km per hour. These speeds are comparable to the highest speeds within the jet stream on Earth. A space probe would have a hard time surviving these extreme conditions.

Mars has the liveliest history of study and debate of all the planets. Although Venus is closer to Earth in mass and distance from the Sun, Mars has provided a greater fascination to the human mind ever since telescopes were pointed at it. When flights of fancy considered the subject of life elsewhere in the universe, the place that usually came to mind was Mars. Telescopic observations of the Martian surface began with Galileo. Observations ever since have provided data from which many speculations arose. As with all the terrestrial planets, the quality of data about Mars was limited by the planet's small, angular size and by turbulence within the Earth's atmosphere. Telescopic observations did show both daily changes due to Mars's rotation and seasonal changes in the colors of certain regions (**Figure 9-3**).

FIVE PHASES OF VENUS

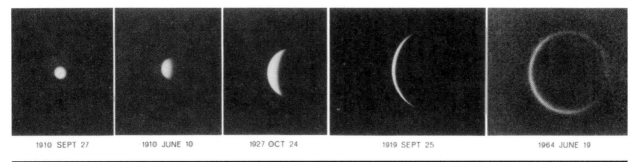

1910 SEPT 27   1910 JUNE 10   1927 OCT 24   1919 SEPT 25   1964 JUNE 19

Figure 9-2   Venus, as seen from Earth, shows only a featureless layer of clouds.

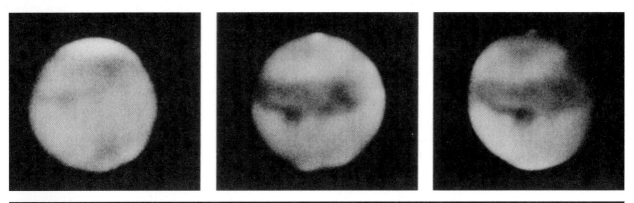

Figure 9-3   Several views of Mars as seen from Earth. The photographs are taken during different Martian seasons.

Figure 9-3 shows some of the better older photographs of **Mars** taken from Earth. (Modern-day electronic detectors allow better-quality images to be produced.) From such photographs we learned that the length of a day on Mars, and the tilt of its rotation axis, are close to those of Earth. The axial tilt produces seasons that last twice as long as earthly seasons, owing to the greater length of the Martian year.

Inquiry 9-1

Can you detect evidence for seasonal changes in the photographs of Figure 9-3? What do you see?

One of the earliest explanations of the observed seasonal changes in **Mars's** coloration from light to dark was that they might be caused by the presence of vegetation. This explanation also led to a more speculative school of thought that quickly caught the public's fancy—the notion that Mars harbored intelligent life. This line of thinking had its origin in nineteenth-century observations by the Italian astronomer Giovanni Schiaparelli, who saw the Martian surface as being crisscrossed by an extensive network of straight lines that he called *canali.* Thus began a controversy that lasted until the era of space exploration. The U.S. astronomer Percival Lowell observed similar features on Mars about 1896 and speculated freely about their origin. Although the Italian word *canali* is somewhat noncommittal and can be employed to describe almost any sort of channels or ditches, Lowell interpreted the markings quite literally as *canals;* that is, artificial ditches constructed by intelligent creatures to bring water from the melting polar caps to the presumably populated temperate regions of Mars. But while Lowell continued to see his "canals," other observers did not, and their reality was debated for more than half a century.

Mars's canals were never photographed, which should be no surprise, because during the time required to make a time exposure Earth's turbulent atmosphere erases fine details. The eye, by contrast, is a highly sensitive instrument and can see details that are visible only in those rare and fleeting moments when the atmosphere is exceptionally calm. A large number of short-exposure images can be taken by modern electronic devices and combined in a computer to mimic the eye's abilities. Such electronic observations have pro-

duced images not dissimilar to some features in the drawings of Lowell, but we now understand them as an effect of differences in contrast rather than real physical objects.

The hypothesis of intelligent life on **Mars** received a setback when astronomers discovered that its atmosphere was predominantly carbon dioxide, as on Venus. Furthermore, Mars's atmospheric pressure is very low, about one hundred times less than that of the Earth. Observations also revealed only the barest traces of water vapor in the planet's atmosphere, and this fact argued against the existence of plant life on Mars.

PLANETARY SPACE EXPLORATION

A variety of spacecraft have provided us with ever more exciting information about our Solar System companions. Although the entire history is fascinating, we will mention only those spacecraft whose specific information will be discussed in this book.

In 1974, the *Mariner 10* space probe[2] arrived in the vicinity of **Mercury** after receiving a boost on its way as it passed close to Venus (**Figure 9-4**). Using one planet's rotational energy to send a probe farther into space than its own rocket could propel it is a commonly used technique, unfortunately misnamed *gravitational assist.* An additional bonus in the case of *Mariner 10* was the fact

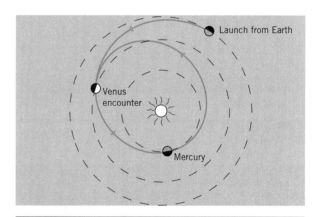

Figure 9-4 The *Mariner 10* spacecraft used the orbital motion of Venus to change the spacecraft's orbit and send it on to Mercury.

---

[2]Because not all the spacecraft that have been sent to the planets will be discussed, they will not appear in an obvious numerical order. For example, while *Mariner 2* was the first probe to Venus and *Mariner 4* was the first to Mars, their results have been superseded by more recent probes.

that the spacecraft was given an orbital period of 176 days, just twice Mercury's orbital period. As a consequence, we were able to obtain images[3] when the vehicle passed close to the planet on three separate occasions in 1974 and 1975. Most of our knowledge of the surface of Mercury, which will be described in detail later in this chapter, came from the *Mariner 10* flybys. No other spacecraft have yet returned to Mercury (although one is currently on its way, as described later).

In 1990, the *Magellan* spacecraft arrived at **Venus**. The primary goal of this satellite was continuous high-resolution radar imaging of the entire planetary surface. After its arrival at Venus, Magellan mapped almost the entire surface with an improved resolution. For the first time, we are able to "see" what the surface really looks like on a global scale. The only landers to study Venus's surface were the *Venera* spacecraft from the former Soviet Union. Results from them will be described later in Section 9.4.

Hoping to resolve some of the perennial questions about **Mars**, the United States launched a number of successful spacecraft visits. The history of Mars studies from space is full of problems that have caused many missions to fail. The successes, however, have been spectacular! The 1971 *Mariner 9* spacecraft took a series of more than 7,000 magnificent images of Mars. Many were of regions that had not been examined before, and *Mariner* revealed unexpected volcanoes, valleys, craters, and channels that considerably revived interest in the planet, while dramatically increasing our knowledge.

The objective of the 1976 Viking mission to **Mars** was to land a probe on the surface of the planet, examine the surface geology from close up, and carry out experiments that might detect the presence of life forms in the Martian soil. After some excitement about a possible detection of life processes, biologists upon further analysis concluded that none had been found. This is discussed in more detail in Chapter 24.

A new era of **Mars** exploration began in 1997 when the Pathfinder mission, wrapped in airbags, bounced to a successful landing. *Pathfinder* carried a small rover, called *Sojourner,* to provide the first mobile spacecraft on a planetary surface. *Sojourner* had the capability of moving from rock to rock and performing a chemical analysis. Other *Pathfinder* instruments were able to study Martian rotational and orbital dynamics, weather, and magnetism. The Mars Global Surveyor also arrived in 1997 and began its work of mapping the surface with higher resolution than ever before (in some cases as fine as 4 meters).

At the time of this writing (spring 2005) Mars is home to two landers (*Spirit* and *Opportunity*) and three orbiters.

Spacecraft views of each of the terrestrial planets are shown in **Figure 9-5**.

9.2 Large-Scale Characteristics

We now look at planetary properties that apply to each body as a whole, including orbital characteristics, rotation, size, mass, density, and seasons.

Orbits

	MERCURY	VENUS	EARTH	MARS
Orbit eccentricity	0.206	0.007	0.017	0.093
Orbit inclination	7°	3°	0°	2°

The elliptical orbits of the terrestrial planets are nearly circular. **Mercury's** orbit differs from circularity the most, and **Venus's** orbit is the most nearly circular of all. Earth's orbital plane, by definition, is the ecliptic plane; it defines what we view as the fundamental geometric plane of the Solar System. The 7° tilt of Mercury's orbital plane to the ecliptic is larger than that of any planet but Pluto (review Figure 6-4). All these planets orbit the Sun in a counterclockwise direction, as observed from above Earth's North Pole.

An important and interesting property of **Mercury's** orbit is that it *precesses*. This means that the major axis of Mercury's orbit slowly rotates over time by 43 seconds of arc per century more than can be accounted for by Newtonian gravity. While admittedly a small difference, this is of great historical importance, because accounting for these 43 seconds of arc per century by Einstein's theory of general relativity was one of relativity's major triumphs. Sometimes small things make a big difference!

---

[3]We use the term *image* rather than *photograph* due to the process used in obtaining it. Because the photographic process is not used here, the term *photograph* would be erroneous.

(a)

(b)

(c)

(d)

Figure 9-5 The terrestrial planets as seen by spacecraft. *(a)* Mercury, *(b)* Venus, *(c)* Earth *(d)* Mars.

Rotation

	MERCURY	VENUS	EARTH	MARS
Rotation period (sidereal)	58.65 days	243 days retrograde	23.9 hours	24.6 hours
Tilt of spin axis to the perpendicular of the orbital plane	0°	3°	23.5°	25°

In 1965 astronomers used Earth-based telescopes (backward!) to send out a radio signal to bounce off **Mercury**. (When used like this, the radio tele-scope becomes a radar telescope—it is both sending and receiving.) In analyzing the structure of the returning radar signals, they found to their surprise that the rotation period of Mercury was only 59 days rather than the 88 days that had been thought. Mercury did *not* present the same face toward the Sun at all times, as had been thought!

This interesting result provides a good lesson for those who argue that the scientific method leads inexorably in a straight line to the truth. In fact, the path is often twisted, with numerous dead ends. Here was a result that had been established by both theoretical and observational evidence and had been accepted for decades, yet it was suddenly found to be completely wrong. There had been one clue that something was wrong with the hypothesis of an 88-day rotation period, because the temperature of Mercury's dark side was surprisingly high. While suggestions were made to explain the anomalously warm dark side of Mercury, none were particularly successful.

Inquiry 9-2

How would a rotation period of 59 rather than 88 days provide an explanation for the dark side being warm?

How could observers have been so wrong about **Mercury**? On Earth, we would see pretty much the same apparent movement of surface features, whether Mercury has a 59-day or an 88-day spin. This is so because we see the planet only when it is far from the Sun, at the extremities of its angular distance from the Sun. The ratio of the orbital period to the rotational period, 88/59, which is almost exactly 3/2, would give us the same view of the surface with either rotation period. Because we were unable to observe Mercury continuously, we had a *stroboscopic* effect in our data, analogous to the effect that causes the wagon wheels in the western movies to appear to rotate in a backward direction. The 3/2 ratio of orbital to spin period results in the solar day on Mercury (the time for the Sun to go from overhead to overhead) lasting two Mercurian years! This situation is shown in **Figure 9-6**, which shows Mercury starting at perihelion (closest to the Sun) with a feature (indicated by the arrow) pointing toward the Sun at point 1. After one rotation of the planet relative to the Sun, its orbital position is at point 2, and so on.

Why does **Mercury** rotate three times for every two orbits? Just as gravitational tidal effects with the Moon forced its rotation to become equal to its revolution, Mercury has gravitational tidal effects with the Sun that produce the 3/2 rotation to revolution relationship. This is a *resonance*, as described in Chapter 7.

Inquiry 9-3

Use Figure 9-6 to determine how many full orbits Mercury completes during three rotations starting from point 1 through point 4. To which point in its orbit has it returned?

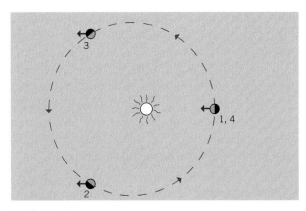

Figure 9-6 The rotation and revolution of Mercury. The arrow, which represents a surface feature, points toward the Sun when Mercury is at its closest approach to the Sun, at point 1. One rotation later, Mercury is at point 2 in its orbit, ⅔ of the way around its orbit; one further rotation later, it is point 3 in its orbit. After 1 more rotation, completing three Mercurian rotations relative to the stars, Mercury returns to its original position in its orbit.

To study **Venus's** surface rotation, astronomers had to find some way to penetrate the thick clouds. This was first done in the 1960s when scientists started bouncing radar beams from Earth-based telescopes off the planet. From such radar studies, astronomers made two surprising discoveries about Venus: It rotates opposite (clockwise, or **retrograde**) in relation to the other terrestrial planets, and it rotates at the remarkably slow rate of once every 243 days. This slow rotation means that solar heat is not evenly distributed around Venus's atmosphere, and thus the side of the atmosphere facing the Sun might be expected to be hot. However, unequal heating is expected to cause high winds (as explained in Chapter 8) that would transport heat around the planet and thereby moderate the temperature distribution. The end result is that while the temperature is highest directly under the Sun, it does not vary greatly around the planet. Two suggestions as to the cause of the retrograde rotation include a collision early in the planet's history, or tidal forces by the Sun on Venus's thick atmosphere and friction between the atmosphere and the surface.

Observations of the surface of **Mars** show it to have an Earthlike rotation of just over 24 hours. Thus, Mercury and Venus both rotate considerably more slowly than Earth and Mars.

With the 20-year time period between the Viking and Pathfinder missions, *Pathfinder* was able for the first time to detect the effects of precession in

Mars's rotational axis (if needed, refer to Chapters 4 and 5 to review precession). The period is a long (and slow) 171,000 years. To complicate matters even more, a variety of observations indicate that the position of Mars's poles has changed considerably over time. In fact, some planetary scientists think Mars's rotation axis may once have been where its equator now is! Thus, Mars is dynamically more complex than anyone had thought.

Basic Properties: Size, Mass, Density

	MERCURY	VENUS	EARTH	MARS
Diameter (Earth units)	0.381	0.951	1	0.531
Mass (Earth units)	0.0558	0.815	1	0.107
Average density (g/cm^3)	5.42	5.25	5.518	3.94
Uncompressed density (g/cm^3)	5.3	4.0	4.1	3.7

Earth is the largest of the terrestrial planets, with Venus a close second. Their diameters are given in **Table 9-1**, which is a general summary table for the terrestrial planets. (These and additional data on the planets are also given in Appendix C.)

The masses of planets having one or more natural moons can be found using Newton's version of Kepler's third law. For example, Mars has two moons, which have been studied to find a mass for Mars only 11 percent that of Earth. Although neither Mercury nor Venus have moons, their masses have been measured by observing the effects of their gravity on passing or orbiting spacecraft. Mercury's mass was found to be only about 6 percent that of Earth, while Venus's was 82 percent.

When we combine planetary masses with their diameters, we can determine average densities. For comparison, remember that Earth's density is 5.5 g/cm^3, while the Moon's is 3.3 g/cm^3. All the terrestrial planets have nearly the same density. Mercury's is almost exactly the same as Earth's, with Venus's slightly lower at 5.3 g/cm^3. Mars has the lowest density at just under 4 g/cm^3. As we shall soon see, these basic numbers provide important information about the interior structure of each planet.

More important than the average density, though, is something called the **uncompressed density**. To see the difference between the average (or compressed) and the uncompressed density,

consider a small rock that is sitting on a table top; its density would be what we call the uncompressed density. If you took that same rock and pushed hard on it (as happens to material far inside the Earth), that pressure would tend to compress the volume, and thus make its density higher. Or, if you heated the rock, it would tend to expand and thus lower its density. Each planet has a different temperature and internal pressure, so to compare the average density is misleading. To determine the uncompressed density from the average density requires a model of the pressure and temperature inside the body. Thus, in comparing the terrestrial planets, we see that Mercury's uncompressed density is significantly higher than that of the other planets, which is telling us something important (but not yet fully understood) about the composition and structure of Mercury.

Once the mass and radius are known, we can also determine the strength of gravity at the surface, and the escape velocity. For Earth and Venus, this is about 10 km/sec, and only half that much for Mercury and Mars. The value for each of the terrestrial planets is given in Table 9-1.

Seasons

	MERCURY	VENUS	EARTH	MARS
Seasons	no	no	yes	yes

Seasons, as we know, are produced by the variation in the concentration of solar heat throughout the year. The primary agent of change in the amount of solar heat is the angle at which the solar energy strikes the planet's surface. For **Earth**, this angle comes from the 23.5° tilt of Earth's axis to the perpendicular of its orbital plane. By contrast, **Venus** has only a 3° tilt (and **Mercury** has 0°), and therefore, they have essentially no seasonal variations.

Because **Mars's** axial tilt is only slightly greater than Earth's, seasons similar to Earth's are expected. However, due to Mars's larger orbital eccentricity, the temperature variations in Mars's southern hemisphere are greater than those in its northern hemisphere or those on Earth. As shown in an exaggerated form in **Figure 9-7**, because Mars is closest to the Sun during summer in its southern hemisphere and farthest from the Sun during summer in its northern hemisphere, the

Table 9-1   The Terrestrial Planets: A Comparison

	MERCURY	VENUS	EARTH	MARS
Orbital data				
Semi-major axis (AU)	0.387	0.723	1.000	1.524
Orbital period (years)	0.241	0.615	1.0	1.881
Orbit shape (eccentricity)	0.206	0.007	0.017	0.093
Orbit inclination (°)	7	3	0	2
Rotation period	59 days	243 days retrograde	23.9 hours	24.6 hours
Axial tilt (°)	0	3	23.5	25
Physical data				
Diameter (Earth units)	0.381	0.951	1	0.531
Mass (Earth units)	0.0558	0.815	1	0.107
Average density (g/cm³)	5.42	5.25	5.518	3.94
Uncompressed density (g/cm³)	5.3	4.0	4.1	3.7
Surface gravity (Earth units)	0.38	0.90	1.00	0.38
Surface pressure (Earth units)	—	~90	1.0	;0.01
Escape velocity (km/s)	4.3	10.3	11.2	5.0
Temperature (surface—equator) (K)	100 to 700	730	260 to 310	150 to 310
Moons	0	0	1	2
Surface features				
Craters	many	few	few	many
Impact basins	yes	?	yes	yes
Continental-sized areas	no	yes	yes	Tharsis?
Mountains	no	yes	yes	yes
Ancient lava flows	yes	yes	?	yes
Active volcanism	no	perhaps	yes	no
High scarps	yes	no	no	no
Plate tectonics	no	probably not	yes	no
Surface winds	no	slight	yes	strong
Polar ice caps	perhaps (H_2O)	no	yes (H_2O)	yes (H_2O, CO_2)
Surface water	no	no	much	no
Surface composition	silicates?	silicates?	silicates	silicates, iron oxide
Other characteristics				
Atmosphere	very slight	carbon dioxide	nitrogen, oxygen	carbon dioxide
Atmospheric water	no	very little	2% (variable)	very little
Atmospheric oxygen	no	no	≈21%	trace
Atmospheric carbon dioxide	no	≈96%	trace	≈95%
Atmospheric nitrogen	no	≈3%	≈78%	≈3%
Clouds	no	sulfuric acid	water	water, dust
Magnetic field	yes	no	yes	no
Differentiated core	yes	unknown	yes	no

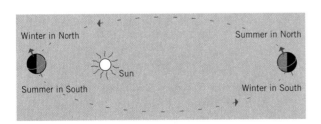

Figure 9-7 Seasons on Mars are particularly extreme for winter in the southern hemisphere due to Mars's greater distance from the Sun. (The ellipticity is exaggerated and not to scale.)

summers are substantially warmer in the southern hemisphere than in the northern. Likewise, Mars is farthest from the Sun in its southern hemisphere winter, producing colder winters there than in the northern hemisphere. A variety of observations indicate that Mars's tilt has been as much as 60° in the distant past. Such large variations produce large seasonal variations that could play a significant role in the planet's history.

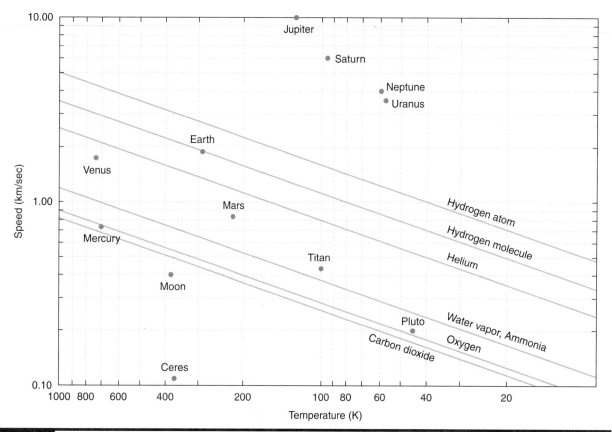

9.3 Atmospheres

The most important properties of planetary atmospheres include their chemical compositions, temperatures, and pressures. Other aspects that we examine include the presence of clouds and lightning.

The atoms and molecules in a planet's atmosphere are in continual motion. The greater the atmospheric temperature, the faster the particles move. At any given temperature, a small number of particles can have a low or a high speed, but the majority will be moving with some intermediate speed. The speed of an atom or molecule will also depend on its mass; the greater the mass, the slower the speed. Particles having a speed greater than about one-sixth the escape velocity of the planet, which is determined by the planet's mass and radius, will escape from the atmosphere and be lost to the planet. Thus, the composition of a planet's atmosphere can change over time due to the loss of atoms and molecules, particularly low-mass ones.

These concepts are brought together in **Figure 9-8**. For example, at Mercury's temperature of ~700 K, an average hydrogen atom is moving with a speed of well in excess of one-sixth Mercury's escape velocity. Thus, over time, hydrogen will be lost to Mercury. We therefore expect it to have no hydrogen, which is what we observe. On Jupiter, by contrast, its 125K temperature produces speeds that are so low in comparison to one-sixth of its escape velocity that hydrogen will be retained. Thus, application of these concepts allows you to have some understanding of why planetary atmospheres are the way they are.

For Earth, the expectation is that neither hydrogen nor helium would be present in significant amounts, which is the case. Most of the helium remaining on Earth is a byproduct of radioactive decay, and we separate it from natural gas deposits.

However, some atmosphere might be gained through **outgassing** by the planet's interior, which

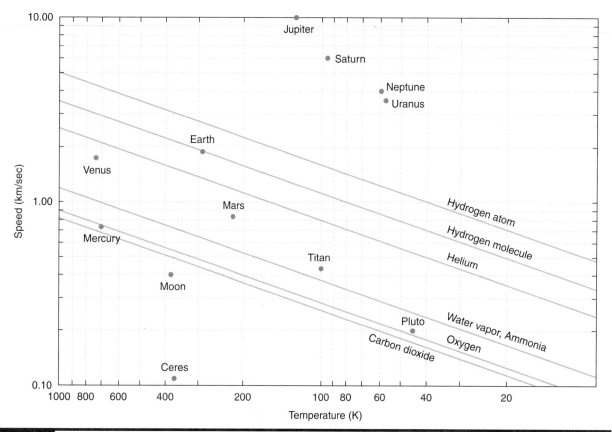

Figure 9-8 A comparison of the speeds that various atoms and molecules have at varying temperatures along with one-sixth the escape speed from some Solar System bodies. Gases below the point representing the planet are held by the planet's gravity.

is the release of material from the inner, hotter regions of a planet. Life processes, if present, would also help to modify the original planetary atmosphere.

Atmospheric Chemical Composition

	MERCURY	VENUS	EARTH	MARS
Nitrogen	no	≈3%	≈78%	≈3%
Oxygen	no	no	≈21%	trace
Carbon dioxide	no	≈96%	trace	≈95%
Water	no	very little	<2% variable	very little

The *Mariner 10* encounter with **Mercury** confirmed ground-based observations that the planet has almost no atmosphere. However, the spacecraft did find a thin atmosphere of surprising complexity. At its surface, Mercury's atmosphere has a density of only about 100 hydrogen atoms and 6,000 helium atoms per cubic centimeter. Ground-based observations have shown that the largest constituent of Mercury's atmosphere is sodium, with a density of 30,000 atoms per cubic centimeter—some 10^{15} times less dense than Earth's sea-level density. The hydrogen and helium undoubtedly come from atoms ejected by the Sun; the other elements are probably released from rocks during the heat of the Mercurian day. This is not surprising, as helium is four times more massive than hydrogen, so it moves more slowly given the same temperature. Sodium is 23 times more massive than hydrogen, so it moves even more slowly.

Venus and **Mars** have similar atmospheric compositions, which have been known for many years to be about 95 percent carbon dioxide. Nitrogen contributes only about 3 percent of the gas in these atmospheres, while it is 78 percent in Earth's atmosphere. Oxygen is not present in Venus's atmosphere, and only a small amount is found on Mars. Water vapor is present only in trace amounts in the Venusian and Martian atmospheres, but it amounts to about 2 percent in Earth's atmosphere. Sulfur dioxide, SO_2, is an important trace element in Venus's atmosphere because it forms clouds. These data are summarized in Table 9-1.

Methane has recently been discovered in the atmosphere of **Mars**. Three independent research groups, using both ground-based and space-based techniques, made the discovery. The discovery is important because methane is rapidly destroyed by a variety of processes under the conditions present

on Mars. Thus, its presence must mean there is a source for replenishment. That source could be volcanic and geothermal outgassing (such as in Yellowstone National Park), but there is no evidence of present-day outgassing. Methane, at least on Earth, also results from biological processes. The observations might be understood in terms of such processes, but as yet there is no evidence of either current, or past, life on Mars.

ATMOSPHERIC TEMPERATURE

Because **Venus** lacks atmospheric oxygen and ozone to absorb solar radiation, the temperature in the atmosphere decreases continuously up to the level of the clouds at 50 km (**Figure 9-9**). A similar situation holds true for **Mars**. The temperature of the Martian atmosphere rarely rises above the freezing temperature of water. The variations of temperature with height are dramatically different from that of Earth because of differences in the chemical makeup of the atmospheres of the two planets.

Atmospheric Pressure

	MERCURY	VENUS	EARTH	MARS
Pressure relative to Earth	v	~90	1	~0.01

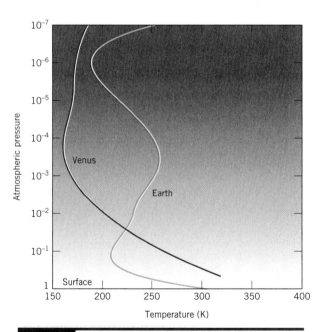

Figure 9-9 The variation of Venus's atmospheric pressure with decreasing atmospheric temperature above its surface. The Earth's variation is given for comparison. Pressure is given in units of the Earth's atmospheric pressure at sea level.

Spacecraft observations have found the atmospheric pressure at **Venus's** surface to be some 90 times greater than Earth's atmospheric pressure. In other words, whereas Earth's atmosphere exerts a force of nearly 15 pounds on each square inch, Venus's atmosphere exerts some 1,300 pounds per square inch. This is the same as being 3,000 feet (nearly 0.6 mile) under the ocean on Earth. Only specially designed research submarines are able to survive to such depths. At the other extreme, **Mars's** average atmospheric pressure is about 100 times less than the pressure on Earth.

Lightning

	MERCURY	VENUS	EARTH	MARS
Presence of lightning	—	?	yes	none observed

The presence of lightning in a planet's atmosphere is of interest because lightning plays a role in the atmosphere's chemistry and provides information on the cloud structure and dynamics. The 1978 *Pioneer* atmospheric probes of Venus confirmed some results that had been suggested by two earlier Soviet Venera missions—that the planet's atmosphere exhibits intense lightning flashes (up to 25 per second) and even occasional crashes of thunder. However, the *Cassini* spacecraft, while passing Venus on its way to Saturn, found no lightning. Researchers conclude either that none exists or it is different from that on Earth. The topic remains highly controversial. No lightning or thunder has been reported for Mars.

Inquiry 9-5

In visible light, Venus appears to be much cooler than the temperature determined for its surface by radio astronomers. What is the probable reason for this?

Clouds

	MERCURY	VENUS	EARTH	MARS
Clouds	no	sulfuric acid	water	water, dust

Those terrestrial planets having substantial atmospheres also have observable clouds. Earth's clouds, of course, consist of water in the form of vapor, droplets, and ices.

Venus's sulfuric-acid clouds are concentrated at three levels about 65, 53, and 48 km above the surface. Below 48 km, the atmosphere is clear of clouds.

Mars shows two different types of clouds, which can be observed from Earth. The white clouds consist of water vapor with some carbon dioxide. Such clouds have been observed by spacecraft to be present in low-lying areas in the early morning, and then to dissipate as the day warms. Extensive cloud systems form over the polar regions as the atmosphere begins to cool in the Martian late summer and early fall. The second type of clouds is yellowish in color; these arise from dust storms that are often so extensive they cover the entire planetary surface. Such storms arise due to high-speed winds moving over the surface, sometimes well in excess of 160 km per hour. Without such high speeds, the thin atmosphere would not be able to lift the dust from the surface.

Atmospheric Circulation

	MERCURY	VENUS	EARTH	MARS
Winds at high elevation	—	320 km/hr	jet stream 400 km/hr	—
Winds at surface	—	average 3 km/hr	0–160 km/hr	140 km/hr

Earth's atmospheric circulation is extremely complicated, as we have seen, because in addition to energy received directly from the Sun and indirectly from the land masses, large amounts of energy are released by the oceans and absorbed by the atmosphere. Coriolis effects (see Chapter 5) also complicate the atmospheric circulation on Earth.

Although **Venus's** clouds appear complex, their structure and motions are simple because atmospheric circulation is determined only by the heating that occurs at the point directly under the Sun. These heated gases rise, then travel around the planet at high speeds (over 300 km per hour), until they cool and descend toward the surface in the cooler nighttime regions. Surprisingly, this great wind speed falls off to nearly zero at the surface. The planet's slow rotation means that coriolis effects, which play an important role in Earth's atmospheric circulation, will be unimportant for Venus.

Mars's atmospheric pressure may vary by as much as 30 percent throughout its year because of variations in solar heating. For this reason, high-speed winds of a hundred miles per hour sometimes move across the surface. These winds produce the Martian dust storms discussed previously. On a smaller scale are dust devils, many of which blew past the *Pathfinder* lander repeatedly.

WHY DO THE ATMOSPHERIC COMPOSITIONS DIFFER SO MUCH?

Given that **Earth** and **Venus** are nearly the same size and located at nearly the same distance from the Sun, why is **Venus's** atmosphere almost entirely carbon dioxide while Earth has nitrogen, oxygen, and water vapor instead?

It seems clear that the present-day atmospheres of Venus and Earth are not the same as their original, primitive atmospheres (see Chapter 8). Those were blown away by the strong solar wind when the Sun was young. It appears certain that the current secondary atmospheres were formed from volcanic outgassing over geological time. Even today, volcanic outgassing is adding to the atmosphere of Earth, and possibly of Venus.

Among the primary constituents of the volcanic gases would have been carbon dioxide and water vapor, with smaller amounts of nitrogen and no oxygen to speak of. There is no reason not to think that the original chemical composition of each terrestrial planet was nearly the same.

What processes might have occurred on **Earth** that would have decreased the amount of carbon dioxide in its atmosphere? As discussed in Chapter 8, over geological history, atmospheric carbon dioxide, catalyzed by liquid water, combined with silicate rocks to form carbonates. This process removed carbon dioxide from the atmosphere and tied it up in rocks. Two additional processes contributed to a decrease of CO_2 in Earth's atmosphere. First, when life evolved on Earth, plant photosynthesis chemically removed carbon dioxide. Second, marine animals used CO_2 to create carbonate shells, which sank to the bottom of the sea when the animals died. The accumulated deposits of minerals such as limestone and chalk that are still extensive today contain much of the original carbon dioxide. Thus, once a full accounting is completed, Earth and Venus are found to have the same *total* amount of CO_2. All these processes are still occurring, but equilibrium was established prior to the industrialization of humanity.

Why was CO_2 not removed from **Venus's** atmosphere? The fact that Venus is closer to the Sun than Earth would have given it a somewhat higher surface temperature initially. There might not have been much liquid water to help catalyze the chemical reactions that, on Earth, helped remove carbon dioxide from the atmosphere. Furthermore, Venus's higher temperature would have placed more water vapor into the atmosphere and would have caused a small greenhouse effect. This greenhouse warming would have helped raise the surface temperature and driven carbon dioxide out of the rocks. As more and more carbon dioxide accumulated, the greenhouse effect would have increased even more, thus releasing more CO_2 and increasing the temperature, and so on. Such a *runaway greenhouse effect* would have effectively prevented the removal of carbon dioxide from the atmosphere and eventually led to the situation we see today. The greenhouse effect was discussed in more detail in Chapter 8.

Oxygen on **Earth** arose as a side benefit of the photosynthetic process that removes CO_2. Only when plants formed on Earth did the slow accumulation of oxygen begin. That oxygen combined with other atoms to form molecules that animals eventually evolved to use as an energy source. The reason **Venus** does not have Earth's large amount of nitrogen in the atmosphere is that a substantial fraction of Earth's nitrogen results from the decay of dead organisms.

The difference in water content between the two planets is also explained by **Venus's** high temperature. Although Venus may have formed as a water-rich world, once its temperature became high enough to vaporize any liquid water present, the vapor rose into the atmosphere, where ultraviolet radiation from the Sun could break up the water molecules into hydrogen and oxygen. Venus's gravity was not strong enough to hold onto the hydrogen, which escaped into space.

Mars had yet a different history. It is colder than Venus or Earth, and much of the water would have been in the form of ice, which does not chemically remove carbon dioxide from the atmosphere. Furthermore, Mars's smaller size both limits the volume of outgassed material and results in a smaller escape velocity, which allows atmospheric gases to escape easily. Finally, Mars's volcanic activity appears to have ceased early in its history, thus restricting the length of time over which outgassing occurred.

This scenario is our current best understanding of the history of the atmospheres of the terrestrial planets. There are some unexplained parts, however. One example is that Venus has a high abundance of argon in comparison to Earth. In searching for an explanation, some astronomers have revived speculations that some volatile and noble elements (such as argon) came from collisions with comets. If this were true for Venus, such collisions would undoubtedly have also occurred for Earth, in which case arguments that comets played an important role in bringing biochemical compounds to Earth may be strengthened.

9.4 Surfaces

The passage of a spacecraft close to a planet greatly increases the resolution of the planet's surface details over that seen from Earth. Studies of planetary surfaces have provided us with detailed information about processes occurring not only on the surfaces but also within the atmosphere and the interior. In this section we look at the surfaces to learn both facts about them and the processes that produced the surfaces we see today.

THE VIEW FROM ORBIT

Nothing was known about the surface of **Mercury** before the *Mariner 10* spacecraft made its closest approach to Mercury in 1974. Scientists were elated as the first pictures arrived on Earth and showed a landscape resembling that of the Moon (Figure 9-5*a*). But as they began to examine the orbital images in more detail, distinct differences from the Moon appeared. For example, Mercury lacks the large mare we see on the side of the Moon facing the Earth. Further differences, along with similarities, will be discussed later in this chapter.

Our knowledge of **Venus** was rewritten in the early 1990s, thanks to the *Magellan* spacecraft. Earlier radar studies had shown two-thirds of Venus's surface to be covered by rolling plains, about one-quarter covered by lowland regions, and less than one-tenth by highland areas. The improved resolution of *Magellan* shows details as small as 100 m across, about the size of a football field. **Figure 9-10** shows the best global radar view of Venus yet produced. Radar images do not necessarily look like what your eyes would see. They show the ability of planetary material to

Figure 9-10 A radar view of Venus from Magellan. Bright regions are those that reflect radar beams well.

reflect radar beams. The reflecting and absorbing properties of a surface depend on a variety of factors, including the composition and sizes of particles. Finely ground materials reflect poorly compared with coarse materials. For these reasons, radar images are difficult to interpret, and great care must be taken to avoid errors.

We can infer what conditions on the surface of **Venus** must be like. With its thick atmosphere, the planet is probably quite dark with only about 1 percent of the sunlight reaching the surface. Red light would be scattered somewhat less than blue, so objects would take on a reddish color. The late astronomer and popularizer of science Carl Sagan, in his book *The Cosmic Connection,* described the surface of Venus in this way:

> . . . *broiling temperatures, crushing pressures, noxious and corrosive gases, sulfurous smells, and a landscape immersed in a ruddy gloom . . . there is a place astonishingly like this in the superstition, folklore and legends of men. We call it Hell.*

The view of the **Martian** surface so long awaited by astronomers is shown in Figure 9-5*d*. Although about 40 percent of the planet's surface consists of plains regions, in this view we see giant volcanoes, a long and deep valley, and craters. These features are far larger than anything compa-

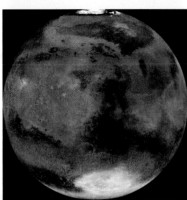

Figure 9-11 Opposite sides of Mars. Part *a* shows the Martian volcanoes and the Mariner Valley. Notice the difference between the two sides as well as the differences between the northern and southern hemispheres.

Figure 9-12 The surface of Venus as photographed by Soviet Venera probes.

rable on Earth. There is no doubt that these mountains are volcanoes, for one can see near the summit the remains of lava flows that took place during past eruptions. The other 60 percent of the Martian surface is cratered. Most of the plains (and their volcanoes) are confined to the northern hemisphere; the more heavily cratered regions are in the southern hemisphere. The differences in the northern and southern hemispheres can be seen on the images of **Figure 9-11**.

But what about the **Martian** canals suggested by Schiaparelli and Lowell? None of the *Mariner* images support the hypothesis that *canali* exist on the surface of Mars. We must therefore conclude that these features were an optical illusion, perhaps due to the ability of the eye and brain to see fine differences in contrast. We might also ask if Lowell and others were not strongly influenced by their *desire* to see canals, because many reputable observers were never able to see them. Sagan commented wryly that there is no question that intelligence was involved in the problem of the Martian

canals, but that the real question was on which end of the telescope the intelligence was located. Clyde Tombaugh, the discoverer of Pluto, once commented that Mars looked "different" to him when he compared the view through Lowell's telescope with the view through other telescopes at the observatory, raising the question of whether there may have been some unsuspected instrumental problem at work.

THE VIEW FROM THE SURFACE

Even though no spacecraft has yet landed on the surface of Mercury, there is much about the surface that can be inferred from the variety of ground-based and orbit-based observations. These will be discussed when appropriate in this section.

The Soviet *Venera 13* and *Venera 14,* which landed in 1982, sent back visible-light images of the surface of **Venus** (**Figure 9-12**) showing it to be a barren, rocky plain. From these pictures, taken with a fish-eye lens and therefore covering a large field of view, planetary astronomers can draw some important conclusions. The sharp-edged rocks show a lack of erosional processes, indicating a lack of wind-blown dust on the surface. Instrumented landers had shown previously that the high-velocity winds in the upper atmosphere cease at the lower levels. The presence of shadows shows that some direct sunlight is able to penetrate the multilevel clouds all the way to the surface. A horizon can be seen in the upper-left corner, indicating the clarity of the dense atmosphere. No further landings on Venus are planned.

The first incredible images from the **Martian** surface were returned by the *Viking* spacecraft

Figure 9-13 The Martian landscape as seen by *Viking 1*.

Figure 9-14 A computer-simulated view of Venus, using Magellan spacecraft altimeter data. In this view, the observer is about 0.5 km above the 8-km-high mountain, Maat Mons. The colors are meant to represent the result of sunlight filtered through Venus's sulfuric-acid clouds. Relief is exaggerated.

(**Figure 9-13**). We see a landscape not unlike a desert on Earth, with windblown sand and weathered rocks. Some of the discoveries of the *Viking* landers are discussed in the coming sections.

Surface Temperatures

	MERCURY	VENUS	EARTH	MARS
Equatorial surface Temperature (K)	100 to 700	726	260 to 310	150 to 310

The surface of **Mercury** has a daytime temperature at the equator of some 700 K (800° F), and a nighttime temperature of 100 K (–279° F). This wide range comes about from the nearness of the Sun and from the lack of a substantial atmosphere to help retain heat.

Venus surprised astronomers, who found unexpectedly large amounts of heat coming from the planet. Analysis of these data showed the average surface temperature to be 730 K (854° F), above the melting point of lead! The fact that Venus is closer to the Sun than the Earth cannot by itself account for this remarkably high temperature. It turns out that Venus exhibits an extreme example of the greenhouse effect that was described in Chapter 8 and mentioned earlier in this chapter. It shows us what can happen if the abundances of greenhouse gases (CO_2, and SO_2) get out of hand.

Venus never obtains any relief from its high temperature. The blanketing by the atmosphere and the movement of heat by atmospheric circulation cause Venus's temperature to be nearly the same over the entire surface.

Venus's high surface temperature would have observable effects. Planetary scientists have conjectured that over billions of years at such high temperature, the surface might behave like Silly Putty. In addition to making it difficult for rigid plates to form, the heat would cause material on mountains to flow downward, or slump, due to gravity. The mountain called Maat Mons shown in **Figure 9-14** has such a slumped appearance, which confirms the temperature measured by other methods.

Mars's surface temperature variations range from a rare extreme high of 280 K (45° F) at the equator during summer to 150 K (–189° F) at the poles in winter. (The air temperature, though, rarely rises above freezing.)

Surface Compositions

	MERCURY	VENUS	EARTH	MARS
Surface composition	silicates?	silicates?	silicates	silicates, iron oxide

From Earth-based observations of **Mercury**, astronomers concluded many years ago that the nature of the surfaces of Mercury and the Moon must be similar. Spacecraft have shown the similarity in more detail. Although no surface samples of Mercury have yet been collected, astronomers infer from the observed planetary density and other data that Mercury's surface materials are most likely basaltic lava.

The only direct measurements we have of **Venusian** surface materials come from the many Soviet *Venera* landers. Various detectors measured gamma ray and X-ray emissions from surface rocks. The results show the rocks are basalts and granites, which are igneous rocks formed by volcanism. *Venera 13* and *Venera 14* revealed that these Venusian basalts contain substantial amounts of calcium, a result of some importance that we will return to later.

The **Mars** *Viking* landers finally answered the question of why Mars is red. Astronomers have long suspected that the red color of Mars was due to iron oxides—rust—in its soil. The soil analysis completed by each *Viking* lander showed the surface to consist mostly of silicate rocks, but with a large fraction (20%) as iron oxide. Another indication of iron is that the soil is magnetic. Such a high surface abundance of iron, combined with its overall density, means that the planet should not have much of an iron core. In other words, Mars should show little, if any, differentiation. However, *Pathfinder* data came to a different conclusion, as discussed in the section on planetary interiors later in the chapter.

An important finding by *Sojourner* about **Mars's** surface is that not only does it have dust and pebbles, but it also has sand. Sand results from weathering processes such as wind and water erosion.

When *Sojourner* determined the soil composition of the landing site, we learned that the composition is the same as at the distant *Viking* landing sites. From this information, we infer that Mars's winds are global and sufficient in strength to distribute material uniformly.

Water

	Mercury	Venus	Earth	Mars
Surface water	no	no	much	none currently

The high temperatures present on **Mercury** and **Venus** preclude the presence of surface water. However, in the case of Mercury, as discussed later in the section on the polar regions, the poles may be frigid enough for water ice to exist.

The question of water on **Mars** has been perhaps the foremost question in planetary astronomy for more than 100 years. Mars is dustier and drier than any desert on Earth. For this reason, the discovery made by the *Mariner 9* orbiter in 1971 that the Martian surface has many features that are clearly the result of fluid flows was surprising. **Figure 9-15** shows two regions of this type. In Figure 9-15*a* is a sinuous, twisting channel that looks as if it were caused by flows of a mudlike slurry. Such slurry could be produced if *permafrost,* which consists of a frozen mixture of soil and ice, were to melt. Such melting could be produced by climatic changes or by geothermal heat. There is clearly nothing artificial about these Martian regions. Indeed, they strongly resemble the beds of ancient streams and rivers on Earth. Figure 9-15*b* shows two sand bars around which flowed a running fluid. Similar formations are seen in rivers all over the Earth.

Our study of the search for a watery history to **Mars** will take an historical approach in which we look first at the tantalizing indications that significant amounts of liquid water were present in the past. This, then, will lead us to the newest evidence provided by the two Mars Explorations Rovers, *Spirit* and *Opportunity,* that studied the Martian surface in 2004.

Figure 9-16 shows two kinds of evidence of past water. First, the rounded boulders and rocks (red arrows) are interpreted as obtaining their shapes when floods knocked off their rough edges. Second, one interpretation of light-colored areas, marked with white arrows, are that they are material left behind after mineral-rich water flooded the area and then later evaporated, leaving the lighter-color material behind. Another interpretation of these same white areas is that they are clumps of material fused together by water.

Sojourner's photos of the hills called Twin Peaks show four or five horizontal layers that are thought to result from significant past water action that occurred more than once. **Figure 9-17*a*** shows features near the top that suggest a continual fluid flow and subsequent erosion. **Figure 9-17*b*** shows crater ejecta thought to result from impact into a surface containing either liquid or frozen water. When the impact occurred, water along with rock and debris were released and the ejecta flowed like mud to produce the observed ejecta blanket.

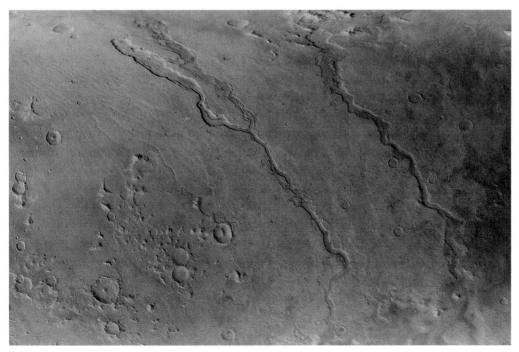

(a)

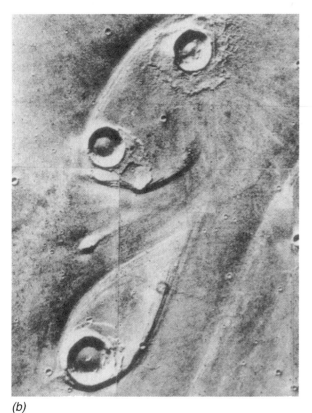

(b)

Figure 9-15 Evidence of the past presence of water on Mars. *(a)* Ancient channels are thought to have been produced by a sudden release of water due to melting of permafrost. *(b)* Similar structures are found in rivers on Earth when there is a barrier to water flow.

Figure 9-16 A *Sojourner* image of Mars. The red arrows point to rocks thought to be rounded by water. The blue arrows point to sharp-edged rocks thought to have resulted either from nearby impact craters and/or volcanic origin. Finally, the white arrows point to deposits thought to have been left behind by evaporating water.

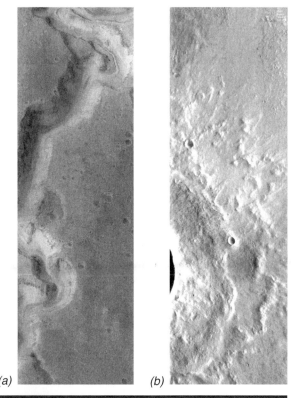

(a)   *(b)*

Figure 9-17 The effects of water on Mars. *(a)* shows the results of fluid flow and erosion, while *(b)* shows what results from an impact into a liquid or frozen water surface.

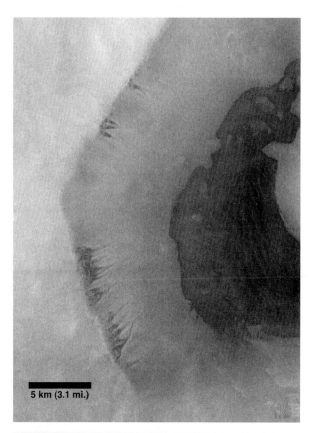

5 km (3.1 mi.)

Figure 9-18 A crater on Mars that suggests the seepage of fluid and the subsequent flooding of the crater floor. The dark material at the bottom may result from sediment deposited by the seeping water. The resolution is about 24 m over the 26 by 32 km region.

Scientists have been looking for years for direct evidence of locations where standing water might have been. They now argue that the *Global Surveyor* image in **Figure 9-18** provides that evidence. The image shows a 50-km-wide crater with channels in the rim suggestive of fluid seepage from the crater walls. Furthermore, at the interface between the supposed once-flooded darker sediment areas and the lighter crater walls are what appear to be small bays and peninsulas. The lack of cratering indicates that the events producing this apparent ponding may have occurred relatively recently.

Inquiry 9-6

What are two mechanisms in addition to those suggested above that could have provided the heating to melt the permafrost?

The exciting observations just discussed provide strong indications of past water flow on Mars. They provide a natural progression in the study of Martian water up to the spring 2004, when the *Spirit* and *Opportunity* rovers moved over the surface and performed a variety of studies, which we now discuss.

Let us cut to the chase: the conclusion from the landers and orbiters most recently studying **Mars** is that significant amounts of water were present on the surface in the past. Numerous lines of evidence lead to this important, and startling, conclusion. The Mars exploration rover *Opportunity* observed large numbers of what they referred to as *blueberries,* shown in **Figure 9-19** as the rover entered the crater named Endurance. Hypotheses to explain these include debris from volcanic eruptions, debris from meteoroid impacts, and *concretions,* which are irregular or rounded rocks formed when water percolates through rocks. Scientists have concluded, on the basis of the triplet blueberry in **Figure 9-20**, that they are concretions made of gray hematite, a mineral that at least on Earth is formed in the presence of water. The bedrock surrounding the landing sites of the two rovers contains small, elongated holes (called *vugs*) that are best explained through the washing away by water of crystals that formed within the sediments (**Figure 9-21**). Rocks have been observed to consist of sulfur, bromine, and chlorine salts; on Earth, such salts form in water and then precipitate out when the water evaporates. The mineral jarosite, which has been found, requires liquid water to form. Some rocks show the presence of fossilized ripples (**Figure 9-22**), which formed when gently flowing water moved across the surface. Finally, the landers observed layers of sedimentary rock at angle to one another—something called crossbedding (**Figure 9-23**)—that can result from either wind or water. Mars's crossbedding resembles that produced by flowing water. The effects of water are seen to depths of centimeters in some locations and to meters in others. All these observations, especially when taken together, lead to a strong conclusion for the presence of significant amounts of liquid water in Mars's past history.

There is no evidence that large quantities of liquid water exist on **Mars** now; in fact, the low atmospheric pressure prevents it. This raises an obvious question: If there were once substantial amounts of water, where has it gone? Various scientists have suggested that the Martian climate may alternate between ice ages, which are cold and dry with

Figure 9-21 Elongated holes, called vugs, in Martian rocks provide indirect evidence of a watery past.

Figure 9-19 "Blueberries" on the surface of Mars.

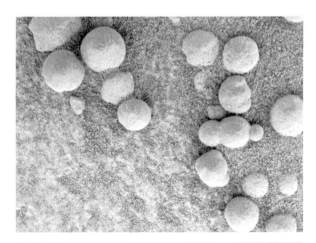

Figure 9-20 The triplet blueberry at the right-center provides evidence of how they formed.

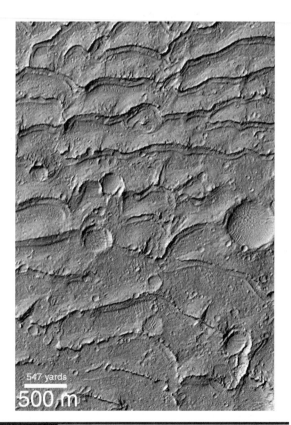

547 yards
500 m

Figure 9-22 Fossilized ripples on the Martian surface.

moisture largely locked in the form of permafrost and ice, and temperate ages, which are warmer and wetter. When subsurface permafrost melts, landslides may result. In **Figure 9-24**. the water produced by the melted permafrost flowed away toward the left.

Craters

	MERCURY	VENUS	EARTH	MARS
Craters	high density	low density	low density	medium density

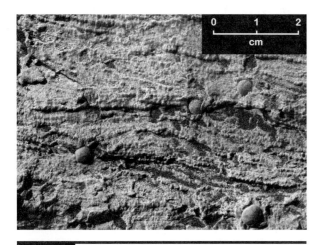

| | | | 0 | 1 | 2 |
| cm |

Figure 9-23 Layering of rocks at various angles, which is called crossbedding, provides evidence for past fluid flow on Mars.

Mercury shows extensive cratering, including numerous bright-rayed impact craters and relatively flat regions between craters (Figure 9-5a). Superficially, Mercury's craters resemble those on the Moon, but scientists have noted one significant difference between the two bodies: The entire lunar highlands are extensively cratered and rolling, whereas the analogous regions on Mercury are relatively smooth. This difference is probably explained by the fact that Mercury has a higher surface gravity than the Moon. When a crater formed on Mercury, the ejected material did not travel very far, and the secondary craters remained close to the primary crater. On the Moon, the ejecta could travel farther, and secondary cratering would have taken place over a much wider area.

Venus, too, shows impact craters (**Figure 9-25**), some with central mountain peaks. With a crater density some thousand times less than that of the Moon, Venus is relatively free of impact craters. Such a low crater density implies that planetwide resurfacing has been occurring during the past 500 million years. Such resurfacing may have been caused by volcanism.

The **Moon**, **Mercury**, and **Mars** all have craters in a wide range of sizes, from hundreds of kilometers downward. **Venus**, by contrast, has no craters less than two kilometers across. Such a lack of small craters is interpreted as a lack of small impacting bodies, which must have been destroyed in passing through Venus's atmosphere. Shock waves that resulted from the passage of meteoroids through the thick atmosphere could have pulverized

Figure 9-24 A landslide on Mars. Melting of the permafrost allowed the surface to slide. The resulting fluid flowed away toward the left.

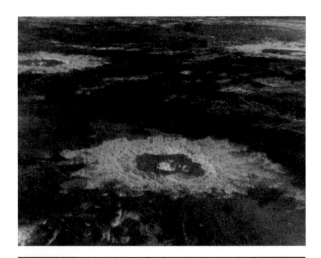

Figure 9-25 Craters on Venus. This is a radar image obtained by the *Magellan* spacecraft. Note the central peak. The crater is about 35 km in diameter.

Venus's surface material into fine dust. This phenomenon might explain the various dark splotches observed to surround some craters on Venus.

Impact Basins

	MERCURY	VENUS	EARTH	MARS
Impact basins	yes	?	a few	yes

We saw in Chapter 8 that the Moon has basins as large as 1,100 km across that resulted from the impact of large bodies. On the one hand, the basins on the Earth-facing side of the Moon are completely filled with lava that oozed from the interior, forming the maria. The Mare Orientale, on the other hand, has not completely filled in but exhibits multiple rings. The impact basins completely on the far side of the Moon are not filled in at all, which makes the two sides appear more different than they are.

Some large impact basins are found on **Earth**, but we assume that subduction by continental drift obliterated most of them long ago. One large impact feature appears to be a crater nearly 200 km in diameter discovered in 1992 off the Yucatan Peninsula in the Gulf of Mexico.

In **Figure 9-26**, you see one of **Mercury's** sizable basins. The **Caloris Basin** (the proper name is Caloris Planitia) is the largest one, with a diameter of 1,300 km. It is a multiringed basin, similar to the Mare Orientale on the Moon (see Figure 8-25).

Figure 9-26 Mercury's giant multi-ringed basin, Caloris Planitia.

Mercury's surface on the side opposite the Caloris Basin contains what planetary scientists labeled *weird terrain* because its jumbled appearance had not been observed elsewhere in the Solar System. Subsequently, similar but less extensive weird terrain was found on the Moon directly opposite the Mare Orientale. One explanation of weird terrain is that it arose in response to the impacts causing the Caloris Basin and Mare Orientale. Seismic waves traveling through and around the surface caused material opposite the impact to be lifted off the surface, and to fall back in the jumble we now observe (**Figure 9-27**).

More than 20 basins have been identified on **Mars**. One of them, called Hellas, is a multiringed basin. Because the basin floor is some 5 km below the surrounding plains, the greater atmospheric pressure at the bottom allows frost and clouds to form there. The high reflectivity of the frost and clouds makes Hellas readily visible from Earth.

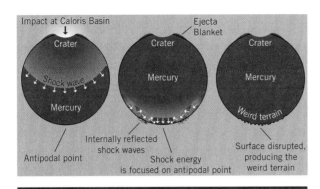

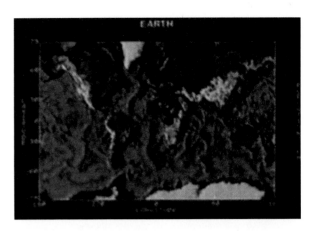

Figure 9-27 The formation of "weird terrain" from a gigantic impact. Seismic waves traveling through the planet and over the surface converge on the opposite side.

Continents, Mountains, and Volcanoes

	MERCURY	VENUS	EARTH	MARS
Continental-sized areas	no	yes	yes	yes
Mountains	no	yes	yes	yes
Ancient lava flows	yes	yes	no	yes
Active volcanism	no	perhaps	yes	no

Relief maps of three of the terrestrial planets are shown in **Figure 9-28**. Mercury does not show any great volcanoes like those on Venus or Mars. (However, because the *Mariner* spacecraft observed only one side of the planet, some could exist on the other side.) The resolution of the images *Mariner* sent to Earth is too low for us to be able to see lunar-type domes and lava flow fronts. Nonetheless, Mercury does have rimless pits similar to those found on the Moon; they are thought to be volcanic vents (see Figure 8-22a). Finally, the plains regions between craters have the general appearance of previously molten material.

While about two-thirds of **Venus's** surface is rolling plains, almost 10 percent of the surface is concentrated in extreme highland regions. For example, the area known as *Ishtar Terra* is comparable in size to Africa and contains the mountains called Maxwell Montes, which have an elevation of more than 11 km (35,000 ft.) above the surrounding plain. This is a mile higher than Mt. Everest! Other continental-sized areas are also present.

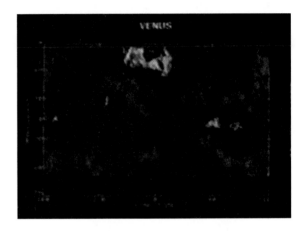

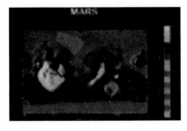

Figure 9-28 Relief maps of Venus, Earth, and Mars. The high regions on each planet are the Himalayas on Earth, Ishtar Terra on Venus, and the Tharsis Plateau on Mars.

Venus shows a range in type and size of volcanically produced features. These include domes, mountains, highland plateaus, and meandering flow channels (**Figure 9-29**). These features can be as large as 500 km across, and one channel runs for 6,800 km. Figure 9-14 shows a computer-generated false-color view of Maat Mons, the youngest-looking volcano on Venus.

On **Mars**, volcanoes are located primarily on the Tharsis plateau, an area covering nearly a quarter

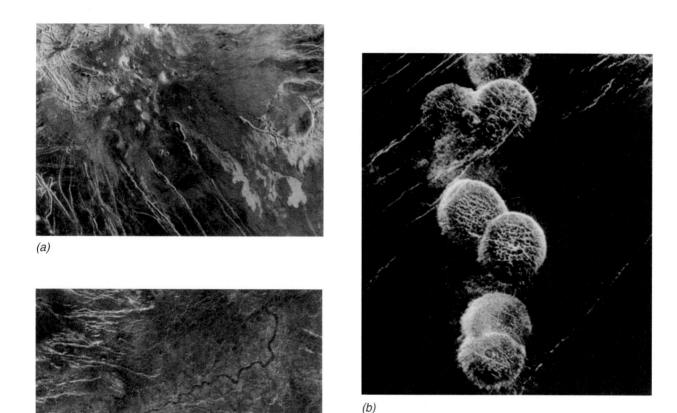

(a)

(b)

(c)

Figure 9-29 Volcanism on Venus, as seen by radar on the Magellan spacecraft. *(a)* Lava flows invading and degrading an impact crater, whose diameter is 65 km. *(b)* Pancake domes, some 25 km across, as observed by the *Magellan* orbiter. *(c)* A meandering channel that could be produced only by flowing lava. This section is 200 km long. Channels as long as 6,800 km have been seen on Venus.

of the Martian land mass. Tharsis is much higher than the surrounding plain (Figure 9-28). These volcanoes are part of the plains region that covers 40 percent of the surface. The plains apparently came about from extensive lava flows. In contrast, the southern hemisphere is rough and covered with craters, making it geologically older than the northern hemisphere. Such geologic asymmetries are common on the Earthlike planets.

Mars has the tallest volcano in the Solar System. **Olympus Mons** (Mount Olympus) rises almost 27 km (nearly 90,000 feet) above the planet's surface (**Figure 9-30**). The central caldera is 80 km across and resulted from collapse when the lava inside suddenly withdrew. The large circular ring surrounding it is a cliff 500 km in diameter. In many places the cliff height is as much as 6 km.

Why are some of the **Martian** volcanoes so much larger than those on Earth? In the case of the

Hawaiian Islands, for example, the volcanoes formed when molten lava from hot spots located on moving continental plates broke through the lithosphere. The plate movement means that new volcanic cones will be continuously formed as the plate moves over the static hot spot (**Figure 9-31**). On Mars, however, where there is no plate movement, volcanic activity occurs time and again at the same location so that lava piles up to great elevations. The presence of such high mountains also tells us the Martian crust is strong enough to hold such a concentrated mass.

Are volcanoes currently active on any terrestrial planets other than Earth? The answer is a definite no for Mars and Mercury (subject to the caveat that we have seen only half of Mercury's surface), but the answer is not so clear for Venus. The *Pioneer Venus* spacecraft detected 50 times the amount of sulfur dioxide (SO_2) usually present in Venus's atmosphere. Because calcium present in

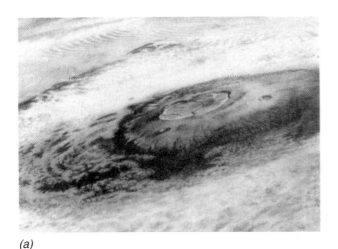
(a)

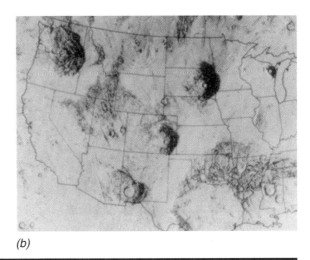
(b)

Figure 9-30 Olympus Mons, the Solar System's highest volcano. *(a)* The summit, at an elevation three times that of Mt. Everest on Earth, is well above the clouds. The large ring is 500 km in diameter. *(b)* Olympus Mons and other Martian volcanoes compared in size with the United States.

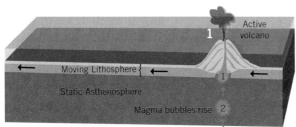

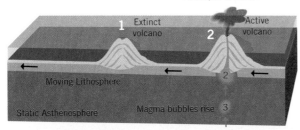

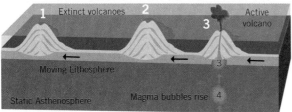

Figure 9-31 Why volcanoes on Mars are larger than those on Earth. Formation on a moving lithosphere limits volcanic sizes on Earth, while the lack of motion on Mars allows them to grow.

Venus's surface material readily combines with sulfur dioxide, thus removing it from the atmosphere, there must be a source of fresh sulfur dioxide available. Since on Earth sulfur dioxide comes from volcanoes, we therefore infer that active volcanoes may be present on Venus.

Faults, Tectonics, Cliffs, and Valleys

	MERCURY	VENUS	EARTH	MARS
High scarps	yes	no	no	no
Plate tectonics	no	probably not	yes	no

Each of the terrestrial planets has its own unique set of surface features caused by its own unique history. In this section, we look briefly at features that on Earth are usually explained in terms of plate motion, but that require different explanations on the other terrestrial planets.

Mercury has cliffs that are several kilometers high and often hundreds of kilometers long (see the darkish arc passing through the center of **Figure 9-32***a*). These cliffs are unique among the Earthlike planets. They are thought to have formed when pressures from compression produced during cooling and shrinking of the crust caused local folding of the crust (**Figure 9-32***b*). As with the Moon, however, there is no ironclad proof of Earthlike plate motion at any time in Mercury's history.

One of **Mars's** surprises awaiting scientists was an enormous canyon slicing across the surface of the planet, (**Figures** 9-5*d* and **9-33**). Named the Valles Marineris (Mariner Valley) for the *Mariner* spacecraft that discovered it, this huge feature is about 5,000 km long, more than 5 km deep, and 80 km wide in places. On Earth, this colossal canyon would stretch from Los Angeles to New York, and Mars is much smaller than Earth! The

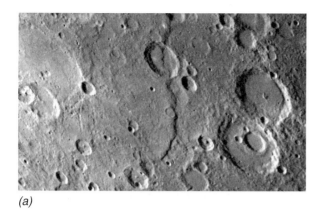

(a)

(b)

Figure 9-32 *(a)* A close-up of Mercury. The smallest visible details in this figure are about 250 meters across, or the length of three football fields. The dark arc passing through the center is a piece of a high cliff (scarp) named Discovery. *(b)* Mercury's system of cliffs is produced when the surface shrinks, creating compression forces that eventually cause one layer to override another.

part of the Valles Marineris called the Candor Chasma appears to have been eroded by both wind and water (Figure 9-33*b*). Within the Martian canyons, the *Mars Global Surveyor* observed layering of materials throughout the depths of the canyons. The scale of the layering is from a few to tens of meters; the layering is observed to extend thousands of meters within the Valles Marineris rock wall. Such layering provides information necessary for understanding the story of Mars's crustal history. Possible interpretations consistent with various observations include volcanic flows, as well as material deposited by wind and/or water. Thus, all we can conclude at this time is that the investigating scientists do not yet understand Mars's crustal history.

Mars is covered by a variety of fault regions, one of which is shown in **Figure 9-34**. This is the most intense fault region on the planet and shows two complex sets of faults: one in a north-south direction, the other running northeast to southwest. Although the cause of the stresses producing the faults is unknown, they could have resulted from convection within the mantle. Not only is there no direct indication of large-scale crustal motion, but, as discussed earlier, the large volcanoes argue against plate motion.

(a)

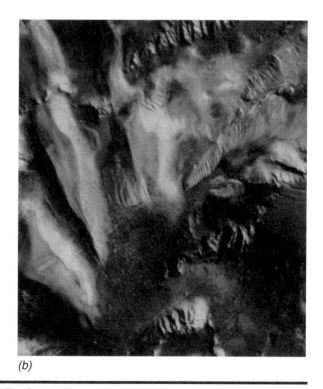

(b)

Figure 9-33 *(a)* Valles Marineris is the size of the United States. *(b)* The Candor Chasm region of the Valles Marineris shows erosion and perhaps material deposited on its floor.

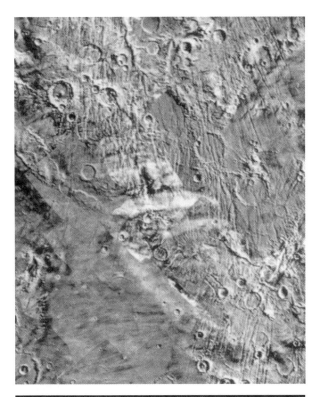

Figure 9-34 Martian faults. The area shown here covers 1,500 km from top to bottom. One set of faults runs north–south, and another runs northeast–southwest. Most of these large fault systems radiate from the region where the volcanoes formed, indicating a relationship.

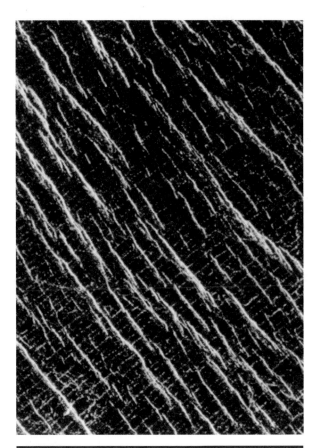

Figure 9-35 The *graph paper region* of Venus. The fainter lines, perpendicular to the brighter ones, occur at regular intervals of nearly a kilometer.

The *Magellan* images of **Venus** confirm that it does not exhibit continental drift. Thus, Earth is the only planet having large-scale horizontal plate motions. Nonetheless, Venus does appear to release internal heat by means of cracks in the mantle through which plumes of material flow to form Venusian mountains. Evidence of such cracking is shown in **Figure 9-35**, which shows a region on Venus known as the *graph paper region*. This area, unique within the Solar System, is thought to be due to an unknown type of subsurface force. Further fault features are visible in Figure 9-29*b*.

Wind Erosion

	MERCURY	VENUS	EARTH	MARS
Wind erosion	no	little/none	yes	yes

Mars has abundant evidence, both within its atmosphere and over its surface, of wind-blown dust.

Winds provide an important erosional process on Mars today because of the lack of competing processes from running water and active volcanism. Such wind provides one connection between observable surface features and the atmosphere. Shifting dust, covering some areas and uncovering others, provides the most likely explanation of the changing appearance of Mars's dark markings. An extensive region of wind-deposited sand dunes appears in **Figure 9-36**. Analysis of images of such regions shows that winds of up to 90 mi/hr (150 km/hr) exist on the planet. Sojourner images show numerous features both in the soil and on specific rocks indicative of wind erosion. **Figure 9-37**, of the rock called Moe, shows the effects of such erosion.

Inquiry 9-7

What previously discussed observations are there that wind erosion is not important on Venus?

Figure 9-36 Wind-blown sand on Mars. A region of Mars covered by sand dunes, evidence of the action of winds in the thin Martian atmosphere.

Figure 9-37 The rock called Moe as observed by the Sojourner rover. The flute-like textures on the rock may have been caused by wind abrasion.

Polar Regions

	MERCURY	VENUS	EARTH	MARS
Polar ice caps	perhaps	no	yes	yes

Study of planetary polar regions is difficult from Earth because they are viewed from a highly oblique angle. Radar observations of **Mercury** made in 1991 gave astronomers a view of the planet never before seen. At the time these observations were made, the planet's north pole was tilted toward Earth. The observations revealed an oval area, measuring 640 by 300 km around Mercury's north pole, which reflected radar waves strongly. The researchers interpret these data as an ice cap on Mercury's north pole. Although surprising, the observations are consistent with recent theoretical calculations indicating Mercury's polar temperatures can be as cold as 125 K ($-235°$ F). Because interpretation of radar images is both difficult and tricky, it is possible the region is bright due to intense cratering or other ragged terrain at the pole. However, most planetary astronomers think such an explanation is unlikely, and that the ice interpretation is correct. Additional observations, such as those to be made by spacecraft missions at the end of the decade, are required before a definitive statement may be made.

Mars's polar ice caps had been extensively observed for many years, but controversy over whether they were water ice or dry ice (frozen carbon dioxide) continued for a long time. Earth-based observations indicated that Martian ice caps completely melted each summer; would higher-resolution close-up images show this to be the case?

Long-term studies of the temperature of **Mars's** ice caps by orbiting spacecraft found ice caps to be composed of water *and* carbon dioxide ices. The more abundant dry ice produces the polar cap viewed from Earth in the views in Figure 9-3; the water ice cap lies underneath it. In the summer, the dry ice in the northern cap completely sublimates and leaves behind a water-ice remnant. Some frozen carbon dioxide remains at the southern polar cap throughout its summer (**Figure 9-38**). These observations are surprising because they are the opposite of what might be expected on the basis of the seasons as illustrated in Figure 9-7. The Mars *Global Surveyor* returned detailed images of the Martian surface underneath the melted polar caps. The terrain is observed to be layered, and a light amount of cratering indicates that the layering is young on the Martian geological time scale. A highly complex set of ridges, whose origin is not

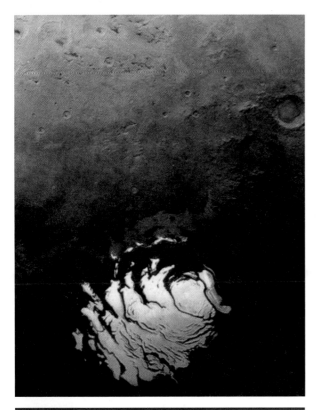

Figure 9-38 | Polar regions. Mars's remnant south polar cap is about 400 km across.

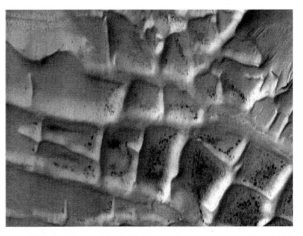

Figure 9-39 | Mars's south polar cap revealed complex intersecting ridges to the *Mars Global Surveyor.* The area shown is 20 by 14 km. The cause of the dark spots in some regions is unknown. Reprinted with permission from *Science Magazine,* Vol. 279, No. 5377, March 13, 1998, p. 1681. Copyright © 1998 American Association for the Advancement of Science.

understood, is shown in **Figure 9-39**. The *Global Surveyor* has found a variation of 1.5 to 2 meters in the CO_2 snow depth throughout the Martian year.

Inquiry 9-8

What might you expect the temperatures of Venus's polar regions to be compared to its equator? Explain.

9.5 Interiors

A planet's interior cannot be observed directly. Therefore, astronomers infer the interior structure from theoretical models that must be consistent with all of the direct observations. The observations include the planet's magnetic field, overall shape, density, and spin, as well as data collected from meteorites found on Earth, and the motion of their moons and visiting spacecraft. In the cases of Venus and Mars, landers have collected data about surface radioactivity, chemical composition, and seismic activity that help constrain the interior models.

Magnetic Fields

	MERCURY	VENUS	EARTH	MARS
Global magnetic field	yes	no	yes	no

Much to the surprise of planetary scientists, *Mariner 10* found that Mercury has a significant magnetic field. With a strength 1 percent that of Earth's magnetic field, Mercury's field is stronger than that of either Venus or Mars. The Mars *Global Surveyor* spacecraft confirmed that no significant global magnetic field exists on Mars but it did find various small-scale magnetic anomalies in the crustal material. *Sojourner* also found the surface dust to be highly magnetic. Although the meaning of all this has not yet been determined, it certainly implies that Mars had a magnetic field when its crust solidified. What can we infer about each planet's interior from its magnetic field? We examine that next.

INTERIOR STRUCTURES

The most persuasive theory for how an overall planetary magnetic field is produced is the dynamo theory, which requires an ordered motion of charged particles in a planet's interior, presumably from the planet spinning about its axis. How then can **Mercury**, which has such a slow rotation speed, show a significant magnetic field? We infer that Mercury has a core occupying a substantially larger fraction of its interior volume than is true for the other terrestrial planets. Models of Mercury's interior show the core to contain 42 percent of the planet's volume, compared with 16 percent for the Earth.

Other observations have not been made for all of the planets, so the following information and, therefore, the models are incomplete—a work in progress.

Venera 8 found that the level of radioactivity in the surface rocks of **Venus** was similar to that of terrestrial rocks. This observation supports those who think that Venus is similar to Earth not only in size and mass but also in its stratification into layers: core, mantle, and crust. One difficulty with this model is the lack of a strong magnetic field. However, even if Venus does have a molten core, dynamo action might not have developed due to the planet's slow rotation.

Mars's interior is, according to the *Pathfinder* lander, more differentiated than previously thought. The early results from *Pathfinder* indicate the presence of a metallic core with an uncertain radius between 1,300 and 2,400 km, or 5 to 36 percent of the planet's volume. These results are consistent with the chemical abundances of the Martian meteorites found on Earth. How the core came about is not yet understood.

Although each of the *Viking* spacecraft visiting **Mars** had a seismometer, only one of them worked, and it showed that Mars does have seismic activity. However, the amount of seismic activity per unit area on Mars is less than that on Earth. The Mars quakes that were observed lasted about a minute, indicating that Mars's internal structure is more similar to that of Earth than it is to the Moon, whose quakes last for hours. This implies that Mars's interior is less solid than the Moon.

Figure 9-40 summarizes our knowledge of planetary interiors through a scaled figure of the interiors of each of the terrestrial planets and the Moon. Mercury's core is unique in its large fractional volume.

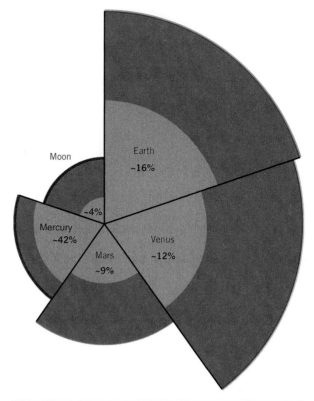

Figure 9-40 The interiors of the terrestrial planets and the Moon, drawn to scale. In a relative sense, Mercury has the largest core. The fraction of the planet's volume occupied by the core is noted.

9.6 Moons

	MERCURY	VENUS	EARTH	MARS
Moons	no	no	1	2

Mars is the only terrestrial planet other than Earth to have any moons, which are natural satellites. *Viking* succeeded in photographing the tiny **Phobos**, whose diameter is some 23 km (**Figure 9-41a**), and **Deimos**, which has a 12 km diameter (**Figure 9-41b**). Their names come from the mythological horses that pulled the war god's chariot and mean *fear* and *panic*, respectively. The two moons turned out to be quite irregular in shape and pitted by craters.

An observer on **Mars** might well be confused by observations of Phobos and Deimos, because Deimos rises in the east and sets in the west, while

(a)

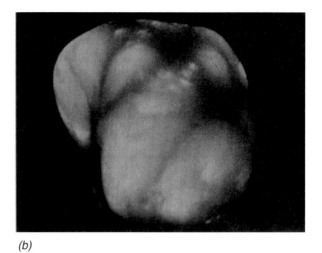
(b)

Figure 9-41 The Martian satellites. *(a)* Phobos *(b)* Deimos.

Phobos, which is closer to Mars, does the opposite! The apparent eastward motion of Phobos results from its orbital period of only 7 hours, much less than Mars's rotation period of slightly more than 24 hours.

Phobos has numerous small impact craters and one large one. Due to the moon's small mass, none of the craters have central peaks or rings. Phobos has strange parallel linear features, many of which are associated with its large crater. Perhaps the parallel lines are indications of a strong impact that might have come close to destroying the moon, producing deep cracks at the time.

Deimos is much smoother than Phobos and has no large craters. Its surface is dotted with numerous bright patches whose nature is unknown.

The *Viking* spacecraft passed close enough to the moons to interact gravitationally and therefore allow masses and densities to be determined. Although the density of Mars is about 4 g/cm³, that of the moons is only 2 g/cm³. They appear to be made of material different from that of Mars. Observations indicate they are similar to large asteroids and the carbonaceous chondrite meteorites. From their chemical makeup, and from their appearance, one might conclude that they are captured asteroids. However, we do not expect randomly captured bodies to be in circular orbits and in the equatorial plane, as both Phobos and Deimos are. However, if they formed by accretion at the same time and place as Mars, why the difference in densities? Although various hypotheses have been made, no one answer is clearly better than any other.

9.7 Evolutionary Comparison

The similarities of the terrestrial planets are assumed to result from their similar evolutionary histories. The differences probably arise from variations in their location in relation to the Sun and Jupiter, mass, possession of a massive moon, rate of meteoroid impact, and so on.

The evolutionary story for **Mercury** serves as the simplest example of terrestrial-planet evolution. An examination of the *Mariner* images suggests that there was an early period of surface melting. Such early melting is required because the surface shows the effects of cratering from several billion years, without much erasure of features having taken place. The picture of Mercurian history that seems to fit the data is roughly as follows:

1. Creation of the planet out of the primitive solar nebula, as discussed in Chapter 6
2. Soon after formation, chemical differentiation that led to a heavy iron core and a lighter silicate mantle
3. Possibly an early smoothing of the surface, perhaps by volcanic action, to erase the scars of the formation itself
4. An episode of meteoritic bombardment similar to that which took place on the Moon, which created large features such as the Caloris Basin
5. Thereafter, minor volcanic activity at most, and a long era of relative quiescence with occasional peppering by meteoroids, even to the present day

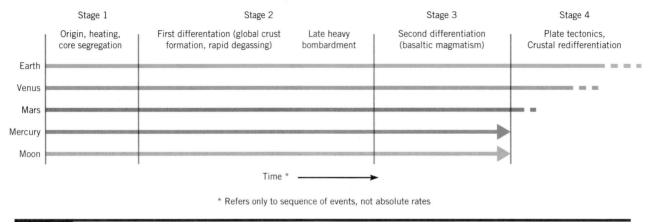

Comparative Crustal Evolution

	Stage 1	Stage 2		Stage 3	Stage 4
	Origin, heating, core segregation	First differentiation (global crust formation, rapid degassing)	Late heavy bombardment	Second differentiation (basaltic magmatism)	Plate tectonics, Crustal redifferentiation

Earth

Venus

Mars

Mercury

Moon

Time * ⟶

* Refers only to sequence of events, not absolute rates

Figure 9-42 A comparison of the crustal evolution of the terrestrial planets and the Moon.

The story of **Mars** is more complicated, while Venus's history is obscured by its relatively young surface.

We have seen that **Mars** has a puzzling mixture of surface features. How are we to interpret them? Mars's crustal history has been more active than that of either Mercury or the Moon. Certain regions of Mars are relatively young; for example, there are fewer impact craters near some of its bigger volcanoes than elsewhere, indicating that these volcanoes formed later. There are also other heavily cratered and presumably very old regions, so Mars has a considerably varied history. The different appearances of the channels indicate that some of them are a mere 100 million years old, whereas others are as much as 2.5 billion years old.

We are now in a position to sum up our new views of the terrestrial planets. It is clear that after many years of thinking of them as places somewhat similar to Earth, we are forced to conclude that they are unique entities, unlike the Earth or the Moon, and each with its own alien soil, chemistry, history, and geology.

Figure 9-42 summarizes the evolution and emphasizes that the early crustal evolution of the terrestrial planets was the same. In Stage 1, they all were formed, heated, and had their cores separate to one extent or another from the remaining material. In Stage 2, the igneous crust formed. The lunar highlands, for example, formed during this stage. Heavy bombardment toward the end of this stage produced basins such as Hellas on Mars, Caloris on Mercury, and the large circular basins on the Moon. Stage 3 involved the production of relatively smooth, level regions. During this

period, the maria basins that were excavated in Stage 2 were filled with lava, and the smooth plains of Mercury were formed. Crustal evolution of the Moon and Mercury ended here. At Stage 4, crustal evolution becomes unique for each object. We know some crustal evolution continued on Mars because the Valles Marineris is most likely a tectonic depression rather than a canyon formed exclusively from the flow of liquids. Data from *Magellan* show that crustal development on Venus continued and may still be in progress. On Earth, erosion and extensive tectonic activity (continental drift and volcanism) show that crustal evolution is continuing even today.

Inquiry 9-9

Considering the successes and failures of the Viking missions, what suggestions can you make for future exploration of Mars?

Inquiry 9-10

How does Viking illustrate the differences between space missions carrying astronauts and those carried out by robot probes?

Inquiry 9-11

Using Figure 9-42, make a statement about how the evolution of crustal evolution of terrestrial planets depends on the planet's mass. How might you explain such a relationship?

9.8 Future Studies

NASA launched the first spacecraft to Mercury since the 1974 *Mariner 10* in mid-summer 2004. Three flybys will occur in 2008 and 2009, and it will go into orbit in 2011. Called *MESSENGER* (MErcury Surface, Space ENvironment, GEochemistry, and Ranging), this NASA mission's goals are to answer questions dealing with understanding why Mercury's density is so high, its geologic history, its core structure and the nature of the magnetic field, the nature of the bright material at Mercury's poles, and the importance of certain volatile gases in the planet's thin atmosphere. Various hypotheses for each of these make specific predictions that *MESSENGER* will test. Thus, the mission is meant to falsify many hypotheses while verifying others.

The European Space Agency (ESA) is planning to launch the *BepiColombo Mercury Orbiter* in 2012. Current plans are for two modules. One, the *Mercury Planetary Orbiter*, will make remote sensing and radio science observations. A second is the *Mercury Magnetospheric Orbiter*, which is meant to better characterize Mercury's magnetic field. (A lander had been planned, but it was cancelled due to budget constraints.) The scientific goals are to study the planet's form, interior structure, surface geology and composition, the planet's magnetic field and atmosphere, test Einstein's theory of general relativity, search for asteroids close to the Sun, and study the planet's origin and evolution.

The Mars *Reconnaissance Orbiter*, to be launched in 2005, will be able to see objects as small as a few feet across. It will search for additional evidence for a long-term past presence of water. In addition, it will analyze minerals and study atmospheric water and dust as a part of giving daily global Martian weather reports. The mission is planned to last to the end of 2010.

A 2007 launch is planned for the next Mars lander, called *Phoenix*. The Mars *Science Laboratory* is planned to launch in 2009 to explore and assess Mars as a potential habitat for past or future life.

Chapter Summary

Observations

- Table 9-1 summarizes our factual knowledge of the terrestrial planets.
- Venus's surface temperature is hot enough to melt lead. Mars's temperature is generally well below the freezing temperature of water. Mercury's surface is hot.
- Mars's surface is red because of large amounts of iron oxide.
- Although no liquid water is present on Mars today, there is evidence that in ancient times water used to flow and pool on the surface in abundance, and indications that water currently exists as **permafrost** below the surface.
- All terrestrial planets exhibit impact craters (although those on Earth have been greatly eroded). Some impacts produced large basins still visible (a number with multiple rings around them) on Mercury and Mars.
- Venus and Mars both have ancient, gigantic volcanoes, and continent-sized raised regions. Neither shows evidence of continental drift. While there is no indication of current volcanism on Mars, there is such evidence for Venus.
- All terrestrial planets show faults that indicate various degrees of surface activity. Mercury has extensive cliffs caused by crustal shrinkage. Venus has a region of crisscrossed faults whose origin is not understood. Mars has a variety of fault structures, including an extensive canyon system whose origin is still unknown.
- Mars has high-speed winds traveling over the surface and producing dust storms and erosion. Venus has high-speed winds at the top of its atmosphere, but not at the surface.
- Mars's polar region contains ice caps made of both frozen carbon dioxide and water that change with the seasons. Mercury appears to have an ice cap.

Theory

- The terrestrial planets have changed over the life of the Solar System. The nature and rate of change depend on the location of the planet in relation to the Sun, Jupiter, and a massive moon (if any), and the original mass and composition of the planet. We can infer what the surfaces used to be by comparing the planets to each other and applying universal laws of nature.

Conclusions

- Venus's surface temperature is hot enough to melt lead due to the presence of a runaway greenhouse effect. Mercury's surface is hot because of its closeness to the Sun.

- Although the total carbon dioxide content of Venus and Earth are similar, Earth's CO_2 is tied up in carbonate rocks. Venus's water evaporated, at which time the water molecules were destroyed; the rapidly moving hydrogen atoms then escaped the gravitational pull of the planet. Oxygen on Earth came about mostly from plant life. Earth's nitrogen came from volcanic outgassing and also from the decay of dead organisms.

- Mercury's surface composition is thought to be basaltic, like the Moon's surface. Venus's surface composition is basically similar to that of Mercury and the Moon but probably with some as yet unknown chemical differences.

- Numerous observations of Mars point to an extensive amount of liquid water existing for a short period of time in the distant past.

- The planetary interiors are similar in that they have cores surrounded by mantles and crusts. Mercury's magnetic field indicates that the planet has a substantially larger core than the other terrestrial planets.

- The characteristics of the terrestrial planets—surface, atmosphere, magnetic fields—continue to change, although at different rates and by different mechanisms.

Summary Questions

1. What are the major facts known about each of the terrestrial planets? Include distances, general orbital characteristics, size, mass, and density relative to the Earth; also include temperature, atmospheric composition, etc.

2. Compare and contrast all the terrestrial planets with one another. In the discussion, include any distinctive features of orbital characteristics, appearance, rotation and revolution, general atmospheric features and composition, interior structure, magnetic fields, and moon systems.

3. What planetary features were Earth-bound astronomers able to see through their telescopes on each of the terrestrial planets? Describe any changes in these features that astronomers could have observed.

4. What do we mean by the *runaway greenhouse effect?* Why do we think the atmospheres of Earth and Venus developed as they did? What is the critical role of life in the maintenance of Earth's atmosphere?

5. Describe the history of Martian canals, paying attention not only to the objective evidence prior to *Mariner* but also to possible nonobjective aspects.

Applying Your Knowledge

1. Refer to Figures 4-27 and 4-29, which show the location of planets in relation to Earth. Make a drawing similar to these figures showing where Venus would have to be, in relation to the Earth, to produce each part of Figure 9-2. Label your drawing with the dates. In your drawing, place the Earth at the top. (Note: You want to be able to reproduce both the phase and the relative size of Venus.)

2. In what ways might Earth and Mars have been different than they currently are if each exchanged places with the other; in other words, if a Mars-sized body were 1 AU from the Sun and an Earth-sized body at 1.5 AU?

3. Examine Figure 9-5a. How do you know that this is a picture of Mercury and not the Moon (other than because the figure caption tells you so)?

4. Why does the fact that Venus has a dense core lead to our expectation that the surface rocks probably have a low iron content?

5. On Earth a solar day is a little longer than a sidereal day, but on Venus it is shorter. Explain why. (*Hint:* Use a *carefully* made drawing.)

6. You have just landed on Mars. Your view is that of Figure 9-13. Describe the view vividly and in detail.

▶ 7. If Mercury's angular diameter is observed to be 0.002 degrees when its distance from Earth is 0.92 AU, what is Mercury's diameter? Express the diameter in kilometers, in units of the Earth's diameter, and in terms of the Moon's diameter.

▶ 8. Use the planetary orbital periods in Table 9-1 or Appendix C to calculate the distances of the terrestrial planets from the Sun in AU.

▶ 9. How much would a 150-pound person weigh on each of the terrestrial planets, as well as on the Moon?

▶ 10. Deimos circles Mars at a distance of 23,500 km in 1.26 (Earth) days. Calculate the mass of Mars, and compare your value with the "true" value. What is the percentage difference?

▶ 11. Under the best observing conditions on Earth, the smallest angle astronomers can discern is about 0.25 seconds of arc. Assuming Martian "canals" must have at least this angular size, compute their true size in kilometers. Is such a size reasonable for channels produced by intelligent beings?

▶ 12. The photographs of Venus in Figure 9-2 are at the same scale. Assuming the first one was taken when Venus was at its greatest distance from Earth, and the last one when it is nearly at its closest approach, determine the ratio of greatest to least distance from Earth. Does your answer make sense in view of what you know about Venus's orbit?

▶ 13. Compute the size in kilometers of the smallest Martian feature on the photographs in Figure 9-3.

▶ 14. Does the failure of Newton's theory of gravity to explain the amount of the observed precession of the orbit of Mercury mean that Newton's theories are incorrect descriptions of nature and should be disregarded? Explain.

Answers to Inquiries

9-1. Two major effects are color changes and variations in the size of the polar caps.

9-2. With a 59-day period, every side of Mercury eventually receives sunlight before it can cool off completely.

9-3. 2/3 orbit for one rotation, 1-1/3 orbits for two rotations, and 2 orbits for three rotations.

9-4. Mercury's axial tilt of 0° means there are no seasons. However, unlike Earth, Mercury's high orbital eccentricity will produce annual temperature variations.

9-5. Because we see primarily the cloud tops in visible light, we are looking at a cooler layer.

9-6. Volcanism and meteoritic impact.

9-7. Rocks on the surface have sharp edges.

9-8. Because the clouds retain heat and the circulation moves it around in a planetwide pattern, you would expect high temperatures at the poles, too.

9-9. This is an open-ended question. One might think of: better-designed experiments that would unambiguously detect life or its lack; an "intelligent" robot that could move about the surface and sample a wider area of Mars; a probe designed so that it can land safely in the Martian highlands or near a polar cap.

9-10. Many questions were left unresolved, because the range of experiments that could be carried out was so limited.

9-11. The greater the mass of a terrestrial planet, the longer the period of time over which crustal changes occur. The reasons are not clear-cut but would involve the consequences of a larger mass. These include greater core pressures and temperature that would affect the overall planetary structure, which would affect the ability of the planet to produce and hold heat, which would determine whether or not the planet would have a structure allowing for continents and their motions.

Mercury

The View from Earth

Mercury is difficult to observe from Earth because it is the nearest planet to the Sun and is never more than about 27° away from it. For this reason, it is always observed near the horizon and can never be viewed in dark skies far from the Sun, but mostly in twilight. Furthermore, its angular size is always small. For these reasons little information, beyond knowing that it is one-third the Earth's size and 5 percent the Earth's mass, has been obtained about the planet from ground-based optical observations.

Mercury's orbit deviates considerably from a circle; it has the second most eccentric planetary orbit in the Solar System after Pluto. Likewise, its orbit has the second greatest tilt to the ecliptic of any planet.

Analysis of radar waves reflected from the surface show the planet to rotate three times on its axis for every two revolutions about the Sun. Mercury has no known moons.

The View from Spacecraft

Spacecraft images of Mercury show its surface to have numerous craters; it appears similar to the Moon. Because Mercury has a low mass and high temperature, the planet cannot maintain a substantial atmosphere. The only atmosphere it has comes from the Sun and from gases emitted by its heated surface material.

Unique Features

Mercury has a surprisingly strong magnetic field. Assuming this magnetic field is caused by a dynamo effect similar to that thought to cause Earth's magnetic field, Mercury must have a large metallic core. Mercury's surface contains a large number of steep cliffs called scarps that may have been caused by stresses in the crust while it was cooling. The planet has a large multiringed basin called Caloris Planitia, most likely caused by an asteroid impact. The surface area between craters appears to be covered by ancient lava flows.

Additional Important Features

Mercury lacks mountain ranges similar to those on the Moon. Recent observations indicate a possible ice cap at its north pole.

---

[1]For detailed information on each of the planets, refer to Table 9-1.

Venus

The View from Earth

Although Venus comes closer to Earth than any other planet, and is nearly the same size and mass as Earth, few surface details have been observed visually because the planet is covered with dense clouds. Furthermore, Venus is never more than about 48° from the Sun. Radar waves from ground-based telescopes penetrate the clouds to the surface. Such data have shown that Venus rotates in a retrograde manner, in the direction opposite to its orbit about the Sun. Because Venus's spin axis is nearly perpendicular to its orbital plane, the planet has almost no seasonal changes.

Venus's atmosphere is mostly carbon dioxide. The CO_2 atmosphere produces a greenhouse effect that holds the Sun's heat and results in an average surface temperature of 730K, high enough to melt lead. There is no water on the surface and only a trace of water vapor in the upper cloud layers. No planetwide magnetic field has been observed. Venus does not have a moon.

The View from Spacecraft

Spacecraft observations show Venus to be encircled by high clouds whose circulation is determined by solar heating. The clouds move at high speeds.

A surface pressure that is about 90 times sea-level pressure on Earth accompanies the high surface temperature. The clouds contain concentrated sulfuric acid droplets. The *Magellan* spacecraft has observed ancient volcanoes, long winding channels, and regions of extensive faulting on the surface. No definite indication of continental motions has yet been observed.

Unique Features

Venus stands out from the other terrestrial planets because of its slow retrograde rotation; its high density, carbon dioxide atmosphere; and its runaway greenhouse effect. Its high surface temperature and pressure are also unique features. There is circumstantial evidence for current volcanism, but further data are needed.

Additional Important Features

Soviet *Venera* landers found the surface to contain sharp-edged rocks, indicating a lack of surface erosional processes. Lightning occurs often in Venus's atmosphere.

Earth

Mars

The View from Earth

The third planet from the Sun has a surface showing a wide diversity of features caused by a variety of processes. These include moving continents that float on underlying, higher-density material. When one continental mass moves beneath another, the crust can be uplifted and mountain ranges produced. Seismic events thus occur along such plate boundaries. These seismic events are used to study the interior structure, which is thought to consist of a solid inner core, a liquid outer core, and a surrounding mantle. Material that sinks into the mantle can be heated to produce volcanoes along the borders between the continental plates. Earth contains some impact craters and basins, but wind, water, freeze/thaw cycles, and continental motions erode them over geologically short periods of time.

The atmosphere contains 78 percent nitrogen and 21 percent oxygen with trace amounts of argon, carbon dioxide, and water. The oxygen on Earth results from plant photosynthesis. The atmosphere's weight yields a surface pressure of 14.7 lb/in^2 at sea level. The temperature variation with height above the surface depends on various factors, including chemical composition. Heat from the Sun, along with heat emitted from the oceans and continents, produces global wind patterns.

A compass, which in the Northern Hemisphere points to the north magnetic pole, shows Earth to have a magnetic field. The magnetic field traps charged particles from the Sun, forming the Van Allen radiation belts and the aurorae.

The View from Spacecraft

Viewed from space, Earth shows changeable clouds and large land masses with surrounding oceans. Spacecraft also detect Earth's magnetic field and Van Allen belts.

Unique Features

Earth is the only planet known to have moving continents and liquid water on its surface, and is the only planet known to currently nurture life.

Additional Important Features

Earth's surface rocks consist of three types: igneous, sedimentary, and metamorphic. These participate in a continuous rock cycle because of continental motions. Earth's oceans are subjected to gravitational tidal effects caused by the Moon and Sun's gravitational pull being different on opposite sides of the planet.

The View from Earth

The fourth planet from the Sun is half the size and one-tenth the mass of Earth. It has a characteristic reddish color that makes it easy to spot during its apparent trek among the background stars. Mars's light and dark areas change appearance throughout its year as winds cover and uncover various surface areas. Polar caps are observed to fade in the Martian summer and grow as winter progresses. Clouds are seen in the Martian atmosphere. Mars has two moons, Phobos and Deimos.

The View from Spacecraft

The northern hemisphere is mostly lava-covered plains, which contain volcanoes as large as many states in the United States. These large volcanoes are considered evidence against continental drift. The southern hemisphere is strongly cratered. A gigantic valley, the Valles Marineris, extends some 5,000 km across the Martian surface.

Mars's surface is red because its surface rocks contain relatively large amounts of iron. To our surprise, the Pathfinder mission found a core, but its size is not well known. This result may be consistent with Mars's lack of a global magnetic field.

Evidence for the presence of liquid water in Mars's distant past abounds. Examples include ancient riverbeds showing debris that was carried along with the water, sandbars shaped by flowing water, and a variety of chemical and physical observations of rock and rock layers. Evidence for water in the form of permafrost comes from layering of surface deposits of sand and ice, as well as from landslides. The polar ice caps consist of two parts: the large caps seen from Earth are made of frozen carbon dioxide, while smaller caps that remain for a while after the CO_2 sublimates contain water ice.

Unique Features

Mars's unique features include the volcanoes, valleys, and ancient riverbeds.

Additional Important Features

The differences between Martian seasons in the northern and southern hemisphere are greater than those on Earth. This difference occurs because the planet has a more eccentric orbit, which produces greater changes in its distance from the Sun throughout its year. These seasonal changes produce winds that sculpt the surface.

10

The Jovian Planets and Pluto

Do there exist many worlds, or is there but a single world?
This is one of the most noble and exalted questions in the
study of Nature.

Albertus Magnus, thirteenth century

The Jovian planets have gross similarities, yet no two are alike. Jupiter and Saturn are liquid giants—liquid throughout except for the very central volume and upper gaseous atmosphere. Uranus and Neptune, by contrast, are often referred to as *ice giants* because a large fraction of their interior appears to be water ice with thick gaseous atmospheres on the outside. Although they all have moon and ring systems, even these differ dramatically from each other. In this chapter you will learn about some of the exciting results of recent explorations, and you will see that processes studied in earlier chapters are applicable to each of the Jovian planets and their moons.

Although Pluto is not a Jovian planet (nor a terrestrial planet, either) we present it in this chapter, because of its great distance from the Sun, and its similarity to some moons of Jovian planets.

10.1 Introduction

We will see what information we have obtained about the Jovian planets with observations from Earth's surface before we look at the data obtained from spacecraft.

AN OVERVIEW OF GROUND-BASED STUDIES

The four Jovian planets, also known as the giant planets, are shown in **Figure 10-1** as observed through Earth-based telescopes. Even a quick glance reveals striking similarities and differences. More detail is observable on Jupiter because it is much closer to Earth than any of the others (about 4 AU at its nearest). Neptune is about 30 AU from Earth at its closest. Jupiter shows distinct, variable, multicolored bands across its face. Markings include a large reddish area in the southern hemisphere, which has become known as the **Great Red Spot**. Its continual presence has been observed ever since first sighted more than 300 years ago. Saturn has bands that are similar but less distinct and less colorful. Some photographs of Uranus and Neptune show a hint of bands, but they are not as obvious. The relative motions of the bands and the different rotation rates at various latitudes show that we are observing complex atmospheres rather than a solid surface.

Moon systems are readily observed around Jupiter and Saturn with even the smallest telescopes. Observations over only a few hours will reveal changes in the locations of the moons in their orbits. Moderate-sized telescopes are required to observe the moons of Uranus and Neptune.

Perhaps the most distinctive property of any of the planets is the beautiful ring surrounding Saturn. Although Galileo observed it as fuzzy protrusions, the ring shape was inferred some 50 years later by the astronomer and physicist Christian Huygens. Because Saturn was the only planet known to have a ring system, theorists scrambled to explain why rings around planets were rare.

As Saturn and Earth orbit the Sun, we are able to view the ring system from different orientations, as shown in **Figure 10-2**. The rings are so thin that when we see them edge-on, they practically disappear from view. The rings are only a few tens to hundreds of meters thick and are composed of many small bodies—ice, dust, rocks, and boulders up to a few meters in diameter. We know that the rings are not solid, because bright stars can sometimes be seen through them. Furthermore, the inner parts of the ring system rotate more rapidly than the outer parts, in accordance with Kepler's third law. The total mass of Saturn's rings is estimated to be perhaps one-fourth that of our own Moon.

Chapter opening photo: Churning clouds in the atmosphere of Jupiter.

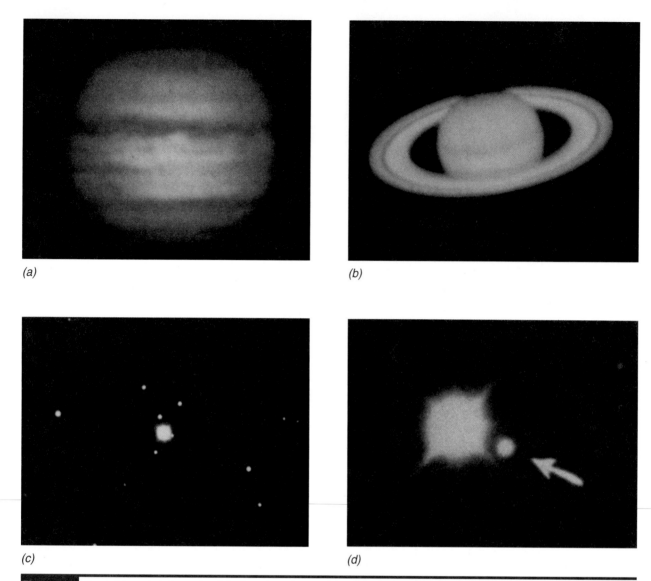

(a)

(b)

(c)

(d)

Figure 10-1 Ground-based photographs of the four giant planets: *(a)* Jupiter, *(b)* Saturn, *(c)* Uranus, and *(d)* Neptune. (Not to the same scale.)

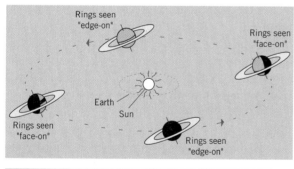

Rings seen "edge-on"

Rings seen "face-on"

Earth

Sun

Rings seen "face-on"

Rings seen "edge-on"

Figure 10-2 The apparent orientation of Saturn's rings as seen from Earth varies with its position in its orbit. The small circle around the Sun represents the orbit of the Earth.

Inquiry 10-1

If Saturn's rings were solid, how would you expect them to rotate? Would the inner or the outer parts of such a ring take longer to go around?

The visible light we receive from each Jovian planet is reflected sunlight. However, because each planetary atmosphere affects the light in ways that depend on its chemical makeup (which will be described in Chapter 13), we have been able to discover, from Earth-based observations, the presence of methane and ammonia in each planet's atmosphere.

Astronomers were surprised in the middle 1950s when they recorded intense bursts of radio energy coming from **Jupiter**. These bursts of energy were found to depend on the orbital location of the moon Io. Energetic discharges were interpreted as being caused by charged particles trapped in an intense Jovian magnetic field, forming Jovian *Van Allen belts,* as described for Earth in Chapter 8. The planetary magnetic fields are discussed further in Section 10.4.

While the ancients knew about Jupiter and Saturn, Uranus and Neptune were not known. **Uranus** was discovered by the English astronomer William Herschel in 1781. He immediately understood that it was not a star, because it moved from night to night in relation to the stars. Furthermore, it was blue-green. The story of the discovery of **Neptune** in 1845 and 1846, and its importance as a verification of Newton's ideas, was discussed in Chapter 5.

SPACE EXPLORATION OF JOVIAN PLANETS

The first spacecraft launched to the outer planets were *Pioneer 10* and *Pioneer 11* in 1973 and 1974. These were small spacecraft carrying instruments to study the charged particle and magnetic environments of Jupiter and Saturn. In addition, they were to produce images of higher resolution than were possible using telescopes on Earth's surface. NASA lost contact with *Pioneer 10* in early 2003, when, at a distance of 82 AU, its power source was no longer able to generate power. Prior to that time, however, *Pioneer 10* continued to send data back to Earth from regions of the Solar System never before studied. A number of important discoveries were made that paved the way for the more sophisticated and better-equipped *Voyager 1* and *Voyager 2* probes launched in 1977 and the *Cassini/Huygens* mission to Saturn that was launched in 1997.

The two *Voyager* missions were incredible testimony to both scientific curiosity and knowledge, and to our willingness to explore the unknown. Originally designed to last only long enough to study Jupiter and Saturn, the spacecraft proved to be hardy and long-lived. *Voyager 1* flew by Jupiter in March 1979 and Saturn in late 1980. *Voyager 2* passed Jupiter in July 1979, Saturn in August 1981, Uranus in January 1986, and Neptune in August 1989. Both are traveling through the outer Solar System; we hope to continue to receive data for the next one to two decades as they exit the Solar System and enter interstellar space. *Voyager 1* passed 90 AU in late 2003. In fact, some scientists argue that the evidence shows it has left the Solar System and is now in interstellar space.

Voyager 2 passed within 80,000 km of Uranus (and within 16 km of its designated target point, a feat that has been compared to bowling from Los Angeles to New York and making a strike) and was able to tell us more about Uranus in a few minutes than we had learned since its 1781 discovery. *Voyager 2* then passed 5000 km above Neptune's clouds.

The *Galileo* spacecraft is the most recent one, with Jupiter as its main mission. The mission involved two spacecraft: an orbiter and a probe that entered the Jovian atmosphere with the purpose of the first direct measurement of the atmosphere of one of the giant planets. The spacecraft arrived at Jupiter in December 1995 after a six-year trip and three gravitational assists: one from Venus and two from Earth. The main mission had the spacecraft circle Jupiter 11 times, each time in different elliptical orbits that allowed detailed studies of Europa, Ganymede, and Callisto, and studies of Io from medium distances. An extended mission to study Europa in more detail, because of the suggestions that Europa could harbor life, continued for 14 months after. Thereafter, it moved closer to Io, reaching there in October 1999.

As you read this book, the *Cassini* spacecraft is circling Saturn. After a seven-year trip, *Cassini* went into orbit about the planet in mid-2004 to study the planet, moons, and ring system for at least four years. You may want to check *Cassini's* Web site for updates: *http://saturn.jpl.nasa.gov/home/index.cfm.*

10.2 Large-Scale Characteristics

In this section we investigate such basic macroscopic properties as orbital and rotational characteristics, mass, size, density, seasonal changes, and energy emission.

Orbits and Rotation

	JUPITER	SATURN	URANUS	NEPTUNE	PLUTO
Orbital eccentricity	0.048	0.056	0.047	0.009	0.248
Orbital inclination (degrees)	1.3	2.5	0.8	1.8	17
Rotation period (hours)	9.92	10.66	17.24	16.11	153.3

The orbital characteristics of all the Jovian planets are similar to one another, but distinctly different from those of Pluto, whose orbit is both eccentric and inclined to the ecliptic.

Continuous observations of spots moving across a planet's disc allow astronomers to determine the rotation period. Spots at **Jupiter's** equator complete their journey around the planet in only 9 hours and 55 minutes, written as 9^h55^m. Regions closer to the pole take longer to complete their cycle. **Saturn**, too, completes its rotation in a short time, 10^h39^m. The small angular size of Uranus and Neptune, along with indistinctness of their features, makes their rotation periods difficult to determine accurately. Prior to the *Voyager* missions, astronomers thought their periods were about 18 hours.

A rotation period of roughly 10 hours for **Jupiter** and **Saturn** means that the equatorial regions are moving at high speeds. In the case of Jupiter, this means nearly 45,000 km/hour. Such high rotation speeds easily explain the observed flattening in the appearance of the planet (Figure 10-1).

Inquiry 10-2

Measure the flattening of Jupiter and Saturn in Figure 10-1 by finding the ratio of polar to equatorial diameters.

Uranus's rotation has been known for many years to differ from the general pattern in the Solar System, in which rotation axes are nearly perpendicular to the orbital plane. For Uranus, the rotation axis is nearly *in* the orbital plane, with a tilt angle of 98°. If current theories are correct, the Solar System was formed from turbulent gas, and it is entirely possible that local irregularities in the motion of this material caused the unusual rotation of Uranus. We have seen that, once established, motions persist until other forces enter the picture. Although this allows us to understand and predict the behavior of bodies moving under the influence of gravity, it does not tell us how they got to be the way they are now observed. By analogy, suppose we see a baseball come crashing through the living room window. Looking outside, we can only conjecture—albeit, in this case reasonably—that the ball was set into motion by the scared kid holding the baseball bat. For the Solar System, the game was played nearly 5 billion years ago, and the players have all gone home.

Pluto's orbit deviates from a circle by more than any other planet. In fact, Pluto's distance from the Sun varies from as little as 29.7 AU to as much as 49.3 AU. When closest to the Sun, it's closer than Neptune. However, with an orbital inclination of 17°, by far the largest in the Solar System, it cannot collide with Neptune. Furthermore, Pluto and Neptune are in resonance (see Chapter 7) with each other: For every three revolutions Neptune makes around the Sun, Pluto completes two. In other words, they have a 3:2 resonance with one another.

Size, Mass, Density

	JUPITER	SATURN	URANUS	NEPTUNE	PLUTO
Diameter (Earth units)	11.19	9.41	4.01	3.89	0.18
Mass (Earth units)	317.9	95.1	14.5	17.1	0.0022
Density (g/cm³)	1.31	0.69	1.29	1.64	2.10

In Chapter 6 we found that the first object a visitor from space would notice in our Solar System, after the Sun, would be Jupiter. **Jupiter** is brighter than the other planets because it is considerably larger. At 11 times the diameter of Earth, Jupiter could contain some 1,300 Earths inside it. Saturn, Uranus, and Neptune each have diameters of nine, four, and four times Earth's diameter, respectively.

The masses of all the Jovian planets are easy to obtain thanks to the presence of their many moons. Jupiter, with a mass more than 300 times Earth's mass, contains more material than all the other planets put together—about 70 percent of the mass of the Solar System that is not bound up in the Sun. Jupiter is so massive that it is perhaps more appropriate to think of it as a binary companion to the Sun rather than just the largest of the planets, even though it is not massive enough to be a star in its own right. The other Jovian planets, while smaller, still contain substantial amounts of material.

The density of each of the Jovian planets is low. In the case of Saturn, which has a mass of 95 Earth masses and a size of 9 Earth radii, the density is only 0.7 g/cm³. This value, which is less than that of water, means that Saturn could float if placed in an appropriately sized tub! The densities of the other Jovian planets are between 1.3 and 1.6 g/cm³, considerably less than the 5.5 g/cm³ for Earth.

Inquiry 10-3

Considering their low densities, what two elements would you expect to make up the bulk of the mass of the Jovian planets?

Seasons

	JUPITER	SATURN	URANUS	NEPTUNE	PLUTO
Axial tilt (Degrees)	3	25	98	30	123
Seasons	minor	yes	extreme	yes	extreme

The existence of seasons is determined by the tilt of the rotation axis to the perpendicular of the orbital plane. The tilts were discussed and shown in Figure 6-5b. Although **Jupiter's** angle of only 3° produces essentially no seasonal variations, **Saturn** and **Neptune** have axis tilts between 25° and 30°. These tilts, therefore, produce seasonal changes in the amount of energy received by every part of the planets' atmospheres.

Uranus is unique among the Jovian planets with its 98° tilt. There is a time in its orbit when the north pole points almost directly at the Sun; half of its 84-year orbital period later, the south pole points at the Sun. Midway between these times, the Sun shines directly onto the equatorial regions.

Inquiry 10-4

For approximately how long would the Sun be above the horizon for an observer at Uranus's north pole? Would you describe Uranus's seasonal changes as mild or harsh? Why?

Excess Energy

	JUPITER	SATURN	URANUS	NEPTUNE	PLUTO
Excess energy	yes	yes	small	yes	no

Ground-based observations of **Jupiter**, later confirmed by the two *Pioneer* spacecraft, indicated that Jupiter emits about twice as much energy as it receives from the Sun. Jupiter is not, therefore, just a passive reflector, but it has an internal energy source of its own. However, it is not a star, because its mass is more than 10 times too small for it to generate energy by means of nuclear fusion, as real stars do.

The source of **Jupiter's** excess energy is heat left over from the time of formation, produced when the gravitational potential energy in the protoplanetary cloud was converted into trapped heat during the collapse. The slow rate of heat escape indicates that Jupiter's interior is probably in a turbulent, boiling state, with much of the internal heat being carried outward by convection.

Saturn, too, radiates approximately twice as much energy as it receives from the Sun. In fact, relative to its mass, it emits more excess energy than Jupiter. The amount of Saturn's excess energy requires an additional energy production process, which seems to be the ongoing gravitational settling of helium, which is heavier than hydrogen, toward the interior. This settling, or differentiation, releases gravitational potential energy. An observation consistent with this idea is Saturn's smaller abundance of atmospheric helium compared to that in Jupiter.

Of the two outer Jovian planets, **Uranus** emits only a small amount of excess energy. **Neptune** radiates more than 2.7 times as much energy as it receives from the Sun. But if these planets are as similar as we have always thought, why does Neptune emit more energy? Models show that, like Jupiter, it may be radiating energy from it formation. The reason for Uranus's lack of excess energy is not yet known. In addition, Uranus and Neptune are not quite as similar as we had thought, as we will see in the next section.

10.3 Atmospheres

Each of the *Voyager* spacecraft returned dramatic new images. Worlds that had never before been seen as anything more than a point of light were turned into unique places. This section begins our more detailed examination of results from the *Voyager* and *Galileo* missions to Jupiter, and *Cassini* to Saturn. The discussion of the moons comes in Section 10.5.

APPEARANCE AND CIRCULATION

Full-disc images of each of the Jovian planets, obtained by the *Voyager* spacecraft, are shown in **Figure 10-3**. Jupiter revealed a varied and

Figure 10-3 Voyager images of the Jovian planets: Jupiter, Saturn, Uranus, and Neptune. The Earth is shown for a size comparison.

continually changing weather pattern of great complexity. The banded structure within Jupiter's atmosphere was once thought to be relatively stationary, but now material has been observed moving between bands. In addition, narrow bands have been seen to consolidate and wide bands to come apart. When the *Cassini* spacecraft flew by Jupiter on its way to Saturn it discovered that, contrary to previous ideas, the darker "belts" are rising and the lighter "zones" are falling.

The spotted regions on **Jupiter** are thought to be Jovian storm systems, not unlike those on Earth, although larger in scale and more persistent. The Great Red Spot, for example, is similar to a terrestrial cyclone (actually an anticyclone) that contains rising gases. The upward motion of the gas is aided by the excess energy flowing from the interior. This motion causes the Great Red Spot to rise to about 8 km above the surrounding gases. Forces generated by the rapid rotation of the planet cause the inflowing gas to swirl around like a whirlpool (the Coriolis effect of Chapters 5 and 8). **Figure 10-4** shows a close-up of the Great Red Spot. About 30,000 km in length, it would easily swallow the Earth!

Jupiter's Great Red Spot rotates, and small spots approaching the storm have been observed to accelerate and be gobbled up by it. This is not a temporary phenomenon; the spot has been observed for more than 300 years, although sometimes its contrast diminishes somewhat. The long duration of this giant storm turns out to be less surprising than it might at first seem, because one consequence of Jupiter's large size and dense

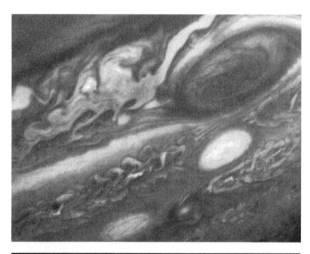

Figure 10-4 Jupiter's Great Red Spot, showing the turbulent gases that trace its wake.

atmosphere is that its weather takes place on time scales much longer than those on Earth. Calculations indicate that 300 years is roughly the length of time one would expect a major storm on Jupiter to last. Some theorists think that the Great Red Spot may be a permanent feature of Jupiter's atmosphere.

Regions of **Jupiter's** atmosphere outside the Great Red Spot (Figure 10-4) exhibit fantastic sawtooth-shaped patterns of gas that are remarkably similar to the turbulence created in air by the passage of an airplane. The diversity of color shown in the convoluted flow patterns, from reds and oranges to yellows and browns and even blues, provides a challenge to chemists and meteorologists. At pres-

Figure 10-5 The results of the collision of Comet Shoemaker-Levy 9 with Jupiter in 1994.

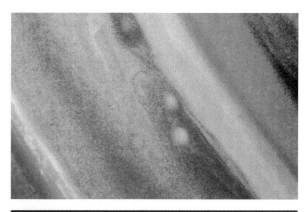

Figure 10-6 Saturn's atmospheric features appear less distinct than those on Jupiter.

ent, at least three major cloud levels have been distinguished on Jupiter: a high layer of frozen ammonia crystals, a middle layer that appears brownish, and regions where our line of sight apparently penetrates deep down into the planet's atmosphere, where there is a blue appearance.

The *Galileo* atmospheric entry probe found winds of 700 km per hour below the clouds. Such high-speed winds cannot be produced by solar heating or the condensation of water, as on Earth. Rather, these high winds appear to be caused by the heat escaping from the planet's deep interior.

A further opportunity to study **Jupiter's** atmosphere arose when nature decided to help astronomers with the collision of a comet with the planet. Comet Shoemaker-Levy 9, shown in Figure 7-7 after it broke into many pieces after a passage past Jupiter, collided with the planet in 1994 and provided astronomers with a unique opportunity to see how a planet's atmosphere reacts to the sudden input of millions of megatons of energy. This event is the first time we have ever witnessed such a collision. **Figure 10-5** shows the result of the collision of the piece of the comet

labeled G. The impact shows a central region 2,500 km across, a thin, dark ring 7,500 km across, and a dark outer ring about the size of Earth. This event provided astronomers with new observational constraints for their models of Jupiter's atmosphere.

Saturn shows many features similar to those on Jupiter. However, Saturn's features have a muted appearance in which the color contrast with the surrounding regions is not as strong (**Figure 10-6**). *Voyager* detected massive electrical discharges coming from a region near the equator, and scientists suspect that these are lightning discharges from a giant storm.

Voyager's view of **Uranus** (Figure 10-3) was expected to give us the same dazzling views of atmospheric features that we saw on Jupiter and Saturn, but such was not to be. Uranus turned out to be enveloped in a featureless haze. There are no stripes, bands, or spots of the sort that we might have expected by analogy with Jupiter, even though Uranus is also rotating rapidly. In spite of careful analysis, only four small clouds were seen. We are apparently witnessing the effects of hydrocarbon compounds in the upper atmosphere, molecules that we would call *smog* if they appeared over Los Angeles! These molecules are highly effective at scattering light; just as in a fog on Earth, detail is lost. The blue regions are indications of methane in the atmosphere, a compound that scatters blue light effectively while absorbing red light.

The absence of details in **Uranus's** cloud structure (see Figure 10-3) may come from the fact that, at the time of the observations, solar radiation was incident directly on the planet's north pole

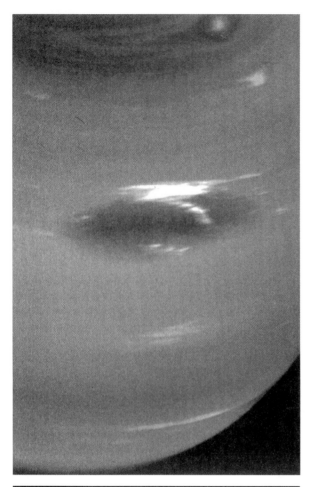

Figure 10-7 Neptune as observed by *Voyager 2*. The full disc of the planet shows high-altitude white clouds in addition to the Great Dark Spot, and a bright, rapidly moving bright spot called Scooter.

One definite statement that can be made about the meteorology of the Jovian planets is that it is extremely complex, and we do not yet understand it.

CHEMICAL COMPOSITION

The low overall density of the Jovian planets indicates a high abundance of the lighter elements such as hydrogen and helium. In fact, *Galileo* found that 92 percent of the gas atoms and molecules in **Jupiter's** atmosphere is hydrogen, while 7 percent is helium. These numbers are close to the abundance in the Sun, but not exactly the same. We also expect nitrogen and carbon atoms to be present because they are present in the Sun. Surprisingly, *Galileo* found that Jupiter contains substantially more carbon and sulfur than the Sun—an observation with important consequences for various theories of Jupiter's formation. The observation of less oxygen than in the Sun is a surprise, because oxygen was expected to be enriched from collisions with comets during the past 4.5 billion years. Finally, there is little evidence for organic molecules. However, at the cool temperatures at the higher levels of the Jovian planetary atmospheres, hydrogen will combine with carbon and nitrogen to form the volatile compounds methane (CH_4) and ammonia (NH_3). Ground-based observations showed these compounds to be present in the Jovian planets' atmospheres. Although observations indicate less ammonia in Saturn, Uranus, and Neptune than in Jupiter, the lack is readily explained: at the lower temperatures of the outermost planets, ammonia condenses into ices that are not readily observed.

Voyager and *Galileo* observations, along with information about temperature and pressure at various depths within the atmospheres, allow astronomers to compute models of the planetary atmospheres. The atmospheric structure of each of the Jovian planets is shown in **Figure 10-8**. The models show that Saturn's cloud structure is different from that of Jupiter. These differences can explain the muted appearance of Saturn's features.

For **Jupiter**, the *Galileo* atmospheric probe obtained new results that complicate the situation. *Galileo* found that the upper atmosphere was hotter and denser than expected, meaning there must be some source of atmospheric heat in addition to the Sun. Furthermore, no thick clouds were found,

(due to the extreme tilt of the planet's spin axis). Perhaps this unusual situation works against the formation of jet streams and other weather patterns familiar on Earth. The lack of substantial excess energy may contribute to the lack of detail in the atmosphere's appearance.

Neptune has a (Figure 10-3 and **Figure 10-7**) blue color, which comes from absorption of red light by methane, and is richer because we see to deeper layers of its atmosphere. Neptune is more dynamic than Uranus, with wispy clouds floating at high levels in the atmosphere and spots similar to Jupiter's Great Red Spot. The largest, dubbed the **Great Dark Spot**, rotates and is undoubtedly a long-lasting storm system. Other spots, such as a white spot called *Scooter,* move rapidly; winds in excess of 2,000 km/hour have been measured.

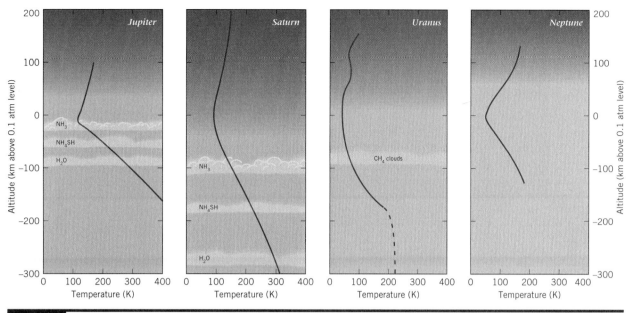

Figure 10-8 A comparison of the cloud structures of the Jovian planets. The thicker clouds on Saturn help mute the appearance of atmospheric structure.

meaning that the diagram for Jupiter in Figure 10-8 may be obsolete and wrong! The atmospheric probe detected methane, neon, argon, krypton, and xenon, but in amounts some three times greater than in the Sun; the meaning of this enhancement is not yet known. A search for water found far less than expected; planetary scientists now think the probe entered the atmosphere in a very dry, atypical, location. If that is the case, the expectation is that there is more water elsewhere, or deeper than the probe was able to go before being destroyed by the high atmospheric pressure it encountered while dropping through the atmosphere. These results show the importance in science of repetition and verification of results.

10.4 Planetary Interiors

The interior structure of a planet is not directly observable but must be inferred from models. The model computations are based on the laws of nature thought to be important in the interior of a gaseous planet and the constraints imposed by a variety of observations. Some of these observations include those previously discussed—density, flattening due to rotation, excess energy, temperature—and the presence of strong magnetic fields, which we discuss next.

Magnetic Fields

	JUPITER	SATURN	URANUS	NEPTUNE	PLUTO
Magnetic field	strong	strong	yes	yes	no?

Unexpected radio emissions were observed in the middle 1950s to be coming from **Jupiter**. The observations, which are unrelated to the excess energy previously described, were inconsistent with energy coming from either a hot or a cool body, but consistent with energy coming from an object having a strong magnetic field in which charged particles were accelerating at speeds near that of light. Such radiation is called **synchrotron radiation**. In other words, the observation of synchrotron radiation told astronomers that Jupiter has an intense magnetic field. (We will see later that synchrotron radiation also appears elsewhere in the universe.)

A detailed study of the charged particle and magnetic environment around each of the Jovian planets was one of the major goals of space missions to the outer planets. The magnetic environments will consist of charged particles fed into the magnetosphere and captured from the solar wind. Additional particles come, in some cases, from the

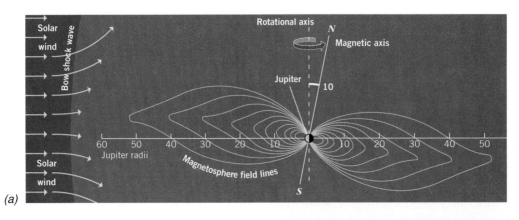

(a)

Figure 10-9 Jupiter's magnetosphere. *(a)* an illustration showing its size and tilt. *(b)* an image of the magnetic field obtained by the passing Cassini spacecraft on its way to Saturn. The black circle shows the size of Jupiter; sketched on either side of the center is the location of the ring of gas called the Io torus (Section 10.5) located in the orbit of Jupiter's moon, Io.

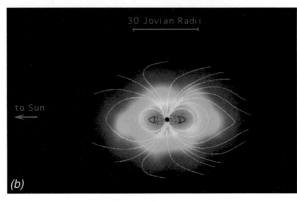

(b)

planet's moons. Just as the Earth's magnetic field is distorted by the solar wind, similar effects are observed for the other planet.

Jupiter's magnetosphere is illustrated in **Figure 10-9a** and is shown in a computer-produced image obtained by the *Cassini* spacecraft in **Figure 10-9b**. Jupiter's magnetic field is physically larger than the Sun. Expressed differently, if you were able to see it visually, its angular extent would appear two to three times larger than the full moon! The magnetic field is flattened by Jupiter's rapid rotation. Furthermore, the magnetic axis is tilted in relation to the rotation axis, which means the magnetosphere oscillates up and down as the planet rotates. Charged particles from the solar wind trapped by the magnetic field form radiation belts, analogous to Earth's Van Allen belts. The *Galileo* atmospheric probe discovered an intense radiation belt, 10 times stronger than Earth's Van Allen belts, between the ring and the upper atmosphere. Some of the particles spiral into the magnetic poles, producing auroras. **Figure 10-10** shows the auroral ovals for Jupiter and Saturn, which are like that for Earth as seen in Figure 8-18b. The Jovian atmosphere also has an electrical environment that produces extensive lightning, which, according to the

Galileo entry probe, is vastly different from lightning produced on Earth in that there is less of it but each bolt is more energetic.

Jupiter's charged particle environment contains particles from its moon Io, which is discussed further later. This moon, through its volcanic activity, supplies the Jovian magnetosphere with additional particles.

The other Jovian planets also have magnetospheres that are also fed and shaped by solar wind particles. Saturn's magnetic field, while not as strong as Jupiter's, is still physically larger than the Sun. For the other two Jovian planets, the magnetic fields are not as strong or as extensive as Jupiter's. For example, Jupiter's magnetosphere extends a distance of more than 50 times the planet's radius, but the Earth's magnetosphere extends only 10 times its radius. Also, the magnetic field at Jupiter's equator is nearly 14 times as strong as it is at the Earth's equator. **Figure 10-11** compares the magnetic field information for each of the Jovian planets. Just as Earth's magnetic axis is tilted with respect to its rotation axis (by 12°), the magnetic fields of the Jovian planets generally are not aligned with their rotation axes. Saturn is the main exception; its rotation axis and magnetic

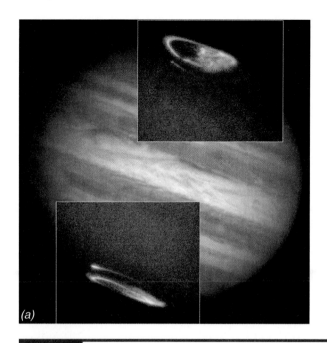

Figure 10-10 Auroral ovals as observed by the Hubble Space Telescope. *(a)* Jupiter and *(b)* Saturn. Courtesy of NASA and the Space Telescope Science Institute

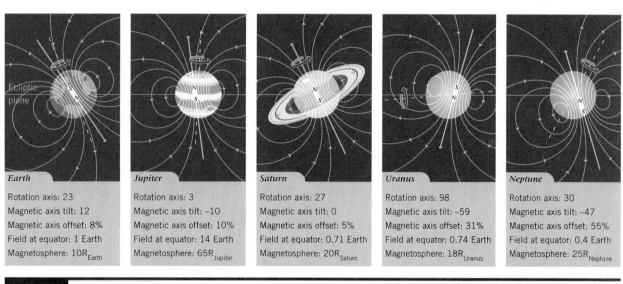

Earth

Rotation axis: 23
Magnetic axis tilt: 12
Magnetic axis offset: 8%
Field at equator: 1 Earth
Magnetosphere: $10R_{Earth}$

Jupiter

Rotation axis: 3
Magnetic axis tilt: −10
Magnetic axis offset: 10%
Field at equator: 14 Earth
Magnetosphere: $65R_{Jupiter}$

Saturn

Rotation axis: 27
Magnetic axis tilt: 0
Magnetic axis offset: 5%
Field at equator: 0.71 Earth
Magnetosphere: $20R_{Saturn}$

Uranus

Rotation axis: 98
Magnetic axis tilt: −59
Magnetic axis offset: 31%
Field at equator: 0.74 Earth
Magnetosphere: $18R_{Uranus}$

Neptune

Rotation axis: 30
Magnetic axis tilt: −47
Magnetic axis offset: 55%
Field at equator: 0.4 Earth
Magnetosphere: $25R_{Neptune}$

Figure 10-11 Summary and comparison of the magnetic environments of each of the Jovian planets. Note the relative strengths of the magnetic field at each planet's equator, the tilt of the magnetic field relative to the rotation axis, and the offset of the field from each planet's geometric center.

field axis are nearly identical. Uranus and Neptune provide a real surprise: in both cases, the magnetic axis is tilted 47° to 60° from the rotation axis. Furthermore, none of the magnetic field centers are at the planets' geometric center. The center of Neptune's field is offset farther from the geometric center than for any other planet. The cause of the field is discussed in the next subsection.

Inquiry 10-5

For review (from Chapter 8), what two conditions are required to produce a planetary magnetic field?

The magnetic fields of the planets rotate at rates that differ from what is observed for the planets' clouds. Because the rotation period of the observable clouds also depends on their latitude, such periods are not necessarily fundamental to the planetary body itself. The magnetic field rotation period is, however, related to the planetary body, and provides the most accurate measure of rotation.

Interior Structure

	JUPITER	SATURN	URANUS	NEPTUNE	PLUTO
Liquid metallic core	yes	yes	no	no	no
Differentiated	yes	yes	yes	yes	?

Magnetic field observations are telling planetary scientists a great deal about planetary interiors. Their job now is to interpret and understand what they are being told.

Magnetic fields are produced by moving materials that are good conductors of electricity, such as a metal in the case of the Earth. The Jovian planets, however, are mostly hydrogen and helium, neither of which are metals. A possible answer to the apparent dilemma appears when models of the interior structures are computed, as shown in Figure 10-12. Both temperature and pressure increase strongly with increasing depth. Underneath the visible clouds of all four Jovian planets, hydrogen is compressed into liquid molecular hydrogen.

Further in, interior models of **Jupiter** show the pressure to be greater than a million times the sea-level pressure on Earth—so high that hydrogen is compressed into an exotic state only recently produced in scientific laboratories on Earth known as **liquid metallic hydrogen**. In this state, hydrogen is a liquid but behaves like a metal, so it becomes an excellent conductor. Its conducting properties, along with the interior's rapid rotation and interior convection, are thought to produce a magnetic dynamo such as we discussed for Earth, and thus to generate the observed magnetic field. Jupiter's central region is thought to consist of a hot (15,000 to 20,000 K), high-pressure solid core (40 million times Earth's surface pressure) consisting of the heavier atoms that sank to the center because of gravity. Thus, the core may actually be rocky. The core's mass is some 10 to 15 Earth masses. A detailed knowledge of Jupiter's chemical make-up is considered to be a crucial missing link in our knowledge of the Solar System's formation.

A similar interior model results for **Saturn**. However, the lesser pressures (10 million times Earth's atmospheric pressure) decrease the size of

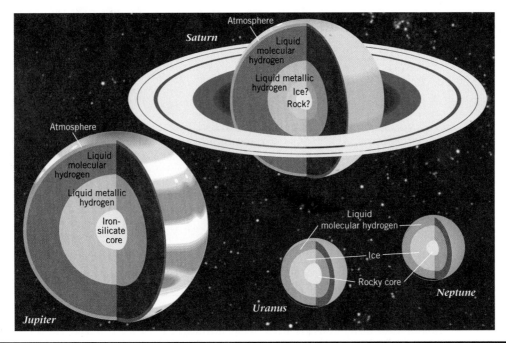

Figure 10-12 The interior structure of each of the Jovian planets. Note in particular the similarities of Jupiter and Saturn, and of Uranus and Neptune, and the differences between these two groupings. (Not to the same scale.)

the region of liquid metallic hydrogen. The core, which may be rocky like that of Jupiter, is thought to be 10 to 15 Earth masses—the same mass as Jupiter's core.

The interior models for **Uranus** and **Neptune** do not include liquid metallic hydrogen; the interior pressures are not great enough. Therefore, the magnetic field may be caused either by high pressure in ice surrounding the core or by as yet unknown conditions in the interior. In any case, the tilts between the magnetic and rotation axes and the offset from the geometric centers remain mysteries. The interior models indicate that most of the interior appears to consist of compressed water ice. It is for this reason that they are often referred to as *ice giants* rather than *gas* or *liquid giants*.

10.5 Moons

	JUPITER	SATURN	URANUS	NEPTUNE	PLUTO
Moons (as of May 2005)	63	47	27	13	1

The views of the outer planets as observed by the *Voyager* spacecraft were remarkable. However, they were almost upstaged by the magnificent sights of their moons. The moons generally fall into two groups: those having nearly circular orbits in the planet's equatorial plane, and those having orbits that are peculiar in one way or another (such as orbital shape, inclination, or direction of motion). Most of them have densities near that of water, although some are higher. The densities indicate a compositional mix of water ice and rock.

Perhaps the most surprising fact discovered about these moons is their lack of similarity in outward appearance. Each seems to have its own dramatic and individual personality. In this section, we will examine each moon system separately, comparing individual bodies when appropriate.

THE MOONS OF JUPITER

While Jupiter has a large complement of known moons (63 as of May 2004), the four largest were first seen by Galileo when he first pointed the telescope at Jupiter and are known in his honor as the **Galilean moons**. Io, Europa, Ganymede, and Callisto have densities that are higher than almost all other moons of Jovian planets, indicating surfaces covered with ices of methane and ammonia but with denser material in the interior.

Io, the innermost, is multicolored, looking very much like a pizza (**Figure 10-13a**)! Its surface appears to be pitted, but it is not cratered, and it gives an impression of having been sanded down by some process. The lack of impact craters makes it appear to be geologically young, just as the lunar maria are geologically young compared with the highlands.

(a)

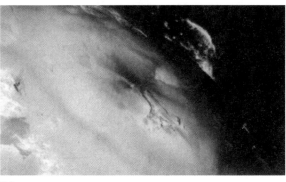

(b)

Figure 10-13 *(a)* Io. *(b)* A volcano during eruption, along with a volcanic vent in the middle of the figure.

Jupiter's gravitational pull on **Io** is strong. When combined with that of the relatively nearby Europa and the more distant but large Ganymede, Io is repeatedly stretched and squeezed and thus heavily stressed. Such stressing causes heating of the interior, just like the constantly changing stresses in a paper clip that is bent back and forth produces heat. From predictions of this **tidal heating**, planetary scientists predicted that Io might have active volcanoes. Imagine their pleasure when active volcanoes were observed on Io! Not only that, but many were observed to be going off simultaneously (see **Figure 10-13b** for an example). This discovery makes Io the most geologically active body in the Solar System.

Is the material the volcanoes spew out liquid sulfur or liquid silicate minerals? Sulfur melts at less than 700 K but silicates require more than twice the temperature. The *Galileo* spacecraft made observations of high temperature material, indicating that silicate materials dominate the volcanic flows.

Io's volcanoes also emit sulfur and oxygen atoms that become part of a doughnut-shaped region of glowing charged sulfur particles known as the Io torus that encompasses its orbit (**Figure 10-14**). A glowing sodium cloud results because sodium is sand blasted from Io's surface by particles caught in Jupiter's intense magnetic field. These charged particles interact with Jupiter's magnetic field and induce a flow of electricity between Jupiter and Io's surface. The charged particles then interact with the Jovian magnetic field to produce the aurorae and radio emissions discussed previously.

Europa, the smallest of the four Galilean moons, has an orangish, off-white color and a high reflectivity indicative of an icy surface (**Figure 10-15**). The false-color image's brown and reddish colors indicate contaminants in the ice. Icy plains are represented here in different shades of blue, which is caused by differences in grain sizes of the ice. Its most distinguishing and intriguing features are lines 3,000 km long and 100 km wide running across the surface. Their depth is unknown. They may be extensive surface cracks in the 100-m-thick surface ice. The moon has few craters and no tall mountains, indicating a surface that is not strong enough to hold a mountain's weight without sinking. Water ice has been observed on its surface. The *Cassini* spacecraft discovered a torus of gas in Europa's orbit. This torus is formed by the bombardment of Europa's icy surface by fast-moving particles from Jupiter.

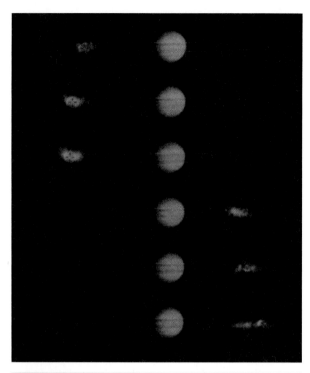

Figure 10-14 Ground-based photographs showing sodium vapor surrounding Io. As the moon circles Jupiter, sodium forms a doughnut-shaped cloud known as the Io torus around the planet. The figure is to scale, in which Io is the dot inside the cross-hair over the cloud.

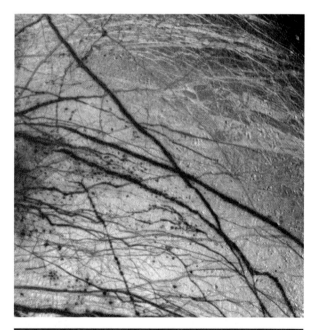

Figure 10-15 The surface of Europa shows icy plains (bluish hues) divided into regions that appear different because of different sizes of the ice particles.

Figure 10-16 A montage of the varied landscapes and surface features on Europa.

Other images of **Europa** show areas having rocky material that either comes from the interior or from impact (or a combination of both). Overall, the surface contains a striking array of exciting features (**Figure 10-16**). These include ridges, fractures, wedge-shaped bands where the surface appears to have been pulled open and then filled with material from underneath, the ice equivalent of lava flows on Earth, large areas of ice that broadly resemble icebergs, smooth puddles indicative of past fluid flow, isolated hills (the largest being more than a kilometer high), and impact basins with rays. The interpretation of many of these features involves a somewhat warmed water or a water-ice mixture, which serves to provide a fluid medium on which other material can float, much like Earth's continental plates float in the aesthenosphere. Many of Europa's surface features can thus be understood in terms of a migration of an icy surface (**Figure 10-17**, which has a resolution of 55 meters).

Does **Europa** have liquid water underneath the crust? If so, might it have conditions under which life might form? To help answer these and other questions, the two-year long Jovian mission was extended an additional two years, mostly to study Europa in more detail because the higher resolution of *Galileo* allows the surface to be seen in great detail. The current consensus is that the evidence is in favor of a 50–100 km deep ocean under perhaps as much as 20 km of ice.

Ganymede

The third Galilean moon, **Ganymede**, is bigger than the planet Mercury and is, in fact, the largest moon in the Solar System (**Figure 10-18**). Whereas the inner two moons have densities similar to those of the terrestrial planets, the outer two have densities only about twice that of water, suggesting that a significant portion of these moons may, in fact, be water. Water ice has, in fact, been observed on it. Ganymede shows evidence of an ice-like surface, along with many features somewhat similar to those on our own Moon, such as maria, highlands, and craters. All these features are in and on an icy surface, as compared to the rocky surface of the Moon. The icy surface may have resulted when ice in the interior melted, worked its way to the surface, and refroze. Although most of the icy surface is frozen methane and ammonia, water ice has also been found. Ganymede has more craters than the lunar maria, but there are no big craters. Perhaps of most interest to geologists is the existence of discontinuously broken lines; these are extensive fault regions, indicating the probable presence of internal upheavals.

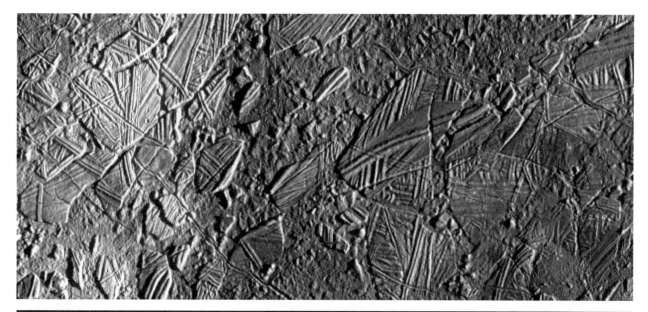

Figure 10-17 The complex icy surface of Europa. At a resolution of about 55 meters, the Galileo spacecraft shows a thin, jumbled icy crust. The white and blue areas are blanketed with fine ice particles from a crater 1000 km toward the bottom. The reddish brown colors lack the fine ice particles and result from minerals spread by water vapor released from below the surface.

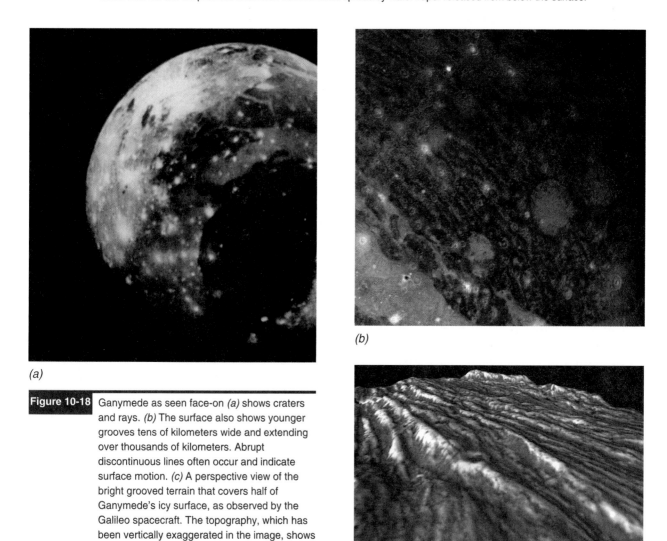

(a)

(b)

(c)

Figure 10-18 Ganymede as seen face-on *(a)* shows craters and rays. *(b)* The surface also shows younger grooves tens of kilometers wide and extending over thousands of kilometers. Abrupt discontinuous lines often occur and indicate surface motion. *(c)* A perspective view of the bright grooved terrain that covers half of Ganymede's icy surface, as observed by the Galileo spacecraft. The topography, which has been vertically exaggerated in the image, shows elevation differences of a few hundred meters between high and low points.

Figure 10-19 Callisto shows the most heavily cratered landscape of all the Galilean moons, indicating an old surface. Callisto also shows a multi-ringed basin called Valhalla.

The high resolution of the *Galileo* spacecraft cameras provided a detailed image of Ganymede's grooves (Figure 10-18c). The image shows ridge crests having bright icy material while the troughs contain dark material whose nature and origin are unknown.

Ganymede has been found to be the only one of Jupiter's moons to have its own magnetic field. The cause of it is uncertain because our knowledge of the details of the interior is too incomplete.

Ganymede may, like Europa, have an ocean of liquid water underneath the surface ice.

Callisto

The outermost Galilean moon, **Callisto**, is heavily cratered (**Figure 10-19**). This appearance could result from the fact that Callisto is the coolest of Jupiter's moons and hence the one whose surface solidified first. However, its smooth-appearing edge and lack of craters larger than 50 km in diameter suggest that it has a rather soft (plastic) surface that may not be able to support features as large as mountains. Such a soft surface is consistent with the observation of water ice, just as found on Europa and Ganymede. Callisto has one major impact basin, Valhalla, (Figure 10-19), which is a multiringed basin like Mare Orientale on the Moon and the Caloris Basin of Mercury. Those structures, however, were formed on rocky surfaces.

When the *Galileo* probe observed Callisto, the heavily cratered moon was observed to show few craters less than 100 meters in diameter. This is a surprise because, like the Earth's Moon, there is a lack of current volcanic activity present to erode craters.

Callisto may join Europa and Ganymede in having a liquid water ocean.

Other Observations About Jupiter's Moons

The other moons of Jupiter are not as large as the Galilean ones. **Amalthea**, one of Jupiter's inner moons, is small (only 130 by 80 km) and irregularly shaped, similar to Figure 7-14. Amaltea, along with others, play crucial roles in forming Jupiter's ring system, as discussed later.

While none of Jupiter's moons have dense atmospheres held by gravity, we now have observations about variable, low-density atmospheres for some Galilean moons. For example, because of **Io's** volcanic activity, *Voyager* found the moon to have sodium, potassium, and sulfur associated with it. *Galileo* observed sulfur dioxide and atomic oxygen. The *Galileo* probe confirmed an observation by the Hubble Space Telescope that **Europa** has an extremely thin, variable atmosphere that includes oxygen. **Callisto** was found to have a thin atmosphere of hydrogen and carbon dioxide. All four Galilean moons have ionospheres, which are atmospheric regions of charged particles; a more complete atmosphere may be inferred from this.

The Galilean moon interiors were studied by observing how each of the moons modified the spacecraft's orbit during each flyby. In addition, Io, Europa, and Ganymede are each magnetically connected to Jupiter and strong electric currents run between them. The high density of Io and Europa, and the magnetic field observed for Ganymede all imply the existence of an iron core in each of these objects. Taking all the information together, astronomers can infer the interior structure; **Figure 10-20** shows the results. The figure shows that Io has a large iron core surrounded by a rocky shell. Europa, by contrast, has a core surrounded by a shell perhaps containing liquid water and/or ice, surrounded by an icy crust. Callisto differs from the others, having a more homogeneous interior without a core.

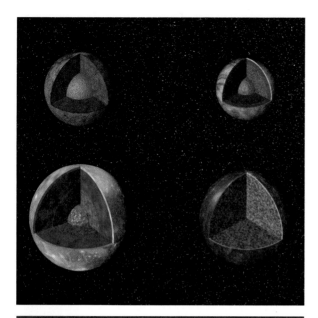

Figure 10-20 Cutaway views of the Galilean moons. Io (top left), Europa (top right), Ganymede (lower left), and Callisto (bottom right). Gray cores are iron and nickel; brown represents rocky shells. Blue and white are water in ice or liquid form. The sizes and interior structures are to scale.

Inquiry 10-6

From the size and shape of Amalthea, what might you conclude about its origin as a Jovian moon?

THE MOONS OF SATURN

The *Voyager* mission also spent considerable time examining the moons of Saturn, which turned out to be every bit as diverse as those of Jupiter.[1] (As of May 2005, Saturn is known to have 47 moons.) Saturn's largest moons are shown in **Figure 10-21** as seen by the *Cassini* spacecraft. Most of these moons have densities between 1.0 and 1.4, indicating interiors of water ice. Other data also indicate the surfaces are water ice rather than the heavier methane and ammonia of the Galilean moons. Each has one side that is more heavily cratered than the other side. This asymmetry is easy to understand: all of them have **synchronous rotation**, so that the same side always faces Saturn. Even more impor-

tant is that the same side always faces toward the direction of the moon's orbital motion, thus receiving more impacts than the trailing side.

Mimas

The smallest of Saturn's major moons at only 390 km in diameter, **Mimas**, looks like a Cyclops with a distinguishing giant (100 km) crater on its surface. The crater, whose diameter is one-third that of the moon itself and 10 km deep, has a central peak 6 km high. The impacting body is estimated to have been 10 km across. Furthermore, Mimas is scarred by troughs 100 km long, 10 km wide, and 1 to 2 km deep—a possible residue of fracturing from past meteoroid impacts. Its density, 1.2 g/cm^3, implies that it contains primarily water ice.

Enceladus

Enceladus is one of the more complex moons in the Solar System. Its surface is geologically young and diverse. Reflecting more than 90 percent of incident sunlight, it is the most reflective object in the Solar System, reflecting more than newly fallen snow! This high reflectivity, along with the body's low density, points toward a water ice surface. Although it has some impact craters 35 km in diameter, it is not densely cratered, implying that continual resurfacing occurs. Voyager did not detect active volcanism or landforms indicative of past volcanic activity. The variety of surface features may imply some unknown source of internal heating.

New results from the *Cassini* spacecraft in 2005 for Enceladus found it to have an atmosphere consisting of charged water vapor. An atmosphere is not expected for Enceladus, because its small size means it cannot hold an atmosphere for long. Therefore, it must be constantly replenished. That replenishment is thought to come from either the surface or the interior via, perhaps, ice volcanoes or geysers. Such an explanation for the atmosphere probably also explains the high reflectivity as well as the youthful appearance of the moon's surface.

Tethys

Tethys is distinguished by a 2000-km-long gorge covering three-fourths of its circumference. Planetary scientists are unsure whether the gorge was caused by an impacting body or came from internal activity. Furthermore, Tethys (**Figure 10-22**) has a 400-km-diameter crater, which is larger than Mimas's crater. Tethys's more densely cratered landscape indicates an old surface.

---

[1] The names of the nine classical moons, in order of distance from Saturn, may be remembered by the mnemonic *MET DR THIP,* for **M**imas, **E**nceladus, **T**ethys, **D**ione, **R**hea, **T**itan, **H**yperion, **I**apetus, and **P**hoebe.

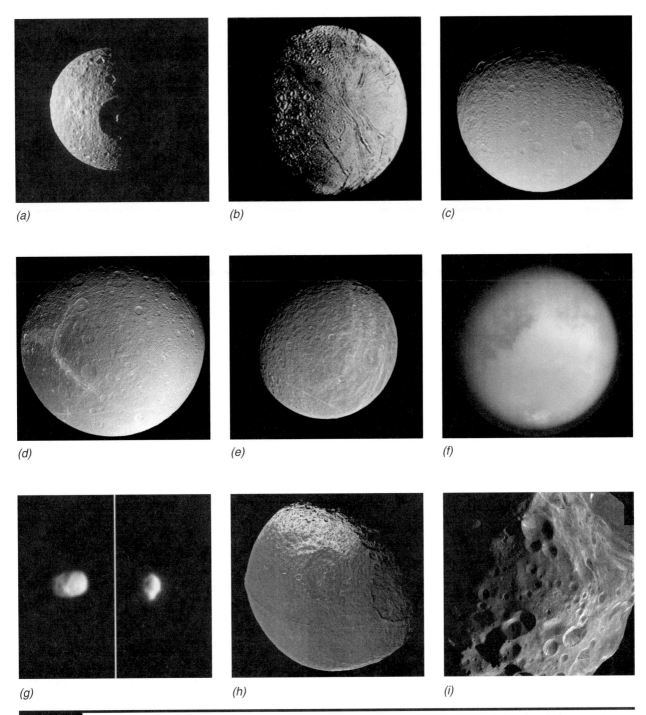

(a)

(b)

(c)

(d)

(e)

(f)

(g)

(h)

(i)

Figure 10-21 Saturn's major moons. *(a)* Mimas *(b)* Enceladus *(c)* Tethys *(d)* Dione *(e)* Rhea *(f)* Titan *(g)* Hyperion *(h)* Iapetus, at nearly full-phase. (i) Phoebe. The images are not to the same scale. All the images were obtained by Cassini except for Enceladus, which is a Voyager image.

The surface of **Dione** shows bright wisps, which are probably (geologically) fresh ice. Its higher-than-average density (1.4 g/cm³) means it contains more rock than the other moons. Its winding valleys are suggestive of internal processes; its plains regions point toward some type of unknown surface renewal. Cratering was more intense on the side facing its direction of orbital motion, while the other side, shown here, is darker and contains the wispy streaks. The *Cassini* spacecraft surprised scientists by finding that the wispy regions were ice cliffs created by tectonic fractures and not thin ice deposits as previously hypothesized.

Figure 10-22 Additional view of Saturn's satellites. Tethys has a crater whose diameter is 400 km, in addition to a long valley.

Rhea, the largest of the innermost moons at 1,530 km in diameter, is nearly a twin of Dione. Its forward-facing hemisphere, which is covered with almost pure water ice, has two sections. One consists of large craters without small ones, while the other has small craters without any large ones. This side also contains the remains of three multi-ringed basins. Rhea's trailing hemisphere, shown in **Figure 10-21e**, also has wisps. Thus, similar processes must be occurring on Dione and Rhea.

Titan received its name when it was thought to be the largest moon in the Solar System. Larger than Mercury but slightly smaller than Ganymede, Titan was studied in some detail because Earth-based telescopic data had shown it to have its own atmosphere. **Figure 10-23**, for example, shows separate haze layers within the atmosphere as observed by the Cassini spacecraft.

As the image also shows, little detail is seen in **Titan's** atmosphere, because the atmospheric haze was far thicker than anticipated. The atmosphere is at least 90 percent nitrogen and contains enough material to make the pressure on the surface 50 percent greater than on Earth. The frigid surface temperature, 94 K, is about right for puddles and small lakes of liquid hydrocarbons to exist on the surface. As a result of these atmospheric conditions, the chemistry on Titan can be expected to be highly complex. Indeed, with a composition rich in methane and other organic molecules, the chemistry may not be too dissimilar from that on the primitive Earth before life arose.

The *Cassini* spacecraft carried a separate probe, called *Huygens* after the astronomer who first described Saturn's rings as rings, that descended

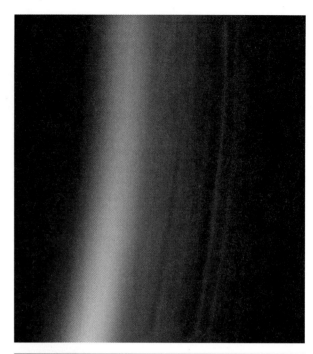

Figure 10-23 Titan, as seen by the *Cassini* spacecraft, shows separate haze layers within the atmosphere.

through **Titan's** atmosphere on January 14, 2005, and landed successfully. At the time the *Huygens* lander probed the upper atmosphere on its way to the surface, it was buffeted by winds of some 400 kilometers per hour. At the surface, however, the winds slowed to a breeze.

Titan's density of 1.90 g/cm^3 implies that it has equal amounts of water ice and rock. Thanks to its atmosphere, the surface remained unseen and mysterious—until the *Cassini* mission, which will make some 45 planned encounters with the moon between 2004 and 2008.

Figure 10-24*a* is an image of Titan from the *Huygens* lander, taken from about 8 kilometers above the surface—the distance of a typical passenger jet above the Earth. Showing surface features as small as 20 m across, the image shows light-colored high-elevation regions and dark-colored lower areas. The higher regions also show what may be drainage channels from the flow of a fluid. *Huygens* appears to have landed in a region having the consistency of mud. **Figure 10-24***b* shows a historic image; it is the first color image ever taken on an extraterrestrial moon. The two objects just below the center are about 6 inches and 1.5 inches across, respectively. Scientists do not yet know if they are rocks or blocks of ice.

Other images of **Titan** show clouds that change size in hours. Such observations may indicate

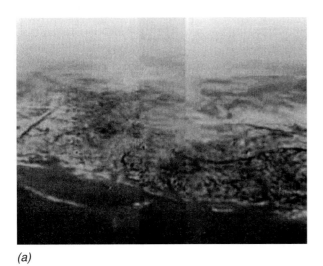

(a)

Figure 10-24 Titan's surface. *(a)* High elevation regions (light colors) may contain channels caused by fluid flow. Lower elevation regions are dark. *(b)* The view from the surface.

(b)

methane rain falling onto the surface. If correct, that would explain the hypothesized lakes of hydrocarbons that were sought by the *Huygens* probe. Further goals for the descent involve studies of the surface by analyzing radar waves reflected by it.

Inquiry 10-7

How would you interpret Titan's lack of an observed magnetic field?

Hyperion and Iapetus

Hyperion is a small, irregularly shaped chunk of material of unknown density (Figure 10-21*g*, which shows separate views). It probably resulted from a collision and was later captured by Saturn.

Iapetus was known from Earth-based studies to have two dramatically different sides—one light and the other dark. This interpretation of the observations was confirmed by *Voyager* and shown by the *Cassini* image in Figure 10-21*h;* the dark-appearing side is in full sunlight. This side faces the direction of orbital motion and contains material that is as dark as black tar. It either is thick or is replenished often, because no bright spots or craters are seen on top. The new *Cassini* observations indicate that the darker material is an externally produced coating of the surface; there is no evidence of erupted fluids from the interior being the source of the material.

Phoebe

Phoebe is small and dark. Its darkness is reminiscent of the dark side of Iapetus. Only low resolution images had been obtained until *Cassini;* Figure 10-21*i* shows its surprising view. It had a battered past that indicates an old surface with no major resurfacing activity. From this image planetary scientists think it is an icy body (like Saturn's other moons) but covered by a thin layer of dark material. Given that the nucleus of Halley's comet was found to be dark and similar to the carbonaceous chondrite meteorites, we can wonder whether there is a relationship between these bodies.

New Moons

Pioneer and *Voyager* discovered a number of new moons. One of these is called **Dione B**, because it occupies Dione's orbit but leads Dione by a 60° angle. Tethys also has two other moons in its orbit—one 60° in front and one 60° behind. These three so-called **co-orbital moons** are all small and irregular and are probably fragments from collisions. The existence of the co-orbital moons is readily understood in terms of the Lagrangian points described in Chapter 7 (Figure 7-16) for the Trojan asteroids. In this case, however, the triangle is formed by Saturn, Dione, and Dione B. Locations 60° ahead and behind are where the net gravitational forces of the planet and the moon cancel, allowing bodies there to remain in stable positions.

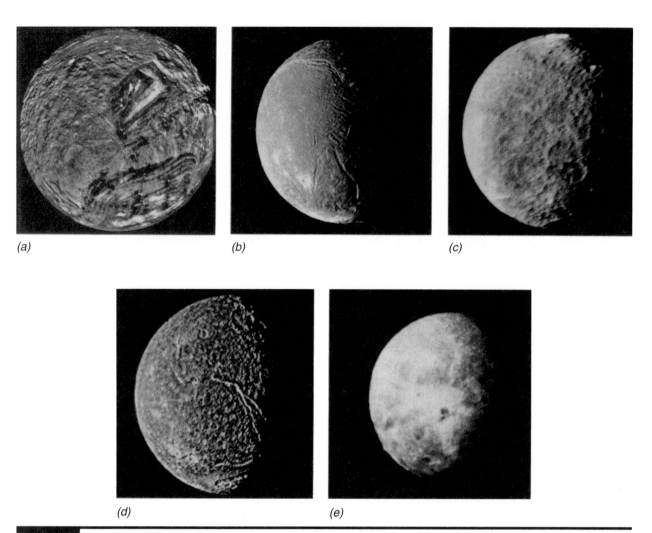

Figure 10-25 Uranus' classically known moons. *(a)* Miranda *(b)* Ariel *(c)* Umbriel *(d)* Titania *(e)* Oberon.

THE MOONS OF URANUS

Prior to *Voyager,* Uranus was known to have 5 moons; now 27 are known.[2] Images of some of these moons (**Figure 10-25**) show the same pattern of heavy bombardment exhibited by Saturn's Mimas. They all have densities slightly higher than that of Uranus, 1.5 to 1.7 g/cm^3, and probably consist of methane and ammonia ices. But their ability to reflect light is lower than that of Saturn's moons, perhaps hinting at the presence of small, dark particles in the outer Solar System.

The moons' orbital planes are in the strongly inclined equatorial plane of the planet, not in the ecliptic plane (Figure 6-6). This finding is significant when it comes to understanding the formation of the Uranian system.

Because **Miranda** is less than 500 km across, astronomers thought it was too small to be particularly interesting. How wrong they were! Its appearance is strikingly odd (Figures 10-25a, **10-26**). Miranda's surface contains 5-km-high cliffs, deep gorges, faults, craters, and valleys. Abrupt changes in the surface occur from one area to another. The light-colored L-shaped feature near the center, called the *chevron,* is located within a well-defined trapezoidal block having 200 km sides. In another area is an elliptical region called the *Circus Maximus* (because it is reminiscent of a Roman racetrack). The 5-km-high cliff is shown in more detail in Figure 10-26.

Planetary scientists have been driven to extreme lengths of speculation to understand Miranda's surface. One hypothesis suggests that sometime in its past, Miranda was actually broken (perhaps by a collision) into a number of pieces, some of rock and some of ice. The pieces re-formed into a hap-

---

[2]The five classical moons can be remembered with the mnemonic *MAUTO,* or Miranda, Ariel, Umbriel, Titania, and Oberon.

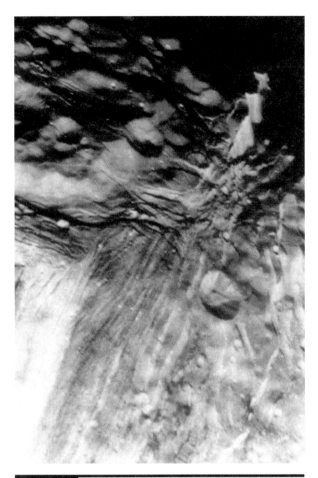

Figure 10-26 Miranda. The satellite's distinct surface shows ridges, valleys, and cliffs indicating a unique history.

hazard assemblage, and there was a tendency for the lower-density ice to rise and the higher-density rock to sink. This process might generate enough heat to melt portions of the crust and also create surface stress fractures. Another hypothesis is that tidal interactions with Uranus heated the interior. Partially molten material in the core underwent convection and rose to the surface, broke it open, and then froze in place.

Ariel contains a complex system of faults, along with large flat regions that resemble mudflows. **Umbriel** has a surface with few craters. It has a large bright surface ring of unknown origin. **Titania**, too, has faults that must indicate past internal activity. Some of its craters show rays, while others appear to contain a black substance that oozed from the crater's bottom. The outermost moon, **Oberon**, shows impact craters with bright rings of associated ejecta. Some of the craters have dark centers, perhaps dirty water erupting from the

icy interior of the moon. Oberon is distinctive in showing a 6-km-high mountain. Its surface structure must be strong to prevent such a mass from sinking.

The mysteries of the Uranian moons are ultimately tied up with the mystery surrounding the origin of the 98° tilt of the spin axis of Uranus. On the one hand, if the planet's tilt was caused by a collision (possibly with a huge asteroid), and the moons were already in existence around the (previous) equatorial plane of the planet, their orbits would have been severely perturbed; they could have repeatedly collided with each other, ultimately re-forming into the bodies we see today orbiting around the new, tilted equatorial plane. But this process would be a slow one, and at least one moon of Uranus (Umbriel) shows very old features that have remained undisturbed for a long time. On the other hand, if Uranus spun off gas to form the moons after its tilt had been changed, then it is difficult to explain where the rocky parts of the moons came from.

THE MOONS OF NEPTUNE

Neptune's number of known moons jumped from two to eight after *Voyager*'s encounter, and as of February 2004 the number is 13. The recently discovered ones are all small; one is 30 km in diameter, but the others are less than 24 km in diameter.

Triton, whose radius is 1,352 km, has been known since the 1970s to have an atmosphere composed of nitrogen and methane. Its surface pressure is low, only 0.001 percent the sea-level pressure on Earth. It seems to contain a haze, perhaps of frozen hydrocarbons.

The thinness of the atmosphere allowed observations of the surface, which is composed mostly of frozen nitrogen. Triton's surface rivals those of Enceladus and Miranda in complexity and wonder (**Figure 10-27a**). One part is smooth and dimpled like a cantaloupe rind, but the other part is heavily cratered and contains grooves and ridges traversing the surface, indicating tectonic processes. Its few craters indicate a relatively young surface. The images show traces of flows; a solidified liquid appears to fill craters (**Figure 10-27b**). Perhaps the most dramatic discovery is the presence of some small dark streaks, which astronomers have interpreted as carbon-rich geysers shooting material 8 km above the surface. The energy to produce the geysers is thought to come from heat generated during changes in the frozen nitrogen under the surface.

Triton shows aurorae in its upper atmosphere. These are caused by an interaction between charged particles in Neptune's radiation belts and Triton's atmosphere.

Voyager provided data for the first accurate determination of Triton's mass, from which a density of 2 g/cm³ was found. It is thus similar to Titan in being about half water ice and half rock.

Triton's orbital motion is retrograde. Why should a large moon be in such an orbit? Was it formed in a retrograde orbit, or did it undergo a collision with one or more bodies? This retrograde orbit, when combined with its density that is similar to that of Pluto, may mean that the two are related in their past histories. Perhaps they are related to the trans-Neptunian objects discussed in Chapter 7. The astronomical community does not yet agree on the explanation.

Nereid, Neptune's other moon known prior to the *Voyager* mission, is small, with a diameter of only 340 km. It moves in a direct (non-retrograde) orbit that is highly eccentric and inclined at a large angle (28°) to the equatorial plane. The orbital characteristics of the moons raise difficult questions about their origin.

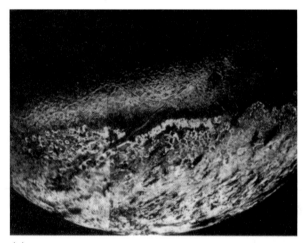

(a)

(b)

Figure 10-27 *(a)* Neptune's satellite Triton. The south polar region of Triton shows a complex and rough terrain with dark regions thought to be ice geysers. There is also an extensive dimpled region containing long cracks in the icy surface. *(b)* A lake of frozen water, methane, and ammonia.

Inquiry 10-8

What orbital characteristics would you expect if these moons were formed from natural processes at the time of Neptune's formation? How might you explain the observed orbital characteristics?

Table 10-1 summarizes the variety of processes that act to modify the surfaces of moons. **Table 10-2** summarizes some of the distinguishing physical characteristics of the giant-planet moons. In this table, the characteristic titled "bi-hemisphere differences" means that characteristics between two hemispheres are vastly different.

10.6 Planetary Rings

	JUPITER	SATURN	URANUS	NEPTUNE	PLUTO
Rings	yes	yes	yes	yes	no?

Table 10-1 Processes of Moon Modification

PROCESSES	EXAMPLES (NOT INCLUSIVE)
External processes	
Tidal forces (heating/stretching)	Io
Impact	
Large body collisions	Callisto, Phoebe
Material accumulation from surroundings	Phoebe, Iapetus
Internal processes	
Radioactive decay	all
Volcanism	Io, Enceladus (?)

All the Jovian planets are now known to have ring systems (**Figure 10-28**). However, only Saturn's rings are readily visible through a telescope. The telescopic view of Saturn's rings on a clear night is a thrill not to be missed. Although pictures and images from spacecraft may show more detail, seeing the planet itself with your own eyes is a fantastic experience! Our discussion will follow the order of the discovery of the planetary rings, after which we will compare and contrast the rings of the four planets.

The telescopic view of **Saturn's rings** (refer back to Figure 10-1) generally shows two rings. (There is a faint inner ring that is sometimes visible.) Between the outer two rings is a region named the **Cassini division**, which appears to lack reflecting particles. The cause of this gap was thought to be similar to the process that produced the Kirkwood gaps in the asteroid belt. The Cassini division is located at a distance from Saturn corresponding to a particle moving with one-half the period of the moon Mimas. In other words, a particle in the Cassini division would be at a 2:1 resonance, orbiting Saturn two times for each one of Mimas. Because such a particle would be directly aligned every two orbits, gravity would provide an extra tug, and the particle would be removed from that orbit, leaving the gap we observe.

Table 10-2 Summary of Giant Planet Moons

Characteristic	Examples (not inclusive)
Volcanism	Io, Enceladus (?)
Ice geysers	Triton, Enceladus (?)
Dark material on surface	Titania, Phoebe
Icy surface	Most/all
Ice flows	Europa
Few craters/mountains	Io, Europa, Enceladus, Triton
Cliffs	Miranda, Phoebe
Dominant crater	Mimas, Tethys
Heavily cratered	Callisto, Phoebe
Troughs/gouges	Mimas, Tethys
Atmosphere	Titan, Triton, Enceladus (only ones)
Bi-hemispheric differences	Mimas, Enceladus, Tethys, Iapetus, Triton

(a)

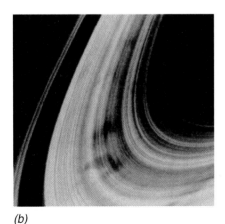

(b)

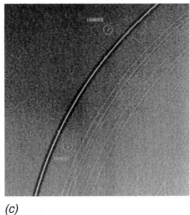

(c)

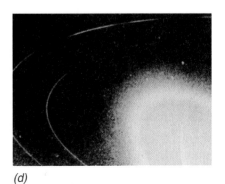

(d)

Figure 10-28 Comparison of ring systems for *(a)* Jupiter, *(b)* Saturn, *(c)* Uranus, and *(d)* Neptune. Note the shepherd satellites for Uranus.

Inquiry 10-9

How might you determine that the ring system is not composed of solid rings? (What would be observed if solid rings passed in front of a star?)

During 1980 and 1981, the two *Voyager* spacecraft passed near **Saturn**, and their sophisticated cameras revealed a wealth of new and surprising information about the rings. The greater resolution of the cameras revealed that the ring system actually consists of many thin ringlets, as the beautiful image shown in **Figure 10-29a** indicates (see also Figure 1-7). Additional rings were found both inside and outside those visible from Earth. *Voyager* confirmed the overall flatness of the rings; the thickness is about 100 m over the 300,000 km radius. There is a range in particle size, from 0.05 mm up to meter-sized boulders.

Saturn's F ring surprised project scientists by its twisting, winding structure and its nonuniform brightness (**Figure 10-29b**). At first, it was thought that gravity alone could not explain this behavior. However, two previously undetected moons were found to straddle the braided F ring (**Figure 10-29c**) and explain the complex structure. These moons are called **shepherd moons** because their combined gravity keeps this ring confined to a narrow band. A shepherd moon found just outside the outermost ring visible from Earth is thought to maintain that ring's sharp outer boundary.

However, most of Saturn's rings do not appear to have such companions, so there must be an additional mechanism to explain the narrowness and sharpness of the rings' edges. That mechanism appears to be a sound-like compression wave caused by gravitational influences of Saturn's moons on the ring particles. Such patterns were found by *Cassini* in Saturn's A ring and are shown in **Figure 10-29d**.

Voyager also observed dark ring structures pointing radially away from Saturn, like spokes of a wheel (**Figure 10-29e**). The spokes rotate with the rings, appearing within a few minutes and remaining for some tens of hours at the most. The material forming the structures appears to be above the ring plane and to be partly guided by electric and magnetic forces. The spokes consist of small particles that have acquired a net positive charge by losing electrons through collisions. This charge makes the particles repel one another and lifts them above and below the ring plane so we can see their presence.

Inquiry 10-10

The ring spokes were a surprise because, according to Kepler's third law, they should not last as long as they do. Why does Kepler's third law indicate the spokes should rapidly dissipate?

Rings around **Uranus** were unexpectedly discovered in 1977; the story of the discovery is an interesting tale of how scientific progress often occurs. In the 1970s, attempts were made to observe occultations of stars by the outer planets on a regular basis. An occultation takes place when a planet passes in front of a star, hiding it from our view. By timing the duration of the occultation, astronomers can obtain both an accurate measure of a planet's diameter and information about its atmosphere's structure. In early 1977, while awaiting an occultation of a star by Uranus, astronomers flying in the *Kuiper Airborne Observatory* were surprised to notice that the light from the star dimmed and brightened several times before the planet occulted the star. At first they suspected clouds, but they quickly concluded that nine thin rings surrounding the planet caused what they observed. Several other observatories around the world independently recorded these events, thus verifying their reality.

After the spectacular view of Saturn's rings by the two *Voyager* missions, astronomers were excited by the prospect of actually seeing **Uranus's** rings. *Voyager 2* confirmed the presence of nine major rings, and found them to be composed of dark particles that reflect only 3 percent of the incident light. Some planetary scientists have suggested that they are coated with dark carbon compounds. We wonder whether their darkness is in any way related to the dark material on Saturn's moons Iapetus and Phoebe, or the dark material in the bottoms of craters on Uranus's Titania and Oberon.

Planetary rings will appear differently when a spacecraft approaches compared with when it leaves, because of the complex way small particles scatter light. From a comparison of different views of the rings, astronomers are able to draw conclu-

(a)

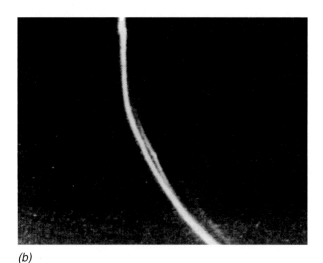

(b)

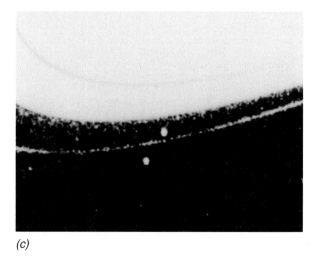

(c)

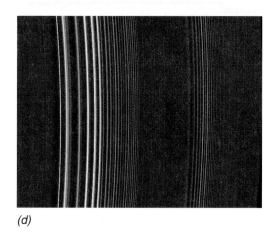

(d)

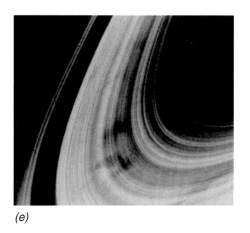

(e)

Figure 10-29 Characteristics of Saturn's ring system. *(a)* The rings consist of hundreds of "ringlets." The Cassini division is shown to contain material not observable from Earth. *(b)* The braided F ring. *(c)* Shepherd satellites on either side of the F ring. *(d)* Waves in the ring in Saturn's A ring *(e)* Saturn's ring "spokes."

sions about the sizes of the particles in them. Surprisingly, the rings of Uranus showed little in the way of tiny dust grains; most of the particles in the ring system appear to be centimeters to meters across or larger, although microscopic particles are also present.

Although narrow, the rings of **Uranus** do have breadth, and the breadth of the outermost ring is variable. The narrowness of the outermost ring was readily explained when *Voyager 2* discovered two shepherd moons on either side (see Figure 10-28*c*). As with Saturn, the lack of shepherd moons near the other rings probably means compression waves have formed from the gravity of the various moons.

With the discovery of Uranus's rings, some astronomers suggested that *all* the Jovian planets

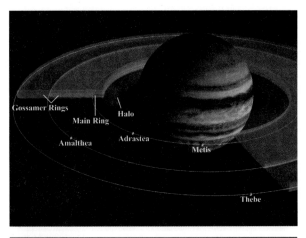

Figure 10-30 Jupiter's moons Thebe, Amalthea, Adrastea, and Metis, which cause the main and gossamer rings. Note the size scale.

Figure 10-31 Jupiter's gossamer ring extends outward from the overexposed flat main ring, which adjoins a toroidal inner halo ring. The gossamer ring results from the continuous grinding of material into smaller pieces. It extends beyond the outer edge of this image.

would have rings. For this reason, the *Voyager* spacecraft made a long-exposure image slightly offset from **Jupiter** itself, specifically looking for a faint ring. A system of three rings was found, and a *Galileo* image is shown in Figure 10-28a. (Evidence for the existence of a faint ring around Jupiter was published as early as 1960 by the Soviet astronomer S. Vsekhsvyatskij, but it had never been confirmed.) Jupiter's ring is at most a few tens of kilometers thick, with little structure. The particles are small, about 0.001 mm in size (less than the thickness of tissue paper!) and dark.

Jupiter's ring particles are subject to a variety of forces and will not last for long. In orbiting the planet, the moons provide a varying gravitational force on the rings. For example, the innermost moons Metis and Adrastea skim along the main ring's outer edge (**Figure 10-30**). Particles ejected from these moons by micrometeoroid bombardment lose energy and move toward the planet to form a ring. Thus, the moons are the source of the ring material. Io's volcanoes may also supply some.

The *Galileo* mission discovered two very low-density rings (referred to as *gossamer rings*) outside the main ring (**Figure 10-31a**). As indicated in Figure 10-30, the material for these rings comes from the small moons Amalthea and Thebe. These rings are thicker than the main ring because the moons' orbits are inclined to the ring plane. Figure 10-30 shows these four sources of material for Jupiter's rings.

By this time (1986), astronomers were in a frenzy trying to observe rings around **Neptune**! Conflicting reports appeared. Some researchers

inferred the presence of ring *arcs,* which are incomplete pieces of rings. These scientists were certainly pleased when the images shown in Figure 10-28d were returned by *Voyager 2* in 1989. Although the images show complete rings, bright concentrations could easily be interpreted as ring arcs. These arcs are now understood in terms of complex gravitational interactions between the ring and the 150-km-diameter moon Galatea. Located 1,000 km inside the outer ring, Galatea acts as a single shepherd of the fine dust particles composing the ring.

Inquiry 10-11

Ring arcs are unexpected because a clump of particles only 15 km long (in a direction away from a planet) will spread completely around a planet in only three years. What concepts discussed in this book explain why clumps of particles will not exist for long?

Inquiry 10-12

What characteristic distinguishes all the planets that so far have been discovered to have rings?

What is the chemical composition of ring particles? Jupiter's ring particles appear to be relatively high-density mineral grains while those of the other planets are low-density ices. Astronomers can explain this difference if the more massive Jupiter had a higher temperature than the other planets.

Inquiry 10-13

Why would a higher temperature at Jupiter produce rings composed of minerals rather than ice?

Why do planets have rings? The law of gravity tells us that if a body more than a few tens of kilometers in diameter were located at the ring's distance from the planet, the planet's large mass would produce tidal forces that would eventually tear the smaller body apart. Outside a critical distance, called the **Roche limit**, break-up would not occur. The major rings are observed to be within the Roche limit for each Jovian planet. This does not *necessarily* mean that the rings are the residue of a broken-up moon, but they could be. The presence of rings could also mean that while the protoplanet was forming into a planet, material inside the Roche limit was prevented from becoming a single, large moon.

An additional mechanism might also be hypothesized: collisions. Two scales of collisions could be envisioned, one in which an entire moon is shattered by a collision with the resulting particles forming a ring, and one in which small impacts continuously erode small inner bodies.

Current thinking is that ring systems do not persist forever but are transient. For example, rings are subject to bombardment by meteoroids. Some of the resulting residue will remain in the ring system, but some will be dispersed or fall into the planet. Interactions between rings, and between rings and shepherd moons, appear to set up waves, which influence the rings' long-term stability. For these and other reasons, astronomers no longer think that ring systems formed at the time the parent planet formed. Rather, we now think that different mechanisms occurred for different planets. For example, Saturn's rings are known to consist primarily of icy particles that range in size from small to boulder-size. These properties are readily understood if the rings resulted from the break-up of a moderate-sized icy moon. However, we cannot say for sure if the break-up resulted from collision or the tidal break-up of a body inside the Roche limit.

The rings of Jupiter and Neptune, however, consist of dark particles that are mostly small. Uranus's rings contain both large and small particles. Such small particles may be understood to result from the steady, long-term erosion of small inner moons as discussed earlier for Jupiter's main and gossamer rings.

Figure 10-32 Clyde W. Tombaugh, the discoverer of Pluto, in a photograph taken when the University of Kansas Observatory was dedicated in his honor in 1980 (the 50th anniversary of his discovery).

10.7 Pluto

Pluto is so small and distant that astronomers have a hard time obtaining even the most basic information about the planet. It is clearly not one of the gas giants, but neither is it a terrestrial planet. The search for it began because of predictions based on irregularities observed in the motion of Uranus. A new planet was discovered during an exhaustive search in 1930 by the 24-year-old **Clyde Tombaugh** (1906–1998) (**Figure 10-32**), whose systematic search method included viewing two photographs at a time with an instrument called a **blink comparator**. This machine uses a mirror that flips between two aligned pictures; an object that has moved between the time the photographs were taken appears to jump back and forth, or *blink*. The photographs on which Tombaugh made his discovery are shown in **Figure 10-33**.

Until recent years, all that was known about Pluto was that its orbit was highly elliptical and inclined to the ecliptic. The ellipticity is such that between about 1979 and 1999, the planet was closer to the Sun than Neptune. There is no chance of Pluto and Neptune colliding, however, because the orbits are inclined to one another. Observations of Pluto's brightness showed it to vary regularly, with a period of 6.39 days.

January 23, 1930

January 29, 1930

Figure 10-33 Pluto's discovery plates obtained by Clyde Tombaugh.

Inquiry 10-14

What are some possible explanations for the regular variations in Pluto's brightness?

Only in late 1992 did astronomers detect frozen ices on Pluto. We now know the surface is composed of 97 percent nitrogen, 1 to 2 percent carbon monoxide, and 1 to 2 percent methane. Chemically, then, Pluto's surface is similar to Neptune's moon Triton. Because Pluto's surface has remained virtually unmodified since its formation, an understanding of its surface chemistry can provide a look at the chemistry of the early Solar System. And if, as some scientists have speculated, some of Earth's atmosphere resulted from collisions with comets formed from the same material as Pluto, we will be better able to understand Earth's primitive atmosphere.

Pluto has a weak atmosphere containing mostly nitrogen with some methane. It may have been produced only in recent years, because Pluto's elliptical orbit has brought it closer to the Sun now than it normally is.

Inquiry 10-15

Hypothesize that Pluto is covered by an extensive atmosphere. In this case, one would expect its disk to be uniform in brightness. Would it be possible for the planet to exhibit the regular variations in brightness that are observed if this were the case?

To everyone's great surprise, tiny Pluto turned out to have a moon. Discovered in 1978, it has been named **Charon**, after the mythical boatman who ferried the dead to the underworld. Charon is small and difficult to observe; nevertheless, astronomers have been able to estimate the mass of Pluto with moderate accuracy from the moon's motion. The mass of Pluto is a low 0.0022 (1/450) Earth masses. **Figure 10-34** shows the best ground-based image of Pluto and Charon, as well as an image taken with the Hubble Space Telescope. (The rings are not real!)

By pure luck for modern-day astronomers, in recent years Charon and Pluto were involved in a series of mutual eclipses of each other. These events allowed a reliable diameter to be determined for each body for the first time. The currently accepted value for Pluto's diameter is between 2,290 km and 2,400 km. With this value, we can finally determine its density, which is 2.13 g/cm^3. This means that Pluto contains somewhat more rock than ice. Nonetheless, Pluto's surface certainly

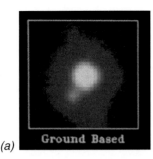

(a) Ground Based

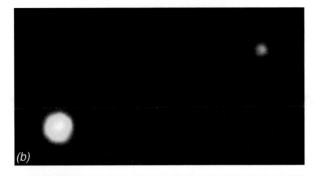

(b)

Figure 10-34 Pluto and Charon. *(a)* The best ground-based image of Pluto and Charon. *(b)* A similar image from the Hubble Space Telescope. (The rings are not real planetary features.)

consists of frozen gases, such as methane and water. Charon's diameter of 1,200 to 1,300 km means its density is only about 1.3 g/cm³. Because Pluto and Charon are of similar size, they can be thought of as a double planet, as is often done for Earth and the Moon.

The mutual eclipses provided astronomers with data to model the appearance of Pluto's surface. This model shows the presence of a bright south polar cap that reflects 95 percent of the incident radiation. The darkest regions reflect less than 15 percent.

Pluto's orbit is the most eccentric and the most inclined to the ecliptic of any planet in the Solar System. The existence of a small planet at the outer edge of the Solar System has led some astronomers in the past to wonder whether Pluto might not be an escaped moon of Neptune. Its properties are, as we have seen, reminiscent of many of the moons of the outer planets. In fact, models have been computed considering the possibility that Pluto escaped from Neptune, leaving Triton in its retrograde orbit and Nereid in its high-eccentricity orbit. Escape models also show that

Pluto would consist mostly of methane and have a low density. In fact, Pluto and Triton have identical densities. However, the dominance of nitrogen and the observed density of 2.13 g/cm³ indicate that Pluto formed directly out of the solar nebula, just like all the other planets. The fact that Pluto has a moon of its own could argue against the escaped-moon hypothesis, but we need to remember that many asteroids have companions, and so the presence of the moon may not be an important argument against escape.

Some people have asked the question, "Is Pluto a planet?" At some level, the question is meaningless. To paraphrase Pluto expert David Tholen, when does a creek become a stream? When does a stream become a river? It's a matter of definition, with the borderlines being fuzzy and unimportant. No firm definition of *planet* is yet available; the existence of extrasolar planets has made it more difficult to develop such a definition. More meaningful questions are whether Pluto is the largest of the trans-Neptunian objects discussed in Chapter 7 and whether it is, in fact, related to the *Plutinos,* which are other objects also in a 3:2 resonance with Neptune. Since astronomers have yet to agree on the definition of *planet,* the International Astronomical Union, the body that names celestial objects, decided to use historical precedent and leave it classified as a planet.

10.8 Future Studies of the Outer Solar System

Future studies of the Solar System beyond the terrestrial planets include space probes that will increase our knowledge of the objects we already know about. There is, too, the question of whether there are planets beyond Pluto.

FUTURE SPACE PROBES

The *Cassini* mission to Saturn, launched in October 1997, went into orbit about Saturn in 2004. To get there, *Cassini* received gravitational boosts from Earth, two from Venus, and one from Jupiter. Following the lead of the *Galileo* mission, *Cassini* has an atmospheric probe that entered the atmosphere of the moon Titan in January 2005. The orbiter will remain circling Saturn and returning data for four years.

Are There Planets beyond Pluto?

Pluto's discovery took so much time and effort that few astronomers are eager to search for planets beyond Pluto. In fact, after discovering Pluto, Tombaugh spent many years searching for other planets. More recently, the catalog of objects found by the *Infrared Astronomy Satellite* has been examined unsuccessfully for possible planets. This is not surprising, because there are theoretical reasons not to expect major planets beyond Neptune. Finally, recent reanalysis of planetary orbits, in which a few discrepant data points are neglected in the orbital calculations, shows no reason to expect the presence of more distant planets. The old argument for there being a Planet X beyond Pluto is no longer considered valid.

Chapter Summary

Observations

- The characteristics of the Jovian planets and Pluto are summarized in **Table 10-3** at the end of the chapter.
- The Jovian planets—Jupiter, Saturn, Uranus, and Neptune—are all gas or ice giants composed of hydrogen, helium, methane, and ammonia. They all have low densities, moon and ring systems, magnetic fields, and rapid rotations. Their nearly circular orbits have low inclinations to the ecliptic.
- The atmospheres of Jupiter, Saturn, and Neptune contain numerous large and small spots, which are storm systems of various sizes. Uranus's atmosphere is featureless.
- The magnetic fields of Jovian planets are often tilted relative to the rotation axis and offset relative to the planet's center.

Table 10-3   The Jovian Planets and Pluto: A Comparison

	JUPITER	SATURN	URANUS	NEPTUNE	PLUTO
Orbital data					
Semi-major axis (AU)	5.203	9.54	19.18	30.06	39.53
Orbital period (years)	11.86	29.46	84.01	164.79	248.68
Orbit shape (eccentricity)	0.048	0.056	0.047	0.009	0.248
Orbit inclination (°)	1.3	2.5	0.8	1.8	17
Rotation period (hours)	9.92	10.66	17.24	16.11	153.3
Axial tilt (°)	3.1	26.7	97.9	29.6	122.5
Physical data					
Diameter (Earth units)	11.19	9.41	4.01	3.89	0.18
Mass (Earth units)	317.9	95.2	14.5	17.15	0.0022
Density (g/cm^3)	1.33	0.69	1.27	1.64	2.13
Surface gravity (Earth units)	2.64	1.13	0.89	1.13	0.06
Escape velocity (km/sec)	60.0	36.0	21.2	23.5	1.2
Temperature (surface) (K)	120	88	59	48	40
Moons (as of spring 2005)	62	35	27	13	1
Visible features					
Banded appearance	strong	muted	none	slight	no
Storm systems	strong	yes	no	yes	no
Ring system	yes	yes	yes	yes	no
Other characteristics					
Clouds	thick	thick	thick	thick	?
Magnetic field	strong	strong	yes	yes	unknown
Differentiated core	yes	yes	yes	yes	no
Excess energy	yes	yes	small amount	yes	no
Atmospheric composition	H, He, NH$_3$, CH$_4$	H, He, NH$_3$, CH$_4$	H, He, NH$_3$, CH$_4$	H, He, NH$_3$, CH$_4$	N$_2$, CH$_4$, CO

- The moons of the Jovian planets have icy surfaces composed of either methane and ammonia, or water ices. Some moons show volcanoes and geysers; others exhibit craters, valleys, and cliffs that suggest all sorts of complex histories. Some moons have retrograde orbits; some have orbits that are highly elliptical.

- Certain moons interact with the ring systems as **shepherd moons**. Other moons form **co-orbital** systems in which groups of moons occupy a single orbit.

- Pluto is in an elliptical orbit that has the greatest inclination to the ecliptic of any planetary orbit. It has a surface of frozen nitrogen and an atmosphere of nitrogen and some methane.

- Some outer-planet moons have surfaces as bright as newly fallen snow, while others are as dark as tar.

- Planetary rings are within a planet's Roche limit. Gaps in the rings are observed to be associated with resonances of moons.

Theory

- Planetary magnetic fields cause charged particles to become trapped. When accelerating near the speed of light, they emit **synchrotron radiation**, which is recorded as intense bursts of radio energy. Observation of such radiation shows the presence of magnetic fields.

Conclusions

- Planetary rings are not solid bodies but are composed of particles ranging in size from small dust grains to meters across. Each individual particle orbits its planet governed by Kepler's laws.

- Complex planetary storms are thought to be caused by an interaction between heating from the Sun, the **Coriolis forces** that result from rapid rotation, and the **excess energy** radiated from the interiors of Jupiter and Saturn.

- The interiors are **differentiated**. High interior pressures in Jupiter and Saturn produce a region in which **liquid metallic hydrogen** exists. These regions, and especially the central cores, are thought to produce the extensive magnetospheres around each of the Jovian planets.

- Rings are thought to form and re-form during periods of 100 million years in which moons break up, either by collisions or by tidal forces, when a body comes inside the planet's **Roche limit**.

- Pluto's moon, Charon, has allowed us to determine Pluto's mass, size, and density. These results indicate that Pluto formed directly from the solar nebula, and that it is not a moon ejected by Neptune.

Summary Questions

1. In what ways are each of the Jovian planets similar to and different from one another? In your discussion, include the orbital characteristics, appearance, rotation and revolution, general atmospheric features, and chemical composition.

2. In what ways are the observed characteristics of the major moons of the Jovian planets similar to and different from one another?

3. In what ways are the interior structures of the Jovian planets similar to and different from one another? Why do the interior structures differ?

4. What are the magnetic field properties of each of the Jovian planets? Where do we think the magnetic fields come from?

5. How do the appearances of the ring systems of the Jovian planets differ from one another?

6. What is meant by shepherd and co-orbital moons?

7. Explain how astronomers are able to determine the interior structure of a Jovian planet.

8. Describe the methods used to discover Uranus, Neptune, and Pluto. In what year was each discovered?

9. Describe the magnetic fields of the Jovian planets in comparison with that of Earth.

10. Make a table of the characteristics of the major moons of the Jovian planets as follows: Along the left edge place the moon names; along the

top, write "craters," "faults or cracks," "mountains," "valleys," "scarps," "atmosphere," "volcanic activity," and "density." Fill in the table with an "X" for each characteristic shown by each moon. If the density is known, place it in the "density" column. What general conclusions about groups of moons and moon systems can you make?

11. In what ways is Pluto similar to and different from both the Jovian and the terrestrial planets?

Applying Your Knowledge

1. The Jovian planets can be described as objects that have changed little since their formation, while the terrestrial planets have changed a great deal. Explain the meaning of this statement, and give examples of its validity.

2. What is the distance in kilometers from Dione to Dione B? From Tethys to its co-orbital moons? (*Note:* You can answer this question without mathematics!)

3. What would you conclude about the formation of the Solar System if Jupiter had the same composition as Earth?

4. Speculate on what the *Voyager* spacecraft might have found if it had gone on to Pluto.

▶ 5. Compute Saturn's density relative to Earth's, given its diameter of 9 Earth radii and 95 Earth masses. What is its density relative to that of water (1 g/cm^3)?

▶ 6. Saturn's Roche limit is located at 2.4 times the planet's radius. Measure the inner and outer radii of Saturn's rings from Figure 10-1, and express them in terms of Saturn's radius. Are these distances inside or outside the Roche limit?

▶ 7. Saturn's innermost ring begins at about $1.11R_s$ (Saturn radii), and the outermost one extends to about $6R_s$. How much longer does it take an outer-ring particle to revolve around the planet relative to a particle at the inside of the innermost ring?

▶ 8. Why would you weigh nearly the same on Saturn as you do on Earth? Make the computation.

▶ 9. How far above Io's surface is the volcano in Figure 10-13*b?* What implicit assumption are you making in solving the problem?

▶ 10. Determine the size of the smallest feature you can see on the Earth-based photograph of Jupiter in Figure 10-1.

▶ 11. Compute the ratio *moon radius/planet radius* for the largest moon in each planetary system throughout the Solar System. Which, if any, stand out from the rest?

Answers to Inquiries

10-1. If they were solid, the inner and outer edges of a ring should go around the planet in the *same time,* just as the hub and end of a clock hand go around a clock in the same time.

10-2. The equatorial diameter is about 5.9 mm, whereas the polar diameter is 5.2 mm. Thus, the ratio is 5.2/5.9 = 0.88, which is a significant deviation from round (where the ratio would be 1.0).

10-3. The two lightest elements, hydrogen and helium.

10-4. Half of Uranus's period of revolution around the Sun, or 42 years. Because a planet whose rotation axis is perpendicular to its orbit will not have seasons, while a planet whose angle is 90° will have the largest possible seasonal changes, Uranus's seasons will be nearly the maximum possible.

10-5. Electrically conducting region in the interior, and rotation.

10-6. It might be a captured asteroid.

10-7. It lacks a metallic core and/or its rotation is too slow.

10-8. The orbits would be in the planet's equatorial plane and would be nearly circular. Collisions of some kind are probably required to explain the observations.

10-9. In addition to the answer given in Inquiry 10-1, stars can in fact be seen through the rings.

10-10. From Kepler's third law, the innermost particles move faster than the outer ones, just as on a track the inside runners advance on the outside ones. Therefore, the particles should be spread evenly around the rings.

10-11. The answer is the same as for Inquiry 10-10.

10-12. They are all members of the Jovian group of planets.

10-13. Ices are volatile. Higher temperature would cause the ice to vaporize and the volatile elements to escape more readily.

10-14. Pluto is rotating once in 6.39 days, and the brightness varies because the surface is not uniformly bright but has bright and dark areas. A moon with a 6.39-day period of revolution will also contribute to such a variation.

10-15. No, *unless* there is a moon or ring system successively adding or blocking light.

Jupiter

The View from Earth

Jupiter is a giant planet by any means of measurement. Located fifth from the Sun, its 318 Earth masses and 11 Earth diameters give it an average density 1.3 times that of water. Its appearance through ground-based telescopes is that of a gas giant flattened from its rapid rotation, with reddish and whitish belts and zones. Located within one of the zones in the southern hemisphere is the Great Red Spot, a feature about three times the size of Earth. Ground-based observations show the planet to be escorted by a retinue of moons. The largest, the Galilean moons, are large enough and bright enough to have been seen by Galileo; hence their name. Their changing positions are readily observed within a period of hours.

Ground-based observations with radio telescopes found synchrotron radiation, which occurs only when charged particles accelerate near the speed of light in a magnetic field. Such radiation points to the presence of a strong magnetic field.

Other observations show that Jupiter emits nearly twice as much energy as it captures from the Sun. This excess energy comes from internal heat left over from the time of Jupiter's formation.

The View from Spacecraft

From the *Pioneer, Voyager,* and *Galileo* probes, features became more highly resolved and showed greater detail within the turbulent and stormy atmosphere. Lightning and aurorae have been observed. These spacecraft confirmed the expected presence of hydrogen and helium as the main constituents of the planet. *Voyager* discovered a ring consisting of small particles while *Galileo* showed the source to be a moon.

The magnetosphere was found to be extensive and to change with time. The magnetic axis is tilted 10° to the rotation axis.

Models of the interior structure show the planet to contain a region composed of liquid metallic hydrogen. Because this material behaves like a metal, its rotational and turbulent motions are thought to produce the planet's magnetic field.

The Galilean moons are each unique and show a variety of geological features. Io has volcanoes. Some of the gases form a cloud around the moon and its orbit. Europa exhibits straight-line features within its icy methane and ammonia surface. Ganymede, the largest moon in the Solar System, is cratered and contains faultlike features indicative of some type of internal motions that move the crust. Callisto is saturated with craters and contains a large multiringed basin.

Saturn

The View from Earth

Saturn's most distinctive feature when observed with ground-based telescopes is its glorious ring system. Although farther from the Sun and smaller in mass and diameter than Jupiter (95 Earth masses and 9 Earth radii, respectively), the planet is observed to be flattened from its rapid rotation and to contain faint bands parallel to its equator. Saturn's axial tilt implies that it has seasons. Moons revolving around the planet can be seen even with small telescopes.

The View from Spacecraft

Saturn's appearance is muted compared to Jupiter's. Its storm systems are smaller. The planet emits energy in excess of that received from the Sun. In fact, its excess energy is greater than that from Jupiter.

Saturn's ring system is complex. Composed of hundreds of narrow ringlets, the particles range in size from dust grains to boulders. They show a variety of features. Not all rings are perfectly symmetrical. They exhibit spokes, which are caused by the charging of dust particles exposed to solar emissions. The F ring shows both an inhomogeneous distribution of material and a transient braiding or twisting. The sharpness of the ring system is caused by a complex interaction with some of Saturn's many moons. The ring system may be a relatively young feature whose existence comes and goes.

Saturn's moons consist of a variety of unique and interesting bodies that rotate synchronously. Their surfaces consist of water ice. Mimas' heavily cratered surface has one crater nearly one-third the planet's diameter. Enceladus's surface is as bright as newly fallen snow and has a region that is geologically young, in that resurfacing appears to be a continuous process; it also has an atmosphere. Tethys has a gouge covering a large part of its total circumference; it also has a gigantic crater covering a large part of its surface. Dione and Rhea both contain wisps of relatively freshly produced ice. Titan has a dense nitrogen atmosphere and may have lakes of hydrocarbons. Iapetus has one side that is highly reflective and another side that is black. Phoebe is as dark as tar.

Two of Saturn's moons, Dione and Tethys, have companion moons orbiting Saturn. These are co-orbital moons that occupy Lagrangian points in Dione's and Tethys' orbits around Saturn. Some of the moons also act as shepherds to the ring by using their gravity to prevent ring particles from escaping.

Saturn has a magnetic field axis aligned with its rotation axis. Models of the interior predict a zone of liquid metallic hydrogen.

Uranus

Neptune

The View from Earth

Discovered in 1781, Uranus is the least massive of the Jovian planets, at nearly 15 times Earth's mass. With a radius of somewhat less than four times the diameter of Earth, its density is lower than that of Jupiter or Neptune. Its great distance gives it a small angular diameter so that few markings had been seen using Earth-based telescopes.

The most significant single piece of information determined from ground-based data was that its rotation axis lies nearly in the ecliptic.

Five moons were discovered using Earth-based telescopes.

Ground-based observations of Uranus's occultation of a star found the planet to have a system of nine thin, dark rings. It was the first planet other than Saturn known to have a ring system.

The View from Spacecraft

The *Voyager 2* mission, which arrived at Uranus in 1986, found a planet devoid of surface markings. The planet emits little excess energy.

Voyager found Uranus to have a magnetic field whose axis is tilted 59° to the rotation axis. Furthermore, the center of the magnetic field is offset from the center of the planet by one-third of the planet's radius. The planet's mass is too low to allow liquid metallic hydrogen to exist in the interior, and the source of the magnetic field is therefore unknown.

Uranus's ring structure is at least partly determined by shepherd moons on either side of its outermost ring. At least one of the rings varies in its width as it circles the planet.

Each of the moons is small, none being larger than 0.08 the size of Earth's Moon. Nevertheless, the larger ones are as interesting as the moons around Jupiter and Saturn. Miranda shows a surface that appears to have been put together from random pieces of a puzzle. It contains a 5-km-high cliff in addition to faults, craters, and valleys. Ariel and Titania have faults. Titania's craters sometimes contain a dark-appearing material in their bottoms. Umbriel has a light-colored ring on the surface that may be (relatively) freshly deposited ice. Oberon has a 6-km-high mountain. *Voyager* discovered 11 new moons.

Unique Features

The placement of the rotation axis near the ecliptic is unique. This angle produces extreme seasons on the planet and its moons.

The View from Earth

Discovered in 1845/1846, Neptune is the smallest of the Jovian planets. With a mass of 17 times the Earth's mass, the density of 1.64 is the largest of all the Jovian planets. Its small angular size allows little to be seen from Earth. The two moons known from Earth-based observations have strongly differing characteristics. Triton, which is nearly as large as Earth's Moon, has a mass of 0.3 times the mass of the Moon and a retrograde orbit. Nereid is small and revolves in a highly elliptical, highly inclined orbit.

Ground-based observations of occultations of stars hinted at a possible ring system consisting of pieces of rings rather than complete entities.

The View from Spacecraft

Voyager 2 revealed clouds and storm systems of various sizes. High-level clouds were observed to move and layering of the cloud structure was apparent. Neptune emits more excess energy than any of the other Jovian planets.

Neptune's magnetic field, which is tilted 47° relative to its rotation axis and offset from the planet's center by half the radius, is similar to that of Uranus. The planet's low mass means that it lacks the liquid metallic hydrogen thought to produce the magnetic fields in Jupiter and Saturn.

The rings are both thin and dark, and thus somewhat similar to Uranus's rings. The ring material is not uniformly distributed; bunched material gives the clear impression of ring arcs. These arcs may be caused by a single moon that acts as a shepherd.

In addition to the discovery of six new moons, *Voyager* obtained detailed images of Triton. With only a thin nitrogen atmosphere covering it, the surface of frozen nitrogen revealed grand complexity. One large area is relatively smooth; an adjoining area around a pole is heavily cratered and contains dark, streaky areas that appear to be geysers of ice emanating from the surface. Its density of 2.0 g/cm⁴ shows Triton to be about half water ice and half rock.

Unique Features

The ice geysers have been seen nowhere else in the Solar System. Triton is one of only a few moons in the Solar System to have an atmosphere. The orbital characteristics of Triton and Nereid are unique in the Solar System.

Pluto

The View from Earth

Pluto was found in 1930 by Clyde Tombaugh after a systematic, exhaustive search. From Earth, Pluto appears as little more than a point of light. The planet's brightness was observed to vary with a 6.39-day period, which was assumed to be caused by variations of brightness across the surface. Photographs of the planet made in 1978 showed the presence of a moon, later named Charon. Its period of 6.39 days partially explains the brightness variations observed for Pluto. Furthermore, both Pluto and Charon rotate with the same 6.39-day periods.

Only in 1980 was a thin atmosphere detected. The atmosphere may have been found only recently because Pluto was closer to the Sun than it usually is in its highly elliptical, highly inclined orbit.

In recent years, Pluto and Charon underwent a series of mutual occultations. Analysis of the brightness changes allowed astronomers to determine an accurate diameter: 0.18 the size of the Earth. Pluto's mass was also determined from an analysis of Charon's motion; Pluto's mass turned out to be 0.0022 Earth masses. The resulting density of 2.3 g/cm^3 indicates that Pluto formed directly out of the solar nebula rather than as an escaped moon from Neptune.

The View from Spacecraft

No spacecraft has yet visited Pluto.

PART THREE

Discovering the Techniques of Astronomy

Because astronomy is an observational science, astronomical data, especially for objects beyond the Solar System, are limited entirely to the diverse types of radiation received from the distant universe. Only through an understanding of the properties of this light and particles can astronomy progress beyond descriptive observations. In Chapter 11 we discuss the observed properties of light, and how these properties lead scientists to models that allow us to describe the behavior of light.

The radiation from planets, stars, and galaxies is collected by means of telescopes and analyzed by a variety of auxiliary instruments. Great advances are being made in telescopes and the electronic instruments used to detect radiations from beyond Earth. These topics are the subject of Chapter 12.

In our final chapter of Part Three, Chapter 13, we break starlight into its component parts as we begin to see how astronomers determine the characteristics of radiating bodies.

The Nature of Light

11

It is indeed wrong to think that the poetry of Nature's moods in all their infinite variety is lost on one who observes them scientifically, for the habit of observation refines our sense of beauty and adds a brighter hue to the richly coloured background against which each separate fact is outlined. The connection between events, the relation of cause and effect in different parts of a landscape, unite harmoniously what would otherwise be merely a series of detached scenes.

M. Minnaert, *The Nature of Light and Colour in the Open Air*, 1954

Until recently, our knowledge of the universe was obtained exclusively from a study of the visible light that happened to arrive on Earth, and this is still a significant source of information. However, since the 1930s, astronomers have been able to study other kinds of light and particles—radio waves, X-rays, gamma rays, cosmic rays, as well as exotic items such as neutrinos and gravitational radiation, as we will see in later chapters. We will now examine three ways to think about light—as a ray, a wave, or a particle—all of which astronomers use to extract detailed information from light of all types.

11.1 Light as a Ray

A ray of light can be thought of as a small piece of a light beam, like the beam from a flashlight or laser pointer. A ray can be simply represented as an arrow that shows the light's direction. The ray model of light can conveniently illustrate the properties of reflection and refraction.

REFLECTION

A basic property of an isolated light beam is that it travels in a straight line. This is why we cannot see around corners under normal conditions—it would certainly be hard to make sense of the world if it were otherwise! However, light *can* change direction under certain conditions, for example, when it is reflected from a surface.

The concept of **reflection** explains in a simple way why we see objects as we do. We see a chair, for example, because it reflects some of the light that falls on it toward our eyes; it acquires form to our eyes because we can distinguish the light rays that come from various directions. We see color by virtue of the fact that the chair does not reflect all colors equally well; a blue chair is blue because it reflects blue well while tending to absorb other colors.

> **You should do Discovery 11-1, Images in a Mirror, at this time.**

REFRACTION

The direction a ray of light travels also changes when it passes from one transparent material, such as air, into another, such as glass. This bending of

light as it passes from one material to another is called **refraction**.

> **You should do Discovery 11-2, Refraction of Light, at this time.**

11.2 Light as a Wave

The ray model allows us to predict extremely well how light will reflect and refract. However, other observations about light cannot be understood with the ray model, and so another model, the wave model, provides a useful description. In describing the behavior of light with the wave model, we will examine waves of different lengths, and discuss how fast they travel.

DIFFRACTION

If light consisted of parallel rays, they could travel through a small pinhole and make a small bright spot on a screen a short distance away (**Figure 11-1a**). When we actually observe the light that passes through such a hole, however, we see the rather unexpected pattern of **Figure 11-1b**. The spot is larger than the pinhole and not uniformly bright. What is happening? The pinhole has done more than just prevent some light from passing through; it has also affected the light that did get through, causing some of it to change direction.

We can observe a similar phenomenon when sound waves or waves of water pass through a narrow opening (**Figure 11-2**). This analogy leads us to conclude that light, too, must have some of the properties of waves. Indeed, the pattern we observe when light passes through a pinhole can

Chapter opening photo: A rainbow exhibiting some properties of light.

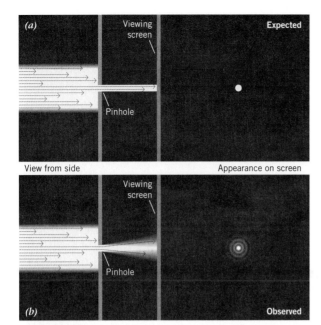

Figure 11-1 Light passing through a pinhole. *(a)* If light passed in straight lines through a pinhole, the expected pattern would have the same size as the pinhole and be sharp-edged. *(b)* The observed pattern is larger than the pinhole, and consists of rings of decreasing intensity.

Figure 11-2 The bending of water waves passing through a slit.

be explained only by assuming wave properties rather than ray properties for light. The bending of a wave as it passes near the edges of a hole or around an obstacle is known as **diffraction**.

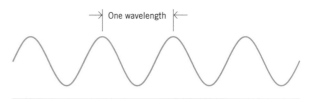

Figure 11-3 The definition of wavelength: the distance between two crests of a wave.

We describe a wave, such as that shown in **Figure 11-3**, in terms of its **wavelength**, the distance between wave crests or its period, the time between the arrival of wave crests. We find that the bending effects of diffraction are greatest when the opening is comparable in size to the wavelength. You can see diffraction by looking at a distant bright streetlight through the space between your fingers held directly in front of your eyes. While looking at the lamp, squeeze your fingers together slightly so as to change the opening size. Diffraction will cause the light to spread out in a direction perpendicular to your fingers. You can also observe diffraction of light by looking at the lamp while squinting with your eyes.

INTERFERENCE

What happens if two waves run into each other? Waves can interact and combine with each other, resulting in a composite form to which both have contributed. For example, from your experience with stereo music systems, you may be familiar with what results when sound waves combine. Stereo systems emit sound waves in which the wave peaks from one speaker can line up with those from the other and add together. Or, the sound waves can have the peaks from one speaker fall on the troughs of the waves from the other speaker and subtract from each other, producing areas of decreased sound throughout the room or car interior. When wave peaks add together, or when wave peaks and troughs subtract, the effect is known as **interference**. Without careful placement and electronic adjustment (e.g., fade) of speakers, your favorite seat could be in a location where interference of waves occurs and the quality of the sound is diminished.

The same things can happen with light waves. In **Figure 11-4a**, two identical waves are shown that are *in phase* with each other—that is, all their crests (the high points of the waves) line up, as do all their troughs (the low points). In this case, the two waves add to produce a wave that is twice as

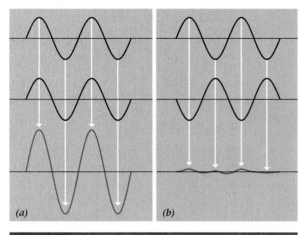

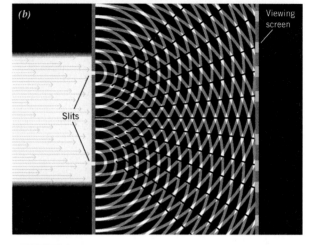

Figure 11-4 *(a)* Constructive interference. The two in-phase waves add up, reinforcing the effect.
(b) Destructive interference. The two waves exactly out-of-phase cancel each other out.

Figure 11-5 *(a)* Observation of diffraction fringes when light passes through a narrow slit. *(b)* Diffraction explained. The wave crests from two symmetrically placed sides of the slit, which are isolated here by being shown as separate slits, interfere constructively where the wave crests meet and destructively in between. Thus, diffraction is really interference.

high as the individual waves. If the two waves combine in this way and reinforce each other, the result is called **constructive interference. Figure 11-4***b* shows the case where the two waves are *out of phase* so that the crests of one line up with the troughs of the other. The two waves cancel each other out, giving **destructive interference.** Interference occurs if a large number of the peaks and troughs line up; having only one out of a million peaks lining up will not produce observable interference. Thus, random waves having no phase relationship to one another will not interfere but will pass through each other unaffected.

Here's another view of the same concept: When light passes through two slits in an opaque sheet, the waves that come from one slit will interfere with the waves from the other slit to produce light and dark bands, seen in **Figure 11-5***a*. The explanation can be seen in **Figure 11-5***b*, which shows the waves from each of two small slits. Notice that in certain directions (marked with solid lines), the crests of one wave align with the crests of the other wave. An observer looking in these directions would see that the light was especially bright, because constructive interference is taking place. By contrast, in other directions, the crests of one wave line up with the troughs of the other, producing destructive interference. An observer looking in these directions would see no light. Because there are several directions in which constructive interference takes place, separated by others with destructive interference, the result is the banded appearance in Figure 11-5*a*.

With an understanding of interference from two slits we can now understand the cause of diffraction through a single slit. A wave passing through one part of a single slit interferes with a wave coming from another part of the same slit. The resulting pattern of bright and dark regions is the diffraction pattern.

The properties of diffraction and interference give astronomers a method other than using a prism for breaking light into its component colors (spectrum). The method involves a glass or plastic surface on which thousands of fine, parallel lines have been cut by a precision machine, producing many extremely narrow scratches, each capable of diffracting light as a slit. The amount of diffraction (i.e., spreading) depends on the wavelength of the light, and the many narrow slits all acting together separate the wavelengths that make up the incoming light. Such a device is called a **diffraction grating.** A complete instrument that uses a diffraction grating or a prism as a component to produce a spectrum is called a **spectrograph.** When

Figure 11-6 Water waves spread outward from the point of contact when an object is dropped into a pool.

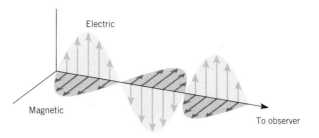

Figure 11-7 Electric and magnetic waves vibrate at 90-degree angles to the direction of motion, and to each other.

astronomers attached a spectrograph to a telescope, they discovered that other stars have the same type of spectra as the Sun, and thus proved that the Sun is a star as well.

In summary, the observations of diffraction and interference cannot be understood with a ray model of light. However, they lead naturally to another model of light—the wave model—that we now explore more completely.

ELECTROMAGNETIC RADIATION

Diffraction and interference demonstrated that light has wavelike properties, but exactly what kind of wave is it?

We can visualize more about the wave model of light through an analogy with water waves. If you drop a rock into a pond, a series of waves spreads from the point of impact, as illustrated in **Figure 11-6**. The crests of the waves spread in circles, but the water itself does *not* move outward with them. This is easily seen if you watch a surfer in the ocean. The surfer simply bobs up and down as the waves pass by but does not move along with them. What is moving through the water is mechanical energy in the shape of a wave.

An experiment shows the appropriateness of the wave model for light. If we put a long wire in the path of light, an electrical current is produced. We can explain this current by a model where the light causes the wire's electrically charged electrons to oscillate as if a wave of light is passing by. The electrons in the wire are bobbing up and down,

just like a surfer bobs up and down in ocean waves. We see this connection between waves interacting with wires and the subsequent production of an electrical current every time we turn on a radio, because electrons in a radio antenna vibrate in response to the radio waves falling on it. (Note that the electric current in the wire must get converted to pressure (sound) waves through the air; it is not possible to hear the waves carrying electrical energy.)

It was soon discovered that light also carried magnetic energy as well. The Danish physicist Hans Oersted (1777–1851) carried out experiments and observed magnets responding to a nearby wire carrying a current. Continuing these observations, he discovered the very close connection between electric currents and magnetic fields. Thus, in summary, we see that light is an electromagnetic wave—electric and magnetic energy moving through space like a wave.

The grand synthesis of all these ideas about electricity and magnetism was made by the Scottish scientist James Clerk Maxwell (1831–1879), who was the Newton of electricity and magnetism. Maxwell's synthesis is in the form of four equations that, along with the concept of forces between charged particles, describe everything we know of electric and magnetic fields. For example, they show us that the electric and magnetic waves vibrate perpendicularly to the direction of motion of the wave (see **Figure 11-7**).

Maxwell's equations show that electric and magnetic fields become unified with each other to produce what we call **electromagnetic radiation**. One example of electromagnetic radiation is what our eyes perceive as light. Other manifestations are what we call gamma rays, X-rays, and ultraviolet, infrared, and radio waves. All these types of radiation taken together form what is called the **electromagnetic spectrum**. The only thing that distinguishes one from another is the wavelength of the particular form of radiation.

The wavelength of what we call *visible light* is extremely short. For example, what the average person calls green is produced by radiation having a wavelength around 5×10^{-5} cm. Because such numbers are not particularly convenient, astronomers often use a special unit of length, the angstrom, to measure them. An angstrom, abbreviated Å, is equal to 10^{-8} centimeters, or a hundred-millionth of a centimeter. The wavelengths of light that appear reddish are roughly 6,500 to 7,000 Å, whereas radiation having wavelengths in the range of 4,000 to 4,500 Å would appear bluish. Many physicists prefer to measure wavelength in nanometers (billionths of a meter), an internation-

ally agreed-upon standard of length. In these units, red light at about 6,500 Å would be 650 nm (1 nm to 10 Å). Angstrom units are used here because most astronomers use them to identify infrared, visible, and ultraviolet light.

White light is a beam of light containing all wavelengths.

The wavelengths of ultraviolet radiation, X-rays, and gamma rays are shorter than those of visible light, where those for radio and television waves, microwaves, and infrared radiation are longer. Yet all are essentially the same in their physical nature—all consisting of traveling waves of electromagnetic energy moving at the speed of light. **Figure 11-8** diagrams the entire electromagnetic spectrum and shows the approximate wavelength range occupied by each type of radiation. Notice that visible light is only a small portion of the entire spectrum.

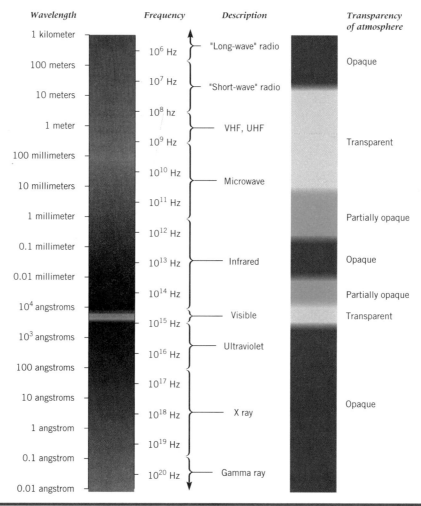

Figure 11-8 The electromagnetic spectrum extends from the very long radio waves to the shortest gamma radiation and includes many kinds of radiation in between. Because the Earth's atmosphere is not equally transparent to all wavelengths, not all of them reach the ground.

Inquiry 11-1

Why are wavelength designations more objective than using color words, such as blue, violet, indigo, and purple?

Not all parts of the electromagnetic spectrum reach the Earth's surface, though, because atoms and molecules in the Earth's atmosphere absorb certain wavelengths and transmit others. At some wavelengths the light travels directly through the Earth's atmosphere so that we can look through these nearly transparent **atmospheric windows** into the universe beyond. Figure 11-8 also indicates the transparency of the Earth's atmosphere to each form of radiation. It is no coincidence that the visible and then the radio regions of the spectrum were the first to be exploited by astronomers.

Inquiry 11-2

If you were in charge of the astronomical program for a space station orbiting Earth above the atmosphere, to what portions of the electromagnetic spectrum would you give highest priority for observing? Why?

THE SPEED OF LIGHT

Even though the wave model gives us a decent understanding of the nature of light, experimental physicists exploring more thoroughly discovered that light had some unexpected properties.

For example, every type of wave that had previously been investigated had required some type of medium to move through. A sound wave is unable to propagate in the vacuum of empty space; it needs something to move through, just as the seismic P waves discussed in Chapter 8 move through the Earth. However, light and other forms of electromagnetic waves need no such assistance; they travel easily through empty space. Your remote control would work even without air in the room—although your voice wouldn't!

Another surprise was that the speed of light turned out to be extremely high, but not infinite. Galileo was the first to attempt to measure it, exchanging lantern signals with a colleague on a distant hill; however, this technique was not fast enough. In the seventeenth century, Danish

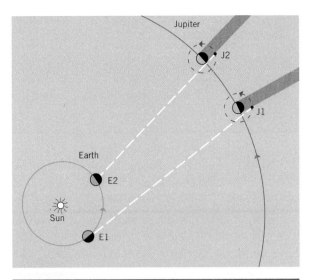

Figure 11-9 The Earth, and one of Jupiter's moons just entering Jupiter's shadow, viewed at different times. The radiation arriving at E1 has a greater distance to travel than that reaching E2.

astronomer Ole Roemer made the first reliable measurement by cleverly using the moons of Jupiter. He paid careful attention to the eclipse and occultation timings of Jupiter's moons and found that the eclipses occurred slightly sooner than expected when the Earth was closer to Jupiter and slightly later than expected when Earth was farther from Jupiter. He correctly interpreted this as being due to the time it takes light to cross the space between the Earth and Jupiter (**Figure 11-9**).

Modern experiments have measured the speed of light to be fast—3×10^5 km/sec (186,000 miles per second). This is often abbreviated by the letter c. Light travels very quickly, but it still takes time for it to get from one place to another. Theorists were able to show that the speed of light is also the ultimate top speed in our universe. Nothing with mass can go faster than—or even as fast as—light.

Inquiry 11-3

How far does light travel through empty space in one year? Express your answer in kilometers.

Inquiry 11-4

How many minutes does light take to reach the Earth from the Sun, which is at a distance of about 150,000,000 km?

Inquiry 11-5

Observations of radar signals reflected from the Moon take about 2.6 seconds to travel from the Earth to the Moon and back. What is the approximate distance to the Moon?

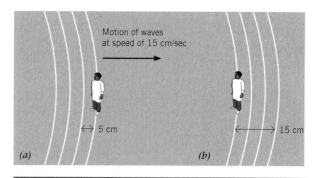

Figure 11-10 The relationship between wavelength, frequency, and speed. *(a)* Initially. *(b)* One second later.

SPEED, FREQUENCY, AND WAVELENGTH

When working with visible light and other short-wavelength electromagnetic radiation, it is normal to speak of the *wavelength* of the radiation as we have been doing. However, astronomers who observe radio radiation from the universe generally find it more convenient to refer to the **frequency** of the radio waves they are studying. By definition, the frequency of a wave is the number of complete wave crests that pass the observer per second. For example, if we throw a rock into water, we would probably count a few crests per second passing by. However, light waves have a much higher frequency, on the order of 10^{15} waves per second, or 10^{15} *hertz,* to use the standard unit of measure for frequency. One hertz means one wave passing per second. If your favorite radio station is 91.5FM, it is broadcasting electromagnetic waves at 91.5 megahertz, or emitting 91.5×10^6 waves per second. (You can't hear a radio without a speaker, the part that creates sound waves by vibrating the air based on the signal from the light waves.)

The wavelength, frequency, and speed of any wave are related to each other in a simple but important way. Consider water waves in a pond, shown in Figure 11-6. Suppose that the wave crests are 5 cm apart and that we count three wave crests per second. **Figure 11-10** shows that in 1 second, the wave must have moved (3 waves/s) × (5 cm/wave), or 15 cm in all. The speed of the wave is therefore 15 cm/s. More generally, *for any type of wave,* this is true:

Wavelength × Frequency = Speed of wave.

Symbolically, if λ (the Greek letter lambda) is the standard symbol to represent the wavelength, f is the frequency, and c is the speed of light, then we can express the relationship like this:

$$\lambda f = c.$$

Replacing the symbols with their units of measurement shows the validity of the equation: On the left side we have length × 1/s = length/s; on the right side, we have velocity, which is length/s.

Inquiry 11-6

Using this relationship, what is the speed of a water wave if its frequency is 2 hertz and its wavelength is 10 cm?

Inquiry 11-7

Think about a wave traveling with a constant speed of 100 m/sec. Would its frequency increase or decrease if the wavelength were to decrease?

Inquiry 11-8

You measure a wave having a wavelength of 5000 Å. Would the frequency increase or decrease if the wave's speed were to decrease?

Inquiry 11-9

What is the wavelength of radio radiation having a frequency of 15 megahertz (1.5×10^7 hertz)?

Inquiry 11-10

What is the frequency of microwave radiation with a wavelength of 3 cm?

In glass, air, or other transparent materials, electromagnetic waves interact with the charged particles—usually electrons—in the matter. Thus, they travel more slowly than in the vacuum of

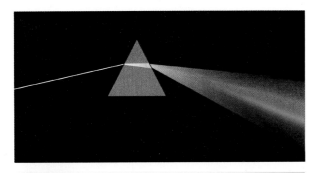

Figure 11-11 The dispersion of light by a prism into its constituent wavelengths. Notice that blue light is most strongly refracted.

Figure 11-12 The light of the daytime sky is polarized. The student is holding two Polaroid filters with their axes perpendicular to each other, as indicated by the arrows. The polarization is shown by the different sky brightness through each filter. Where the filters intersect, the light is fully polarized by the first filter so that no light passes through the second one.

empty space. For example, in glass, visible light travels about two-thirds its speed in empty space; in water, light waves travel at only three-fourths of their speed in empty space. Because air has low density, the speed of a light wave through air is only slightly less than its speed in empty space. If, in going from one material to another, a light wave strikes the transparent material *at an angle,* the part of the wave that hits first changes speed first, causing the light wave to bend as it goes through the interface between the two materials. This uneven slowing of the wave in the denser material, or uneven speeding up in less dense material, causes it to refract. If the entire wave hits the new surface at the same time, however, it does not refract.

Furthermore, because the speed of an electromagnetic wave in a substance changes with wavelength, the amount of refraction depends on wavelength in the sense that the shorter the wavelength, the more the wave bends. This spreads out the colors of light, an effect called **dispersion**. The dispersion of light passing through a prism is illustrated in **Figure 11-11**. Dispersion will be examined in detail in Chapter 13.

POLARIZATION

If you were able to examine the electric part of an electromagnetic wave, you would notice that the wave was vibrating in a single plane, say up and down (see the electric wave in Figure 11-7).[1] A photograph of the next wave taken 10^{-8} seconds later would usually find this wave vibrating in a

different plane, say sideways. A series of such photos would show successive waves vibrating in different (random) directions. Such is the case for what we think of as *normal* light. However, situations often arise in nature that cause the wave to vibrate in one plane over a long interval of time; when this happens, we have **polarized light**. This commonly happens to light reflecting off the surface of a pool or the hood of your car. Polarized sunglasses block out the portion of light that reflected off the surface so you can better see light coming from other directions, like the other cars around you or swimmers under the water.

Polarization of light demonstrates two important properties. First, it shows that the wave model of light is valid, because the wave model accurately predicts polarization phenomena, whereas the ray model cannot. Second, it shows that light waves are fundamentally different than sound waves. (For readers who have read Chapter 8, polarization shows that light waves are similar to the seismic S waves (Figure 8-1), not the compression P waves, which cannot be polarized.)

The study of the polarization of light is important to astronomy because many processes in the universe polarize light. For example, when molecules in Earth's atmosphere scatter light, polarization results (**Figure 11-12**). Similar effects occur

---

[1] The same is true of the magnetic part. We consider the electric part here only for convenience.

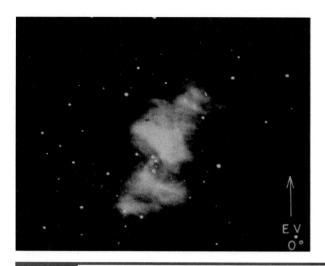
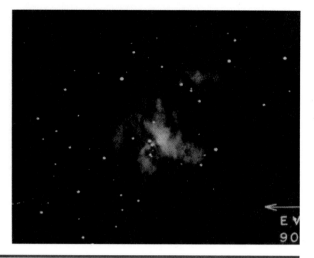

Figure 11-13 The Crab Nebula photographed in polarized light. The arrows in the lower right corners show the orientations of the polaroid filters.

for light scattered from the atmospheres of the other planets, and give information about the surface characteristics. Processes that emit polarized radiation occur in a variety of astronomical objects, such as the Crab Nebula shown in polarized light in **Figure 11-13**. The difference in the appearance of the two photographs provides astronomers with information about the processes taking place within or around the object. Further examples of objects emitting polarized light are discussed later in this book.

11.3 Light as a Stream of Particles

In the previous sections we examined two models describing the behavior of light: rays and waves. We have described experiments involving reflection, refraction, diffraction, interference, and polarization, for which these models provide predictions of past and future behavior. There are other experiments, however, for which neither the ray model nor the wave model provides satisfactory explanations.

At the beginning of the twentieth century, physics had reached some apparently insurmountable obstacles to further progress. Although experiments showed the intensity of electromagnetic energy from hot bodies to rise to a peak and then fall off toward shorter wavelengths, the theoretical understanding of the day predicted that such a body ought to radiate more and more energy at shorter and shorter wavelengths (**Figure 11-14**).

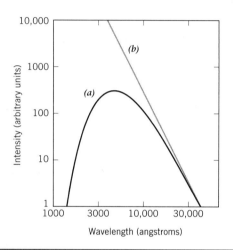

Figure 11-14 Radiation from a hot body. *(a)* As observed. *(b)* According to early theory.

This *ultraviolet disaster* obviously meant that there was something wrong with the theoretical understanding—but what?

The physicist Max Planck ushered in a new era of physics in 1899 when he showed that agreement between observation and theoretical predictions could be obtained by assuming a model in which electromagnetic radiation is always emitted in discrete amounts of energy, called **photons**. Photons are bundles of energy and can loosely be pictured as particles that have no mass. In addition, Planck's photon model suggested that the amount of energy contained by each photon is proportional to the frequency of the radiation. Expressed mathematically, the sentence says this:

$$E = hf$$

where E is the energy of a photon corresponding to a wave of frequency f, and h is a quantity called *Planck's constant* (given in Appendix A2). This equation links the wave and photon models of light by linking the wave property of frequency with the photon property of energy. The constant h, relating the photon energy to the wave property of frequency, is one of the more important fundamental constants of physics. Owing to the relation between frequency and wavelength ($\lambda f = c$), we can also express this equation as follows:

$$E = hc/\lambda.$$

This concept tells us that the higher the frequency of the radiation (i.e., the shorter the wavelength), the more energetic are the individual photons composing the beam of radiation. Planck's proposal was the first step toward developing the twentieth-century theory called *quantum mechanics*, which assumes that most aspects of the natural world are discrete bits and pieces when viewed on the smallest scales. Once the photon nature of radiation was understood, theory and the observations in Figure 11-14 matched perfectly.

In the photon model, we describe the **intensity** of light in terms of the number of photons striking a detector (your eye) each second. Intensity is not the same as the energy of the individual photons in the beam. For example, you can have a high-intensity beam of low-energy (e.g., infrared) photons, and you could have a low-intensity beam of high-energy (UV, X-ray, or gamma-ray) photons. However, a century ago the idea that light could be modeled as a particle was, to put it mildly, not well received by the physics community. It took someone not immersed in the mainstream of physics thinking to prove its value.

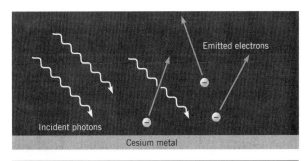

Figure 11-15 The photoelectric effect. When light falls on certain metals, electrons are dislodged from the metal, causing an electrical current to flow.

light falling on certain substances forces electrons from the surface, thus making an electric current flow. This phenomenon is the basis of a number of devices in common use (e.g., the exposure meter in modern cameras, and the automatic door openers in many business establishments).

The accepted theory of the time predicted that an electric current ought to be given off by the photoelectrically active substance no matter what the wavelength of the light, provided that the light's intensity was large enough (**Figure 11-15**). Instead, experiments found that the electrons were not knocked loose unless the wavelength of the light was less than some critical value, *no matter how intense the light*. In addition, light that had a sufficiently short wavelength caused an electric current to flow, *no matter how feeble the intensity of the light*.

Einstein reasoned as follows. If the energy of each photon depends on its wavelength, then only a photon that has a sufficiently great energy will be capable of knocking an electron loose from the surface. If the electron acquires only a small amount of energy, it will not have enough energy to leave the surface. Because shorter-wavelength photons have the larger amount of energy, only they will generate an electric current. Although just one photon could knock an electron loose if it had sufficient energy, photons with too little energy could *never* knock the electron loose, regardless of how many there were. How much energy is required depends on the substance. The photon nature of light, therefore, explains all observations concerning the photoelectric effect. Einstein won his only Nobel Prize for this work, not for his better-known theories of special and general relativity.

Inquiry 11-11

What do you conclude about the nature of light now that observations and theory described in Figure 11-14 are in agreement?

THE PHOTOELECTRIC EFFECT

In 1905, the young Albert Einstein applied this daring idea of photons to explain another phenomenon that had been puzzling physicists for years: the photoelectric effect. In the photoelectric effect,

Inquiry 11-12

As the wavelength of the incident photon gets shorter, what should happen to the speed (kinetic energy) of the ejected electron?

Inquiry 11-13

For the following types of radiation, what is the order of increasing wavelength? Of increasing photon energy? Gamma rays, infrared radiation, visible light, X-rays, radio radiation, ultraviolet radiation.

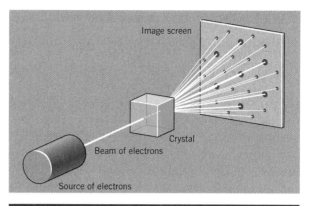

Figure 11-16 Electrons passing through a crystal are diffracted, demonstrating that electrons also exhibit some wave properties.

MODELS OF LIGHT

At this point, you may be asking, "What *is* light, really?" We have talked about it as a ray, as a wave, and now as a stream of particles. It seems contradictory that light could behave both as a wave, which cannot be localized to a single point, and as a particle, which by definition is localized. However, the evidence tells us that light can be either a wave or a particle, depending on the experiment. Scientists must accept the evidence, even if it seems counterintuitive. When we describe light as a wave or as a particle, we are saying that under certain circumstances it can behave as a wave and under other circumstances as a particle. The models we call *waves* or *particles* have a certain meaning in our everyday experience, but we are applying them to describe a submicroscopic phenomenon that is far removed from that experience. In fact, *nothing* in our macroscopic world is, or even can be, a perfect analogy to explain the nature of light. The best models explaining the phenomenon of light are abstract mathematical models that enable phenomena to be predicted but have no easily visualized analogies such as rays, waves, or particles.

This **dual nature**, or **wave-particle duality**, is not confined to light, but is a property of all matter as well. Material that we are accustomed to thinking of as only particles, such as electrons and neutrons, also exhibit wavelike behavior. For example, if a beam of electrons passes through the closely spaced atoms of a crystal and then falls on a detector, the images of the electrons will not form a single spot in the center but, rather, a pattern of spots similar to a diffraction pattern for light (**Figure 11-16**), thus showing that electrons have wavelike characteristics, too.

We will apply the photon nature of light to an understanding of spectra in Chapters 13 and 14. First, however, in Chapter 12 we will look at the telescopes that astronomers use to collect the electromagnetic radiation we studied in this chapter.

Images in a Mirror

After completing this activity, you should be able to do the following:

- State the relationship between incident and reflected rays and use it to trace rays reflected from a mirror.
- Define *virtual image* and identify situations in which a virtual image occurs.
- Diagram and explain the formation of a virtual image by a mirror.

Take a mirror and observe the reflection of a printed page in it. What do you see? The familiar fact is that the mirror reverses the page and the writing. In the "looking-glass world," everything is reversed: left-handed shoes become right-handed, people hold out their left hand for a handshake, and bottle tops screw on and off in the wrong direction.

To study this phenomenon, shine a light (flashlight or desk lamp) along a piece of paper so that the beam from the light makes a long streak on the paper (**Discovery Figure 11-1-1**). This will work better if you make a mask (perhaps two pieces of black tape or an index card with a slit cut out of it) for the flashlight or lamp so that you allow only a thin slit (rectangle) of light to pass, as shown in part *c* of the figure.

- **Discovery Inquiry 11-1a** What model of light is illustrated by the fact that the beam of light from the flashlight makes a long streak on the paper?

Now place a small mirror at the edge of the paper so that it intercepts the beam and sends it off in another direction. Tilt the mirror so that the outgoing beam can also be seen on the paper, as shown in Discovery Figure 11-1-1*b*. By varying the angle *A* between the incoming beam and the line perpendicular to the mirror, you can vary the direction in which the reflected beam will go. For various incoming angles, *A*, roughly estimate the size of angle *B* between the reflected beam and the line perpendicular to the mirror. (In the language of physics, angle *A* is called the **angle of incidence** and angle *B* the **angle of reflection**.)

You can also do this activity with a flashlight and a full-sized mirror. Place a removable mark or a piece of tape on the mirror. Move straight toward the wall (at an angle to the spot as shown in **Discovery Figure 11-1-2**) as you continue to shine your flashlight on the marked spot. As you approach the wall, you are constantly increasing the angle of incidence. Watch the reflected beam.

- **Discovery Inquiry 11-1b** From your observations, what can you infer is the relationship between the angle of incidence and the angle of reflection?

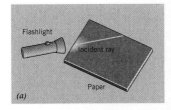 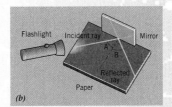 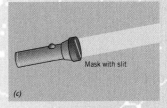

Discovery Figure 11-1-1 Making a ray of light reflect off a piece of paper *(a)* without and *(b)* with a mirror in place to reflect the beam. Angle *A* is the angle of incidence, while angle *B* is the angle of reflection. Placing a slit over the end of the flashlight *(c)* helps to define the light beam better.

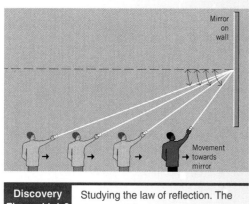

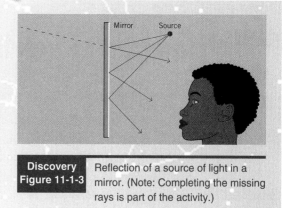

Discovery Figure 11-1-2 Studying the law of reflection. The observer shines a flashlight at a point on a mirror while walking toward the wall.

Discovery Figure 11-1-3 Reflection of a source of light in a mirror. (Note: Completing the missing rays is part of the activity.)

The rays of light behave as if they consisted of a stream of particles all traveling in the same direction. When hitting a surface, they bounce away in another direction determined by the direction from which they came. Billiards players will recognize that the answer to the previous question is used to estimate the angle at which a ball will rebound from the cushion. Thus, you have found the law of reflection: the angle of reflection always equals the angle of incidence.

We are now in a position to understand the formation of an image in a mirror. **Discovery Figure 11-1-3** shows a source of light in front of a mirror. Several rays from the source are shown, and their paths are sketched, using the relationship just discussed. One of these rays, denoted by a dashed line, is extended backward from the mirror surface. *Trace or photocopy the figure;* on the copy, take a straightedge and extend the other two lines straight back to where all three lines intersect. This point of intersection is the point from which the rays *appear* to come—from behind the mirror at a distance equal to the distance of the actual object in front. An observer in front of the mirror has no way of knowing what has happened to the rays along the way and can only sense the *direction* from which the rays *appear* to come.

- **Discovery Inquiry 11-1c** Describe the location behind the mirror from which the light appears to come.
- **Discovery Inquiry 11-1d** Explain why the eye is fooled into thinking that the source of light is *behind* the mirror.

If you were to place a camera at the point where the image appears to be located and then tried to photograph it, no image would result because there are no actual light beams there. For this reason, the kind of image we are discussing is called a **virtual image**. This means that the rays of light that enter our eyes project from an apparent source, as if they have traveled in straight lines with no deflections. It is an apparent image only, which is what puzzles a kitten trying to "catch" its own reflection in a mirror. Virtual images are to be distinguished from the so-called **real images** we will study later.

Refraction of Light

After completing this activity, you should be able to do the following:

- Diagram and explain the formation of a virtual image by refraction at the surface of a glass of water.
- Diagram and describe the effect of refraction on a ray of light as it passes from one medium to another.

Fill a clear glass with water. Put a pencil into the water at an angle and observe the appearance when you look directly along the part of the pencil that is out of the water. Draw the appearance of the pencil. (*Important:* Look directly *along* the part of the pencil that sticks out of the water, not from the side.)

- **Discovery Inquiry 11-2a** What appears to have happened to the pencil?
- **Discovery Inquiry 11-2b** Where does the tip of the pencil appear to be, relative to its actual position?

When light rays pass from one medium to another, as from air to water or from water to air, they bend. The rays of light bend *toward* the line perpendicular to the surface when they enter a denser medium (water) from a less-dense medium (air). They bend *away* from this line when they leave the denser medium, as shown in **Discovery Figure 11-2-1**.

Discovery Figure 11-2-1 shows the rays of light leaving the tip of a pencil and entering the air from the water. Trace or copy the figure and extend the rays in the air backward along straight lines to the point from which they appear to come in the water, as illustrated by the dashed line.

- **Discovery Inquiry 11-2c** Is the apparent source of the rays of light from the tip of the pencil in agreement with your answer to the previous inquiry? Where is the apparent source relative to the actual position of the pencil tip? (Show your figure.)
- **Discovery Inquiry 11-2d** What kind of image (real or virtual) are you observing when you look at the pencil tip in the water?

Now imagine a ray of light that enters a piece of glass with parallel sides, as illustrated in **Discovery Figure 11-2-2**. The path of the light ray inside the glass is determined using the rule stated above that light entering a denser medium is bent toward the perpendicular. Using a pencil, draw the path of the ray as it leaves the glass in the

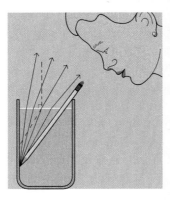

Discovery Figure 11-2-1 The pencil experiment, with rays partly drawn in. The dashed line shows the apparent origin of one of the rays. The drawing does not show what you will see if you view the glass from the side.

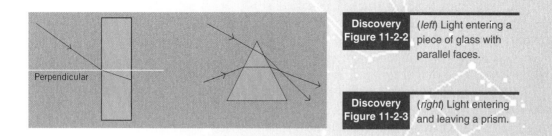

Discovery Figure 11-2-2 (*left*) Light entering a piece of glass with parallel faces.

Discovery Figure 11-2-3 (*right*) Light entering and leaving a prism.

figure. (Note that the beam that strikes the surface perpendicular to the surface is not refracted.)

- **Discovery Inquiry 11-2e** After the ray leaves the glass, is it traveling in the same direction it was originally? Explain.

The answer to Discovery Inquiry 11-2e depends on the fact that the two sides of the piece of glass are parallel. For example, in a prism (**Discovery Figure 11-2-3**) the two sides are not parallel. A ray of light that enters the first side perpendicularly will not be bent at that side but will be bent at the other one. Even if it enters the first side at an angle, there will be a net change in the direction of travel of the ray.

- **Discovery Inquiry 11-2f** In **Discovery Figure 11-2-4**, light strikes the circular surface of the glass perpendicularly. Draw the path of the light through the glass and show how it exits the bottom surface.

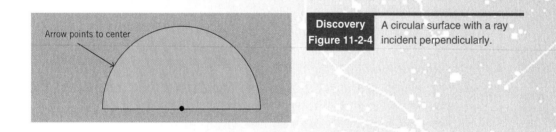

Arrow points to center

Discovery Figure 11-2-4 A circular surface with a ray incident perpendicularly.

Chapter Summary

Observation

- Light exhibits properties of rays through **reflection** and **refraction**, and properties of waves through **diffraction**, **interference**, and **polarization** (in addition to reflection and refraction).

- All types of electromagnetic waves in the vacuum of empty space move with the speed of light and are characterized by a **wavelength** and **frequency**. These three quantities are related to one another by $\lambda f = c$, where λ is the wavelength, f is the frequency, and c is the **speed of light**.

- Light, gamma rays, X-rays, ultraviolet and infrared radiation, and radio waves all travel at the speed of light and differ only in their wavelengths, frequency, and energy per photon.

- Not all types of **electromagnetic radiation** pass equally well through the Earth's atmosphere.

- Light can exhibit the properties of particles (**photons**), while particles of matter can show the properties of waves, thus giving a dual nature to light and particles.

- The **photoelectric** effect is a phenomenon in which electrons are ejected from the surface of a material when photons having high-enough energy strike the material.

Theory

- We describe light by means of various models, none of which say what light *is*.
- Maxwell's theory of electromagnetism predicts the behavior of electromagnetic waves. It predicts that these waves vibrate perpendicularly to their direction of motion (and can therefore be polarized), that they move with the speed of light, and that they show the properties of reflection, refraction, diffraction, and interference. It also predicts that waves carry energy.

- Nothing can move faster than the speed of light in empty space.
- Planck showed theoretically that the energy possessed by a photon is directly related to the frequency, f, associated with it, and inversely related to the wavelength, λ: $E = hf = hc/\lambda$, where E is the photon energy and h is Planck's constant.

Conclusions

- Both theory and observation indicate the wave–particle duality of light and matter.

Summary Questions

1. What is meant by *reflection* and *refraction?*
2. What is meant by *diffraction?* In what situations does it occur? How does diffraction vary with the wavelength of the light and the width of the slit through which it travels?
3. What is the meaning of *wavelength* and *frequency?* What is the relationship among wavelength, frequency, and speed of a wave?
4. What do the words *electromagnetic spectrum* mean?
5. What is the importance of the concept of atmospheric *windows* for astronomy?

6. What is the meaning of *interference?* How does the concept explain the slit experiment in Figure 11-5?
7. What is meant by the polarization of light? Why is it important to astronomy?
8. What is meant by the wave–particle duality of both light and matter?
9. In what way does the energy of a photon depend on its wavelength? Arrange radiations of various types in order of increasing photon energy.

Applying Your Knowledge

1. Summarize the models we use to understand light. What evidence is used to validate each model?
2. Explain the differences between the bending of light by refraction and the bending by diffraction.
3. What is the importance of the photoelectric effect in providing evidence in favor of the photon description of light?

▶ 4. Suppose your favorite radio station's frequency is 91.5 megahertz (millions of cycles per second). What is the wavelength of the signal?

▶ 5. How much more energy does a 1000 Å photon have than one having a wavelength of 10,000 Å?

▶ 6. What are the frequency and the energy of a photon having a wavelength of 21 cm?

Answers to Inquiries

11-1. Individuals differ in the way they perceive color and the words they use to describe them.

11-2. Very high priority should be given to observing wavelengths that are absorbed by the atmosphere and hence inaccessible at the Earth's surface: gamma ray, X-ray, ultraviolet, infrared. Observations that are possible from the Earth's surface should be done from there, because space operations are extremely expensive.

11-3. 3×10^5 km/s $\times$ 1 year $\times$ 365 days/year $\times$ 24 hours/day $\times$ 60 minutes/hour $\times$ 60 s/minute = 9.5×10^{12} km.

11-4. 150,000,000 km = 1.5×10^8 km / 3×10^5 km/s = 500 s = 8.3 minutes.

11-5. 3×10^6 km/sec $\times$ 2.6 seconds / 2 (round trip) = 3.9×10^5 km.

11-6. Speed = 2 hertz $\times$ 10 cm = 20 cm/s.

11-7. Decreasing wavelength means increasing frequency.

11-8. Decreasing speed means decreasing frequency.

11-9. $\lambda f = c$; $\lambda = 3 \times 10^{10}$ waves/s 1.5×10^7 waves/s = 2×10^3 cm, or 20 m (0.02 km).

11-10. $\lambda f = c$; $f = 3 \times 10^{10}$ cm/s / 3 cm = 10^{10} hertz or 10^4 megacycles/s or 10 gigacycles/s = 10 gigahertz.

11-11. Because the theory depended on the assumption of discrete photons, you should conclude that the description of light in terms of photons has some degree of validity.

11-12. The speed (kinetic energy) of the ejected electron should increase as the wavelength of the incident photon decreases (i.e., as the energy of the incident photon increases).

11-13. Increasing wavelength: Gamma rays, X-rays, ultraviolet radiation, visible light, infrared radiation, radio radiation. Increasing energy is the reverse order.

Telescopes: Our Eyes of Discovery

12

O telescope, instrument of much knowledge, more precious than any sceptre!

Is not he who holds thee in his hand made king and lord of the works of God?

Galileo Galilei, 1610

Telescopes provide humanity with a view of the universe not imagined by our ancestors. The telescope is the most basic and fundamental instrument that astronomers use, indispensable since light spreads out and gets dimmer the farther it travels (the inverse square law). It has existed as a research tool since the time of Galileo, almost 400 years ago. A well-constructed telescope is a versatile instrument with perhaps the longest useful lifetime of any major piece of scientific equipment. The detectors and analyzers of light mounted on telescopes are constantly being improved by the rapid progress of modern technology, but a soundly constructed telescope will continue to perform its basic functions for many decades.

There are, of course, many astronomers who never use telescopes. Theorists regard large, powerful computers as their principal research tools, but even they are ultimately dependent on the data collected by observational astronomers to test their theories and hypotheses.

The main function of a telescope is to gather light from a celestial object and bring it to a focus for further study. Secondary functions are to resolve the detail in the image, and to magnify the angular scale of celestial objects. The telescope is placed on a mounting that can be moved, thereby enabling it to follow the apparent paths of objects across the sky.

Astronomers have used many different kinds of telescopes since Galileo first turned his small instrument skyward in 1609. Modern astronomers use telescopes of all types and sizes. Some are sent above the atmosphere to capture light that cannot get through the Earth's atmosphere (Figure 11-8), and to get above its distortions. In this chapter, you will study how lenses and mirrors form images, and how they can be combined to form telescopes. In addition you will read about the auxiliary instruments that are attached to telescopes to detect the light. Finally, the observatories at which telescopes are located are discussed.

12.1 The Formation of Images

A simple device that illustrates the basic process of image formation is the pinhole camera. **Figure 12-1** shows the principle of its operation. It forms an image on a screen that can be observed directly with the eye. Light travels from a particular part of the scene being viewed through the pinhole and then (except for diffraction effects) travels onto the screen, where it is observed. The image is upside down, which is easily understood by looking at the figure.

If a piece of photographic film is used instead of a screen, we can take a picture of the object.

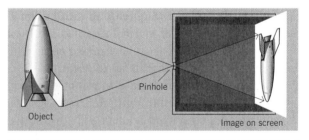

Figure 12-1 The principle of the pinhole camera. An inverted image is formed because light travels in straight lines through the pinhole.

Pinhole cameras are still used today to take very wide-angle photographs that would be difficult for conventional lenses. But the pinhole camera suffers from some obvious drawbacks. The image is faint and hard to see, because not much light gets through the pinhole, and enlarging the pinhole to

---

Chapter opening illustration: An artist's rendition of the European Southern Observatory's Very Large Telescope in Chile. The sky shows the Large Magellanic Cloud.

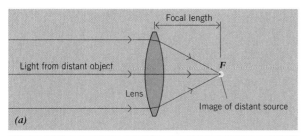

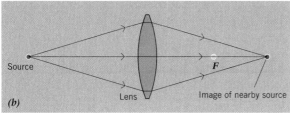

Figure 12-2 Image formation of a point source of light using the same lens but from (a) a distant source and (b) a nearby source.

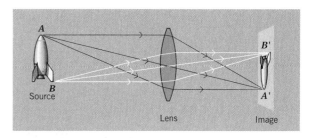

Figure 12-3 Formation of an image of an extended source by a lens.

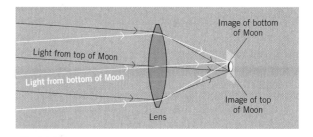

Figure 12-4 Formation of an image of the Moon by a lens.

let in more light would blur the image. Therefore, small holes and long time exposures are necessary to produce good pictures. To photograph faint celestial objects, astronomers need something that has the ability to collect a great deal of light but that can still concentrate the light to a focus, so that sharp, bright pictures can be made. These goals can be accomplished using either lenses or curved mirrors.

We will describe how lenses image a *point source* of radiation, which is an object whose distance is so much greater than its size that it looks like a point of light. A star is an example of such a source. **Figure 12-2a** shows parallel light rays from a distant point source brought to a focus by a lens. The distance from the lens to the focus is the most important quantity used to describe a lens and is called the **focal length**. The lens brings the light to a focus because each ray encounters the glass at a different angle, and each is bent by a different amount. **Figure 12-2b** is similar to Figure 12-2a except that it shows several rays of light from a *nearby* point source entering the same lens; in this case, the rays are brought to a focus farther from the lens than the distance defined by the focal length. In both cases, the net result is that all the rays are focused to a single point. Such an image is called a **real image**, because the light rays actually go through this point; you can record it by placing film there. The

real image is different from the virtual image discussed for mirrors in Discovery 11-1.[1]

Any object that is not a point source is called an *extended source* (**Figure 12-3**). Light from various parts of an extended object is focused onto different parts of the observing screen. Light from the rocket's nose at point *A* is imaged onto the screen at point *A'*, whereas light from the rocket's fin at *B* is imaged at *B'*. The resulting image is inverted.

When an object is at a great distance, light rays from any part of the object will arrive at the lens in almost parallel rays. To visualize why this is true, imagine the source of light in Figure 12-2b moving away from the lens until it was many miles away. The rays of light would clearly be almost perfectly parallel and images would form exactly one focal length behind the lens. If we were to put a piece of film there, we could photograph the image.

Figure 12-4 shows what happens when a lens focuses light from a distant extended source. In the case of the Moon, one set of parallel rays comes from one side of the Moon and another set from

---

[1]For those of you who did not do Discovery 11-1, an example of what we call a virtual image is the image you see when you look at yourself in a mirror. It is called a virtual image because if you were to place a piece of film at the location where your image appears to be, no photographic image would form because there are no light rays passing through that point.

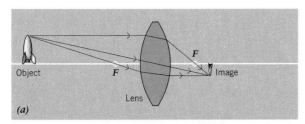

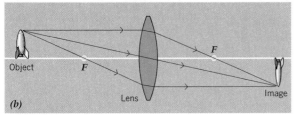

Figure 12-5 Image size depends upon the focal length. Each lens shown here has a different focal length and therefore produces images of different size.

the other side. Each set of parallel rays will be focused to a particular point by the lens. The image is, of course, inverted.

The image sizes of the rocket in Figure 12-3 and the Moon in Figure 12-4 are directly proportional to the focal length of the lens. **Figure 12-5** shows why. The symmetrical lenses shown here have foci at equal distances from the center of the lens. The lens in Figure 12-5a has more sharply curved surfaces than the lens in Figure 12-5b. This bends the light more, and 12-5a's lens therefore has a shorter focal length. The result is that the longer-focal-length lens produces the larger image.

Figure 12-5 tracks light rays from one part of the rocket to the corresponding position in the image. Note that the ray through the lens's center is not refracted. Furthermore, the rays entering the lens parallel to the axis exit the lens traveling through the focal point. These principles of optics determine the locations at which an image forms.

Inquiry 12-1

If a lens having a 50-mm focal length produces a 1-cm-high image of a person, what will be the image size when a 150-mm focal length lens is used?

12.2 Telescopes

Lenses have been in use for a long time to correct vision. Although probably not the originator of the telescope, Hans Lippershey, a Dutch spectacle

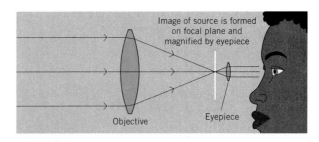

Figure 12-6 How a telescope works.

maker, is usually given credit for making the principle of the telescope widely known in the early 1600s.

The basic telescope has two lenses, an objective and an eyepiece lens, physically arranged as illustrated in **Figure 12-6**. The objective lens or mirror (often referred to as the **objective**) is a lens or mirror of longer focal length and (usually) larger diameter than the second lens so it can produce a large real image with as much light as possible. Telescopes with objective lenses are called **refractors**, while those with mirrors are called **reflectors**. The real, inverted image can be photographed or inspected through the second part of the telescope, the **eyepiece**. The eyepiece, which magnifies the image, has a shorter focal length than the objective.

THE LIGHT-GATHERING POWER OF A TELESCOPE

Inquiry 12-2

How many times more light is collected by a 10-cm-diameter objective than by a 1-cm-diameter objective?

The telescope's objective collects light. The larger the area of the objective, the greater the telescope's **light-gathering power**. In the Inquiry, the objective diameter has increased 10 times and the area 10^2 times. For this reason, 100 times as much light is collected by the larger objective.

The diameter of a telescope's objective is its most important single characteristic as far as an astronomer is concerned, because the diameter determines the light-gathering power. Even a 1 cm lens will collect about four times as much light as the human eye. One of the world's largest operating optical telescopes, each of the two 10 m Keck telescopes on Mauna Kea in Hawaii, has a diameter approximately 2,000 times that of the human

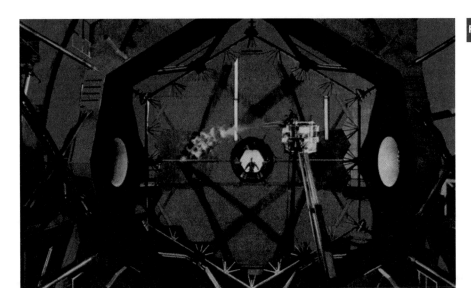

Figure 12-7 The 10-meter Keck telescope.

pupil. As a consequence, it is able to gather roughly 2000^2 or 4,000,000 times as much light as the eye (**Figure 12-7**).

    Table 12-1 lists some of the largest telescopes in the world.

Inquiry 12-3

If the dark-adapted eye has a diameter of 0.5-cm, how much more light can a lens of diameter D cm collect than the human eye?

Table 12-1   Some Large Telescopes

TELESCOPE (DIAMETER OF OBJECTIVE)	FUNDING COUNTRY (COUNTRIES)	OBSERVATORY (LOCATION)
ESO Very Large Telescope Observatory (four 8.2 m)	Europe	La Silla (Paranol, Atacama, Chile)
Hobby-Eberly Telescope (11 m to 9 m effective)	USA/Germany	McDonald Observatory (Mt. Fowlkes, Texas)
Gran Telescopio Canarias (10.4 m)	Spain/Mexico/USA	Observatorio del Roque de los Muchachos (La Palma, Canary Islands)
The Keck Telescopes (two 10 m)	USA	W.M Keck Observatory (Mauna Kea, Hawaii)
Southern Africa Large Telescope (10 m)	South Africa/Germany/ Poland/USA, New Zealand/UK	South African Astronomical Observatory (Sutherland, South Africa)
Large Binocular Telescope (two 8.4 m mirrors on one mounting)	USA/Germany/Italy	Mt. Graham International Observatory (Mt. Graham, Arizona)
Subaru Telescope (8.3 m)	Japan	National Astronomical Observatory of Japan (Mauna Kea, Hawaii)
Gemini North (8.1 m)	USA	National Optical Astronomical Observatory (Mauna Kea, Hawaii)
Gemini South (8.1 m)	USA	National Optical Astronomical Observatory (Cerro Pachon, Chile)
MMT (6.5 m)	USA	MMT Observatory (Mt. Hopkins, Arizona)
Magellan Project (two 6.3 m)	USA	Observatories of the Carnegie Inst. of Washington (Las Campanas, Chile)
Bol'shoi Telesop Azimutal'nyi (6 m)	Russia	Special Astrophysical Observatory (Mt. Pastukhov, Russia)
Large Zenith Telescope (6 m liquid mirror)	Canada/France/USA	UBC Malcom Knapp Research Forest Maple Ridge (British Columbia)
George Ellery Hale Telescope (5.08 m)	USA	Palomar Observatory (Palomar Mountain, California)

The advantage of big telescopes over the human eye is even more dramatic than just light-gathering power, because at the focus of the telescope the astronomer can mount a camera or sensitive electronic detector and make an observation over an extended length of time (e.g., a time exposure on film). The longer the observation, the fainter the objects the detector can record. The human eye, by contrast, is adapted to be efficient in detecting motion, so it clears itself many times a second; as a consequence, it cannot accumulate light as a photographic plate can. Our large telescopes are capable of detecting objects more than 100 million times fainter than the faintest stars detectable by the naked eye.

RESOLVING POWER

The edges of a telescope objective diffract the light passing around it. Diffraction makes the edges of images fuzzy, and for this reason no image can ever be perfectly sharp. (Remember, diffraction from a hole produces an image with a blob at the center surrounded by rings.) The larger the diameter of the lens, however, the less important the diffraction and sharper the image will be. Sharpness is measured in terms of the **resolving power**, or **resolution**, of the telescope, which is the angular size of the smallest detail that can be observed with it (**Figure 12-8**). The smaller the number, the better the resolving power. Thus, whatever makes the number smaller improves the resolving power. Because short wavelengths diffract less than longer wavelengths, resolution is better at shorter wavelengths. The resolving power therefore depends on both the

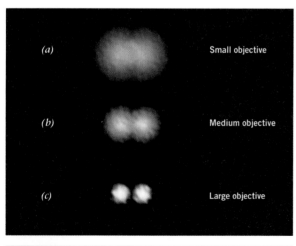

Figure 12-8 A double star as seen through telescopes having different resolving powers. (a) Small objective. (b) Medium objective. (c) Large objective.

wavelength of the incoming radiation and the diameter of the telescope objective:

$$\text{Resolving power (in degrees)} = 57.3°$$
$$\times \frac{\text{Wavelength of light}}{\text{Diameter of objective}},$$

where the wavelength and diameter are measured in the same units.

For optical wavelengths (at about 5,000 Å), a practical rule of thumb that measures resolving power in the more convenient unit of seconds of arc is given by this formula:

$$\text{Resolving power (in seconds of arc)}$$
$$= \frac{10}{\text{Diameter of objective in cm}}.$$

Note that resolving power is expressed as an angle. The human eye is able to distinguish objects separated by an angle of about 1 minute of arc. A 1-inch (2.5 cm) telescope can distinguish objects separated by an angle of 4 seconds of arc as two distinct images. We say that these objects are *resolved*. Objects whose angular separation is less than 4 seconds of arc will not be resolved by a 2.5 cm telescope. In looking at a planet with this same telescope, only details separated by at least 4 seconds of arc will be distinguishable from one another. Using this formula, the 10 m Keck telescopes are each theoretically capable of resolving objects 0.01 second of arc apart, about four powers of ten better than the unaided human eye.

Unfortunately, the Earth's atmosphere plays a crucial, limiting role in ground-based observations under most situations. The atmosphere behaves like a lens in that it refracts the light passing through it. However, this lens is constantly moving. You have probably seen the effects of this: If you have ever observed stars carefully at night, even with the naked eye, you will have noticed that they twinkle. This is due to moving masses of air in the atmosphere through which light passes. Observed with a high-magnification telescope, the image of the star dances rapidly about and becomes blurred. A similar effect is seen when you look at scenery beyond a hot grill or through heat waves rising from a hot road.

The steadiness of the atmosphere can vary from minute to minute during the night, and from night to night. Astronomers refer to the blurring of the image as **seeing**. The seeing is expressed as an

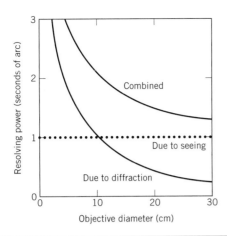

Figure 12-9 The blurring effects of seeing and diffraction in telescopes of different diameter objectives. The top curve gives the actual resolution achieved.

angular size and typically is 1 to 2 seconds of arc. On a night of good seeing, however, the atmosphere is stable enough that the blurring may be as small as a quarter of a second of arc from the best observatory sites.

The formulae just given for resolving power are the best theoretically possible values. **Figure 12-9** compares the blurring effects of diffraction and seeing by showing the theoretically possible resolving power of telescopes of different objective sizes, and the seeing for a typical night. You will note that even modest-sized telescopes are more limited by seeing than by diffraction, and that only the very smallest telescopes are limited in their performance by diffraction. Seeing limitations are eliminated, however, when telescopes are placed above the Earth's atmosphere. They are then able to operate at their theoretically possible resolution limit.

Inquiry 12-4

At roughly what objective diameter does *seeing* rather than *objective size* become the limiting factor in determining actual telescopic resolution? Assume the best seeing is between 0.25 and 0.5 seconds of arc.

Magnification

We have said that a telescope forms a real image, and if a screen were placed at the focus, we would see an image. But that image would not be very large. For this reason, an *eyepiece* lens is used to magnify the real image produced by the objective (see Figure 12-6). Because we are using the eyepiece as a magnifier, we actually observe a virtual image.

The overall **magnification** of a telescope depends on both the size of the real image formed by the objective and the amount by which the eyepiece magnifies this real image. We said previously that the size of the image produced by the objective lens depends on its focal length. For this reason, the overall magnification of the telescope must depend on the focal lengths of *both* the objective and eyepiece. In fact, the magnification is expressed simply:

$$\text{Magnification} = \frac{\text{Focal length of objective}}{\text{Focal length of eyepiece}},$$

where the focal lengths are in the same units.

Inquiry 12-5

If an amateur astronomer's telescope having an objective diameter of 15 cm and a 150 cm focal length is used with an eyepiece of 2 cm focal length, what would its magnification be?

Inquiry 12-6

Would the ability of this amateur telescope to resolve detail in an image be limited by diffraction or by seeing when the seeing is 0.5 seconds of arc?

It is possible in theory to make a telescope with as much magnification as desired simply by choosing eyepieces of shorter and shorter focal length. However, there are practical limits to the amount of magnification that can be used, because of diffraction in the telescope and unsteadiness in the atmosphere, both of which tend to blur the image. Beyond a certain point, one only magnifies the blur without seeing anything new. Astronomers rarely work at a magnification of more than about 500, and it would be an unusually good night when a small, amateur telescope could be used at a magnification greater than 10 times the diameter of the objective measured in centimeters (e.g., a magnification of 100 for a 10 cm objective).

Inquiry 12-7

Suppose you see a catalog that offers to sell you a 5-cm-diameter telescope with a magnification of 300. How do you respond? What would be a more reasonable magnification?

Inquiry 12-8

In what three ways can we talk about the *power* of a telescope? Is it meaningful to ask someone what the power of his or her telescope is?

THE MOUNTING OF A TELESCOPE

Astronomical telescopes are built with special mountings that enable them to rotate along with the moving sky and continuously observe a celestial object as long as it is above the horizon. **Figure 12-10** shows the polar axis of the University of Texas 82-inch (2.1 m) telescope running between the south and north piers that support the telescope. The **polar axis** always points toward the North Celestial Pole while the telescope rotates around it to follow a star. This arrangement is called an **equatorial mount**.

Mountings that allow the telescope to move in altitude and azimuth are much easier to construct

and less expensive than equatorial mountings. However, such telescopes are more difficult to use because keeping an object in the center of the field requires moving the telescope in two directions instead of one. Because computers can perform these operations easily and without error, most radio telescopes and large modern reflectors are built with such mounts.

12.3 Comparing Reflecting and Refracting Telescopes

Refracting telescopes suffer from numerous disadvantages. In a refractor, the light passes through the glass. Because, as we saw in the previous chapter, the amount of refraction depends on the wavelength, different colors are refracted in slightly different directions. This means that in a simple refracting telescope consisting of only two lenses, red and blue light focus at different points and produce colored halos around star images. This blurring effect, known as *chromatic aberration*, is shown in **Figure 12-11a**, Chromatic aberration can be corrected by making compound lenses containing several lens elements of different materials and having different shapes (**Figure 12-11b**). Good lenses with excellent correction of various aberrations may have many lens elements—the standard 55 mm lens on a good camera may have, for example, from five to nine elements. Such lenses are expensive and difficult to design. Furthermore, they lose light at each lens surface due to reflection, which can seriously degrade their performance. By contrast, as Sir

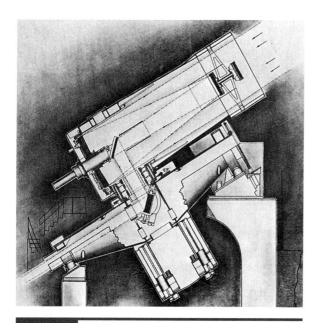

Figure 12-10 The polar axis and the coudé focus of the 2.7m telescope at the University of Texas and McDonald Observatory.

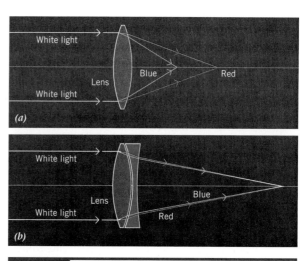

Figure 12-11 *(a)* Chromatic aberration. *(b)* The correction for chromatic aberration using a compound lens.

Figure 12-12 The 40-inch refractor of Yerkes Observatory, Williams Bay, Wisconsin.

Isaac Newton first pointed out, mirrors are inherently free of color effects.

Another drawback of lenses results from the fact that a lens is made up of a lot of material through which the light must pass. When the telescope moves to follow a heavenly object, the lens changes its orientation with respect to the center of the Earth and bends a little under its own weight. Because astronomical lenses and mirrors are ground to close tolerances—a millionth of a centimeter or so—even a miniscule amount of flexure can seriously distort the image. The lens can be made thicker, but casting a large, thick piece of glass that is so perfect it does not contain many bubbles and other imperfections is difficult. In addition, thicker lenses weigh more (requiring stronger, more expensive mountings to hold them) and absorb more light. The practical limit for refracting telescopes was attained in 1897 with the 40-inch (1.02 m) lens at Yerkes Observatory, Williams Bay, Wisconsin (**Figure 12-12**). A mirror, by contrast, does not suffer these limitations. Because it is supported from the back, it can be made quite stiff and can be cast with reinforcing ribs inside it.

Another drawback to refractors is that all lens surfaces must be perfect; in the case of a compound lens, that means at least four surfaces. Only the front surface of a mirror needs to be ground to high tolerances, and the presence of imperfections inside the mirror material does not degrade the performance because light does not pass through it. For these reasons, per centimeter of objective diameter, reflectors are also much less expensive to build.

Most of the cost of many telescopes is the dome that houses it. As can be seen in Figure 12-12, the Yerkes refractor has an enormous dome compared to the diameter of the telescope. This is necessary to accommodate the long tube, due to the long focal length of the lenses. Reflectors can fold the light path and be shorter for their diameter. Their domes can also be much smaller and cheaper.

12.4 Reflecting Telescopes of Various Types

Reflecting telescopes are made with concave mirrors. While there are various possible curved shapes, *spherical* shapes are desirable because they are easy to manufacture. **Figure 12-13***a* shows a section of a spherical mirror with numerous parallel light rays from a distant source of light falling on it. For each incident ray, the reflected ray is shown, following the law of reflection discussed in Chapter 11. The fact that rays of light from a star will *not* be focused into a single point by this mirror is a problem called *spherical aberration*. One

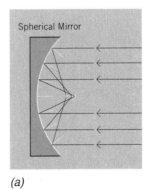
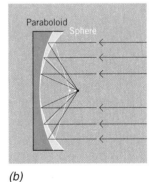

(a)   (b)

Figure 12-13 (*a*, above left) Spherical aberration, in which a spherical mirror will not bring parallel rays of light from a distant star to a single focus; (*b*, above right) Spherical aberration, and the correction provided by changing the mirror's shape into a parabola.

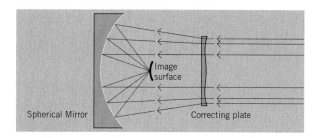

Figure 12-14 The Schmidt telescope. The aberrations of the spherical mirror are corrected by the thin correcting plate, producing an excellent wide-field image.

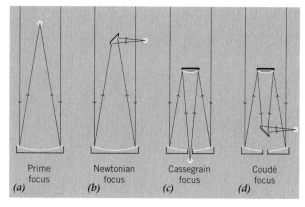

Figure 12-15 A variety of focal configurations are available with reflecting telescopes. *(a)* Prime focus. *(b)* Newtonian focus. *(c)* Cassegrain focus. *(d)* Coudé focus.

way to achieve a proper focus is to give the mirror a surface in the shape of a parabola, in which the surface (shown in **Figure 12-13*b***) is not as steep as a sphere. In this case, the rays striking the outer part of the mirror focus at the same place as those closer to the mirror's center. Most professional and many amateur telescopes therefore have parabolic mirrors.

Another method of overcoming spherical aberration is found in the **Schmidt telescope**. This system (**Figure 12-14**) uses a spherical mirror because of its ease of manufacture. To overcome spherical aberration, a specially shaped, thin correcting lens is mounted in front of the mirror. It is thicker in the middle and around the edges and thinner in a ring-shaped region in between. The amount of spherical aberration correction required is small, and as a result the telescope is not nearly as sensitive to chromatic aberrations as a refractor would be.

The Schmidt design combines both refracting and reflecting elements in the objective. Its advantage is that it combines a wide field of view with a short focal length, so that short exposure times can be used. For this reason the Schmidt telescope is ideal for making surveys of large regions of the sky.

Because large telescopes are expensive, astronomers prefer them to be as versatile as possible. Astronomical telescopes are usually built so that they can focus light in a variety of ways, which means a variety of observational goals can be accomplished with a single telescope.

One problem with a mirror is that the image forms in front of it. A method is required for observing the image without blocking most of the incoming light. If the mirror is large, this is not a major problem, and the telescope can be used at the mirror's focus, which is also called the **prime focus** (**Figure 12-15*a***). A camera or other instrument can be mounted at this position. Although some light is blocked, most passes by and is focused by the main mirror. Until the 1980s, in the largest telescopes the astronomer would actually ride for hours in a cage placed at the prime focus. During the long period of observation throughout the night, the astronomer would keep the telescope focused and the object being observed centered. The observer's cage blocks only a small fraction of the light falling on the mirror. Nowadays, however, advances in electronics have moved the astronomer out of the cage and into a room where observing is done in front of a computer terminal.

In smaller telescopes, the amount of light blocked can be minimized by inserting a small, flat mirror into the optical path to redirect the image to a focus outside the telescope tube. This arrangement is common in amateur telescopes. **Figure 12-15*b*** shows a common way of accomplishing this in the **Newtonian telescope**, named for its inventor, Sir Isaac Newton. The focal position in such an arrangement is called the Newtonian focus and is used in the least-expensive amateur telescopes.

In another arrangement, a convex mirror reflects the light back through a hole in the objective mirror. No additional light is lost on account of the hole, because it is in the shadow of the secondary mirror. Because the secondary mirror has some curvature, it increases the focal length of the system over either the prime or Newtonian focus. This arrangement is called the **Cassegrain focus** (**Figure 12-15*c***). (The Cassegrain focus can also be seen in Figure 12-10 at the back end of the telescope.) High-quality amateur telescopes with a Cassegrain focus and a Schmidt correcting plate to alleviate spherical aberration are common.

Some telescopes can be set up to direct the light to yet a third focus, the *coudé* (from the French verb meaning "to bend"), and the word describes the rather complicated path taken by the light in this case (**Figure 12-15d**). A sequence of mirrors brings the light down the hollow axis of the telescope mounting and into a room below. With this arrangement, no matter how the telescope moves, the light is brought to a focus at the same point in the room. The advantage of the coudé focus is that large and heavy instruments, which cannot be mounted on the moving telescope itself, can be placed in the coudé room. Figure 12-10 shows the path of light taken to the coudé focus located off the figure to the bottom left.

Large telescopes are usually set up so that by moving mirrors, any one of the foci—prime, Cassegrain, or coudé—can be selected. Why would an astronomer choose one focus over another? The focal length increases from prime focus to Cassegrain to coudé. Because, as we have seen, the image size depends on the focal length, the coudé focus produces the largest image of any of the foci. However, the larger the image size, the more spread out the photons are, making the image fainter and increasing the time required to obtain precise observations. Therefore, faint objects cannot be observed with the coudé focus. However, the short focal length of the prime focus is ideal for the faintest objects. The Cassegrain focus is in between. Thus, the most appropriate focus to use for an observation depends on the brightness of the object and exactly what the astronomer is hoping to learn.

12.5 The Telescopes of the Future

As astronomers tried to increase the size of their telescopes, they found themselves confronted with astronomically rising costs. Roughly, the cost of a traditional observatory rises as the third power of the mirror size. Thus, to double the size of a telescope, the cost was eight times as much.

Not only that, but astronomers working with enormous pieces of heavy glass were running into some fundamental limits. For example, the 6 m Pyrex mirror of the Bol'shoi telescope is too mas-sive to reach temperature equilibrium during the night. As its temperature slowly changes, the mirror flexes and produces a blurred image.

During the last few decades, astronomers and engineers have experimented with radically new telescope designs to achieve reductions in the weight and cost of building large instruments. An important development has been the segmented mirror—a mirror that is ground in sections and then assembled into one giant mirror, like pieces of a puzzle. This type of assembly takes state-of-the-art computers and optics technology to keep all the various segments properly aligned. Sensitive instruments continuously monitor the shape of the mirror segments and rapid computer-controlled motors work to keep them properly aligned. In other words, thick and heavy mirrors have given way to thin mirrors whose shape can be computer controlled. The goal is to concentrate all of the light collected by all the segments at one point in space. Figure 12-7a illustrates the type of telescope that results. The example shown here is one of the Keck telescopes in Hawaii, built with a segmented mirror by the University of California and California Institute of Technology. The mirror is 10 m in diameter.

Another creative method of producing large, lightweight mirrors has been developed by Roger Angel at the University of Arizona. The method uses a rotating oven containing melted glass. The rotation produces a concave parabolic curvature that remains when the mirror is cooled. With this technique, significantly less time is required in grinding and polishing the surface to reach the final mirror shape. The mirror lightness comes from an internal honeycomb structure that reduces the quantity of glass required. Mirrors of 8-meter diameter are routinely produced with this technique.

One of the latest developments in mirror design is also one of the oldest—the liquid mirror. The idea was proposed by Newton. Various problems prevented working liquid telescopes from being made. However, a variety of technological advances now makes them practical. A rotating liquid will form a parabola, which, as discussed earlier, is the perfect shape for a telescope mirror. Of course, the mirror must always be pointed straight up, but that's not a problem for many types of projects. Liquid mirrors as large as 6 m across have been made that rotated about five times per minute to achieve the right shape. The

Figure 12-16 The Overwhelmingly Large (OWL) telescope design. Notice the people and the truck for scale.

cost savings is significant. This is the design for Canada's Large Zenith Telescope.

A partnership among astronomers and engineers at the University of Kansas, Dartmouth College, San Diego State University, and private business is designing, building and testing an ultra-lightweight mirror and telescope made of carbon composite materials. Such designs, if successfully proven to work well, should significantly reduce the cost of building large telescopes in the future.

Astronomers worldwide are currently making feasibility studies for the design and construction of substantially larger telescopes of sizes from 30 to 100 m. The most ambitious is the OWL (Overwhelmingly Large Telescope) project of the European Southern Observatory, which is studying a new design to produce a 100 m optical and infrared telescope. The OWL concept is illustrated in **Figure 12-16**. Other studies are looking at 30 m to 50 m telescopes.

EFFORTS TO OVERCOME THE UNSTEADY ATMOSPHERE

The Chapter 12 opening photograph shows the Very Large Telescope (VLT), currently the world's largest telescope system; the four separate 8 m telescopes can be used separately or together, giving the equivalent of a 16 m mirror. It was built in the high altitude of the Chilean Andes by the European Southern Observatory, far from any interfering lights and above much of the turbulent air in the atmosphere that causes the incoming light to distort.

This arrangement of separate telescopes allows the instrument to be used as an interferometer, combining the incoming light from the various telescopes to produce an interference pattern. When these patterns are analyzed by modern techniques, they produce greatly improved spatial resolution. When fully implemented, these telescopes should be able to produce a resolution of 0.025 second of arc at near-infrared wavelengths when the separation of the telescopes is 130 m. Such possibilities provide a dramatic improvement over the limits usually imposed by seeing.

Inquiry 12-9

How large a single objective would be required *theoretically* to give the same possible resolution as the VLT at a wavelength of 10,000 Å?

Another development that astronomers are now using is *active optics* (also known as adaptive optics). The idea of active optics is to change the shape of the mirror in response to changes in the Earth's atmosphere. By monitoring a star, or a laser beam reflected from the atmosphere, astronomers can make rapid small changes in the shape of the mirror to compensate for changes in the atmosphere. In this way, the effects of the atmosphere can be dramatically reduced. This can also be used to correct for flexure due to changing positions of the mirror.

12.6 Detectors and Instruments

Astronomers no longer use their eyes at the focus of a telescope. The eye is not as sensitive a detector of radiation as film or modern electronic detectors. Furthermore, the eye is connected neither to a perfect storage medium nor to one that allows for perfect duplication of the data. For these reasons, astronomers use a variety of detectors.

CAMERAS AND FILM

Just as important as the telescope itself is the radiation detector that is placed at the focus. We cannot overemphasize the dramatic boost given to astronomy by the invention of the photographic emulsion. The ability to take a time exposure at the telescope enabled astronomers to penetrate much more deeply into space, and it also created a permanent record for other scientists to study. If you realize that a photograph is composed of many microscopic picture elements called **pixels**, you can see that the data storage capacity of film is enormous. An 8- by-10-inch photograph has some hundred million pixels.

The major problem astronomers have always had with photographic emulsion is that its response to light intensity is not linear: Twice the brightness does not result in twice as much exposure on the film. This nonlinearity of photographic film makes it difficult to determine the intensity of the light falling on it. Furthermore, if you bombard a piece of film with too much light, eventually it stops responding at all. Another problem with film is that it is insensitive to most infrared radiation.

PHOTOELECTRIC PHOTOMETERS

The desire of astronomers for a more sensitive detector was answered by the discovery of the photoelectric effect discussed in Chapter 11. When incident light of an appropriate wavelength falls on certain metals (such as cesium), they respond by ejecting electrons; that is, they produce an electrical current. Such photoelectric surfaces are highly sensitive, and the amount of current given off is strictly proportional to the intensity of the incident light.

Astronomers found that such detectors could be used to measure the brightness of stars to a high degree of precision. The light from a star is passed through a colored filter before being measured, so that by changing filters we can measure the amount of light the star gives off at various wavelengths. For example, by putting a blue filter (one that passes only blue light) in the light path, we can measure the amount of energy the star gives off at blue wavelengths only.

The photoelectric photometer has one dramatic disadvantage: It can measure only one star at a time because you must pass the light from the star through a small hole that excludes light from other stars that might be confused with it. The system then gives an accurate reading of the brightness. So although a photographic emulsion can register many stars at once, but with low accuracy, photoelectric systems can measure only one star at a time, but with high precision.

MODERN ELECTRONIC DETECTORS

Astronomers have always longed for an instrument that combines the sensitivity of the photoelectric detector with the two-dimensional data storage capacity of the photographic emulsion.[2] For a few years they experimented with TV cameras mounted at the focus of the telescope. Although not ideal detectors, they were used successfully in the International Ultraviolet Explorer satellite. In the 1980s, a new type of detector called a **charge-coupled device (CCD)** was invented. When light strikes the detector surface, electric charges are produced in small regions of the material (called pixels, as in a photograph). Because the charge is directly proportional to the amount of light falling on the material, the amount of incident light can be determined by finding the electrical charge within each pixel. Displaying all the pixel data at once can produce an image. A CCD image of one of the authors is shown in **Figure 12-17a;** part *b* shows the pixel structure around the eye. The CCD is

---

[2]Film-based photography has no major role in professional astronomy today. In fact, the last photograph taken with the Hale 200-inch telescope was made by Sidney van den Bergh on September 29, 1989.

(a)

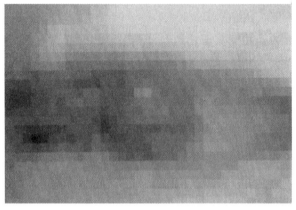

(b)

Figure 12-17 *(a)* A CCD image appears to be continuous and smooth but, as the enlargement of the eye in part *(b)* shows, the image is made up of distinct pixels.

found in home video, digital, and cell phone cameras and makes use of the same kind of technology that underlies transistors and microchips.

Because the CCD provides us with a number proportional to the brightness of the radiation incident on each pixel, we describe the image in terms of numbers, referred to as *digital data.* Analysis of the millions of pixels in an image has given rise to what is called **image processing**. Image processing is the mathematical manipulation of digital data to extract the maximum amount of information contained in the image. Such processing allows astronomers to view the universe in ways not previously possible. **Figure 12-18**, for example, shows a processed view of a globular cluster (see Figure 1-14 for an unprocessed example). Numerous other examples will be presented throughout this book.

Such improvements come at a price, however. These new instruments collect data so rapidly that

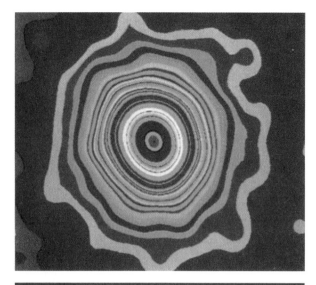

Figure 12-18 An image of a globular cluster after various techniques of image processing have been applied.

a researcher can become completely buried under it. Astronomers are having to devote considerable effort to developing efficient methods of data reduction, ways to get computers to take all the raw numbers, do all the corrections (e.g., for the absorption of the Earth's atmosphere), and complete the image processing automatically. Data reduction itself has become a career specialty for astronomers.

12.7 Radio Astronomy

In the early 1930s, while studying certain kinds of interference that affected transatlantic radiotelephone transmissions, Karl Jansky of Bell Telephone Laboratories discovered that some of the interference was coming from a region in the sky that moved in the same way the stars do. His antenna was rigidly mounted and could not compensate for the rotation of the Earth. A source of radiation in the sky would move across his antenna about once a day as a result of the apparent rotation of the sky. In fact, because of the difference between the solar and sidereal days, this source among the stars would be observed about four minutes earlier each day.

Jansky had discovered radio emissions from the center of our own galaxy. Strangely enough, his discovery did not generate much interest among professional astronomers. One of the few people to become seriously interested was an amateur astronomer and professional radio technician named Grote Reber, who set up a small radio telescope—the first one designed for the purpose—in his backyard in Illinois. He began to make systematic observations and was the first to make a map of the radio sky.

During World War II, Reber was probably the world's only radio astronomer. The rapid growth of radar technology and the enormous amount of surplus radar equipment left over after the war soon triggered an explosion of postwar astronomical activity.

Radio astronomy became important because it enables astronomers to study an entirely different portion of the electromagnetic spectrum than had been available to astronomy. Many objects that appeared to be uninteresting faint stars when examined optically turned out to be powerful emitters of radio radiation. An excellent example is a subset of the class of objects known as *quasars*. Although they are insignificant in optical telescopes, some quasars are among the most luminous objects in the sky when viewed by radio telescope. The *pulsar*, another type of object that has attracted great attention from astronomers and has provided much insight into the final stages of the lives of stars, might never have been detected using only optical light.

Radio telescopes have other advantages as well. Clouds of interstellar dust in the plane of our galaxy obscure most of our own stellar system from the view of optical telescopes. But radio waves pass through the dust without much dimming. For this reason radio astronomy has become an important tool for the study of the Milky Way. Because the daytime sky does not appear as bright to radio telescopes as it does to optical telescopes, radio astronomers can make many observations during the day. Bad weather is not as much of a problem for the radio astronomers either, because radio waves readily penetrate Earth's clouds.

However, individual radio telescopes do have some limitations. Most serious is that their resolving power is generally poor. We saw earlier (Section 12.2) that the resolving power of any telescope depends on the wavelength of radiation divided by the diameter of the objective. Therefore, the long wavelength of radio waves will produce poor resolution unless the objective diameter is correspondingly large. Even the world's largest radio *dish*, the 1,000 foot (300 m) telescope at Arecibo, Puerto Rico (**Figure 12-19**), can be

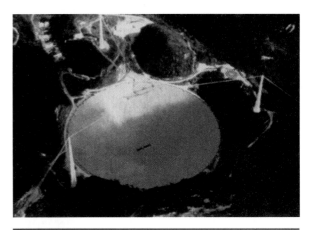

Figure 12-19 The 1000-foot radio telescope at Arecibo, Puerto Rico.

seriously affected by diffraction and produce poorly resolved data.

Until recent years radio astronomy data were presented only as *contours*—lines that connect regions of constant intensity. **Figure 12-20** shows intensity numbers from the image in Figure 12-17*b* and contour lines connecting locations of constant intensity. **Figure 12-21** shows a series of such radio contours superimposed over an optical photograph. Because the cross-hatched oval indicates the degree of resolution of the telescope that obtained the radio data, apparent structure smaller than the oval is not fully resolved. As shown next, modern techniques of image processing allow radio data to be presented as images rather than numbers and contours.

Inquiry 12-10

Estimate the resolving power of the 1,000 feet (300 m) dish if it is operating at a wavelength of 3 feet (1 m). How does this compare with your eye (with a resolution of about 1/60°)?

Figure 12-22 shows the basic components of a radio telescope. An antenna collects the electromagnetic energy and converts it to a weak electrical current. The antenna can be a dish, as shown here, or it could resemble a television antenna. If it is a dish antenna, it need not be polished to the great accuracy that is required for an optical telescope. Imperfections in the surface of the dish need be an order of magnitude smaller than the

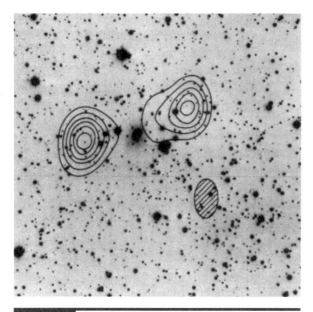

```
100  99  99  95  90  83  84  85  82  84  83  81  77  81  82  81  83  85  88  88  79  81  80  81  81  85  88  94  97  90  87  87  77  76  79  81  84
111 112 112 105  93  85  92  88  90  87  91  91  83  83  89  83  85  83  87  74  85  82  80  86  94  97 102  99  91  84  85  78  83  81  80
108 111 115 111 108 101 101 100  96 100  93  92  94  87  88  91  84  90  85  89  91  88  87  86  88  89 100 107 104 101  93  89  84  88  81  89
105 118 117 115 112 113 109 108 112 110 108 101  99  93  92  90  95  92  93  94  95  97  88  95 101  96  96 105 112 103  99  96  91  88  90  94  91
105 104 110 125 124 123 124 127 128 132 118 116 108 103 105  99 100  99  94  97 100 104 104 100 111 109 113 112 106 101  97  92  98  99 100 100
 94 115 112 121 124 134 132 147 143 143 140 130 132 113 107 104 104  99 102 103 109 112 114 118 114 118 121 116 107 104  98  91  94 100 108 104 106
 95 109 117 128 129 142 140 151 154 161 157 150 138 128 114 113 108 111 113 117 117 123 122 131 134 134 136 120 115 107  96  90  94  99 106 106 108
100 104 119 133 139 142 147 162 161 162 161 152 147 136 128 122 123 119 120 125 128 134 140 149 151 153 147 134 117 121 101  98  93  95  97 101  89
 92 108 112 128 130 136 133 146 150 157 157 157 147 136 135 130 132 129 135 132 141 158 153 163 166 168 167 142 128 116 101  96  92  92  94  90  91
 83 101 111 117 119 124 125 125 143 140 147 152 148 147 140 133 143 142 144 161 146 159 160 171 168 173 167 143 135 117 107 100  96  92  89  92  85
 83  92  94 108 110 118 116 125 126 129 143 140 151 147 154 153 156 163 162 166 167 169 171 167 167 168 163 145 136 122 116 100 100  97  92  83  77
 81  89  94  95 102 108 109 116 114 122 127 139 146 154 160 168 178 173 185 183 187 175 172 158 167 161 154 140 133 119 117 106 106 100  91  90  85
 89  83  90  90  91 101 106 112 121 121 133 142 152 156 169 183 190 201 203 199 191 184 173 165 156 154 147 135 128 124 107 100 104  98  91  91  80
 83  84  90  83  91 101 106 113 115 124 135 140 153 164 184 180 200 205 205 205 200 187 179 160 158 145 142 130 123 115 113 109 106  99  93  90  90
 81  81  88  96  97 101 104 114 121 131 138 151 158 172 172 188 192 209 204 203 194 172 171 163 150 141 131 124 119 117 104 110 105  94  92  88  83
 79  83  88  96  97 103 108 110 129 135 156 154 158 171 174 181 185 183 188 184 174 170 162 148 145 139 129 121 113 114 110 108  98  92  84  79
 78  84  88  85  96 102 109 115 126 132 146 153 160 159 164 166 170 179 169 166 155 151 145 141 138 127 126 117 116 111 112  95 101  93  92  79  78
 79  77  84  88  91 104 104 116 120 125 141 146 156 153 153 152 156 158 153 147 144 136 135 129 125 120 116 114 116 104 106  96  93  89  90  81  86
 82  85  84  89  91  99 104 109 119 123 133 135 140 138 135 141 139 137 133 132 126 127 121 123 117 114 114 109 106 103 100  92  92  89  83  76  83
 80  82  77  88  89 100 105 109 109 121 116 125 122 125 127 125 128 128 126 123 123 117 114 122 122 113 123 115 112  99  99  92  87  81  84  75  80
 74  83  89  82  92  90  99 100 107 110 110 117 114 116 124 119 119 121 116 121 114 111 115 120 121 123 116 114 109 105  95  89  87  83  78  77  75
 83  82  81  77  89  90  89  98 105 104 111 111 109 110 109 118 115 114 116 112 112 112 109 112 110 113 118 116 113 109  99  95  87  82  81  76  67  70
 83  88  82  82  78  90  91  92  99 104 105 115 113 112 111 113 115 109 112 112 112 109 112 110 113 118 116 113 109  99  95  87  82  81  76  67  70
 89  83  82  81  84  79  88  87  88  98 103 111 117 107 108 118 114 107 103 105 104 106 103 109 109 104 100  97  98  79  72  66  65
 86  84  78  82  83  80  77  84  94  90 115 112 114 118 113 101 112 109 106 100  97  99 100 106  97  98  99 102  94  97  83  84  84  68  67  59
 82  79  81  81  76  75  80  82  90  84 101 109 120 111 110 105 105 103 104  93  96  99  96  95  93  96  97  98  90  97  89  86  77  73  70  74  66
 78  80  73  74  77  77  75  76  80  87  94 103 107 111 102 104  97  96  99 101  90  92  88  96  90  91  85  90  89  89  87  84  78  75  70  71  62
 74  74  72  70  69  71  70  75  76  82  90  91  96 101  99  96  97  93  86  94  95  90  87  88  82  89  86  91  88  88  84  78  80  69  79  66  67
```

Figure 12-20 Raw data from part of Figure 12-18b is composed of numbers representing intensities. Contour lines connect regions of equal intensity.

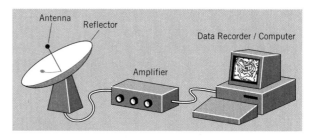

Figure 12-22 The basic components of a radio telescope are an antenna to collect the radiation, an amplifier to amplify it to detectable levels, and a recorder to make a permanent record.

Figure 12-21 A galaxy as seen both optically and with a radio telescope. The radio contours have been superimposed on an optical photograph. Reprinted, with permission, from the *Annual Review of Astronomy and Astrophysics*, Volume 5, © 1967, by Annual Reviews, www.annualreviews.org

shortest wavelength at which the dish is to be used. Deviations of a few inches can often be tolerated.

The second major component of a radio telescope is a highly sensitive amplifier, which multiplies the strength of the signal millions of times. Finally, a device to record the amplified signal for further study is required. In the early days this was usually a chart recorder that simply drew a wiggly line whose height was proportional to the signal

strength; modern radio telescopes record the intensity data (which are simply numbers) directly into a computer and save it for analysis later.

INTERFEROMETERS

A method to overcome the problem of poor resolving power of an individual radio telescope is to combine signals from two separated antennas, as shown in **Figure 12-23**. When this is done, a much-increased resolving power is obtained, equivalent to a radio dish whose diameter is equal to the distance between the two antennas. Such an arrangement is called a **radio interferometer**. For example, two 25-m-diameter radio dishes separated by 3 km will provide data having a resolution equal to a single telescope 3 km in diameter. Radio astronomers have even used antennas on opposite sides of the Earth as an interferometer to obtain highly precise information. In this case, the separate telescopes are not physically connected, but the data from each are stored, along with accurate

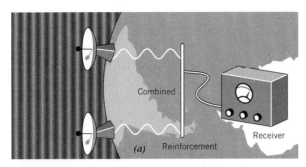

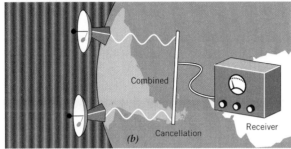

Figure 12-23 Principle of the interferometer. *(a)* When the waves reaching the two antennas are in phase, the output of the receiver is at a maximum. *(b)* After the Earth has turned a bit, the waves are out-of-phase, and destructive interference takes place. The output of the receiver is at a minimum.

Figure 12-24 The Very Large Array (VLA) in New Mexico.

time signals from state-of-the-art atomic clocks, and then a computer combines the two signals. This cannot be modeled using the ray model of light; the wave model has to be used to require the light reaching both dishes to be identical. By putting antennas on opposite sides of the Earth, international collaborations of astronomers are able to simulate the resolving power of a telescope as large as the Earth (although the radiation gathering power is not enhanced to a comparable degree, of course). This field of astronomy is known as **very-long-baseline interferometry**, or VLBI. Ironically, after struggling with the notoriously low resolution of single-dish telescopes, radio astronomers using VLBI now obtain higher resolution than any optical equipment under construction at this time.

One of the most powerful radio interferometers in the world, called the Very Large Array (VLA), has been constructed on the desert plains near Socorro, New Mexico (see **Figure 12-24**). It consists of 27 parabolic reflectors, each of which is 25 m in diameter. The telescopes can be arranged along a giant Y pattern covering an area 17 miles in extent. Using this system, a radio astronomer can see radio-image details comparable to the resolution obtained by optical astronomers.

A large step forward for radio astronomy was the 1992 completion of the Very Large Baseline Array (VLBA), a system of ten 25 m dishes spaced out over the United States and the Caribbean. This system provides both radiation-gathering power and resolution, with a resolution some 100 times better than any existing telescope system. Other important radio observatories are located in England, the Netherlands, Australia, and Germany.

Orbiting Very Long Baseline Interferometry (OVLBI) provides capabilities to extend resolution significantly. A single space-based radio telescope was launched by the Japanese Space Agency in 1997 to work in conjunction with ground-based radio telescopes. With an orbit varying from 300 to 13,000 miles (500 to 21,000 km) above the Earth, the resolution of the interferometer is that of a virtual telescope three times the size of the Earth. A Russian project, called RadioAstron, is planning to launch a 10 m radio telescope in 2006. The orbit will extend out to about 220,000 miles (350,000 km) and will provide resolution equal to a single telescope nearly 30 times the size of Earth. That will be similar to resolving a quarter from a distance of 40,000 miles (65,000 km). NASA is studying the Advanced Radio Interferometry between Space and Earth (ARISE) system. The plans for it involve one or two 25 m radio telescopes in a highly elliptical orbit that will provide resolution some 3,000 times better than the optical *Hubble Space Telescope.*

Once the data have been obtained by each of the component telescopes in the interferometer array,

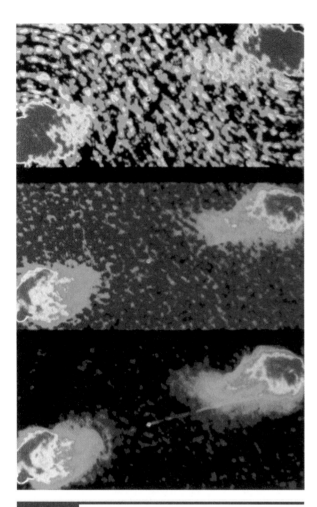

Figure 12-25 An image in various stages of image processing, improving from top to bottom.

the data are combined into one set of numbers containing the total information. The raw data are represented in an image in the top part of **Figure 12-25**. To convert these data into a final image requires extensive image processing. The middle section shows an intermediate stage in the production of a final image (shown in the bottom part of the figure) of what is called a double-lobed radio galaxy. Production of such images requires the largest, most sophisticated computers available.

Neither radio astronomy nor optical astronomy can stand by itself in modern research, because each technique is capable of providing information that the other cannot. Used together, the two techniques reinforce each other and give us a much clearer idea of the nature of celestial objects. In a real sense, the whole becomes greater than the sum of its parts.

12.8 Optical and Radio Observatory Sites

Optical astronomical observatories are located on remote mountain tops for a variety of reasons. As discussed previously, the Earth's atmosphere absorbs radiation incident upon it. Even in the optical region where the transparency is relatively high, only some 80 percent of the incident optical radiation reaches sea level. This amount increases to 90 percent at 7,000 feet (1,600 m). Not only does the atmosphere absorb radiation, but it also emits its own faint glow. With less atmosphere above an observer on a high mountain, there is both less atmospheric emission and less turbulence to distort the image. Cloud cover above the observatory must be relatively infrequent because optical telescopes cannot see through clouds. In addition to these requirements, an observatory site must also be accessible to personnel and have power, water, and other utilities.

Because infrared radiation is absorbed by the Earth's atmosphere, ground-based infrared observations require a location also having the lowest possible amount of infrared-absorbing water vapor overhead. Such locations are found on high mountaintops in the arid regions of Arizona, Hawaii, and Chile.

The final and perhaps most important requirement for locating an optical observatory is to be away from city lights. Just as you see few stars when the full moon is up, city lights cause important details to be lost in astronomical images. This problem is known as **light pollution**. Encroachment of cities, both large and small, into once-dark areas can drown out the view of the universe not only to astronomers but also to humanity as a whole. As an example, **Figure 12-26** shows the growth of lighting in the Tucson area as observed from Kitt Peak National Observatory between the years 1959 and 2003. Lights from cities as far as 100 miles from an observatory measurably increase sky brightness. The number of areas on Earth suitable for astronomy is small and getting smaller; the only remaining sites in the continental United States include limited parts of Arizona and New Mexico. **Figure 12-27** shows a satellite mosaic at night on which cities even as small as Lawrence, Kansas, appear. For this reason, astronomers help communities around observatories find ways to provide adequate lighting

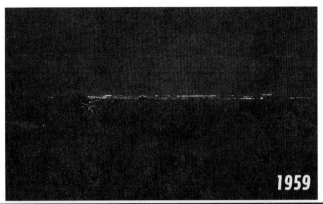

Figure 12-26 The growth of light pollution around Tucson, Arizona, between the years 1959 (upper) to 2003 (lower).

Figure 12-27 A photomosaic of the United States at night as seen from a satellite.

while still protecting the astronomical environment. Low-pressure sodium lamps, for example, use less energy and cost less to operate than incandescent, mercury vapor, or high-pressure sodium lamps, yet provide enough light to illuminate streets without washing out a large amount of the optical spectrum. Well-designed shielding around light fixtures directs light downward where it is useful, thereby reducing sky illumination.

Proponents of bright lighting argue that astronomy can be done from space. However, the expense of making all astronomical observations from orbit is prohibitive and impractical. Perhaps even more important is the loss of the night sky to citizens, for the night sky is a treasure everyone should be able to enjoy. It's a basic human right.

Radio astronomy also has a pollution problem; there are a limited number of radio frequencies and innumerable uses for them. Powerful radio and TV transmitters, along with lower-power CB radios, cellular phones, telecommunications via microwave, satellites, and so on illuminate the sky at radio wavelengths and make observations of faint astronomical sources impossible at certain wavelengths. Although some astrophysically important radio frequencies are currently protected by international agreement, some others have been lost.

12.9 Beyond Optical and Radio Astronomy

Especially since 1960, astronomers have learned to exploit the entire electromagnetic spectrum, from long radio waves to very short gamma rays. The impetus has been the tremendous growth of technological capability, including improvements in detectors, computers, and the ability to send instruments above the turbulent, obscuring atmosphere of Earth. To the astronomer, the atmosphere is professionally a nuisance because, as illustrated in Figure 11-6, it absorbs most of the radiation coming into it. Indeed, the most important reason for building observatories on mountaintops is to get above some of the atmosphere. Even mountaintops are not high enough to allow undisturbed observation of X-rays, gamma rays, ultraviolet, and infrared radiation. While the technology required to make instruments in these wavelength regions differs from that in the optical region, the fundamental principles of optics discussed previously still apply.

A surprising amount of successful research has been performed by instruments carried up in balloon gondolas, which can reach altitudes in excess of 20 miles (30 km). Prior to the use of satellites equipped with telescopes, such balloons provided the only technique to obtain some ultraviolet and infrared data. In 2005, the *Stratospheric Observatory for Infrared Astronomy (SOFIA),* a 747SP jet carrying a 2.5 m infrared telescope, will start providing astronomers with an unimpaired view of the infrared sky. Because the most serious obstacle to astronomical observations at infrared wavelengths is the absorption by atmospheric water vapor, *SOFIA* rises above about 99 percent of the water vapor by flying at about 41,000 feet (12.5 km).

Our ultimate look at the infrared part of the spectrum will be with the *Spitzer Space Telescope,* a fully equipped space infrared telescope facility that was launched in 2003; it is discussed further later in this chapter.

The *Infrared Astronomy Satellite (IRAS),* which operated 1983—1984, was the first major satellite to have a (small) telescope and detectors that operated at infrared wavelengths. Observing in the infrared region presents unique problems because telescopes themselves are warm and radiate at infrared wavelengths. Therefore, to obtain useful data at these wavelengths the entire telescope must be cooled to 4K in a bath of liquid helium. In the case of IRAS, once the liquid helium evaporated (after about nine months), the instrument became inoperable. However, during its short period of operation, IRAS made numerous important discoveries concerning star formation, galaxies, dust within the Solar System, and many other subjects discussed elsewhere in this book. The **Infrared Space Observatory** (ISO) satellite ran out of coolant after 2.5 years. A new design will allow the Spitzer Space Telescope (discussed later in this section) to remain operational for some 5 years.

Important ultraviolet observations were made by instrumented rockets, but they were aloft for only a few minutes per flight. Longer opportunities to observe were presented by the *Copernicus* satellite during the years 1972–1980. The *International Ultraviolet Explorer (IUE)* satellite, with a 45 cm mirror, was launched in 1978. Although its design lifetime was only a few years, it sent back data for 20 years!

The ability to observe in the ultraviolet was one of the original justifications for building the *Hubble Space Telescope (HST).* With its larger diameter and more modern and sensitive detectors, *HST* is providing a more detailed view of the ultraviolet universe than the earlier satellites.

The hottest and most energetic objects in the universe emit energy in the X-ray and gamma ray regions of the spectrum and the pursuit of understanding them has moved astronomy into a culture of international collaborations. NASA's *Einstein Observatory,* an X-ray telescope that operated from 1978–1981, the Japanese *Ginga* satellite (1987–1991), the German/U.S.A./U.K. *Roentgen Satellite* (1990–1999), and the Italian/Dutch *BeppoSaX* satellite (1996–2002), provided crucial information on a variety of objects and processes. They monitored X-rays from such objects as neutron stars, quasars, possible black holes, and galaxy clusters. NASA's *Swift Gamma Ray Burst Explorer* was launched in 2004 to capture light from gamma ray bursters.

The satellite systems just discussed, and many others not mentioned, were all necessary precursors to the most sophisticated systems in orbit today and in recent years, which we now discuss.

12.10 NASA's Great Observatories

A grand plan for astronomical exploration was developed by NASA in the 1980s. The plan was to build and launch telescopes to study electromagnetic radiation from celestial objects in all parts of the electromagnetic spectrum other than radio—gamma ray, X-ray, ultraviolet, optical, and infrared. A variety of problems delayed these cutting-edge and expensive projects, but the last one was finally launched August 25, 2003.

Some people have suggested that astronomers will no longer need to build any new telescopes on Earth now that we have orbiting telescopes. Those making such suggestions are wrong for at least four reasons. First, space-based observations are extremely expensive to make compared to those from the ground. The rule has always been that only projects that cannot be done successfully from the ground are to be done from space. Second, orbiting telescopes cannot be as large as those on the ground. Third, these telescopes have limited capability and only a small number of researchers are able to use the space-based observatories. The last reason is that the space observatories have produced an even stronger need for ground-based telescope than before. For example, in the case of the *Hubble Space Telescope (HST)* there has been a natural symbiosis between *HST* and the new (and upcoming) generation of giant ground-based telescopes. The *HST*, with its sharp and unblurred view of much of the electromagnetic spectrum, has proven to be the *discovery* instrument that finds new and surprising objects and results that jerk science forward, sometimes in spite of itself. When a new object is discovered, then the giant ground-based telescopes, with their enormous light-gathering power and versatility,

take over and provide the detailed studies necessary to understand the new object.

The following sections will look at each of the Great Observatories, whose results are spread throughout the rest of this book.

THE HUBBLE SPACE TELESCOPE

The **Hubble Space Telescope (HST)**, the first of the Great Observatories, was launched in 1990. In the years since first reaching orbit, it has continued to provide astronomers with a long-desired optical, ultraviolet and near-infrared observatory in space (**Figure 12-28**). This telescope has the largest diameter of any thus far sent into space—2.4 m. More important, however, is that for the first time astronomers have a complete observatory in space with the full array of observing instruments available in ground-based observatories. All the previous telescopes that have been sent into space have not only been smaller but also, more fundamentally, they have been special-purpose instruments designed for only a few specific types of observations. The *HST* was designed to be able to do complete studies of objects many times fainter and almost 10 times deeper into space than our best ground-based systems; it explores a volume of space almost 1,000 times larger than before. In addition, without the degrading effects of the Earth's atmosphere, the telescope obtained more highly resolved images than possible from the

Figure 12-28 The Hubble Space Telescope (HST) attached to the bay of the Space Shuttle.

ground. In addition, the telescope is well outside the atmosphere and its attenuating effects on incoming radiation. It is able to observe across the electromagnetic spectrum, from the energetic ultraviolet radiation to the long-wavelength infrared.

In recent years advances in the technology of active optics, detectors, and software have allowed dramatic improvements in ground-based astronomy. Therefore, the advances in optical astronomy provided by *HST*, although certainly strong, are not as great relative to ground-based work as was achieved when it was first launched. In any case, the telescope is the largest ultraviolet telescope in orbit today, and is producing the highest-resolution images ever made.

The telescope has compartments for five main instruments. The instruments have been changed over the years as detector technology has advanced. The second Wide-Field Planetary Camera is the main instrument for obtaining images of planets, galaxies, clusters of galaxies, and quasars. The advanced camera for surveys (ACS) consists of three cameras and provides the capability of obtaining wide-field, high resolution, and ultraviolet images. It has the capability to detect the polarization of light from astronomical objects. The ACS can also place an occulting disk in front of a star to block the light and permit observations of any fainter gas and dust near the star. The Near Infrared and Multi-Object Spectrograph does much the same as the ACS but in the near-infrared spectral region. The fine-guidance sensors used to point the telescope are designed to determine positions of stars more precisely than ever before. They provide additional capabilities in positional astronomy. Two upgraded new detectors, the third Wide-Field Planetary Camera and the Cosmic Origins Spectrometer, were built, but the Space Shuttle mission to install them has been cancelled.

THE COMPTON GAMMA-RAY OBSERVATORY

The *Compton Gamma-ray Observatory* (*CGRO;* **Figure 12-29**) was named after Arthur Holly Compton, the Nobel Prize–winning physicist whose work was pivotal in understanding the interaction of gamma rays with matter. It was this understanding that provided the knowledge required for the detection of these highest-energy electromagnetic radiations. Gamma rays contain so much energy that materials can neither refract

Figure 12-29 The *Compton Gamma-Ray Observatory* (CGRO) being deployed.

nor reflect them. Thus, the telescope contains neither lenses nor mirrors. Rather, the observatory consists of large, massive instruments that respond to gamma rays when they pass through them. The different onboard instruments contained different materials for detecting gamma rays of varying energy. For these reasons, *CGRO* was by far the heaviest of the Great Observatories when launched in 1991 (17 tons compared to *HST*'s 12 tons).

The *CGRO* successfully detected and located gamma-ray bursters, which are discussed in Chapter 19. The phenomena that produce these very intense, sudden outbursts of gamma rays are still not well understood. Other types of objects, which also emit gamma rays, also needed more observations to be fully characterized. One of its instruments, the Energetic Gamma Ray Experiment Telescope, was 10 to 20 times larger and more sensitive than any that had been used in space before. After eight productive years of service, the *CGRO* was deorbited into a watery grave in the Pacific Ocean in 2000.

THE CHANDRA X-RAY OBSERVATORY

The *Chandra X-Ray Observatory* (*Chandra;* **Figure 12-30**) was named for astrophysicist Subrahmanyan Chandrasekhar. Of Indian descent, he immigrated to the United States in 1937 to the University of Chicago. His contributions to many fields of theoretical astrophysics are legendary. You will learn about one of his important contributions in Chapter 18 when you study white dwarf stars. His work on the physical processes occurring in stars earned him the Nobel Prize in 1983.

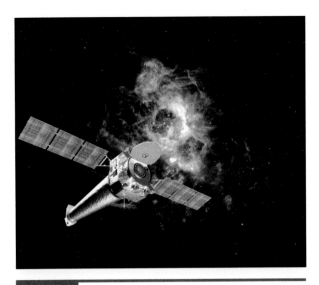

Figure 12-30 The Chandra X-Ray Observatory.

Figure 12-31 The Spitzer Infrared Space Telescope.

X-rays from astronomical sources are difficult to study, although not as difficult as gamma rays! As with gamma rays, X-ray photons have too much energy to be gathered by a concave mirror or a lens. However, X-rays incident nearly edge-on can be reflected, brought to a focus, and then sent to one of the four detectors on board. Two of the detectors are designed to provide information on the number of X-rays, their position, energy, and time of arrival. Two other instruments record the spectrum of the source being observed.

Chandra is designed to study the most energetic phenomena in the universe, and so far has obtained the spectrum of the powerful wind shooting out of a pair of violently interacting binary stars, searched the center of our galaxy for a supermassive black hole, and explored the intergalactic medium.

THE SPITZER SPACE TELESCOPE

The *Spitzer Space Telescope* (**Figure 12-31**) was named after astronomer Lyman Spitzer, who was a driving force behind the development of *HST* and a major contributor to a wide variety of fields of astronomy and astrophysics. The *Spitzer Space Telescope* was the last of the Great Observatories to be launched (2003). Although the telescope itself is small, only 0.85 m (compared to *HST*'s 2.4 m), it is the largest infrared telescope ever launched. Working with longer-wavelength, less-energetic light, it is concentrating its studies on cool objects, such as brown dwarfs and extra-solar planets, planet and star formation, and ultralumi-

nous galaxies. It has discovered two populations of very red galaxies in the early universes, one type that is shrouded in dust and thus invisible in optical light, and the other that, surprisingly, consists of old stars.

The telescope system consists of three instruments that are cooled to a temperature of about 5 K. The Infrared Array Camera obtains high-resolution infrared images. The Infrared Spectrograph provides researchers with spectra that will allow for a better understanding of the materials and physical conditions of the target objects. Finally, the Multiband Imaging Photometer produces both images and brightness measurements over a more extended range of the infrared.

THE FUTURE

The success of the *HST* has generated enthusiasm for an even more powerful system, originally called the *Next Generation Space Telescope* but now officially known as the *James Webb Space Telescope (JWST)* after the NASA administrator from 1961 to 1968 during the build-up to the Apollo moon landings. The telescope, currently planned for launch in 2011, will have a mirror composed of 18 segments, for a total diameter of 6.5 m. Whereas the *HST* that it will replace was optimized for the UV, optical, and near-infrared wavelengths, the *JWST* will be optimized for the near-infrared. The near-infrared camera is designed to obtain images of the first luminous objects to have formed after the big bang, as well as studies of star formation and planetary systems in our galaxy. The near-infrared spectrograph will provide information on chemical abundances and motions of the variety of objects *JWST* will observe. The final instrument, a mid-infrared instrument, provides both imaging and spectroscopy at longer infrared wavelength. The telescope is designed for a 5-year lifetime, although the goal is for a 10-year mission. Given that its orbit will place it one million miles (1.6 million km) from Earth, it will not be able to be visited by astronauts to repair any problems or to change instruments.

The scientific goals of this telescope include answering questions about the geometry of the universe, the evolution of planets, stars, and galaxies, the chemical evolution of the universe, and the nature of dark matter.

Although this project is an obvious next step, two controversies surround it. First is the naming. Since the telescope is an international collaboration among a large number of countries, naming it after the U.S. NASA administrator rather than something more international is seen by some as improper, even though Webb's contributions to

space exploration during the *Apollo* moon-landing era are seen as pivotal. The second controversy stems from the fact that the funding of this telescope may require turning off the *HST*, which could be upgraded and producing first-rate science until at least the year 2020, given the political will. If the *HST* is turned off before *JWST* is operational, there will be no large UV-optical space telescope in operation during that interim period. In addition, if there were any problems with the *JWST*, then there would be no large UV-optical observatory in space because it cannot be reached to be serviced. These are seen as a problem because the hiatus will slow down, and perhaps halt, progress in the field. In addition, it may be a negative sign to students considering entering science and, without new people, fields stagnate. In addition, given that *JWST* is optimized for the near-IR, without *HST* there will be no significant UV-optical telescope in space. The international community of astronomers will have to come together to solve these problems.

Chapter Summary

Observations

- The main purpose of a telescope is to gather light and bring it to a focus. Secondary purposes are to resolve and magnify the images.
- The **light-gathering power** of a telescope depends on the area of the objective lens or mirror. The **resolving power** of a telescope depends inversely on the objective diameter. The **magnification** depends on the ratio of the objective focal length to the eyepiece focal length.
- **Image size** is determined by the focal length. The greater the focal length, the larger the image produced.
- **Seeing** measures the degree of motion of the Earth's atmosphere above an observatory site.
- **Refracting telescopes** suffer from chromatic aberration. Spherical aberration is corrected by shaping a mirror into a nonspherical shape or by using a correcting lens (Schmidt telescope).

- The properties of **interference** may be used to improve the resolution of a telescope. **Long-base-line interferometry** is necessary to obtain usable resolution with radio telescopes because of the long wavelengths of radio waves.
- Detectors used by astronomers include photographic film and various electronic devices, including the charge-coupled device (CCD). These detectors are used with cameras, photometers, and spectrographs.
- Astronomical studies have been made from balloons, airplanes, rockets, and satellites.
- The Great Observatories are four satellite telescopes covering all wavelengths except the radio. They include the *Hubble Space Telescope* (UV/optical/near-IR wavelengths), the *Compton Gamma-Ray Observatory,* the *Chandra X-Ray Telescope,* and the *Spitzer Space Telescope* (infrared wavelengths).
- The *Webb Space Telescope,* a 6.5 m reflector, is planned for launch into solar orbit in 2011.

Summary Questions

1. How do both a pinhole camera and a lens form a real image of an object? Use a diagram in your explanation.

2. What effects do spherical and chromatic aberration have on an image?

3. What are the basic parts of a telescope? What is the function of each part? Use diagrams in your answers.

4. What is meant by resolving power as it applies to a telescope?

5. How does seeing affect the image produced by a telescope? How does the effect of seeing compare with the effects of diffraction?

6. What is the principal advantage that telescopes of large diameter have over those with small diameter?

7. What are the advantages and disadvantages of each of the principal types of reflecting telescopes?

8. What are the advantages and disadvantages of radio telescopes as compared with optical telescopes?

9. Why are optical and radio telescopes discussed separately from telescopes in the higher-energy parts of the light spectrum?

Applying Your Knowledge

1. What are the pros and cons of placing an optical observatory on the roof of a university building?

2. Why do astronomers need large telescopes on Earth? In space? Why is the word *large* important?

3. Why are radio telescopes so much larger than optical telescopes?

4. Why must the shape of the mirror of an optical telescope be more accurate than the shape of the antenna of a radio telescope?

5. Given the choice of a reflecting telescope at the prime focus, Cassegrain focus, or coudé focus, which would you choose to use to photograph a faint galaxy? To obtain a detailed picture of a planet? Explain your reasoning.

▶ 6. Giovanni Schiaparelli, who claimed to observe *canali* (later translated as *canals*) on Mars, used a telescope having a 22-cm-diameter objective. What is the resolving power of such a telescope? What would the true size of Martian canals have to be for Schiaparelli to have observed them? Assume Mars is 0.5 AU from Earth at the time of observation.

▶ 7. A 20 cm (8 in) amateur telescope with a focal length of 2,000 mm produces an image of the moon 1.7 cm in diameter. How large an image of the Moon does the 5 m (200 in) telescope (focal length = 16.7 m) produce? What would be the image size for a 1.25 m (50 in) telescope having the same focal length as the 5 m (200 in) telescope?

▶ 8. Suppose you want the Moon to appear 45° across when viewed through a telescope of 125 cm focal length. What focal length eyepiece must you use?

▶ 9. Compare the theoretical light-gathering and resolving powers of telescopes having the following objective diameters to a telescope having a 2.5 cm (1 in) objective: 20 cm (8 in), 127 cm (50 in), 2.54 m (100 in), 5.08 m (200 in), 10 m (400 in).

▶ 10. Compare the resolving power of one 10 m Keck telescope with what will be possible along with the second Keck telescope 85 m away, when the combination is used for interferometry at a wavelength of 2 microns (1 micron = 0.0001 cm).

▶ 11. The European Southern Observatory's VLT consists of four separate telescopes, each having 8-m-diameter mirrors. How large would a single mirror have to be to have the same light-gathering power? How much more light than the human eye will it collect?

12. Present arguments on both sides of the controversy involving the *HST* and *JWST*. What do you think should be done?

Answers to Inquiries

12-1. The size of the image is directly related to the focal length, so if the focal length is three times longer, the image is three times taller. Using ratios, 1 cm/ 50 mm = image size/150 mm; therefore, image size is 3 cm.

12-2. The explanation is in the following paragraph.

12-3. $(D/0.5)^2 = 4D^2$ if D is in centimeters.

12-4. Roughly 20 to 30 cm.

12-5. The magnification would be 150/2 = 75; the diameter is unimportant for magnification.

12-6. The resolving power for a 15 cm telescope is 10/15 = 2/3 second of arc, which is greater than the seeing of 0.5 seconds of arc. Thus, the limiting factor will be diffraction

12-7. A 5 cm telescope has such small light-gathering and resolving power as to be not worth purchasing for astronomical purposes. From the rule of thumb given in the text, the maximum magnification you would want to use with such a telescope would be about 50.

12-8. Light-gathering power, resolving power, and magnification. No, because the word *power* has no single meaning.

12-9. To get a resolution of 0.025 seconds of arc, which is 6.9×10^{-6} degrees, at a wavelength of 1 micron (10^{-4} cm), requires a telescope of diameter $D = 57.3 \, \lambda/$ resolving power = 57.3×10^{-4} cm / 6.9×10^{-6} = 8.3 m. (Remember, the VLT will be limited by the Earth's atmosphere (seeing), not the limits of diffraction.)

12-10. The resolving power is $57.3° \times 1/300$ = $0.2°$. The naked eye can resolve detail more than 10 times finer than this.

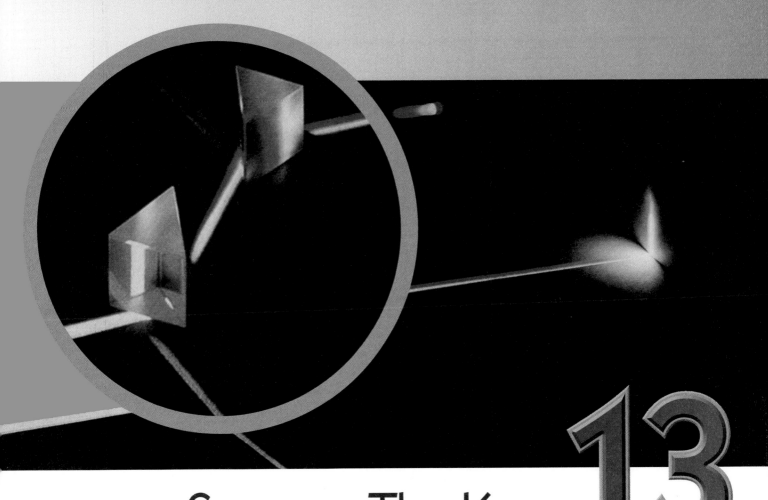

Spectra: The Key to 13 Understanding the Universe

We understand the possibility of determining their [celestial bodies'] shapes, their distances, their sizes and motions, whereas never, by any means, will we be able to study their chemical composition.

Auguste Comte, Cours de philosophe positive, *1830*

The techniques used to understand light derive from investigations begun by Newton in the mid-1600s. Newton used a prism to break up a beam of white light into a band of colors called a **spectrum** (the plural is **spectra**). He also showed that the colors could not be broken down further, and that they could be recombined to give white light. The existence of radiation beyond the two ends of the visible spectrum was unheard of at that time.

Because spectra of starlight do, in fact, reveal their chemical makeup, the chapter introductory statement by Auguste Comte is a classic example of the risk inherent in trying to set a priori limits to human knowledge. Although Comte could conceive of no method that could span the vast distances between the stars and determine their composition, a century later astronomers would carry out such analyses routinely.

During the nineteenth century, laboratory experiments by physicists showed that the spectra of light emitted from various luminous substances differed greatly from each other. Finally in the twentieth century, it became possible to obtain highly detailed information about the stars by studying their spectra. Without exaggeration we can say that the overwhelming majority of modern astrophysical knowledge comes from a study of the spectra of celestial bodies. For this reason, this chapter is of fundamental importance.

In this chapter we investigate the three principal types of spectra and learn their significance. We will also examine the principles by which the chemical compositions of and physical conditions in stars can be measured. Finally, we apply these principles to analyzing various sources of radiation both in the laboratory and in the outside world.

13.1 Observations of Spectra

Scientists made tremendous advances in understanding spectra toward the end of the nineteenth century. Many of these advances occurred because of scientists like Gustav Kirchhoff, who analyzed light by passing it through a prism and breaking it into its component parts. From his studies, he observed that all light sources could be described in terms of three types of spectra. One of these had a continuous spread of color from violet to red (**Figure 13-1**) and is called a **continuous spectrum**. Such a spectrum, when photographed with black-and-white film, appears as continuously changing shades of gray. A second type of spectrum showed discrete, bright lines, and is called an **emission spectrum** or **bright-line spectrum** (**Figure 13-2**). The third type has discrete, dark lines superimposed on a continuous spectrum

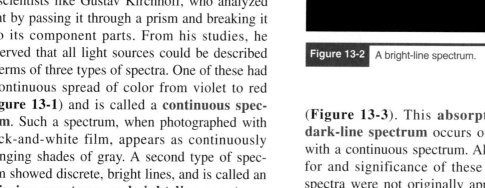

Figure 13-1   A continuous spectrum.

Figure 13-2   A bright-line spectrum.

(**Figure 13-3**). This **absorption spectrum** or **dark-line spectrum** occurs only in conjunction with a continuous spectrum. Although the reasons for and significance of these different types of spectra were not originally appreciated by scientists, Kirchhoff's research provided important clues, as will be discussed throughout this chapter.

The spectrum of light is studied with an instrument called a **spectrometer**. We detour briefly to examine this instrument before trying to understand astronomical spectra.

Chapter Opening Photo: Breaking light into its component colors is how astronomers study the universe.

Figure 13-3 | A dark-line spectrum.

THE PRINCIPLE OF THE SPECTROMETER

A spectrometer is a device to view the spectrum with the eye. A **spectrograph**, by contrast, is a device for recording a spectrum; basically, it is a spectrometer, but with the eye replaced by another type of detector (a camera or some type of electronic device that makes a permanent record of the spectrum). You could make a simple spectrograph by taping a diffraction grating over the lens of a digital camera or a film camera using high-speed color film to photograph street lamps and other outdoor sources of radiation at night.

Astronomers sometimes use a similar technique to record the spectra of stars. They place a large prism over the end of a telescope and photograph stars directly (**Figure 13-4a**). The arrangement is called an *objective prism spectrograph,* and the advantage it offers is that spectra of many stars can be recorded on a single photograph. Instead of a photograph full of star images, one obtains a photograph full of stellar spectra (**Figure 13-4b**).

Although the objective prism spectrograph is especially useful for making surveys of many stars rapidly, the spectra it produces show relatively little detail and cannot be made at all if the object is faint or is an extended source of radiation, such as a planet or galaxy. The typical spectrograph, therefore, has a number of additional parts that maximize both the amount of radiation that falls on the detector and the amount of detail in the spectrum. **Figure 13-5** illustrates how the astronomical spectrograph works. Light entering the telescope from distant stars is gathered by the telescope objective and brought to focus at the slit. The slit passes only the light of the star being studied. As the figure shows, when the light passes through the slit, it is diverging. For this reason the system must have another lens—the collimator—to make the rays of light parallel for the prism. The prism produces a

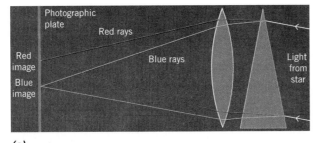

(a)

(b)

Figure 13-4 | (a) An objective prism spectrograph can be used to obtain the spectra of many stars at the same time. (b) An objective prism plate showing the spectra of three stars.

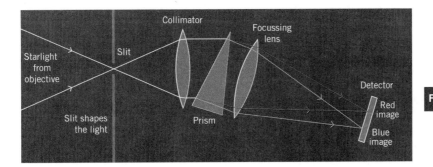

Figure 13-5 | A diagram showing the main parts of an astronomical spectrograph—one of the astronomer's most valuable tools.

spectrum that can then be detected and studied in detail. Either a grating or a prism can be used to disperse the light into its component colors.

THE CONTINUOUS SPECTRUM: KIRCHHOFF'S FIRST LAW

We will now formulate the laws first elaborated by Kirchhoff. **Kirchhoff's first law** addressed the continuous emission spectrum.

> *Radiation given off by luminous solids, luminous liquids, and hot opaque gases produces a continuous spectrum.*

An incandescent light bulb is an example of a solid giving off radiation; it contains a tungsten filament through which an electrical current flows, causing it to heat up, glow, and produce a continuous spectrum. The hot, molten lava that flows from an erupting volcano is an example of a hot liquid that, if observed with a spectrometer, would show a continuous spectrum. By an opaque gas, we mean a gas you cannot see through, such as the Sun or a star.

A precise and quantitative description of a continuous spectrum is obtained by measuring the amount of energy emitted at each wavelength. This is done by passing the light through a diffraction grating or prism, which sends each wavelength in a slightly different direction. A photoelectric cell (e.g., like a camera's exposure meter) placed at each wavelength in the spectrum then measures the amount of radiation at each wavelength. A graph of the resulting readings is then drawn—the horizontal axis giving the wavelength and the vertical axis the amount of energy at that wavelength. The resulting energy distribution curve for a typical continuous spectrum might look like **Figure 13-6**. Thus, the word *spectrum* can refer either

to a graph as in Figure 13-6 or an image as in Figures 13-1 through 13-3.

THE DEPENDENCE OF THE CONTINUOUS SPECTRUM ON TEMPERATURE

You probably have noticed that as an object is heated it begins to glow, and that the color of the glow changes slightly as the temperature of the object rises. For example, when you turn on an electric stove, you feel the heat, which is infrared radiation, before you see any visible changes. After a while, the burner glows red and then may take on an orange hue. The changing color as the temperature rises is caused by a change in the continuous distribution of energy with wavelength. In other words, as the object's temperature increases, the relative strength of its radiation shifts, from relatively low in the visual spectral region and high in the infrared to relatively high in the visual and low in the infrared.

Inquiry 13-1

The hotter an object gets, the bluer the radiation from it becomes. Does a hotter object emit its most intense radiation at shorter wavelengths than a cooler object?

A graph of the variation of radiation intensity with wavelength—that is, the spectrum of radiation—between wavelengths 0 Å and 300,000 Å is shown in **Figure 13-7a** for bodies of temperature 3,000 K and 6,000 K. Because both the wavelength and the intensity ranges in the graph are large, the graph is highly compressed. In particular, no detail can be seen in the 3,000 K spectrum.

Figure 13-7b graphs the identical data as in part *a*. However, the scales of the axis are drawn differently so that the large ranges of wavelength and intensity can be shown on the one graph, while making visible the differences between the curves for the two temperatures. Whereas Figure 13-7a has *linear* axes in which the interval between tick marks changes by a constant number of angstroms or intensity, the axes in part *b* are *logarithmic*; that is, each tick mark increases by the same *multiplicative* factor. For example, in Figure 13-7a, the interval between each wavelength tick mark is 50,000 Å, while in part *b* each tick mark corresponds to an increase of wavelength by a *factor* of 10. Note that the lengths on the axis between the wavelengths of 1,000 Å,

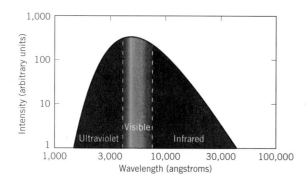

Figure 13-6 Energy distribution curve for a continuous spectrum.

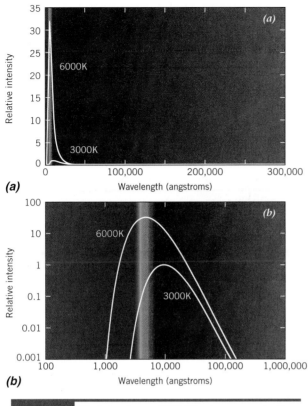

(a)

(b)

Figure 13-7 | Radiation from a 3000-K body compared with that from a 6000-K body. *(a)* The scales are linear. *(b)* The scales are logarithmic.

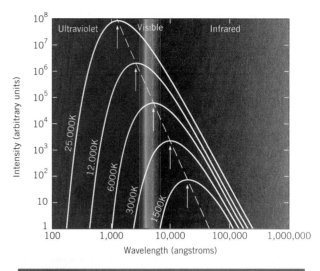

Figure 13-8 | Radiation from bodies of various temperatures is compared. The variation of the wavelength of maximum intensity in relation to the temperature is indicated.

10,000 Å, and 100,000 Å are all the same. A similar logarithmic scale is used for the intensity axis. These scales allow more detail to be seen on one graph than could otherwise be seen. Because of the large range in numbers used in astronomy, astronomers often use logarithmic scales, as we will see in many chapters in this book.

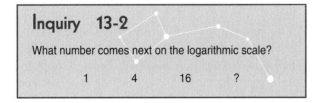

Inquiry 13-2

What number comes next on the logarithmic scale?

1   4   16   ?

The one disadvantage of the logarithmic scale is that interpolating between divisions on the axis becomes more difficult. On a linear scale, half the distance between 2 and 4 is 3; however, on a logarithmic scale, halfway between the 10 and 100 tick marks is 31.62.

Any object whose continuous spectrum has the shape shown in Figure 13-7 is known as a **blackbody**. A blackbody (not to be confused with the term *black hole*) is a hypothetical object physicists find convenient to use when describing radiation emitted from any body whose temperature is greater than 0 K. Stars roughly approximate blackbodies. For this reason, blackbody characteristics can be used to determine temperatures of stars. This works because the spectrum of a blackbody depends only on temperature and not on chemical composition.

Figure 13-8 compares spectra of blackbodies having temperatures between 1,500 K and 25,000 K. The higher-temperature bodies emit more of their energy in the ultraviolet part of the electromagnetic spectrum. In fact, because the eye is insensitive to radiation whose wavelength is shorter than about 4,000 Å, much of the radiation emitted by a body whose temperature is 12,000 K or more is not visible. The predominant visible radiation from this body is in the shorter wavelengths and appears bluish to the eye.

Notice how, in Figure 13-8 in the visual region of the spectrum, the slope of each curve changes. That is, for the 3,000 K body the curve increases from short to long wavelengths through the visual spectral region. The curve is nearly flat for the 6,000 K body, it decreases somewhat for the 12,000 K and decreases strongly for the 25,000 K body. In other words, the slope of the blackbody curve is uniquely determined by temperature. Reversing the argument, by *observing* the slope of the continuous spectrum, the astronomer can readily determine the body's temperature.

If we measure the total energy radiated, we find that the 3,000 K blackbody radiates most of its energy in the red and infrared part of the spectrum, whereas the 6,000 K blackbody has its maximum energy output in the yellow-green part of the spectrum. In addition, we see that all blackbodies of temperatures greater than 0 K radiate at least some energy at all wavelengths, from the shortest to the longest. From the spectral energy curves we see that hot bodies radiate more energy at each wavelength than cooler bodies.

THE LAWS OF BLACKBODY RADIATION

Each spectrum shown in Figure 13-8 exhibits a maximum intensity at some specific wavelength. We denote the wavelength of the maximum intensity as λ_{max}. Laboratory experiments with approximate blackbodies show a relationship between λ_{max} and the temperature of the object. This relationship, first given by the German physicist Wilhelm Wien, is as follows:

$$\lambda_{max}T = \text{constant} = 3 \times 10^7 \text{ Å K.}$$

For the product to be a constant, as one variable (e.g., temperature) increases, the other one must decrease. In other words, the greater the temperature, the smaller is the wavelength of the maximum intensity. More energy is emitted at λ_{max} than at any other wavelength. Because a hot body emits most of its energy in the ultraviolet, the body will emit more blue light than red light in the visual spectral region and thus appear to be blue. The same type of argument for a cool body, which emits most of its energy at long wavelengths, shows that it will appear to be red. For these reasons, the color of an emitting body is determined by λ_{max}. However, because the wavelength of maximum emission is much more objectively and precisely defined than the subjective notion of color, we will specify the wavelength. To give an example of the use of this formula, let's ask a question: What will be the wavelength of maximum emission for a body whose temperature is 6,000 K? The answer can be computed:

$$\lambda_{max} = \frac{3 \times 10^7 \text{ Å K}}{6,000 \text{ K}} = 5,000 \text{ Å.}$$

This wavelength is in the green part of the electromagnetic spectrum. Usually, the question is turned around, because the wavelength of maximum emission is the observed quantity and the temperature is the desired (computed) result.

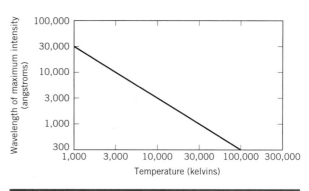

Figure 13-9 A graph showing Wien's law, the relationship between temperature and the wavelength of maximum intensity.

A graph of the relationship between λ_{max} and temperature is shown in **Figure 13-9**. Instead of using the formula, you can read values directly from the graph if you prefer. Notice, however, that the scales are logarithmic rather than linear, so interpolation is not easy.

Inquiry 13-3

What is the temperature of a blackbody whose wavelength of maximum emission is observed to be 2500 Å? Check your result on the graph in Figure 13-9 and compare it with the curve in Figure 13-8.

Inquiry 13-4

Although the temperature on the surface of Mars varies, let us assume that it is at approximately 300 K (about room temperature). At what wavelength should you look for Mars's peak emission of radiation? (This is nearly the same as for your body.)

The curves shown in Figure 13-8 actually represent the amount of energy per second radiated by *each square meter*[1] of a blackbody having a temperature greater than 0 K. Notice that the area between the wavelength axis and the curve itself is greater for the hotter body; this means that the hotter body radiates more energy per second from each square meter than the cooler one does. How much more? Two Austrian scientists, Josef Stefan and Ludwig

---

[1] A more general way of saying energy radiated by each square meter is to say energy radiated per unit area. A unit area can be one square meter, one square inch, or one square kilometer. However, the energy per square meter will not be the same as the energy per square kilometer.

Boltzmann, answered this question when they found that the energy radiated per second from one square meter of a body is proportional to the fourth power of the temperature. Symbolically, the **Stefan-Boltzmann law** is as follows:

$$F \propto T^4$$

where T is the temperature in kelvins, $\propto$ means *proportional to* and F is the *energy flux* with units of Joules/s/m^2. Thus, a body twice as hot as the Sun will emit $2^4 = 16$ times as much energy per unit area per second as the Sun; a star 10 times as hot as the Sun will emit 10^4 times as much energy per unit area per second as the Sun emits.

Inquiry 13-5

How much more energy does each square meter of a 50,000 K star radiate than a 5,000 K star?

In summary, we have four significant points about blackbodies. First, Wien's law shows that the wavelength of maximum emission is shorter for hot objects than for cool ones. Second, the Stefan-Boltzmann law shows that *every* object with a temperature above absolute zero emits radiation. Third, the amount of continuous radiation depends only on temperature, not on chemical composition. Fourth, hotter objects emit more energy at every wavelength than do cooler objects, such that hotter objects emit more energy per second per unit area in an amount that depends on the temperature to the fourth power.

OBJECTS CAN BE SEEN BY THEIR EMITTED LIGHT OR THEIR REFLECTED LIGHT

The energy *emitted* by Mars and the other planets that results from their temperature is *not* the radiation by which we see these objects with our eyes. The reason is that such cool bodies radiate their energy at wavelengths to which human eyes are insensitive. We *see* the planets, and everyday objects such as tables and chairs, by light *reflected* from them, not by radiation resulting from their temperature. The visible colors of these objects are determined by their reflective properties. If a chair appears to be red, it is because it reflects light of longer, redder wavelengths more efficiently than shorter, bluer wavelengths, not because it has been heated to a temperature of several thousand

degrees Kelvin. Similarly, the visible-light spectrum of Mars closely resembles that of sunlight because it *is* reflected sunlight.

Most astronomical objects, the stars in particular, have sufficiently high temperatures that we see them by the radiation they *emit* as a result of their high temperatures; therefore, their temperatures can be determined from their colors or from the wavelength at which they emit the largest amount of energy, $\lambda_{\max}$, provided they radiate like a blackbody. There are some exceptions to this, which we will study later.

Inquiry 13-6

Stars in the process of forming (we call these objects *protostars*) emit most of their radiation in the infrared part of the spectrum. What would be the surface temperature of such a protostar if the wavelength at which it emits the greatest amount of energy were observed to be 30,000 Å?

Inquiry 13-7

In the 1960s, astronomers discovered a new type of object called an *X-ray source*. These objects show a continuous spectrum with the peak emission at very short wavelengths. What is the temperature of an X-ray source whose peak radiation falls at 0.5 Å?

EMISSION SPECTRA: KIRCHHOFF'S SECOND LAW

Kirchhoff's second law concerns the second type of spectral pattern, the **emission or bright-line spectrum**.

An emission (bright-line) spectrum is characteristic of transparent (i.e., low density) gases that have been excited to glow by being heated or by other means.

The energy distribution curve of an emission spectrum can also be graphed, as in **Figure 13-10a**, which shows the emission spectrum of low-density mercury vapors. Because energy is emitted only at discrete wavelengths, the energy distribution curve consists of spikes at several wavelengths. A different amount of energy may be radiated in each of the spectral lines, so the heights of the spikes will, in general, be different. (When we speak of the strength of a bright spectral line, we mean the height of the peak.) Because bright spectral lines of mercury are in the blue/ultraviolet

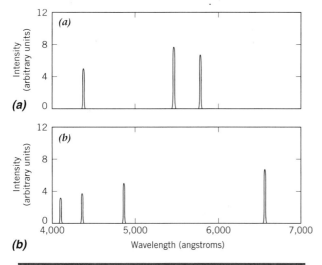

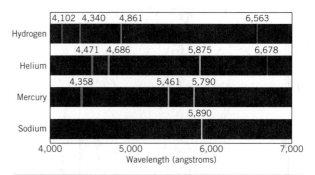

Figure 13-11 Discrete-emission spectra of selected elements in the visible-wavelength region. The colored lines marked at certain wavelengths are places where energy is being emitted.

Figure 13-10 Energy distribution curve for a discrete-emission spectrum. *(a)* Mercury. *(b)* Hydrogen. Compare these two diagrams with Figure 13-11.

region of the spectrum, the mercury lamp appears bluish in color. Similarly, **Figure 13-10*b*** shows the spectrum for hydrogen. The light from hydrogen appears red because the structure of the hydrogen atom causes the brightest spectral line to appear in the red part of the spectrum. (The dark lines in Figure 13-3 are also caused by hydrogen but at a shorter wavelength.) Notice that the separation of the hydrogen emission lines decreases toward shorter wavelengths; any useful model of spectra must explain this observed convergence of the spectral lines.

The properties of these transparent gases are not the same as those of denser, opaque gases that cause a continuous spectrum. For a denser gas, the spectrum of the emitted radiation is continuous and depends only on the body's temperature. None of the laws governing radiation from dense bodies holds for transparent gases. In fact, a transparent gas does not need to be hot to emit energy; low-temperature gases can also produce emission lines, as we will see in Section 13.2.

THE SPECTRAL FINGERPRINTS OF THE CHEMICAL ELEMENTS

Early researchers noticed that each chemical element emitted a different pattern of wavelengths in the spectrum. Consequently, this pattern could be used as a guide to identify the chemical element producing the spectrum. If several elements were mixed together in the same gas, the resulting spectrum would contain a mix of the emission features of each of them, so the presence of each element

could be confirmed. Furthermore, the strengths of the bright spectral lines would give a measure of the proportions of the elements in the mixture.

In other words, the presence of discrete features in the spectrum of a celestial object, as opposed to continuous features, allows us to determine its chemical composition. We can identify not only the kinds of elements present in the stars but also their amounts. It is clear that Auguste Comte was wrong in stating that we would never know the composition of celestial objects!

Figure 13-11 shows emission spectra of several other elements found in common light sources of radiation. Readers who have access to a diffraction grating or prism can compare these spectra with the spectra of city lights where they live and attempt to identify the elements involved.

The reasons spectra are unique for each chemical element will become clear in Section 13.2.

THE SUN'S ABSORPTION SPECTRUM: KIRCHHOFF'S THIRD LAW

Kirchhoff observed a third type of spectrum when light having a continuous spectrum, characteristic of a blackbody at some temperature, was passed through a low-density, cooler gas, as shown in **Figure 13-12**. He found that under such circum-

Figure 13-12 An absorption spectrum results when light from a source of continuous radiation passes through a cooler gas.

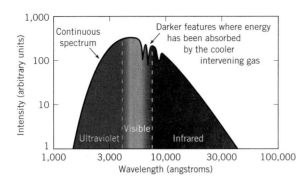

Figure 13-13 Energy distribution curve of a continuous emission spectrum with energy removed at certain wavelengths to produce an absorption-line spectrum.

stances the rarefied cool gas absorbed radiation at discrete wavelengths. Most of the continuous spectrum passed through the cool gas unhindered, but at certain wavelengths, energy was removed from the beam. Significantly, the wavelengths absorbed by the cool gas were the same as those that would have been emitted if the gas had been heated. Such a spectrum is called an absorption or dark-line spectrum.

The energy distribution curve for the resulting beam of radiation would resemble **Figure 13-13**, a continuous distribution of energy whose basic features, determined by the star's temperature, are preserved, but with dips in the energy at discrete wavelengths. The continuous part of the spectrum can still be studied to learn about the temperature of its source, but the presence of the cooler, low-density gas reveals itself by the absorption it produces. Note that when we say *cool* gas we mean cooler than the object that produced the continuous spectrum that flows through it.

Figure 13-14 shows a spectrum of the Sun, with the elements that produced some of the dark lines identified. Each dark line is at the same wavelength as would be emitted by the same element if it were vaporized into a transparent gas at a high temperature. In other words, each chemical element also exhibits its characteristic fingerprint pattern of lines, even when it appears in dark-line form. This means that we can determine the chemical composition of the Sun and other stars, because they all show an absorption spectrum.

An absorption spectrum cannot exist without a background continuous spectrum. For this reason, the Sun's absorption spectrum has a continuous emission of energy between the absorption lines. We can extract a great deal of information from its total spectrum, namely, its temperature *and* its chemical composition. We will see in later chapters that when we examine the lines in a spectrum in magnified detail, we can determine even more properties of the object that emitted that spectrum—for example, its density, motion through space, rotation, turbulence, and even the presence and strength of a magnetic field.

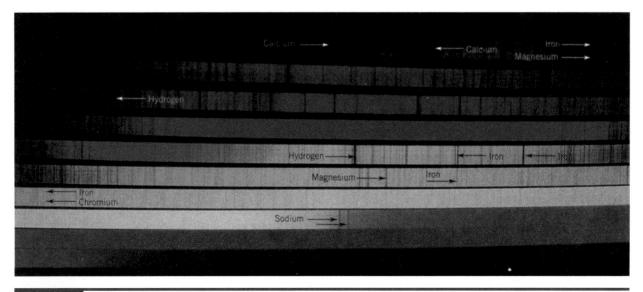

Figure 13-14 The spectrum of the Sun, showing a few of the elements in the solar atmosphere.

Inquiry 13-8

The Sun is so hot that water vapor cannot exist there; it would be broken up into atoms of hydrogen and oxygen. Yet when astronomers observe the solar spectrum, they find absorption lines produced by water molecules. How can this be?

Inquiry 13-9

What kind of spectrum would you expect when observing the Moon? What additional features might you expect to see in the spectrum of the planet Venus that you would not see in the Moon's spectrum?

Kirchhoff's third law on absorption spectra explains these observations, and can be summarized as follows:

An absorption (dark-line) spectrum forms whenever radiation with a continuous distribution of energy, characteristic of a blackbody of some temperature, is passed through a cooler, low-density gas. Energy is removed from the beam at the same wavelengths that the gas would have emitted had it been heated to a higher temperature or otherwise excited into emission.

Taken together, the elements are capable of emitting and absorbing at probably millions of discrete wavelengths. With such a large number of possibilities, scientists have developed extensive tables that give the wavelengths and characteristic strengths of the various spectral lines of all elements. Astronomers refer to such tables when analyzing the spectra of celestial bodies.

13.2 Understanding Spectra

In Section 13.1 we examined observations of spectra with no consideration of their cause beyond Kirchhoff's nineteenth-century laws. In the early part of the twentieth century, scientists began to lay the groundwork for understanding spectra as they began to learn more about the structure of atoms. About all they knew was that atoms were composed of both positively and negatively charged particles. In fact, Ernest Rutherford proposed a model for an atom that had a heavy, positively charged **nucleus** surrounded by orbiting negatively charged **electrons**.

On the basis of experiments, Rutherford constructed a model for hydrogen, the simplest atom, that consisted of a single positively charged particle, called the **proton**, at the center of the atom with an electron orbiting around it. Because the mass of the proton was found to be about 1,800 times the mass of the electron, virtually all the mass of the atom resided in its nucleus. However, Rutherford showed that the nucleus was very small (approximately 10^{-13} cm in size), and hence the *size* of the atom depended on how far away the electron was from the nucleus. Because an average electron stays approximately 10^{-8} cm (1 Å) away from the nucleus, the diameter of the atom is about 10^5 times the diameter of the nucleus itself. Atoms, therefore, are mostly empty space.

The most common form of the helium atom has a nucleus that consists of two protons and two **neutrons** (chargeless particles of approximately the same mass as the proton). Its total **atomic weight**, the number of protons plus neutrons in the nucleus, is four. However, the electrical charge of the nucleus is only two (because of the two protons), so only two electrons are needed to orbit outside the nucleus to give the helium atom overall electrical neutrality.

The number of protons in the nucleus determines what chemical element a nucleus is. Two nuclei having a different number of protons are different elements. In general, the nuclei of the chemical elements are designated with a concise notation in which the *charge* of the nucleus—the number of protons in it—is written as a subscript before an abbreviation of the element's name, and the total atomic weight is written as a superscript before the element name. Thus we designate hydrogen as $_1^1$H and helium as $_2^4$He.

As a final example, consider the element carbon, which is designated by $_6^{12}$C. The subscript of 6 tells us that the carbon nucleus has 6 protons, whereas the superscript of 12 tells us that the nucleus has a total atomic weight of 12, so it must have 6 neutrons in addition to the 6 protons. To electrically balance the 6 protons, it has 6 electrons as well. The atom $_6^{13}$C is a carbon atom having 7 neutrons rather than 6, while $_6^{14}$C is a radioactive form of carbon used in carbon dating. Atoms having the same number of protons but different numbers of neutrons are called **isotopes** of an atom.

Inquiry 13-10

While the most common isotope of oxygen has 8 neutrons, isotopes having 7, 9, and 10 neutrons also exist. What is the chemical designation for each of these isotopes? Why are they all oxygen?

Rutherford's simple model agreed well with experiments in which atoms scattered fast-moving helium nuclei incident upon them, but it ran into difficulty when it was used to explain the observed spectra of atoms. The reason for this is that, according to the theory of electromagnetism, accelerated electrons ought to radiate away energy in the form of electromagnetic waves so rapidly that in less than a millionth of a second the electrons should collapse into the nucleus. This prediction of the model was in violent disagreement with experiments.

In 1913, the Danish physicist Niels Bohr proposed a radical modification to Rutherford's model. In the older model, electrons could orbit the nucleus at *any* distance; all possible orbital distances from the nucleus were allowed. Bohr's radical proposal was that electrons could exist only in discrete, well-defined orbits around the nucleus. He put forth a model for the hydrogen atom that consisted of a single, heavy, positively charged particle, a proton, with a negatively charged electron in orbit around it. The electron could occupy orbits only at discrete radii—other orbits were forbidden to it. This is analogous to a ladder: You are allowed to step only on the discrete rungs but not in between them. In the Bohr model there was an orbit having minimum radius and energy; if the electron were in that orbit, it could not radiate away any more energy. This idea ruled out the catastrophic collapse of atoms that was marring the earlier model. The resulting model resembles a miniature solar system in which the more massive proton plays the role of the Sun and the less massive electron the role of a single planet (**Figure 13-15**). Planets, however, can occupy any orbit; electrons, only certain specific (allowed) ones. In addition, the electron and proton are bound together by electromagnetic rather than the gravitational forces that bind the planets to the Sun.

The Bohr model now makes it possible to explain the spectrum of a radiating atom. Bohr realized that the closer the electron was to the nucleus, the more energy would be required to remove it from the atom because of the strong electromagnetic attraction between the positively charged nucleus and the negative orbiting electron. In this way, he was able to assign to each orbit a precise amount of energy equal to that required to remove an electron in that orbit from the atom. Bohr then proposed a revolutionary idea; he said that an atom emits and absorbs energy by emitting and absorbing photons having *exactly* the energy required to move the electron from one orbit to another. Because the energy and wavelength of photons are related as previously discussed ($E = hc/\lambda$), once Bohr knew the energy needed to move electrons between orbits he could calculate the possible wavelengths that could be emitted or absorbed by a hydrogen atom. When he did so, the agreement of his new model with the observed spectrum of hydrogen was excellent.

The absorption and emission of photons in the Bohr model are illustrated in **Figure 13-16**. Symbolically, we can write absorption as

$$\text{Atom in inner orbit} + \text{Energy} \rightarrow$$
$$\text{Atom in outer orbit}$$

and emission as

$$\text{Atom in outer orbit} \rightarrow$$
$$\text{Energy} + \text{Atom in inner orbit}$$

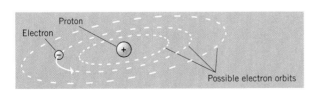

Figure 13-15 The Bohr model pictured the atom as a miniature solar system, with electrons orbiting the nucleus as planets orbit the Sun.

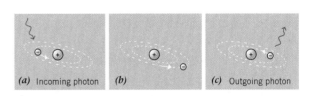

(a) Incoming photon (b) (c) Outgoing photon

Figure 13-16 Emission and absorption of light by a hydrogen atom. *(a)* A photon impinging on an atom whose electron is close to the nucleus. The photon is absorbed and the electron becomes excited as shown in *(b)*. After a while, the electron drops to an inner orbit, with the emission of a photon, as in part *(c)*.

In the case where an atom absorbs energy and moves to an excited (outer) orbit, the energy can come from two places:

1. The atom can absorb a photon (referred to as *photoexcitation*).
2. The atom can collide with an electron or another atom, a process called *collisional excitation*. The greater the temperature, the greater the speed of the colliding particles and the greater the excitation can be.

The transition either to inner and outer orbit must always involve an amount of energy exactly equal to the energy difference between the two orbits of an atom. In the case of emission, an atom *spontaneously* emits its "excess" energy by emitting a photon in a very short time (10^{-8} second) and then moves to an inner orbit. In the case of photoexcitation, an atom must encounter a photon of just the right amount of energy to absorb it and reach an outer orbit. In the case of collisional excitation, the colliding particle must have enough kinetic energy (energy of motion) to be able to raise the atom to a particular orbits.

Successful as it was in explaining the hydrogen spectrum, the Bohr model is too simple to explain the detailed spectrum of even hydrogen, much less the more complex spectra of the other elements, which have more electrons. The theory of *quantum mechanics* (quantum theory, for short), which replaced the Bohr theory in the 1920s, shares some of its main features, but recognizes that because of its wave nature the electron cannot be localized to simple circular orbits as in the Bohr theory. **Figure 13-17** shows a schematic representation of the electron cloud of an excited atom. Rather than

being represented by a thin line as in Figure 13-16, the electron orbit is represented as a fuzzy cloud in which the darkest regions represent places where the electron has the highest probability of being at any given moment. Like the Bohr theory, quantum theory predicts that electrons can exist only in discrete orbits (now referred to as *energy states*) and that to move from one energy state to another, the atom must emit or absorb a photon of *precisely* the right energy or wavelength. If the energy is not right, no absorption occurs. Mathematically, the relationship between the energies of the two states and the energy of the photon is as follows:

Energy of photon = Energy of higher energy state
– Energy of lower energy state

In symbols, if E is the energy, then

$$E_{\text{photon}} = E_{\text{higher state}} - E_{\text{lower state}}$$

Inquiry 13-11

How does the relationship between the energies of the two states and the energy of the photon lead to the conclusion that the total amount of energy in the universe is not changed when an electron jumps from one energy state to another?

Inquiry 13-12

What physical principle is embodied in Inquiry 13-11?

ENERGY-LEVEL DIAGRAM OF THE HYDROGEN ATOM

Atoms having electrons in different orbits are said to be in different energy states or **energy levels** because each electron orbit corresponds to a different energy. These energy levels are best understood in terms of the **energy-level diagram** (shown in **Figure 13-18** for the hydrogen atom). Each numbered energy level corresponds directly to an allowed electron orbit in the Bohr model. An electron in the innermost orbit is said to be in the **ground state**. Any other orbit is called an **excited state**. Because an electron in the ground state feels such a strong attraction to the nucleus, it takes considerable energy to excite that electron to the first excited state. Thus, the size of the interval between the energy levels labeled 1 and 2 in Figure 13-18 represents the amount of energy required to move the electron from its lowest

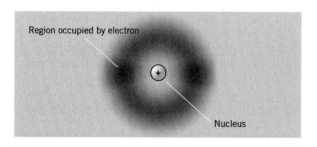

Figure 13-17 The electron cloud for an excited hydrogen atom. The darkness of the "cloud" in the figure represents the probability of where the electron can be. According to quantum mechanics, the exact position of an electron in an atom cannot be precisely located.

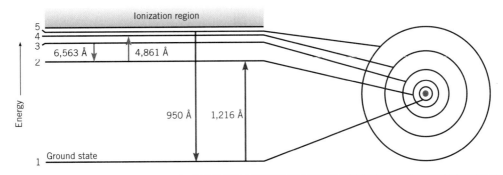

Figure 13-18 Energy-level diagram for the hydrogen atom showing several transitions. The wavelengths of the transitions in angstroms are given alongside the transition. The electron orbits are highly schematic and are drawn to scale, or the correct spacing.

allowed level (the ground state) to the level labeled 2 (an excited state). The figure shows that a smaller amount of energy is required to move the electron from level 2 to level 3; still less is required from 3 to 4, and so on. Going to successively higher enerßgy levels requires successively less energy because the electron occupies more and more distant orbits, where the attractive force between the nucleus and the electron becomes weaker. For this reason, the energy levels appear to converge toward the top of the diagram; this convergence of energy levels predicts the convergence of spectral lines observed in Figure 13-3.

The arrows in Figure 13-18 represent **transitions**, in which the electron jumps from one energy level to another, with the simultaneous absorption or emission of energy. The amount of energy absorbed or emitted is equal to the difference between the energies of the two levels. Several of the transitions that are responsible for producing lines in the hydrogen spectrum are shown. For example, when the electron makes a transition between levels 2 and 3, a spectral line at a wavelength of 6563 Å is produced; the 4861 Å line is produced when the transition is between levels 2 and 4. Notice that *energy levels* do not have wavelengths; only *transitions* between any two of them do.

Inquiry 13-13

The arrows in the diagram point from the initial state of the atom toward the final state. Which direction represents emission, an upward-pointing arrow or a downward-pointing arrow? Which represents absorption? (*Hint:* For each case, absorption and emission, which state has the higher energy—the initial or the final?)

The spectral lines at 6563 and 4861 Å belong to a series of lines called the **Balmer series**. These are lines formed when the hydrogen atom makes a transition between the second energy level and *any* higher energy level. A transition from level 2 to 4 will produce a Balmer *absorption* line at 4861 Å; a transition from level 4 to 2 will produce a Balmer *emission* line at the identical wavelength. The Balmer series is important because its wavelengths are in the visual spectral region and are thus observable in spectra taken by telescopes located on the Earth's surface.

In a similar manner, **Figure 13-19** shows that the ultraviolet **Lyman series** of hydrogen spectral lines is generated by transitions between the

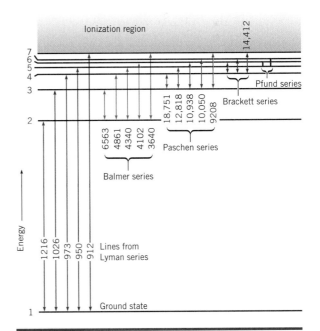

Figure 13-19 Energy-level diagram for the hydrogen atom, showing the principle transition in the Lyman, Balmer, Paschen, Brackett, and Pfund series. The wavelengths in angstroms are given.

ground state and any higher level; the Paschen series by transitions between the third level and a higher level; the Brackett series between the fourth level and a higher level; and the Pfund series between the fifth level and a higher level. Series involving the sixth, seventh, eighth, and higher levels also exist but do not have special names. The wavelengths of some of these high-numbered series are so long that they lie in the radio part of the electromagnetic spectrum. Astronomers have detected some of these lines.

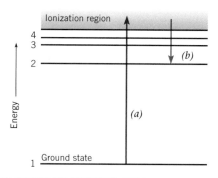

Figure 13-20 Energy-level diagram for hydrogen, showing (a) ionization and (b) recombination.

Inquiry 13-14

In which regions of the electromagnetic spectrum will each of the hydrogen series produce spectral lines? Which can be observed from the Earth's surface? Which require instruments above the Earth's atmosphere? (*Hint:* Look at Figure 13-19.)

Inquiry 13-15

Each of the following transitions gives rise to a line in one of the previously named series. Which series does each line belong to? Does the transition give rise to an emission or absorption line?

a. From level 4 to level 2
b. From level 1 to level 2
c. From level 5 to level 3
d. From level 5 to level 1
e. From level 5 to level 28

The shaded region at the top of the energy-level diagram represents what happens when the electron is given so much energy that it can no longer be held by the electrical attraction of the proton. The electron is removed from the atom, and the electron and proton go their separate ways. In such a case, we say that the atom has been **ionized**. We may represent this process, called **ionization**, by the upward arrow labeled *a* in **Figure 13-20**, and by the formula

Hydrogen atom + Sufficiently energetic photon → Proton + Free electron.

As with excitation, an atom can also be ionized in two ways: (1) photoionization (absorbing an energetic photon) and (2) collisional ionization, colliding with a particle having sufficient energy to ionize the atom (that is, move the electron to the ionization level).

Any energy the photon has above that required to ionize the atom goes into the electron's kinetic energy of movement. Because the free electron can move with any speed (i.e., can have any energy), there is a continuous range of ionized states that are distinguished from the discrete energy states of the bound electron. After ionization, the proton is all that remains of this simplest of atoms. We often designate an ionized atom with a plus to emphasize that the charge is no longer balanced.

The reverse process can also occur, in which the proton recombines with a free electron and gives off a photon. This reaction, which is called **recombination**, may be represented as a downward transition from the shaded region of Figure 13-20 into the lower region:

Proton + Free electron → Hydrogen atom + Emitted photon.

Transitions into or out of the ionization region of the energy-level diagram do not produce discrete spectral lines. The reason is that the energies in the ionization regions are not discrete (as electron orbits are) but continuous. Thus, absorptions and emissions involving the ionization region cover a continuous range of wavelength and thus contribute to the continuous, rather than to the discrete line, spectrum.

As represented in Figure 13-19, the only difference between emission and absorption is the direction of the arrow. It would thus appear that there is complete symmetry between emission and absorption. In fact, this is not so, because if an atom is in an excited energy state, it will spontaneously give off its excess energy in a short time (about 10^{-8} seconds) and make a transition into a lower energy state. But an atom in a lower energy state cannot make a transition to a higher state unless a photon of exactly the right energy arrives and is absorbed.

For example, referring to Figure 13-19, if the hydrogen atom is in its ground state (level 1) and a photon of 1,210 Å passes by, nothing will happen because the atom cannot absorb a photon of this energy. Similarly, a photon of wavelength 1,220 Å will also be ignored. Only if the photon has just the right energy to lift the electron to a higher state—for example, if its wavelength is 1,216 or 950 Å—can it be absorbed. Similar considerations hold for absorption to higher levels if the atom is already in one of the excited energy states.

Inquiry 13-16

Under which of the following situations is it possible for the photon to be absorbed by the hydrogen atom and produce a discrete absorption line? If the photon is absorbed, in what state will the atom find itself? Use Figure 13-19.

a. Atom in ground state (level 1); the photon wavelength is 1,026 Å

b. Atom in excited state (level 2); the photon wavelength is 1,026 Å

c. Atom in excited state (level 2); the photon wavelength is 4,861 Å

d. Atom in excited state (level 3); the photon wavelength is 6,563 Å

Once the electron is in the excited state, however, it will have excess energy and will rapidly and spontaneously make a transition into any lower state, if another absorption has not occurred first. If this lower state is also an excited state, then the electron will again make another transition into an even lower state, and so on. In other words, after *each* transition, the electron will cascade to lower energy levels with the emission of a photon having an appropriate wavelength.

Inquiry 13-17

What wavelength or wavelengths can a hydrogen atom in level 3 emit by spontaneous emission?

Inquiry 13-18

What are all the possible wavelengths a hydrogen atom in level 4 could possibly emit by spontaneous emission as the electron cascades to the ground state?

WHY DIFFERENT ELEMENTS HAVE DIFFERENT SPECTRA

The spectra of most elements are much more complicated than that of hydrogen. Although the simple Bohr model of the atom was adequate to explain the wavelengths of the hydrogen spectrum, the development of quantum theory was required to give a satisfactory account of the other elements. The reason for this is that in the Bohr model the electrons are imagined to be localized as points orbiting the nucleus, whereas in fact the wave nature of the electron makes this picture inadequate if more than one electron is present. Thus, although Bohr's model was conceptually wrong and was replaced by later advances, it was an important first step in understanding spectra.

For example, consider the neutral helium atom. It has two protons in its nucleus and therefore two electrons orbiting it to make the atom electrically neutral. In addition, there are two neutrons in the nucleus. The two electrons are located somewhere in a hazy cloud surrounding the nucleus, because their wave nature makes them impossible to localize, as illustrated in **Figure 13-21**. Because the two electrons occupy roughly the same region of space and are both negatively charged, they tend to repel one another, which distorts the regions each can occupy and affects the energy states that the atom can have. The resulting energy-level diagram is considerably more complex than the one for hydrogen.

Similarly, as one progresses through heavier and heavier elements, the number of protons in the nucleus increases. The underlying pattern of allowed orbits is determined primarily by the number of positive charges (protons) in the nucleus and only secondarily by the number of neutrons. It follows that the number of electrons surrounding the nucleus must also increase to maintain electrical neutrality. The electrons tend to arrange themselves

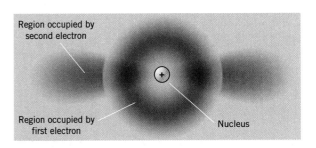

Figure 13-21 In a helium atom, the two electrons occupy overlapping regions.

into shell-like regions, each containing a certain number of electrons; the electrons in the outermost shell make the transitions that are responsible for the visible spectrum of the particular element. But it is primarily the number of protons that determines the amount of energy required to make any given transition. With more and more electrons, the mutual repulsion between the various electrons becomes more and more complex, with a resulting increase in the complexity of the energy-level diagram and spectrum. Thus, the spectra of two elements with almost the same number of protons will be completely different.

Inquiry 13-19

Considering everything you have learned, would you expect the spectrum of singly ionized helium to more closely resemble the spectrum of neutral helium, neutral hydrogen, or ionized hydrogen? (By *more closely resemble*, we mean the relative positions of the spectral lines.)

ATOMIC EXPLANATION OF KIRCHHOFF'S LAWS

Let's not forget Kirchhoff's laws, which provided us with information on the conditions under which spectra of various types are formed. We now need to merge his ideas with those of atomic structure.

You might well wonder at this point why a dark-line spectrum is ever seen. It might seem that the absorption of a photon will usually be followed by reemission of a photon of the same wavelength so that they cancel one another and *nothing* would be observed. If this is so, then why is a dark line seen?

Consider 100 photons at each of the wavelengths, as noted in **Figure 13-22a**. These photons, coming from the right, encounter a cloud of hydrogen gas. All the photons of each wavelength other than the 1,216 Å pass through the cloud unimpeded. The photons of wavelength 1,216 Å, however, correspond exactly to the energy required to excite an electron in the hydrogen atom from the ground state to the first excited state. The 1,216 Å photons are therefore absorbed by the atom, and the electrons become excited to a higher energy level. They remain excited for only a hundred millionth of a second, however, after which they drop back down to the ground state, emitting photons of wavelength 1,216 Å *in random directions*. The observer on the left, then, sees a spectrum of intensity 100 at all wavelengths except 1,216 Å, where, in this example, only eight photons are detected. Thus, a dark line is observed at that wavelength (**Figure 13-22b**).

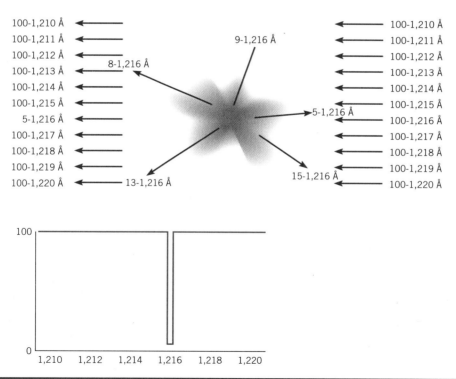

Figure 13-22 Formation of absorption lines in a cool, low density hydrogen gas cloud. Only some emitted photons are shown.

Figure 13-23 summarizes the concept. The figure shows light that has a continuous spectrum passing through a rarefied, cool gas. The atoms in the gas absorb photons of certain wavelengths from the passing beam of radiation. The atoms then re-emit photons at wavelengths characteristic of the gas. However, the emitted photons are radiated randomly into *all* directions in space. At the particular wavelengths at which absorption is taking place, there will be a net decrease in the number of photons in the beam moving toward spectrometer A, because most of the photons that are emitted at the wavelength will go in other directions. Observer A will therefore see a continuous spectrum with energy missing at the wavelengths that the gas is capable of absorbing. In other words, there will be a continuous spectrum with a superimposed dark-line spectrum.

Inquiry 13-20

Suppose an observer were to study the spectrum of the radiation received by a spectrometer pointed in the direction of spectrometer B in Figure 13-23. What kind of spectrum would be seen? Why?

This inquiry shows that the type of spectrum seen depends not only on the physical conditions in the source of light but also on the geometrical relationship between the source and the observer. We illustrate spectra further in the next section.

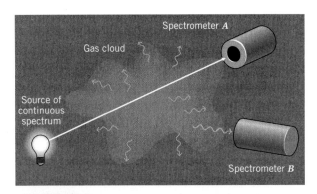

Figure 13-23 | Light with a continuous spectrum passing through a low density gas. Both absorption *(A)* and emission *(B)* spectra can be seen, depending on the location of the observer.

13.3 Applications of Spectroscopy

In this section we present examples of spectra formed under a variety of physical conditions. These examples, which include the Sun, gas clouds found near hot stars, and the pervasive cool interstellar gas, contain no new fundamental concepts but will enhance your understanding of the concepts presented in this chapter. They also serve to emphasize the overriding importance of spectra to the study of the universe. Because space contains extremes in the physical conditions of density, temperature, pressure, mass, and energy that have never been duplicated in a laboratory on Earth, these illustrations also reveal extraterrestrial space as a place that provides a unique laboratory for the study of nature.

A MODEL OF THE SUN BASED ON ITS SPECTRUM

Observations of the solar spectrum, along with Kirchhoff's laws of radiation, can be used to construct a first approximation of a model for the Sun. First, the fact that the Sun exhibits a continuous spectrum with absorption features shows that the light-emitting portion of the Sun must be either a hot solid, a hot liquid, or a hot, dense, opaque gas. But we also know that the Sun is yellow in color, corresponding to a peak emission of energy at about 5,000 Å or a bit less.

Inquiry 13-21

Adopting 5000 Å as the wavelength of peak emission for the Sun, roughly what is the temperature (in K) of the Sun's surface?

At such a high temperature (6,000 K, or almost 11,000° F), matter cannot remain as a solid or liquid because it is quickly vaporized. It is clear, then, that the Sun must be composed of a hot, dense, opaque gas. This part of the Sun is called the **photosphere** (Greek for *sphere of light*).

The Sun also has dark lines in its spectrum (see Figure 13-14) indicating that there must be a layer of relatively cool, lower density gas above the photosphere. So, in the first approximation, our model

Cooler gases above photosphere

Photosphere ≈ 5500K

Figure 13-24 A simple model of the Sun. Light from the center of the Sun's disk comes from a deeper, hotter layer and appears brighter than the light from the Sun's edge. The figure is not to scale; the thin layer of cooler gases would not be visible if drawn to scale.

of the Sun is as follows: A ball of hot, dense, opaque gas with a temperature around 6,000 K surrounded by a cooler, lower density outer atmosphere (by cooler we mean perhaps 4,500 K) that produces absorption lines in the spectrum. Such a model of the Sun is shown in **Figure 13-24.** As we will see in Chapter 14, other stars, too, exhibit dark-line spectra and surface temperatures comparable to the Sun's; as a result, similar models can be used to explain the principal observed features of the stars.

Given the basic model, astronomers can refine it by careful measurement of the wavelengths and strengths of all the dark lines in the spectrum. By comparing these measurements with laboratory measurements, astronomers can determine not only the chemical composition of the Sun but also the *abundances* of the various chemical elements in the Sun and stars. In addition, other details about the structure of the stars can be determined. These aspects of spectral analysis will be developed further in Chapter 14.

THE SUN'S FLASH SPECTRUM

Astronomers expected the temperature of the Sun to decrease with distance from its center. They therefore were surprised to find a spectacular con-

firmation of Kirchhoff's laws during a total eclipse of the Sun. As the Moon moves in front of the Sun and progressively blocks more and more of the Sun's light, the spectrum remains a dark-line spectrum on top of a continuous spectrum but of decreasing intensity. Eventually the Moon completely covers the Sun's photosphere. Now, only those atmospheric gases surrounding the photosphere of the Sun and beyond the edge of the Moon are visible to us, and in an instant the spectrum changes from a dark-line to a bright-line spectrum. This change occurs so suddenly that the new emission-line spectrum has been named the **flash spectrum** (**Figure 13-25**). (The bright-line spectrum is, of course, always there, but is washed out by the much brighter photosphere.) Because the emission lines appear pinkish for this part of the Sun, the emitting region is called the **chromosphere** (color sphere).

Inquiry 13-22

Accepting the idea that the Sun is composed primarily of hydrogen, how can you explain the pink color of the flash spectrum?

From the observed flash spectrum, and from Kirchhoff's second law, we are forced to conclude that there is a region surrounding the photosphere that is hotter than the photosphere itself.

In the middle of a total eclipse, when the Sun's disk is completely covered by the Moon, the solar **corona** appears (see Figure 4-26*b*). This normally invisible tenuous envelope of the Sun emits an emission-line spectrum from highly ionized gases, indicating a temperature of more than one million degrees.

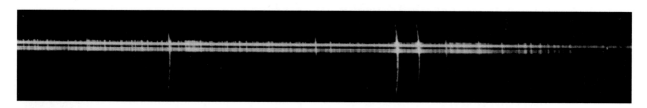

Figure 13-25 The Sun's flash spectrum. In this case we are positioned like the observer at point *B* of Figure 13-23. The lines are curved because the source of light is the thin crescent of radiation coming from the eclipsed Sun.

Inquiry 13-23

The spectra of a certain class of stars (known as Wolf-Rayet stars) possess not only characteristics of the absorption spectra found in ordinary stars but also bright-line emission features. Suggest a model of a Wolf-Rayet star that might explain the observed spectrum.

Inquiry 13-24

The great galaxy in Andromeda, M31, has an absorption-line spectrum. What does this indicate about the types of objects that make up M31?

Figure 13-26 The Lagoon Nebula in Sagittarius is a good example of an emission nebula—a region where stars, gas, and dust interact.

IONIZED GAS BETWEEN THE STARS

Our next application of spectroscopy involves colorful clouds of gas found throughout our galaxy, the Milky Way. These clouds play an important role in star formation, the subject of Chapter 17. First, we look at the spectrum of hydrogen from these clouds, then we see the spectrum from heavier elements.

THE NATURE OF NEBULAE

The term **nebula** comes from a root meaning *cloud,* and historically it referred to any fuzzy, indistinct, nonstellar object in the sky. Even as recently as the twentieth century a controversy about the nature of the *nebulae* was still very much alive. The first progress in understanding nebulae was made when the science of spectroscopy began to develop. Astronomers found that the spectra of *some* nebulae looked like the spectra of stars, indicating that these nebulae contained stars so distant that individual ones could not be recognized (just as the Milky Way is simply a glow to the naked eye rather than distinct, individual stars). Some prominent nebulae, by contrast, showed bright-line spectra due to the emission of energy only at discrete wavelengths.

Inquiry 13-25

What conclusion about the nature of nebulae could be drawn from those that showed bright-line spectra?

A good example of a nebula having a bright-line spectrum is the Lagoon Nebula, M8, in Sagittarius (**Figure 13-26**). M8, which is just visible to the naked eye and quite satisfying to view in a small telescope, was the eighth object on a list of "fuzzy" objects compiled by the eighteenth-century French comet hunter Charles Messier. An **emission nebula** (or **diffuse nebula**) such as M8 is a cloud of rarefied gas found in association with a hot star or group of stars. The gas is luminous not because it is at a high temperature but because ultraviolet radiation from the associated hot star ionizes the gas. The process can be described symbolically as follows:

Hydrogen atom + Ultraviolet photon → Proton + free Electron

To explain further, we can see from Figure 13-19 that ionization can occur only when photons are present having energies greater than that corresponding to a wavelength 912 Å. **Figure 13-27** illustrates the continuous spectrum of a hot star, whose peak emission is at short wavelengths. Only those stellar photons having wavelengths shorter than 912 Å have sufficient energy to ionize the hydrogen atom. The ionization of hydrogen atoms creates a gas cloud surrounding the star consisting of protons and electrons at a temperature of about 10,000 K. The surrounding region of ionized gas is known as an **H II region** (hydrogen that is not ionized is called H I).

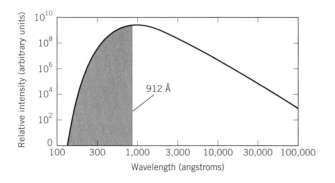

Figure 13-27 The spectrum of a star of surface temperature 30,000 K. The wavelength region containing photons energetic enough to ionize hydrogen atoms is shaded. Photons at wavelengths greater than 912 Å may excite, but not ionize, electrons in hydrogen.

Inquiry 13-26

Will a star like the Sun, whose temperature is about 6,000 K, produce an emission nebula? Explain why or why not.

The speed that particles in a gas move depends on temperature. In fact, temperature is often defined in terms of how fast the particles are moving. Even though the high temperatures cause the particles to move about at high speeds, a proton will occasionally recombine with an electron. When this occurs, the electron will cascade from high energy levels to lower and lower ones and emit photons, as described previously. Quantum mechanics shows us that an electron will almost always make the transition from the second to the first excited state, and in so doing emit a photon of wavelength 6,563 Å. Because this wavelength is in the red end of the visual spectral region, nebulae generally appear reddish. These photons resulting from recombination make the nebula visible to us and to be reddish in color.

Recombination does not simply re-create a spectrum like that of the original star. The stellar spectrum in Figure 13-27 is a continuous spectrum, whereas recombination produces primarily an emission-line spectrum in the ultraviolet, visible, and infrared parts of the spectrum from the cascade process described earlier (in Figure 13-19) for the hydrogen atom.

FORBIDDEN-LINE EMISSION

Even after astronomers understood how the ionized gas in nebulae and the hydrogen emission lines were created, questions remained, because there were many emission lines in the spectra of nebulae besides those of hydrogen, as **Figure 13-28** shows. In fact, in the early part of the twentieth century researchers had no success in identifying the elements producing many of these emission features. Because helium was first observed in the solar spectrum and only subsequently found on Earth, frustrated astronomers looked for a new chemical element—dubbed *nebulium*—that did not exist on Earth. The explanation of these mysterious emission lines was provided by the astronomer Ira S. Bowen in 1927. He showed that the emission lines were due to common chemical elements, such as oxygen, in unusual energy states called *metastable states*. Metastable levels are unlike normal energy levels in that once an atom finds itself in a metastable level it will stay there for a relatively long time before radiating. Although an ordinary transition downward typically takes place within one hundred millionth of a second after the electron arrives in the excited state, an atom may remain in a metastable level for a second or even more before radiating and dropping to a less excited state. As a result, emission lines not otherwise possible can be formed. Emission lines resulting from transitions from metastable states are called *forbidden lines*, not because they disobey laws of physics but because they had not been observed previously and were not understood originally.

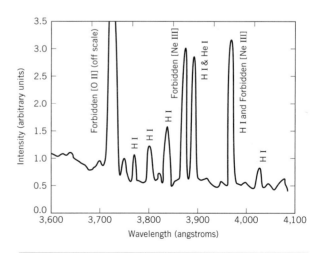

Figure 13-28 The spectrum of the Orion Nebula, showing the hydrogen lines and the forbidden-line emission.

Two forbidden lines that often occur in nebulae are at wavelengths of 4,959 and 5,007 Å. These lines are caused by transitions in oxygen that has had two electrons removed. Because these strong lines are in the green part of the electromagnetic spectrum, nebulae often have a greenish tint to them.

But even granted that these lines are unusual, why hadn't such transitions been seen in laboratories on Earth? The answer is that forbidden lines can form only when the density of gas is extremely low. When the density is high, the atom will experience a high rate of collisions with other gas particles, and it will most likely lose its excess energy in a collisional de-excitation before it has time to radiate. Even the best vacuum that can be created on Earth is four to seven orders of magnitude more dense than the gas in diffuse nebulae. The observation of forbidden lines in nebulae immediately tells us that the density of the nebular gas is low. Thus the study of nebulae provided scientists with objects under never-before-seen physical conditions.

Perhaps the most important forbidden line for astronomers is one found at a wavelength of 21 cm that is produced by cold clouds of atomic hydrogen. It has allowed astronomers to map gas clouds in our galaxy and probe the rotation of other galaxies. Because the mechanism that produces this spectral line is different from the previous discussion, we leave a discussion of the 21-cm line until Chapter 20.

Inquiry 13-27

What conclusion about nebulae results from observations of forbidden lines in the spectrum?

NEUTRAL GAS BETWEEN THE STARS: INTERSTELLAR ABSORPTION LINES

The ionized gas in the emission nebulae is the easiest component of the gas between stars to observe because the hot star excites the gas cloud to radiate. Detecting interstellar gas that is predominantly *neutral*—that is, not ionized—is more difficult. Such gas is cool and does not emit energy efficiently; in fact, it was first found by virtue of its absorbing powers.

Shortly after the year 1900, the astronomer Johannes Hartmann was studying the spectrum of the binary star delta (δ) Orionis (one of the stars in Orion's belt). He discovered that some spectral lines changed wavelength with time while others did not. (The reason why some lines change position will be discussed in Chapter 14.) He also noticed that the unchanging lines were extremely narrow in wavelength, distinctly different in appearance from the broad and constantly shifting absorption lines of the stars. He reasoned—correctly—that these absorption features were caused by a cloud of material between him and the star (**Figure 13-29**). In other words, from an analysis of spectra he accidentally discovered **interstellar absorption lines** and the fact that the space between stars is filled with gas. The narrowness of the lines also indicated to him that the cloud consisted of very cold material, with a temperature of only about 100 K (−173° C).

Interstellar absorption lines are difficult to study because there are not many of them in the visible part of the spectrum. The reason for this is worth noting, because it also explains much of the great interest astronomers have in making observations from space.

The schematic energy-level diagram of a typical atom is shown in **Figure 13-30**. Most atoms in cool interstellar space have their electrons in the ground state. For this reason, transitions from the ground level are the strongest ones and produce the most important spectral lines. Most atoms have energy levels spaced so that transitions from the ground state to an excited state involve a considerable amount of energy. Such transitions occur at wavelengths in the ultraviolet part of the spectrum, typically in the range from 500 to 2,000 Å. Unfortunately, these ultraviolet wavelengths are also the ones that are hardest to observe, because ozone in the Earth's atmosphere absorbs these

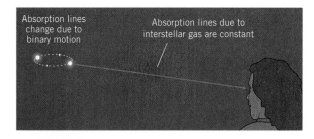

Figure 13-29 Interstellar gas clouds were first discovered by examining the spectra of binary stars.

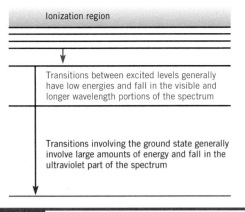

Ionization region

Transitions between excited levels generally have low energies and fall in the visible and longer wavelength portions of the spectrum

Transitions involving the ground state generally involve large amounts of energy and fall in the ultraviolet part of the spectrum

Figure 13-30 An energy-level diagram for a typical atom.

ultraviolet wavelengths. We can therefore see them only with telescopes placed above the Earth's atmosphere.

Inquiry 13-28

Some 70 percent of the mass of the observable matter in the universe is hydrogen; presumably, this applies also to interstellar gas. Consult the energy-level diagram of the hydrogen atom. Would we be likely to observe interstellar absorption lines of hydrogen gas in the visible part of the spectrum? Explain your reasoning.

The ultraviolet spectrograph on the *Copernicus* satellite, which gathered data between the years 1972 and 1980, surveyed the absorption-line transitions found in the ultraviolet region of the spectrum for a number of bright stars. Because it was able to observe the strong transitions beginning in the ground state, the satellite returned a wealth of data on the abundances of elements in interstellar gases. It found that many determinations of the elemental abundances agreed well with the values that had been discovered for most stars, but there turned out to be a set of chemical elements whose abundances were dramatically *depleted*. Elements found from both ultraviolet and optical observations to be depleted include carbon, nitrogen, calcium, potassium, nickel, aluminum, and iron. We will argue later in the book that these depleted elements have probably gone into forming dust grains in space, which play an important role in the formation of stars and planets.

Chapter Summary

Observations

- The solar spectrum contains a continuous spectrum with dark absorption lines superimposed on top of it. Just as the Sun reaches totality during a total eclipse, the spectrum becomes an emission-line spectrum. The spectrum of the solar corona is also an emission spectrum of highly ionized atoms.

- The spectrum of a gaseous nebula shows emission lines.

- Cool gas in the space between stars absorbs background starlight and produces interstellar absorption lines.

Theory

- **Kirchhoff's first law** tells us that radiation from an opaque solid, liquid, or dense gas will emit a **continuous spectrum**. His **second law** is that a low-density hot gas will emit an **emission (bright-line) spectrum**. His **third law** says that white light passing through a cooler gas will be absorbed and produce an **absorption (dark-line) spectrum**.

- The (continuous) spectrum of a **blackbody** depends on temperature but not on chemical composition.

- **Wien's law** relates a blackbody's temperature to the wavelength at which the body emits its most intense radiation: $\lambda_{\max}T = 3 \times 10^7$ Å K where $\lambda_{\max}$ is the wavelength of maximum radiation and T is the temperature in kelvins. The **Stefan-Boltzmann law** relates the amount of energy radiated by each unit area of a body to its temperature: $F \propto T^4$ where T is temperature.

- The number of **protons** in the nucleus determines what element a nucleus is. Differing numbers of neutrons produce different isotopes of an element. To achieve electrical neutrality, there are as many electrons as there are protons.

- The temperature of a gas is a measure of the average speed of particles within the gas.

- The **energy-level diagram** describes the amount of energy required to excite an electron from one discrete orbit to another. Electrons in lower energy states may absorb a photon and move to more energetic, **excited states**. Electrons in excited states move to less excited states (closer to the nucleus) by emitting a photon. The frequency of the absorbed or emitted photon is determined by the energy difference between the two orbits. An electron in the **ground state** cannot emit a photon. An atom can become **ionized** if the electron is given enough energy to be removed from the atom.

- Each atom and isotope has a unique energy-level diagram, meaning that each atom's absorption-line and emission-line spectrum is unique.

- Quantum mechanics provides scientists with a model of the atom that offers a more complete description of the workings of nature on the atomic level.

Conclusions

- From the observed continuous spectrum, we infer that the Sun's interior consists of a hot,

opaque gas. From the observed absorption lines, we infer the presence of a lower-temperature atmosphere surrounding the photosphere. From the observed flash spectrum we infer the presence of a higher-temperature chromosphere. From observations of the highly ionized corona, we infer temperatures in excess of 10^6 K.

- The chemical composition of a gas can be determined by comparing an observed emission-line or absorption-line spectrum with that of a gas in a laboratory.

- The **Balmer series** of hydrogen occurs in the visual region of the spectrum as a result of transitions involving the first excited state. The **Lyman series** involving the ground state appears in the ultraviolet, while all other hydrogen series occur at infrared or longer wavelengths.

- The emission-line spectrum of a gaseous nebula comes from ionization of hydrogen gas by high-energy photons from a hot star near the nebula. Some strong emission lines come from transitions involving metastable states, states whose lifetimes are longer than normal.

- From absorption lines observed in the spectra of some stars, astronomers infer the presence of cool gas in the regions between stars.

Summary Questions

1. What are the three principal types of spectra? Explain the conditions under which each is produced.

2. What effect does temperature have on the color and energy distribution of the radiation emitted by a hot object?

3. How can you describe the formation of the absorption spectrum of the Sun in terms of a simple solar model?

4. How do the interactions between photons and atoms produce emission and absorption spectra?

5. Explain what is represented in an energy-level diagram. Draw the energy-level diagram for a

hydrogen atom, and indicate the principal transitions.

6. Why do different chemical elements have different characteristic spectral lines?

7. What are Kirchhoff's laws for absorption and emission spectra? How does atomic theory explain the Sun's flash spectrum?

8. How can you use the spectra of "nebulae" to show that some are composed of stars and others are composed of hot gases?

9. How is the spectrum of an emission nebula formed?

Applying Your Knowledge

1. Consider the following experiment. A clear glass tube filled with sodium gas has light from an electric sodium lamp focused into it. (A sodium lamp is yellow in color.) The inside of the tube is observed to contain a cone of yellow light in the region in which the sodium lamp is focused. However, when the electric sodium lamp is replaced with an electric lamp filled with mercury gas, no cone of light is observed from inside the tube. Explain the experiment.

2. List all possible *downward* transitions an electron in the sixth energy level in the hydrogen atom can make. List their wavelengths when known.

3. Why does the Lyman series occur at shorter wavelengths than the Balmer series?

4. Given that most of the hydrogen atoms in the Sun's photosphere are in their lowest energy state, what spectral series of hydrogen would be produced most strongly in the Sun?

5. Suppose you observed the spectrum of some astronomical object and found it to have both absorption and emission lines. What might be a reasonable explanation for the nature of the object?

6. If you are star-gazing on a clear night and notice two stars, one of which appears bluish and the other reddish, what could you conclude about the relative characteristics of the two stars?

▶ 7. What is the wavelength of the peak emission of energy for a body of temperature 3 K? (As you will see later in the book, this temperature is the current temperature of radiation still around from shortly after the universe began!)

▶ 8. What are the frequency and the energy of a photon having a wavelength of 21 cm?

▶ 9. Relative to the Sun, how much more energy per unit area does a star with a temperature of 22,000 K radiate?

▶ 10. Relative to the Sun, how much more energy per second comes from the entire surface of a star that has a radius one-third that of the Sun and a temperature of 22,000 K? (*Hint:* Remember that the energy per second *per unit area* depends on T^4.)

▶ 11. Consider the following logarithmic scale:

$$3 \quad 15 \quad ? \quad 375$$

What number should be in the place of the question mark? Explain your reasoning.

▶ 12. Compute the temperature of a star that appears to be red. Do the same for a star that appears to be blue.

Answers to Inquiries

13-1. Yes.

13-2. In this example each succeeding tick mark is four times greater than the previous one. Thus, the next number is 64.

13-3. 3×10^7 Å K / 2,500 Å = 12,000 K

13-4. 3×10^7 Å K / 300 K = 10^5 Å

13-5. (50,000 Å K / 5,000 K)^4 = 10^4

13-6. 3×10^7 Å K / 30,000 Å = 10^3 K

13-7. 3×10^7 Å K / 0.5 Å = 6×10^7 K

13-8. There is water vapor in the Earth's atmosphere through which the stars are observed.

13-9. The Moon's visible spectrum should be similar to the Sun's, because the Moon shines by reflected sunlight. We would also see CO_2 in Venus's spectrum.

13-10. $^{15}_{8}O, ^{17}_{8}O, ^{18}_{8}O$. They are all oxygen because they all have eight protons.

13-11. If the relationship is rewritten, it becomes

Energy of higher energy state = Energy of photon + Energy of lower energy state.

Therefore, the total energy after the transition is the same as the total energy before the transition.

13-12. The principle of conservation of energy.

13-13. Emission is represented by a downward-pointing arrow; absorption by an upward-pointing one.

13-14. Lyman: ultraviolet; Balmer: optical; Paschen, Brackett, Pfund: infrared. Only the Balmer lines can be observed from the Earth's surface; all the others require instruments above the Earth's atmosphere.

13-15. (a) Balmer emission; (b) Lyman absorption; (c) Paschen emission; (d) Lyman emission; (e) Pfund absorption

13-16. (a) Photon may be absorbed and the atom will be in the third energy state. (b) Photon cannot be absorbed to produce an absorption line. (c) Photon may be absorbed and the atom will be in the fourth energy state. (d) Photon will be emitted, not absorbed, and the atom will be in the second energy state.

13-17. From Figure 13-19, the only downward (emission) transitions are at 6,563 Å and 1,026 Å.

13-18. From Figure 13-19, we find: 973 Å, 4,861 Å followed by 1,026 Å, 18,751 Å followed by either 1,026 Å or both 6,563 Å and 1,216 Å.

13-19. Neutral hydrogen, because both have a small nucleus surrounded by a region containing one electron. But because the nuclear charge of helium is greater, all the energy levels will be higher, and the wavelengths will be shorter than the corresponding hydrogen wavelengths.

13-20. An emission spectrum. The observer would see the light being emitted in random directions by the atoms after being absorbed by the beam. There is no continuous spectrum because there is no higher temperature source behind the cloud.

13-21. 3×10^7 Å K / 5,000 Å = 6,000 K

13-22. High temperature will ionize hydrogen. Recombination of protons with electrons will produce transitions between levels 3 and 2, giving a strong emission line at 6,563 Å, in the red.

13-23. This could be produced by a star (which gives the absorption spectrum) surrounded by a shell or ring of gas that is hotter than the outer surface of the star.

13-24. Because the stars have absorption spectra, we are presumably seeing the composite effect of a large number of stellar spectra superimposed when we observe the galaxy.

13-25. They must consist of rarefied gases (by Kirchhoff's laws) at some temperature.

13-26. A Sunlike star is too cool to emit a significant number of UV photons. Therefore, such a star will not produce an emission nebula.

13-27. The density of the gas is very low.

13-28. We would not be likely to observe interstellar absorption lines of hydrogen gas in the visible part of the spectrum because most of the hydrogen would be in the ground state. For this reason, the only transitions that occur involve photons with wavelengths in the ultraviolet part of the spectrum.

PART FOUR

Discovering the Nature and Evolution of Stars

When we view galaxies we first see light from stars, which appear to be the fundamental building blocks of galaxies. However, looking further, we find that stars are formed from clouds containing atoms, molecules, and dust. Some people, therefore, might describe these as the fundamental components of galaxies. The purpose of Part 4 is to describe these components and to obtain a basic understanding of them—their characteristics, formation, and evolution.

Most of what we know about the universe comes from the study of spectra; thus, we open Part 4 with a study of stellar spectra. In Chapter 14, we study what astronomers can learn about stars from an analysis of their spectra.

In Chapter 15 we look at a large population of stars to see what general conclusions we can form about them. It is here we find that stars range in size from that of the Earth up to the diameter of Saturn's orbit, with masses ranging from a small fraction of the Sun's mass to 100–150 times that of the Sun. We also begin to determine the distances to stars. We also look at binary stars, because more than half of the stars in our solar neighborhood are in such systems.

Chapter 16 contains our study of the Sun, along with the beginning of our discussion of stellar evolution. We look at possible sources of energy in the Sun and examine the conditions required for it to generate the energy we observe. We examine the conditions of equilibrium that must occur within a star for it to survive for long periods of time. These conditions, along with the sources of energy, determine how long a star will live. The final section of the chapter examines the Sun as a typical star. We learn about its various layers, and the range of processes that occur both within it and on its surface. Finally, we see how astronomers are studying the interior of the Sun by observing oscillations on the surface.

The evolution of stars is the subject of Chapters 17, 18, and 19. In Chapter 17 we learn about the formation of stars—where and how they form. In Chapter 18, we look at middle age and the approach to old age for stars of all masses. The death of low-mass stars is discussed here. Finally, in Chapter 19, comes our discussion of the deaths of massive stars, those that explode in the fury of a supernova explosion, leaving behind either a neutron star, a black hole, or just a new cloud of gas and dust. All along the way, we pay particular attention to observational evidence for or against our theoretical picture of the life cycle of stars, especially observations of star clusters, as well as clues given by detailed examination of stellar spectra.

Understanding Stellar Spectra

14

OH BE A FINE GIRL, KISS ME!
(THE ORIGINAL MNEMONIC)
OH Be A Fine Girl/Guy, Kiss Me Right Now, Sweetheart!
Oh Boy, An F Grade Kills Me!
Oven Baked Ants, Fried Gently, Kept Moist,
Retain Natural Succulence
Our Best Aid For Gnat Killing—My Reindeer's Nose. Snort.
Only Boys Accepting Feminism Get Kissed Meaningfully

Mnemonics for spectral types of stars

When faced with large quantities of seemingly random information, scientists respond by first classifying the data. For example, biologists classify animals into kingdom, phylum, subphylum, class, order, family, genus, species, and subspecies; geologists classify minerals in terms of form, cleavage, and hardness. Classifications are like boxes into which similar objects are placed. Such classifications provide scientists with opportunities to see similarities and differences among objects. In astronomy, we classify types of galaxies (see Chapter 21) as well as types of stars. As we shall see, stars can be classified on the basis of their spectrum (this chapter), their variation in light output (Chapter 18), and their membership in clusters of stars (Chapters 15 and 18). Because so much of what we know about the universe comes from an analysis of spectra, we now study stellar spectra in some detail.

14.1 Classification of Stellar Spectra

In the late nineteenth century, photographic emulsion became sensitive enough for astronomers to obtain stellar spectra, and they started to obtain spectra for a large number of stars. Faced with all these spectra, astronomers set out to classify them. Of course, no one understood what caused the spectra to appear the way they did, but at this initial stage of their study, understanding was not important. The purpose of classification is to place all objects having similar characteristics together in an attempt to make meaningful order out of apparent chaos. Then, hopefully, the classifications will suggest possible underlying reasons for why the spectra are different.

The classification scheme astronomers set up for spectra was based on the appearance of the hydrogen Balmer lines that showed up on photographs. The types of spectra were simply designated by letters of the alphabet. Thus, there were stars of **spectral type** A, B, C, and so on. But around 1900, when our story begins, the American astronomer Annie Jump Cannon of Harvard College Observatory (**Figure 14-1**) began a project to classify the spectra of nearly 400,000 stars on the basis of their spectral appearance and the relative strengths of the dark-line features in them. Her classifications were published between the years 1918 and 1924 as the famous *Henry Draper Catalog,*[1] which is still in use by astronomers.

Figure 14-1   Annie Jump Cannon (1863–1941), classifier of stellar spectra.

At the time Cannon did her work, astronomers did not know why the spectra of stars varied so greatly in appearance—that understanding had to wait until the full development of quantum physics in the 1920s. However, the classification scheme Cannon established led the way toward a proper understanding of stellar spectra, and it is still used today in modified form (and is currently being updated by Nancy Houk at the University of Michigan).

MODERN CLASSIFICATION

The classification scheme now used by astronomers is shown in **Figure 14-2**. To the side of each spectrum is the name of the star; each also has a letter designation for its **spectral class**. Each major class has been divided into 10 subclasses using the numbers 0 through 9. Thus, within the class of stars of type B, astronomers distinguish subtypes B0, B1, . . . B9. Within each major class, the spectra are similar; yet the various

---

Chapter opening photo: The spectrum of the Sun.

[1] The catalog is named for the person whose estate funded its publication. Henry Draper was the first astronomer to photograph the spectrum of a star.

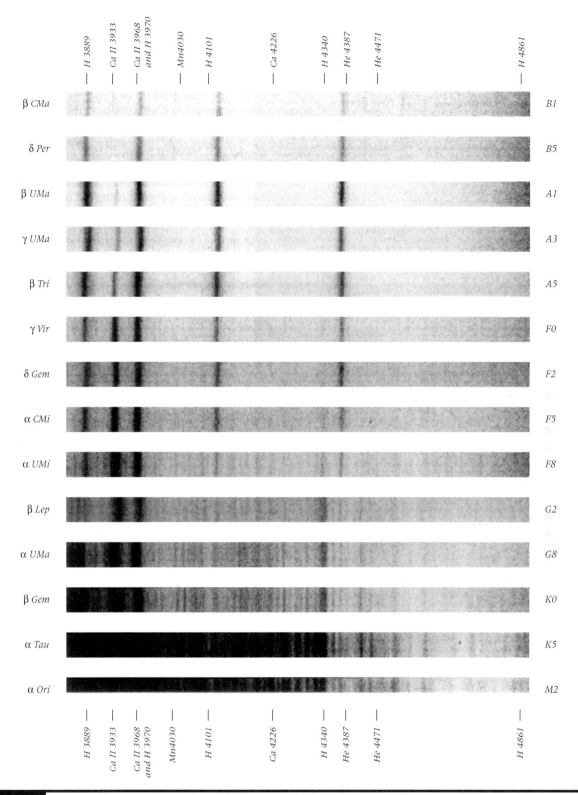

The following spectral lines are marked at the top and bottom:

H 3889 · Ca II 3933 · Ca II 3968 and H 3970 · Mn 4030 · H 4101 · Ca 4226 · H 4340 · He 4387 · He 4471 · H 4861

Star	Type
β CMa	B1
δ Per	B5
β UMa	A1
γ UMa	A3
β Tri	A5
γ Vir	F0
δ Gem	F2
α CMi	F5
α UMi	F8
β Lep	G2
α UMa	G8
β Gem	K0
α Tau	K5
α Ori	M2

Figure 14-2 The sequence of stellar spectra. Some important spectral lines are marked at the top and bottom.

classes merge smoothly into each other, so that a B9 star and an A0 star are more similar to each other than a B9 star and a B0 star, for example. At the top and bottom of Figure 14-2, the major spectral lines are identified. If you follow a particular spectral line down the figure, you will notice that it varies in strength as the spectral type varies. For example, examine the line of ionized calcium, denoted as Ca II, at 3,933 Å.[2] You will notice that as you go through the A stars toward the F stars, the strength of this line continually increases.

We should clarify our notation of atoms and ions. Using iron as an example, we refer to neutral iron as Fe I; that is, the Roman numeral I is attached to the element name. Singly ionized iron, which has lost one electron, is designated as Fe II, doubly ionized iron as Fe III and so forth. The spectrum of an iron atom with 15 electrons removed (as actually seen in the solar corona) would be designated as Fe XVI. This convention will be used throughout the text where appropriate.

In the modern classification scheme, the letter types do not follow in alphabetical order. In the original Harvard classification of Annie Cannon, the most important criterion was the strength of the Balmer lines of hydrogen. Notice that they are strongest in type A, but they are also strong in types B and F and faintly visible in most of the other types. It also turned out that there were some duplicate categories that had been given different letters. When this was understood, certain letters disappeared from the list (e.g., C, D, E, H, I, J, and L).

More important, when the colors of the stars in various classes were examined, astronomers found that stars of a given spectral type had similar colors. For example, O and B stars were very blue, whereas M stars were quite red. We have already seen that the color of a hot, glowing object, such as a star, is an indication of its temperature (Wien's law in Chapter 13). When these variations in color were studied, astronomers came to realize that each of the different spectral types belonged to stars of different temperature; type O stars had temperatures greater than about 25,000 K, type M stars had temperatures below 3,500 K, and the other types were in between (**Figure 14-3**). Temperature, then, must be the underlying reason for the variation of stellar spectra. Variations in the surface temperature of a star cause some spectral lines to grow stronger and others to become weaker. Later in this chapter we will see how this can happen.

Once these temperature variations were appreciated, the spectral classification sequence was rearranged in order of *decreasing* temperature, and the resulting sequence came out as OBAFGKM, omitting the spurious and duplicate classes. Three rarer categories (R, N, and S) include somewhat unusual spectra and are frequently named at the end of the list (although they do not follow the temperature sequence). Generations of astronomy students have remembered the correct order of the spectral sequence using the mnemonic phrase first suggested in less enlightened times by Henry Norris Russell, "Oh Be A Fine Girl, Kiss Me," to which some add the phrase "Right Now, Smack" to account for the special classes R, N, and S. You can substitute "guy" for "girl" if you wish, but over the years various astronomers and students have had fun constructing their own pet mnemonics for the spectral sequence. A few that we have heard are given as the chapter opening quotation.

The variation in the observed strengths of the spectral lines for some elements is summarized in **Figure 14-4**. We can readily see from this figure what spectral features dominate at each spectral type.

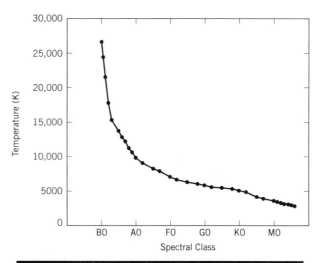

Figure 14-3 The relationship between spectral type and surface temperature.

---

[2]When discussing stellar spectra in the visible region, we will use angstroms (Å) because of historical and user preference, although some scientists prefer nanometers, where 3,933 Å = 393.3 nm.

Although the current spectral classification system comes from the early 1900s, it is far from stagnant. In recent years, a class L has been added to include the recently discovered brown dwarf stars (discussed in Chapter 17) and a class T that designates very cool stars called *methane dwarfs* because of absorption lines in the spectrum caused by the methane molecule.

The basic characteristics of stars having various observed spectral types are summarized in **Table 14-1**. Notice that the surface temperatures of these "normal" stars vary from roughly 50,000 K down to as little as 2,000 K, and that the spectral features vary in a continuous manner. Also included in the table is the approximate percentage of normal stars in each spectral class. You see that the vast majority of stars are cool. Take the time to compare the descriptions in Table 14-1 with the spectra themselves shown in Figure 14-2. Remember that the classification scheme is based only on the *observed* appearance of the spectrum, especially the presence and relative strength of the spectral lines. Only after the spectra have been classified on the basis of their appearance do we attempt to understand the physical meaning of the classes—which we do in the next subsection.

What does an astronomer mean when talking about the temperature of a star? That question is not easy to answer because a star does not have a well-defined physical surface. Rather, temperature varies throughout the star. For example, as will be discussed in later chapters, the temperature at the center of a star is usually very high—meaning tens of millions of degrees. That temperature decreases significantly toward the outer regions.

In looking at a star, the line of sight first passes through very low density, cool gases. Such gases are transparent and are generally referred to as the **stellar atmosphere**. A highly transparent gas allows one to see to depths of thousands or hundreds of kilometers. In viewing deeper into the atmosphere, both density and temperature rise and

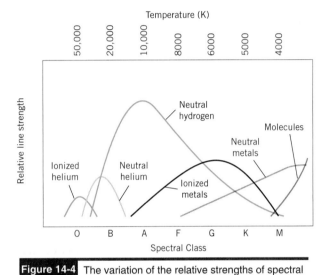

Figure 14-4 The variation of the relative strengths of spectral lines from different elements with spectral type.

Table 14-1 Characteristics of Spectral Classes for Normal Stars[3]

Type	Temperature	Color	Prominent Spectral Features	Percentage	Examples
O	25,000–50,000 K	Blue	Ionized helium; multiply ionized metals; Balmer lines of hydrogen weak	0	Mintaka
B	11,000–25,000 K	Blue	Neutral helium; ionized metals; Balmer lines moderately strong	0.1	Rigel Regulus Spica
A	7,500–11,000 K	Blue-white	Balmer lines very strong; other features very weak or absent	1	Vega, Sirius, Altair, Deneb
F	6,000–7,500 K	White	Balmer lines moderately strong; singly ionized metals		Polaris Procyon A
G	5,000–6,000 K	Yellow-white	Balmer lines weaker; neutral metals strong; easily ionized metals strong	4	Sun Capella
K	3,500–5,000 K	Red-orange	Neutral metals strongest; singly ionized metals	14	Arcturus Aldebaran
M	2,000–3,500 K	Red	Balmer lines weak; many lines of neutral metals; molecules prominent	72	Betelgeuse Antares

[3]The normal stars in this table account for about 91 percent of stars. The other 9 percent of stars are abnormal types like spectral types R, N or S.

the depth to which one can see becomes shallower. Eventually, one reaches a region where the gases become opaque, meaning you can see to depths of only centimeters. Where the gases become opaque can be defined as the *surface* of a star. More correctly this region is called the **photosphere**, but astronomers refer to this as the surface. The temperature at the photosphere is the temperature that determines the color of a star; this is basically the temperature referred to in Wien's law.

Absorption lines are formed throughout the cool, low-density regions of the stellar atmosphere. Thus, spectral line formation is complex because it occurs in stratified layers of the star's atmosphere. For simplicity, however, we will simply refer to the *surface* and the *surface temperature*.

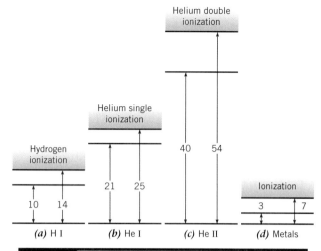

Figure 14-5 Energy-level diagrams for hydrogen, neutral and ionized helium, and a typical metal. The amount of energy needed to excite to the first excited state, and the amount needed for ionization, are given.

Inquiry 14-1

The spectra at the top of Figure 14-2 are brighter at short wavelengths, while those at the bottom are brighter at long wavelengths. Why is there this difference in appearance?

WHY SHOULD SPECTRA VARY IN APPEARANCE?

Most stars show a continuous spectrum with absorption lines superimposed. This means that there is a hot, opaque region (in the Sun, its photosphere) that emits a continuous spectrum, and above it a cooler, low-density, and more transparent layer where the absorption lines are formed. This cooler layer is the star's atmosphere, and it is the part of the star on which we now focus our attention. (Many stars have additional very-low-density regions farther out with yet higher temperatures, but for now we will neglect these stellar chromospheres and coronae.)

Scientists were able to understand the different appearances of stellar spectra once they understood two ideas. First, as discussed further in Section 14.2, most stars are chemically similar to the Sun, so that for most stars differences in chemical composition are unimportant. Stars consist of some 71 percent of their mass in hydrogen, 27 percent helium, and only 2 percent everything else, which astronomers will refer to as *metals*.[4] (Although 71 percent of the *mass* of a star is hydrogen, 90 percent of its *atoms* are hydrogen.)

The second important idea is the understanding of the structure of the atom and the energy-level diagrams that we discussed in Chapter 13. Because of stars' similar chemical compositions of hydrogen, helium, and metals, we can understand spectra by examining simplified energy-level diagrams for these three types of atoms.

Figure 14-5 shows the ground state, first excited state, and ionization level for hydrogen, neutral and ionized helium, and a typical metal. (We are simplifying by ignoring all the other energy levels.) In part *a* we see that it takes 10 units of energy to excite hydrogen to its first excited state. If 14 units of energy are available, the electron is given enough energy to leave the atom, and the result is a hydrogen ion. The energy to excite and ionize comes primarily from the photons producing the continuous spectrum that are then available for being absorbed by the overlying photosphere. Collisions between particles can also play a role. The greater the temperature, the more energy the average photon has and the greater the collision speed between the particles in the gas. Thus high levels of excitation, including ionization, occur only when the temperature is sufficiently high. In the case of hydrogen, the temperature must be about 6,000 K to reach the first excited state, and 12,000 K to ionize.

Figure 14-5*b* shows a similar diagram for neutral helium. Because helium has two protons in its nucleus, the allowed orbits are different from those of hydrogen. Helium's greater nuclear charge also means that more energy is required to excite an electron. In fact, it takes more energy to *excite*

---

[4]Only astronomers use the term *metals* in this way. When chemists talk about metals, they mean something else.

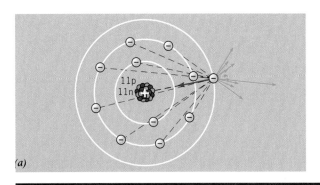
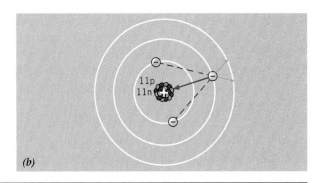

(a)   (b)

Figure 14-6 *(a)* The atomic structure of sodium (a typical metal), showing that the outermost electron feels a smaller attractive force from the positively charged nucleus than do the electrons closer to the nucleus. Furthermore, the distant electron feels a repulsive force from all the inner electrons. *(b)* When ionized, the outer electron is closer to the nucleus, and there is a smaller repulsive force from the few remaining electrons. (Drawing is not to scale.)

helium (21 units of energy) than to *ionize* hydrogen. Helium ionization requires 25 units of energy. Such ionization releases one of helium's two electrons, thus producing a helium ion.

The one electron remaining in the helium ion can also undergo transitions. However, due to the loss of the first electron, the locations of the allowed orbits are different from that of the neutral atom. Therefore, as indicated in Figure 14-5c, the transitions that occur in ionized helium require different energies than was the case for neutral helium. From the figure we see that ionized helium becomes excited if there are at least 40 units of energy, and that the second electron will be dislodged only if 54 units of energy are available.

Lastly, we have the energy level diagram of a typical metal *(d)*. Although metals have many electrons, only the outermost one(s) undergo transitions and produce spectral lines that are important to us at this time. As indicated in **Figure 14-6a** for sodium, the outermost electron is at a large enough distance from the nucleus that it feels only slightly attracted to the positive nuclear charge. In fact, the many electrons closer to the nucleus repel the outer electron, which makes it even less strongly held. For these reasons, Figure 14-5d indicates that a small amount (only some three units) is required to excite a typical metal. A total of seven units of energy is sufficient to remove one electron from many elements.

With this understanding of the structure of the important atoms, we can attempt to predict what stellar spectra should look like for stars of different temperatures. We will also be able to understand why Figure 14-4 looks the way it does. We will confine ourselves to the visual region of the spectrum, because historically astronomers were limited to that region due to Earth's atmosphere.

Low-Temperature Stars

Low-temperature stars (spectral types K and M) have temperatures less than about 5,000 K. Low temperatures are sufficient to excite electrons in most elements other than hydrogen and helium. Excitation readily occurs because the outer electron, which is the most important one in producing spectral lines, is only weakly held to the nucleus, for reasons already discussed. Therefore, electron transitions occur easily. The large variety of atoms present in a star, combined with the ease with which transitions can occur, produces a huge number of possible absorption lines. Therefore, cool stars are expected to have a plethora of lines, and to have highly complex spectra.

In cool stars, hydrogen and helium exist almost entirely in their ground states.

Inquiry 14-2

Do you expect the Balmer lines to be strong, weak, or absent from the spectrum of a low-temperature star? Explain why.

Inquiry 14-3

Do you expect absorption lines of helium to be strong, weak, or absent from the spectrum of a cool star? Explain why.

Medium-Temperature Stars

Medium-temperature stars (spectral types A, F, and G) are those having temperatures between 5,000 and about 12,000 K. Metals in a medium-temperature star are readily excited and sometimes ionized. The removal of electrons has two important effects on the spectrum. First, with distant electrons removed, the new outermost electrons are now closer to the nucleus, where the attractive electrical force is greater (**Figure 14-6b**). Second, there are fewer electrons repelling the now-outer electron. The result is that the now-outer electron is held somewhat more tightly than before, and more energy is required to excite it. The greater energy means spectral lines from ions will generally occur at a higher frequency (because, as we saw in Chapter 11 and 13, $E = hf$), or shorter wavelength, than for a neutral atom. Therefore, spectral lines for ions are expected to be primarily in the ultraviolet spectral region. The result is that medium-temperature stars are not expected to have as many spectral lines in the visible region of the spectrum as lower-temperature stars.

In Chapter 13 we learned that the speed of a particle in a gas depends on the temperature. The greater the particle's speed, the greater its kinetic energy. This means that in the medium-temperature range, a large fraction of the hydrogen atoms present are in the second energy level (the first excited state). Therefore, the spectral lines belong to the Balmer series of hydrogen (Chapter 13), which will be strong in the spectrum of such a star.

Inquiry 14-4

Do you expect absorption lines of helium to be strong, weak, or absent from the spectrum of a medium-temperature star? Explain why.

High-Temperature Stars

In a hot star (one of spectral type O and the hottest of the B stars, with temperatures greater than about 20,000 K), the high temperature causes ionization of the metals. The situation is similar to but more extreme than that for medium-temperature stars, because more than one electron can be removed from each metal atom. For this reason,

few if any spectral lines from metals are expected to be present in the visible part of the spectrum because they will mostly be in the ultraviolet.

Inquiry 14-5

In a hot star, hydrogen will be mostly ionized. What would be the appearance of the *absorption-line spectrum* from hydrogen in such a star? (*Hint:* There is a slight trick to the question!)

A large fraction of the hydrogen atoms will be ionized in a hot star. Remember, spectral lines are formed by electron transitions. Because there are few hydrogen atoms still having electrons to make transitions, spectral lines from hydrogen atoms will be weak.

Helium in a hot star can be excited to higher energy levels. Therefore, absorption lines from neutral helium will be present. In the hottest stars, helium will be ionized. The one remaining electron will be able to undergo transitions and produce spectral lines at wavelengths characteristic of ionized helium.

A summary of the discussion of what is expected in spectra is in **Table 14-2**.

Molecules in Cool Stars

The spectra of cool stars may be expected to have one additional attribute. In such stars, the low speeds of collisions allow two or more atoms to stay combined as a molecule. Most of the absorption observed in the visible part of a cool star's spectrum comes from molecules of titanium oxide (TiO) (**Figure 14-7**), but in certain stars there is also noticeable absorption from molecules of C_2 (two carbon atoms bound together) and CN (cyanogen). The TiO, C_2, and CN molecules are exceptionally effective at absorbing visible light. Although a typical atom has only a relatively small

Table 14-2 Expected Spectral Lines from Metals, Helium, and Hydrogen in Stars of Various Temperatures

Temperature	Metals	Helium	Hydrogen
Cool (T < 3,500 K)	Metal lines dominate; TiO molecule	Ground state none	Ground state weak to none
Medium	Some lines present	None	First excited state strong
Hot (T > 20,000 K)	Ionized—none in visual	Yes, He I; He II in hottest	Ionized—yes, but weak

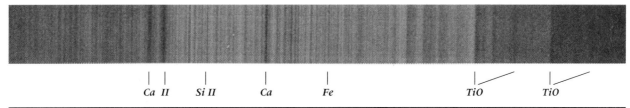

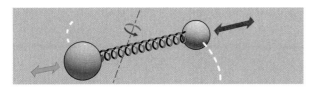

| Ca II | Si II | Ca | Fe | TiO | TiO |

Figure 14-7 The spectrum of the M-type star α Hercules, with absorption bands from molecules indicated.

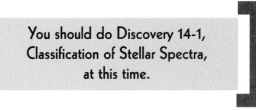

Figure 14-8 Atoms in molecules vibrate along the line of force holding them together. The molecule also rotates.

number of absorption lines in the visible part of the spectrum, molecules are able to absorb at so many wavelengths that in certain parts of the spectrum they seem to block nearly all the light.

The reason molecules are so effective at absorbing light is not hard to understand. A simple molecule made up of two atoms can be modeled as two balls held together with a spring (see **Figure 14-8**) in which the spring represents the electrical forces holding the atoms together. This structure, resembling a flexible dumbbell, can vibrate along the line between the atoms in the molecule; the atoms can vibrate back and forth, toward and away from each other. The entire dumbbell can also rotate. Just as the electrons in the atom are allowed to occupy only discrete energy states, not all molecular vibrations and rotations are allowed. Those vibrations and rotations that are allowed exhibit slightly different amounts of energy and give rise to separate energy levels. Because each vibrational and rotational energy level is close to the next one, when transitions do occur, the lines are close together and blend to form a band of absorption rather than a line. (The individual lines can be seen if a high-resolution spectrograph is used.)

The forces that hold a molecule together are weak compared with those that bind an electron to the nucleus of an atom. At the high temperatures found in stellar atmospheres, most molecules are decomposed into their component atoms, a process called **dissociation**. Dissociation can be caused either by the absorption of a photon or by the colli-

sion of a molecule with a high-speed atom, electron, or other molecule. The titanium oxide molecule is one of the hardiest, and enough of it survives at the temperatures found in the atmospheres of M stars for it to be prominent in their spectra.

> ## Inquiry 14-6
>
> On the basis of the explanations in the previous paragraphs, if you found a star whose spectrum contained only the absorption features of titanium oxide, would you be justified in concluding that the star was composed entirely of titanium oxide? Explain.

OBSERVATIONS OF STELLAR SPECTRA

Now that we know what to expect on the basis of atomic theory, we will examine what astronomers actually observe.

> **You should do Discovery 14-1, Classification of Stellar Spectra, at this time.**

The M stars, shown at the bottom of Figure 14-2 and in Figure 14-7, have strong TiO bands as well as numerous lines from metals. On the basis of our previous discussion, we conclude that they are low-temperature stars.

Moving upward in the sequence, we now consider the spectra of K and G stars. As Figure 14-2 illustrates, most of the spectral features in a K-type star are the absorption lines of neutral metals (such as iron and calcium), whereas only weak molecular absorption bands (chiefly CH) are seen. Evidently, the atmospheres of K stars are hot enough that most of the molecules have been decomposed into their constituent elements.

Inquiry 14-7

Raising the temperature of a gas causes the average velocity of the molecules and atoms in the gas to increase. How does this explain the smaller number of molecules in the atmosphere of a K-type star as compared to an M-type star?

The G stars have temperatures in the range of 5,000 K to 6,000 K. The Sun, for example, is a typical G2 star. In G stars, absorption lines due to ionized elements begin to appear. As we saw before, the spectrum of an ionized atom is expected to be different from that of a neutral atom, because the energy levels are completely rearranged by the loss of an electron. Therefore, the different pattern of lines makes it possible to distinguish between lines due to ionized atoms and lines due to neutral atoms of the same element.

For example, the strongest single feature in the entire solar spectrum is the absorption line of ionized calcium (Ca II) at 3,933 Å. This is easy to understand, because not only is the outer electron in the atom relatively far from the nucleus, but also the attractive force of the nucleus on this electron is nearly neutralized by the repulsive force of the inner electrons. This means that the outer electron is only weakly bound to the atom and can easily be dislodged from it either by absorbing a photon of sufficient energy or by colliding with another atom.

Inquiry 14-8

Raising the temperature of a star's surface increases the average energy of the photons it emits. How does this affect the ionization of atoms in the star's atmosphere?

With F-type stars, few absorption features due to neutral elements are seen. From the earlier discussion, these stars must have higher temperatures than G-type stars. At the 7,000 K temperatures of F-type stars, most elements have lost at least one electron and many lines of multiply ionized elements are seen. For example, the absorption features of Ca have faded and absorption by Ca II is seen instead.

The main observed spectral characteristics of A-type stars are the dominating Balmer lines. These strong lines arise because the temperature is high enough that the electrons in the hydrogen atoms are mostly in the first excited state as previously discussed—the state from which the Balmer lines of hydrogen form. The temperatures of these stars are, therefore, about 10,000 K.

B-type stars show lines of neutral helium and a decreased strength of the hydrogen Balmer lines. According to our arguments based on atomic structure, the temperatures must be greater yet, in the range of 11,000 to 25,000 K.

The hottest stars—the O stars—show a high degree of ionization. Absorption lines due to ionized helium (He II) are observed in the spectra of O stars, indicating that a significant amount of ionized helium does exist in their atmospheres, and that the temperatures must be high, in excess of 25,000 K.

Inquiry 14-9

Absorption features of neutral silicon (Si I) are seen in G stars. In F stars, these lines are weak, but lines of ionized silicon (Si II) are strong. In B stars, the only strong silicon lines are due to silicon atoms that have lost two electrons and are said to be doubly ionized silicon (Si III). Why is this so?

Inquiry 14-10

What are two reasons why it is more difficult to remove a second electron from an ionized atom than it was to ionize the atom in the first place?

Inquiry 14-11

If you photograph the spectrum of the Sun from Earth, you will find certain strong absorption bands due to molecular oxygen. How can this observation be explained?

Inquiry 14-12

Ionized helium has a Lyman and a Balmer spectral series like hydrogen, except that all the wavelengths are shorter. Which of these two series, the helium Lyman or Balmer series, would probably appear first in the ultraviolet spectrum of stars as the temperature is raised?

14.2 The Cosmic Abundance of the Chemical Elements

The abundance of a chemical element in a star is not necessarily determined from its presence or absence in the star's spectrum. To understand what an observed spectrum is telling us, we must first know the star's surface temperature. For example, we have seen that the strengths of the hydrogen Balmer lines change with temperature, as do the line strengths for ionized calcium. Furthermore, the *relative* strength of the hydrogen line compared to the calcium line depends on the temperature. Once the temperature is known, it is possible (with much work) to interpret the spectrum and deduce the proportions of each of the chemical species appearing in the spectrum. This is not an easy task, because some substances (such as ionized calcium) exhibit very strong lines even though there may be only a small amount of the substance in the star. Hydrogen spectral lines are strong only because 90 percent of the number of atoms in stars (71 percent of the mass) is hydrogen.

Another complication has made it difficult to deduce precise chemical abundances. Some substances have no strong electron transitions (and therefore no strong lines) in the visible part of the spectrum. For example, neutral carbon has almost all its significant transitions in the ultraviolet, at wavelengths shorter than 3,000 Å. Earth's atmosphere, which absorbs ultraviolet light, has prevented us from observing these lines from Earth's surface. In fact, many important chemical elements have their strongest transitions, which involve the ground state, in the ultraviolet. Astronomers who wish to study these elements at visible wavelengths are forced to observe transitions between two excited states, even though only a tiny fraction of the atoms are in those states. The resulting spectral lines are weak and difficult to observe.

Some other substances have all their lines in the infrared, microwave, and radio regions of the spectrum. For example, molecular hydrogen (H_2) and many other molecular species fall into this category. One benefit of the new astronomical observing techniques, particularly observing from above Earth's atmosphere, has been to open these previously inaccessible regions of the spectrum for study.

Inquiry 14-13

The visible-light spectrum of Saturn is dominated by absorption bands of ammonia (NH_3) and methane (CH_4), yet we saw in Chapter 10 that the low density of Saturn can be explained only if the planet's atmosphere is composed of at least 90 percent hydrogen. How can so much hydrogen be present and not be seen in the visual spectrum?

Starting in the 1930s, when the structure of atoms had become reasonably well understood, astronomers finally began to draw some broad inferences about the abundances of various chemical elements in stars. A surprising finding was made by Cecilia Payne-Gaposchkin, the first woman ever to receive a doctorate from Harvard, and one of the more prolific astronomers of the century (**Figure 14-9**). She found that most of the stars in the Galaxy had similar chemical compositions—in each star, 90 percent of the atoms are hydrogen,[5] nearly 9 percent are helium, and everything else in the periodic table accounts for less than 1 percent of the atoms in a star! In other words, only a few atoms in a thousand are the

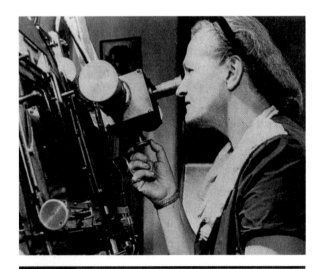

Figure 14-9 Cecilia Payne-Gaposchkin (1900–1979), who showed that stars are primarily hydrogen.

[5]Although 90 percent of the atoms are hydrogen and 9 percent are helium, about 71 percent of the *mass* is hydrogen and 27 percent of the *mass* is helium.

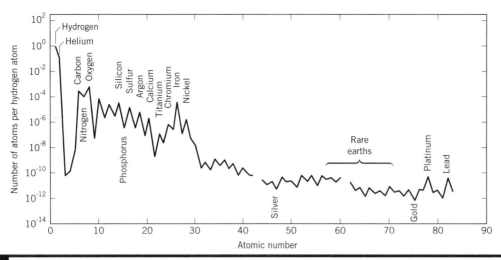

Figure 14-10 The relative abundances of the chemical elements in the universe. Note the peak at iron, due to the stability of the iron nucleus.

heavier elements we think of as being common on Earth—carbon, nitrogen, oxygen, silicon, and the rest. The small amount of heavy elements in the universe explains why we earlier considered spectra of only hydrogen, helium, and a typical metal.

The interstellar gas also turned out to have approximately the same chemical composition as did individual stars in other galaxies. In fact, the only parts of the universe that did *not* have such a composition seemed to be the terrestrial planets in the Solar System, meteorites, and interstellar dust grains. These three items make up only a small part of the universe as a whole. Therefore, it appears that Earth is an atypical sample of the material found in the universe.

Overall, the chemical composition of the universe has a broadly uniform **cosmic abundance**. **Figure 14-10** illustrates the relative abundances of the chemical elements in the universe from studies of all available sources, such as the Sun, stars, interstellar gas, and meteoritic rubble that has landed on Earth. Note that the abundance scale is a logarithmic scale, where each tick mark represents a factor-of-100 change in abundance. The attempt to understand the reasons for the detailed shape of this element-abundance curve spawned the field of nuclear astrophysics and has proven to be key to our current understanding of the evolution of stars (which we will discuss in Chapters 16–19).

The existence of a common elemental composition for almost all the objects in the universe is a spectacular idea, and certainly something that a successful theory of the origin and evolution of the universe must explain. A universal elemental abundance may also be important when it comes to discussing the possibilities of life elsewhere in the universe (Chapter 24).

Inquiry 14-14

We have just seen that meteorites are *not* typical of the universe as a whole—they have little hydrogen and helium. How, then, can studies of meteorites help us understand the compositions of the stars, which are *mostly* hydrogen and helium?

14.3 The Doppler Effect

Stellar spectra can reveal more about stars than just their surface temperature and composition. For example, there turns out to be a straightforward way to measure how fast a star is moving either directly toward or away from us, a component of the total velocity at which a star is moving through space. Motion along the line of sight is known as the **radial velocity**. The physical principle behind the measurement is based on the wave nature of light and will be introduced through an analogy to the familiar behavior of sound waves.

If you have ever been on a highway as an ambulance moved past blasting its siren, or near a railroad track as a train moved past while blowing its horn, you have probably noticed a change in the

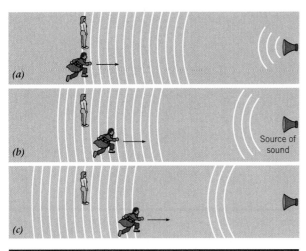

Figure 14-11 Sound wave passing by two observers *(a)* initially, *(b)* later, and *(c)* still later. More wave crests per second pass the moving observer, who detects a higher pitch than the stationary observer hears.

pitch of the siren or horn as the vehicle moves by. The pitch is higher than normal when the vehicle is approaching, but after it passes the pitch suddenly lowers. The changing frequency is caused by the relative motion of the sound source and the observer, and is called the **Doppler shift** or **Doppler effect**. It is due to the wave nature of sound. Notice the *relative* motion is what matters; the source or the observer or both may be moving.

A sound wave is a disturbance that moves through the air at a speed of about 343 m/s (the speed of sound in air). The disturbance consists of waves of high and low pressure that follow each other in rapid succession, causing the eardrum to vibrate. (For those readers who read Chapter 8, they are similar to the seismic P waves discussed there.) The more rapidly the eardrum vibrates, the higher will be the pitch of the sound we perceive.

In **Figure 14-11**, a sound wave is passing two observers, one standing still and the other moving toward the source of sound. In the figure, the lines correspond to the high-pressure areas while the low-pressure regions are halfway between. At any instant of time, a larger number of high-pressure and low-pressure regions will pass the moving observer than will pass the stationary observer. As a result, the moving observer will hear a higher-pitched note than the stationary observer. Had the moving observer been going *away* from the sound source, the effect would be the opposite: fewer regions of high and

low pressure would pass by each second, and a lower-pitched note would be heard. It does not matter if it is the source or the observer that is moving; only the relative motion along the line between the source and the observer counts.

Think about sound waves emitted in all directions from a stationary source. Such waves will be spherical. After, say, one second the peak of the wave will be as shown (**Figure 14-12a**). Now another wave is emitted, producing wave number 2. During this time, the first wave has moved outward. Now, two more waves are produced until four waves are at the positions shown in parts *a–d*. The length between each wave is the wavelength.

Now, repeat what was done in the previous paragraph but let the source move toward a stationary listener at the bottom of the page. Figure 14-12*e* shows the first wave, which was produced with the source at position 1. Before the next wave is emitted, however, the wave source moves to position 2, as shown in part *f*. Each wave produced is circular about the position of the source at the time when the wave was emitted. The process of wave emission continues each second. The end result, in Figure 14-12*h*, shows how the regions of high pressure—called *wavefronts*—get bunched up in front of the moving source of sound and spread out behind it. The listener in the direction shown will hear a higher-pitched sound than a listener located in the opposite direction (toward the top of the figure). This change in frequency is the Doppler effect.

From the discussion, you can see that a Doppler shift occurs only if there is motion toward or away from the observer—that is, along the line of sight. If the motion is across the line of sight between the object and the observer, there is no Doppler shift. If the motion is at an angle to the line of sight, only the *component* of motion toward or away from the observer contributes to the Doppler shift.

Inquiry 14-15

What will be the pitch of sound perceived by a listener toward the right of the diagram in Figure 14-12, relative to observers at the top and bottom of the diagram? (Note that, at the instant shown, the source of sound is moving perpendicular to the line between it and the listener, so it is moving neither toward nor away from the listener.)

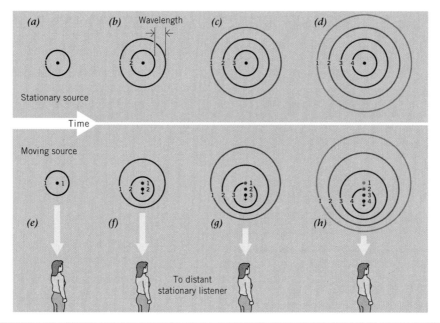

Figure 14-12 Waves from a stationary source at four times, each of which is separated by one second *(a–d)*. Waves from a source moving toward an observer at the bottom of the page (not drawn to scale) at four times, each separated by one second *(e–h)*. The waves bunch up in front of a moving source and are streched out behind it, producing the Doppler effect.

Light also shows a Doppler effect. If an observer moves toward a source of light, or if the source moves toward the observer, an electromagnetic wave of a higher frequency (shorter wavelength) will be detected than if the relative motion between source and observer were zero. If the source and observer are moving away from each other, then the frequency detected will be lower and the wavelength longer. From Figure 14-12, it makes sense that the greater the relative speed between the source and the observer, the greater will be the resulting shift in wavelength. Thus, by *observing* the wavelength shift, we should be able to determine the relative speed between the source and the observer (us).

The amount of change in the wavelength produced by the relative motion of the source and the observer is easy to compute, using the *Doppler shift formula:*

$$\frac{\left(\begin{array}{c}\text{Relative speed between}\\\text{the source and observer}\end{array}\right)}{\text{Speed of light}} = \frac{\text{Change in wavelength}}{\text{Rest wavelength}}.$$

Expressed symbolically, we have

$$\frac{v}{c} = \frac{\Delta\lambda}{\lambda},$$

where the symbol $\Delta\lambda$ stands for the *change* in the wavelength (*not* some quantity Δ times λ), λ is the rest wavelength (the wavelength the wave would have if there were no motion), c is the speed of light, and v is the relative radial velocity of the source and the observer. The change in the wavelength, $\Delta\lambda$, is defined as the *observed* wavelength minus the *rest* wavelength. If the observed wavelength is greater (longer) than the rest wavelength, $\Delta\lambda$ is positive and the source and observer are separating. If the rest wavelength is in the visible spectral region, the observed wavelength has been shifted toward longer wavelength—redward—and is called a **redshift**. This same term is used even if the rest wavelength is in, say, the radio spectral region because the shift is, again, toward longer wavelengths. If the observed wavelength is less than the rest wavelength, $\Delta\lambda$ is negative and the source and observer are approaching each other, and the result is called a **blueshift**.

Inquiry 14-16

Why does a firetruck going past us have a noticeable change in the siren's pitch but not in the color of the headlights?

The Doppler effect in stellar spectra is illustrated in **Figure 14-13**. The figure shows the spectrum of

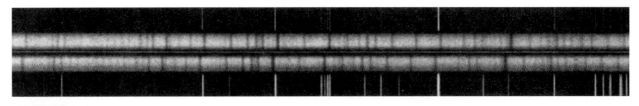

Figure 14-13 Spectra of α^1 Geminorum at two different times. The spectra are Doppler shifted relative to one another because of the orbital motion of the star.

a star orbiting about another, fainter, star. The figure is divided into four parts: the central parts include two spectra of the star taken at different times, while the top and bottom quarters include emission spectra from a lamp attached to the telescope. These spectral emission lines are at known, fixed positions and therefore define fixed reference positions. The upper stellar spectrum shows absorption lines to the red (right) of the brightest reference lines, while the lower stellar spectrum shows the same lines blueshifted. The changed positions result from the star's movement away from the observer when the upper spectrum was obtained and toward the observer at the time the second one was taken. (The observed positions of the spectral lines include the motion of the Earth about the Sun, and astronomers need to account for that to obtain the actual motion of the star itself.)

This Doppler shift formula is accurate, provided the speeds are not too close to the speed of light. Stars, in fact, move at only a small fraction of the speed of light. In words, the formula for the Doppler shift states that the fractional change in wavelength ($\Delta\lambda/\lambda$) is the same as the ratio of the relative speed between source and observer to the speed of light (v/c). For example, if the observed change in the wavelength were 1 percent of the rest wavelength, the relative motion of the objects would be 1 percent of the speed of light, or 3,000 km/s.

Using the Doppler shift formula to find speed requires knowing the rest wavelength, which astrophysicists have determined through measurements in a laboratory. Once the identity of the element causing a particular spectral line is known, its rest wavelength can be found in a database of wavelengths.

Let's look at an example. In the spectrum of M31, the great galaxy in Andromeda (see Figure 1-15), the line of hydrogen whose rest wavelength is 4,861 Å is observed at a wavelength of 4,856 Å. What is the speed of this galaxy relative to Earth? Is it moving toward or away from us?

Because the observed wavelength is shorter than the rest wavelength, the light has been blueshifted, and the Andromeda galaxy must be moving toward us. The change in wavelength is 4,856 Å – 4,861 Å or –5 Å. Therefore, using the Doppler shift formula we find that

$$\frac{\text{Relative speed between M31 and Earth}}{\text{Speed of light}} = \frac{-5\text{Å}}{4,861\text{Å}}$$

$$= -0.001.$$

The galaxy's speed relative to Earth is therefore –0.001 times the speed of light, or –0.001 × 300,000 km/s = –300 km/s, which is toward each other. At that speed, one could travel around the Earth in two and a quarter minutes! (In fact, it is expected that Andromeda and the Milky Way will collide and merge in a few billion years!) Yet Andromeda is a slowpoke compared with most of the galaxies in the universe. By contrast, nearby stars are found to move somewhat more slowly than this. For example, the α Centauri system is moving toward the Earth at about 22 km/s. Note that the wavelength change produced by stars is so small that no noticeable change in the star's color results. A further numerical example using the Doppler shift formula is in Appendix A9.

The radial velocity we have just discussed is only one component of a star's motion. The other is perpendicular (across the line of sight to the star) and is called the *transverse motion*.

Inquiry 14-17

In the spectra of certain stars, the wavelengths of all the lines shift from longer to shorter wavelengths and back to longer wavelengths in a regular fashion over the course of a few days. Can you suggest a possible explanation for this?

14.4 What We Can Learn from Spectral Lines

We have seen that the strengths of lines in a stellar spectrum are affected by the temperature and chemical composition of the star, and how these physical parameters can be measured, in principle, by studying line strengths. We also have seen how the observed positions of lines can be used to measure the speed of the object toward or away from us. Not all spectral lines appear the same: Some lines appear strong while others appear weak; some appear narrow while others appear broad. Because the appearance of a spectral line depends on a variety of a star's physical characteristics, astronomers learn much about the physical characteristics of stars through a detailed study of the *shape* of the spectral line. Such a detailed shape is known as a **line profile** (**Figure 14-14**). A careful study of the line profile yields information about a variety of stellar properties: temperature, rotation, atmospheric density, atmospheric turbulence, and magnetic field. We discuss each of these next.

THERMAL BROADENING

A stationary atom would produce a very narrow spectral line because it comes from transitions between two precisely defined energy levels. However, because gases in a star are hot, the atoms are always moving; some are moving toward an observer while others are moving away. And, some move slowly while others move rapidly. These relative motions cause transitions to occur at very slightly different wavelengths, giving spectral lines a slightly wider line profile that depends on the star's temperature.

ROTATION OF STARS

Although stars are too far away for us to see different parts of the actual surface, let us consider the light coming from different parts of the stellar surface. **Figure 14-15a** shows a spectral line formed by a nonrotating star. In **Figure 14-15b**, one side of a rotating star is approaching us; the Doppler effect shifts the radiation we receive from this side to shorter wavelengths. The light from the star's other side must be similarly shifted to longer wavelengths, because that part of the star is moving away from us. Because the part of the star in the center moves perpendicularly to our line of sight, the Doppler effect will not shift that light at all (unless, of course, the entire star is moving along the line of sight). When we observe the distant pointlike star, we observe a mixture of light from all parts of the star, each shifted by its own amount. The net effect is to broaden the spectral lines compared with that for a nonrotating star. Thus, the greater the rotation speed, the broader the line. In practice, we reverse the process; an observation of the width of the spectral line provides information on the rotation speed (**Figure 14-16**).

DENSITY IN A STELLAR ATMOSPHERE

An atom in a real gas (such as one in a stellar atmosphere) will continually interact with neighboring atoms. These collisions disturb the energy levels in an atom in random, unpredictable ways, partly as a result of the force of the collision itself and partly as a result of the electromagnetic effects of the charged particles in the neighboring atoms. **Figure 14-17** shows the effect of collisions on the atom. The energy level diagram on the left is for

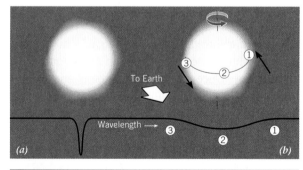

Figure 14-15 The non-rotating star on the left produces a narrow spectral line, while the rotating star on the right emits radiation that is Doppler shifted by different amounts across the stellar surface. The resulting spectral line is broadened.

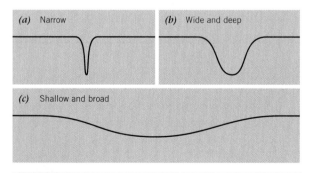

(a) Narrow   (b) Wide and deep   (c) Shallow and broad

Figure 14-14 Line profiles of different characteristics.

an isolated atom; it has sharp energy levels, and transitions between them produce a very narrow line. The diagram on the right corresponds to an atom that is strongly affected by collisions with its neighbors. Because of collisions, at any time the energy levels of some atoms might be slightly farther apart than normal; at another time, slightly closer. With large numbers of atoms all colliding and all undergoing transitions, the effect is the same as replacing the discrete energy level of the isolated atom with the broadened energy levels, as shown in Figure 14-17b. Transitions between the two broadened (disturbed) levels shown can involve photons of slightly greater or slightly lesser energies and wavelengths than for undisturbed energy levels. Because a spectral line in a star is made up of transitions from a large number of atoms, instead of absorbing only photons of a wavelength close to the wavelength of an isolated atom (called λ_0 in the figure, pronounced lambda-naught or lambda-zero), the atoms can absorb photons of many slightly different wavelengths in a range around λ_0. The result is a broadened line centered on λ_0 as a central wavelength.

This effect primarily depends on the *density* of a star's atmosphere, or how close the atoms are together. The density is determined by the strength of the star's gravity, which depends on the star's mass and radius. The higher the density, the more likely photons will be absorbed and collisions between atoms will occur. If the density is high, there will be many collisions between atoms and the energy levels will be much perturbed. In addition, the denser the material, the closer together the atoms are and the greater the effect of one atom's electrical forces on another. We call this **collisional broadening**.

What makes this factor particularly useful is that the density of a star's atmosphere is determined by the strength of gravity at the surface of the star. If the mass of the star can be estimated, then using the surface gravity and Newton's law of gravitation yields an estimate of the radius of the star. Since most stars are so far away that their size cannot be imaged by current technology, this is a powerful tool indeed!

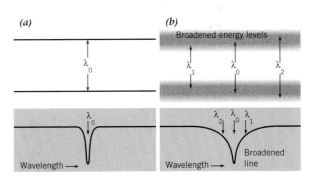

Inquiry 14-18

Figure 14-18 shows the spectra of two stars of the same temperature. One is for a dwarf star and the other is for a supergiant star, where the density of the atmospheric gases is thousands of times less than that in a dwarf star. Which is which? Explain.

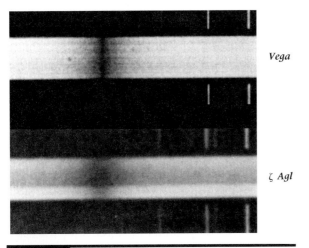

Figure 14-16 The spectrum of Zeta Aquilae, a rapidly rotating star (345 km/s). The slowly rotating star Vega (15 km/s) is shown for comparison.

Figure 14-17 Energy levels and line profiles of (a) an isolated atom and (b) an atom affected by its neighbors.

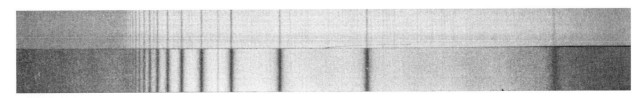

Figure 14-18 Spectra of two stars that have the same surface temperature but different atmospheric densities. See Inquiry 14-18

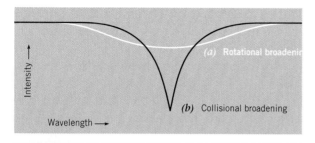

Figure 14-19 Line profiles for *(a)* rotational broadening and *(b)* collisional broadening. Note the sharper peak in the case of collisional broadening.

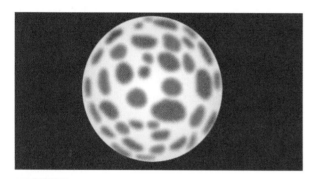

Figure 14-20 Turbulence in the atmosphere of a star is observed by Doppler shifts of rising (blue) and falling (red) regions.

Astronomers can distinguish between collisional broadening of spectral lines and the broadening caused by rotation discussed earlier because the detailed shape of the resulting line is different in the two cases. **Figure 14-19** shows the theoretical shape of the line profile in each situation. Notice that the rotational broadening produces a smooth profile, whereas collisional broadening produces a profile with a sharp center.

ATMOSPHERIC TURBULENCE

A pot of boiling water is turbulent in that hot material rises and cooler material falls. The same occurs for some stars. **Figure 14-20** is a diagram that illustrates parts of a stellar atmosphere rising while others are falling. With such atmospheric turbulence, rising elements are moving toward the observer while falling ones are moving away. The result is that rising elements emit radiation that will appear at shorter wavelengths than normal, while falling ones will be at longer wavelengths. The end result of a large number of rising and falling elements, each moving with slightly different velocities, is a broadened spectral line. Because turbulent motions are relatively small, they are important only in highly detailed analyses.

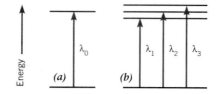

Figure 14-21 Zeeman effect. *(a)* No magnetic field. *(b)* Magnetic field present. Note the splitting of the upper energy level, giving rise to several closely spaced transitions.

MAGNETIC FIELDS IN STARS

If an atom is placed in a strong magnetic field, the energy levels in the atom will be split into several different, closely spaced levels (**Figure 14-21**). The larger the magnetic field, the greater the amount of splitting of the energy levels. Transitions can occur between the various levels, as shown in Figure 14-21*b,* which produces a corresponding splitting of the spectral lines. **Figure 14-22** shows the spectrum of a gas (the Sun) in which the top and bottom areas show no magnetic field and no splitting, while the central part shows the absorption line split into three parts. By measuring the amount of the line splitting, astronomers can determine the strength of a star's magnetic field. The splitting of spectral lines in a magnetic field is called the **Zeeman effect**.

Except for those few bright stars that have strong magnetic fields and are observed with a high-resolution spectrograph, it is unusual to see the lines actually split. Instead, they usually merge together and appear broadened. Because the light from a Zeeman-split line is polarized (see Chapter 11), this kind of broadening can be distinguished from the types previously mentioned.

The Sun has a magnetic field that is analyzed daily from observations of the sunlight's Zeeman effect and polarization. The observations show that the Sun's magnetic field changes with time and that regions of intense magnetic fields are associated with sunspots.

BINARY STARS

Only about one-third of stars are isolated and exist by themselves. Most stars are in gravitationally bound systems are called **multiple star systems**. If there are only two stars, they form a **binary star system**. Here we will show how stellar spectra reveal the existence of one type of a binary system. Chapter 15 will discuss what we can learn from the study of binary stars.

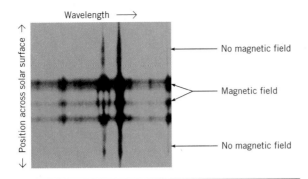

Figure 14-22 The spectrum of the photosphere and a sunspot, showing Zeeman splitting of radiation from the sunspot. Each point on the spectrum in the vertical direction corresponds to a unique point of light on the Sun coming through the slit.

Consider two stars that are bound together by gravity. Suppose, as will always be the case, that the stars are revolving around one another. Suppose, further, that the stars are equally bright. The observed spectrum, then, will be the sum of the light of both stars. If one star is, say, a B-star and the other is a K-star, the observed spectrum will be the sum of the two spectra.

Unless the orbits of the stars are seen from the top, there will always be a component to their motion along the line between the star and the observer. At one time, one star's spectrum will be redshifted while the other is blueshifted. Half a period later, they reverse. Thus, by observing the periodic shifting of spectral lines, astronomers can "see" the presence of binary and multiple star systems. This technique has been refined in recent years to discover and study planets around other stars. Figure 14-13 shows the line shifts from the Doppler effect for stars in a binary system.

A Final Example of Stellar Spectra

Let's pull many of the ideas of this chapter together. Suppose a star is rapidly rotating. An example might be the star β Lyrae, whose equatorial regions are moving at hundreds of kilometers per second—so rapidly that material is actually flying away from the star's equator (**Figure 14-23**). Let's consider the spectrum that such an object might have.

Inquiry 14-19

What type of spectrum comes from the star itself?

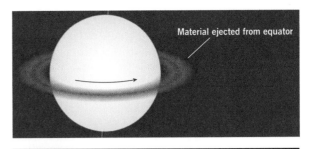

Figure 14-23 Artist's conception of β Lyrae.

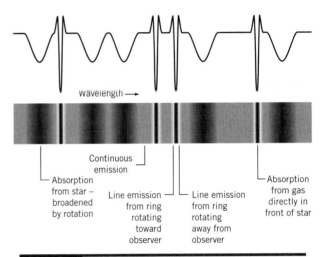

Figure 14-24 A schematic spectrum resulting from a star surrounded by a rotating, nonexpanding shell.

Inquiry 14-20

What type of spectrum would you expect to see if you consider only the light from the edges of the gaseous ring?

Inquiry 14-21

What type of spectrum would you expect to see if you consider only that part of the ring directly between the observer and the star?

The resulting spectrum is shown in **Figure 14-24**. Such a star and ring system would have a continuous spectrum with absorption lines from the star itself. Rapid rotation of the star would broaden the star's spectral lines. The radiation from the edges of the ring would be in emission, because the ring edge is hot compared with its background of cold space. Because of the ring's rotation, the emission line will be broader than if the ring were not rotating. Finally, the gas directly in front of the star will produce an absorption line in the middle of the emission line because this gas will be cool compared with the stellar photosphere behind it.

The photographic spectrum, shown in shaded colors, can also be represented as a diagram. The translation of the complex photographic spectrum into a graphical representation is shown directly above the "photograph" in Figure 14-24. This gives a more precise, measurable picture of the spectrum, and is usually the only representation available from modern digital detectors.

Classification of Stellar Spectra

After completing this activity you will be able to:

- Classify an unknown spectrum to within one-half a spectral class, given a sequence of known comparison spectra.
- Discuss the characteristics of spectra of stars having different temperatures.

Compare each of the unknown spectra in **Discovery Figure 14-1-1** with the spectra of Figure 14-2. None of the unknown spectra will be exactly like the comparison spectra, because they are for different stars; some will be somewhat similar and others will be quite different. Also, the scales of the figure are close but not identical. Attempt to classify each spectrum by finding the standard spectrum that most closely resembles the unknown spectrum, feature for feature. Write down the spectral type of the spectrum that gives you the best match. If the unknown spectrum clearly lies about halfway

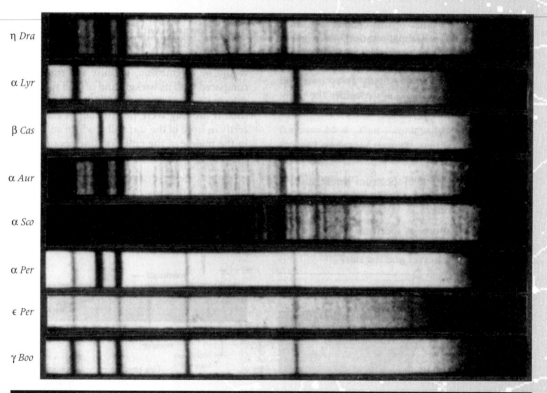

η Dra

α Lyr

β Cas

α Aur

α Sco

α Per

ε Per

γ Boo

Discovery Figure 14-1-1   Unknown spectra for classification, using the spectra in Figure 14-2 for comparison.

between two known spectra, write down the spectral type that is about halfway between the two known spectra (e.g., if your unknown spectrum lies halfway between A0 and A7, then you could write down either A3 or A4).

Remember that you are looking for specific spectral features that characterize each spectral type. Are lines of neutral or ionized helium present when the hydrogen Balmer lines are weak or absent? Then the spectrum is an O or B type, if you identified the helium lines correctly. Are the hydrogen lines the dominant feature? Then the spectrum is probably type A. How about metals? Molecular bands? Then the star is cooler, probably type F, G, K, or M, depending on what spectral lines are present and how strong they are. In this way you can make a rough classification.

To match the spectra more exactly, you will need to look at more subtle characteristics. For example, in stars of types A to G, you can compare the strengths of the hydrogen Balmer lines at 4,340 Å and the ionized calcium line at 3,933 Å. The hotter the star, the weaker the calcium line will be in comparison to the hydrogen line. At spectral type A7, they are about equal in strength, whereas in F and G stars, the line of ionized calcium is stronger.

In a similar way, you can compare the strength of the neutral calcium line at 4,226 Å and the group of iron lines at 4,271 Å with the hydrogen line at 4,340 Å. The iron and calcium lines will become more and more prominent as the star gets cooler. Similarly, the titanium oxide (TiO) bands near 4,761 Å and 4,944 Å can be compared with the calcium and iron lines if the star is very cool. In O and B stars, you can compare the strengths of the hydrogen, helium, and ionized helium lines.

- **Discovery Inquiry 14-1a** Give the spectral type and numeric subtype you determined for each star. Include information about what features you found and the relative strengths of the various features that guided you to the classification you made.
- **Discovery Inquiry 14-1b** Summarize in your own words the characteristic features of each different spectral type, referring only to Figure 14-2. Then compare your answers with **Table 14-1** to check your understanding.
- **Discovery Inquiry 14-1c** What is the contradiction if a stellar spectrum exhibits both molecular lines and helium lines? How might such a contradiction be resolved?

Chapter Summary

Observation

- Stars are observed to have a variety of different types of spectra. The types of spectra are given alphabetic names that we refer to as the **spectral type** or **spectral class** (OBAFGKM).
- Stars of spectral type O and B show lines of helium; type A stars have strong hydrogen lines; type F, G, and K stars have ever-stronger metal lines and ever-decreasing hydrogen lines. The M stars are dominated by absorption from

molecules along with a large number of lines from neutral metals.

- Stars of spectral type O and B are observed to be hot stars, while stars of spectral type K and M are cool, as are the newly found L and T stars.
- Large stars with low atmospheric densities are observed to have lines that are narrow, while small stars with higher atmospheric densities have lines that are much broader. The broader lines in small stars happen because the atoms collide more frequently and disturb the energy levels.

Theory

- The variety of spectra are interpreted with the help of atomic energy-level diagrams. The higher the temperature and density, the more electrons can be excited to higher atomic orbits, and the greater the **ionization** can be.
- The **Doppler effect** results from the wave nature of light; it is a shift in the wavelength of the observed radiation caused by the relative motion of a source and the observer. The motion that results in the wavelength shift is along a line connecting the source and observer; motion perpendicular to this line does not contribute to the observed shifting of the spectrum.
- Rotating stars produce broadened spectral lines.
- The random and turbulent motions of gases within the atmosphere of a star cause the line profile to broaden.
- Magnetic fields cause energy levels to split, thus splitting spectral lines into multiple lines.

Conclusions

- From a detailed analysis of the spectra of stars and galaxies, astronomers conclude that hydrogen and helium are the most abundant elements in the universe. All other elements in the universe, the metals, comprise only about 2 percent of the elements present.
- From a careful analysis of spectral **line profiles** and comparison with theoretical models, we can deduce detailed information on the elemental abundances in stellar atmospheres.
- From an analysis of a star's spectrum, astronomers can infer many of the star's characteristics: its atmospheric pressure, whether or not it rotates and how rapidly, whether or not the atmosphere is turbulent, whether or not it has a magnetic field and its strength, and whether or not the star has a shell of gas surrounding it.

Summary Questions

1. What are the standard spectral types in order, from hottest to coolest? What are the main characteristics of each class (temperature range, color, prominent spectral features)?
2. Why do the principal features of spectra change from one star to another?
3. Explain why relative motion between a source and an observer results in a change in the observed wavelength of light.

4. What are three effects that alter the observed profile of a spectral line? Explain why and how each one changes the line profile.
5. The prominent lines in stellar spectra such as α Hercules (Figure 14-7) do not include hydrogen and helium. How, then, do astronomers claim that this star is made up mostly of hydrogen and helium?

Applying Your Knowledge

1. When lines of *ionized* helium are present in a spectrum, will the lines of *neutral* helium have the same or different strength when compared to a spectrum in which no ionized helium lines are present? Explain.
2. In what energy level must a hydrogen atom be for it to absorb light at a wavelength of 4,861 Å? Describe how the atom might get to such an energy level.
3. What will happen to a photon of wavelength 1,220 Å when it encounters a hydrogen atom?

4. Stars known as *white dwarfs* (studied in Chapter 18) have very high densities. How do you think spectral lines of a white dwarf might differ from that of a normal, less-dense star?
5. Consider a star with an expanding shell of hydrogen surrounding it. Make a drawing of the system, and then describe and draw the resulting spectrum by considering the radiation coming from the system's various parts.
6. It has been found that the lines of hydrogen in the spectra of galaxies tend to be observed at

longer and longer wavelengths the farther the galaxy is from us. Interpret this observation in terms of what you have learned.

7. How rapidly would an object have to be moving for its color to change from blue (4,000 Å) to red (7,000 Å) because of the Doppler effect? What (incorrect!) assumption are you implicitly making?

8. What would be the maximum contribution (in angstroms) of the Earth's orbital motion to the shift of a stellar absorption line? Assume a rest wavelength of 4,000 Å, and that the Earth travels at some 30 km/s around the Sun. What would be the maximum contribution from the Earth's rotational motion, which is 0.46 km/s at the equator?

9. At what wavelengths will the D lines of sodium at a rest wavelength of about 5,900 Å be observed in the spectrum of a star that is moving away from us at 90 km/s? If it is moving toward us at 150 km/s?

10. What would be the approximate spectral type of a star whose wavelength of maximum intensity was (a) 1,000 Å, and (b) 12,000 Å?

11. A galaxy is observed to have two prominent lines in its spectrum at wavelengths of 5,104 and 6,891 Å. Assuming these are redshifted Balmer lines, what is the speed of the galaxy relative to Earth? (*Hint:* If you have identified the lines correctly, you should get the same speed regardless of which line you see.)

Answers to Inquiries

14-1. The stars at the top are hot and thus emit more strongly in the blue; the stars at the bottom are cool, and thus emit more strongly in the red and emit less blue radiation.

14-2. The Balmer lines involve the first excited state. In a low-temperature star, because few atoms will be in the first excited state, the Balmer lines will be weak if present at all.

14-3. There are no helium absorption lines. The temperature is too low to excite the electrons.

14-4. The temperature is too low for helium to be excited.

14-5. Because absorption lines are caused by transitions of electrons, and because there are no electrons remaining attached to the nucleus, there will be *no* hydrogen absorption lines in this situation.

14-6. No. The presence of TiO means the temperature is very low; therefore, other elements may not become excited at such low temperatures.

▶ 14-7. The higher the temperature, the more frequent and violent are their collisions, which dissociate them.

▶ 14-8. It will increase the level of ionization, because more frequent and energetic collisions will ionize more particles.

▶ 14-9. The argument here is similar to the changing of the strengths of hydrogen lines explained in the text. The higher the temperature, the greater the amount of ionization and the stronger the lines due to ionized atoms.

▶ 14-10. (a) The second electron is closer to the nucleus and therefore more strongly bound. (b) There are fewer electrons to repel the second electron than to repel the first one.

▶ 14-11. These are lines from molecular oxygen in the Earth's atmosphere.

14-12. The helium Lyman lines should appear before the helium Balmer lines.

14-13. Saturn is cool, so most of the hydrogen is in molecular form. Because molecular hydrogen does not have a spectrum in the visual spectral region, we don't detect it.

14-14. Even though hydrogen and helium are lacking, the meteorites may give information about the proportions of the heavier elements in the universe.

14-15. There is no change in pitch because there is no radial component of motion toward or away from the observer.

14-16. The wavelengths of visible light are so much shorter than the wavelengths of sound we hear that the change in wavelength isn't detectable in visible light, but it is in sound.

14-17. Such stars are the *spectroscopic binaries,* which consist of two stars in orbit around each other. The period is short enough and the relative velocities high enough that the change in Doppler shift is readily seen.

14-18. The spectrum on the bottom has broader lines and comes from the dwarf star. Denser stars result in broader lines because the atoms interact often and slightly modify the electron energy levels.

14-19. A continuous spectrum with absorption lines from the stellar atmosphere; a normal stellar spectrum.

14-20. Because the gas would be a low-density gas that is hot compared with whatever is behind it, the spectrum would be an emission-line spectrum.

14-21. There would an absorption line from the ring superimposed on the stellar continuous spectrum.

Discovery Inquiry 14-1c. The presence of molecules requires a star with a low surface temperature, whereas helium lines require a star with high surface temperature. A single star cannot have both characteristics, so this apparently single star must be an unresolved binary system.

The Observed Properties of Normal Stars

15

Twinkle, twinkle little star
How I wonder what you are
Up above the world so high
Like a diamond in the sky.

Nursery Rhyme

In this chapter we will learn how astronomers go about finding more of the fundamental properties of stars. We begin by learning how astronomers determine directly the distance to the nearest stars. Distance, while fundamental and important to understand, is an external property that is independent of the internal properties of stars. The most fundamental of a star's internal properties is the mass, and the only direct way to find stellar mass is to observe the gravitational effects one star has on another. Such effects can be observed for stars in a binary system, which are two stars gravitationally bound to one another. Binary systems can also be used to obtain direct information on stellar radii, as can other methods we will discuss. Once we have both mass and radius, we can compute the density, which is the mass divided by the volume.

Combining the information about stellar mass, radius, and surface temperature, along with energy output for a large number of stars, we are then able to discover important relationships between these stellar properties. One of the most powerful graphs in astronomy is of energy output versus surface temperature. First suggested early in the twentieth century by Danish astronomer Ejnar Hertzsprung and American astronomer Henry Norris Russell, it is now called the *Hertzsprung-Russell, or H-R, diagram*. Similarities and differences among the various types of stars become apparent when a large number of stars are plotted on an H-R diagram. We will make constant reference to H-R diagrams from this point on.

Finally, we will use the Hertzsprung-Russell diagram to infer the distances to stars too far away to be found by the direct methods used for nearby stars. We will refer to such inferred distances as *indirectly* determined distances.

15.1 Distance Measurements

Our fundamental knowledge of the distances to the stars comes from an application of stellar parallax discussed in Chapter 5. The parallax angle observed by astronomers for stars is an extremely small angle, thus implying that the stars are far away. How far does the small observed parallax imply? To find out, **Figure 15-1** shows a right triangle formed by the Earth, the Sun, and a nearby star. The side labeled A is the distance from the Earth to the Sun, which is 1 AU. The angle p is called the **parallax** of

the star; it is one-half the total parallactic shift shown in Figure 5-26*a*. Because the triangle is so long and skinny, the distance D from the Sun to the star can be computed from the angular size formula introduced and discussed in detail in Chapter 3;

$$D = 57.3° \times \frac{A}{p°}.$$

Parallax $p°$ is expressed in degrees. For example, the nearest star to Earth beyond the Sun, Proxima Centauri, has a measured parallax of 0.00021°, an extremely small angle. The distance is then found to be

$$D = 57.3° \times \frac{1}{0.00021} = 2.6 \times 10^5 \text{ AU.}$$

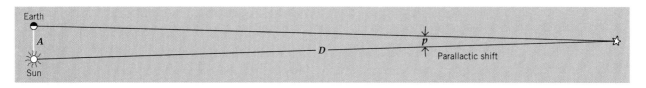

Earth

A

Sun

D

p

Parallactic shift

Figure 15-1 | The relationship between the distance of a star and its parallax. The greater the distance, the smaller the parallax angle.

Because the angles are such a small fraction of a degree, astronomers have chosen to use angles measured in seconds of arc rather than degrees. The angular size formula then becomes

$$D = 2.1 \times 10^5 \times \frac{A}{p''}$$

where p'' is the angle measured in seconds of arc; whatever units are used for A will be the units for D. Substituting this angle and the length of an astronomical unit in kilometers (A) into the formula, we find the distance to be

$$D = 2.1 \times 10^5 \times \frac{150,000,000 \text{ km}}{0.76''}.$$

$$= 41,000,000,000,000 \text{ km}$$

$$= 4.1 \times 10^{13} \text{ km}.$$

Just as you would not express the distance from San Francisco to New York in millimeters, we see that kilometers, miles, and even astronomical units are too short to be appropriate units for measuring distances to stars. A more suitable unit is the light-year (ly), because one light-year is about 9.5×10^{12} km or 5.9×10^{12} miles. The distance of Proxima Centauri, then, is 4.3 ly.

Measurements to distant objects were discussed previously in units of light-years. While the light-year is a valuable unit of measure, astronomers generally use a unit of distance called the **parsec**, equal to 3.26 light-years. Expressed in kilometers, 1 parsec = 3.1×10^{13} km. The reason for defining the parsec, which we abbreviate pc, is that it is the distance to a star whose parallax angle is exactly 1 second of arc. (Rather than always saying parallax angle, astronomers generally speak of the parallax.) With this definition, the formula relating distance and parallax is simplified:

$$D \text{ (in parsecs)} = \frac{1 \text{ AU}}{p'' \text{ (in seconds of arc)}}.$$

For those preferring to express distance in light-years, we have

$$D \text{ (in light years)} = \frac{3.26}{p'' \text{ (in seconds of arc)}}.$$

As an example, astronomers have measured the parallax of Alpha Centauri to be 0.76 second of arc. The distance is then $1/0.76 = 1.32$ pc. Expressed in light-years we have 1.32 pc times 3.26 ly/pc = 4.3 ly.

Inquiry 15-1

If the smallest angle an instrument on Earth's surface can accurately measure is 0.01 second of arc, what is the largest distance (in parsecs) for which this method can be applied? What is it in light years?

Inquiry 15-2

If the instrument in the previous question were placed on one of Saturn's satellites, which is 10 AU from the Sun, to what distance could we accurately determine the positions of stars?

From Earth's surface, we are able to determine distances using the parallax technique only for stars closer than about 100 parsecs (300 light-years). In addition, the farther away a star, the more unreliable is the observed parallax. For these reasons, in 1989 the European Space Agency launched a satellite whose purpose was to improve dramatically our knowledge of stellar distances. Called the **Hi**gh **P**recision **Par**allax **Co**llecting **S**atellite, or *HIPPARCOS* (to honor Hipparchus for his pioneering work in mapping 1080 stars in the sky), the satellite's primary mission was to measure parallaxes for 120,000 stars to an accuracy of 0.001 second of arc. Its secondary mission was to observe more than one million stars with 0.020–0.030 second of arc accuracy. Because most fields of astronomy benefited from the parallaxes obtained by *HIPPARCOS*, the results are spread throughout the rest of the book.

Observations of stellar parallax provide the only *direct* method of finding stellar distances. It is a direct method because the parallax, and thus the distance calculated from it, is directly observed and not inferred from anything else. Nothing other than the observed parallax and the angular size formula are needed. Since the angular size formula is from trigonometry, parallaxes determined in this way are often referred to as **trigonometric parallaxes**. Trigonometric parallax measurements provide the most accurate distances possible. All other methods are *indirect methods* and are less accurate.

From our discussion in the previous section, you saw that the trigonometric parallax method could be used only for stars closer than about 300 light-years for observations made from the surface of the Earth. That distance is only about 1 percent of the distance to the center of our galaxy, which shows the importance of the HIPPARCOS satellite that provided accurate and precise *direct* distances to stars 10 times further away. The European Space Agency is working on project GAIA, which in 2010 will begin to obtain parallaxes for a billion stars to an accuracy two to three powers of ten better than HIPPARCOS! We will then have accurate distances to stars 100 to 1,000 times farther yet— truly a remarkable feat that will improve our understanding of the universe in significant ways.

For now, however, we have to work with what we have, and most stellar distances must be determined by other means, which we will discuss later in this chapter and in subsequent ones.

VARIATION OF STELLAR BRIGHTNESS WITH DISTANCE

As a distant car comes closer, its headlights appear to get brighter. Of course, you know that the actual light output of the headlights is not changing. Thus, because your experience gives you an idea of how bright the headlights of a car actually are, you intrinsically know there is a relationship between how bright the headlights appear and the distance to the car. However, when looking at the night sky, you cannot tell if a particular star you see is relatively nearby or distant because you do not know the star's actual energy output.

A star's actual energy output, or power, is called **luminosity**. Luminosity is similar to the wattage of a light bulb. For example, a 1,000-watt light bulb is much more luminous than a 10-watt bulb.

We will now use **Figure 15-2** to help us see how apparent brightness varies with distance. By **apparent brightness** we mean *the number of photons passing through a unit area each second.* Consider only those photons coming from the source that are confined between the straight lines in the figure. No photons will enter or leave the confined region unless they come from the source. A certain number of photons from the light source pass through an area at a distance of 1 unit from the source, as shown. At a distance of 2 units, the same number of photons passes through an area that is now twice as large on each side, or four times greater in area than at location *1*. Thus the apparent brightness is $(1/2)^2 = 1/4$ what it was before. At a distance of 3 units, the apparent brightness will be $(1/3)^2 = 1/9$ the apparent brightness at location *1*.

We can now generalize our result. We have found that the apparent brightness of a light source

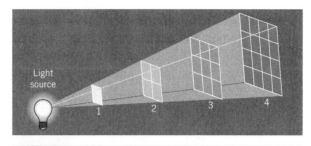

Figure 15-2 The inverse square law of light. The apparent brightness is the energy per unit area, which decreases with the square of the distance.

decreases with the distance squared. This relationship, called the **inverse square law of light**, tells us that the amount of light received from a distant source decreases in proportion to the square of its distance from us. Suppose a star of luminosity *L* is at a distance *d* from Earth. The star will have an apparent brightness *B*. The inverse square law then says

$$B \propto \frac{L}{d^2}.$$

As an example of the inverse square law, suppose we observe two identical stars (which, because they are identical, have the same luminosity), one of which is five times closer to us than the other. The inverse square law of light tells us that the nearer star will *appear* to be 5^2 or 25 times brighter than the farther one. Or, to turn this around, again suppose we know from a detailed study of their spectra that the two stars have the same luminosity, but we *observe* that one star appears 100 times brighter than the other. We then *deduce* that the apparently fainter star is actually 10 times farther away than the brighter one, because $\sqrt{100} = 10$.

Luminosities obtained from trigonometric parallaxes, which are *directly* determined quantities, are themselves directly determined quantities. Thus, astronomers can determine the luminosity for any star for which a parallax has been observed. Later, we will see how to determine luminosity using indirect means.

Inquiry 15-3

Suppose two stars of equal luminosity are observed and one of them appears 50 times brighter than the other. If the nearer star is at a distance of 12 light-years, how far away is the distant star?

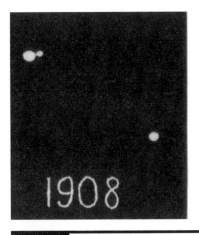

Figure 15-3   The visual binary Krüger 60, photographed in 1908, 1915, and 1920. The star to the lower right is not associated with the binary system; it is 30 seconds of arc from the primary (brighter) star of the binary system.

Inquiry 15-4

Suppose one star is 8 light-years from us and another 48 light-years away. If the two stars have the same luminosity, how many times brighter does the nearer star appear?

15.2 Binary Stars

William Herschel was the first astronomer to demonstrate from continued observations that there were such things as stars orbiting other stars under their mutual gravitational attraction. In fact, we now know that more than half the stars in the vicinity of the Sun are members of gravitationally bound multiple-star systems. Binary systems are particularly valuable to astronomy because it is from them that virtually all *direct* measurements of stellar masses, and many stellar-diameter measurements, are made. For these reasons, we now examine three types of binary-star systems.

VISUAL BINARIES

Binary systems in which both components can be seen as two separate stars are called **visual binary stars (Figure 15-3)**. The discovery of such pairs gave astronomers their first clue that stars had a large range in luminosity. For example, because both of the stars on the left side of each photograph in Figure 15-3 are at the same distance, the brightness difference they exhibit is caused by their actual difference in luminosities. Not all stars that appear close together are gravitationally connected, however. Some pairs of stars, which we call *optical doubles,* appear to lie near each other in the sky but are actually at different distances from Earth and are not binaries.

Visual binaries give astronomers some of their best information about the masses of stars. In fact, they provide a *direct* method of determining the mass of a star. Kepler's third law (Chapter 5), as modified by Newton to account for the masses, is used for this purpose:

$$(M_1 + M_2)P^2 = A^3.$$

Remember, the masses are in solar mass units; P is the orbital period in years, and A is the semi-major axis of the elliptical orbit in astronomical units. Thus, an observational determination of the period and the semi-major axis will allow us to compute the sum of the masses of the stars in the binary system.

The period of the two stars in their orbit is *observed* directly, and the average separation between the stars can be calculated from the *observed* angular size of the orbit and the *observed* distance to the star system.[1] Therefore, if the distance to the binary is known, we can compute the sum of the masses of the two stars. **Figure 15-4** summarizes this procedure for obtaining information about stellar masses from visual binary stars.

---

[1]The separation of the stars in astronomical units is related to the observed angular separation and the distance through the angular-size-formula introduced in Chapter 3. An extremely useful form of this equation is angular separation (seconds of arc) $= \dfrac{\text{linear separation (AU)}}{\text{distance (parsecs)}}$.

Finally, observations of the binary system over many years can, in some cases, provide the ratio of the masses of the two stars. The mass ratio, when combined with the sum of the masses from Kepler's third law, allows us to solve for the mass of each star.

Inquiry 15-5

If the sum of the masses of two stars is 5 solar masses and the ratio of their masses is 4, what are the masses of the individual stars?

You should do Discovery 15-1, Visual Binary Stars, at this time.

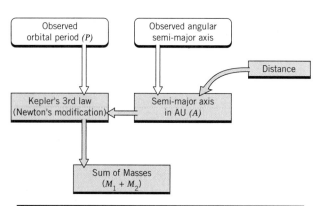

Figure 15-4 The process of obtaining information on stellar masses using observations of visual binary stars.

SPECTROSCOPIC BINARIES

In a binary system that is too distant, or in which the stars are too close together to be resolved, direct viewing of each star is not possible. The presence of a binary system may be inferred, however, from a periodic Doppler shift observed for the spectral lines as the stars orbit one another. Such a system is known as a **spectroscopic binary**. The spectrum of the star Mizar, which appears on photographs to be a single star, shows it to be a spectroscopic binary. Mizar's spectrum at two different times is shown in **Figure 15-5**.

To understand these spectra, we consider two such stars in various positions in circular orbits as shown in **Figure 15-6a**. Each star revolves around a point between the two stars. That point, called the **center of mass**, is determined by the relative masses of the two stars. If the masses are equal, the center of mass is at the midpoint between the stars, just as the balance point of a barbell is at the center of the bar. If unequal, the center of mass is closer to the more massive star. **Figure 15-6b** contains schematic spectra at each position. At the bottoms of Figures 15-6b are comparison spectra, which are emission spectra of a lamp attached to the telescope to provide lines of known wavelength for reference. In the stellar spectra, some spectral lines for each star are noted.

At time 1, the stars are moving along the line of sight from the observer to the stars. Because star A is moving away from the observer, its spectrum will be Doppler shifted toward longer wavelength while the lines for star B will be blueshifted, as shown in position 1 of Figure 15-6b. Because the stars must remain on opposite sides of the center

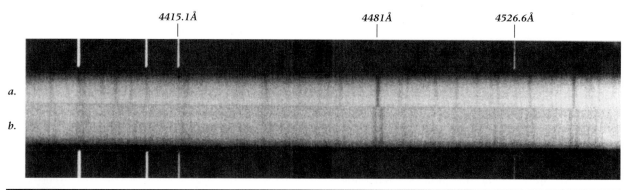

Figure 15-5 Spectra of the spectroscopic binary Mizar. *(a)* The spectral lines are superimposed when the stars are moving across the line of sight. *(b)* The spectral lines are separated when the stars move toward or away from the observer. Note the comparison spectra above and below the stellar spectra.

of mass at all times, both stars have the same period. For this reason, because the outermost star has the greater distance to go during the orbital period, its velocity is larger. Therefore, as shown, star B's lines are shifted more than star A's lines.

Continuing on, when the stars have moved in their orbits to position 2, the stars are moving across the line of sight; neither star shows a Doppler shift, and the spectral lines from both stars coincide. At position 3, the situation is the reverse of position 1; star A shows a blueshift and star B, a redshift. Finally, the stars will again move across the line of sight (time 4), and there will again be no Doppler shift of the spectral lines.

A star system like the one just described is called a *double-lined spectroscopic binary*—one in which the spectral lines of both components can be observed. From the observed Doppler shift, each star's radial velocity at every position in the orbit is computed using the Doppler shift formula. The results are then graphed, producing what astronomers call a **radial velocity curve (Figure 15-6c)**. Furthermore, the less massive star moves faster than the more massive one. From these considerations we find that the ratio of the observed velocities of the stars equals the ratio of the masses of the stars. Thus spectroscopic binaries are important to astronomers because of the information they provide about masses.

Inquiry 15-6

Refer to Figure 15-6c. Suppose that this radial velocity curve showed one star (call it star A) to have a maximum radial velocity of 25 km/s while the other star (B) has a maximum velocity of 125 km/s. What is the mass of star A relative to that of star B?

Whenever one star in a spectroscopic binary system is more than about 15 times fainter than the other, only one spectrum will be visible. However, because both the visible and invisible objects still move about the center of mass, the spectral lines of the brighter star may be seen to shift back and forth periodically. In such a *single-lined spectroscopic binary*, detailed analysis allows us to infer the mass of the *unseen* star. Astronomers use such techniques to search for and discover black holes.

You should do Discovery 15-2, Spectroscopic Binary Stars, at this time.

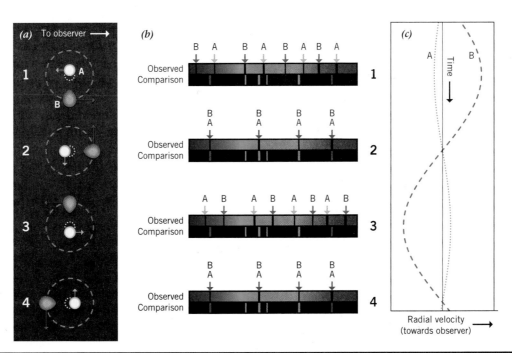

Figure 15-6 (a) Two stars in circular orbits in a binary system at different times. (b) Schematic spectra of the stars. The symbols for the spectral lines for each star are the same as for the stars in part a. (c) The radial velocity curves resulting from the spectra in part b.

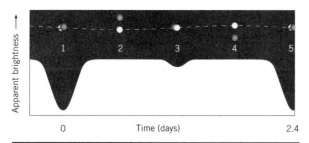

Figure 15-7 The light curve of an eclipsing binary. Notice that the light curve is constant except during the eclipses.

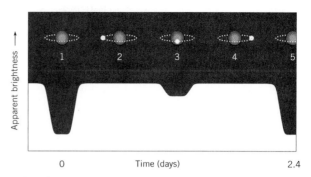

Figure 15-8 The cause of light variations in an eclipsing binary.

ECLIPSING BINARIES

Sometimes, one star will pass in front of the other, producing what we call an **eclipsing binary**. The archetype of this kind of star is β Persei, or Algol, which means "the demon star" in Arabic. The brightness of an Algol-like eclipsing binary is normally quite constant, but every few days it suddenly dims for a short time and then brightens again (**Figure 15-7**). The variation of the brightness with time produces a **light curve**. The distinctive shape of an eclipsing binary light curve is easily explained by the fact that periodically one of the stars moves in front of the other, partially or completely cutting off the light of the second star (**Figure 15-8**). Approximately half an orbit later, the second star eclipses the first one in a similar fashion, producing a secondary drop in brightness.

From a detailed study of the light curve of an eclipsing binary, we can deduce many details about the orbit and the eclipsing stars themselves. In particular, we can determine the diameters of the stars in terms of the average distance between them.[2] Stellar diameters determined from nearby eclipsing binary star observations are *direct* determinations because they result from observations without the need for using any intermediary relationships for inferring anything about them. Stellar radii measured in eclipsing binary systems are in good agreement with those inferred from the formula that relates radius, luminosity, and temperature; this agreement gives us confidence that the radii determined by indirect means are reliable.

For stars that are both spectroscopic *and* eclipsing binaries, astronomers are able to find the mass of each star. Because the mass of a star is the single most important thing we need to know to understand it, the binary systems that allow us to determine stellar masses are prized finds in astronomy. The resulting masses are as accurate as we can get because these methods are *direct* methods.

Bringing together all the observational and theoretical information we have obtained from stellar mass determinations, we find the stars range in mass from about 0.1 to perhaps as much as 100 to 150 times the mass of the Sun.

15.3 Other Studies of Stellar Radii

Knowledge of the radii of stars is necessary for complete knowledge of stellar properties. We have seen that some direct information on stellar radii can be obtained from eclipsing binary stars. Technology has provided some additional direct methods, two of which we now mention.

LUNAR OCCULTATIONS

Developments in instrumentation for observing stars have opened up new channels for obtaining *direct* information on stellar radii. In the 1970s, considerable effort was put into the study of events called *lunar occultations* in which the Moon passes in front of a star. From observations of these occultations, direct measurements of stellar diameters are sometimes possible.

For years it was assumed that when the sharp edge of the Moon passed between a star and us, the light from the star would extinguish instantly. But the advent of modern detectors and electronics allows us to take an accurate reading of stellar

---

[2]Sharp-eyed readers may have noticed the differences between the light curves in Figures 15-7 and 15-8. These differences are due to different relative radii of the two stars in the systems that produced the light curves shown. The stars that produced the light curve in Figure 15-7 are of nearly equal radius whereas there is a substantial difference for the stars in Figure 15-8.

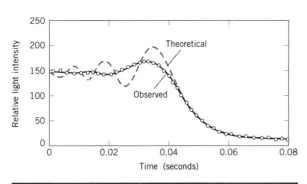

Figure 15-9 | Occultation light curve of λ Aquarii. Shown are a theoretical curve (dashed line) for an infinitesimally small point source, and the observed light curve (solid line).

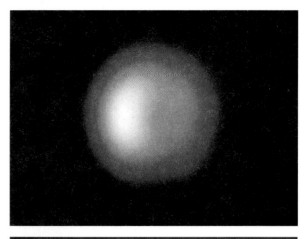

Figure 15-10 | The resolved disc of Betelgeuse, showing a hot spot.

brightness every thousandth of a second or so. As a consequence, we can measure the short time it takes for the Moon's edge to move across a tiny stellar disk. Because the Moon is outside our atmosphere and has no atmosphere of its own, there is no blurring of the data involved anywhere along the line of sight.

Figure 15-9 illustrates what was seen when astronomers at the University of Texas applied these super-fast observing techniques to a lunar occultation of the star λ Aquarii. The dotted curve is what we would expect to see if the star really were an infinitesimally small point source; the ups and downs are the *diffraction pattern* that occurs when a light wave from the star passes near a straight obstacle (the lunar edge). This is the effect of diffraction discussed in Chapter 11.

In Figure 15-9, notice that the curve actually seen for the star λ Aquarii is somewhat smoothed and broadened. The light does not fall off as rapidly as it would for a point source. In fact, the stellar trace is exactly what we would expect to see from an occultation of a star with an angular radius of 0.0037 second of arc—the diameter of a penny seen from 300 km away. An independent determination of the distance of the star (say, measuring a parallax for it) would allow us to convert this apparent radius into a linear radius in kilometers.

OPTICAL INTERFEROMETRY

We explained the principles of radio interferometry in Chapter 12; these same principles can be applied at optical wavelengths, although the instruments are more difficult to use and the data harder to interpret. For a few very large, nearby stars interferometric measurements have been successful. With an angular radius of 0.044 seconds of arc, which is an actual radius 530 times that of the Sun, Betelgeuse is one of the largest stars that has been measured in this way. The large angular size made it possible for the *Hubble Space Telescope* to obtain a remarkable image of Betelgeuse (**Figure 15-10**), which is the first star other than the Sun on which surface markings have been seen. A small, hot companion star is hypothesized to cause the "hot spot" or bright area that is seen in the figure. The smallest star for which the radius has been determined using interferometry, is Altair, with a radius of only 1.6 times that of the Sun. The only reason Altair could be measured in this way is that it is relatively near the Sun (only 17 light-years), so that its angular radius is relatively large.

Technological advances in recent years, and those expected in the near future, will allow optical interferometric determination of the diameters of many more stars. Anticipation of these advances is the reason for the layout of the European Southern Observatory's *Very Large Telescope* shown in the illustration at the opening of Chapter 12, and for construction of the 10-meter Keck 2 telescope 85 meters away from the Keck 1 telescope in Hawaii. Not only will these telescopes directly measure radii for many more stars, but changes in stellar radius will be observable for certain stars whose radius varies with time.

Studies of stellar radii show us that typical stars range in radius from about 0.1 to 1,000 times the radius of the Sun.

DENSITIES OF STARS

Using the mass determined from analyzing observations of binary stars and the stellar radius obtained from studies of eclipsing binary stars, lunar occultations, or interferometers the average density (the amount of mass inside a volume) can be found. Although stars are massive, they are also large; the stellar material can be spread out into a huge volume. Thus the average density of a typical star is not necessarily high. For example, the average density of the Sun is just slightly greater than that of water (it is nearly 160 times that of water in the center but extremely low in the outer parts). Many stars are so large that their average densities are millions of times less than that of the Sun. Cases of extremely high average density will be mentioned later in this chapter when we talk about stars called *white dwarfs* and then later in Chapter 19 when we learn about other stars called *neutron stars.*

15.4 The Hertzsprung-Russell Diagram

One of the major goals of science is to make sense out of apparent chaos. Given the large number of stars observable with telescopes, astronomers work to understand them by searching for relationships among the various stellar properties. This section examines perhaps the most important of all such relationships for stars, the one between luminosity and temperature.

A GRAPHING EXPERIMENT

Let us start with a graphing experiment that is analogous to the idea behind the Hertzsprung-Russell diagram and that is straightforward to interpret. In 1992, one of the authors requested all students in introductory astronomy courses at the University of Kansas to anonymously submit to him their sex, height, weight, and number of letters in their last name. In **Figure 15-11a** we have plotted the height of each person versus the number of letters in his or her last name. In the graph, each person is represented by a single point. There is clearly no systematic relationship between the two quantities plotted; this particular diagram provides no useful information. In **Figure 15-11b**, we have plotted height versus weight. In this case, there is evidence of a systematic relationship: the taller a person is, the heavier that person tends to be. There is, of course, some scatter; given any height, there will always be people of various weights. But the variations do not obscure the existence of the relationship shown in the diagram. In other words, the points do not simply scatter randomly around the diagram but form a sequence of points within the graph that can provide useful information about people.

Furthermore, in Figure 15-11b we can even detect evidence of two different relationships, one

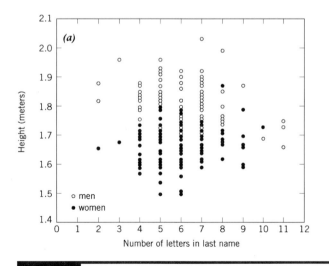

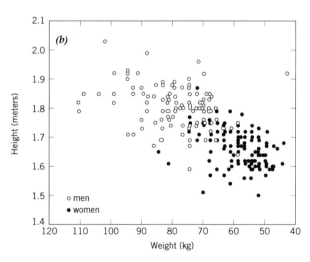

Figure 15-11 *(a)* Graph comparing the heights of a group of people with the number of letters in their last name.

(b) Graph of the heights of the same group of people compared with their weights.

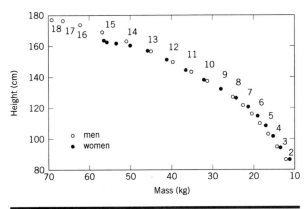

Figure 15-12 | Graph that shows the "growth" path of children's height and weight from age 2 to 18.

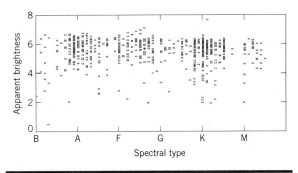

Figure 15-13 | Graph showing an expected lack of correlation between apparent brightness and spectral type for a large number of bright stars.

for males and one for females. If you did not realize that there were any differences between the body structures of males and females, you could have discovered it by looking at this graph. (Where do you fall within the two graphs in Figure 15-11*a* and *b*?)

More detailed analysis of such data shows a tendency for people's location in the diagram to change with time from areas of low weight (for a given height) to regions of higher weight. A graph of height versus weight for children from the ages of 2 to 18 is shown in **Figure 15-12**. This height-weight diagram demonstrates that children change as they age. We can say that as people change, they "move" in the height-weight diagram. Of course, the person does not move, but the point that represents the person's height and weight moves. This notion of a systematic change in the position within a diagram will be used for the discussion of stellar evolution in coming chapters.

INTRODUCTION TO THE HERTZSPRUNG-RUSSELL DIAGRAM

You have already learned quite a bit about stars. You have seen that stars have a temperature that can be determined from the spectral type, a rotation and perhaps turbulence that can be determined from the Doppler effect, and perhaps a magnetic field that can be studied using the Zeeman effect. Furthermore, stars have radial velocities toward or away from the observer that can be studied with the Doppler effect. From the inverse square law, we can find a star's luminosity once its apparent brightness and distance are known. Are any of these characteristics related to any other? For example, **Figure 15-13** shows a graph of apparent

brightness plotted against spectral type. No relation is seen, and none is expected because a star's apparent brightness (which is an extrinsic property) is not expected to be determined primarily by its temperature (which is an intrinsic property).

Early in the twentieth century, Hertzsprung and Russell independently proposed a powerful method for comparing stars by plotting their luminosity against their surface temperatures. Now known as the **Hertzsprung-Russell diagram**, or simply the **H-R diagram**, this graph, which is perhaps the most important one in the field of astronomy, pulls together fundamental information on stellar temperatures, luminosities, radii, and masses. We will use the H-R diagram to find the distances to stars. Historically the H-R diagram has been the basis for breakthroughs in understanding both stellar and galactic evolution.

We saw previously that the average surface temperature of a star can be determined from Wien's law by measuring the star's peak wavelength (as discussed in Chapter 13) or by determining its spectral type (as discussed in Chapter 14). Astronomers frequently use the spectral type rather than the temperature determined from the peak wavelength, because spectral type is what they determine when classifying a spectrum, and the spectrum they have may not include the peak wavelength. On the other hand, temperature (not spectral type) is the quantity needed for calculations, so that we might also plot the H-R diagram using temperature. To emphasize the strong relationship between spectral type and temperature, in the diagrams that follow, temperature is on the bottom axis and approximate spectral type along the top.

We can calculate a star's luminosity from its apparent brightness when its distance is known.

We often express it in terms of the Sun's luminosity. For example, the star Arcturus radiates approximately 100 times more energy than the Sun. Arcturus' luminosity in solar units is therefore equal to 100 $L_\odot$. where the symbol $L_\odot$ represents the Sun's luminosity. Because the Sun's luminosity in metric units is 3.8×10^{26} watts, it follows that the luminosity of Arcturus in metric units is 3.8×10^{28} watts.

Inquiry 15-7

What would be the luminosity, in metric units, of a star of 5 $L_\odot$ Of a star of 0.2 $L_\odot$

Inquiry 15-8

What would be the luminosity, relative to the Sun, of a star whose luminosity in metric units was 10^{24} watts? Of a star whose luminosity was 8×10^{27} watts?

THE NEAREST STARS

The nearest visible stars have all had their parallaxes measured. Because the distance in parsecs is given by $1/p$ (seconds of arc), the distances to these stars are known. Therefore, we can find their luminosities by applying the inverse square law. **Figure 15-14** is the H-R diagram for the 125 stars *nearest* the Sun; the diagram is a survey of all known stars closer than about 22 ly. Many of these stars are double or triple star systems (i.e., several stars bound together by gravity), and in such cases each member of the system is plotted as a separate point. For each star, the surface temperature (in

kelvins) and luminosity (relative to that of the Sun) are plotted. Notice the point representing the Sun.

Notice, too, that in the H-R diagram the temperatures *decrease* from left to right. Furthermore, because this is a *logarithmic* graph (discussed in Chapter 13), the divisions shown are not equally spaced. Equal *intervals* on the graph correspond to equal *ratios* of the quantity being plotted. For example, the interval between 3,000 and 6,000 K is equal to the intervals 10,000–20,000 K and 20,000–40,000 K, because the temperature ratio in each case is two. Logarithmic graphs are required here because the luminosity axis in Figure 15-14 has a large range (10^{12}), as does the temperature axis (2,000 K–40,000 K).

Inquiry 15-9

Where on Figure 15-14 are most of the points, which represents stars close to the Sun, found?

Inquiry 15-10

How might you describe the second group of stars that is distinctly separate from the first?

THE BRIGHTEST STARS

Most of the stars that appear brightest to the naked eye at night are close enough to the Sun to have observable parallaxes, and thus have known distances and luminosities. An H-R diagram for these stars is shown in **Figure 15-15**. In cases of multiple-star systems, we have plotted only the brightest star of the group.

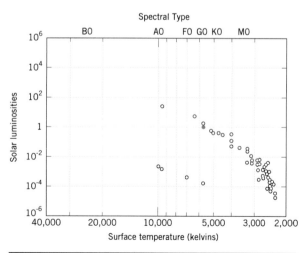

Figure 15-14 | The Hertzsprung-Russell diagram for the stars closest to the Sun. The Sun is denoted as an orange point.

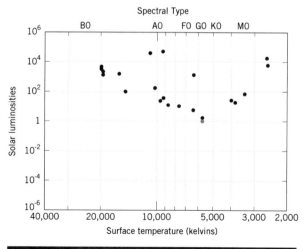

Figure 15-15 | The Hertzsprung-Russell diagram for the brightest stars in the night sky. The Sun is denoted as an orange point.

Inquiry 15-11

Describe the distinctly separate groups of stars plotted in Figure 15-15.

Inquiry 15-12

Can you identify one of these groups as being a natural extension of a group you discovered among the nearby stars?

Figure 15-16 combines the plots of both the nearest and the brightest stars.

Inquiry 15-13

Considering *all* the stars in Figure 15-16, approximately what is the ratio of luminosity between the most luminous and least luminous stars plotted?

Inquiry 15-14

Are the stars nearest the Sun the same ones that appear brightest in the sky? How do you explain this?

Inquiry 15-15

Compare the luminosities of the stars having temperatures near 3,000 K. Within this group, approximately what is the ratio of luminosity between the most luminous and least luminous stars?

INTERPRETING THE HERTZSPRUNG-RUSSELL DIAGRAM

The diagrams you have been inspecting contain a wealth of information, and now is a good time to demonstrate the power of the H-R diagram. Comparing the H-R diagrams makes it immediately apparent that the two groups of stars, near and bright, are different in nature; only three objects (Sirius, Procyon, and alpha (α) Centauri) appear on both graphs.

Inquiry 15-16

Are the brightest stars we see in the sky that way because they are the same as the group of nearest stars?

The brightest stars in the sky are definitely *not* the nearest stars—the bright ones are, instead, extremely luminous objects, so intrinsically bright that even though they are many times farther away than the nearest stars they are still the most conspicuous objects in the night sky. From the diagrams you can see that the ratio of the luminosity of the brightest to the faintest star shown here is greater than 10^{10}—ten billion—and the actual ratio is even greater, because there are stars even more and less luminous than those that happen to be close to the Earth.

Because most of the brightest stars in the sky are highly luminous objects fairly far away, the diagram of the *nearest* stars is more representative of the *average* population of stars. The plot of the nearest stars includes the statistics on *all* the known stars within a certain volume of space—a sphere centered on the Sun whose radius is 22 ly. Although it is possible that a few extremely faint stars may yet be discovered, the group of stars is a nearly complete sample.

By analogy, you would probably get a better idea of the characteristics of U.S. citizens by surveying everyone in your hometown rather than those people whose names appear in the headlines of the newspaper. The people in the headlines are generally atypical in some respect, which is how they got there in the first place.

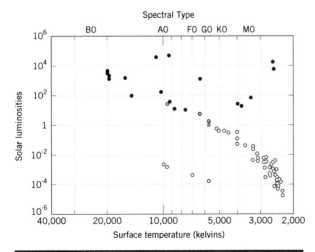

Figure 15-16 The data from Figures 15-14 and 15-15 have been plotted together in this diagram. Note the separation into distinct groups.

Inquiry 15-17

There may be stars within a 15-ly radius that have not yet been detected, so the plot of Figure 15-14 may not be complete. What do you think would be the most likely reason for such a star not to have been detected?

Inquiry 15-18

In arguing that the stars near the Sun are probably typical of the average stars in our galaxy, what is the fundamental (post-Copernican) assumption being made by astronomers?

Inquiry 15-19

Make this assumption, and compare the number of stars in Figure 15-14 that are fainter than the Sun to the total number of stars. Roughly what proportion of stars in our galaxy is fainter than the Sun? Hypothesize about what fundamental property of stars might produce this result.

The H-R diagram clearly indicates that superluminous stars are relatively rare objects and that the vast majority of stars fall in the lower (less luminous) part of the diagram.

15.5 Main-Sequence, Giant, Supergiant, and White Dwarf Stars

In examining these H-R diagrams, we clearly see that some parts of the graph are heavily populated with stars, while other parts are empty. The most obvious feature of the H-R diagram is the band of stars running diagonally across it from the upper left to the lower right—from high temperature, high luminosity to low temperature, low luminosity. This band has come to be known as the **main sequence**. The majority of stars are found in this region of the diagram and are thus called **main-sequence stars**. Indeed, because the nearby stars are representative of average stars in our galaxy, and most of them fall on the main sequence, the obvious conclusion is that most stars in our galaxy are main-sequence stars. Later, when we consider the evidence that stars evolve and change their characteristics with time,

we will also see an additional significance to this clustering in the diagram—namely, that the main sequence is where stars spend most of their lives.

There is a second group of stars among those closest to the Sun that makes up a fair proportion of the nearby stars. These stars therefore must be a fairly common type of star in the universe. This is the group of stars below the main sequence that consist of what are called **white dwarf** stars (for reasons that will soon become apparent). These stars are the most important and populous category of non–main-sequence stars.

Inquiry 15-20

What percentage of stars in our galaxy do you estimate are white dwarf stars? Use information in Figure 15-14.

When we consider the *brightest* stars, we find that many of them are *not* main-sequence stars. Notice that among the stars with temperatures near 4,000 K, there are some with luminosities around 100 times the Sun's luminosity. We need to understand how this range of 100 in luminosities among the brightest stars comes about.

The rate at which energy is radiated by a hot surface is related to its temperature, as we saw in Chapter 13. For a given stellar temperature, we can use the Stefan-Boltzmann law to calculate how much energy per second the star will radiate from each square centimeter of its surface. If two stars have the same surface temperature, then a square centimeter on the surface of each star will radiate the same amount of energy per second. So it follows that if one of these two stars radiates more *total* energy per second than the other, the star with the higher energy output must have a greater surface area. To consider a more "earthly" analogy, compare two hot plates that are each at the same temperature; that is, equally red. The only way one hot plate will radiate more total energy than the other is if it is larger.

As an example, in comparing the Sun and a more luminous star having the same 6,000 K temperatures, each square centimeter on the surface of each star radiates the same amount of energy per second. The more luminous star must simply be much larger than the Sun. A star 100 times as luminous as the Sun must have a surface area 100 times greater. Its radius, then, is $\sqrt{100} = 10$ times that of the Sun.

This same argument can demonstrate the existence of even larger stars in the H-R diagram. Notice that there are some stars in Figure 15-16 with surface temperatures about 3,000 K that are over a million times more luminous than main-sequence stars of this same temperature. Betelgeuse and Antares are two of them. These stars must therefore have more than a million times the surface area of the main-sequence stars, or $\sqrt{10^6} = 1,000$ the main-sequence star's radius.

The stars in the luminous group, whose radii are ~10 solar radii, have come to be called **giants** to distinguish them from the more common stars on the lower part of the main sequence. Arcturus is a familiar example of a giant star. The more luminous stars, whose radius is ~100 to 1,000 times the Sun's radius, have come to be called **supergiants**. If the supergiant Antares were to replace the Sun in our solar system, it would swallow up the Earth and extend all the way out to the orbit of Saturn!

Stars on the lower main sequence are also generally called **dwarfs**. Dwarf stars should not be confused with the dramatically different white dwarfs below the main sequence. Thus, we have divided stars into three **luminosity classes**: main-sequence, giants, and supergiants. These distinctions will help us later to find the distances to stars.

The relationship between the luminosity, temperature, and radius of a star that we have been discussing qualitatively can be expressed quantitatively in a simple way: The luminosity is given by the formula

Luminosity = (Energy/s radiated by a cm²
of the star's surface)
× (Total surface area of the star).

We have noted that the energy per second radiated by a square centimeter depends on the temperature; in fact, it depends on the *fourth power* of the temperature, as discussed in Chapter 13 (it is known as the Stefan-Boltzmann law). For a spherical star, the second part of the expression is the surface area of a sphere the radius of the star. Because the area of a sphere depends on the square of its radius, we find that

Luminosity of the star ∝ (Stellar radius)²
× (Surface temperature)⁴.

In symbols, we have

$$L \propto R^2 T^4.$$

where R is the radius of the star and T is the temperature in kelvins.

Inquiry 15-21

If two stars have the same temperature, but the radius of one is twice the radius of the other, how much more luminous will the brighter star be?

Inquiry 15-22

If two stars have the same radius but differ in temperature by a factor of two, how much more luminous will the hotter star be?

Inquiry 15-23

The stars Antares and CD −46°11540 have the same surface temperature (2,700 K), but Antares is two million times more luminous than CD −46°11540. How much larger is it?

We conclude that we must consider both the radius and surface temperature of a star when discussing its luminosity. We have seen from the H-R diagram that there are stars ranging from nearly 10^5 times more luminous than the Sun to stars nearly 10^5 times less luminous, a range of 10^9 to 10^{10} (1–10 billion). Some of this difference is due to differences in surface temperature, because normal stars range from approximately 3,000 to 30,000 K. Because luminosity depends on T^4, this factor of 10 in the temperature can account for a factor of 10^4 in the luminosity. The remaining factor of 10^5 to 10^6 in the luminosity range must be accounted for by different stellar radii.

Figure 15-17 shows the general location on the Hertzsprung-Russell diagram of the principal groups of stars we have mentioned. In addition, the relationship among temperature, radius, and luminosity has been used to plot *lines of constant stellar radius* on the diagram. This means that all the stars on a given line have the same radius (the lines turn out to be straight lines only because the plot is on a logarithmic graph). Lines are plotted for various multiples of the Sun's radius. For example, the topmost line is for stars whose radius is 1,000 times the Sun's, the next for those whose radius is 100 times the Sun's, and so forth.

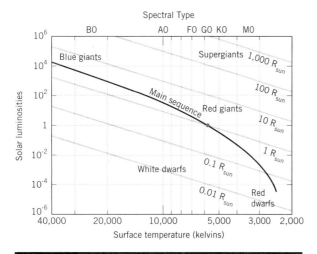

Spectral Type

Figure 15-17   The Hertzsprung-Russell diagram, showing the locations of the main groups of stars and several lines of constant radius.

Inquiry 15-24

Approximately how many times bigger than the Sun are the largest stars on the diagram? How does this compare with the Earth's orbital radius? About how much smaller than the Sun are the smallest stars on the diagram? What solar system object is this size close to?

Inquiry 15-25

The Sun is roughly 1.5 million km in diameter. From Figure 15-17, roughly what is the diameter of a typical white dwarf? Of a supergiant star?

We can readily see that the observed range of stellar radii is indeed large and sufficient to explain the remaining range of stellar luminosities not explained by temperature differences. As for the Sun, it is intermediate in radius between the smallest and largest stars (excluding the exceptional neutron stars, which will be discussed in Chapter 19). For this reason, the Sun is often referred to as a *typical* star. Further justification for this terminology is that its mass lies approximately in the middle of the range for all stars. However, as we saw from our examination of the H-R diagram, the Sun is actually more luminous and larger than the great majority of stars in the galaxy, so the word *typical* must be used with care.

Figure 15-18 is a more detailed and complete Hertzsprung-Russell diagram. It shows not only the location of many well-known stars, but also the locations of various types of objects you will read about later in this book.

15.6 Explanation of the Main Sequence

The H-R diagram is a graph that compares luminosity and temperature; however, these quantities are determined in a complex way by a star's mass. We therefore look first at how mass and luminosity are related and then at the average density of a star.

THE MASS–LUMINOSITY RELATION

The H-R diagrams indicate that most stars are found on the main sequence. The farther down the main sequence we go, the more stars there are, so that even on the main sequence the more luminous stars are rare and the less luminous ones are common. In fact, only about 3 percent of stars in our galaxy on the main sequence are more massive than the Sun, and stars having about half the mass of the Sun or less make up about 85 percent of the stars in our galaxy.

Astronomers have obtained further insight into the properties of stars by including their masses on the main sequence, as is done in **Figure 15-19**. (Remember, the mass comes from studies of binary star systems.) From this figure, you can see that the main sequence is actually a *mass* sequence, with the most massive stars at the upper left and the least massive at the lower right. The most massive star known with an accurately determined mass is Plaskett's star, at more than 70 solar masses. Although confirmation is needed, the star known as LBV 1806-20 is thought to be 150 solar masses; it may, in fact, be a binary system consisting of two very high-mass stars. The least massive star that has been measured is about 0.1 solar masses. We might well ask whether stars can exist that are substantially more or less massive than these limits. Surprisingly enough, theoretical reasons indicate that neither case is possible. In particular, an object much more massive than the upper limit would be unstable; it would then eject enough matter to bring it within the observed range. An object much less massive than 0.1 solar masses is not massive enough to produce sufficient energy to become a star. Given some uncertainties in the measurement of both the highest and lowest masses, we can say that the range of stellar masses runs from approximately 0.1 to 150 solar masses.

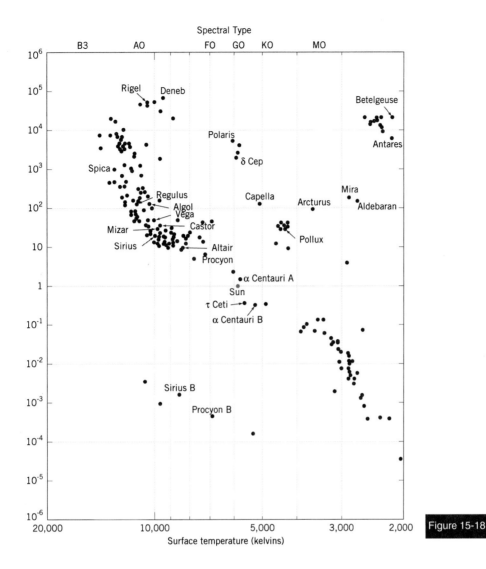

Figure 15-18
A more detailed Hertzsprung-Russell diagram.

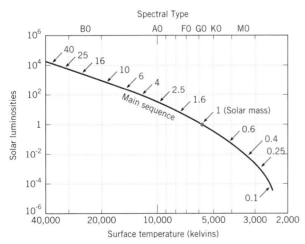

Figure 15-19
The main sequence, showing the locations of stars of various masses.

There is a sound physical reason for the existence of the main sequence. For normal main-sequence stars, mass determines a star's temperature and radius, and thus its luminosity: The more massive the main-sequence star, the more luminous it is. In Chapter 16, where we consider the internal structure of stars, we will see that this relationship exists because of nuclear reactions taking place in the centers of stars. In the meantime, we will simply take it as an observed fact. If we use the masses and luminosities from Figure 15-19 and plot a graph of these quantities, the result is **Figure 15-20**, which is called the **mass-luminosity relationship**. Like the Hertzsprung-Russell diagram, it is a logarithmic graph, and it shows that the increase of luminosity with mass is rapid. For example, a main-sequence star only twice as massive as the Sun is roughly 10 times as luminous. Mathematically, astronomers find very

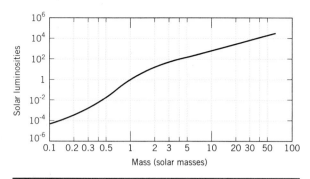

Figure 15-20 The mass-luminosity relationship for main-sequence stars.

roughly that $L \propto M^4$ where the mass and luminosities are expressed in units of the Sun.

It follows that the stars in the upper part of the main sequence are more luminous than the Sun because they are more massive. Their radii are all roughly three times the Sun's, so they lie along a constant-radius line just above the line that runs through the Sun. As we travel down the main sequence, the stars become less and less massive, and hence less luminous. At roughly the mass of the Sun, the decrease of luminosity with mass becomes more gradual; we will see the reason for this in a later chapter. At the same time, the radii of the stars on the main sequence start to decrease, so the main sequence dips below the line for one solar radius and into the region of cool stars called **red dwarfs**.

The mass-luminosity relation can be used to infer one quantity, given the other, indirectly. For example, suppose you observed the trigonometric parallax of a star; from this, you calculate a distance. That distance was determined using a direct method. The luminosity, which then can be calculated from the inverse square law, is a direct luminosity determination. If the star is a main sequence star, we can then use the mass-luminosity relation to infer, indirectly, the mass. That mass will be more uncertain than determinations made from nearby binary stars.

Inquiry 15-26

Classify each of the following quantities, some of which are from the previous chapters, as either directly or indirectly determined:

a. Distance to a car from the observed headlight brightness
b. The outside temperature from reading a thermometer
c. The outside temperature from an observation of ice on a tree
d. The radial velocity of a star
e. The luminosity class of a star found from measurements of the width of a spectral line
f. The stellar rotation velocity from the measured width of a spectral line
g. The presence and strength of a magnetic field

Inquiry 15-27

About how many times more massive is Plaskett's star than the least massive observed stars?

Inquiry 15-28

Which are most numerous in our galaxy: stars of high mass or stars of low mass?

DENSITIES OF STARS: A REPRISE

We have seen that the range in the masses of stars is relatively modest, whereas the range in their diameters is large. It follows that the range in stellar *densities* (i.e., mass per unit volume) is enormous. For example, the diameter of a typical white dwarf is roughly 1/100 that of the Sun's (that is, the diameter of the Earth), yet the star may have the same mass as the Sun. Because the volume of a spherical star is proportional to the *cube* of its diameter, the volume occupied by a white dwarf is only $(1/100)^3$, or one-millionth that occupied by the Sun. If the same amount of mass is packed into only one-millionth the volume, the density must be a million times as great. Because the average density of the material in the Sun is about that of water, the density of the material in a white dwarf must therefore be roughly one million times that of water. A cubic centimeter, if brought to Earth, would weigh over a ton! Similarly, the material in a red supergiant is, on the average, only one-millionth as dense as water.

15.7 Stellar Distances and the H-R Diagram

Hertzsprung and Russell took data supplied by others and drew conclusions about stars from the data. Their work was possible only because accurate parallaxes using photographic techniques were becoming available, and Annie Cannon was publishing her spectral types in the *Henry Draper Catalog* of stellar spectra.

One of the first applications of the Hertzsprung-Russell diagram was to infer the distance to stars that were too distant to exhibit a directly observable parallax. Although parallaxes (and therefore distances) can be directly observed only for the nearest stars, if it were possible to infer a star's luminosity by some means, then from its observed brightness one could infer the distance using the inverse square law of light.

We can see from the formula in Section 15.5 that to determine the luminosity of a star we must first be able to estimate both its surface temperature and its radius.

Inquiry 15-29

Assuming a star is a main-sequence star and its temperature has been measured to be 10,000 K, approximately what would be its luminosity relative to the Sun? Use one of the H-R diagrams discussed earlier to determine your answer.

We cannot realistically assume, as done in Inquiry 15-29, that a given star is necessarily a main-sequence star. Fortunately, it is possible to distinguish among luminosity classes (main-sequence, giant, and supergiant stars) by studying their spectra. Distinguishing between the groups is possible because of the enormous range in the densities of stellar atmospheres, which is accompanied by a similar range in the atmospheric pressures. Thus, as we saw in the previous chapter, the low-density stars that we now call supergiants have narrow spectral lines, while the high-density ones, the dwarfs, have broadened lines (see Figure 14-18). This technique works well when hydrogen lines are present; the situation is more complex in stars having weak or no hydrogen lines in their spectra.

A star's spectrum thus allows us to infer both the star's surface temperature and its luminosity class. With these two pieces of information, we can tell approximately where the star falls on the Hertzsprung-Russell diagram. From its location on the H-R diagram, we infer the star's luminosity. While it is true that there will be some uncertainty in this inference, an imprecise result is better than none at all. If this process is done for a large number of stars, the overall average for the population as a whole is precise. Then, by comparing its inferred luminosity to its observed brightness (which is a quantity that is relatively easy to obtain), we can estimate the distance to the star from the inverse square law of light. A distance determined in this way, by using the *observed* spectral type and luminosity class, and the H-R diagram to infer the luminosity, is known as a **spectroscopic parallax**. The name is something of a misnomer, because parallax is an angle, but the word is used here as a synonym for distance. The spectroscopic-parallax technique is summarized in **Figure 15-21**.

The spectroscopic parallax technique is an example of an *indirect* method of finding distance. Indirect methods use relationships—in this case, the H-R diagram—to infer information (in this case, *L*) needed to obtain the final result. Such methods have greater uncertainties than direct methods.

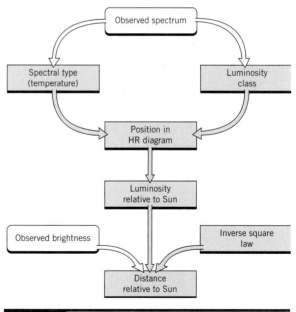

Figure 15-21 The process of determining stellar distances using the spectroscopic-parallax method.

Spectroscopic examination of a star and the use of the H-R diagram to infer its luminosity are the keys to this method. In using the H-R diagram, we are assuming that the properties of the star whose distance we wish to find are similar to those of the vast majority of other stars. Although direct parallax measurements from the ground are valid only to, at most, 300 light-years from the Sun, the spectroscopic method can be used on any star for which a good spectrum can be obtained and classified with accuracy, excluding those stars that are "peculiar" in some way. Astronomers are therefore able to determine distances indirectly to a large number of stars. Of course, the GAIA parallax project referred to previously will move all of astronomy to a new level of knowledge that astronomers have thought of previously only in their wildest dreams!

We illustrate the spectroscopic-parallax method of finding distance using the star α Centauri. Spectroscopic observations of its absorption lines show that it is similar to the Sun (a main-sequence, G-type star) and, in fact, is only slightly more luminous. Measurements, however, show that the much-closer Sun *appears* much brighter—almost 9×10^{10} times brighter than α Centauri. From the inverse square law, the distance is given by the square root of this brightness ratio. Therefore, we find that α Centauri is nearly 3×10^5 times farther away from us than the Sun. This distance of 3×10^5 AU may be expressed in other units as follows:

$$\frac{3 \times 10^5 \ \text{AU}}{2 \times 10^5 \ \text{AU/pc}} = 1.5 \ \text{parsecs}$$

$$(1.5 \ \text{parsecs}) \times (3 \ \text{ly/parsec}) = 4.5 \ \text{ly}$$

$$(3 \times 10^5 \ \text{AU}) \times (1.5 \times 10^8 \ \text{km/AU}) = 4.5 \times 10^{13} \ \text{km}.$$

Inquiry 15-30

How many powers of ten are there between the numbers used to express a distance in parsecs and the same distance expressed in astronomical units?

By contrast, the star Canopus appears to be just about as bright as α Centauri, and yet its location in the Hertzsprung-Russell diagram tells us it is about 1,200 times as luminous. Again, we use the inverse square law: because the square root of 1,200 is about 35, it follows that Canopus is about 35 times farther from us than α Centauri, or about 160 light-years. (Readers desiring a more general mathematical treatment of spectroscopic parallax may refer to Appendix A10.)

Visual Binary Stars

DISCOVERY 15-1

When you have completed this Discovery, you should be able to do the following:

- Determine the period of stars in a visual-binary system from a series of photographs.
- Determine the scale of a photograph, given the angular separation of two objects in the photograph.
- Determine the separation of two stars in a binary system given their observed angular separation and the distance of the stars from the Sun.
- Determine the sum of the masses of the stars in a visual binary system.

Figure 15-3 contains photographs of the binary-star system Krüger 60 taken in 1908, 1915, and 1920. We will refer to the brighter star in the binary system as the *primary,* and the fainter one as the *secondary.* The star in the lower-right corner is simply a star within this field of view. The motion of the secondary relative to the line between the primary and the field star is clear.

Note: This activity calls on many pieces of knowledge from earlier in the book. You may want to review Section 3.3, the end of Section 5.4, and the discussion of distance at the beginning of this chapter. A drawing may help you visualize the relationships of the stars in the system and yourself.

You should make a photocopy of Figure 15-3 or use tracing paper rather than drawing over the original.

You will first determine the period of the binary system. Start by drawing a line through the primary star parallel to the left edge of the photograph. This line is your reference line. Repeat for each photograph. Place a pencil mark at the center of each primary and secondary star image and draw lines connecting the centers of the primary and secondary stars. Your lines should extend well past the secondary star. Use a protractor to determine the angle at the primary between your reference line and the line to the secondary. Tabulate the angle for each of the dates. Determine the angle through which the star has moved between each date.

- **Discovery Inquiry 15-1a** In a complete cycle the secondary star will travel through an angle of 360°. Using the information about the motion of the secondary derived previously, determine the period of the binary star system. (We are making the valid assumption that all the photographs are from the same period.)

A photograph is similar to a map. All maps have a scale that relates the distance on the ground to a distance on the map. For example, a map scale might be 1 inch on the map equals 50 miles on the ground.

Astronomers cannot directly measure miles in space between objects but they can measure angles. Angles between stars become distances measured on a photograph. For instance, a two-inch photograph of the Moon, whose angular diameter is 0.5°, would have a scale of 1° per inch.

To determine the scale of each photograph of the visual binary star system for this Discovery use the fact that in 1908 the field star to the lower right was 30 seconds of arc from the primary. (Although the scale in each photograph is the same, it appears different because the stars themselves have moved.)

- **Discovery Inquiry 15-1b** Using this number and a ruler, determine the angular separation *in seconds of arc* between the center of the primary and secondary components for each photograph.

We will assume that the average of your measured angular separations equals the semi-major axis of the elliptical orbit of the secondary around the primary. Although this assumption is not necessarily true, it is sufficient for our purposes here.
  Krüger 60 has a parallax of 0.254 second of arc.

- **Discovery Inquiry 15-1c** What is the distance of Krüger 60 from the Earth in parsecs? In AU?

- **Discovery Inquiry 15-1d** Use the distance you just computed, along with the average angular separation you found, to determine the length of the semi-major axis of the orbit in both astronomical units and kilometers. A drawing that includes the observer and the two stars may help; then use this form of the angular-size formula:

$$\text{Angular separation (seconds of arc)} = \frac{\text{Linear separation (AU)}}{\text{Distance (parsecs)}}.$$

- **Discovery Inquiry 15-1e** Using all the information you now have, determine the sum of the masses of the stars in the Krüger 60 binary system.

Although we have not determined the masses of the individual stars in this system, knowledge of their sum is useful. In some systems, the motion of the primary and the secondary in relation to the background stars can be determined. In these cases, we can determine the *ratio* of their masses. Then, given the ratio, and the sum as determined above, we can obtain the masses of the individual stars.

Spectroscopic Binary Stars

When you have completed this Discovery, you should be able to do the following:

- Determine the scale of a spectrum, given the wavelengths of two lines in a comparison spectrum.
- Measure the shift between spectral lines in millimeters and determine their shift in Å.
- Apply the Doppler shift formula to these measurements to determine the relative radial velocity of the two stars.

If a binary system is composed of two stars that have about the same brightness, the spectrum of the system will consist of the spectrum of each star superimposed on one another. If we see the orbital plane edge-on, there will be times at which one star approaches the observer and the other moves away. At these times the Doppler shift will cause two spectral lines rather than one to be seen. From the separation of the lines, we can determine the relative speeds of the stars in the system.

The spectrum of ζ Ursae Majoris (also known as Mizar) is shown in Figure 15-5. The top spectrum *(a)* was taken when the stars were moving across the line of sight so there was no Doppler shift of the stellar spectrum; the bottom one *(b)* was taken when one star was moving toward the observer and the other was moving away, thus producing the observed change in spectral-line position.

Above and below the two stellar spectra are emission spectra of a standard lamp that is attached to the spectrograph. Because this lamp is motionless with respect to the spectrograph, its lines are at their rest wavelengths and may be used as reference markers.

Use a millimeter ruler to determine the scale of the photograph by finding the number of angstrom units in each millimeter measured on the figure. Next, use your ruler to measure the separation of the two lines centered on 4481 Å in the bottom spectrum; then, use the value of the scale you computed to find the separation of the stellar lines in Å. This separation is due to the relative motion of the stars.

- **Discovery Inquiry 15-2a** What is the value of the scale of the spectrum in Å/mm?
- **Discovery Inquiry 15-2b** What is the value of the separation of the lines in mm? In Å?
- **Discovery Inquiry 15-2c** What is the value of the relative radial velocity of the two stars?

In practice, measurements such as this are made at many times throughout a complete orbit. When this is done we have a graph of the radial velocity versus time—a radial velocity curve as shown in Figure 15-6c. Further analysis of this graph gives detailed information about the orbit and, finally and most important, the masses of the stars.

Chapter Summary

Observations

- The **Hertzsprung-Russell diagram**, the most important graph in astronomy, plots **luminosity** and **spectral type** (which is determined by the star's temperature). The majority of stars lie on the **main sequence**, with higher-luminosity groupings called **giants** and **supergiants**, and lower-luminosity **white dwarfs**.

- The brightest stars in the sky appear bright because they have high luminosities, not because they are close to the Sun.

- For stars on the main sequence there is a **mass-luminosity relation** showing that stellar luminosity increases strongly with stellar mass.

- There are three types of binary star systems: **visual binaries**, **spectroscopic binaries**, and **eclipsing binaries**. Such systems are most important in determining stellar masses and sometimes their diameters.

Theory

- A star's average surface temperature can be calculated from its spectral type.

- Luminosity is proportional to the fourth power of the surface temperature and to the star's surface area (which is proportional to the square of the radius).

- The **inverse square law** of light states that the apparent brightness varies inversely with the distance squared.

Conclusions

- Most stars in our galaxy are main-sequence stars.

- The **spectroscopic parallax** method of distance determination requires the star's spectral type and **luminosity class**, both of which are readily obtained from the spectrum. The star's luminosity is then inferred from the H-R diagram. When inferring the luminosity, we implicitly assume the star whose distance is to be found is similar to the other stars in the H-R diagram. The determination of luminosity from the H-R diagram is uncertain because we do not know if the star in question actually has the average luminosity of its luminosity class, or if it is more or less luminous. The distances determined using these luminosities, therefore, contain uncertainties.

- Stellar masses are inferred from observations of the period and semi-major axes of stars in binary systems. If a star's luminosity is known, and if the star is a main-sequence star, its mass can be inferred from the mass-luminosity relation.

- Stellar diameters may sometimes be obtained from analysis of binary stars. The techniques of interferometry and occultations also give astronomers information on stellar radii.

Summary Questions

1. Sketch and label the Hertzsprung-Russell diagram for all types of stars and identify each of the various groups of stars on it. Explain the terminology main sequence, red dwarf, white dwarf, giant, and supergiant.

2. Why is the Hertzsprung-Russell diagram for the nearest stars different from that for the brightest stars in the sky?

3. How and why are a star's radius, temperature, and luminosity interrelated?

4. How do the various groups of stars on the Hertzsprung-Russell diagram differ in terms of their radii?

5. Where on the main sequence do stars of different masses lie? Draw a schematic H-R diagram and indicate such locations.

6. What is the range of orders of magnitude for stellar mass, luminosity, radius, surface temperature, and density?

7. What is the method of spectroscopic parallax? What are its advantages and disadvantages?

8. What are the three main types of binary star systems? Describe them.

9. How can astronomers determine the masses and radii of stars directly?

Applying Your Knowledge

1. Suppose you have an eclipsing system in which the stars have the same diameter. Assuming the eclipses are central—that is, during the eclipses the centers of the stars coincide—draw the resulting light curve.

2. Discuss the process by which astronomers determine a spectroscopic parallax. In your discussion be sure to specify what is observed and what is inferred.

3. Make up a table that shows the maximum and minimum observed values for the following properties of stars: temperature, mass, diameter, density, and luminosity. Express your results both in absolute numbers and in terms of the Sun's value.

▶ 4. Rewrite the formula relating distance in parsecs and the parallax angle to be valid for astronauts observing from an orbit whose average distance from the Sun is A astronomical units.

▶ 5. What would be the distance to a star identical to the Sun but whose apparent brightness is 10^{16} times less than that of the Sun?

▶ 6. What would be the approximate period of revolution for a binary system composed of two stars like the Sun for which the semi-major axis is (a) 1 AU, (b) 10 AU, and (c) 40 AU?

▶ 7. To explain a *possible* period of 26 million years in the extinction of some biological species, scientists have suggested that the Sun might have a companion whose period is the interval between extinctions. What would be the distance to such a hypothetical star (which is known as Nemesis)? How does this distance compare with the distance to the nearest star?

▶ 8. The star α Canis Majoris (Sirius) is a binary of period 50 years. Its semimajor axis is 7.5 seconds of arc, and the system's parallax is 0.378 second of arc. What is the sum of the masses of the two stars in solar mass units?

▶ 9. What is the distance to a blue supergiant star whose apparent brightness is 10^{14} times less than the Sun's brightness, assuming a true luminosity of 10^4 the luminosity of the Sun? As seen in the H-R diagram in Figure 15-18, such supergiants do not all have the same luminosity. If the true luminosity is actually 2×10^4 that of the Sun, what is the star's distance? Compare the two distance estimates, and discuss how uncertainties in the inferred luminosity produce uncertainties in the derived distance. (*Hint:* Appendix A10 may be helpful.)

▶ 10. How many times less massive is Jupiter than the least massive star shown in the mass-luminosity relation in Figure 15-20?

▶ 11. What would be the luminosity (in solar units) of a star whose temperature was 3,000 K and radius was 1,000 times the radius of the Sun?

▶ 12. Consider the star Betelgeuse, which is located in the H-R diagram of Figure 15-18. Use the data for the star from the graph to determine the radius of Betelgeuse in terms of the Sun's radius.

▶ 13. What would be the parallax of Proxima Centauri if it were observed from a telescope on one of Saturn's satellites?

14. Write an essay discussing direct and indirect methods for finding properties of stars. Give specific examples.

15. Draw a flow diagram, as in Figure 15-21, showing how to find the mass, distance, and luminosity, as well as the mass–luminosity relation.

Answers to Inquiries

15-1. Distance = 1/parallax angle = 1/0.01 = 100 parsecs, or 326 ly.

15-2. The distance is directly proportional to the length of the baseline. Because Saturn is 10 AU away, the distance will increase to 1,000 parsecs.

15-3. From the inverse square law, $\sqrt{50} \sim 7$, so the more distant star is about $7 \times 12 = 84$ ly away.

15-4. The ratio of the distances is 6. From the inverse square law, the luminosity depends on the distance squared, or 36 times brighter.

15-5. If $M_1 + M_2 = 5$, and, assuming that M_1 is the more massive star, $M_1/M_2 = 4$, then $M_2 = 1$ and $M_1 = 4$. The next Inquiry addresses one method to observe which is the more massive star.

15-6. The mass ratio equals the velocity ratio, with the more massive star moving more slowly. Therefore, because $V_B = 5V_A$, we have $M_A = 5M_B$.

15-7. 2×10^{27} watts; 8×10^{25} watts.

15-8. Approximately $1/400 \ L_\odot$; $20 \ L_\odot$

15-9. On a rough diagonal from the center of the plot to the lower-right corner. That is, the vast majority of stars near the Sun are both cooler and fainter than the Sun.

15-10. Significantly fainter than the majority of the stars. (As we will see, these are called white dwarfs.)

15-11. The bright stars divide into two main groups: those along a diagonal line in the upper-left part of the diagram, and a group to the right of the first set of points.

15-12. Those on a rough diagonal from the upper left corner to the center of the plot.

15-13. About 10^{10}.

15-14. The nearest and brightest stars form two distinct groups. From the H-R diagram we see that the nearest stars are intrinsically faint, while the brightest-appearing ones are also intrinsically bright. The nearest ones are typical of the vast majority of stars, while the brightest ones are seen only because they are so luminous. The latter group is atypical.

15-15. The ratio between the brightest and faintest stars on the graph approaches 10^6 to 10^7.

15-16. The brightest stars in the sky are that way because they are intrinsically very luminous, not because they are nearby.

15-17. Such a star would be intrinsically faint. Given the high sensitivity of recent sky surveys, it is thought that almost all such stars have been identified.

15-18. We are located in a part of the Galaxy that is typical of the Galaxy as a whole.

15-19. It is often stated that the Sun is a "typical" star. Insofar as it is outshined by many other stars, this is true; however, only 6 of the 125 stars in the graph of nearby stars are more luminous than the Sun. Therefore, the Sun is more luminous than most of the stars in our galaxy. This is because most stars are less massive than the Sun.

15-20. The data in the figure indicate that roughly 6% (8/125) of the stars are white dwarfs.

15-21. Since the stars have the same temperature, the luminosity must depend on the area of the star; the area depends on the radius squared. Thus, with $L \propto R^2$, we have $2^2 = 4$ times as luminous.

15-22. Since the stars have the same radius, the luminosity must depend on the temperature to the fourth power. Thus, with $L \propto T^4$, we have $2^4 = 16$ times as luminous.

15-23. The luminosity ratio is 2×10^6. Because luminosity depends on the square of the radius ($L \propto R^2$), the radius depends on the square root of the luminosity. Antares is $\sqrt{2} \times 10^6 = 1.4 \times 10^3$ times, or 3 orders of magnitude larger in radius than CD $-46°11540$.

15-24. More than 1,000 times as large, about 10 times Earth's orbital radius. Only 1/100 as large, or about the radius of Earth.

15-25. White dwarfs are 100 times smaller or roughly 15,000 km (about the diameter of Earth). Supergiants are roughly 300 times greater or 5×10^8 km in diameter (a few astronomical units).

15-26. a. indirect, because you must infer it after making an implicit assumption about the headlight's intrinsic brightness.

b. direct.

c. indirect, as the temperature could be well below the temperature at which water freezes.

d. direct.

e. indirect, as you are making an inference from a relationship.

f. direct measurement from the Doppler shift.

g. direct measurement of line splitting.

15-27. Plaskett's star is roughly 700 times more massive than the least massive star.

15-28. Stars of low mass.

15-29. From Figure 15-17, a main sequence star with a temperature of 10,000 K would have a luminosity of 50–60 $L_\odot$. (Remember, the luminosity scale is a logarithmic scale, so interpolation is not easy.)

15-30. The *number* representing the distance in parsecs is 5 powers of ten smaller than the number of astronomical units.

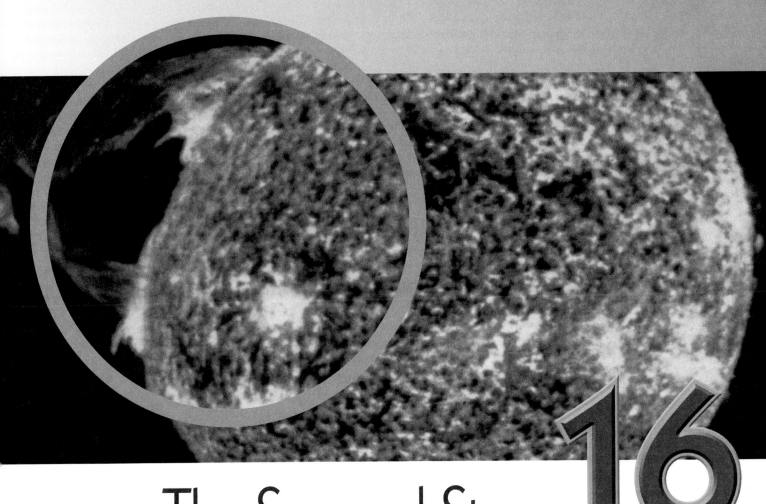

The Sun and Stars: Their Energy Sources and Structure

16

We had the sky, up there, all speckled with stars, and we used to lay on our backs and look up at them, and discuss about whether they was made, or only just happened.

Mark Twain, Huckleberry Finn

When scientific research depends heavily on government support, as it now does, debate over funding priorities is inevitable. Many feel that research should be directed primarily toward the solution of practical problems such as improvement of crop yields, discovery of new ways to produce energy, cures for disease, and the like. Others point out that many of the most important practical discoveries of science have come from basic research conducted solely for the purpose of unlocking nature's secrets, without any practical objectives in mind. In fact, both kinds of research are important.

The discovery of nuclear fusion is one of the most dramatic examples of the importance of basic research. For good or ill, few discoveries have had more practical implications for the survival of humanity, yet the basic concepts were first discovered by astronomers and physicists doing research that appeared to be utterly without any practical significance. Their question was: What is the source of power for the Sun and other stars?

16.1 The Power Produced by the Sun

As soon as the approximate distance to the Sun was known, astronomers could determine the Sun's power, or luminosity. Measurements show that 1,400 watts—enough power to light 14 standard 100-watt light bulbs—falls on a square meter at the Earth's distance from the Sun. Averaging cloud cover and atmospheric characteristics throughout the year, some 13 percent of this prodigious amount of energy strikes the ground in the United States. For example, in the course of a year approximately 10^{14} kilowatt-hours of sunlight fall onto the top of the atmosphere above the state of Kansas. If it were possible to recover it all, at seven cents per kilowatt-hour it would be worth about $7 *trillion* and would amount to roughly 13 times the world's entire electricity consumption.

The Sun supplies abundant energy to Earth. Unfortunately, even with the best technology only a small portion of this power could actually be converted into electricity. Any energy crisis might well be called an energy *cost* crisis, because at present it is not *economically* feasible to harness most of this solar energy. But if and when the cost of conventional types of energy rises high enough, there will come a time when solar power will indeed be economically competitive, and we can look forward to a day when much of the energy we use will be derived directly from the Sun.

From observations of the power of sunlight received on Earth, and knowing the Sun–Earth distance, we can calculate that the luminosity of the Sun is about 4×10^{26} watts.[1]

> **Inquiry 16-1**
>
> What principle or law would you use to compute the energy radiated by the Sun, given an observation of the amount of energy falling on Earth?

The amount of power radiated by the Sun is enormous, and it was obvious to nineteenth-century astronomers that ordinary burning simply could not supply this amount of energy for more than a few thousand years. At the same time, because geologists were providing convincing evidence that Earth was millions of years old, it was clear that some other method of producing energy in the Sun must exist.

16.2 Energy Sources

In our discussion of historical candidates for the source of stellar energy, we consider a variety of mechanisms, some that work and some that do not. We begin with the conversion of chemical energy

---

Chapter opening photo: The Sun as seen from space during an eruption of a prominence.

[1] A watt is the amount of power produced by 1 joule of energy in a second. In other words, 1 watt = 1 joule/s.

into radiation and then consider gravitational potential energy and a medley of mechanisms involving the atomic nucleus.

CHEMICAL REACTIONS

An early proposal for the source of energy in the Sun involved chemical reactions. Some familiar chemical reactions are those that occur during the burning of a candle flame or those that occur within the internal combustion engine of your car. Those reactions involve changes in the arrangement of the bonds that hold atoms together to form molecules. The bonds involve the sharing between atoms of the outermost electrons located around the atomic nucleus; the nucleus itself is not involved. Energy may be given off whenever the atoms or molecules rearrange themselves or combine with other atoms or molecules. The amount of energy given off varies with the chemical reaction.

To check the hypothesis of chemical reactions as the Sun's energy source, let us see how long the Sun could last if its luminosity came from chemical reactions. Just as the length of time for a water-filled bucket to empty will depend on how much water is initially there and the rate at which water flows out, the lifetime of a star depends on the amount of energy available, and the luminosity or rate at which the star emits this energy:

$$\text{Lifetime} = \frac{\text{Energy available}}{\text{Rate of loss of energy}} = \frac{\text{Energy}}{\text{Energy/s}}.$$

The Sun contains some 10^{38} joules of *chemical* energy and radiates 4×10^{26} watts. Thus, the lifetime over which chemical reactions could supply the observed luminosity would be as follows:

$$\text{Lifetime} = \frac{10^{38} \text{ joules}}{4 \times 10^{26} \text{ joules/s}}.$$

$$\text{Lifetime} = 2.5 \times 10^{11} \text{ s}$$

$$\text{Lifetime} = 8,000 \text{ years}.$$

Clearly, the Sun has insufficient chemical energy to provide the observed luminosity over geological and biological time scales.

GRAVITATIONAL COLLAPSE

The physicists Lord Kelvin and Hermann von Helmholtz proposed one answer to the mystery of the Sun's energy source. They pointed out that a large mass such as the Sun would, under the action of its own gravity, arrange itself into a spherical

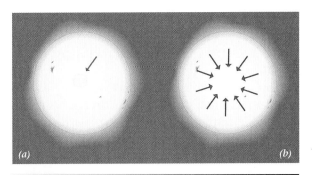

Figure 16-1 *(a)* Gravity acting on a volume of material pulls the matter toward the center. *(b)* Gravity causes the gas from which the Sun formed to collapse slowly into a sphere.

shape. Because each particle of matter will tend to be drawn toward the Sun's center (**Figure 16-1a**), the effect of gravity will be to cause the Sun to contract and shrink in size (**Figure 16-1b**).

The arguments of Kelvin and Helmholtz are easy to understand. From earlier discussions in Chapters 5 and 7, you may recall that bodies in motion have kinetic energy, while bodies having the *possibility* of motion have potential energy. A star has potential energy by virtue of its gravity. Gravity's pull on the outer parts of a star causes those parts to move towards its center. This slight decrease in the size of the star decreases the amount of potential energy present. The law of conservation of energy, however, requires that any decrease in potential energy result in an increase in the kinetic energy within the star. The end result is that half the loss of potential energy produces an increase in the temperature of the star, while the other half is released as light and is radiated away.

Let us hypothesize that the energy of the Sun comes from a continuous but slow decrease in its size through the conversion of gravitational potential energy into radiation. Although this *possibility* makes good physical sense, is it *reasonable* as the energy source in the Sun?

To get to its current density, the Sun had to have lost about 10^{42} joules of gravitational potential energy since its birth. Gravitational collapse would need to supply the observed luminosity at the rate of 4×10^{26} watts. This means that this energy source could power the Sun for 100 million years:

$$\text{Lifetime} = \frac{10^{42} \text{ joules}}{4 \times 10^{26} \text{ joules/s}}$$

$$= 2.5 \times 10^{15} \text{ s}$$

$$= 10^8 \text{ years}.$$

This is 10,000 times longer than chemical energy could last, but still short by a factor of 50 to match the age of Earth's oldest rocks, just over 4 billion years.

Throughout most of the nineteenth century this **Kelvin-Helmholtz contraction time** appeared to be a sufficiently long time, and the question of the source of the Sun's power was not considered troublesome.

However, by the end of the nineteenth century, geological evidence had increased the estimated age of Earth to *several* hundred millions of years. Further, the discovery of radioactivity at the close of the century made it possible to measure Earth's age with even greater certainty, to where the accepted value today is 4.54 *billion* years. Because it is hard to imagine how Earth could be much older than the Sun, the source of the Sun's energy, which can be neither chemical nor gravitational, once again became one of the great puzzles of astronomy.

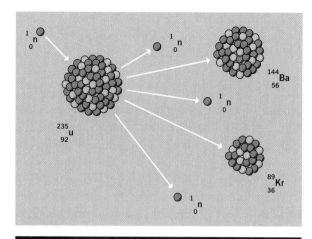

Figure 16-2 The fission of uranium into barium, and krypton.

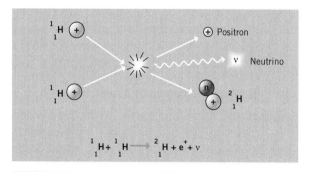

Figure 16-3 The fusion of two hydrogen nuclei to form a deuterium nucleus releases energy.

Inquiry 16-2

The conclusion that gravitational contraction is not the source of the Sun's energy depends on what assumptions about the constancy of the Sun's energy output over time? Explain.

Although gravitational collapse does not account for most of the luminosity of the Sun, starting in the next chapter we will see that this mechanism does play an important role for certain types of objects at a number of points in their lives.

NUCLEAR REACTIONS: FISSION VERSUS FUSION

The attention of astronomers and physicists was drawn to the nucleus of the atom as a possible source of power for the stars and the Sun. Theoretical investigations showed that there were two basic types of nuclear reactions that might be a source of large quantities of energy. In the first of these, **fission** reactions, heavy nuclei split into lighter ones. An example is the fission of uranium into nuclei of krypton and barium, along with neutrons, as illustrated in **Figure 16-2**. The other type of reaction, **fusion**, takes place when several light nuclei collide and combine (that is, fuse) into a heavier nucleus, releasing energy in the process. Because the Sun is composed almost entirely of light elements, the British physicist Sir Arthur

Eddington in around 1920 reasoned that fission would not be likely since spectroscopic studies of the Sun show elements other than hydrogen and helium to be exceedingly rare. Calculations show that even if the Sun were composed entirely of uranium, this would produce only half the observed luminosity. However, fusion might well provide the Sun's energy. We now look more closely at fusion reactions.

THE PROTON-PROTON CHAIN

It was not immediately obvious that fusion processes were possible in the core of the Sun. One possible fusion reaction was shown in the 1930s to be workable. A young graduate student, Charles Critchfield, who worked for the physicist George Gamow, calculated that at the temperatures and pressures expected to exist at the center of the Sun, two hydrogen nuclei (which are protons) could collide and form a nucleus of deuterium, or heavy hydrogen (**Figure 16-3**). This is the first reaction in a series called the **proton-proton**

chain. Hans Bethe, who was awarded the 1967 Nobel Prize in physics for his important work, made additions and refinements. We refer to such reactions as **hydrogen burning**, although it is not *burning* in the usual chemical sense but, as we will see, in the sense of consuming fuel. Symbolically, the reaction can be written as follows:

$$^1_1H + ^1_1H \rightarrow ^2_1H + e^+ + \nu.$$

In this nuclear reaction, the subscript represents the number of protons (positive charges) in the nucleus. The superscript represents the number of protons plus neutrons in the nucleus. Thus, the particle ^1_1H is a hydrogen nucleus with one proton and no neutron; in other words, ^1H is a protein. The deuterium nucleus, called a *deuteron* and written ^2_1H, consists of one proton and one neutron; therefore, its charge is one unit and its atomic mass number is two units. In addition, two other particles are produced: a positron, a positively charged electron written e^+, and a very low mass, chargeless particle known as a **neutrino** and designated by the Greek letter ν (nu). We will have more to say about these particles in Section 16.4.

Note that the number of positive charges is the same on both sides of the reaction equation. This is called *conservation of charge* and must be obeyed for a reaction to occur.

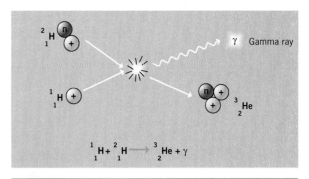

Figure 16-4 The fusion of hydrogen and a deuterium nucleus to form a nucleus of helium-3.

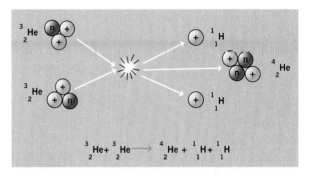

Figure 16-5 The fusion of two helium-3 nuclei to form a helium-4 nucleus. To get the two helium-3 nuclei requires cycling through the previous steps twice.

Inquiry 16-3

Assuming the mass of the positron is negligible in comparison with that of a proton or a neutron, is the total atomic mass number conserved in the fusion of two protons? The electrical charge? Justify your answer.

Inquiry 16-5

Nuclear reactions must conserve charge and total atomic mass number. On the basis of such considerations, which of the following reactions is impossible, and why?

$$(a)\ ^2_1H + ^1_1H \rightarrow ^3_1H + e^+ + \gamma$$

$$(b)\ ^2_1H + ^1_1H \rightarrow ^3_1H + \gamma$$

Once created, the deuteron can collide with yet another proton, producing a light isotope of helium that has two protons and one neutron (^3_2He). At the same time, a high-energy photon, or gamma ray (symbolized by the Greek letter γ), is produced (**Figure 16-4**). Symbolically, the reaction is written

$$^2_1H + ^1_1H \rightarrow ^3_2He + \gamma.$$

The final step in the proton-proton chain takes place when a second nucleus of light helium, created in the same way as the first, collides with light helium nucleus, forming a nucleus of ordinary helium (^4_2He) and ejecting two protons (**Figure 16-5**). In symbols,

$$^3_2He + ^3_2He \rightarrow ^4_2He + 2^1_1H.$$

Inquiry 16-4

Because the photon is massless and chargeless, show that this reaction conserves atomic mass number and charge.

Inquiry 16-6

Considering the entire sequence of the three reactions in the proton-proton chain, how many protons are involved in the production of one nucleus of helium-4 (^4_2He)? How many are left over?

The first two parts of the cycle occur twice for every one occurrence of the final step. Counting all the protons that go into the proton-proton chain, we find that the net result is to transform four hydrogen nuclei into one nucleus of helium-4. But if we compare the *masses* of the four protons with the resulting helium nucleus, we find that mass appears to be missing because the mass of four protons is slightly more than the mass of a helium nucleus. In fact, the mass of a proton is 1.67252×10^{-27} kg, whereas the mass of a helium nucleus is 6.64258×10^{-27} kg. If we subtract the mass of one helium nucleus from that of four hydrogen nuclei, we find the following:

Mass of four hydrogen nuclei = 6.69008×10^{-27} kg

Mass of one helium nucleus = 6.64258×10^{-27} kg

Difference in mass = 0.04750×10^{-27} kg.

It appears that a small fraction of the original mass ($0.04750/6.69008 = 0.007$) is missing. What has happened to it? Albert Einstein (**Figure 16-6**) came up with the answer in 1905. As a consequence of his theory of special relativity, Einstein showed that mass and energy are different forms of the same thing and that each can be converted into the other. His famous formula $E = Mc^2$ gives the relationship between the amount of mass and an equivalent amount of energy. In words,

Energy = Mass × (Speed of light)^2.

The speed of light is 3×10^8 m/s and its square is 9×10^{16} m^2/s^2—a large number. This means that

Figure 16-6 Albert Einstein (1879–1955). This great physicist was also a humanitarian and an accomplished violinist.

even a small amount of mass is equivalent to a large amount of energy. Just 10^{-3} kg of matter (about five paper clips), if converted entirely into energy, is equivalent to nearly 10^{17} joules, or about 3×10^7 kilowatt-hours. At a cost of seven cents per kilowatt-hour, this would be worth about $2 million! And so, the missing mass was not missing but was converted into another form, namely energy.

Inquiry 16-7

How much energy (in joules) is produced by fusion of four hydrogen nuclei into one helium nucleus? Assume that 0.05×10^{-27} kg is entirely converted into energy.

Inquiry 16-8

Each second the Sun radiates about 4×10^{30} joules of energy into space. How many kilograms per second of matter is this equivalent to, assuming that the source of the Sun's energy is the conversion of mass into energy?

The rate at which mass is converted to energy in the Sun, as computed in Inquiry 16-8, is huge by Earth's standards. How long would it take the Sun to convert only 10 percent of its mass of hydrogen into helium plus energy at the current rate? Remember from earlier in the chapter that the lifetime of an energy source is given by the energy available divided by the rate of loss of energy. Accounting for the fact that in hydrogen burning only 0.007 of the mass becomes energy, we can compute the lifetime as follows:

$$t = 0.007 \frac{\text{(Mass to be converted)} \times \text{(Speed of light)}^2}{\text{Energy radiated per second}}$$

$$t = 0.007 \frac{(2 \times 10^{30}\,\text{kg}) \times (3 \times 10^8\,\text{m/s})^2}{4 \times 10^{26}\,\text{joules/s}}$$

$t \sim 10^{10}$ years.

Even at the prodigious rate of conversion of mass to energy implied by Inquiry 16-8, the Sun would be able to shine for 10 billion years and still have radiated away less than one-tenth of 1 percent of its mass. So nuclear reactions do not conserve mass (because a small amount is used up), but they do conserve mass *plus* energy. The energy from the nuclear furnace in the core of the Sun works its way out through the enormous mass of the Sun, eventually emerging from its outer surface as its visible luminosity and 3 percent as neutrinos.

Does the fact that fusion of hydrogen to helium is physically plausible *prove* its existence in the Sun? No. A discussion of observations concerning fusion in the Sun will come in Section 16.4.

In summary, a chemical reaction involves only the outermost electrons of an atom. The atoms retain their identity because their nuclei are not affected. Nuclear reactions involve nuclei and produce a change in the identity of the reacting nuclei. Gravity releases four orders of magnitude more energy than chemical reactions, and nuclear reactions are two to three orders of magnitude more energetic than gravitational collapse.

THE CARBON–NITROGEN–OXYGEN CYCLE

Astrophysicists think that the proton-proton chain is the main power source for the Sun and for stars less massive than the Sun (which means the majority of stars) because it is the most likely reaction at the core temperature of these stars. But there is another important set of hydrogen-burning reactions called the **carbon–nitrogen–oxygen cycle** that occurs at higher temperatures. Although it contributes only a small amount to the Sun's luminosity, it dominates in stars that are more massive than a few times the Sun's mass.

In the carbon–nitrogen–oxygen, or **CNO**, cycle, hydrogen is also converted into helium. Nuclei of the elements carbon, nitrogen, and oxygen are also involved. The sequence of nuclear reactions in the CNO cycle follows:

$$^{12}_{6}C + ^{1}_{1}H \rightarrow ^{13}_{7}N + \gamma$$
$$^{13}_{7}N \rightarrow ^{13}_{6}C + e^{+} + \nu$$
$$^{13}_{6}C + ^{1}_{1}H \rightarrow ^{14}_{7}N + \gamma$$
$$^{14}_{7}N + ^{1}_{1}H \rightarrow ^{15}_{8}O + \gamma$$
$$^{15}_{8}O \rightarrow ^{15}_{7}N + e^{+} + \nu$$
$$^{15}_{7}N + ^{1}_{1}H \rightarrow ^{12}_{6}C + ^{4}_{2}He.$$

You should be able to verify the conservation of charge and atomic mass number in each of these reactions. Notice that, at the reaction's end, the original carbon-12 nucleus is returned unchanged. Here, carbon functions as a catalyst—a substance that helps a reaction to take place without itself ultimately being changed. If a star was made from material containing no carbon atoms, it would not be able to generate energy using this cycle even if it had enough mass and therefore a high enough core temperature to do so.

The CNO cycle, just like the proton-proton chain, also produces gamma rays (γ), positrons (e^{+}), and neutrinos (ν) in addition to helium.

Inquiry 16-9

How many hydrogen nuclei enter the CNO cycle? How many helium nuclei and positrons are produced? How is the net effect of the CNO cycle similar to that of the proton-proton chain?

16.3 The Conditions Required for Fusion

Fusion will occur only where the physical conditions are right. In this section we will discuss what these conditions are and where within a star they are met.

TEMPERATURES AND DENSITIES

Fusion will happen only when the atoms are stripped of their electrons and collide with each other frequently. This happens when the temperature is high and the gas is dense. Unfortunately, the required conditions are difficult to produce on Earth, which is why nuclear power plants use only fission and never fusion reactions.

In the Sun, the temperatures and densities are high compared with those on Earth. Models tell us that the Sun's central temperature is about 15 *million* kelvins, and that the central density is about 150 times that of water. Under such extreme conditions, atoms are flying about at such great speeds and colliding so frequently and violently that their electrons are stripped away from the atomic nuclei. A gas such as this, composed of charged particles, is known as a **plasma**. Although a somewhat uncommon state of matter on Earth, plasmas are extremely common in the cosmos. In fact, most of the luminous matter in the universe probably exists in the form of plasma.

High temperatures and high densities are necessary because the particles that fuse together in the various reactions are all positively charged. The reacting particles therefore repel one another. The repulsive electromagnetic force is a strong barrier for incoming particles to overcome. Yet, once they get sufficiently close a stronger and attractive nuclear force that binds protons and neutrons together in the nucleus becomes dominant and pulls the incoming particles together. This nuclear force, which is one of the fundamental forces of nature, is called the **strong nuclear force**. With high temperature, the particles collide with sufficient energy to overcome the repulsive electromagnetic forces between the two positively charged nuclei. Then, the two particles are close enough together for the strong nuclear force to act, and the particles will be pulled together and fusion will take place. (This discussion has been over simplified, because a concept called *quantum mechanical tunneling,* which is beyond the scope of this book, is a necessary part of the process. Nevertheless, high temperature and density are a prerequisite for fusion reactions to occur.)

Elements heavier than hydrogen have many protons in the nucleus. The mutual repulsion of the positively charged colliding nuclei is, therefore, much greater than for two colliding hydrogen nuclei. Therefore, nuclear reactions involving heavier nuclei require yet higher temperatures.

High density also helps the fusion process. Although some reactions certainly take place at low densities if the temperature is high enough, only at a sufficiently high density will collisions be frequent enough for significant amounts of energy to be released. **Figure 16-7** summarizes this discussion.

Inquiry 16-10

Would you expect the rate of energy generation to *increase* or *decrease* as the temperature is raised? How about if the density is raised?

The rate of energy generation per gram of material depends strongly on the temperature. For the proton-proton reaction, the energy generation rate depends on T^4. This means that a star with a central temperature twice that of the Sun would generate 16 times as much energy. The CNO cycle is even more strongly dependent on the temperature; the energy generation rate is proportional to T^{17}. This means that even a small increase or decrease in the temperature will produce a large change in the amount of energy produced: a 1 percent change in temperature will produce nearly a 20 percent change in energy. Because more massive stars will, in general, have higher central temperatures, it is apparent why the CNO cycle will dominate in their energy generation.

16.4 Antimatter and Neutrinos

Recall that the first reaction in the proton-proton chain produced two additional particles, the positron and the neutrino. The positron, denoted by the symbol e$^+$, is a type of **antimatter** predicted theoretically by the physicist Paul Dirac several years before it was actually produced in a laboratory experiment. A positron is an antielectron—the same as an ordinary electron except with an opposite electrical charge. If a positron and an electron meet, they will annihilate each other totally, their mass being converted into energy in the form of two gamma rays. It can be represented in symbols as follows:

$$e^+ + e^- \rightarrow \text{energy}.$$

The positron was the first of a whole range of antiparticles to be detected. Antimatter, which consists of these various antiparticles, is the mirror image of ordinary matter. Every particle has a cor-

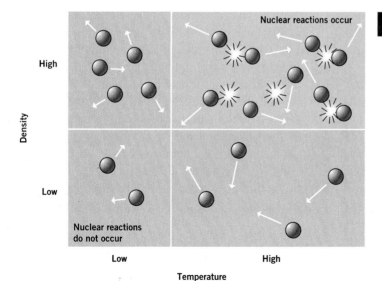

Figure 16-7 Particle interactions will result in nuclear reactions only where temperature is high. The higher the density, the more interactions will occur per second.

responding antiparticle. The negatively charged antiproton corresponds to the proton; the antineutron corresponds to the neutron, and so on. Whenever a particle meets its own antiparticle, the two annihilate each other in a flash of energy. (The photon is somewhat special; it is its own antiparticle, and photons do not annihilate each other.) Thus, antimatter is not science fiction but something that actually does exist!

Like the positron, the neutrino was also first predicted theoretically as an explanation for an experiment and only later observed. The neutrino is chargeless and was originally thought to be massless like the photon, and to move at the speed of light. Unlike the photon, however, it interacts with matter only rarely. This makes it difficult to detect. The neutrinos produced in the center of the Sun, for example, stream freely outward. When they reach the Earth, most of them pass right through as if it were not there.

Inquiry 16-11

Why would it be of great value to detect neutrinos that have traveled from the center of the Sun? (*Hint:* Only gamma ray photons are created in the center of the Sun; do we receive any of them?)

Although the probability of a given neutrino being absorbed by any intervening matter is extremely small, we expect that so many neutrinos are produced in the Sun that a certain number must inevitably be absorbed when passing through matter. Raymond Davis, a U.S. chemist and 2002 Nobel Prize winner for the work we will now describe, began trying to detect solar neutrinos in the late 1960s. His apparatus, located deep underground in a gold mine to shield it from as much stray radiation as possible, consists of a 100,000-gallon tank filled with perchloroethylene (C_2Cl_4), a common commercial dry-cleaning fluid (**Figure 16-8**). The idea behind the experiment is that there is an extremely small but nonzero probability that a solar neutrino will be absorbed by one of the chlorine atoms in the tank and be turned into an atom of radioactive argon. The reaction can be written symbolically (using an asterisk to indicate the radioactive argon) as follows:

$$_{17}^{37}\text{Cl} + \text{n} \rightarrow (_{18}^{37}\text{Ar})^* + \text{e}^-.$$

Inquiry 16-12

Is charge conserved in this equation? Show that your answer is correct.

Periodically, helium gas is bubbled through the tank and any argon atoms that have been created are swept up. When the radioactive argon decays after a time, the number of decays is counted and the number of captured neutrinos is determined.

It would be nice to be able to say that Davis detected enough solar neutrinos to confirm from observations that nuclear fusion is the source of the Sun's power. Some neutrinos have been detected, but less than a third as many as detailed computer models of the Sun predict.

Davis's experiment could not detect the low-energy neutrinos produced by the most common reaction in the p-p chain. Rather, it could detect only some higher-energy neutrinos from a rather rare reaction that accounts for only 2 percent of the Sun's energy. Furthermore, it could detect only one of the three known types of neutrinos—the electron neutrino, which is the type produced in the Sun.

Figure 16-8 The Davis solar neutrino experiment in the Homestake Gold Mine in South Dakota.

Understanding the discrepancy from the expected amount is important. Is the experiment wrong? Is there something wrong with the models of the Sun, or is there something about neutrinos that we do not yet understand? To test the experimental result, many more sensitive experiments to search for solar neutrinos have been designed. One uses a large quantity of the element gallium. However, gallium is so expensive that large international collaborations are required to finance such experiments. One project, known as **SAGE** (the **S**oviet-**A**merican **G**allium **E**xperiment), is located in the Caucasus Mountains. When a neutrino interacts with an atom in the 30 tons of liquid gallium, the element germanium is produced. Periodically scientists search the tank for germanium to determine the number of neutrinos incident on the tank since the last time it was searched. Another project, known as GALLEX, involves Italy, Germany, France, the United States, and Israel. Its gallium detector is located inside a mountain at Gran Saso, Italy. Both these experiments are in agreement with each other and have detected about 60 percent of the number of expected neutrinos. The Japanese Super-Kamiokande detector uses a tank containing 50,000 tons of pure water (**Figure 16-9**). Located 1 kilometer underground, when a neutrino strikes a particle in the water, light is given off, which is then detected by some of the 11,146 light detectors surrounding the tank. At present, this experiment has detected only 36 percent (about one-third) of the expected number of solar neutrinos.

All of these neutrino experiments supply independent verification that less than the expected number of neutrinos is being observed. Why? How can we explain it?

Some scientists have suggested that our models of the Sun might be wrong, and that the temperature at the Sun's center might be less than we calculate, although theorists feel that an error of this magnitude seems unlikely.

A more likely explanation for the lack of agreement between the expected and observed neutrino counts comes from experiments indicating that the neutrino does interact with the particles inside the Sun. It would do this if it had mass, even a tiny bit, much smaller than the electron's mass. The Japanese Super-Kamiokande neutrino experiment announced in 1998 that they had definitive evidence that neutrinos have mass. Confirmation of this result has been reported by the Canadian/British/American Sudbury neutrino experiment and by the Japanese/American KamLAND experiment.

Neutrinos are known to come in three different types: the electron neutrino, the tau neutrino, and the muon neutrino. If neutrinos have a small mass, they would not be stable as they moved through the Sun but instead they would change from one of the neutrino types into another; these changes are called **neutrino oscillations**. The missing neutrinos are the ones that have oscillated into types not detected by the experiments. Thus, the 40-year-old solar neutrino problem has been solved. It represents an excellent example of how science works at its best. Experiment and theory interact; disagreements and new experiments result in a new understanding of fundamental properties of matter. Astrophysicists are gratified that the solution showed that the solar models, which are described in the next section, are correct and that the problem, as it was, was in our understanding of the nature of the neutrino and not in our understanding of the Sun. What appeared for years to be a failure of a model was instead a step in the advancement of scientific knowledge.

16.5 Stellar Structure

In addition to having a source of energy, a star must obey other physical concepts to remain stable. In this section we discuss a number of processes that enable stars to live long lives. We will then put these processes together to produce a scientific model of the interior structure of a star.

Figure 16-9 The Super-Kamiokande neutrino experiment, while being filled with water. For a scale, notice the small raft containing people.

PRESSURE AND ENERGY EQUILIBRIUM

At the high temperatures found in the interior of a star like the Sun, all matter is gaseous. The gaseous nature of matter greatly simplifies the study of stars because, unlike the physical theories that describe the behavior of solids and liquids, those that govern gases are quite simple.

At first it might appear impossible for hydrogen to exist as a gas at the densities found in the interior of the Sun. The hydrogen *atom* (as opposed to the hydrogen nucleus) is mostly empty space because the electrons are far from the nucleus. Therefore, if hydrogen *atoms* were packed as closely as possible, they would form metallic hydrogen, with a density of about 1 gram per cubic centimeter. Yet the density estimated for the Sun's interior is much greater—about 150 grams per cubic centimeter. How can this be?

The answer is that because of its high temperatures the Sun is composed of a plasma, which, as we have seen, means that the electrons have been stripped away from the positively charged nucleus. The diameter of a hydrogen *nucleus* is much less than that of a hydrogen *atom,* and the smaller nuclear particles can be compressed further than can a gas composed of the larger atoms. For this reason, the density of plasma consisting of protons and electrons can be much greater than that of an equivalent mass of atomic hydrogen. This plasma behaves as if it is composed of two gases, one positively and one negatively charged.

The force of gravity acting on each atom in the Sun tends to collapse it. Just as gravity tends to pull the Flying Wallendas to the floor (**Figure 16-10**), the opposing tension in the wire provides a force that counteracts gravity. The wire, which feels the weight of all those above, must provide the largest counter-force. The base of the pyramid is supporting a lesser weight and needs provide a lesser force; the next layer need provide only enough force to hold the light-weight top person. In a star, two principal factors prevent collapse. The first is that any gas, such as that composing the star, has an internal **gas pressure** caused by the motions of atoms. This particle motion resists compression just as the gases in an inflated balloon resist compression. The hotter or denser the gas is, the greater this pressure will be:

$$\text{Gas pressure} \propto \text{Temperature} \times \text{Density.}$$

This relationship between pressure, temperature, and density is known as the **perfect** or **ideal gas law**. **Figure 16-11** demonstrates the relationship between temperature and pressure by showing how the lowered temperature of the gas inside the balloon decreases the outward pressure by the molecules so that the air pressure outside causes it to collapse. The balloon at room temperature is at about 300 K, whereas the balloon cooled by liquid nitrogen at a temperature of 78 K has shrunk dramatically. Inside a star, as the interior becomes hotter and more compressed, the gas law tells us that the outward pressure increases until the star is capable of resisting further collapse.

The second contributor to preventing collapse is a consequence of the radiation that streams outward from the center of a star, where vast quantities of energy are being released by the fusion of hydrogen into helium. Photons have momentum, and as they impinge on the gaseous material above it they tend to push the gas outward. This effect is called **radiation pressure**.

Stars of low to moderate mass, such as the Sun, are supported primarily by gas pressure, whereas in stars of high mass, radiation pressure is more

Figure 16-10 The Flying Wallendas demonstrate the balance between the downward force of gravity and the opposite force caused by the tension in the wire.

Figure 16-11 The perfect gas law. *(a)* The balloons are filled with air, which is mostly nitrogen. Liquid nitrogen, at a temperature of 78 K (-321°F), is in the beaker between them. *(b)* When cooled to the temperature of liquid nitrogen the air in the blue balloon decreases its pressure in proportion to its temperature and collapses.

important because of their higher temperature and higher amount of radiation. In fact, we calculate that no stars larger than about 120 solar masses can exist because in a larger star the radiation pressure would be so great as to blow away the outer layer of the star until its mass is reduced to less than 120 solar masses.

A star such as the Sun is said to be in a state of **equilibrium** if it returns to a balanced condition if pushed slightly away from its original condition. Thus, changes in a star in equilibrium take place slowly. In such a state, at every point within the star the inward force of gravity is balanced by the outward force of gas and radiation pressure, so that the gas does not rise or fall (**Figure 16-12**). Without this **gravity-pressure equilibrium** (the technical term astronomers use is **hydrostatic equilibrium**) the Sun would change rapidly. For example, if the outward pressure maintaining the Sun in equilibrium were suddenly to disappear, the Sun would collapse in roughly 1,000 seconds!

To summarize: The deeper the layer, the more weight must be supported, which means the pressure must be higher. Therefore, to maintain the equilibrium the temperature must be higher as you move deeper into the interior of the star.

Another aspect of equilibrium is that the amount of energy lost at the surface of a star is precisely compensated by an equivalent amount of energy released in its interior. We refer to this aspect of equilibrium as **thermal equilibrium**.

What would happen if a star were not precisely in equilibrium (**Figure 16-13**)? For example, suppose the interior temperature were to increase slightly. Due to the ideal gas law, the pressure in

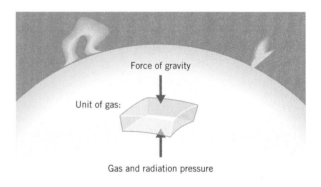

Figure 16-12 Equilibrium between the inward force of gravity and the outward force of gas and radiation pressure in a star.

the interior of the star would then increase slightly and become larger than the inward force of gravity. The star would expand a little. This slight expansion would lower the temperature and the density. Because, as we have seen, the star's energy-generation rate depends strongly on temperature, the slightly lowered temperature would lower the rate at which the star produces energy. This small decrease in energy production would cause the temperature to decrease slightly and, due to the ideal gas law, cause the pressure to decrease slightly. The star would then return to a *new* state of equilibrium in which it is slightly larger and cooler than before. This natural *thermostatic* action is constantly in operation within a star, always tending to bring it back into a state of equilibrium should any minor imbalances occur.

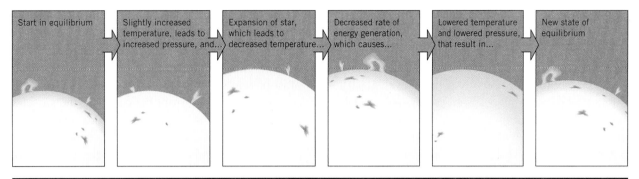

Figure 16-13 How a star adjusts from one equilibrium state to another.

| Start in equilibrium | Slightly increased temperature, leads to increased pressure, and... | Expansion of star, which leads to decreased temperature... | Decreased rate of energy generation, which causes... | Lowered temperature and lowered pressure, that result in... | New state of equilibrium |

Inquiry 16-13

Suppose the energy-generation rate at the center of a star were to become slightly smaller than its equilibrium value. How would the star regain its state of equilibrium?

ENERGY TRANSPORT

Once energy has been generated in the core of a star, it must get to the surface to be radiated away. The high-energy photons produced by the fusion of hydrogen into helium travel only a short distance (typically 1 cm) before they interact with nearby particles such as electrons, nuclei, and highly charged ions. In that interaction, energy is exchanged and photons of somewhat lower energy and thus longer wavelength result. This process continues until the energy finally reaches the surface in the form of mostly visible and infrared radiation, where it is then radiated into space. Because in the core the photon travels only a short distance before interacting with matter, and because the photon can be scattered in *any* direction, it takes energy hundreds of thousands of years to escape (**Figure 16-14**). The process just described, known as **radiative transport**, occurs throughout most of the Sun's interior.

In the outer regions of the Sun, radiative transport does not work well because the cooler gases are more efficient at absorbing radiation than they are at releasing it. The energy trapped by the highly absorbing gas heats it more than would otherwise be the case, much as a blanket keeps us warm by trapping the heat of our bodies. Due to this effect, a blob of warmed gas may become slightly lighter than the surrounding material and tend to rise, like a hot air balloon. After rising a

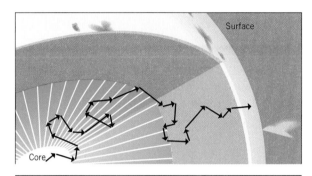

Figure 16-14 The random path followed by photons from production in the solar interior until they reach the surface. The original gamma ray photon is not the one that is emitted from the surface; different photons are constantly absorbed and reemitted.

certain distance, it gives off its heat to the cooler material that was above it. Cooled off, it sinks again to the region below, where the process starts all over again. This process, called **convection**, is familiar in everyday life. For example, in a pot of boiling water, the energy is transported from the bottom of the pot to the surface by convection. Another example is the waves of heat that are seen over a hot fire or rising from a highway on a hot summer day; these actually consist of rising currents of hot air. Convection is highly efficient at moving energy from one point to another.

STELLAR MODELS

The processes we have been discussing—gravitational and thermal equilibrium, the perfect-gas law, energy generation, and transport mechanisms—are the physical concepts and theories that we think govern the structure of a star. They can be expressed in complex mathematical terms. The equations include such quantities as temperature, pressure, density, chemical composition, luminosity, and mass. Solution of these equations, which requires a modern, fast computer, yields values of temperature, pressure, density, chemical composition, luminosity, and mass at each point throughout a star in equilibrium. These quantities, as they vary throughout the star, provide astronomers with a **stellar model**. Such a table of numbers, or a graph, gives us a picture of the conditions throughout the interior of a star.

The results of one such calculation for the Sun are displayed in the three-part **Figure 16-15**. This figure shows how various properties of the Sun change with distance from the center to the surface. For example, the lower panel shows that the temperature (black curve and axis on the left side) decreases rapidly from the center outward. Similarly, the density (brown curve and axis on the right side) drops strongly away from the center. This rapid decrease in temperature and density explains why energy generation occurs only in the core region.

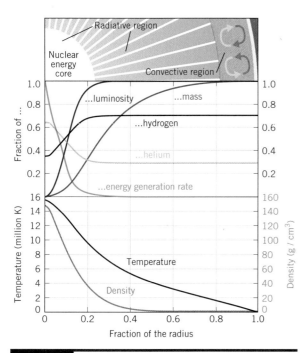

Figure 16-15 A theoretical model of the interior of the Sun. The bottom panel shows the variation of temperature (axis on the left) and density (axis on the right) from the center to the surface. The middle panel shows the variation of the energy generation rate (green) from its maximum at the center; the variation of the abundance of hydrogen (red); the fraction of the Sun's luminosity produced inside a given fractional distance; and the fraction of the Sun's mass that exists inside a given fractional distance. The top panel illustrates the various physical regions of the Sun's interior.

The rate of energy generation throughout the star is shown in the middle panel as a green curve; it shows that the rate at which energy generation occurs in the solar interior drops to about 5 percent of its central value at a distance of less than 20 percent of the distance from the center to the surface. In other words, the curve demonstrates that energy generation occurs solely in the Sun's central core.

In a similar way, the red curve shows that more than 30 percent of the Sun's mass is in the innermost 20 percent of the Sun's radius. In other words, the Sun is centrally condensed. Finally, the results show that the variation of luminosity throughout the Sun is the mirror image of the energy generation rate. Thus, the Sun's luminosity is formed only in the core.

The upper panel illustrates various physical regions of the Sun, with the energy generating nuclear core in the center, a surrounding region that transports energy by means of radiation transport, and finally an outer region that carries energy outward by convection.

Inquiry 16-14

What fraction of the mass of the Sun is in the inner 50 percent of the Sun's radius? The inner 70 percent? What fraction of the Sun's mass is in the outer 30 percent of the radius?

Inquiry 16-15

What fraction of the luminosity of the Sun is produced in the inner 30 percent of the radius?

Inquiry 16-16

What fraction of the original amount of hydrogen in the Sun's core has now been converted into helium?

We might expect stars of different masses to reach different states of equilibrium, with different central temperatures and densities, and this is indeed so. In fact, we can go further and state that if a star obtains its energy from the conversion of hydrogen into helium, as do main-sequence stars, then the star's equilibrium state is determined uniquely by its mass, its chemical composition, and its age. This idea is known as the **Russell-Vogt theorem**. It tells us that for stars of the same chemical composition and age, the different equilibrium states that we see along the main sequence of the Hertzsprung-Russell diagram exist because the stars have different masses. Therefore, because stellar spectroscopy tells us that most stars have the same chemical composition, the main sequence is actually a *mass* sequence, as we saw in Figure 15-19.

Scientists use these stellar models only because they are consistent with the observed properties of real stars.

Inquiry 16-17

The energy-generation rate increases rapidly with temperature and density. If you were to increase the mass of a star, how would you expect its luminosity to change? (*Hint:* A greater mass pushing down on the central regions of the star produces higher pressures, temperatures, and densities there. This is the explanation for the mass–luminosity relationship discussed in Chapter 15.)

16.6 The Lifetimes of Stars

The more massive a star, the more hydrogen it has to burn. The greater the luminosity, the more rapidly the star loses energy. As we saw earlier in the chapter, the lifetime of a star (*t*) depends on the amount of energy available, divided by the rate at which energy is lost. Because the amount of energy available depends on the star's mass (through $E = Mc^2$), and the rate of energy loss is the luminosity, symbolically we have the following:

$$t \propto \frac{M}{L}.$$

The existence of the mass–luminosity relationship has important consequences for the lifetime of stars of differing masses. Both theory and observation indicate that a star of about the Sun's mass will remain on the main sequence as a hydrogen-burning star for about 10 billion years. However, a star that is twice as massive as the Sun has twice as much fuel to burn, but it consumes it at a rate nearly 16 times as fast as the Sun, because its luminosity is almost 16 times greater than the Sun's. As a consequence, its larger supply of fuel does not last nearly as long as the Sun's supply. This is the same idea as the fact that 10 gallons of fuel will take you more miles (last a lot longer) in a car rather than in a truck.

Inquiry 16-18

Using the information just given, how long do you estimate a star of two solar masses to remain on the main sequence?

The variation of stellar lifetime with mass is dramatic. From the previous discussions, astronomers determine that the lifetime in years of a star on the main sequence is given by

$$t = 10^{10} \frac{M}{L},$$

where M and L are in units of the Sun's mass and luminosity. In the previous chapter, we saw that the mass–luminosity relation, which is valid only for main sequence stars, tells us that $L \propto M^4$. Substituting for luminosity in the equation, we find the following:

$$t = 10^{10} \frac{M}{M^4} \propto \frac{10^{10}}{M^3}.$$

This expression says that the lifetime of a star on the main sequence varies inversely with the cube of the mass. Thus, the greater the star's mass, the very much shorter is its main-sequence lifetime. Although the most massive stars can be expected to live for only a few million years or so, the least massive red dwarfs consume their fuel so slowly that they can be expected to live for thousands of billions of years—so long, in fact, that even a red dwarf created at the dawn of the universe has not yet had enough time to move off the main sequence. **Table 16-1** gives the **main-sequence lifetimes** (i.e., the time spent on the main sequence burning hydrogen) for stars of various masses.

Table 6-1 Main-Sequence Lifetimes

MASS (IN SOLAR MASSES)	MAIN-SEQUENCE LIFETIME (IN YEARS)
60	2 million
30	5 million
10	25 million
3	350 million
1.5	1.6 billion
1.0	9 billion
0.1	thousands of billions

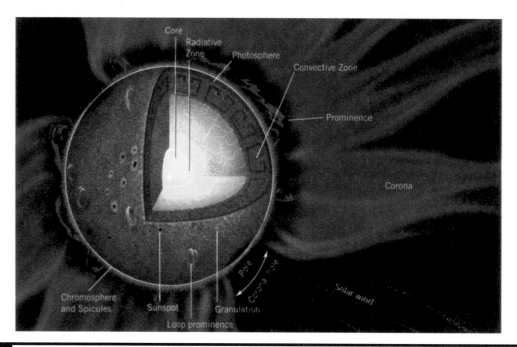

Figure 16-16 A cross-sectional drawing of the Sun, showing the energy-generating core, the surrounding radiative and convective zones, and the solar atmosphere.

The variation of stellar lifetime with mass has many important consequences. To mention just one, we can seek out regions of active star formation in our galaxy by looking for luminous blue stars, which are the high-mass O and B stars at the top of the main sequence. On the one hand, because these stars have such short lifetimes compared to the age of the Galaxy itself, the fact that they exist at all means they must have formed relatively recently and are probably not far from the place where they were born. On the other hand, when we observe a low mass star—say, one-tenth of the Sun's mass—we cannot tell whether it formed recently or not, because its main-sequence lifetime is so long that it could have formed at almost any time in the history of the Galaxy.

Our discussion up to this point has examined the theoretical concepts and processes governing the structure and evolution of stars. These ideas are summarized in **Figure 16-16**, which illustrates the interior structure of the Sun as determined from our models. The illustration shows other phenomena on the Sun's surface, which we are able to investigate only because of its nearness. Detailed observations then allow us to make comparisons with theory. We study the Sun's observable outer regions for the remainder of this chapter.

16.7 The Sun—A Typical Star

Much of what we know about stars comes from our study of the Sun, the only star in our solar system. Because it is so close, we know its distance, and thus its mass and luminosity, accurately. We are able to see details of its surface and surrounding gases that we cannot see for other stars. For these reasons, we have been able to learn more about it than about any other star. We can expect that if models of the Sun can be constructed that agree with the detailed observations we have obtained, then similar models for stars not much different from the Sun are probably not far off the mark. Of course, the more different a star is from the Sun, the more uncertain our models will be. The rest of this chapter looks at some of our detailed knowledge of our star. **Table 16-2** summarizes some basic data about the Sun.

THE PHOTOSPHERE

The bright, visible surface of the Sun that we see is called the **photosphere**. More precisely, it is the shell of hot, opaque gas several hundred kilometers thick, within which the photons making up the Sun's continuous spectrum are produced. The photosphere is the visible part of the Sun's atmosphere. **Figure 16-17** is a photograph of the Sun showing

Table 6-2   Properties of the Sun

Diameter	1,391,980 km (109.3 $D_\oplus$)
Mass	1.99×10^{30} kg (332,943 $M_\oplus$)
Average density	1.41 g/cm^3
Luminosity	3.90×10^{26} watts
Photospheric temperature	5,780 K
Equatorial rotation period	25.04 (Earth) days

Figure 16-18 The cause of limb darkening. The observer's line of sight penetrates to a deeper, hotter layer when looking at the center of the Sun's disk than when observing the limb.

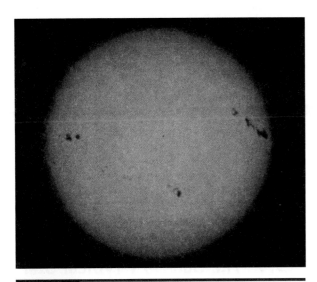

Figure 16-17 A photograph of the Sun showing limb darkening.

the appearance of the visible surface, and illustrating that the center appears considerably brighter than its edge or limb. This phenomenon, known as *limb darkening,* is easily observed by projecting the Sun's image with a small telescope. The reason for limb darkening is that when we look at the edge of the Sun, we are looking at the upper, cooler, and hence fainter levels of the Sun's atmosphere. But when we look toward the center of the disk of the Sun, our line of sight actually reaches to a deeper, hotter, and brighter region (**Figure 16-18**). Even this simple observation provides direct evidence for the fact that the temperature of the Sun increases toward its center, confirming numerous theoretical calculations.

Above the photosphere is the **solar atmosphere**, a layer of cooler gas a few thousand kilometers thick. When light from the photosphere passes through this layer, the photons are absorbed by the atoms and ions making up the solar atmosphere. These atoms and ions absorb at wavelengths dependent on which element and ion they are, and the energy level of their electrons. This produces

the observed absorption spectrum of the Sun. The boundary between the photosphere and the various atmospheric layers is not well defined and depends on the part of the spectrum being observed; it is difficult to say exactly where one ends and the other begins. For this reason, astronomers do not attempt to define a surface of the Sun.

Inquiry 16-19

Observations of the Sun show that its radiation is most intense at a wavelength of about 5000 Å. What is the photospheric temperature?

In addition to finding the temperature from Wien's law, we can also use the Sun's G2 spectral type to determine the temperature from the degree of excitation and ionization, as discussed in Chapter 14. From these considerations we obtain a temperature for the absorbing gases of 5,870 K.

THE CONVECTIVE ZONE

Other interesting details about the photosphere are revealed in images made with telescopes above much of the Earth's turbulent atmosphere. **Figure 16-19** is a photograph taken with Project Stratoscope, for which a telescope was suspended from a balloon about 100,000 feet above Earth's surface. At this altitude, much sharper pictures revealing great detail can be made.

In this photograph, you will notice a pattern of light and dark areas known as **granulation**. The bright areas are slightly hotter than the darker ones. Photographs taken in rapid succession show that this pattern changes with time, with a typical bright area lasting for only a few minutes. If we look at a moving picture of this granulation, the impression we get is of a region of boiling gas, with material bubbling out of the bright areas and falling back into the darker ones.

Inquiry 16-20

Figure 16-20a illustrates photospheric granulation with the slit of the spectrograph across a number of the bright and dark areas. **Figure 16-20b** shows one spectral line at a wavelength of 8,542 Å. Wavelength, increasing to the right as shown, indicates that light from the bright areas is blueshifted whereas that from the dark areas is redshifted. What is the most likely interpretation of this observation?

Spectra of the granulation show the individual granules to be in motion. The bright regions are blueshifted and thus rising, while the darker regions in between are redshifted and thus falling. The observed Doppler shifts of the granules tell us that the observed granulation is the visible tops of the convection cells. Thus we have observational evidence that the energy transport mechanism of convection occurs in the outer regions of the Sun, just as predicted by theory.

Further characteristics of the outer convection regions are revealed by detailed computer models of the Sun. This convection occurs in a shell that extends from the Sun's surface down to a depth of nearly 200,000 km, or about one-quarter of its radius (Figure 16-16). In the interior, the energy is transported by the radiative transport mechanism.

A star less massive than the Sun is similar, except that the region of convection is proportionately larger; in low-mass stars, in fact, the entire star is convective. Calculations indicate a different story for high-mass stars; in these, models show that the envelope is transporting energy by radiative transport, whereas the core is convective.

> You should do Discovery 16-1, Solar Granulation, at this time.

SUNSPOTS

Chinese and Japanese astronomers occasionally observed sunspots thousands of years ago, but systematic observations have been made only for several hundred years. Large sunspots sometimes

Spectrograph slit allows light through

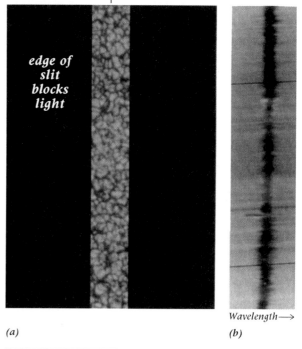

Wavelength⟶

(a)   *(b)*

Figure 16-20 *(a)* A spectrograph slit over solar granulation; only that light coming through the slit forms the spectrum. *(b)* The calcium line at 8542 Å of the solar spectrum examined in great detail. Parts of the solar surface are rising (blueshifted) and parts are falling (redshifted portions of the line), giving rise to the wiggly appearance of the line as the slit samples different points on the Sun. (The straight, nearly horizontal lines are caused by dust on the spectrograph slit.) Courtesy of Project Stratoscope, copyright 2005, The Trustees of Princeton University, supported by NSF, NASA, and ONR

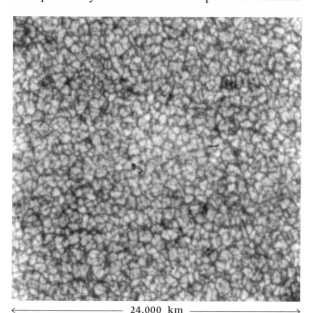

←———————— 24,000 km ————————→

Figure 16-19 Solar granulation, as seen from Project Stratoscope. The region shown is some 24,000 km on a side. Courtesy of Project Stratoscope, copyright 2005, The Trustees of Princeton University, supported by NSF, NASA, and ONR

appear that are visible to the naked eye, especially when the Sun's glare has been dimmed near sunrise or sunset. With a small telescope, the easiest (and safest!) way to observe sunspots is to project the image on a screen. **Do not look directly at the Sun with a telescope unless it has a solar filter**.

Sunspots are complex entities. For example, they generally occur in elaborate groups (see Figure 16-17). Furthermore, they are not uniformly dark. **Figure 16-21** shows a high-resolution view of a sunspot. The dark central region of the spot is known as the *umbra* (shadow) and the lighter region around it is called the *penumbra*. Although they seem dark on the face of the Sun, sunspots are actually quite bright; their dark appearance results from the contrast with the even brighter solar disk.

Inquiry 16-21

The spectrum of a typical sunspot resembles that of a star of spectral type K2. From this information, what is the temperature of such a sunspot? By how many degrees does it differ from the surrounding 5,800-K photosphere? How much less energy does the sunspot radiate per unit area?

The number of sunspots visible at any given time, and their location on the solar disk, varies. Astronomers have been counting sunspots for many years; a graph of the number of sunspots with time is shown in **Figure 16-22**. You can see an obvious periodicity of 11 years, more or less. The number of spots decreases from a strong maximum until, at the end of the cycle, few if any sunspots exist. The number of spots at any given maximum is not the same. In the late 1600s, the number of sunspots was lower than at other times. This so-called **Maunder minimum** indicates a possible important change in the Sun's activity. The time of the Maunder minimum corresponds to a cold period known in Europe as the *Little Ice Age*. Although still controversial, a correlation between sunspot activity and climactic conditions on Earth is inferred by some scientists. The possibility of an important connection provides practical motivation for understanding the Sun so that accurate predictions of future solar behavior can be made.

Any given sunspot has a finite lifetime; a spot may form, only to dissolve a few days or weeks later. Spots continually form as others disappear over the course of the sunspot cycle. As time passes, however, definite trends become apparent. Early in the sunspot cycle, spots are formed in modest numbers at high latitudes, which are far

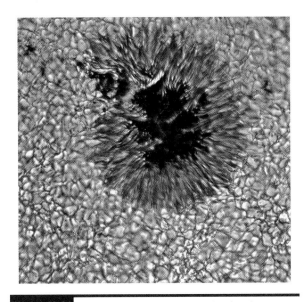

Figure 16-21 High-resolution photograph showing the structure of a sunspot. Solar granulation is also visible.

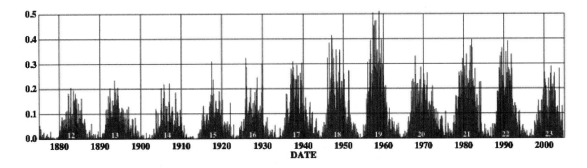

Figure 16-22 Variations of the sunspot numbers with time.

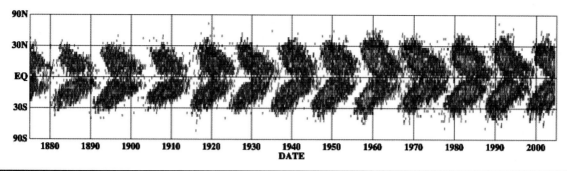

Figure 16-23 The butterfly diagram, showing how the location of newly formed sunspots on the solar disc changes throughout the 11-year cycle. Courtesy David Hathaway, NASA Marchal Space Flight Center

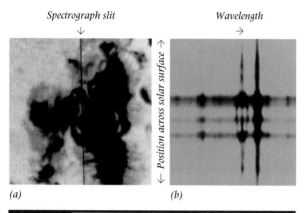

(a)   *(b)*

Figure 16-24 *(a)* The solar photosphere with a spectrograph slit across the photosphere and a sunspot. *(b)* The spectrum of the photosphere and a sunspot, showing Zeeman splitting of radiation from the sunspot. Each point on the spectrum in the vertical direction corresponds to a unique point of light on the Sun coming through the slit.

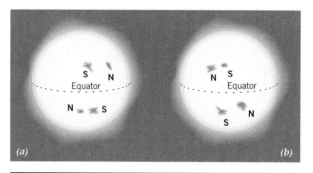

Figure 16-25 Sunspot polarities in the northern and southern hemispheres. *(a)* Notice that the polarities of the leading and trailing spots are reversed in the northern and southern hemispheres. *(b)* Eleven years later, the polarities in each hemisphere are reversed.

from the Sun's equator in both hemispheres. As the years within a cycle pass, spots form and dissolve at lower latitudes, while the number of spots visible at any time increases to a maximum. **Figure 16-23** shows the positions at which sunspots form throughout the 11-year cycles; because of its shape, this figure is known as the **butterfly diagram**. A given spot does not change latitude during its short lifetime; it is the location of formation that changes throughout an 11-year cycle.

THE SUN'S MAGNETIC FIELD

When astronomers place the slit of a spectrograph across a sunspot (**Figure 16-24a**), they observe the spectral lines to be split into two or more parts inside the sunspot but not outside it (**Figure 16-24b**). Such a splitting is caused by the Zeeman effect, discussed in Chapter 14, produced by magnetic zones in the

photosphere. The Zeeman effect is used in making images of the Sun that are used to study the Sun's magnetic field over the Sun's surface and over time.

Sunspots and magnetic fields on the Sun's surface are intimately related. Sunspots have long been known to generally come in pairs, with one leading spot and one trailing. Observations of their magnetic fields revealed that the two spots in a pair have opposite polarities; that is, if the leading spot is a north magnetic pole, the trailing spot is a south magnetic pole, and vice versa. Furthermore, the polarities of the leading and trailing spots in the southern hemisphere are reversed from those in the northern hemisphere (**Figure 16-25**). After an 11-year cycle has run its course, the new spots that appear at high latitudes in each hemisphere will have the opposite polarities from the previous cycle. If this reversal in magnetic polarities is counted, the actual sunspot cycle averages 22 years in length rather than 11. The sunspot cycle is simply a visible manifestation of a more fundamental 22-year magnetic cycle. More recent research suggests that the most fundamental cycle

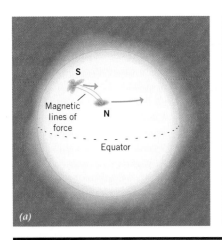
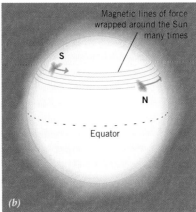
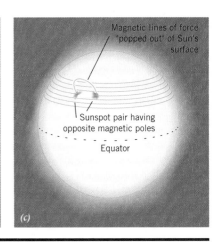

(a) (b) (c)

Figure 16-26 Differential rotation of the Sun tends to stretch out the magnetic lines of force *(a)* until they are wrapped many times around the Sun, beneath its surface *(b)*. If a loop of magnetic lines of force breaks through the surface, as shown in *(c)*, a sunspot pair with the opposite polarities is formed.

may instead be 44 years. Verification and further study are required before the final answer is in.

Astronomers think that the mechanism that produces sunspots works something like this: Intertwined within the plasma of the Sun's surface is a magnetic field. In any plasma there is a tendency for the magnetic field to become frozen in—that is, to be dragged along by the charged particles of the plasma. However, the Sun shows **differential rotation**. With a period of about 25 days at the equator, 28 days at mid-latitudes, and even longer at higher latitudes, the equator of the Sun rotates more rapidly than the higher latitudes, as shown in **Figure 16-26**. As a result, magnetic lines of force near the Sun's surface are stretched out by the differential rotation, eventually becoming wrapped around the Sun many times. This stretching increases the strength of the magnetic field, much as the tension in a rubber band increases when it is stretched. Astronomers think sunspots and other solar activity (to be discussed shortly) occur when the lines of magnetic force erupt through the surface and disturb the normally quiescent atmosphere. Exactly why the magnetic polarities reverse themselves every 11 years is more complex and less well understood.

THE CHROMOSPHERE

One of the applications of spectroscopy that we examined in Chapter 13 was the solar spectrum at the moment during a solar eclipse when the Moon exactly covered the photosphere and revealed a reddish glow peeking around the rim of the Moon. This region was referred to as the chromosphere, or color sphere. It is also a part of the Sun's atmo-

sphere. We saw that the spectrum, referred to as the flash spectrum because it appears suddenly, was an emission-line spectrum characteristic of a hot, rarefied gas. We also saw that the red color comes from the excited hydrogen gas in this layer, which emits much of its energy at 6,563 Å, the red (Hα) line of the Balmer series.

The chromosphere also plays a role in the formation of the solar absorption-line spectrum. In the lower portions of the chromosphere, where the gases are still cooler than those that cause the continuous spectrum, most of the absorption lines in the solar spectrum are formed (although some are formed in the upper photosphere).

Although the chromosphere is best studied during an eclipse, it is also studied at any time by using an instrument called a *coronagraph,* which creates an artificial eclipse by blocking out the light from the Sun's bright photosphere with an occulting disk inside the instrument. This permits the fainter chromospheric emissions to be examined. **Figure 16-27** is an image of the edge of the Sun. It displays finer details of the chromospheric region. Notice particularly the needlelike filamentary structures known as *spicules.* Detailed study has shown that the material in these regions is rising and falling at speeds of about 30 km/s, carrying energy upward. The spicules rise to heights of 5,000 to 20,000 km above the photosphere.

Inquiry 16-22

How does the typical height of a spicule above the photosphere compare with the diameter of the Earth?

Figure 16-27 An image of the Sun showing spicules.

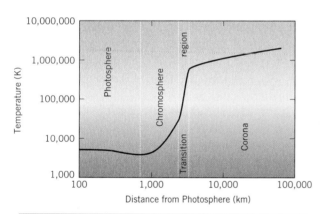

Figure 16-28 The variation of temperature with height above the photosphere.

Figure 16-29 The solar corona at *(a)* sunspot maximum and *(b)* sunspot minimum.

The atmosphere of the Sun falls to its lowest temperature, perhaps as low as 3,000 to 3,500 K, at the base of the chromosphere some 500 km above the photosphere. In spite of the lower chromospheric temperature, absorption lines are not formed here because the density is too low. Through the 1,500-km-thick chromosphere, the temperature rises to some 30,000 K. Thereafter, temperature rises rapidly through a thin transition region until it eventually merges with the solar corona above it (**Figure 16-28**).

THE CORONA

Above the chromosphere is the beautiful and tenuous **corona**, from the Latin word meaning *crown*. The corona is the outermost part of the Sun's atmosphere. **Figure 16-29**, taken during two eclipses, shows the corona as a pearly light extending far beyond the solar disk. The total brightness of the corona is about a millionth the brightness of the Sun itself—about as bright as the full Moon.

The corona changes from month to month; sometimes it exhibits long streamers and at other times is virtually featureless. Detailed studies show that it consists of several components. Part of the light from the corona is nothing more than sunlight scattered in our direction by electrons and dust particles, as illustrated in **Figure 16-30**. In addition, there are a few bright emission lines, such as might be expected from a region of hot, rarefied gas. These emission lines were a source of much mystery for many years after their discovery because they did not appear to be the result of any known element.

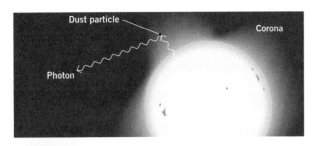

Figure 16-30 Scattering of light toward Earth by particles in the solar corona.

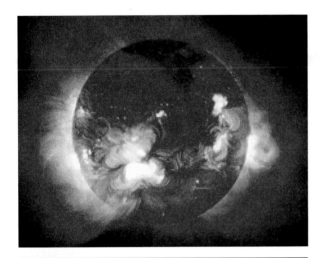

Figure 16-31 X-ray image of the Sun.

It seemed natural to speculate about the existence of a new element, because the element helium was first detected in the Sun in 1868 (the name comes from the Greek word for Sun) before it was found on Earth. But in 1941 astronomers discovered that two of the most prominent emission lines were really due to highly ionized iron atoms. For example, the strongest line, at 5,303 Å, was shown to be the result of iron atoms that had lost 13 of their electrons. Because the amount of energy required to remove 13 electrons from iron is large, the conclusion was inescapable: the temperature of the corona must be much higher than that of the surface of the Sun, exceeding 1 million kelvins.

Figure 16-31 shows the Sun imaged in the short wavelengths of X-rays. The fact that the Sun emits strongly at these short wavelengths provides additional evidence for the high temperature of its outer regions. The bright regions in this figure are the areas of highest temperature and density; the dark areas are regions known as **coronal holes** where the density is much less. Because the Sun's magnetic field lines point away from the Sun within the coronal holes, these holes may be the main escape route for the solar wind particles that are discussed in the next section. X-ray images taken over a long span of time provide astronomers with data that can be used to verify or falsify various hypotheses about the workings of the Sun.

The corona's high temperature was a great surprise to astronomers. Why didn't the Sun's outer atmosphere just grow cooler and cooler as it moved progressively outward, as might be expected? A hypothesis accepted as reasonable since the 1940s was overthrown in about 1980 on the basis of new ultraviolet and X-ray data from spacecraft. The corona shows a structure dependent on the degree of sunspot activity and, thus, the Sun's magnetic field. Figure 16-29*a* shows the corona at sunspot minimum, while Figure 16-27*b* shows its appearance at maximum. Our understanding of these differences is still slight. Newer data from the *SOHO* and *TRACE* spacecraft have provided further evidence that the coronal heating mechanism is related to the magnetic fields, which are continuously churning and twisting. The magnetic fields are produced through internal motions and manifest themselves though regions of opposite magnetic polarity that form what is knows as the *magnetic carpet*. Energy released by these magnetic fields into the corona seems to be the basis of coronal heating (**Figure 16-32**). The details are yet to be understood.

Inquiry 16-23

What are two pieces of evidence that the corona is hot?

THE SOLAR WIND

When the first satellites were launched into orbit in the late 1950s, in addition to discovering the Earth's Van Allen belts (Chapter 8), they also confirmed an earlier hypothesis, based on the changing appearance of comet tails, of a flow of particles away from the Sun. Called the **solar wind**, this flow of particles consists of protons, electrons, and helium nuclei. Somehow, the outer atmosphere of the Sun is ejecting these particles and accelerating them to high velocities, several hundred kilometers per second at the distance of the

Figure 16-32 Thin magnetic loops in the 1 million Kelvins lower solar corona, as viewed in UV light by the TRACE spacecraft. Hot gases within the loops help heat the corona.

Earth. It would not be inaccurate to think of them as having simply boiled off of the hot corona. In a way, the solar wind is really the outermost extension of the Sun's atmosphere. Although the actual amount of mass lost by the Sun is small, there is evidence that some stars have strong stellar winds, which in some cases are much stronger than the Sun's and involve the ejection of much larger and significant amounts of mass.

The particles in the solar wind do not come equally from all parts of the Sun. For example, the *Ulysses* spacecraft, which was the first spacecraft to orbit the solar poles, found that a fast component of the solar wind (moving at some 750 km/s) comes from the polar regions through coronal holes. The fast wind appears to be generated at the boundary intersections of the Sun's magnetic network, which consists of honeycomb-shaped regions of magnetism. A slower solar wind component, moving only half as fast, comes from the equatorial regions. The particles in the solar wind carve out a cavity in space that is called the **heliosphere**. The heliosphere defines the realm of influence of the Sun.

How far into space does the solar wind extend? How will we know where the Solar System ends and interstellar space begins? The place where the solar wind and magnetic field ends is considered to be the boundary between the Solar System and interstellar space; this is known as the **heliopause**. The *Voyager 1* spacecraft, which in 2003 was at a distance of about 90 AU from the Sun, is now about at the heliopause. As the solar wind particles move through space, they eventually run into particles from interstellar space. A shock wave, called the *termination shock,* then forms on the inside of the heliopause. In looking for the termination shock, whose location in space changes as solar activity changes, we are trying to find a moving target. At the termination shock, the solar wind should slow from a speed of 1 million miles per hour to a quarter of that. Various pieces of evidence indicate that *Voyager 1* entered the region of the termination shock in August 2002, but the scientific community has not yet reached consensus on this. *Voyager 2* will reach the same distance from the Sun in about 2008, so there will be a sec-

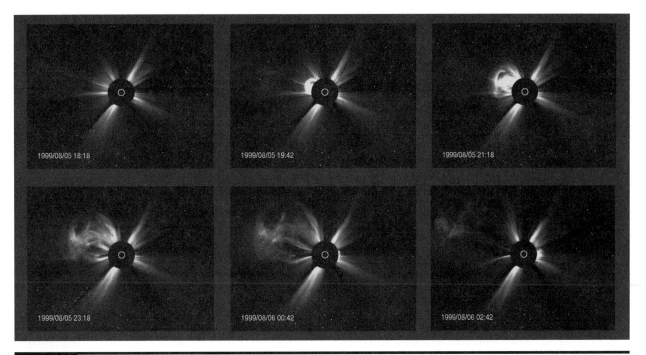

Figure 16-33 A coronal mass ejection as observed by the SOHO satellite. The white circle on the dark occulting disc represents the visible image of the Sun.

ond data point. Since *Voyager 1* is expected to continue providing data at least through the year 2020, the next few years will be providing new information on the outer reaches of the Solar System.

FLARES AND OTHER SURFACE ACTIVITY ON THE SUN

Sunspots are not isolated phenomena; they are often associated with other surface activity, such as **flares** and **prominences**. Solar flares are sudden brightenings on the surface that are thought to occur when broken magnetic field lines reconnect with a subsequent burst of energy. High-energy flares emit powerful X-rays as well as energetic particles such as electrons and protons. Charged particles produced during these bursts produce changes in Earth's ionosphere, thus disrupting radio communications, which depend on the reflective power of the ionosphere. A giant solar flare caused a major power outage in Canada in 1989; other flares have damaged satellites. A major flare could be deadly to astronauts who are outside Earth's protective magnetic field. Solar flares have also been thought to play a major role in producing aurorae on Earth. However, recent work questions this paradigm and instead relates aurorae and communications problems to another phenomenon called **coronal mass ejections** (**Figure 16-33**). These coronal mass ejections dis-

charge substantial amounts of material from the corona at average speeds of 400 km/sec. Each of these ejections, which occur on the average of once a day, ejects a mass equivalent to the amount of material in Mt. Everest! Our understanding of these phenomena and their complex effects on Earth are undergoing a fundamental change.

Prominences, like the ones shown in **Figure 16-34**, are among the most dramatic phenomena seen on the Sun. They are enormous filaments of excited gas arching above the surface and usually stretching hundreds of thousands of kilometers into the corona. They frequently run between active sunspot regions, and astronomers have generally assumed that their form and structure are strongly influenced by the Sun's magnetic field. The loop prominence in Figure 16-32*b* has a shape similar to that of a horseshoe-shaped magnet. Like a horseshoe magnet, for which each side has opposite polarity, sunspots at the foot of each side of the loop are observed to have opposite polarity. When seen at the edge of the Sun, prominences can be spectacular, as Figure 16-32 attests. When seen against the hotter solar disk they appear as dark, filament-like structures. At times they appear to move rapidly; in time-lapse photography they seem to blow away from the Sun at high speed and fly off into space. However, they are not nearly as energetic as solar flares. The detailed origin of

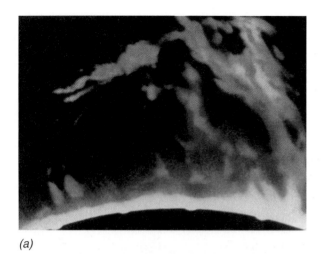

(a)

(b)

Figure 16-34 Solar prominences. *(a)* A large prominence. *(b)* A loop prominence.

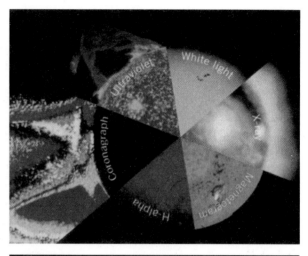

Figure 16-35 Mosaic of the Sun at various wavelengths and in a magnetogram.

prominences is unclear beyond saying they are certainly caused by magnetic effects.

Although the Sun is the only star on which we can observe surface phenomena in detail, the presence of starspots on other stars has been inferred from observations. In addition, stars called **flare stars** are known to show flares that are much more energetic than those on the Sun; these stars are located at the lower right end of the main sequence. Because other stars are so far away, direct observations of surface details that probably are present has not been possible. Spectroscopic evidence that some stars have a strong **stellar wind**, analogous to the solar wind exists; yet other stars show evidence of very large spots on their surface, proportionately much larger than those of the Sun.

Figure 16-35 summarizes the appearance of the Sun in a number of wavelength regions. The various sections of this mosaic look different because the radiations shown emanate from different lay-

ers, each having its own temperature and density. This illustration emphasizes the importance of viewing the universe in all spectral regions.

SOLAR-TERRESTRIAL CONNECTIONS

The Sun, through its energy output, the solar wind, the heliosphere, sunspots, and solar flares, has a varied and important impact on the Earth. As just discussed, solar flares and coronal mass ejections shoot significant amounts of matter into space. When large solar flares occur, the energy input to the Earth can be sufficient to modify the atmosphere and its environment and cause aurorae. Sunspots are visual indicators of deeper solar phenomena; as such, their presence or absence affects Earth. Thus, correlations between solar and terrestrial events provide a powerful motivation for studying and understanding the Sun.

SOLAR OSCILLATIONS

One of the more interesting and important solar observations in recent decades is that some spectral lines move back and forth periodically. Interpreting these cyclical wavelength shifts as periodic Doppler shifts suggests that the solar photosphere oscillates rhythmically. One important oscillation has a five-minute period and is known as the *five-minute oscillation*. Oscillations with other periods have also been measured. These observations have given rise to the research field of *helioseismology,* the study of oscillations of the solar surface. These studies have spawned the field of *astroseismology,* which applies the same ideas to other stars.

The oscillations are like those on the surface of a ringing bell. The details of the oscillations, such as period and amplitude, are determined by the variations of temperature and density in the solar interior. For example, the greater the temperature, the greater is the speed of sound in the gas. Thus, from an analysis of the various frequencies of oscillation, the interior sound speed, and thus the interior temperature, can be determined. These determinations are independent of the solar interior models described earlier in the chapter and thus independently verified those results.

Furthermore, the rotation of the Sun's interior influences the photospheric oscillations. Calculations based on observations of the solar oscillations indicate that the Sun's differential rotation continues from the surface to about one-third of the distance to the center; further in, the rotation is uniform. The boundary between the differentially rotating convection zone and the uniformly rotating radiative zone is where the solar magnetic field may form.

Theoretical calculations that include a range of physical conditions and internal rotations produce models such as that shown in **Figure 16-36**, which shows regions participating in the photospheric oscillation. The redder a region is, the faster that material is moving away from you—the observer. The bluer the region's color in Figure 16-36, the faster it approaches the observer. Figures such as this provide a new method of studying the solar interior by making observations of surface oscillations and comparing them with theoretical models.

Figure 16-36 Theoretical calculations show what solar oscillations would look like on the surface of the Sun.

Such comparisons provide information on the solar interior that is not otherwise obtainable, and will be pivotal in choosing between competing hypotheses of the solar structure.

To understand the Sun better requires that it be monitored continuously. For this reason, solar telescopes have been placed at the South Pole so that six-month-long continuous observations can be made. The U.S. National Solar Observatory has established a worldwide network of specially designed solar telescopes, known as the Global Oscillation Network Group (GONG). The GONG permits year-round observations of the Sun.

Solar Granulation

In this activity you will determine some of the basic properties of solar granulation.

- **Discovery Inquiry 16-1a** If the region of the Sun shown in Figure 16-19 is some 24,000 km on a side, determine the average physical size of a single granule in kilometers. Find the average size by measuring five granules and finding their average. Determine the uncertainty in your result.

- **Discovery Inquiry 16-1b** If a typical granule lasts for about eight minutes, and if it disappears after traveling a distance approximately equal to its diameter, how fast in km/sec does a typical granule move?

- **Discovery Inquiry 16-1c** Refer to Figure 16-20*b*. In this figure, there is approximately 0.023 Å for each millimeter on the photograph. Because the motion of the solar granules produces the width of the line, you can determine the speed of the granulation by measuring the width of the spectral line. Measure the spectral line width in millimeters in five places along the length of the line, find the average and the uncertainty, and then convert the line width to angstroms. Then use the Doppler shift formula to determine the velocity with which the granules move up and down. Is this result consistent with the answer you computed previously?

Chapter Summary

Observations

- The observable outer regions of the Sun include the visible **photosphere**, the **chromosphere**, and the transition region into the **corona**.

- Convection is observable in the Sun as **granulation** in the solar **photosphere**.

- **Sunspots** are cooler regions on the photosphere that are strongly associated with solar magne-tism, which varies with a 22-year cycle. Solar **flares** and **prominences** are other phenomena observed to be associated with solar magnetic activity.

- The **solar wind** is made up of charged particles emitted by the Sun through **coronal holes**.

- The Sun is observed to oscillate with a variety of periods.

Theory

- Sources of energy in stars include gravitational collapse and nuclear fusion at various times in a star's lifetime. Gravitational collapse produces energy from the conversion of potential energy into kinetic energy. Two mechanisms can produce nuclear energy. **Fusion** is when two nuclei collide with enough energy to overcome the electrostatic repulsion of the positive charges of the colliding particles; as a result the nuclei merge. **Fission** is when one massive nucleus breaks into two or more less massive ones. The amount of energy released in a nuclear reaction is given by the difference between the total mass of all the fusing particles and the mass of the resulting particle, multiplied by the speed of light squared ($E = Mc^2$).

- In the **proton-proton chain**, four hydrogen nuclei are transformed into a helium nucleus plus energy plus additional particles. Higher-mass stars on the main sequence obtain their energy from the **carbon–nitrogen–oxygen (CNO) cycle**, which also converts four hydrogen nuclei to helium, with carbon acting as a catalyst.

- The proton-proton reaction produces both **antimatter** and **neutrinos**. Neutrinos are thought to have a very small mass. Because neutrinos interact only weakly with matter, they exit the Sun without absorption. The experimental search for neutrinos from the Sun in the amount predicted by theory has the capability of providing a test of theories of the solar interior. With a new understanding of neutrinos, they have now been observed to come from the Sun in the expected amount.

- The rate at which fusion proceeds depends strongly on temperature, which determines the particles' ability to overcome the repulsive forces, and density, which determines how often particles will interact.

- A star is in **equilibrium** if its gravity and outward pressure (**gas pressure** and **radiation pressure**) are in balance (gravity-pressure or hydrostatic equilibrium), and if the amount of energy radiated equals the amount of energy generated in the core (**thermal equilibrium**).

- Energy produced in a star's core by fusion moves toward the surface by radiative transport and/or convective transport. In the Sun, radiation is important throughout most of the volume, but convection is important in the outer regions.

- The complex equations describing equilibrium conditions along with energy production and transport laws are solved by a computer to give a **stellar model**, which describes how temperature, pressure, density, chemical composition, mass, and luminosity vary throughout a stellar interior.

- The **Russell-Vogt theorem** says that a star in equilibrium has an internal structure determined by its mass, chemical composition, and age.

- The lifetime of a star on the main sequence depends on its mass and luminosity. From the **mass–luminosity relation**, we find the lifetime on the main sequence is shorter for more massive stars.

- The period of solar oscillations depends on the variation of temperature and pressure throughout the Sun, as well as on the details of the rotation of the core and the differential rotation of the outer layers.

Conclusions

- The Sun is too old for gravitational collapse to be the main source of energy over its lifetime. There is sufficient nuclear energy available within the Sun, however, to provide the observed energy for an adequately long time.

- The proton-proton chain occurs in the Sun and other stars of similar and lower mass.

- We infer that the corona is hot from the presence of spectral lines of highly ionized atoms and from the emission of X-rays.

- Although not understood in detail, the complex magnetic field structure observed in the Sun apparently result at least partly from **differential rotation**.

- The details of the Sun's oscillations depend on the interior structure and rotation. Astronomers have used observations of oscillations to model the Sun's interior rotation and have found that the convective zone rotates differentially, whereas the radiative zone rotates uniformly. The Sun's magnetic dynamo produces the solar magnetic field at this interface.

Summary Questions

1. What is meant by Kelvin-Helmholtz contraction? What is its significance insofar as providing the Sun's energy?

2. What do we mean by nuclear fission and by nuclear fusion? Why is nuclear fusion a viable source for the Sun's power, while fission is not?

3. What are the steps involved in the proton-proton chain? What is the net result in terms of what goes in and what comes out?

4. What are the general features of the CNO cycle? What is the end result in terms of particles going into the reaction and those coming out?

5. What are the conditions of density and temperature required for fusion to take place? Why is each variable important? What would happen to the total amount of energy if the variables were altered?

6. What is antimatter? What would be the result if antimatter and matter came into contact with one another?

7. What do we mean by a neutrino? What is the importance of the solar neutrino experiment for astronomy, and what are the current results?

8. What is equilibrium? What forces in a typical star maintain a state of equilibrium? What happens to a star if it deviates only slightly from equilibrium?

9. What is meant by a stellar model? How does the study of a stellar model allow astronomers to understand the interior structure of a star?

10. Why does the mass–luminosity relationship exist? What are its implications for the lifetime of a star?

11. What are the various layers of the Sun? What type of spectrum does each layer produce? Explain.

12. What is the meaning of the term convection? What direct evidence is there that convection occurs in the Sun?

13. How is it known that the solar corona is very hot?

14. What is the sunspot cycle? How are sunspots thought to arise?

15. What can astronomers learn about the Sun from the observed surface oscillations?

Applying Your Knowledge

1. Hypothesis: The Sun's energy comes from a slow contraction and conversion of gravitational potential energy. Question: Present evidence for and against the hypothesis.

2. Suppose the energy generated in the core of a star were to increase suddenly, and that a new state of equilibrium could not be attained rapidly. What would occur, and why?

3. Why is radiation pressure more important in hot stars than in cool ones?

4. Why is it not possible for you to compress an inflated balloon with your hands? (Assume the balloon is small enough for your hands to encompass it.) Explain the source of the forces involved.

5. Why is blueshifted granulation brighter than the surrounding darker regions that are redshifted?

6. Present at least two arguments—one based on observation and one based on theory—that the temperature of the Sun increases from the surface inward.

7. Suppose a rotating star had a large starspot on one side. How might an observer on Earth detect the presence of the starspot?

8. Verify that charge and atomic mass number are both conserved in each of the reactions in the CNO cycle.

▶ 9. How large is the largest sunspot visible in Figure 16-17? How large is the smallest? Express your answers in kilometers, and compare your answers with the diameter of the Earth.

▶ 10. Use Figure 16-22 to determine the percentage variation of sunspot number from cycle to cycle.

▶ 11. Determine the height above the photosphere of the prominence shown in Figure 16-34a in both kilometers and Earth diameters. Similarly, estimate the apparent length of the prominence.

▶ 12. Use Figure 16-22 to predict when the next sunspot maximum will occur. Use the figure to predict the part of the cycle the Sun is currently in.

13. If the typical human eye were able to resolve one minute of arc, how large would a sunspot have to be for it to be seen with the naked eye? Express your answer both in kilometers and in Earth diameters.

Answers to Inquiries

16-1. The inverse square law.

16-2. The assumption is made that the Sun's energy output has not been significantly lower in roughly 5 billion years. This is reasonable, in view of the fact that the earliest known precursors to life on Earth are roughly 4 billion years old. For life to survive, Earth's temperature could not have varied greatly since that time.

16-3. *Atomic mass number:* There are two protons on the left and one proton and one neutron on the right, for a total atomic mass number of two (as shown by the superscript 2 for deuteron). *Charge:* On the right, the deuteron and the positron each have one positive unit of charge.

16-4. *Atomic mass number:* The sum of the superscripts—i.e., the atomic mass number—is 3 on each side. *Charge:* Each hydrogen nucleus has 1 charge for a total of 2. The helium nucleus has a charge of 2.

16-5. Reaction (b) does not conserve charge. Reaction (a) actually does take place in the Sun, although only infrequently. The nucleus designated 13H is that of an extra-heavy form of hydrogen with two neutrons known as tritium.

16-6. Six protons are involved in the creation of each helium nucleus, with two protons left over.

16-7. 0.05×10^{-27} kg $\times (3 \times 10^8$ m/s$)^2 = 5 \times 10^{-12}$ joules, roughly.

16-8. $M = E/c^2 = 4 \times 10^{30}$ watts/ $(3 \times 10^8$ m/s$)^2 = 4.4 \times 10^9$ kg, or almost 4 million tons each second.

16-9. Four hydrogen nuclei enter. One helium nucleus and two positrons are produced. Both convert four protons to helium plus energy.

16-10. Raising temperature increases the rate of energy generation by increasing the speed and thus frequency of collisions between particles. Higher speeds also prevent electrons from recombining with the nuclei. Raising density will increase the rate of energy generation by shortening the distance between nuclei and thus making collisions more frequent.

16-11. Detection of neutrinos allows us to look into the solar interior because, unlike photons, they do not strongly react with the particles in the Sun. They confirm the theory of nuclear fusion as the Sun's energy source, and offer a tool for measuring the temperature and pressure at the Sun's center by providing a method of inferring the reaction rate from the number of neutrinos produced per second.

16-12. The charge of the chlorine atom is 17; that of the argon atom is 18, plus the negative 1 charge of the electron, gives a total charge on the right of 17.

16-13. If the energy-generation rate were too low, the interior of the star would not be hot enough to support its weight. The star would collapse slightly, increasing the internal density, temperature, and pressure. The energy-generation rate would increase, and equilibrium would be regained

16-14. Using Figure 16-15, we see that the inner 50 percent of the radius has 88 percent of the mass; 97 percent of the mass is inside 70 percent of the radius. Only 3 percent of the mass is in the outer 30 percent.

16-15. From Figure 16-15, 99 percent of the luminosity comes from the inner 30 percent of the radius.

16-16. From Figure 16-15, the original amount of hydrogen in the core is the same as currently present in its outer layers. Then, given an original amount of 70 percent hydrogen and a current amount of only 35 percent, (70 − 35)/70 = 50 percent of the original hydrogen in the core has changed to helium.

16-17. We would expect more massive stars to attain higher internal temperatures, densities, and pressures. Therefore, their energy-generation rate would be higher. Thus, more massive stars have greater luminosities (at least on the main sequence).

16-18. It burns twice as much fuel at 16 times the rate of the Sun, so it can be expected to live only about one-eighth as long—about a billion years.

16-19. From Wien's law, 3×10^7 Å K/5,000 Å = 6,000 K.

16-20. The bright areas are rising and the dark areas are falling. This is direct evidence for the existence of convection in the outer layers of the Sun.

16-21. A K2 star has a temperature of about 4,800 K (Chapter 14), so the sunspot is 1,000 K cooler. Because the energy per unit area depends on T^4, we would have $(4,800/5,780)^4 = 0.48$. The sunspot radiates about half as much energy per unit area.

16-22. A height of 5,000 to 20,000 km is 0.4 to 1.6 times the diameter of the Earth.

16-23. Observation of X-rays, and the presence of emission from highly ionized elements.

Star Formation and Evolution to the Main Sequence

17

When we reflect . . . we may conceive that, perhaps in progress of time, these nebulae which are already in such a state of compression, may still be farther condensed so as to actually become stars.

Sir William Herschel (1738–1822)

The study of star formation is a lively subject in which accepted ideas change rapidly. Stars form in cool, dusty molecular clouds that are opaque to visible light, and which hid the details of their formation history from us until the last twenty years. Much of the improved information about star formation results from new instrumentation that is sensitive to light in regions other than the visible spectrum. Ground-based observations at millimeter, centimeter, and meter wavelengths have been crucial, and observations from space in the ultraviolet and infrared parts of the spectrum have added important pieces to the puzzle.

In this chapter, we first study the clouds of dust and gas, with particular attention to the molecules necessary to the process by which stars form inside these clouds. Then we will look at the currently accepted theory of star formation. Finally, we will examine a variety of observations that show that the theoretical picture we have painted is correct in its general outline.

17.1 Matter for Star Formation

Stars must form from *something!* The only material available is what we call the **interstellar medium**, which consists of a mixture of atoms, mostly neutral and ionized hydrogen gas, as well as molecules and dust particles located between the stars in the Milky Way. **Figure 17-1** is a visual-light photograph showing ionized hydrogen (the reddish regions) and dust particles (the black regions). Some aspects of the ionized and neutral components of the interstellar medium were presented in Chapter 13 as examples of what we can learn about the universe with the use of spectroscopy. The roles played by neutral and ionized gas, and dust, in star formation will be presented in this chapter. (Further aspects of dust will be described in Chapter 20.) We also discuss the important role molecules play in forming stars and in providing us with observational tools with which we can better understand star formation.

A little background on the interstellar medium is useful here. In Chapter 13 we learned that cool, low-density neutral gas produces interstellar absorption lines in the spectra of stars lying beyond the absorbing gas. Often, the interstellar lines consist of many components, thus indicating that the interstellar medium consists of numerous **diffuse interstellar clouds** of neutral atoms (mostly hydrogen) and some ions (e.g., calcium) along any given line of

Figure 17-1 The Eagle Nebula, which shows ionized hydrogen in red and dark dust clouds. The smallest, darkest clouds are Bok globules. The brightest appearing stars are part of a young association of stars that provides the energy to ionize the hydrogen.
© Anglo-Austrailian Observ. Photo by David Malin

sight. A typical diffuse gas cloud is 50 light-years across and contains some 400 solar masses of material. This means it has a density of some 10 atoms per cubic centimeter—an extremely good vacuum—at a low temperature of 100 K. (For comparison, the density of air at sea level on Earth is some 10^{19} atoms/cm^3; a good vacuum in a laboratory on Earth still has 10^5 atoms/cm^3.) These properties of the average interstellar cloud are useful for comparison with the different regions in which stars form.

Chapter opening photo: Vigorous star birth observed by the *Hubble Space Telescope* in the nearby galaxy NGC 1569.

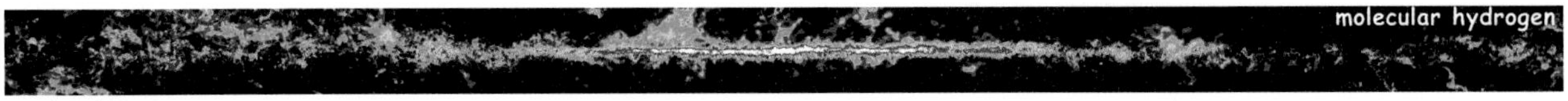

Figure 17-2 The extensive and non-uniform distribution of molecular hydrogen in the plane of the Milky Way.

MOLECULES IN THE INTERSTELLAR MEDIUM

On the one hand, astronomers have known for many years that molecules exist in the atmospheres of some stars. For example, as discussed in Chapter 14, M-type stars show titanium oxide (TiO) in their spectra. These molecular lines are attributed to absorption by cool gases within the extended and diffuse stellar atmosphere. On the other hand, astronomers used to argue that the interstellar medium was an unfavorable location for the formation and presence of molecules, and thus their discovery was unexpected, surprising, and extremely important. As we are about to see, to form stars, the material must be dense enough and cold enough for gravity to collapse the cloud. Thus, there is a clear link between the presence of interstellar molecules in clouds and subsequent star formation.

Inquiry 17-1

The densities and temperatures of many interstellar clouds are quite low. Why might these facts have led astronomers to think that the formation of molecules in such clouds would be highly unlikely? (*Hint:* How would the collision rate between single atoms be affected by these conditions?)

A further reason beyond that in Inquiry 17-1 exists as to why molecules were not expected in interstellar space. Even if a molecule were to form, high-energy ultraviolet radiation flooding space from the many hot stars in the Galaxy would soon destroy it, because only a small amount of energy is needed to break most molecules apart.

In Chapter 14 we found that molecules are capable of radiating energy in many different parts of the spectrum. For example, when a molecule goes from one state of rotation to a state of slightly slower rotation, it will give off the excess energy in the form of a low-energy photon with a wavelength on the order of a millimeter or so—in the radio region of the spectrum. When a molecule goes from one vibrational state to another, it typically emits infrared light. Finally, when electrons within a molecule make transitions, the emitted photons may appear in the ultraviolet, visible, or near-infrared part of the spectrum. Astronomers are thus able to observe emissions from molecules over a range of wavelengths.

So why didn't astronomers find interstellar molecules, the "red flags" of star formation, a long time ago? Radiation at infrared wavelengths is difficult to observe because the Earth's atmosphere blocks much of it. Millimeter-wavelength radiation, however, does penetrate Earth's atmosphere, but only in recent decades did technology develop to the point where it could be detected easily. Using the new technology, the radio-wavelength transitions of the hydroxyl (OH) molecule were first observed in 1963, proving that the early predictions that molecules could not survive in between the stars were entirely wrong. Thus, technological innovations are critical to the advancement of astronomical knowledge. In fact, a survey of emission from molecular hydrogen (H_2) shows its distribution around the Milky Way to be both extensive and clumpy, as **Figure 17-2** indicates. Thus, observations show molecules to be the rule rather than the exception in interstellar space.

New interstellar molecules continue to be discovered. **Table 17-1** shows a partial list of some of the more than 120 interstellar molecules that have been identified. They range from simple molecules such as NH_3 (ammonia), H_2O (water), and H_2CO (formaldehyde, or embalming fluid) to more complex compounds such as CH_3CH_2OH (ethyl alcohol, the stuff of hangovers). In addition, infrared spectra of many objects have the signature of polycyclic aromatic hydrocarbons (PAHs) such as C_6H_6 (benzene). These ring-shaped, carbon-based molecules can be produced by biological processes, and they are stable enough to survive ultraviolet light. Why carbon-based life is the only life found on Earth is explored in Chapter 24.

These interstellar molecules play two important roles. First, molecules provide a mechanism for the clouds to cool off and stay cool so that star formation can occur; they are a necessary part of the process. Second, molecules provide direct observational information about the sites where star formation occurs and the processes involved.

Table 17-1 Some Observed Interstellar Molecules

H_2	Molecular hydrogen
CN	Cyanogen
CO	Carbon monoxide
CS	Carbon monosulfide
SO, SO_2	Sulfur monoxide and dioxide
SiO	Silicon monoxide
OH	Hydroxyl
CH, CH+	——
H_2O	Water
H_2S	Hydrogen sulfide
HCN	Hydrogen cyanide
$HOCH_2CH_2OH$	Ethylene glycol (antifreeze)
NH_3	Ammonia
H_2CO	Formaldehyde
H_2CS	Thioformaldehyde
HNCO	Hydrocyanic acid
H_2CNH	Methylenimine
HC_3N	Cyanoacetylene
H_2CO_2	Formic acid
CH_3OH	Methyl alcohol
CH_3CN	Methyl cyanide
CH_3CH_2OH	Ethyl alcohol
CH_3CH_2CN	Ethyl cyanide

Interstellar molecules are found to exist in clouds with poorly defined edges and structures throughout the flattened plane of the Milky Way galaxy. Most are located within **giant molecular clouds (GMCs)**, which can be as large as 300 to 3,000 ly across, as massive as 10^6 to 10^7 solar masses and as cold as 10 K. They typically have 200 to 300 particles/cm^3, making them 20 to 30 times as dense as the more common neutral hydrogen diffuse interstellar clouds that are spread throughout the galactic plane. Thus, the giant molecular clouds are the largest structures in size and mass in the Galaxy. These largest clouds, and especially the smaller cloud condensations within them, are the places where most star formation is observed.

DUST IN THE INTERSTELLAR MEDIUM

Dust in interstellar space consists of tiny particles made up of anywhere from a few through many molecules. Interstellar dust is more like the particles in smoke or soot than the dust bunnies under your bed. **Figure 17-3** is a picture that shows what interstellar particles look like.

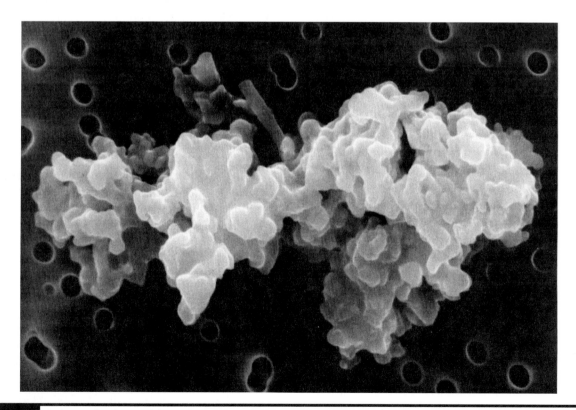

Figure 17-3 An interstellar dust particle.

The amount of dust in the interstellar medium is small—only 1 percent of the mass of hydrogen. However, a small amount of dust significantly affects light passing through it. Because the dust particles are so small, they are able to absorb and scatter effectively the short-wavelength light passing through them. In blocking the light from distant objects, the interstellar cloud is observed in silhouette against the background, as shown in **Figure 17-4**. Further discussion of the role of dust in the Galaxy is in Chapter 20 on the Milky Way.

Dust is distributed throughout the flat plane of the Milky Way on various size scales. At the largest scales are **dark cloud complexes** that are typically 30 light-years across. The dark cloud complexes are made up of individual **dark clouds** some 2 light-years across. The dark clouds are observed to contain **dark cloud cores** that are less than a light-year in size. Each dark cloud component—from the dark cloud complex to the dark cloud to the dark cloud core—increases in the density of dust particles, from some 400 particles per cubic centimeter to some 40,000 particles per cubic centimeter. Compared to the density of Earth's atmosphere at sea level, 3×10^{19} particles per cm^3, this may not seem very dense, but the size of these clouds make up for their low density.

Figure 17-4 The Horsehead nebula is dust that can be seen in silhouette against a bright gaseous background.

THE MIXTURE OF DUST AND MOLECULES

The interstellar medium is a turbulent mixture of neutral gas, ionized gas, molecules, and dust. The components of that mixture are not all observable by the human eye. In fact, only dust and ionized gas can be seen at visible light wavelengths. For example, ionized gas becomes visible when electrons and ions recombine and then emit visible light, as shown in Figure 17-20. The dust is detectable because it blocks the background light, as in Figure 17-4.

Inquiry 17-2

Why is neutral hydrogen gas at a low temperature unable to radiate at visible light wavelengths?

Neutral hydrogen gas at a low temperature is unable to radiate at visual light wavelength because the hydrogen is in the ground state; the temperature is too low to excite the electron to higher energy levels. Similarly, the structure of the interstellar molecules does not allow for emission at visible light wavelengths. Thus, these components are unobservable on photographs but can be detected at radio wavelengths.

Because dust and molecules are well mixed, a light-extinguishing dust cloud is also a molecular cloud. It is these clouds that condense to become stars.

THE FORMATION OF MOLECULES IN DUST CLOUDS

Why are interstellar molecules so abundant in space when preliminary knowledge suggested just the opposite? The important observation was the intimate connection of the dust and molecules previously discussed. Thus, the dark clouds visible in Figure 17-4 as silhouettes are the visible manifestation of a molecular cloud. We can identify molecular clouds in visible light only because the dust is mixed in with the molecules.

The strengths of molecular spectral features of the clouds allow astronomers to determine the density and temperature of the molecular clouds. We have discovered that dark clouds are, in general, more massive, denser, and cooler than other interstellar clouds, and provide the key for understanding how most molecules form.

Because dust is highly effective at absorbing optical and ultraviolet radiation, molecules deep within a dust cloud are almost totally shielded from the ultraviolet radiation that tends to break up molecules. Once a molecule in such a cloud forms, there are not many processes to destroy it, and a concentration of molecules can build up and eventually enough mass clumps together to become vulnerable to rapid gravitational collapse into a star.

Dust does pose serious difficulties when we attempt to observe the visible radiation emitted by any stars forming inside the clouds. The scattering and absorption of visible light by the dust can be so great that less than one-millionth of the visible light escapes! We are more fortunate with the long-wavelength light: because dust is not effective in absorbing or scattering long-wavelength radiation, the millimeter-wave emission given off by molecules embedded in the dusty cloud can freely escape and be observed. Thus, the same dust that is such a villain to the optical astronomer is a hero to the millimeter-wave astronomer. By shielding the centers of the dense clouds from stray radiation, the dust particles allow the concentration of molecules to build up. The long-wavelength radiation emitted by these molecules passes through the dust unhindered on its way to a distant astronomer interested in understanding star formation.

THE HYDROGEN MOLECULE

The most abundant molecule in a molecular cloud is the hydrogen molecule, H_2. Because of its symmetrical shape, the hydrogen molecule cannot be detected at millimeter wavelengths like most molecules. We must instead rely on observing it in the ultraviolet part of the spectrum. The *Copernicus* satellite carried an ultraviolet spectrometer that was extremely successful in detecting molecular hydrogen in the Galaxy. *Copernicus* looked at bright stars and detected H_2 in their absorption spectra from intervening gas clouds. **Figure 17-5** shows a high-resolution spectrum of the star zeta (ζ) Ophiuchi. The cloud in front of zeta Ophiuchi has a relatively low density for a dark cloud; to some extent, radiation can reach the interior regions and destroy some molecules there. The observed presence of H_2 in this relatively low-density dark cloud suggests that in extremely dense and dusty molecular clouds there must be a good deal of H_2. From this we infer that most of the mass of many clouds is probably molecular, rather than atomic, hydrogen.

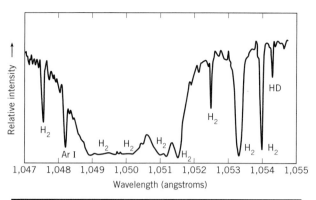

Figure 17-5 Measurements from the *Copernicus* satellite of the ultraviolet spectrum of the H_2 molecule, as seen in absorption in the star ζ Ophiuchi.

The relationship between H_2 and dust particles is especially intimate, because dust grains are responsible for the formation of the hydrogen molecule in the cold interstellar medium. If two hydrogen atoms happen to collide, it is unlikely that they will combine into a molecule. However, if a hydrogen atom collides with a dust grain, there is a high probability of the hydrogen atom sticking there. If two hydrogen atoms have become stuck to a particular grain, they will migrate slowly and randomly around the grain until they eventually encounter each other, and they will be able to combine. When the molecule forms, there is a small amount of energy released that ejects the H_2 molecule from the grain and back into the cloud. Thus the grains first allow the molecules to form, and then protect them from energetic photons that might destroy them.

CLOUDS AND CLOUD COMPLEXES

Collapsing giant molecular clouds fragment into smaller pieces, leaving behind hundreds or even thousands of molecular cloud cores. Thus, giant molecular clouds are the large-scale structures where star formation occurs, while the molecular cloud cores are what will eventually become stars.

Furthermore, only the relatively warm and massive giant molecular clouds form high-mass prestellar cores, while the cooler and less-massive clouds form low-mass cores. The giant molecular clouds produce both large-mass and small-mass stars, but the small clouds form low-mass stars almost entirely.

Table 17-2 Properties of Molecular Clouds

Object	Size (ly)	Density	T(K)
Giant molecular cloud complex	150	100	10
Giant molecular cloud	12	1,000	25
Core of molecular cloud	3	4,000	40
Clump within molecular cloud	1.5	$>10^5$	100
Dark cloud complex	30	400	10
Dark cloud	1.5	10^4	12
Dark cloud core	0.6	4×10^4	10

Table 17-2 summarizes the properties of the various molecular cloud components and the various dark cloud components. The main thing to remember is that they are physically connected.

A specific example of a giant molecular cloud is Sagittarius B2, which lies in the direction of the center of our galaxy. **Figure 17-6** shows a radio contour map[1] of this enormous cloud, which may have a mass greater than one million times the Sun's. With temperatures lower than anywhere on Earth, 10 K to 50 K, it contains every molecule listed in Table 17-1.

Their high mass, relatively high density, and cool temperatures make giant molecular clouds a likely place for star formation to occur. Their large masses provide sufficient material from which stars can form and produce strong gravitational forces that work to collapse the cloud. Resisting such collapse, however, are the pressures caused by magnetic fields, random motions of the gas (even though the gases' temperature is low), and turbulent gas motions. Thus, we infer that the clouds are stable over long periods of time. Therefore, for star formation to occur this state of equilibrium must be broken so that the cloud can collapse. We will discuss how that occurs later in this chapter.

> **Inquiry 17-3**
>
> Some clouds are observed to disperse. Is it more likely that gravity decreased or one of the other, opposing internal pressures increased?

---

[1]Contour maps were introduced in Chapter 12.

Now that we know where the matter for star formation exists, we will examine some of the theoretical questions concerning the collapse of a molecular cloud into one or more stars.

17.2 Star-Formation: Theory

Beyond the general assertion that star formation is taking place, we would like to know more about *how* it actually proceeds. Many questions need answering. For example, what causes stars to form from gas clouds? What do they look like before they become stars? How long does it take them to form? Did star formation proceed more rapidly in the past, or has it occurred at a uniform rate over the history of the galaxy? Does low-mass and high-mass star formation occur in the same way, or are there significant differences? How do the temperature and luminosity change as the star evolves (in other words, what path in the H-R diagram does the point representing an evolving star take)? What observational evidence is there that stars are born, live a long life, and then die out? Thanks to technological advances in detectors and the capability to place telescopes in orbit, astronomers are now making great progress in answering some of these difficult questions.

> **Inquiry 17-4**
>
> Dark clouds and hot, luminous, massive (therefore young) main sequence stars are associated with each other far more often than chance would predict. What conclusion do you draw from this fact?

INITIATION OF STAR FORMATION

Star formation requires not only the presence of material from which the stars will form, but also a mechanism to cause cloud collapse to begin. The physical association of luminous stars with molecular clouds (what we see visually as dark clouds) provides evidence of a close physical relationship between star-forming material and the product of the star-forming process—hot, luminous, young stars.

A gas cloud typically has a great deal of mass. And, as we learned in Chapter 5, all mass has the property of gravity. A gas cloud will begin spontaneously to collapse if its gravitational pull upon itself (often referred to as **self-gravity**) is greater

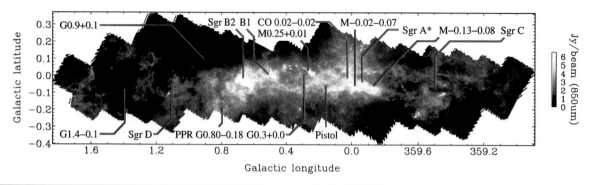

Figure 17-6 An infrared map of the galactic center. The infrared contours show concentrations that coincide with the strong radio sources Sgr A, B2, C, and D. End to end is about 1000 ly.

than the resisting outward pressure. The resisting pressure is caused by a combination of the thermal motions of the particles (determined by the density and temperature of the gas through the ideal gas law discussed in Chapter 16), magnetism, and turbulent motions within the cloud.

To form a star from a cloud requires a triggering mechanism to move the cloud away from equilibrium toward collapse. A triggering mechanism will result in a net force that could squeeze a cloud to the point where its gravitational attraction for itself could take over and continue the collapse. For example, if a cloud moves near a hot star (**Figure 17-7a**), the star's flood of photons will provide a pressure (called *radiation pressure*) that will tend to compress the cloud. Or, as shown in **Figure 17-7b,** gas clouds might collide with one another to initiate a collapse.

Another collapse trigger occurs when a gas cloud is given a "push" by the supersonic blast waves (shock waves) produced by a supernova explosion. Such an event occurs, as we will see in more detail in Chapter 19, when a massive star explodes, not only spewing mass and energy into space, but also producing a supersonic compression, or *shock,* wave. If such a shock wave encounters a molecular cloud, it may induce the cloud's collapse. Because we now know that interstellar space is filled with remnants of supernovae explosions, large numbers of such supernova-induced collapses must have happened in the past. In Chapter 7 we saw observational evidence from the abundance of certain elements in meteorites that the Solar System may have begun in this way; new data are, however, challenging these particular observations.

There is one additional mechanism astronomers consider that might trigger cloud collapse. As we will see in Chapter 20, our galaxy's spiral-shaped

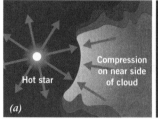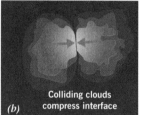

Figure 17-7 Two mechanisms that have been suggested for starting the collapse of interstellar clouds. (a) Radiation pressure from a nearby hot star. (b) The collision of two clouds.

structure may be caused by a wave that rotates relative to the gas, dust, and stars. When a rapidly moving gas cloud overtakes such a slower-moving wave, compression of the cloud occurs. This compression might initiate cloud collapse so gravity can take over and star formation can begin.

To summarize: A giant molecular cloud contains a huge amount of mass, with some parts randomly denser than other parts. As the giant molecular cloud collapses, it fragments into smaller pieces that, in turn, also collapse, forming molecular cores that will become the protostellar core that will eventually become a star. Astronomers do not yet understand the details of this process. We are sure that, in the end, stars do form!

Inquiry 17-5

What are five mechanisms that astronomers think can initiate the collapse of a cloud to form a star? Explain how each causes collapse to occur.

Once collapse begins and gravity takes over, the cloud has begun an irreversible process. What happens to this embryonic star, which we call a **protostar**, until the time it becomes a fully developed star is only partially understood because some processes—such as rotation, magnetic fields and turbulence—are difficult to include properly in the computations. For these reasons, our upcoming description will be simplified and far from complete.

DISK FORMATION AND ACCRETION

A cold molecular cloud core, whose size is on the order of 10^5 AU across, will have at least some rotation. As it collapses, the rotation speed will increase and the gas will begin to flatten—all because angular momentum must be conserved (Chapter 5). As the process continues, material from the protostellar envelope forms into a dusty donutlike disk, which is thick in the equatorial region with holes along the rotation axes. Matter from the collapsing molecular cloud will tend to fall onto the large disk and cause it to become more massive. Matter from the disk is then fed into the core, which is how the protostellar core gains mass. The process is called **accretion**, and we say that the star is accreting matter.

As the protostar forms, it enters a phase where it produces an outward blowing jet-like wind of gas from its poles. Thus, the protostar simultaneously accretes matter from the equatorial disk and loses matter through the polar holes in the disk. This mass loss occurs in opposite directions along the rotation axes, in what is called a **bipolar flow** (*bi* meaning *two, polar* meaning *from opposite extremes*) that forms a **bipolar nebula**.

THEORETICAL MODELS OF EARLY STELLAR EVOLUTION

Our approach to understanding stellar evolution will be to examine how the location of the point representing the star in the H-R diagram changes as time progresses. Our theoretical models predict changes in the star's luminosity and surface temperature at each stage throughout its lifetime. Connecting the resulting successive points in the H-R diagram will produce an evolutionary path in the diagram, much as we did in Chapter 15 for aging children in the height-weight diagram. We will then be able to compare the results of these theoretical models with observations. Agreement is taken to mean that we probably have some understanding of the physical conditions and processes that occur. Lack of agreement means

that the researcher would probably discuss it with colleagues, change the computer program to incorporate the new ideas, compute a new model, and again compare the model with observations.

During the earliest stages of protostar formation, the objects are cool and thus located to the far right of the H-R diagram. During collapse, part of the gravitational potential energy will be converted into raising the temperature and part into infrared radiation. The density of the gas is low enough that any radiation produced will escape from the forming protostar rather than be absorbed. This radiation loss is an energy loss, and a loss of energy cools the interior. Thus, in these early stages the collapse is able to continue because the interior temperature and pressure are prevented from rising. In addition, the processes of dissociation of molecular hydrogen and ionization of atomic hydrogen absorb energy, which also prevents the temperature from rising much.

What follows is the history of the birth of a star. Unfortunately, much as we do know, even the historians (i.e., the astronomers who research this topic) do not yet know all the details. Thus, you should follow the general flow of the story, while keeping in mind that the details will probably change as more information is obtained.

We begin our story after the giant molecular cloud has fragmented into smaller pieces 10^4 to 10^5 AU in size. While the entire cloud fragment collapses, the inner region collapses more rapidly than the outer ones (**Figure 17-8a**). A prestellar core is thus formed first (**Figure 17-8b**). (Note that since the scale of sizes will change from 10^4 to 10^5 AU to a few times the size of the Sun, the parts of Figure 17-8 are not to scale.) Eventually, the collapsing core builds up to such a pressure (**Figure 17-8c**) that the contraction stops and reverses direction (**Figure 17-8d**)—just as happens to a person who jumps on a trampoline. With the reversal of direction, the pre-stellar core is now expanding. This phase is often referred to as the core *bounce*. The expanding core then collides with the still-collapsing outer regions (**Figure 17-8e**) until the entire mass is once again collapsing (**Figure 17-8f**). The core density now increases (**Figure 17-8g**) until photons are no longer free to escape and the core becomes opaque. The renewed collapse process continues again but with increasing strength (**Figure 17-8h**) until there is yet another bounce of the core (Figure 17-8i). After this second core bounce, the

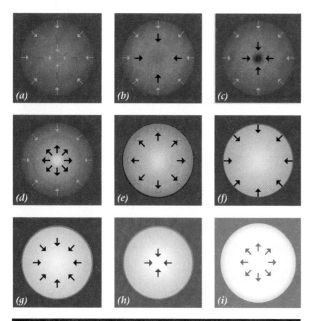

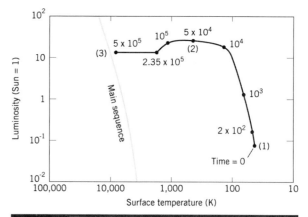

Figure 17-9 The path in the H-R diagram of a protostar during dynamical collapse. The numbers shown are the years since cloud fragmentation.

Figure 17-8 A series of diagrams showing the dynamical collapse to a protostar. Given that the initial cloud in part *(a)* is one million times the size of the final star, the figure is not drawn to scale but rather is to illustrate processes that occur.

temperature and luminosity place the object in an observable part of the H-R diagram. Because the entire process occurs relatively rapidly, in only thousands of years for higher mass stars, astronomers refer to it as the *dynamical phase* of pre-main-sequence evolution. The final bounce (**Figure 17-8*i***) and increase in luminosity are thought to take only some 200 days!

The variation in temperature and luminosity in the H-R diagram during the rapidly changing dynamical phase is shown in **Figure 17-9**. The protostar evolves from a prestellar core at point 1 to point 2 in 5×10^4 years. Still contracting, it reaches point 3 in an additional 4.5×10^5 years. All stars, no matter what their mass, begin in this general region of the H-R diagram.[2]

Our protostar at point 3 is not yet in equilibrium because it is still collapsing and producing energy by the slow conversion of gravitational potential energy into electromagnetic radiation. This energy is transported from one place to another entirely by convection, so the protostar is completely tur-

bulent at this point. We do not observe visible radiation during this phase of collapse because the dust grains mixed in with the gas absorb the radiation. Such slightly warmed grains will radiate in the infrared part of the spectrum. In fact, because the dust in the outer part of the collapsing cloud fragment will usually hide the protostar from view, this infrared radiation is expected to provide the main method of observing star formation in progress within dusty regions of space. We will compare these predictions with observations in the next section of this chapter.

Eventually, ultraviolet light from the protostar will destroy the dust, and visible light from the protostar will break out of its dusty cocoon and be seen. Even now the still-contracting star is probably accreting hydrogen and helium from its surroundings. The situation is complex because, while mass **accretion** must certainly occur, astronomers have not yet observed it for sure. However, extensive mass *loss* is observed to occur through an intense outflow of gas. What is dominant, mass loss by the outflow, or mass accretion? How do the amounts of mass loss and mass accretion vary with time? We do not yet know the answers.

THE APPROACH TO THE MAIN SEQUENCE

Our story of star formation now continues from point 3 in Figure 17-9. Although the following discussion is for a star of one solar mass, we will see that it is basically correct for stars of other masses, also. To portray the evolution in the H-R diagram properly, we begin a new diagram in **Figure 17-10**. The protostar is still collapsing and still releasing

---

[2]The reason is that temperature depends on the ratio of mass to radius. Since the ratio remains nearly constant during this part of the evolution, all stars will have nearly the same temperature.

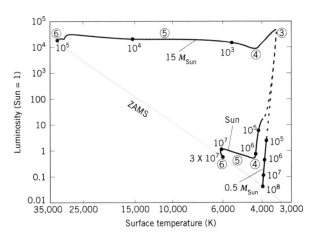

Figure 17-10 The path in the H-R diagram of protostars of different masses. The numbers shown are the times since time = zero in the previous figure.

Hydrogen-burning ignition
↓ produces
Energy generation
↓ which produces
Higher temperature
↓ which produces
Higher pressure
↓ which produces
Slight expansion of core
↓ which produces
Slight cooling of core
↓ which results in
Thermal equilibrium
↓ which means
Collapse ceases

Figure 17-11 The chain of events after the onset of hydrogen fusion in a stellar core.

energy from this gravitational collapse. The shrinking size, along with a nearly constant surface temperature, causes its luminosity to decrease (points 3 to 4 in Figure 17-10), so the evolutionary path in the diagram is nearly vertical. During this time, the changing temperatures and densities cause the energy transport to change from convection to radiation. Eventually, at point 4, all the energy transport is by radiation, which halts the decrease in luminosity. The surface temperature begins to increase, changing the direction of the evolutionary path in the H-R diagram. All the while, of course, the star has been collapsing and the core temperature has been increasing. At point 5, the core temperature reaches the 10 million K required for the proton–proton reaction to begin.

Inquiry 17-6

The protostar is still collapsing. What changes in physical conditions must occur to halt the collapse and bring it into equilibrium?

The onset of nuclear fusion near point 5 has an important consequence that results from the events shown in **Figure 17-11**. The onset of fusion releases energy; this energy, which is slow to leave the core, increases its temperature slightly (recall the ideal gas law). The increased temperature increases the outward pressure in the collapsing core, thus halting the collapse. During this time, in which nuclear fusion gradually increases in impor-

tance and gravitational collapse decreases, the point representing the star moves from point 5 to point 6 in Figure 17-9. The protostar has become a stable star.[3]

This description is basically correct for stars of all masses; details differ depending on the mass. For example, for a more massive star, the changes will occur more rapidly. A general rule in stellar evolution is that the more massive the star, the more rapidly changes occur. The path in the H-R diagram for a 15-solar-mass star is shown in Figure 17-10. The more massive star reaches equilibrium on the main sequence at both higher temperature and luminosity, as shown. For a 0.5-solar-mass star, the vertical distance moved in the diagram is longer, and the time is considerably lengthened, as the star achieves equilibrium on the main sequence at low temperature and luminosity. The time scales for collapse to the main sequence for stars of different masses are shown in **Table 17-3**.

When stars of intermediate mass reach equilibrium, they are located at intermediate points on the H-R diagram. Connecting those points for stars in equilibrium that cover the range of observed stellar masses determines a line known as the **zero-age main sequence**, or **ZAMS**. The

[3]There is no agreed-upon definition of when a protostar becomes a star. We will take it to be around the time when core hydrogen burning becomes dominant.

Table 17-3 Time Scales
for Pre-Main-Sequence Evolution

Mass (Solar Masses)	Spectral Type	Time to Reach ZAMS (in Years)
30	O6	30,000
10	B3	300,000
4	B8	1,000,000
2	A4	8,000,000
1	G2	30,000,000
0.5	K8	100,000,000
0.2	M5	1,000,000,000

ZAMS is the initial main sequence along which stars converting hydrogen to helium fall. Because the pre-main-sequence lifetime is short compared with the main sequence lifetime, the star's age is taken as zero at this point—just as your age is the time since birth, not the time from conception.

The results just discussed came from the calculation of theoretical stellar models, as described in Chapter 16. These models provide us with the theoretical basis for concluding that the main sequence is a mass sequence for stars in equilibrium and having hydrogen fusion as the source of energy. These models also provide a theoretical reason for the observed main sequence mass–luminosity relation presented in Chapter 15.

What about objects less massive than those in Table 17-3? In a protostar less massive than about 0.07 to 0.08 solar masses, the temperature never reaches the 10^7 K required to ignite and sustain nuclear fusion through the proton-proton reaction. Therefore, such an object never really becomes a star, but instead becomes a **brown dwarf**. A very-low-mass object in a binary system may be a planet (depending on how you define *planet,* which is often debated). Such low-mass objects are often referred to as **sub-stellar objects**.

17.3 Star-Formation: Observations

Astronomers have observed a variety of different types of objects that appear to play some role in the overall story of star formation. These are often referred to as **young stellar objects**, and are the subjects of the following sections.

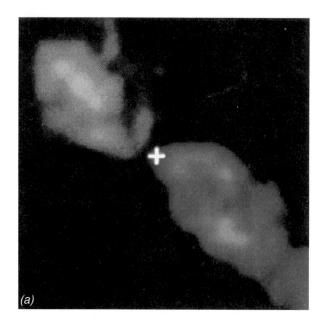

(a)

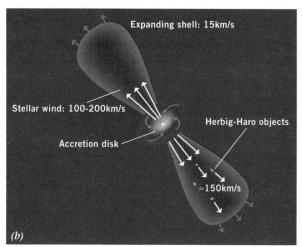

Expanding shell: 15km/s

Stellar wind: 100-200km/s

Accretion disk

Herbig-Haro objects

≈150km/s

(b)

Figure 17-12 *(a)* The bipolar molecular cloud L1551 as observed by a radio telescope in radiation from the CO molecule. *(b)* A model to explain the observations in part a. (Not to scale.)

Bipolar Nebulae and Accretion Disks

Radio astronomers have observed young stellar objects of all kinds to be surrounded by symmetrically placed molecular clouds (**Figure 17-12a**). This figure shows a molecular cloud with blueshifted molecular gas on the lower right and redshifted gas on the upper left. These gas flows, which are an intense stellar wind flowing outward from the protostar at 100 to 200 km/sec, are the **bipolar flows** discussed earlier. These flows are indirect evidence for the presence of an accretion

Figure 17-13 Globules (small, more or less spherical dark spots) in the diffuse nebula NGC 2244 are believed to be the cradles of newly forming stars. Note, too, the luminous O and B stars associated with the nebula. © Anglo-Aust. Observ. Photo by David Malin

disk surrounding the star. The bipolar nebulae are symmetrically placed with respect to a bright source of infrared radiation centered between them. **Figure 17-12b** shows a model that explains the observations in Figure 17-12a. The accretion disk itself has been detected for a nearby massive (about seven solar masses) and young stellar object.

BOK GLOBULES

As we have seen, star formation begins in the coldest, dustiest regions of the Galaxy, and the earliest stages of star formation are hidden from view by dust absorption. Astronomers have long been fascinated by the existence of small, dark regions often seen projected against bright nebulae, as in **Figure 17-13**. These regions are known as **Bok globules**, after astronomer Bart Bok, who first studied them in detail and speculated they might be protostars. Observations show that the smaller globules are the cores of the small, cold molecular clouds discussed earlier, and a quarter or more of them are observed to have a young star or multiple

star system inside. The small dark regions around Figure 17-13 are Bok globules. However, it should be noted that there are regions in our galaxy and in other galaxies where star formation seems to be taking place *without* globules.

INFRARED SOURCES

Suppose a condensation does form and grow within a dusty interstellar cloud, and it increases in temperature and energy output. Calculations show that the condensation will develop a denser, hotter core surrounded by a cooler, more rarefied envelope. Although the hotter core temperatures may vaporize nearby dust particles, the surrounding *envelope* will remain heavily laced with obscuring grains. However, even though the dust grains absorb the energy produced by the central object, that energy must reappear somewhere—the law of conservation of energy says that the energy cannot be created or destroyed. Therefore, the dust grains are heated and reradiate the absorbed stellar energy at infrared wavelengths. This infrared energy is observed as an excess energy over and above what a star would normally have at that wavelength.

In the 1970s, technological advances allowed accurate observations of moderately bright point sources of infrared radiation. At the same time, astronomers learned that these infrared point sources are frequently associated with star-forming regions. A good example of such an object was found in the Orion nebula by Eric Becklin and Gerry Neugebauer, and named the BN source after them. Spectra of this object indicate the presence of dust grains composed of water, ice, and silicates. Because BN is so small that it appears as a point source of radiation, it is now thought to be a protostar in a dust cloud visible only at infrared wavelengths.

Inquiry 17-7

The temperature of the dust grains in many interstellar clouds is about 300 K. At about what wavelength would one expect most of the radiation from such dust grains to be emitted? In what portion of the electromagnetic spectrum does this wavelength lie? (*Hint:* Use Wien's Law.)

Inquiry 17-8

What is one reason why astronomers are interested in studying the infrared portion of the spectrum?

Figure 17-14 An infrared view by the *Spitzer Space Telescope* of the Tarantula nebula.

Our knowledge of star formation is on the verge of a leap forward because of the *Spitzer Space Telescope* (Chapter 12) that was launched in 2003. With its larger size and more advanced infrared detectors, it will build on the wealth of data provided during the one-year lifetime of the 1983 *Infrared Astronomy Satellite (IRAS)*. An example of *Spitzer*'s infrared view of star formation is revealed in **Figure 17-14**, which is a large star-forming region called the Tarantula nebula. This image reveals many previously unknown areas of star formation—unknown because they were hidden from view in visible light by dust that infrared radiation is able to penetrate.

T Tauri Stars

A number of stars called **T Tauri stars** are located in the H-R diagram above the ZAMS. The light output of these stars varies irregularly, in a way that is consistent with accretion from a surrounding protostellar disk. They often emit excess radiation in the infrared, indicating the presence of the dust particles that seem to be necessary for star formation to occur. T Tauri stars also have spectra showing lines of the element lithium. Lithium is not generally present in older stars because it is easily destroyed at the temperatures present in main sequence stars. Thus lithium is often taken as a sign of stellar youth. In addition, T Tauri stars show signs of intense mass loss from a stellar wind in a bipolar flow, just as expected for a protostar

surrounded by a disk. It is this stellar wind that eventually disperses the remaining gas and dust from the region immediately around the fully formed star.

A dark cloud complex important to star formation studies is the Taurus-Auriga complex, which contains a large number of T Tauri stars (including T Tauri itself). Their locations in the H-R diagram are superimposed over theoretical evolutionary paths in **Figure 17-15**. The stars are located above the ZAMS, just as we would expect for stars that are still contracting toward it. Not only that, but the stars are mostly on or below a theoretical *birthline,* which shows from model calculations where stars should first appear on the H-R diagram. The open circles are the classical T Tauri stars, while the filled circles are the naked T Tauri stars—those not having dust surrounding them.

Comparisons of photographs taken a few years apart of the Orion star-forming region revealed a star to have suddenly appeared where no bright star previously existed. The star was named FU Orionis (pronounced F, U Or-ee-onis). Other examples of this phenomenon have also been found and are referred to as **FU Orionis stars**. The event is thought to be associated with a T Tauri star (or an extremely close binary system) whose brightness increases by 10 to 100 times due to accretion of material from the surrounding protostellar disk onto the star. Such outbursts appear to be repetitive and may occur as many as 100 times for a given T Tauri star.

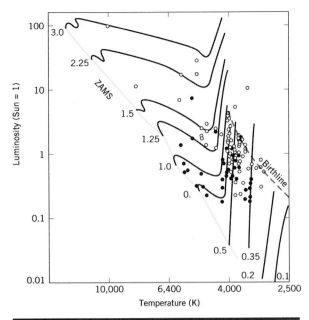

Figure 17-15 H-R diagram showing T Tauri stars in the Taurus-Auriga dark cloud. Also included are theoretical evolution tracks for stars of various masses. Note that most of the stars fall on or below the theoretical birthline.

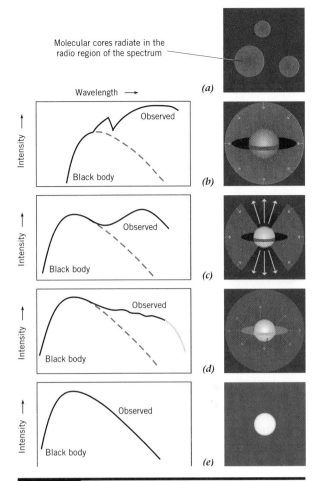

Molecular cores radiate in the radio region of the spectrum

Figure 17-16 Various stages in the formation of a low-mass star; the predicted infrared spectrum at each stage is also shown.

All these data taken together show that T Tauri stars are consistent with low-mass stars still collapsing on their way to the zero-age main sequence. Thus, their energy is not yet coming from fusion reactions in the stellar interior. They provide an important observational confirmation of the entire star formation story thus far presented.

A summary of the formation of these low-mass T Tauri stars is presented in **Figure 17-16**. In Figure 17-16a, cores form within the molecular cloud because collapse is more rapid in the central regions. The core, which has some rotation, accretes material onto the protostar and its disk (Figure 17-16b). Eventually, the strong stellar wind from the protostar punches holes in the surrounding material along the rotation axes, producing a bipolar flow of gas away from the system (Figure 17-16c). Next, the infall of material ceases and the protostellar system becomes a T Tauri star

(Figure 17-16d). Finally, there is a naked T Tauri or post-T Tauri phase in which there is no dust left within the system (Figure 17-16e).

This figure also summarizes how observations and theoretical models are used to provide an understanding of phenomena. Each of the final four phases of evolution presented in Figure 17-16 produces a differently shaped infrared spectrum. The results of calculations of these infrared spectra are shown in the left side of the figure. Thus, by comparing the observed spectrum with a series of model spectra, astronomers can learn where a given pre-main-sequence star is in the evolutionary cycle.

All T Tauri stars have masses less than about three solar masses. More massive pre-main-sequence stars (from 3 up to about 10 solar masses) form somewhat hotter and more luminous stars called Herbig Ae/Be stars. Stars with yet more mass evolve so rapidly that they reach the main sequence before their stellar

Inquiry 17-9

How would you expect intense mass loss in T Tauri stars to make its presence observable?

winds have been able to disperse the surrounding gas and dust. This, along with their rarity, has made the study of pre-main-sequence evolution of the most massive stars more difficult.

HERBIG-HARO OBJECTS

In the early 1950s, several objects whose spectra showed Balmer emission as well as emission of several other elements were discovered. Due to their location within or near dark clouds, their association with star formation was natural. Named after their discoverers, astronomers George Herbig and Guillermo Haro, these **Herbig-Haro objects** are produced by the proto-star outflows, and are not part of the initial star-formation process itself. **Figure 17-17** shows Herbig-Haro objects within a Bok globule.

Recent observations reveal Herbig-Haro (HH) objects moving away from some young stellar objects, thus indicating a physical connection. For example, the jet in Figure 17-17 is moving some 300 km/sec away from an embedded infrared source marked in the figure. **Figure 17-18** shows a more detailed *Hubble* image. The jet is also associated with a molecular cloud. The spectrum of HH 46 shows us that it is actually light from a T Tauri star reflected from dust; this is called a **reflection nebula**. Along the axis of the jet we find not only the bright HH 47A and 47B, but also fainter pieces

labeled 47C and D in Figure 17-17. HH 47C is just emerging from the Bok globule, after having been ejected from the embedded infrared object. Doppler shift measurements show HH 47D to be blueshifted and HH 47C to be redshifted. Thus, all these components are clearly parts of bipolar ejections, as discussed earlier, and provide a coherent and consistent picture of star formation events.

HH 46/47 is complex and appears to indicate that multiple mass ejections have occurred at 500- to 1,000-year intervals. Such episodes of mass ejection might be related to the FU Orionis events described previously. The leading surfaces of all components have shapes (and emission lines) indicative of shock waves that formed when the ejected material encountered a denser medium.

OB ASSOCIATIONS

The giant molecular clouds are the birthplaces of massive stars, those with more than three solar masses. Such massive stars have high surface temperatures and are therefore of spectral types O and B. Because as many as several hundred such stars will form from a single cloud and are observed to be associated with each other, such stellar groupings are called **OB associations**. Remembering that the very existence of luminous O and B stars reveals their youth, we know the stars cannot have strayed far from their birthplace. There are also many low-

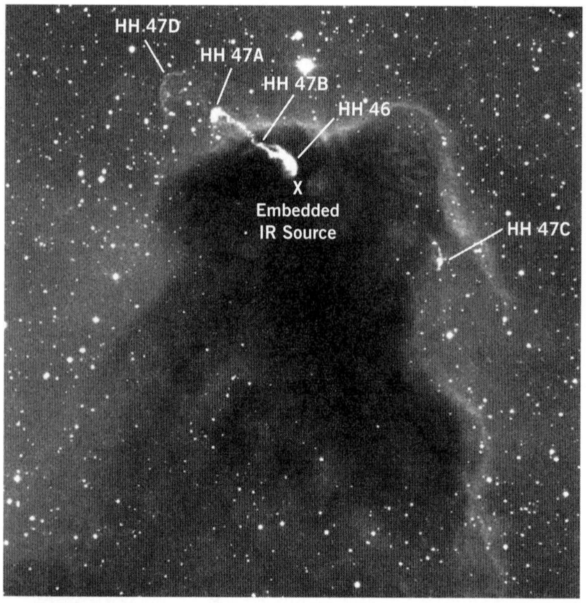

Figure 17-17 Herbig-Haro objects within a Bok globule (which is the entire dark cloud shown). The various labeled objects are all related to one another.

Figure 17-18 A high resolution *Hubble Space Telescope* view of the Herbig-Haro objects shown in lower resolution in Figure 17-17.

Figure 17-19 The formation of an expanding OB association from a giant molecular cloud.

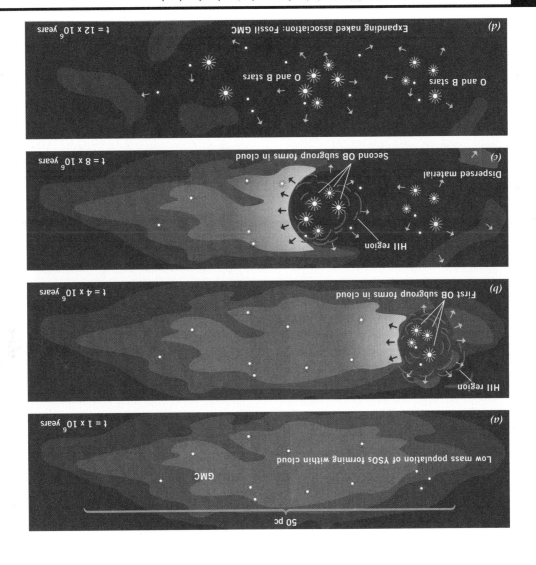

(a) t = 1 × 10^6 years

GMC

Low mass population of YSOs forming within cloud

50 pc

(b) t = 4 × 10^6 years

First OB subgroup forms in cloud

HII region

(c) t = 8 × 10^6 years

Second OB subgroup forms in cloud

Dispersed material

HII region

(d) t = 12 × 10^6 years

Expanding naked association: Fossil GMC

O and B stars

O and B stars

mass stars associated with the visually dominant O and B stars. The presence of nearby remnant gas within OB associations is consistent with their status as groupings of newly formed stars.

One of the early observations made of the OB associations was that they were expanding—that is, gravity was clearly insufficient to hold the association together. In fact, the observed stellar motions were so large that the associations would remain together for only a few million years. But how can all of the OB associations be falling apart when the giant molecular clouds from which they formed were gravitationally bound together? The solution to this long-standing problem was only recently discovered. We describe this problem here because it presents us with a beautiful, consistent story of star formation involving giant molecular clouds and the triggering of star formation.

The first panel in **Figure 17-19** shows a giant molecular cloud some 50 parsecs (150 ly) across in which typically 3 percent of the mass forms low-mass young stellar objects while 97 percent is molecular gas. Each of these young stellar objects has its own random motion relative to every other object. However, the huge mass of the giant molecular cloud provides enough gravity to prevent the young stellar objects from escaping. Eventually, some massive stars form. These massive stars, which make up a small subgroup of OB stars (Figure 17-19b), emit large amounts of ultraviolet photons that dissociate the H$_2$ molecules and ionize the resulting atoms of hydrogen, producing an expanding region of ionized hydrogen called an **H II region**. Some of the dissociated and ionized gas then disperses and is lost to the giant molecular cloud (Figure 17-19c). The

What is happening is that some of the mass of the giant molecular cloud is being compressed into stars, and some is being dispersed into space. In a relatively short time, 90 to 97 percent of the original gas that gravitationally held the giant molecular cloud and stars together in the first place is lost. Because of the loss of mass in the molecular cloud, the speed required to escape the cloud, the *escape velocity*, is lower than before. The random motions of the low-mass stars formed earlier are now greater than the cloud's escape velocity, and these stars are now able to leave the site of their birth and join the general stellar disk population. Any newly formed OB stars, too, will have speeds great enough to escape, but since they are so young and still near where they were born, we see them as an expanding OB association that will eventually dissipate into the general background of stars. The process described can occur a number of times in a given giant molecular cloud and produce a number of separate but related OB associations, some of which are older than others (**Figure 17-19d**). This model fits the observations well and is an example of how a useful model can provide a complete, consistent picture.

The expanding H II region may trigger another episode of star formation farther inside the cloud. As before, more gas will be dissociated, ionized, and lost to the cloud, and yet another subgroup of OB stars may form and then create its own expanding H II region.

The Eagle Nebula shown in **Figure 17-1** provides an excellent summation up to this point of observations having a bearing on star formation. In it we observe H II regions, an OB association, Bok globules, and dust—all physically associated with one another. The Hubble Space Telescope's more detailed image of the dust pillars is one of the most visually exciting images taken (**Figure 17-20a**). These pillars contain not only dust clouds but also significant amounts of molecular hydrogen, as the contour lines on **Figure 17-20b** indicate.

Inquiry 17-10

Stars of type O and B are often seen in regions where star formation is thought to be taking place. How do we know that such OB associations must be young?

The *Hubble* image shows many small nodules that have been given the name EGGs, standing for Evaporating Gaseous Globules. The EGGs protect the gas behind them from the energy of the hot illuminating stars some three light-years away. It is the energy from these stars that not only ionizes the nebula and makes it glow but also evaporates molecular gas on either side of the pillars. At least some of the EGGs may be embryonic stars that will eventually emerge from the gas.

STAR CLUSTERS

We are unable to observe the aging of a single star due to the long time scale for change relative to our short human lifetime. Even the hundreds of years since Galileo first pointed a telescope toward the sky is much too short to allow us to observe the normal evolution of even the most massive stars. The way for us to make headway, observationally, is to study a large number of stars at different stages of their lifetimes. But how can we decide which star is where in its history? Here is where star clusters play an important role.

The Milky Way has thousands of star clusters, gravitationally bound groups of stars, ranging in age from just-born to as old as the Galaxy itself. They have more stars than an OB association, so they stay gravitationally bound longer. One of the more famous is the easily observable Pleiades, or Seven Sisters, in the constellation of Taurus (**Figure 17-21**). Notice the reflection nebulae around them, reflecting the blue light of the stars and parallel lines of density due to their embedded magnetic fields. (The Pleiades long ago shed their natal cloud, and are coincidentally moving through another, diffuse galactic cloud.) Star clusters are important to us because of three (assumed) properties: (1) all the stars are at the same distance, (2) all the stars formed out of the same material; (3) all the stars formed at the

---

Inquiry 17-11

It has been observed that the youngest OB associations are also the smallest in size. What conclusion can you draw from this fact?

Figure 17-20 (a) *Hubble Space Telescope* view of the Eagle Nebula's gaseous pillars. (b) Same as (a) but with contours of infrared emission from molecular gas within the pillars superimposed. Space Telescope Science Institute

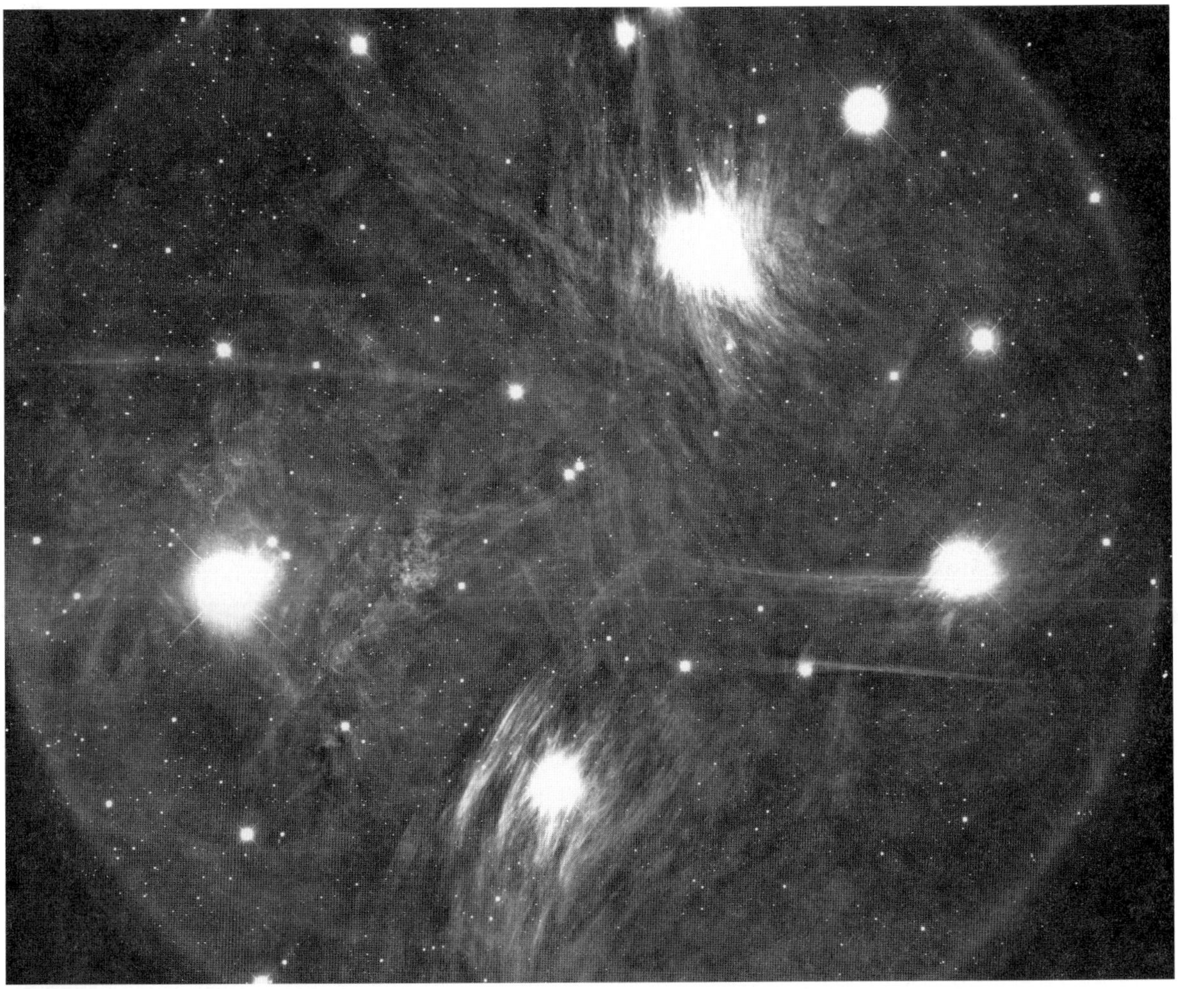

Figure 17-21 The Pleiades star cluster.

same time. Equal distance means that observed differences in star brightness are real differences, not apparent differences due to their unequal distances. From the second assumption of forming from the same material comes the premise of similar chemical composition for each star in the cluster. Finally, formation at one time means that all stars in the cluster are the same age.

Inquiry 17-12

Given the three assumed properties of star clusters what, then, is the difference between the stars in a cluster that allow them to be spread around the H-R diagram?

The Russell-Vogt theorem (Chapter 16) says that the structure of stars in equilibrium is determined by mass, chemical composition, and age. Because of assumptions 2 and 3, the only difference between stars in a cluster is the mass. But we know that stars of different masses evolve at different rates, so the high-mass stars will have evolved more than the low-mass stars in the same cluster.

For example, **Figure 17-22** shows the theoretical H-R diagram for clusters of different ages, all 17-22a, which occurs at an age of zero years. In Figure 17-22a, which occurs at an age of zero years, all the stars are just reaching the beginning of their protostellar stage at the end of the dynamical phase of pre-main-sequence evolution; the more massive stars are somewhat more luminous. After 5,000 years (Figure 17-22b), the more massive stars have evolved toward the left, toward the ZAMS. The less massive stars have not made any significant changes. At 10^5 years (Figure 17-22c), the process continues, with the more massive stars evolving more quickly than the less massive ones. The appearance of the diagram at this point is something we can look for observationally: an H-R diagram for a cluster in which the upper main sequence is populated with stars, but with the less bright (less massive) stars above the projected ZAMS.

Some real clusters present evidence of stars that are not yet completely formed. **Figure 17-23** shows the cluster NGC 2264, which contains large amounts of gas and dust, and a number of young O and B stars. This cluster is still partially embedded within the molecular cloud from which it formed. The association with the gas and dust is excellent circumstantial evidence of their youth. The H-R diagram of the stars in NGC 2264, shown in **Figure 17-24**, matches the theoretical H-R diagram and is therefore an example of a young cluster. Furthermore, symbols outlined in red represent T Tauri stars, and are just where they are expected to be. The sequence of points representing these pre-main-sequence stars in the H-R diagram is broad at least in part because there was a spread in the birth dates of the cluster members. This matching of theory to observations also verifies that our stellar models are fundamentally correct.

A final and important observation related to star formation is that astronomers have never found an example of an *isolated* young star. All the stars that are recognizably young are in associations or young clusters. We also see *moving groups* of stars that appear to be a cluster in the process of breaking up, and *tails* in the orbits of stars in the more massive clusters. This strongly suggests that all stars form initially in clusters, and that the isolated

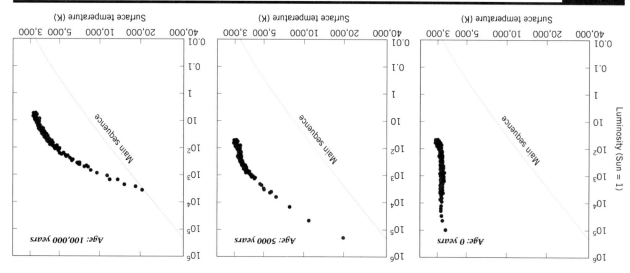

Figure 17-22 The evolution of an imaginary cluster in the H-R diagram.

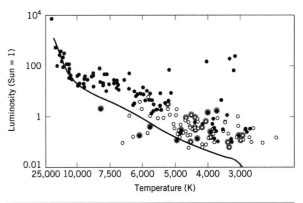

Figure 17-24 The Hertzsprung-Russell diagram for NGC 2264, a very young cluster, showing collapsing stars that have not yet reached the main sequence. Adapted from M. Walker, The Astrophysical Journal (University of Chicago Press 1956)

Figure 17-23 The region around the star cluster NGC 2264. Note the large amounts of both gas and dust. © Anglo-Aust. Observ. Photo by David Malin

stars we see in the Galaxy today were created by the disruption or evaporation of star clusters over time.

Inquiry 17-13

How would you explain the fact that only the stars in the lower part of the main sequence in Figure 17-24 have not yet reached the main sequence, whereas the stars in the upper part of the diagram have already reached the main sequence?

Inquiry 17-14

Draw an H-R diagram showing two young clusters of different ages. Which one is older, and why do your cluster diagrams differ?

Brown Dwarfs

A brown dwarf is an object less massive than a fusion-producing star. The observational confirmation of brown dwarfs required more than a decade of unsuccessful searching. However, persistence

Figure 17-25 The first confirmed brown dwarf, Gliese 229B, as observed by the *Hubble* telescope. The brighter star to the left is Gliese 229, a red dwarf. The spike of light is caused by diffraction within the *Hubble* telescope.

paid off, and the first one, Gliese 229B, was discovered in late 1995 (**Figure 17-25**). Today, dozens more have been found, and it is projected that there are as many brown dwarfs as there are stars. An important distinguishing feature of a brown dwarf from a low-mass main sequence star (a red dwarf) is the presence of lithium, which is destroyed by high stellar temperatures present in red dwarf stars. In addition, the coolest brown dwarfs show methane molecules (the T dwarf stars mentioned in Chapter 14).

Brown dwarfs have been found in star-forming regions, star clusters, and in isolation. Most important is a discovery in 2004 of a brown dwarf in a binary system, since only then can a direct determination of the star's mass be made. This binary system, named 2MASSW J0746425+2000321, contains a red dwarf of mass 0.085 solar masses (just above the limit for hydrogen fusion) and the brown dwarf, with a mass of 0.066 solar masses. Indirect methods applied to two other brown dwarf binaries have provided a consistent estimate of their masses. Thus, the predictions of the theory are in excellent agreement with the observations. Brown dwarfs are about the size of Jupiter, but are at least 80 times more massive and denser.

17.4 A Prominent Region of Star Formation: The Orion Molecular Cloud Complex and Nebula

To illustrate and pull together many of the ideas we have discussed, we conclude with a close look at one of the most prominent star-forming regions in our galaxy—the Orion nebula. **Figure 17-26** shows an image of the Orion nebula, M42, the most famous diffuse nebula in the sky. (See Figure 1-11 for a different view.) It is relatively close to the Sun, at a distance of approximately 1,500 light-years, and is bright enough to be seen with the naked eye. You may know it as the middle star of the sword in the constellation Orion.

At the center of Figure 17-26 we see four bright, high-mass stars known as the Trapezium. The brightest of them is the powerful star $\theta^1 C$ Orionis (pronounced theta one C Or-ee-onis), a main-sequence star of about 33 solar masses. Its spectral type is O6, which means that it is extremely hot, with a spectrum rich in ultraviolet radiation. The energy from this star is almost entirely responsible for the bright, ionized H II region we see as the Orion nebula.

$\theta^1 C$ Orionis is the brightest star in an extensive cluster of stars in the vicinity of the nebula. This cluster contains hundreds of stars, all of them young. Other stars in the Orion nebula have extremely high velocities, which indicates that they are rapidly leaving the region where they

Figure 17-26 The Orion Nebula, M42, the nearest and brightest of the diffuse nebulae. This picture shows both the bright Trapezium stars in the center and the surrounding nebula.

were born. Although it is not obvious why they are moving so fast, their very existence gives another means of estimating the age of the Orion nebula, because we can easily determine how long ago these stars were in the center of the cluster. Of course, they did not all start out from exactly the same place, but their *average* former location is a good estimate of the center.

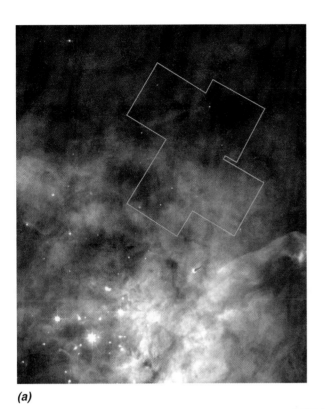

(a)

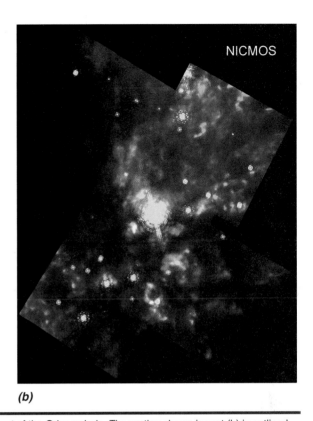

NICMOS

(b)

Figure 17-27 (a) A visible light *Hubble Space Telescope* image of part of the Orion nebula. The portion shown in part (b) is outlined. (b) An infrared image of part of the Orion Nebula showing infrared-emitting objects hidden by dust.

Inquiry 17-15

If you observed a star in Orion to be located 0.1 degree from the center and to have a velocity of 100 km/sec, how long ago would it have been ejected from the center?

The visual-light and infrared views of the Orion nebula are compared in **Figure 17-27**; both images were obtained with the *Hubble* telescope. The brightest object in the infrared image on the right is known as the Becklin-Neugebauer (BN) object, which is a massive young star. The blue filaments are molecular hydrogen clouds. To the north of the BN object is a U-shaped shock front caused by high-speed gas flowing out of this young star.

The region around the Trapezium stars also contains infrared-bright objects at these wavelengths, with more than 3500 low-mass pre-main-sequence stars identified. **Figure 17-28** is an H-R diagram showing 934 stars in relation to the ZAMS. Also shown are lines of constant ages, obtained from theoretical evolutionary model calculations, for

stars of ages of 10^5, 10^6, 10^7, and 10^8 years. From these, an average age near one million years is derived for the Orion stars but with an age spread of some two million years.

Astronomer Bob O'Dell first observed protoplanetary disks, now known as *proplyds,* in the Orion nebula using the *Hubble Space Telescope* (**Figure 17-29**). Thus, the Orion nebula appears to be the place where astronomers are able to observe not only the formation of stars but also the making of planetary systems.

Thus, the Orion nebula that we observe optically is just the visible tip of the Orion iceberg. Surrounding the Orion nebula is a giant molecular cloud known as the Orion molecular cloud (OMC). The OMC contains a wide variety of the molecules discussed previously. The visible Orion nebula, caused by the ionizing radiation from the hot Trapezium stars on the nearside of the cloud, is like a blister of ionized gas protruding from the surface of the molecular cloud behind it (**Figure 17-30**). Most of the action is hidden from sight at visible wavelengths by the dust that is interspersed with the molecular gas. There may be

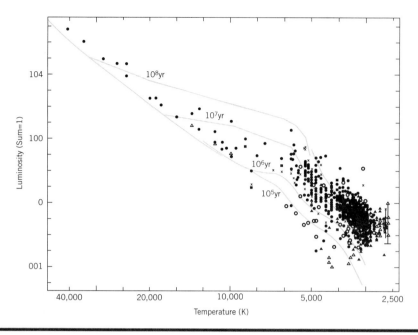

Figure 17-28 The H-R diagram for 934 low-mass stars in Orion. Lines of constant age for 10^6, 10^6, 10^7, and 10^8 years are also shown for comparison.

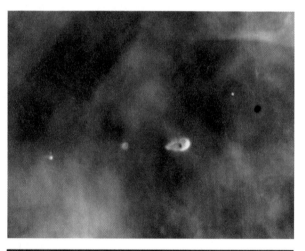

Figure 17-29 Protoplanetary disks (proplyds) in the Orion nebula.

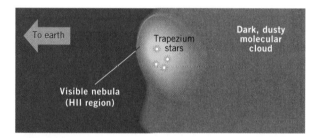

Figure 17-30 A model of the visible Orion nebula and its surroundings.

many other stars in the region embedded deeply in the cloud and invisible optically. All evidence suggests that an impressive new cluster of stars is forming in this region. One day, when the dust is dispersed and we see this cluster in all its glory, the Orion region will be even more spectacular and beautiful to the naked eye than it is now.

Chapter Summary

Observations

- A variety of molecules is found in interstellar clouds located inside dust clouds. The largest molecular clouds are the **giant molecular clouds**, which have high mass, relatively high density, and low temperature. The most abun-

dant molecule is molecular hydrogen, which is best observed in the ultraviolet spectral region.

- Many unusual objects and phenomena are thought to be young stellar objects. For example, young stellar objects exhibit **bipolar flows**. **Bok globules** are small, dark regions in dark clouds with embedded infrared sources.

- **T Tauri stars** vary irregularly, contain the element lithium, and are located in the H-R diagram above the ZAMS. Many show bipolar flows of gas away from the star. **FU Orionis stars** are associated with T Tauri stars that suddenly flare up. **Herbig-Haro objects** are emission regions observed to move away from young stellar objects in a bipolar flow.
- **OB associations** are expanding groups of O and B stars.
- H-R diagrams of young clusters show T Tauri stars still collapsing onto the zero-age main sequence.
- The Orion molecular cloud (OMC) contains molecules, gas, and dust from which stars form. It contains infrared sources as well as hot, blue stars and clusters that have recently formed on the edge of the OMC. Many objects showing protoplanetary disks (proplyds) are observed here.

Theory

- Molecular clouds are supported against gravitational collapse by their internal magnetic fields, random motions, and turbulence.
- A region of a GMC or an entire smaller cloud can collapse under its own weight if it is triggered by an external push from radiation pressure from nearby massive hot stars, by shock waves produced in a supernova explosion, or by clouds colliding with density waves in the Galaxy.
- Each protostar-sized bit of the cloud collapses faster in the central regions than in the outer ones. As the pressure builds up, there is a core bounce. Dust surrounding protostars is expected to radiate in the infrared spectral region. Protostars in different stages of their evolution are theorized to produce differently shaped infrared spectra, which can then be compared with observations.
- Rotation causes collapsing clouds further from the protostellar core to form an accretion disk. Bipolar flows are perpendicular to the disk.

- Eventually the protostar becomes visible in the H-R diagram, at which time energy is transported by convection. Continued collapse increases central pressures and temperatures until hydrogen burning ignites in the core, and the star settles onto the zero-age main sequence.
- High-mass stars evolve more rapidly than low-mass stars. Model stars of different masses that are burning hydrogen in the core array themselves in the H-R diagram in order of mass along the **zero-age main sequence (ZAMS)**. The arrangement has the least-massive stars at the bottom of the ZAMS and thus produces a relationship between mass and luminosity.

Conclusions

- From observations of bipolar flows we infer the presence of a **protostellar disk** surrounding the protostar.
- From observations of infrared sources astronomers infer the presence of heated dust around protostars.
- **Bok globules** are deduced to be the cores of small, cold molecular clouds hiding young stars, often binary systems. From the observations of **T Tauri stars** we conclude that they are young stars still collapsing toward the zero-age main sequence. **FU Orionis stars** are inferred to be T Tauri stars that suddenly accrete matter from their protostellar disk. Astronomers conclude that **Herbig-Haro objects** result from bipolar flows during the star-formation process.
- Comparison of observations of stars in the Orion nebula with theoretical evolutionary calculations finds the Orion stars to be only about a million years old.
- Although there are a number of uncertain details, a variety of consistent observations lend credence to the overall theoretical picture of star formation.

Summary Questions

1. Why are the interiors of dust clouds favorable for the formation and continued existence of interstellar molecules?

2. Why are the conditions of density, temperature, and mass that exist in giant molecular clouds favorable for star formation?

3. What is the picture that astronomers have of the formation of stars? Include in your answer a description of the conditions under which stars are thought to form, and the process that takes a protostar to the main sequence.

4. What is the significance of the fact that O and B stars are often found associated with gas and dust? How does this relate to O and B star's position on the H-R diagram?

5. What is the observational evidence that the variety of objects discussed in the chapter are indeed the precursors to stars?

6. How do observations of one young star cluster differ from observations of a somewhat older cluster? Explain your reasoning.

Applying Your Knowledge

▶ 1. Figure 17-6 is an infrared map of the central region of our galaxy. If the distance from the Sun to the center of our galaxy is 8,500 parsecs, what is the size in kilometers of the infrared region Sgr B2 shown? (*Hint:* Think about the angular size formula.)

▶ 2. Compute what temperature our own protosun had when its luminosity and radius were each 1,000 times the present values. (*Hint:* In Chapter 15 we saw a relationship between the luminosity, temperature, and radius.)

3. How would the spectrum of a protostar having both substantial mass loss from a strong wind and substantial accretion from surrounding material differ from that of a star having neither of these traits?

4. What characteristic of the objects shown in Figure 17-15 causes the spectrum in the infrared to deviate from that of a blackbody?

5. How does the fact that a high-mass star remains on the main sequence for a shorter time than a low-mass star affect the

appearance of an OB association compared to an older star cluster?

6. Why might it be possible to simply *look* at Figure 17-21 and deduce that the bright stars in it are young?

7. In what ways do observations of molecular clouds, bipolar flows, OB associations, Bok globules, T Tauri stars, FU Orionis stars, and Herbig-Haro objects all provide data relating to star formation?

▶ 8. Consider an object similar to that shown in Figure 17-18 in which two Herbig-Haro objects are ejected from a central object. Suppose the central source has a spectral line at 6,000 Å. Suppose further that one component has the same spectral line at 6,002 Å, while the other component has the same line at 5,997 Å. How fast is each component moving with respect to the source, and with respect to each other? What implicit assumption about the motion are you making?

Answers to Inquiries

17-1. Both low densities and low temperatures tend to decrease the number of collisions between single atoms. This would decrease the rate at which one might expect molecules to form.

17-2. This inquiry is answered in the paragraph after the inquiry.

17-3. Gravity would only decrease if the amount of mass decreased; this is not observed. It is more likely that the temperature or turbulence increased. It is also possible that a collision pulled on the cloud rather than pushed on it, lowering its density and thus its internal gravity.

17-4. Their association cannot be due to chance, so it must be causal; that is, the stars must form from the clouds.

17-5. Self-gravity, radiation pressure from a nearby hot star, cloud collision, shock wave compression from a supernova explosion, or cloud collision with a wave within the galaxy.

17-6. To halt the collapse requires increasing the outward pressure. This can be done by increasing the temperature or the density, according to the ideal gas law.

17-7. $\lambda_{max} = 3 \times 10^7$ Å K/3 $\times 10^2$ K = 10^5 Å, which is in the infrared spectral region.

17-8. Infrared radiation can pass through dust, so observing it allows astronomers to look into the interiors of dust-gas clouds and acquire information about star formation.

Infrared radiation also allows the study of cool objects that radiate very little in the visible portion of the spectrum.

17-9. Blueshifted absorption lines would be present.

17-10. The O and B stars have a very short lifetime because they burn their fuel so rapidly.

17-11. The idea of young associations being smaller than older ones is consistent with their observed expansion.

17-12. The fundamental difference between the stars is their mass.

17-13. The stars in the upper part of the main sequence are more massive than the stars in its lower part. Because they are more massive, the force of gravity is stronger, and they tend to contract more rapidly.

17-14. Your drawing might look like Figure 17-24 for one cluster (the younger); the other cluster would be similar but with more stars on the lower main sequence.

17-15. We are given a velocity and angle; the distance is given in the text as 1,500 light-years. From the angular-size relation in Chapter 3, we find the distance of the star from the center of Orion is 0.1° $\times$ 1,500 ly/57.3° = 2.6 ly, or 2.5 $\times 10^{13}$ km. Because time is distance/velocity, we find $t = (2.5 \times 10^{13}$ km)
/[(100 km/s)/(3 $\times 10^7$ s/year)] = 8,300 years.

18

Stellar Evolution After the Main Sequence

This is the way the world ends
Not with a bang but a whimper.

T. S. Eliot, The Hollow Men, *1925*

The development of computers has made it possible for astronomers to calculate the changes expected to take place in stars as they age. Although the actual evolution of a star is slow by human standards, taking millions or even billions of years, a modern computer, in only a few hours, can do the calculations necessary to follow the evolution of a star over most of its lifetime. In this chapter, we will first describe the theoretical picture of stellar evolution that results from extensive computer modeling, and then compare it to a variety of observations.

18.1 The Mid-Life Evolution of Sunlike Stars

Stars cannot remain on the main sequence forever because they run out of fuel. Thereafter, we will examine evolution up to the giant phase and beyond for stars more than 0.2 and up to eight times the mass of the Sun. While some nitty-gritty details are given, remember that the overall evolution of a star is governed by the balance—or imbalance—between gravity and outward pressures.

WHY STARS LEAVE THE MAIN SEQUENCE

We have seen that a stable main-sequence star constantly maintains a state of equilibrium between the inward force of gravity and the outward force of gas and radiation pressure. The energy required to maintain the outward pressure is generated by the fusion of hydrogen in the star's core, the only place where the temperature and density are high enough for fusion to take place.

The star cannot remain on the zero-age main sequence indefinitely, because eventually most of the hydrogen in the star's core will be converted into helium. At this time the star's structure will be as shown in **Figure 18-1**. The changed chemical composition within the core will have changed the star's internal structure. With the core's hydrogen mostly depleted, fusion there must cease because, although the core is a hot 15 million K, it is not hot enough to fuse helium, a process that requires temperatures of about 100 million (10^8) K. At least temporarily, therefore, the star will have lost its source of energy.

Chapter opening photo: The object V838 Monocerotis has a red supergiant at the center that sent out a pulse of light in 2002 that illuminates surrounding dust.

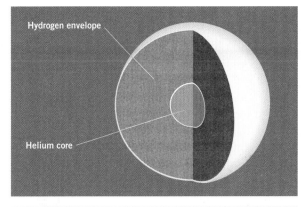

Figure 18-1 Once a star converts its core hydrogen into helium, as shown here, it is ready to leave the main sequence. This figure, and the succeeding ones, are not drawn to scale.

Inquiry 18-1

With its internal energy source gone, what will happen to the core of the star?

How long does it take for a star to convert most of its core hydrogen to helium? We saw in Chapter 16 that the lifetime of a star on the main sequence is determined by 10^{10} M/L, where M and L are the mass and luminosity in solar units. Thus, a star of the Sun's mass will remain quite stable in diameter and luminosity for around 10 billion years. Thereafter, it will find itself undergoing relatively rapid changes, as we now discuss.

The star does not darken immediately when fusion stops, for two reasons. First, as discussed in Chapter 16, it takes a considerable length of time for the photons in the core to leak out and reach the surface; the outward diffusion of energy that has already been produced will take tens of thousands of years. Even more important, the gravitational

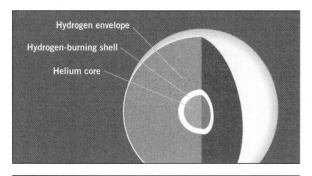

Figure 18-2 Hydrogen burning in a shell around an inert helium core in an evolved star of one solar mass.

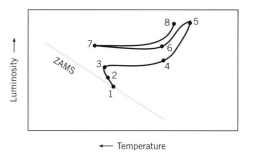

← Temperature

Figure 18-3 A schematic diagram of the evolutionary track of a star of one solar mass in the Hertzsprung-Russell diagram. The events occurring at each numbered point are discussed in the text.

contraction that must set in after fusion ceases will tend to heat the star further by releasing gravitational potential energy. Paradoxically, it seems, the cessation of fusion actually causes the interior of the star to heat up! In a star of one solar mass, such as the ones we are considering, the temperature of the region immediately surrounding the helium core will rather quickly reach 15 million K and cause the hydrogen around the core to fuse. The interior structure will look like that in **Figure 18-2**. At this point the star is said to have a **hydrogen-burning shell**.

Why do the changes just described occur? The basic reason is that as hydrogen fuses into helium in the hot core, the chemical composition of the core changes. You might think that the increasing proportion of helium in the star would be detected in its spectrum, but it is not. The reason is that the convective zone in the outer stellar envelope does not extend deep enough into the core to scoop up helium and move it outward. (The helium we see in the spectrum was either formed during the big bang or in an earlier generation of stars and subsequently ejected into space and later formed into new stars.) However, we are confident that a helium core exists because the composition change from the fusion produces a structural change in the star that modifies the star's surface temperature and luminosity, and thus its location in the H-R diagram.

The changes that result are shown in **Figure 18-3**. This highly schematic figure is not meant to be correct in detail but to provide a general picture of events. The one-solar-mass star's initial position is on the ZAMS at point 1. As the core helium abundance increases during the first 4.5 billion years, the point representing the star's

position moves to number 2, which corresponds to the location of the Sun in the H-R diagram today. Further hydrogen burning moves the location to point 3, at which the core hydrogen is depleted. The core is now helium 'ash' where once there was mostly hydrogen.

Once the core hydrogen is depleted, the star's energy source is gone. The star is still losing energy although no longer releasing any. Thus, thermal equilibrium no longer holds, which means the star must make an adjustment. That adjustment comes about in the following way: because the star is losing energy, the core temperature must decrease. From the ideal gas law, the lowered temperature must produce a lowered pressure. And, from hydrostatic equilibrium, that lowered gas pressure will no longer be sufficient to withstand the downward pull of gravity on the overlying layers, and the star will shrink. But, a shrinking star is losing potential energy and converting it to heat energy, which then increases the core temperature. That temperature increase raises the pressure just enough to balance the downward pull of gravity, and the star reaches a new state of equilibrium.

Meanwhile, as the core shrinks and heats up, it heats the hydrogen surrounding the core. Eventually that hydrogen reaches the temperature needed for hydrogen burning to occur in a thin shell surrounding the core. This, then, is the start of the phase of **hydrogen shell burning**.

Recall from Chapter 15 that the luminosity of a star (the energy per second) is determined by its surface temperature and surface area, so that an increase in the star's luminosity must be due to an increase in its surface temperature, its surface area (or, equivalently, its diameter), or both.

Inquiry 18-2

You will notice that at point 3 in Figure 18-3 the star has a slightly larger luminosity and nearly the same temperature as at its ZAMS location at point 1. What does this imply about the star's diameter?

As hydrogen fuses in the shell, the helium that is produced is added to the central helium core, which slowly shrinks and increases in density. However, the temperature of the core does not increase much at this point, because it is able to radiate away most of the heat liberated by the compression of its gases.

The photons generated in the hydrogen-burning shell flow slowly from the star because they are absorbed by the overlying gases. The absorbed energy causes the outer layers to expand, even though the core is contracting. Because the hydrogen burning shell's photons are absorbed and not getting out, the star's luminosity does not increase but remains approximately constant as the atmosphere expands and cools, and the star's point moves toward number 4 in the H-R diagram.

BECOMING A RED GIANT

As the star expands and its point approaches number 4 in Figure 18-3, models show that the entire envelope of the star outside the hydrogen-burning shell becomes convective. Energy generated in the shell will be transported outward by the turbulent motions of the envelope gases. Convection is much more efficient than radiation in transporting the energy of the star, and the star becomes more luminous. The shell's high temperature causes the reactions to proceed rapidly and thus supply the additional energy required to maintain equilibrium. The increased flood of energy, in turn, causes the envelope of the star to expand rapidly; as the envelope expands, it cools—just as the gases expanding out of an aerosol can cool (making the nozzle cold to the touch). Our star that started out with one solar mass has become a **red giant** star at point 5.

The red giant Aldebaran in Taurus is one of the brightest stars in the winter sky; Arcturus in Boötes is a bright red giant star in the summer sky.

Inquiry 18-3

The decreased surface temperature and greatly increased luminosity of the star move its position in the Hertzsprung-Russell diagram to point 5. What kind of star has resulted?

Inquiry 18-4

At point 5, the Sun will be about 100 times more luminous and 30 times larger than it is now. If the Earth is now about 4.6 billion years old, how long will it be before this event takes place? What consequences will it have for life on Earth?

EVOLUTION DURING THE RED GIANT PHASE

Even though most stars eventually become red giants, relatively few stars are red giants at any one time. This is so because the evolution of a star in this stage is rapid compared to its lifetime on the main sequence. (For an analogy, return to the images taken by the visiting astronaut in Chapter 2. These images would show relatively few people getting their hair cut at any given instant, even though nearly everyone has their hair cut sooner or later.) The percentage of red giants will increase in the future, as the Milky Way gets old enough for its low-mass stars to have evolved to the red giant stage.

We have seen that even as the stellar core contracts, the star's envelope expands. The two parts of the star are moving toward opposite extremes—the core becoming hotter and denser as the envelope becomes cooler and more rarefied. As more and more helium is added to the contracting core, its temperature will rise until at a temperature of 10^8 K helium fusion will commence. The process consists of two basic steps, which must take place in rapid succession (**Figure 18-4**):

$$\frac{4}{2}\text{He} + \frac{4}{2}\text{He} \rightarrow (\frac{8}{4}\text{Be})*$$

$$\frac{8}{4}\text{Be} + \frac{4}{2}\text{He} \rightarrow \frac{12}{6}\text{C} + \gamma.$$

Because the helium fusion reaction involves *three* helium nuclei, or **alpha particles**, it is often called the **triple-alpha reaction**. (The term *alpha particle* for a helium nucleus goes back to the early days of nuclear physics.) Three helium nuclei have

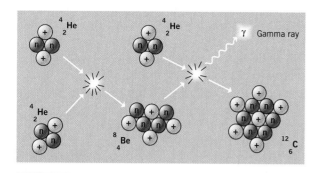

Figure 18-4 The triple-alpha process, which burns helium into carbon in the interior of stars.

been converted into a single carbon nucleus, with an excited beryllium nucleus as an intermediate product. The star is producing new elements, in a general process called stellar **nucleosynthesis**. The reason the two reactions must take place in rapid succession is that the beryllium nucleus is unstable (indicated by the asterisk) and will quickly break down into two helium nuclei if the density is not great enough to make the second reaction take place rapidly.

Inquiry 18-5

The three helium nuclei together have a greater mass than the resulting carbon nucleus. What has happened to the rest of the mass?

An interesting thing now happens. In a normal gas, the energy released by helium fusion would heat the gas, and because the pressure in the gas depends on its temperature, the core would expand. The core would then cool, slowing down the rate of energy production. This process was described in Figure 17-11 in the case of hydrogen burning, in which a natural thermostatic safety valve regulates the temperature and energy-production rate and prevents it from running out of control.

Here, because the core is now extremely dense, the gas at the center of our hypothetical star is no longer an ideal gas. Instead it is now in a special state in which the electrons are packed together so closely that the ideal gas law no longer applies. Physicists call such a gas a **degenerate gas**. Remember, in an ideal gas the pressure comes from the motions of the gas particles, which depends on their temperature; the higher the temperature, the faster the average particle moves and the greater the pressure. Now, however, the rules that govern the

particle motion mean that the pressure is independent of the temperature. The result is that a degenerate core is unable to be compressed, which means the degenerate electrons provide the pressure to withstand the weight of the overlying layers and thus to prevent the star from collapsing.

How is degeneracy important? In the core of a star whose electrons are degenerate, increases in the temperature do not increase the pressure, and so no subsequent expansion takes place. This means that when the helium core heats to the point where the triple-alpha reaction starts, the energy generated by the fusion raises the temperature a little, which increases the energy-generation rate a lot, which raises the temperature further, which increases the energy-generation rate even more, and so on. Without expansion to cool the central core, the star becomes involved in what we call a **thermal runaway**. The process we have just discussed is summarized in **Figure 18-5**, where we compare the events after nuclear burning in an ideal gas with those after burning in a degenerate one.

The thermal runaway occurs rapidly because the energy generation depends so strongly on temperature; the energy-generation rate depends on T^{40}! With such a strong dependence on temperature, a 1 percent increase in temperature produces nearly a 50 percent increase in energy generation. Changes occur so quickly that the star cannot adjust and cannot reach equilibrium. The energy

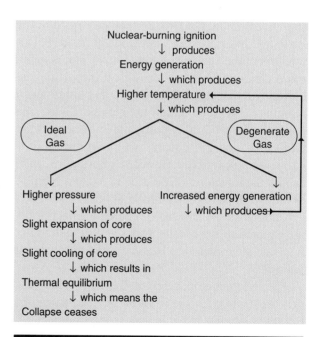

Figure 18-5 Events after the onset of fusion in an ideal gas and in a degenerate gas.

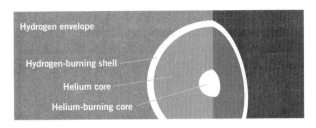

Hydrogen envelope

Hydrogen-burning shell

Helium core

Helium-burning core

Figure 18-6 The core of a star of one solar mass, after core helium burning is established. The hydrogen envelope cannot be illustrated to scale here, as it extends out to many times the diameter of the core region.

within the core rapidly becomes so great that the core undergoes a violent explosion, an event known as the **helium flash**. During the few seconds of the helium flash, an incredible amount of energy—equal to the visible radiation from an entire galaxy—is produced! However, this explosion is hidden deep inside the core of the star.

Following the helium flash, a number of important events occur: (1) the core rapidly expands and cools, soaking up a great deal of the energy and thus keeping the stellar luminosity from increasing greatly; (2) the core's degeneracy ends so it becomes an ideal gas again; (3) the hydrogen-burning shell is disrupted so that its importance as a source of energy decreases, thus causing the star's luminosity to decrease (4) the envelope of the star contracts. These events cause the location of the star in the H-R diagram to move from point 5 down to point 6 (Figure 18-3), where the structure is like that shown in **Figure 18-6**.

By the time a star's point reaches number 5 in Figure 18-3 and becomes a red giant, its structure is already rather complex, and it becomes more difficult to model with a computer. The evolutionary track indicated in the figure depends on a number of assumptions about the star, including its chemical composition and how rapidly the star begins to lose mass, which is a major uncertainty. The solar wind of the main-sequence star has become a solar gale, and up to 40 percent of the mass may be lost before helium-core burning (number 6) is reached. Another uncertainty is the poorly understood role of neutrinos in red giant evolution. These uncertainties, though, are important in understanding details and not our general understanding of stellar evolution. The track indicated in the figure is only one of many possibilities that the Sun might follow in its movement across

the Hertzsprung-Russell diagram, but we think the overall shape of the path is correct.

Eventually, gravity causes the core's expansion to halt and core contraction to ensue. Because the core is again an ideal gas, core contraction causes the temperature to increase. Once the temperature reaches 10^8 K, the triple-alpha reaction begins again (point 6 in Figure 18-3). This time, however, helium ignition is nonviolent, because the core is not degenerate. The situation is similar to that when hydrogen burning first began: the core temperature increases, the core expands and settles into an equilibrium state, and core expansion ceases. From points 6 to 7, then, the star is burning helium into carbon. This is sometimes called the *helium main sequence,* and for stars like the Sun it lasts only about 100 million years.

The red giant Pollux, which can be seen with the apparently nearby twin star Castor in the constellation Gemini, is similar to what our Sun will be when it is a helium-core-burning red giant. It will be about 10 times more luminous than it is now, be 10 times the current diameter, and have an average surface temperature of about 5,000 K.

At point 7 in Figure 18-3 the core is an ash of carbon, just as it was an ash of helium at point 3. As before, because the temperature is not high enough for nuclear fusion to begin, the core begins to contract and heat up. Schematically, the point representing the star then moves in the direction of point 8. A solar-type star will never reach temperatures sufficient for further nuclear fusion to occur. This, then, marks the end of middle-aged period for stars having masses near that of the Sun. In the next step the star enters its death stages, discussed later in this chapter.

18.2 The Mid-Life Evolution of Stars Less Massive Than the Sun

As we saw in the previous chapter, if an object contains less than about 0.07 to 0.08 solar masses, it will never become a star at all; that is, it will not have a main-sequence phase. The force of gravity impelling contraction is insufficient to generate enough central heat to fire up a hydrogen fusion reaction. It will become a degenerate brown dwarf and ultimately (over an extremely long time period) just cool off to become a dead object.

However, stars of mass greater than about 0.1 solar masses will burn hydrogen during a main-sequence phase. Such stars will go through some of the processes described for the Sun. In addition, however, the stars are nearly fully convective, which means that fresh hydrogen is continuously transported into the hydrogen-burning core. The effect of this is to lengthen even further the hydrogen-burning phase for these low-mass stars.

Helium burning will occur only for stars having masses greater than about 0.2 solar masses. Stars of very low mass will not produce red giants. In fact, they are not able to evolve upward from the main sequence at all. If a star has less than about 0.2 solar mass of material, the gravitational contraction that begins after the core has used up all its hydrogen will not be able to heat up the central regions of the star enough to fire up any other nuclear reactions. After the main-sequence phase, these stars will then go directly to a long phase of cooling off, fading away from their main-sequence position down and to the lower right in the H-R diagram.

Although these stars may sound exotic, there are more objects with less than 0.2 solar masses than with more than 0.2 solar masses. This is because more of them are created, and because their lifetimes are so long.

18.3 The Mid-Life Evolution of Stars More Massive Than Eight Solar Masses

Although there is some uncertainty about where the line should be drawn between lower mass and higher mass stars, we adopt eight solar masses as a convenient dividing point. The structure of more massive stars is considerably different from that of stars lower down on the main sequence. In the more massive stars, in which most of their energy production is by means of the carbon-nitrogen-oxygen (CNO) cycle rather than the proton-proton chain, the core transports energy by convection while their outer regions transport energy by radiation. In other words, the convective and radiative regions are switched compared to their locations in low-mass stars. In general, the more massive the star, the greater the fraction of its energy that is carried by radiation transport.

Massive stars, which are highly luminous, generate a strong outward radiation pressure from the vast number of emitted photons. The greater the mass, the greater the temperature, and the greater the radiation pressure. That is why, as we noted earlier, a star more massive than about 100 times the Sun's mass would generate so much radiation that its pressure would cause mass in excess of about 100 solar masses to be rapidly ejected. In further contrast to the less massive stars, the radiation pressure in massive stars is much more important than gas pressure in supporting the star against the inward force of gravity.

In spite of the structural differences we have enumerated, the early stages in the life of a massive star are similar to that of the less massive ones. A massive star spends a relatively quiet and stable life on the main sequence (**Figure 18-7** between points 1 and 3), though for a much shorter time owing to its rapid consumption of fuel. When its core has been depleted of hydrogen (Figure 18-7, point 3), a massive star begins to contract. Just as for the lower-mass stars, a point is rapidly reached where hydrogen burning begins in a shell around the helium core. This helium core eventually contracts to the point at which helium burning (by the triple-alpha process) can begin.

There is a crucial difference between the initiation of helium burning in the more massive stars and in the less massive stars. In the more massive stars the material is *not* degenerate, so there is no explosive burning of helium and thus no helium flash. When the helium ignites, the star is at about position 6 in Figure 18-7. The core expands, because it is not

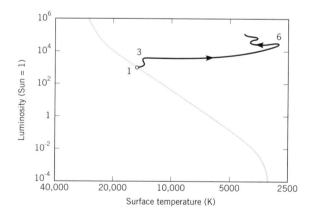

Figure 18-7 The evolutionary track of a massive star on the Hertzsprung-Russell diagram. The numbered locations have the same meanings as in Figure 18-3. That is, the locations refer to physical events in the life of a star, such as the ZAMS (1), depletion of hydrogen in the core (3), and helium core ignition (6).

degenerate; this pushes the hydrogen-burning shell outward, cooling it off. Helium burning then becomes the source of most of the star's energy.

As carbon accumulates in the core of the star, further nucleosynthesis can take place to produce oxygen from a collision between carbon and an alpha particle:

$$^{12}_{6}C + ^{4}_{2}He \rightarrow ^{16}_{8}O + \gamma.$$

As a result, the star forms a carbon-oxygen core. Eventually, of course, most of the helium is burned up; when this happens, the fusion reactions in the core stop and the core begins to collapse once again. Soon, the temperature in the helium surrounding the carbon-oxygen core becomes high enough for the formation of a **helium-burning shell**. At this point the star's energy comes from a **double-shell source** (**Figure 18-8**). One shell is the new helium-burning shell, and the other is the old hydrogen-burning shell that surrounds it and is slowly disappearing. The surface temperature of the star decreases, and the star moves to the right in the Hertzsprung-Russell diagram, along a roughly horizontal path, and into the red giant region (Figure 18-7). Because at point 6 the star's position in the diagram is close to that occupied by low-mass stars during parts of their evolution, distinguishing among red giants of various masses from their positions in the diagram is difficult.

The subsequent evolution of massive stars is not understood as well as the evolution described thus far. In contrast to stars like the Sun, the more massive stars eventually generate interior temperatures and pressures sufficiently high for carbon and oxygen burning to take place with the synthesis of heavier elements. Such burning involves the addition of helium nuclei in a series of **alpha-capture reactions**, in which helium nuclei react with successively heavier nuclei. Some examples are

$$^{12}_{6}C + ^{4}_{2}He \rightarrow ^{16}_{8}O + \gamma$$

$$^{16}_{8}O + ^{4}_{2}He \rightarrow ^{20}_{10}Ne + \gamma$$

$$^{20}_{10}Ne + ^{4}_{2}He \rightarrow ^{24}_{12}Mg + \gamma.$$

In the most massive stars, even the burning of elements such as magnesium, silicon, sulfur, and so on, up to but not including the heavier element iron, may occur. As each heavier nucleus begins to fuse into still heavier nuclei, a shell is formed where this reaction takes place. The temperature at which various nuclear reactions occur, and the time it takes for each energy source to become exhausted, is shown in **Table 18-1** for stars of about 20 solar masses.

The structure of the star becomes more and more complex. The core consists of many shells, much like an onion, each burning the elements in it into still heavier elements (**Figure 18-9**). At the same time, the star shuttles back and forth on the Hertzsprung-Russell diagram, following a path that is difficult to calculate. The final evolutionary phases of massive stars are complex and varied,

Table 18-1 Temperature Required for Nuclear Burning and Time for Exhaustion of Nuclear Fuel in Massive Stars (about 20 Solar Masses)

Energy Source	Temperature (K)	Time for Exhaustion (Years)
Hydrogen burning	1.5×10^7	10^7 years
Helium burning	1.7×10^8	10^6 years
Carbon burning	7×10^8	10^3 years
Neon burning	1.4×10^9	3 years
Oxygen burning	1.9×10^9	1 year
Silicon burning	3.3×10^9	1 day

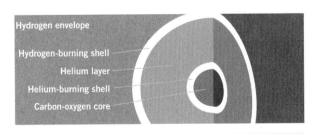

Figure 18-8 The double-shell structure of the core of a massive evolved star. The hydrogen envelope extends out to thousands of times the diameter of the core.

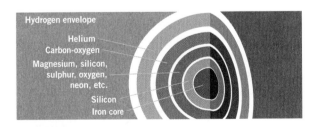

Figure 18-9 A multi-shelled star. Each shell converts the element in it into the heavier elements in the shell beneath. Again, the hydrogen envelope is too extensive to illustrate.

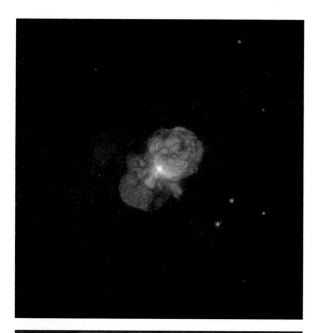

Figure 18-10 Eta Carinae, either the most massive star known or a binary system of two extremely massive stars, is surrounded by bipolar gas expanding at some 2 million miles per hour.

and it will take an entire chapter (Chapter 19) to do the subject justice.

The most massive star currently known is Eta Carinae (**Figure 18-10**). The object is highly unstable, highly variable, and therefore difficult to classify. It most certainly is violently throwing off mass while attempting to reach equilibrium. An outburst in 1841 made it temporarily the second brightest star in the southern sky, even at its distance of 7,500 ly. It ejected at least five solar masses of material, which is seen in the bipolar shells of gas that are perpendicular to a dusty disk, as shown in Figure 18-10 obtained by the *Hubble Space Telescope*.

The conclusion of Eta Carinae's high mass comes from its extraordinarily high luminosity. If Eta Carinae is a single star, its mass would need to be 100 to 150 times that of the Sun—a value that is well above the maximum mass traditionally thought to be possible for a star. For this reason, astronomers have been searching for evidence that Eta Carinae is a binary system, in which case smaller masses, perhaps near the maximum allowable, would make up the system. Such observations have been made and the presence of a binary system is inferred. The current idea, then, is that it is a binary system composed of two very high-mass stars. The 1841 explosion may mean that Eta

Carinae is just a small step away from an explosive death, as discussed in the following chapter.

These observations have also caused a reexamination of the theoretical mass limits. Although the mass limit for high-mass stars is more complicated than previously thought and the answer is not yet in, it does appear that stars up to 150 solar masses can exist after all.

18.4 Pulsating Stars

The Hertzsprung-Russell diagram shown in **Figure 18-11** includes a shaded region known as the **Cepheid instability strip** because stars in this part of the diagram vary in brightness over periods of hours, days, or weeks. When a star enters this region of the diagram, its internal structure oscillates between two states, and it begins to pulsate—actually expanding and contracting in a more or less regular way. At the same time, the surface temperature varies. The star becomes a **variable star**.

The properties of variable stars are determined by their internal structure. For this reason, by observing the properties of variable stars and relating the observed properties to theoretical models describing the pulsation, astronomers can attempt to understand the internal structure of the stars and predict their future behavior.

Inquiry 18-6

What happens to the luminosity of a star whose diameter and temperature vary regularly?

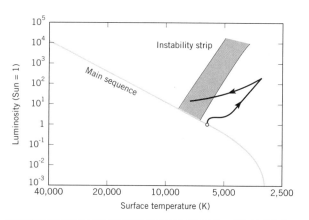

Figure 18-11 The instability strip in the Hertzsprung-Russell diagram, showing the evolutionary track of a star through it.

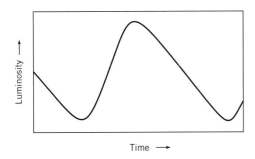

Figure 18-12 A schematic light curve of a Cepheid variable. The shape of the light curve is distinctly different from that of an eclipsing variable, as can be seen by comparing this figure with Figures 15-15 and 15-16.

A number of different types of variable stars are known, the differences depending primarily on mass and their stage of evolution. Usually they are named for the first star of their type to be discovered. Thus **Cepheid variables** (pronounced sef-e-ed) are named for delta (δ) Cephei, a bright star in the constellation of Cepheus. Cepheids are distinguished from other variable stars by their periods, which range from about a day to some weeks in length, and a distinctive brightness variation or **light curve** (**Figure 18-12**), in which the star rises to maximum brightness more rapidly than it fades to minimum brightness. The Cepheid stars are fairly massive (3–10 solar masses) and therefore luminous; some of them are as much as 10,000 times as luminous as the Sun. Their surface temperatures range between 4,000 K and 8,000 K, so they would appear white to us, like the Sun. They evolve through the Cepheid stage in tens of thousands of years.

Stars having masses similar to the Sun also cross into the Cepheid instability strip. One such group is the **RR Lyrae stars**, named for a variable star in the constellation Lyra. These stars are somewhat similar to the more massive Cepheids, but are observed to have pulsation periods of less than a day. In addition to their lower mass, they are less luminous than the Cepheids and all have the same average luminosity. At still lower luminosity is a group of variable stars known as δ Scuti stars. These variables, which have low amplitudes, have periods of only about two hours and are found in the area where the instability strip and the main sequence intersect. Thus, all these types of variable stars form a sequence within the instability strip and provide important information for stellar evolution studies.

A highly distinctive group is the **long-period variable stars**, also called Mira variables after the first one discovered. They vary in visual light by hundreds of times over periods of 100 to 1,000 days. Mode calculations show that these stars have the double-shell sources of hydrogen and helium discussed previously.

We will return to Cepheid and RR Lyrae variables when we discuss stellar and galactic distances in Chapters 20 and 21.

18.5 Mass Loss, Binary Stars, and Stellar Evolution

Our discussion thus far has considered primarily isolated stars having a constant mass. Reality, however, is more complex. Observations indicate that many stars lose substantial amounts of mass while they are red giants (e.g., Eta Carinae in Figure 18-10). Our lack of understanding of the details of this mass loss is the biggest obstacle in the way of our further understanding of stellar evolution.

Inquiry 18-7

What would you expect to happen to the rate at which a star evolves if the star were to lose a substantial amount of its mass?

Inquiry 18-8

How might mass loss be observable in the spectrum of a star?

Mass loss can be detected with a variety of techniques. One is through its effects on a star's spectrum. For example, we observe absorption lines from elements such as hydrogen and calcium that are blueshifted relative to their usual wavelengths (**Figure 18-13**). The blueshifted gas is interpreted as gas expelled from the star, traveling with a considerable velocity toward us. Such observations have been made around the supergiant star Betelgeuse. Consistent with these Doppler shift observations is the infrared image of the star Y Canum Venaticorum that shows an extensive detached envelope of dust surrounding the star (**Figure 18-14a**). Betelgeuse's shows rapid mass loss that has been estimated to take place at the rate of 10^{-5} to 10^{-6} solar masses per year. (In

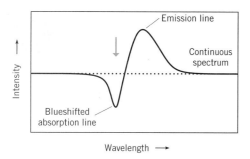

Figure 18-13
Spectrum of a star that is losing mass. The arrow points to an absorption feature on the shorter (blue) side of a line in the star itself, indicating the rapid motion of mass from the star toward us.

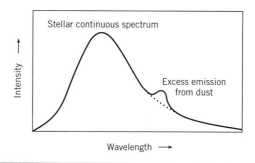

Figure 18-15
Infrared emission from dust adds to the continuous spectrum of a star to produce a hump at infrared wavelengths.

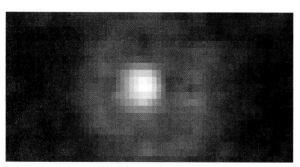

(a)

(b)

Figure 18-14
Two methods of seeing mass loss from stars. (a) Infrared observations of a dust shell around the star Y Canum Venaticorum. (b) Gas flowing away from Mira toward its binary companion.

other words, such a star will lose a solar mass of material in 10^5 to 10^6 years—a relatively short time period.) **Figure 18-14***b* shows another method of detecting mass loss: an ultraviolet Hubble image of the giant variable star Mira shows gas directed toward its binary companion.

The greater the initial mass of a star, the more the total amount of mass lost. For example, a star initially as massive as six solar masses may lose as much as four to five solar masses through a strong

stellar wind before dying. This is different than less massive stars like the Sun, which are thought to lose less than half their initial mass.

The loss of gas from a red giant is often accompanied by the ejection of dust particles into a **circumstellar shell** surrounding the star. These dust particles, which are heated by radiation from the star, emit in the infrared spectral regions and are detected as an excess emission over and above that from the star itself (**Figure 18-15**). Sometimes the circumstellar shell is so thick and opaque that the star can only be detected in the infrared. Furthermore, if the circumstellar shell is a disk or an elliptical cloud, the light may be polarized. Studies of how the polarization changes with time and wavelength provide information on the size and composition of the dust particles, and on the nature of the circumstellar shell itself.

Our discussion has considered mass loss from individual stars. Binary star systems exist in which the component stars are so close to each other that they strongly interact. The gravity of a single isolated star may be represented as shown in **Figure 18-16***a*, in which each circle represents a three-dimensional surface on which gravity is constant. (Gravity's strength decreases with the inverse square of the distance.) Gas inside each of these surfaces is bound to the star. However, when the star has a nearby companion the resulting gravitational surface will be more complex and appear as in **Figure 18-16***b*. One particular gravitational surface appears like a figure 8; each part of the figure 8 is called a **Roche lobe**. Because gravity is the same all along the surface, gas inside one Roche lobe can readily transfer to the other lobe through the crossing point. Gas inside more distant surfaces forms a common envelope surrounding both stars.

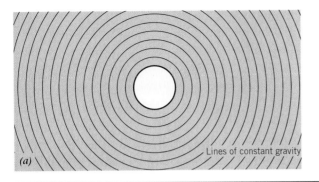

Figure 18-16 *(a)* The gravitational surface of a single spherical star. Along each surface, gravity is constant. *(b)* In a binary system, the gravitational surfaces are distorted. The surface forming a figure 8 defines the two Roche lobes.

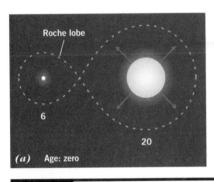
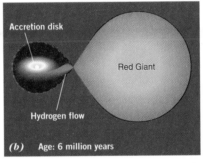
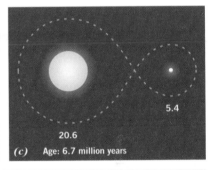

(a) Age: zero   *(b)* Age: 6 million years   *(c)* Age: 6.7 million years

Figure 18-17 The evolution of a close binary system composed of two massive stars. As the 20-solar-mass star evolves rapidly, it becomes a red giant and expands to fill its Roche lobe, transferring material to the less massive star.

Suppose one star in the binary system has 6 solar masses, while the other has 20 solar masses (**Figure 18-17**). Because the more massive star will evolve more rapidly, it will become a red giant before the 6-solar-mass star has left the main sequence. The red giant may completely fill its Roche lobe. As the giant star continues to try to expand, less energy is required for matter to pass through the crossing point to the other side than it would take to expand beyond the Roche lobe. Therefore, the matter from the more massive star comes under the gravitational influence of the less massive one and transfers to it.

and eventually it may be possible for mass to transfer back to the original star. Such a process can continue back and forth for a while. **Figure 18-18** shows an example of a system that has gone through some of the processes just described. R Aquarii is a bloated red giant with a white dwarf

Inquiry 18-9

What happens to the evolution of each star due to the transfer of mass between them?

Depending on the exact situation, it is possible for the less massive star to gain enough mass so it becomes more massive than its companion. The originally less massive star speeds up its evolution,

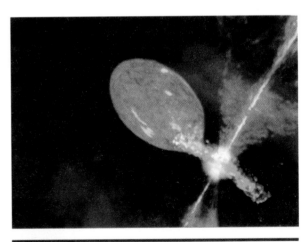

Figure 18-18 An artist's conception of the binary system R Aquarii, consisting of a cool giant and a white dwarf.

companion. Matter from the red giant moves through the Roche lobe and accumulates in an accretion disk surrounding the hot white dwarf before detonating in a nuclear explosion. The process just described occurs in a variety of stellar systems, some of which are so important we will discuss them in further detail in Chapter 19.

18.6 The Death of Stars Less Massive Than Eight Solar Masses

While astronomers have confidence in their observations of *what* happens to low-mass stars after their red giant phase, the theoretical details of *why* are not firmly understood. Remember that at the end of the red giant stage the star contains a burned-out core of carbon along with a helium-burning shell. The star is so bloated that its radius approaches 1 A.U.. A series of explosive events within the helium-burning shell produce instability in the form of pulses of energy in the star's outer regions. These pulses appear to provide an outward push to the extended stellar envelope with the end result that the star's outer envelope is ejected.

Once the star sheds its outer envelope, the hotter inner regions are exposed. The resulting object is called a **planetary nebula** (**Figure 18-19**). Another example, known as the *Ring Nebula* in the constellation Lyra, is shown in Figure 1-12. Note the star in the center; it ejected the observed nebula. The central star eventually cools to the point where we recognize it as a **white dwarf** star. White dwarf stars are discussed further later in this section.

PLANETARY NEBULAE

Planetary nebulae received their name because in a small telescope they somewhat resemble the pale disk of a planet. (You can easily see the Ring Nebula using an amateur-sized telescope.) Actually, they are not at all like planets. Their spectra show a series of bright emission features at individual, discrete wavelengths.

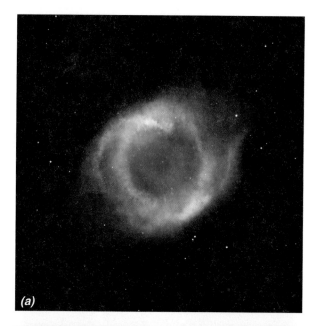

(a)

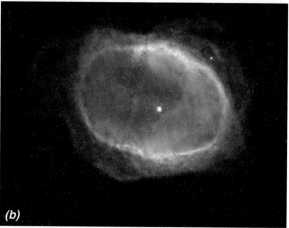

(b)

(c)

Figure 18-19 Three planetary nebulae. *(a)* NGC 6852 the "dumbell" nebula *(b)* NGC 3132 *(c)* Menzel 3, also known as the "ant nebula."

The temperature of the gas in these objects is typically about 10,000 K, whereas their densities are extremely low, ranging from a few hundred to a few thousand atoms of gas per cubic centimeter.

Inquiry 18-10

What specifically can you conclude about the properties of planetary nebulae given that they are observed to have discrete emission lines?

(For comparison, a cubic centimeter of air at sea level has nearly 3×10^{19} atoms.) These physical conditions are similar to the chromosphere of the Sun. The center of a planetary nebula contains a very hot star, and it is the ultraviolet radiation streaming from this star that excites the surrounding gas to glow.

Inquiry 18-11

Why does the previous sentence specify ultraviolet radiation rather than, say, visible or infrared radiation as the cause of the nebular glow?

The ring-like appearance of the nebula is only an illusion. In fact, planetary nebulae exhibit a great variety of shapes. Recent observations show, at least in the case of the Ring Nebula, that the star has ejected bipolar lobes of gas, and that the nebula looks the way it does because we are viewing it along the lobe axis, looking at the pole of one lobe (**Figure 18-20**). A bipolar lobe model helps to explain the variety of shapes observed for planetary nebulae as coming from differing viewing angles. At least some observed features occur because the ejected material may run into interstellar gas and dust as it expands away from the star. Multiple shells have been observed for several planetary nebulas, and some estimates are that more than 50% have multiple shells. Some observations seem to indicate that the central stars of all planetary nebulae are binary stars, which complicates our understanding considerably. If that is

true, then the stars that result in planetary nebulae must be binary, but that is something we do not yet know.

When the light from the nebular ring is analyzed with a spectrograph, the shell of gas is found to be expanding, as expected for a shell of gas that has been ejected from a central star. Typically, it moves away from the central star at speeds of around 25 to 30 km/s. Though this may seem fast, it is actually quite slow when compared to the more violently ejected gases observed around exploded stars, as we will see in the next chapter. Because of the brisk rate of expansion, a planetary nebula is recognizable as such for only some tens of thousands of years.

Because the gas in a planetary nebula is excited to glow by the large numbers of ultraviolet photons emitted by the hot central star, the nebula can be used to study the central star itself. These stars are somewhat small, but they have high surface temperatures and are luminous. In fact, several have surface temperatures in excess of 100,000 K. When the positions of the central stars are plotted on the Hertzsprung-Russell diagram, as shown in **Figure 18-21**, they are found to lie just below the upper end of the main sequence.

Inquiry 18-12

What do the small size of the central stars in planetary nebulae, their location in the Hertzsprung-Russell diagram, and the fact that they have recently shed considerable amounts of mass indicate about the probable place of planetary nebulae in the story of stellar evolution?

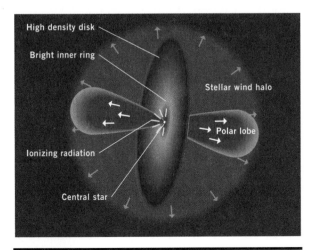

Figure 18-20 Explanation of the ring-like appearance of the shell of gas surrounding the central star of a planetary nebula.

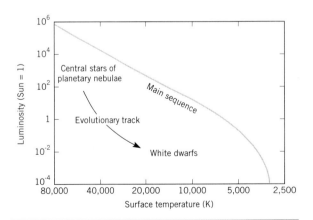

Figure 18-21 The positions of the central stars of planetary nebulae plotted on the Hertzsprung-Russell diagram. The evolutionary path to the white dwarf region is also indicated.

Astronomers have discovered a number of objects that are candidates for planetary nebulae caught in the act of formation. There are many good reasons to suppose that the precursor star was a red giant (actually a long-period variable star consisting of both helium- and hydrogen-burning shells) before entering the planetary nebula phase. We know that red giant stars have a dense core surrounded by a tenuous atmosphere barely held by the star's gravity. Such an object is consistent with theoretical predictions about the death of low-mass stars. The Egg Nebula, shown in **Figures 22a and 22b** is thought to be on its way to becoming a planetary nebulea. This object contains a disk that was observable in infrared radiation by the Hubble telescope. The disk forces outflowing gas into the two bipolar lobes. Gomez's Hamburger, in **Figure 18-22c**, is also on its way to becoming a planetary nebula.

Evidence in favor of the scenario of red giant to planetary nebula to white dwarf comes from a comparison of the number of planetary nebulae existing at any time and their estimated lifetimes. Astronomers estimate that approximately two to three planetary nebulae are born each year in our galaxy. Because each of these nebulae is supposed to produce a white dwarf, about two to three white dwarfs would be expected to form in the Milky Way galaxy each year. Nearly 10 percent (about) of the stars in the Galaxy (10 billion stars) are white dwarfs. Because our Galaxy is about 13 billion years old, this means that an average of a little less than one white dwarf per year has been produced in the Galaxy over its entire history; this number agrees tolerably well (to the same order of magnitude) with the other estimate. This agreement has convinced astronomers that the majority of stars do indeed end their lives as white dwarfs.

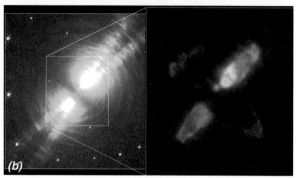

Figure 18-22 The Egg Nebula and Gomez's Hamburger, possible planetary nebulae in formation. *(a)* Optical Hubble image shows oppositely ejected beams in the Egg Nebula. *(b)* The Hubble infrared image of the Egg Nebula shows a dust disk perpendicular to the beams in the optical image. It is the disk that causes the beams to be in opposite directions. *(c)* A Hubble visible light image of Gomez's Hamburger, showing a thick dust disk.

Inquiry 18-13

The estimate of the present production rate for white dwarfs, obtained from the number and lifetimes of planetary nebulae, is about three times the average rate over the entire history of the Galaxy. Assuming that this difference is real, and not just due to inaccuracies in our estimates of the numbers, what would it imply about the present white dwarf production rate compared to the rate in the past?

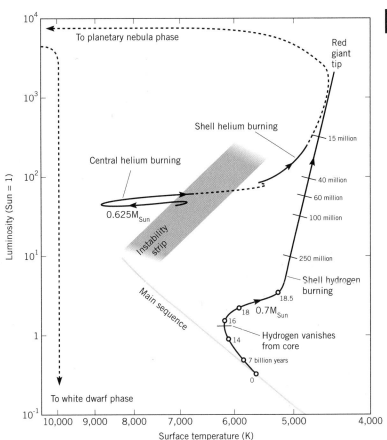

Figure 18-23 Evolutionary track for a star of 0.7 solar masses that contain fewer metals than the Sun. The times shown are the times since the ZAMS. The times along the giant branch are the times from the point indicated until the star reaches the red giant tip. The evolutionary path marked 0.625 solar masses is the path followed by a more massive star once it loses mass as a red giant.

The post-main-sequence evolution of stars is summarized in **Figure 18-23**, which shows the theoretical evolutionary tracks for a star starting with 0.7 solar masses. As the star evolves away from the main sequence, the times since the ZAMS is indicated. The times along the giant branch are the times from that point to the red giant tip.

The evolutionary path in the H-R diagram after the red giant tip depends critically on the mass and the chemical composition. To produce valid models requires that the star's mass be less than its main sequence value. Thus, in the example given, while the model star was a red giant, enough mass loss occurred to bring it down to 0.625 solar masses. Its evolution then proceeded as shown.

OBSERVATIONAL PROPERTIES OF WHITE DWARFS

Many details of stellar evolution are uncertain, but we are certain that a white dwarf is an object that no longer has nuclear reactions occurring in the core. Its energy comes from the cooling of its hot, dead core. To finish our story of the evolution of low-mass stars, therefore, we will replace our computer models with evidence based on observations.

One way to proceed would be to look for stars that *appear* to be either dead or dying. If we wanted to make a list of the characteristics that such a star might have, we might make the following inquiries.

Inquiry 18-14

Would a dying star have a larger or smaller proportion of hydrogen, relative to heavier elements such as helium and carbon, than main-sequence stars?

Inquiry 18-15

Considering that it would have no source of internal energy to provide the high temperatures and pressures required to support it against the force of gravity, would such a star be large or small?

Inquiry 18-16

Assuming that the mass of the star has not changed much, would the star's density be low or high? How might this density affect the appearance of the spectrum?

Inquiry 18-17

In view of the size and lack of energy source for the star, would it have a high or a low luminosity?

A class of stars that appears to have all the expected characteristics of dying stars is the group of white dwarf stars lying below the main sequence. In the 1830s, the German astronomer Friedrich Wilhelm Bessel (who also successfully measured the parallax of a star) discovered that the bright star Sirius exhibited a wobble in its apparent motion across the sky. Sirius is a bright star, primarily because it is close to Earth; as a result, its motion through space can be measured more easily than the motions of most stars. The wobbles in the path of Sirius indicated the presence of a faint, unseen companion orbiting around it that was affecting its motion; without a companion, no wobble would be present. In 1862, the famous telescope maker Alvan Clark actually saw the faint companion, now known as Sirius B, in a newly built telescope. (Even amateur astronomers can sometimes catch a glimpse of it with a 10-inch telescope.) Sirius B was the first white dwarf to be discovered and is noted in **Figure 18-24**, in which the difference in image sizes between Sirius A and the white dwarf is caused by its brightness being some 10,000 times less than that of Sirius A. Over a period of time, the orbit of the object was determined and the mass of the star was calculated to be similar to that of the Sun.

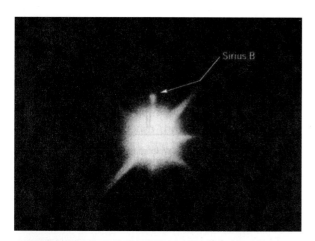

Figure 18-24 Sirius A and its white dwarf companion, Sirius B. The difference in the apparent sizes of the images of the two stars is caused by their great difference in luminosity.

Inquiry 18-18

Describe what specific information astronomers had to have to determine the mass of Sirius B.

Inquiry 18-19

Does the companion of Sirius A obey the mass-luminosity relationship discussed in Chapter 15? Suggest a reason for this.

From the inferred surface temperature and luminosity of this object, the radius was simple to calculate (from $L \propto R^2 T^4$). Imagine the astonishment of astronomers when they found that the diameter of Sirius B is only a hundredth that of the Sun: a mass equal to the entire Sun crammed into a volume similar to that of Earth (**Figure 18-25**)! Its density must be extraordinarily high—because its diameter is only a hundredth of the Sun's, its volume must be only a millionth $[(10^{-2})^3 = 10^{-6}]$ of the Sun's, making its density about a million times greater than the Sun's. In earthly terms, a *cubic centimeter* of white dwarf material, brought here and weighed, would weigh as much as a pickup truck! Sirius B is indeed an extraordinary object.

The spectra of white dwarfs add additional details to the picture we are constructing of small, dense objects. For one thing, the absorption lines are observed to be very broad (**Figure 18-26**). This breadth can be attributed to the high density of the star's atmosphere. (Remember from Chapter 15 why the spectral lines in main sequence stars are broader than those in supergiants?)

Because the white dwarf is small, the inverse square law of gravity tells us that the strength of gravity on the star's surface must be large. A photon leaving the strong gravitational field at the sur-

Figure 18-25 A comparison between the sizes of the Earth, the Sun, and Sirius B.

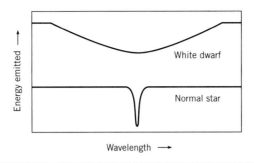

Figure 18-26 The broad absorption lines of a white dwarf compared with the absorption spectrum of a main-sequence star of the same spectral type.

face of a white dwarf does work in leaving the surface and thus loses considerable energy. This energy loss means that the photon's wavelength must become longer than usual. In other words, light leaving a strong gravitational field will have its wavelength shifted toward redder wavelengths. This redshift is not caused by the Doppler shift, but by the gravitational field itself. This effect, known as a **gravitational redshift**, was first predicted by Albert Einstein as a consequence of his theory of general relativity. The observed gravitational redshift of light from a white dwarf is one of many pieces of data that show the theory of relativity is an accurate description of nature.

Inquiry 18-20

To be detected, the gravitational redshift of a white dwarf must be distinguished from any *velocity* redshift (Doppler shift) due to the motion of the star in space. How might this be done in the case of Sirius B? (Hint: Sirius B is a member of a binary system.)

If the Sun were collapsed to the size of a white dwarf, its magnetic field would also collapse and strengthen. White dwarfs are predicted, therefore, to have intense magnetic fields. Such strong fields are capable of polarizing the emitted radiation. Therefore, observations of polarized light from magnetic white dwarfs tell astronomers about the magnetic conditions in the star and the properties of the compressed white dwarf material itself.

How do our observations of white dwarfs compare with our hypotheses about them? We expect a white dwarf to be deficient in hydrogen because it is the remnant of a star whose core hydrogen has been depleted. Making observations that confirm the expected hydrogen deficiency is difficult, however. On the one hand, unless material from the core were to become mixed with the surface layers of the star, the products of nuclear fusion would not be detectable in the spectrum. Furthermore, the extreme broadening of the spectral lines makes them harder to detect. On the other hand, models explaining the production of the so-called type Ia supernova, which are discussed in Chapter 19, require the presence of a white dwarf with a carbon-oxygen core. Observations with the Chandra X-ray telescope of the gases from the Type Ia supernova DEM L71 are interpreted as providing convincing evidence that the supernova explosion involved such a white dwarf.

THE STRUCTURE OF WHITE DWARFS

When the laws governing degenerate gases were discovered, theorists became seriously interested in white dwarfs, because it soon became evident that these were stars composed almost entirely of degenerate material, discussed earlier in this chapter in relation to the helium flash. The outward degenerate electron pressure supports both the helium ash core of a star evolving off the main sequence and a white dwarf. It is not a coincidence that the core of a star passes through a degenerate stage, and that white dwarfs are also degenerate.

Temperature and pressure in a degenerate gas do not depend on each other. For this reason a star composed of degenerate gas cannot shrink, even if the temperature inside it becomes very low. White dwarfs are therefore deprived not only of a source of nuclear energy but also of gravitational energy; with no source of new energy, they can do nothing but slowly cool off, radiating into space the remnants of energy left over from their younger days. Because of their small surface area, cooling takes billions to trillions of years.

Inquiry 18-21

If white dwarfs have no nuclear reactions, and if they cannot contract, what is their source of energy?

The laws of degenerate gases are surprisingly simple, and the structure of stars composed of degenerate material can be worked out with pencil and paper—a computer is not needed. In the early 1930s, the Indian-American astrophysicist

Figure 18-27 S. Chandrasekhar, who discovered the mass-radius relationship for white dwarfs. He was awarded the Nobel Prize in physics in 1983 for his work related to studies of stellar evolution.

S. Chandrasekhar (**Figure 18-27**), after whom the Chandra X-ray telescope was named, made a remarkable theoretical discovery about degenerate stars: He calculated the relationship between the mass and the radius of white dwarfs (**Figure 18-28**). The white dwarf **mass-radius relation** shows that the more massive the star, the *smaller* it is. Not only that, and even more surprising, white dwarfs of mass greater than about 1.4 times the mass of the Sun cannot exist because they will collapse. Note, however, that the effects of stellar

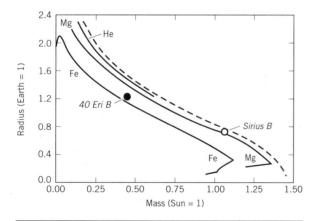

Figure 18-28 The mass-radius relationship for white dwarfs. The particular curve depends on the chemical composition. The observed locations of two white dwarfs are noted. The uncertainties in mass and radius are only slightly larger than the point sizes.

mass-loss discussed above mean that a star that was as massive as 4 solar masses on the main sequence must lose sufficient mass to become a white dwarf.

The significance of this 1.4 solar mass limit, called the **Chandrasekhar limiting mass**, will be seen in the next chapter. Figure 18-28 shows that the mass-radius relation differs for stars of different chemical composition. Thus, accurate observations will allow astronomers to determine the overall chemical composition of a white dwarf. Knowing the chemical composition allows astronomers to deduce the mass of the star that formed the white dwarf, assuming mass loss is understood. The figure shows the locations in the mass-radius diagram of two white dwarfs.

Inquiry 18-22

Statistics on the proportion of stars of various masses and the number of white dwarfs that are seen today indicate that most stars end up as white dwarfs, including all stars up to at least several times the Sun's mass. What must happen during the evolution of these more massive stars for them to end up as white dwarfs?

18.7 The Observational Evidence for Stellar Evolution

Most of our discussion thus far has looked at the evolution of stars on the basis of the predictions from calculations and computer models, because a compact and logical presentation can be made in this way. Actually, some of the important stages in stellar evolution had been worked out from observational evidence before large computers became available. It is now time to see what the observations have to say about stellar evolution, and whether any of the theory discussed in this chapter has anything to do with reality.

THE HERTZSPRUNG-RUSSELL DIAGRAMS OF STAR CLUSTERS

Much of the most dramatic and conclusive evidence for stellar evolution comes from a study of the Hertzsprung-Russell diagrams of the stars in star clusters. Astronomers have long been aware of the existence of stars that are close to each other

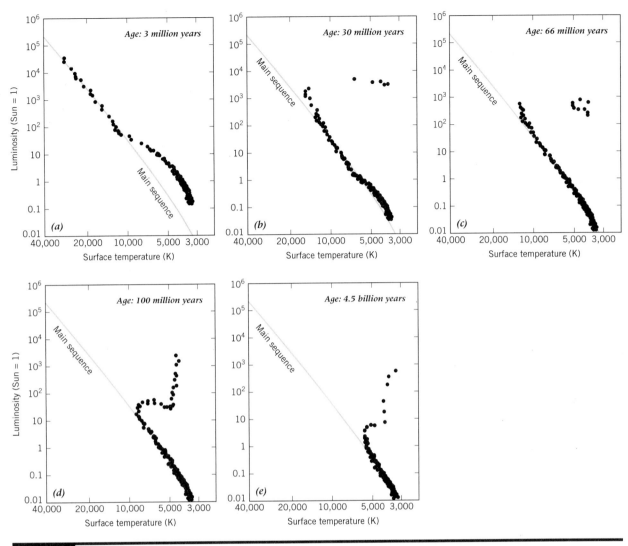

Figure 18-29 The rest of the evolution of an imaginary open cluster. The earlier evolution is shown in Figure 17-20.

and move with a common motion through space. Two examples are the Pleiades (Figure 17-21) and the Hyades, both of which can be seen with the naked eye in the constellation of Taurus, the Bull. The Pleiades is especially pretty when observed with a good pair of binoculars or a small telescope. Because the stars in such clusters are far apart from each other they are often called **open clusters**.

When the cloud from which the cluster forms collapses and fragments into stars, different stars will have different masses and, therefore, different luminosities. In Chapter 17 we saw that such stars will have the same chemical composition and age; we also saw how the H-R diagram of an imaginary but typical cluster changes as the stars evolve onto the main sequence (Figure 17-22) during the first 10^5 years. The rest of the evolutionary picture is

shown in **Figure 18-29**. Between 10^5 and 3 million years (Figure 18-29a), the more massive stars evolve onto the upper end of the ZAMS. The less massive stars evolve toward the bottom end of the ZAMS but at a significantly slower pace. Many star clusters are observed to have Hertzsprung-Russell diagrams of this form, with stars on the upper end of the ZAMS and stars above the lower ZAMS. Such clusters have formed recently.

After only a few tens of millions of years, the most massive stars in the cluster will begin to exhaust the hydrogen fuel in their cores and evolve, moving away from the main sequence and into the red giant region, as shown in Figure 18-29b. As more time passes, stars of lower and lower mass farther down the main sequence will also have time to complete their hydrogen-burning phase, and they will also begin to leave

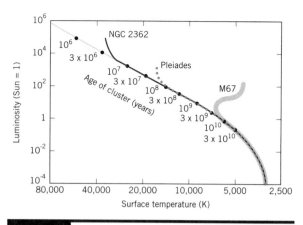

Figure 18-30 The Hertzsprung-Russell diagram of three clusters. The dots on the main sequence indicate the position of the turnoff point for clusters of the indicated age.

Figure 18-31 The globular cluster 47 Tucanae in the southern hemisphere. © Anglo-Australian Observ. Photo by David Malin

the main sequence and move toward the giant region (Figure 18-29*c–e*).

We call the top of the main sequence, from which stars have begun to evolve, the **main-sequence turnoff point**. The older the cluster, the fainter and redder is its turnoff location, Astronomers can therefore determine the age of the cluster from the location of the turnoff. The ages corresponding to the locations of the main-sequence turnoff point are shown in **Figure 18-30**.

Inquiry 18-23

How is the age of a cluster related to the position of its turnoff point on the Hertzsprung-Russell diagram?

Inquiry 18-24

The turnoff point provides a method for estimating the age of a cluster using a plot of its Hertzsprung-Russell diagram. What crucial assumption is being made about the formation of the cluster for this to be a valid method?

Inquiry 18-25

Figure 18-30 shows the Hertzsprung-Russell diagrams of three clusters plotted on the same graph. Estimate the ages of (a) the Pleiades, (b) NGC 2362, and (c) M67.

For isolated stars not found in clusters, ages cannot be estimated with any degree of certainty.

However, in some circumstances we can draw useful conclusions. When we see a single star of one solar mass in the Galaxy, it is in general difficult to tell whether it formed recently or billions of years ago, because such stars spend about 10 billion years in their stable, main-sequence phase. But if we see a massive, luminous star near the top of the main sequence, the short lifetime of such stars assures us that it must have formed recently.

There is yet another type of star cluster, which has an appearance dramatically different from that of the open clusters: the so-called **globular cluster** (named for its spherical, globe-like shape). Figure 1-14 shows the bright cluster known as M13, while **Figure 18-31** shows the bright southern hemisphere cluster 47 Tucanae.

Globular clusters provide us with further confirmation of the evolutionary picture we have been constructing. A composite H-R diagram showing the characteristics of a typical Milky Way globular cluster is shown in **Figure 18-32**. Whereas open clusters exhibit a wide variety to their diagrams, *all* globular clusters in our galaxy have roughly similar-appearing main sequence and giant branches. This is because the turnoff point moves more and more slowly as a cluster ages and its main-sequence stars stay longer on the main sequence. The oldest clusters could be 2 or 3 billions of years different in age and still look almost the same. Ancient star clusters have a short main sequence with a turnoff point at both a low luminosity and a low temperature. In fact, these clusters are all old—more than 10 billion years old. All the stars in a globular cluster more massive than the Sun have had plenty of time to go through their entire life-

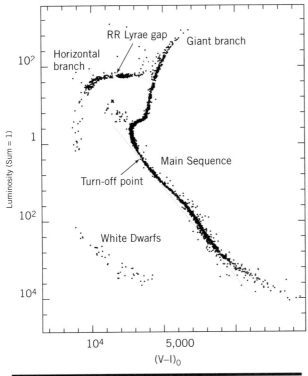

Figure 18-32 The globular cluster H-R diagram. This diagram is not for a specific cluster but is a composite of actual data for 5 clusters. Only RR Lyrae stars inhabit the region noted.

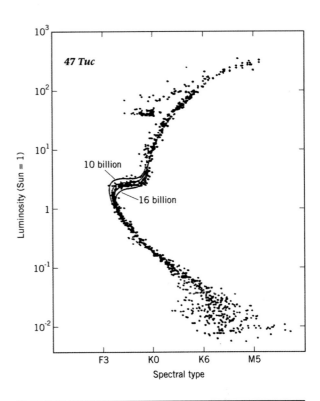

Figure 18-33 The H-R diagram of the globular cluster 47 Tucanae along with evolutionary tracks for clusters of age 10, 12, 14, and 16 billion years. The spread at the bottom of the main sequence is caused mostly by increasing observational uncertainty as the stars become fainter.

times. By now they have evolved into dead or dying stars. In such an old cluster, even stars with as little mass as the Sun will have died and become faint white dwarfs. Such stars have now been observed by the *Hubble Space Telescope.*

The stars we see near the turnoff point in Figure 18-32 are just beginning to evolve away from the main sequence. These stars have a mass of approximately 0.85 to 0.90 solar masses. The stars that are a little farther along in their evolution into the giant region will have masses slightly above this value (having left the main sequence slightly earlier).

In addition to the shortened main sequence and the giant branch, globular cluster diagrams also show a **horizontal branch** of stars (Figure 18-32). These stars are burning helium in the core and have either single or double shell sources. The variable RR Lyrae stars mentioned previously are located along the horizontal branch within the Cepheid instability strip. Because there are no stable stars located here, this part of the horizontal branch is known as the **RR Lyrae gap.**

Figure 18-33 shows some of the best modern data, obtained by astronomer Jim Hesser, for the cluster 47 Tucanae, a cluster located in the southern

hemisphere and pictured in Figure 18-31. (The horizontal branch in Figure 18-33 differs from that in Figure 18-32 because the clusters differ significantly in their heavy-element abundance.) The figure also shows theoretical evolutionary tracks for clusters of ages 10, 12, 14, and 16 billion years. The cluster models are made up of stars having a variety of stellar masses. The observations and the models correlate well, but not perfectly because we still do not fully understand all aspects of stars. But the coincidence between the calculated track and the observations of stars in the cluster's H-R diagram gives us confidence that our theory of stellar evolution as we have presented it is basically correct.

Inquiry 18-26

From a comparison of the observations and models in Figure 18-33, what do you conclude about the age of 47 Tuc?

EVIDENCE FROM SPECTRA PERTAINING TO STELLAR EVOLUTION

In the 1940s, it was discovered that absorption lines due to a newly discovered and rather exotic heavy element, **technetium**, existed in the spectra of some stars. An example is the star ST Hercules, which shows lines of the technetium isotope having 99 nucleons (protons plus neutrons) (^{99}Tc) (**Figure 18-34**). This isotope of technetium is radioactive with a half-life of only about 200,000 years. Its presence in a stellar atmosphere tells us that the nuclear processes required to produce it were operating when the light we see left the star. Thus, the observation of technetium in the spectra of certain stars provides a *direct* observational verification that nuclear processes actually take place in stars and that the resulting elements are mixed into the material of the star's surface, where the spectrum is formed.

About 1 percent of red giant stars show the element barium in their spectra. However, these red giants were too young to have completed the barium-making process in their interiors. How are these **barium stars** formed? This question plagued astronomers for many years until it was found that all stars showing large abundances of barium are, in fact, binary stars with a white dwarf companion. The barium observed in the red giant spectrum is now understood to have formed in a white dwarf companion *when it was a red giant*. The barium arrived at the surface by convection, and then was transported to the binary companion by transfer through the crossing points of the Roche lobes, as described earlier. This finding shows another way in which an understanding of binary star evolution can be so useful to astronomers.

A DIRECT DETECTION OF STELLAR EVOLUTION

Because stars live out their lives on such long time scales, making observational progress in stellar evolution studies means finding stars at different phases of their evolutionary cycles and fitting these various pieces together in sequence. New technology has allowed direct observational verification of the fact that stars do change their characteristics with time.

Computer models predict that a white dwarf will cool and decrease in luminosity along the path shown in Figure 18-21. As the star cools, its surface oscillates. The pattern of these small oscillations will change in a well-determined way as the star cools. In studies since 1979 of a hot white dwarf star called PG 1159–035 and about 30 other *pre–white dwarfs* similar to it, astronomers were able to observe a changing pulsation period in the star, just as predicted (see **Figure 18-35** for a sample of their data).

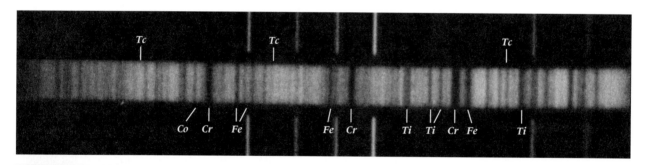

Figure 18-34 The spectrum of ST Hercules shows the element technetium, indicating current nuclear reactions in the star's interior.

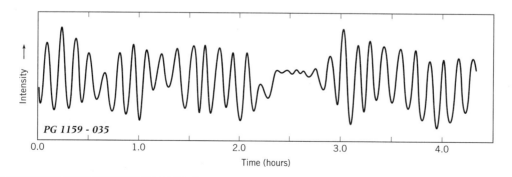

Figure 18-35 The variation in brightness with time of the pulsating white dwarf PG1159–035.

Enough data have been obtained to show that the oscillation is speeding up about 1 second in 1,270 years, about the rate that white dwarf cooling theory predicts. In other words, the predicted movement of a white dwarf star along the cooling track illustrated in Figure 18-21 has been detected.

Chapter Summary

Observations

- The H-R diagrams of the thousands of **OB associations** and **open clusters** in the Milky Way show them to exhibit a variety of shapes. The H-R diagrams of Milky Way **globular clusters'** do not show much variety. All types of clusters have a main sequence of stars truncated at the top left and complete at the bottom right.

- A variety of types of stars are observed to vary periodically in their light output. Such variable stars include the **Cepheids**, **RR Lyrae stars,** delta (δ) Scuti stars, and long-period variables. Observationally, these stars differ in their period and amount of light variation. Light variation occurs when evolving stars pass through a part of the H-R diagram known as the Cepheid instability strip.

- Luminous stars exhibit strong stellar winds that cause the stars to lose a large fraction of their mass in a relatively short interval of time.

- The central star in a planetary nebula is observed to be hot, often up to 100,000 K. Planetary nebulae are shells of gas expanding with speeds of about 25 km/s.

- **White dwarf** spectra contain highly broadened, gravitationally redshifted hydrogen lines. A group of 30 white dwarfs have been observed to change their pulsation period.

- The spectra of some stars have unusual chemical compositions different from that observed in our Sun.

Theory

- The most important factors determining stellar evolution of a particular star are initial mass, environment (including close companions if any), and initial chemical composition.

- **Figure 18-36** summarizes the post–main-sequence evolution of a star of one solar mass

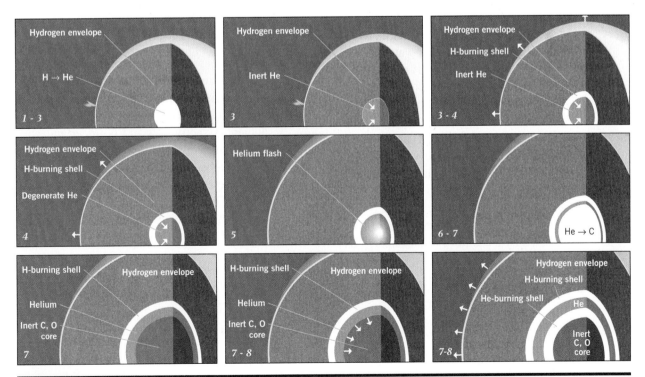

Figure 18-36 A summary of the structure of a star of one solar mass during each part of its midlife phases. The numbers refer to the positions in Figure 18-3. The drawings are not to scale, since the overall size of this star changes a great deal over its life.

by showing the internal structure at each stage of its evolution, which is shown in Figure 18-3.

- The changing chemical composition within a star's interior provides the impetus for change in its structure. These structural changes lead to temperature and luminosity changes that trace an evolutionary path in the H-R diagram.

- Stars similar in mass to the Sun have their energy produced by the proton-proton reaction, transported out of the core by radiation, and through the envelope by convection. In more massive stars, energy is produced by the CNO cycle, transported through the core by convection, and through the envelope by radiation transfer.

- Once the core reaches a temperature of 10^8 K, **helium burning** in the **triple-alpha reaction** begins.

- A **degenerate** gas is one in which the pressure is independent of the temperature. When helium begins to burn in such a degenerate gas, a **thermal runaway** results in a **helium flash**.

- A star like the Sun, after completing its red giant phase, ejects its outer envelope to form a **planetary nebula**. The core part of the star eventually cools to become a white dwarf. The spectrum of the nebula shows emission lines characteristic of a low-density gas.

- **White dwarfs** are hot, low-luminosity stellar remnants containing a solar mass of material squeezed into a volume the size of Earth. With a density a million times that of water, white dwarfs cannot be more massive than the **Chandrasekhar limiting mass** of 1.4 solar masses. Main-sequence stars above this mass limit must lose substantial quantities of mass if they are to become white dwarfs. The degenerate electron gases that compose white dwarfs are subject to a **mass–radius relation** that says that the greater the mass (up to the 1.4-solar-mass limit), the smaller the radius. White dwarf spectra often show a gravitational redshift, which is caused by a photon's loss of energy as it escapes from the strong gravitational field at the white dwarf's surface. All these theoretical ideas are consistent with observations.

- Depending on the relative masses of each star in a binary system, they may transfer mass back and forth between the **Roche lobes**. The star gaining mass speeds up its evolution, while the star losing mass slows down its evolution.

- Objects whose masses are less than about 0.1 solar masses are expected to produce **brown dwarfs** rather than stars.

Conclusions

- From the agreement between the theory of stellar evolution and observations, astronomers conclude that elements heavier than hydrogen are synthesized inside stars. Observations of the presence of certain elements in stellar spectra (technetium and barium) provide observational evidence that nuclear reactions occur in stars.

- A planetary nebula forms from a red giant. The central star that remains eventually becomes a white dwarf.

- Oscillations of the light emitted by white dwarfs are observed to change in the manner predicted by theory, thus providing observational evidence that stellar evolution actually occurs.

- From the variety of types of H-R diagrams exhibited by open clusters, astronomers conclude that they have a range of ages. Cluster ages are determined from **main-sequence turnoff points**. The similar H-R diagrams for globular clusters in our galaxy show them all to be ancient. Exactly how old is uncertain, with ages estimated at some 14 billion years. Such diagrams agree well, but not perfectly, with those predicted by theoretical stellar model calculations. These observations provide evidence that our general understanding of stellar evolution is valid.

Summary Questions

1. Why must a star inevitably leave the main sequence? How do such things as the star's mass and luminosity affect the length of time a star spends on the main sequence?

2. What are the internal and visible changes that take place in a star after its main-sequence evolution? Answer in terms of the physical processes that are taking place. Why does the

internal temperature of a star rise when its core nuclear fuel is exhausted?

3. What does the evolutionary track of a 1-solar-mass star on the Hertzsprung-Russell diagram look like? Relate the luminosity, surface temperature, radius of the star, internal structure, and physical processes taking place in the star to its position along the evolutionary track. Do the same for a 10-solar-mass star.

4. What unusual properties distinguish a degenerate gas from a normal gas? How are these properties responsible for the helium flash? How are they important for the structure of white dwarf stars?

5. What are the nuclear reactions in the triple-alpha reaction? What are the physical conditions under which the reaction takes place? When in the lifetime of a star is the triple-alpha reaction important?

6. Where is the Cepheid instability strip in the Hertzsprung-Russell diagram? What is its significance?

7. What characteristics might a dying star be expected to exhibit? How do these compare with the known properties of white dwarfs? What is the role of the Chandrasekhar mass limit in the theory of white dwarfs?

8. What evidence is there that planetary nebulae are an intermediate stage between red giants and white dwarfs? How does mass loss affect the possibility that a given star may become a white dwarf? What is the approximate maximum initial mass of stars that eventually become white dwarfs?

9. What observational evidence is there for stellar evolution? How would the Hertzsprung-Russell diagram of a star cluster be expected to change over time?

10. What observations of stellar chemical compositions have a bearing on studies of stellar evolution?

Applying Your Knowledge

1. Hypothesis: The Sun is hollow. Use all your astronomical knowledge to argue both for and against this hypothesis.

2. How do observations of the presence of certain elements in a star's atmosphere provide information about stellar evolution?

3. Why is the observed main sequence not an infinitely thin line on the H-R diagram? (Two main reasons are sought: one observational and one theoretical.)

4. Main-sequence stars do *not* evolve up or down the main sequence. From what you know of the properties of stars on the main sequence, what would have to happen to a star to make it evolve up the main sequence?

5. Summarize the events occurring at each point in Figure 18-3.

▶ 6. What percentage of its mass would a 10-solar-mass star lose in 10^5 years if it were losing mass at a rate of 10^{-5} solar mass per year?

▶ 7. Use the formula for the main-sequence lifetime, along with the mass-luminosity relation, to compare the lifetimes of stars of 0.1, 1, 10, and 20 solar masses.

Answers to Inquiries

18-1. The core of the star must start to contract, because the source of thermal energy that was supporting the star against gravity.

18-2. The star's diameter is bigger than it was.

18-3. The star has evolved into a red giant.

18-4. It will be another 4.4 billion years or so before this happens. But in about a billion years the consequences of the Sun's process of becoming a red giant will be catastrophic for life as we know it, which will no longer be able to exist on our planet.

18-5. The remaining mass is converted into energy.

18-6. The luminosity of the star would also vary in a regular fashion.

18-7. The speed of evolution would slow down.

18-8. Emission lines, probably Doppler shifted, would be observable.

18-9. The star gaining mass speeds up its evolution while the one losing mass slows its evolution.

18-10. From Kirchhoff's laws planetary nebulae must consist of a rarefied gas at a high temperature.

18-11. The gas glows because of energy given off as electrons cascade from highly excited energy levels downward toward the ground state. Only ultraviolet photons have enough energy to excite the hydrogen atoms to these upper levels, whereas neither visible nor infrared radiation does.

18-12. It is thought that they are intermediate stages between red giants and white dwarfs.

18-13. The current rate of white dwarf production is greater than that in the past; perhaps more stars are now reaching the stage in their lifetimes when they become white dwarfs. This is reasonable, because the kind of stars that become white dwarfs are low-mass stars, which take a long time to evolve. The other possibility is that the rate of star formation in the Galaxy is not constant (e.g., a burst of star formation 6 to 8 billion years ago would result in a higher rate of current white dwarf formation).

18-14. It would have a deficiency of hydrogen, because it has converted much of it to helium and carbon.

18-15. Small; it would collapse to a point where some other force—specifically, degenerate electron pressure—would be necessary to support the star.

18-16. Because its diameter must be small (see the previous Inquiry), its density must be high. High density would produce highly broadened spectral lines.

18-17. Its luminosity would be low because it has only a small surface area from which to radiate energy.

18-18. Because the mass would come from Newton's modification of Kepler's third law, the period and semi-major axis were needed. Determination of the semi-major axis required the observed angular semi-major axis and the star's distance.

18-19. No; the mass–luminosity relationship is valid only for main-sequence stars.

18-20. The other star is a main-sequence star and should have very little gravitational redshift. Because the two stars are moving together through space, the Doppler shift due to their common motion through space would be the same. Any difference between the redshifts of the two stars, therefore, must be the result of orbital motion (which can be calculated and allowed for) and the gravitational redshift.

18-21. The source of energy is the motion of its individual particles, measured by its high temperature, left over from earlier nuclear reactions. The energy emitted by the star simply causes the white dwarf to cool slowly and continuously.

18-22. The stars must lose mass; otherwise they would be more massive than the Chandrasekhar limit of 1.4 solar masses and could not become white dwarfs.

18-23. The lower the turnoff point, the older the cluster.

18-24. It is assumed that all the stars in the cluster were formed at the same time, so they are the same age. With today's precise Hertzsprung-Russell diagrams, many clusters do show an internal range of age, but the spread in age for open clusters is usually small compared to the age of the cluster.

18-25. According to the diagram, (a) the Pleiades is a little less than 10^8 years old, (b) NGC 2362 is about 5×10^6 years old, and (c) M67 is about 4×10^9 years old.

Stellar Death and Catastrophes 19

A leaf of grass is no less than the journey work of the stars.

Walt Whitman, Song of Myself

In Chapter 18, we studied the evolution of the overwhelming majority of stars—those of modest to moderate mass. We have seen how they evolve from the main sequence into the giant region, become variable stars, and then eject planetary nebulae. As the nebula dissipates, the central star becomes a white dwarf and slowly cools.

The previous chapter left several important questions unanswered. In particular, we did not finish discussing the life cycle of stars of more than a few solar masses because the evolution of these stars is different from that of low-mass stars. In addition to a discussion of the death of high-mass stars, we will also further examine the evolution of binary stars. Such stars often experience rapid and sometimes violent changes as they age, and they play an important role in our discussion. This chapter, then, is a discussion of perhaps the wildest objects studied by astronomers: neutron stars and black holes.

19.1 Novae

Since ancient times people have observed stars to brighten suddenly and then fade slowly. Such events were called *nova stella* (Latin for *new star*) or **nova** (plural **novae**) for short. Only in the twentieth century has it become apparent that these objects are actually old stars passing through an explosive phase of their evolution, and that the term *nova* is therefore a misnomer.

OBSERVATIONS

In late August of 1975, when the sky had darkened sufficiently, many amateur and professional stargazers noticed that a bright nova had suddenly appeared in the constellation of Cygnus, approaching the brightness of the blue giant Deneb at the tip of the Northern Cross. Novae are given names that include both the constellation and the year in which they occur. Therefore, this nova is known as Nova Cygni 1975.

Nova Cygni 1975 faded rapidly in only a few days. By contrast, Nova Delphini 1967, after reaching its maximum brightness, declined in luminosity very slowly, taking about a decade to return to its former state. These two novae, the fastest and slowest of modern times, represent the extremes for most novae. Some examples of nova light curves are shown in **Figure 19-1**.

Several lines of evidence suggest that novae eject mass.

> ## Inquiry 19-1
>
> In what ways might the loss of material in a nova explosion make its presence known?

During its early phases, the spectrum of a nova shows normal stellar absorption lines, but displaced to shorter wavelengths. **Figure 19-2** illustrates that this occurs because the near side of an expanding shell of gas around a star consists of absorbing material moving toward the observer, so the light is Doppler shifted toward shorter wavelengths. The speed with which the gas moves may also be found from observations of the widths of the broad emission lines. **Figure 19-3** shows a star with an expanding ring about it, and a schematic diagram of the spectrum it forms. The greater the expansion speed, the greater the blueshift of the gas moving toward the observer. Similarly, the greater the expansion speed, the greater the redshift of the gas moving away from the observer. The result, then, is a greater width of the spectral line, and the expansion speed can be found from the line's observed width. The expansion velocities of the ejected gas are considerably greater than those found in planetary nebulae; the speeds typically range from a few hundred to more than a thousand kilometers per second.

Chapter opening photo: A *Hubble Space Telescope* image of the LMCN49—the remains of a massive star exploding in a supernova explosion in the Large Magellanic Cloud.

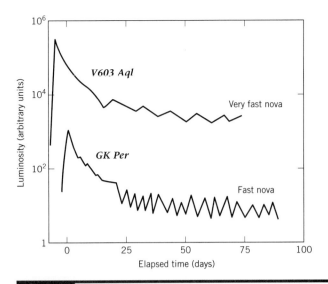

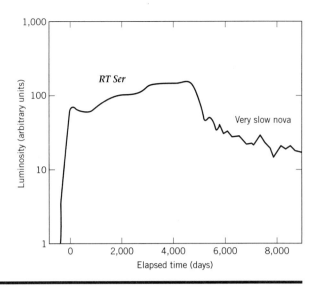

Figure 19-1   The change in brightness over time of fast novae and slow novae.

Figure 19-2   The early spectrum of a nova exhibits blueshifted features because the expanding gases on the near side of the star are moving toward the observer.

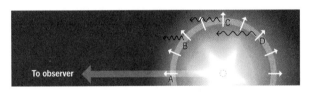

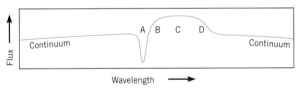

Figure 19-3   A schematic spectrum produced by a cloud of gas expanding around a nova.

Figure 19-4   A photograph of Nova Persei 1901 taken many years later, showing the expanding nebula around the star.

Some time after the nova outburst, the ejected material may become visible on images as an expanding nebula (**Figure 19-4**).

A great deal of energy is involved in a nova outburst. Increases of more than 100,000 times in a day or two have been found by comparing the star's maximum brightness with its pre-explosion value. At their maximum brightness, novae are temporarily among the most luminous objects in the Galaxy—up to 1 million times more luminous than the Sun.

We cannot yet predict which stars are going to "go nova"; we can only hope that when one does, it has been observed previously so that we can study the changes in its properties. From such observations astronomers have concluded that during the early rise to maximum brightness, the surface temperature of the star does not change by much. The increase in luminosity, therefore, primarily comes from the increased surface area of the rapidly swelling star immediately after the explosion.

Inquiry 19-2

Suppose a star were to increase its radius by 10 times so that its surface area increased by a factor of 100 times. Suppose, also, that the temperature of the star did not change. How many times brighter would the star become?

Inquiry 19-3

The situation in Figure 19-2 must eventually be followed by a stage at which some material escapes from the object. What kind of spectral changes would you expect to witness in this ejected material with time? How would the expansion of the gases affect the spectrum?

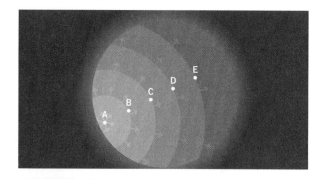

Figure 19-5 If a change in brightness occurs at point *A*, a coordinated change at *B* cannot occur before light from *A* can reach point *B*. Extending the argument across the object means it must be smaller than the light-travel time across it.

THE PLACE OF NOVAE IN STELLAR EVOLUTION

When we plot the positions on a Hertzsprung-Russell diagram of stars that undergo nova explosions, we find that they fall into the same region of the diagram as the hot central stars of planetary nebulae. Apparently, both types of objects are late stages in the lifetime of a star that is trying to adjust its mass to reach a stable condition. But while the nova outburst is much more vigorous, ejecting material at high speeds and increasing the star's luminosity greatly, a typical nova will lose only about 10^{-4} solar masses, about 1,000 times less than that lost by a planetary nebula. So exactly what is the difference between the two? To answer this, let us briefly consider the current theories about novae, and the evidence that led to the development of these theories.

DO NOVAE INVOLVE LARGE OR SMALL STARS?

Observations of the brightness of novae, made with accurate equipment that can determine stellar brightness every thousandth of a second, show that small and rapid brightness fluctuations continually take place. The rapidity with which a light source can vary its intensity is related to the size of the light-emitting source. Specifically, the light-emitting region can be no larger than the time it takes light to travel across the object. For example, since the light variation for a nova is significant in one day, the light-emitting region cannot be larger than one light day. For this reason, we conclude that the nova event occurs in a small space.

To understand why, consider **Figure 19-5**, which shows an object with various points on it noted. If the object is uniformly bright across its surface, an increase in brightness at point *A* will have little effect on the object's overall observed brightness. Furthermore, if the change were random, an increase at point *A* is as likely not to occur during a decrease at point *B*. To have a substantial change in brightness requires that a change at point *A* be followed by changes at points *B, C, D,* and so on across the object. In other words, it is necessary for a change at one point to be causally related to changes at other points. But a change at point *B* cannot be caused by a change at point *A* sooner than the time light would take to travel from *A* to *B* because nothing can travel faster than light. Carrying this argument across the entire object, we conclude that an object must be smaller than the light-travel time across it.[1]

HOW A NOVA EXPLODES

In 1954 the astronomer Merle Walker discovered that the nova DQ Herculis is actually a close binary system with an orbital period of 4 hours and 39 minutes. In subsequent years it has become apparent that novae exist in binary systems where the stars are perhaps a solar diameter apart. The current models of a nova explosion depend on this fact.

Spectra of novae may sometimes reveal the presence of more than one star, but most often the

---

[1]Just as we can speak of distance in terms of light-years, it makes sense to speak of the size of an object in terms of the time it takes light to travel across it. This is the light-travel time.

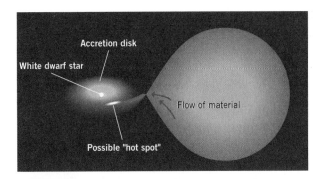

Figure 19-6 Currently accepted models of novae involve the interaction between stars in a binary system.

brighter of the pair so dominates the light from the other that the existence of the second star is not easy to detect.

The model of a nova that best fits most observations consists of a white dwarf star and a low-luminosity main-sequence star (a red dwarf) in orbit around each other. If the two stars are relatively close to each other, mass may transfer from one to the other as discussed in the previous chapter. Such a system is shown in **Figure 19-6**. Material from the outer part of the main-sequence star is forcibly removed by a shrinking Roche lobe,[2] just as toothpaste is forced from a tube whose volume shrinks when it is squeezed. The white dwarf star, which in spite of its small diameter may have a mass equal to or greater than that of the Sun, has a strong gravitational field because it is so small and will attract material from the atmosphere of the main-sequence star. As material flows between the stars, the rotation of the system forces it to spiral around the star, forming an **accretion disc** around the white dwarf. As time goes on and the particles in the accretion disc interact with one another, they heat up (by friction), and the particles lose angular momentum and fall onto the surface of the white dwarf. Thus, the accretion disc is actually continuous all the way to the white dwarf's surface. The friction in the accretion disc heats it to temperatures that are high enough to emit ultraviolet and X-ray photons. Such observations also suggest that there is a moving "hot spot" where the spiraling stream of mate-rial from the main-sequence star collides with the gaseous disk around the white dwarf, liberating large amounts of energy.

The white dwarf has a high surface temperature, and if large amounts of gas from the main-sequence star are dumped into the accretion disc and from there eventually accumulate on the white dwarf, the infall of material will generate even more heat. The infalling gas is rich in hydrogen because it comes from the surface of a main-sequence star. As time progresses, a layer of hydrogen builds up on the white dwarf's surface. Furthermore, some of this hydrogen will become degenerate because a white dwarf is a degenerate star. Eventually, the temperature at the base of the layer becomes sufficiently high for explosive hydrogen fusion to take place; it is similar to the core's earlier helium flash but now it occurs with hydrogen on the surface. In effect, the surface of the white dwarf becomes a giant hydrogen bomb and the blast blows off some of the mass that had fallen onto the white dwarf. Thus, the explosion does not catastrophically disrupt the white dwarf but is merely superficial.

Many novae are observed to be recurrent—that is, they undergo nova explosions multiple times. This observation was crucial to the development of this model because it showed astronomers that a nova explosion was not a one-shot, catastrophic event. It showed that if a nova results from some sort of instability, the unstable conditions are able to recur some time later. Further evidence for a recurring event comes from observations that after the ejected material has been blown off the white dwarf and dissipated, an object very similar to the prenova system remains. Not only has it survived the explosive event, in many respects it seems virtually unscathed and ready to begin the process leading to a new explosion.

Inquiry 19-4

After an outburst as just described, the companion star may start to dump mass onto the white dwarf all over again. How might this fact be used to explain the existence of recurrent novae?

---

[2]The reason why the Roche lobe is shrinking is unclear, but it may be due to magnetic braking, discussed for the Sun in Chapter 6, which helps the system lose angular momentum.

Inquiry 19-5

The answer to Inquiry 19-4 indicates that we may expect many and perhaps all novae to be recurrent. If this is so, how would it affect our conclusions as to how many stars in the Galaxy are involved in the nova phenomenon?

To interpret the significance of novae in stellar evolution, we should correct the observed nova frequency for the fraction that are recurrent. But this fraction is not known; nor is it known how often or how many times a typical recurrent nova explodes. Thus, it is difficult to determine if novae are a common stage in binary star evolution. Current evidence suggests that most normal stars having masses less than several times the Sun's mass eject planetary nebulae if they are single stars or widely separated binaries, or become novae if they are close binaries.

Recent observations indicate that there is a difference in chemical composition between *fast* and *slow* novae, with fast novae showing an unusually high abundance of carbon, nitrogen, and oxygen, and slow novae showing abundances closer to that of the Sun. This suggests that perhaps these two types of objects have a different stellar history, that their predecessors were stars of different internal structure, and that the mechanisms causing their explosions are different.

Inquiry 19-6

In a nova system, a companion star dumps mass onto the surface of a white dwarf star. But the white dwarf must have originally been a normal star itself, possibly more massive than the companion star, because it arrived at the white dwarf stage sooner. Might there have been a stage at which the current white dwarf star was a giant, dumping mass onto its companion? What effect might this have had on the evolution of the system?

Novae and recurrent novae belong to a class of objects often referred to as **cataclysmic variables**. Because the variety of possible stellar systems is so large (due to variations in mass, radius, magnetic field, mass loss rate, and stage in evolution)

these stars can produce bewilderingly complex systems.

19.2 Supernovae

As their name suggests, **supernovae** are even more dramatic and energetic than novae. The supernova shown in **Figure 19-7** is nearly as bright as all the rest of the galaxy in which it is found. Since ancient times, observers have occasionally noticed that a star would suddenly appear in the sky, becoming so bright that it might even be visible during the day. Tycho Brahe observed a supernova in 1572 in Cassiopeia; he studied it carefully and demonstrated that it must be at a greater distance from Earth than the Moon, thus showing the universe beyond Earth to be changeable. Johannes Kepler observed another supernova in the constellation of Ophiuchus only 32 years later. Since this 1604 event, no supernovae have been observed to occur within our galaxy, a fact that makes the study of these enigmatic objects more difficult. Until recently, astronomers had to make do with evidence from supernova events that happened either thousands of years ago or in dis-

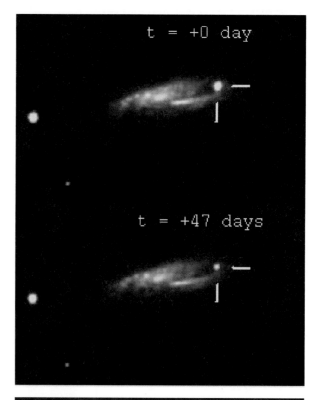

Figure 19-7 A supernova explosion in a distant galaxy, at maximum brightness and 47 days later.

Figure 19-8 The Crab nebula, the filamentary remnant of a supernova observed in 1054 C.E.

Figure 19-9 The Veil nebula in Cygnus, a piece of a remnant of an exploded star.

tant galaxies, from which much less (and less accurate) information is available.

OBSERVATIONS

There are nebulae that give a visual impression of a violent explosion, and detailed observations confirm this. **Figure 19-8** shows a photograph of the famous Crab Nebula in Taurus (see also Figure 1-13). This object is the remnant of a supernova explosion that was witnessed by Chinese astronomers in the year 1054 C.E., when it was at least as bright as Venus. Doppler-effect measurements of the filaments of gas have shown that they are moving at a high rate of speed, on the order of 1,450 km/s. When studied spectroscopically, the gas itself exhibits a high degree of excitation, indicating that there is a considerable amount of energy in these expanding filaments. Images taken of the Crab Nebula over many years clearly reveal the outward motion of the gases. In fact, this expansion, combined with the measured Doppler shifts, provides us with the means to measure the Crab's distance, about 6,000 light-years. If we take the presently observed motion and extrapolate it backward in time, it confirms the Chinese observation that the original explosion occurred almost 1,000 years ago.

The Crab Nebula is not unique; **Figure 19-9** shows the Veil Nebula in Cygnus. Optical studies of the motion and excitation of these gases indicate that they were involved in a violent explosion thousands of years ago. Radio astronomers have also discovered a large number of supernova remnants, because radio telescopes can more effectively penetrate dust in the galactic disk and make observations of more distant objects. **Figure 19-10** shows a radio image of the first radio source to be discovered in the region of the constellation Cassiopeia, Cas A, one of the brightest radio sources in the sky. Its shape, along with the strong excitation of its gases, suggests an origin in a stellar explosion. A considerable amount of mass is present in the gaseous filaments of these nebulae, and it seems as if a star obliterated itself in these explosions. Could it be that these are remnants of stars that became so unstable that they simply blew themselves apart?

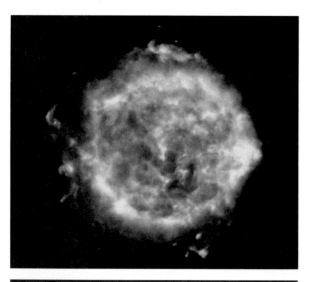

Figure 19-10 A radio image of Cassiopeia A, the remnant of a supernova explosion calculated to have taken place in the year 1667.

Inquiry 19-7

Figure 19-8 shows several stars that appear to be in the nebula. Most of these stars are not associated with the nebula at all, but just lie along our line of sight to it. How might an astronomer determine which, if any, of these stars is actually the star that underwent the explosion?

During a supernova explosion, the brightness of a star suddenly increases billions of times. Such objects in external galaxies have been observed to become as bright as the entire galaxy (Figure 19-7), giving off over time as much energy as 10^{11} stars. Our most reliable information about supernovae has come from studies of those in other galaxies. For this reason, several research groups have built telescopes specially designed to observe thousands of galaxies each night and discover supernovae before they reach their maximum brightness. Occasionally a supernova will be detected, and approximately 3,000 have been discovered either with older photographic techniques or the modern automated searches. However, often the only information that can be obtained for these is the change in brightness over time, their *light curve*. Therefore, if a supernova were to explode in our galaxy anywhere near the Sun, we could collect a great deal of new information.

Judging by how often we see supernovae in galaxies like our own, it appears that we are overdue for such an occurrence, but there is no way to predict when or where it might take place. (One or more could have occurred in our galaxy but have been obscured by dust.) It would, of course, be convenient for astronomers if it were nearby so it could be studied in detail, but a supernova that was too close could have catastrophic significance for life on Earth.

TYPE I SUPERNOVAE

Studies of supernovae in external galaxies show there to be at least two distinct types based on the observed spectral characteristics. **Type I supernovae** are characterized by spectra showing little if any hydrogen (**Figure 19-11**). The stars all have the same characteristic light curve (**Figure 19-12**), with the brightness increasing steadily for several weeks and then slowly declining. From many lines of evidence astronomers infer that the precursors are very old and not very massive.

The best model to date for Type I supernovae suggests that they result from the explosion of a white dwarf star, rich in carbon and oxygen, that has had matter from a companion red giant dumped onto it, a

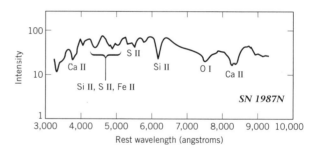

Figure 19-11 Type I supernova spectrum; note the lack of hydrogen lines.

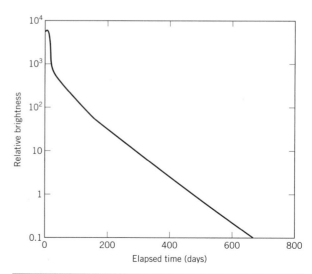

Figure 19-12 Type I supernova light curve.

picture that bears some similarities to the model for a nova explosion. Here, however, the white dwarf does not nova but increases its mass until it surpasses the Chandrasekhar limit of 1.4 solar masses. When this happens, the degenerate electrons in the core are unable to withstand the weight of the overlying layers and collapse ensues. The carbon and oxygen in its interior undergo fusion. Because the white dwarf is in a degenerate state—like the core of a star undergoing a helium flash—large amounts of energy are released suddenly and a thermal runaway occurs, triggering a violent explosion. According to this model, the carbon-oxygen burning produces large amounts of the radioactive isotope nickel-56. This decays rapidly to cobalt-56, which in turn decays to iron-56. The radioactive decay releases prodigious amounts of energy in the form of gamma rays and positrons, which heat the growing nebula surrounding the object.

For many years, astronomers had noted the similarities between the light curves of Type I supernovae and the curve of energy emitted by the decay of radioactive material. But it seemed somehow unlikely that an entire star would transform a significant portion of itself into one radioactive chemical element (nickel-56), and so this explanation never gained widespread acceptance. Currently, however, computer models suggest that this is indeed what takes place, and that our ideas of what was *likely* were conditioned more by habits of thinking than by solid evidence. Much of the energy released by Type I supernovae during their decline from maximum luminosity does appear to be due to radioactive decay processes.

Type I supernovae (and especially a subclass called Type Ia) appear to be similar to one another. In fact, it appears that their peak luminosities are all nearly the same. If true, and all the observational evidence indicate that it is, then these objects are extremely useful for measuring the distances to the galaxies in which they are found. In fact, as you will learn when you read about cosmology in Chapter 23, the current interpretation of observations of Type Ia supernovae have revolutionized our understanding of the universe.

The Type Ia supernova model of a binary star system in which one star feeds matter to the other makes some testable predictions. If valid, once the supernova explodes, one might expect that the companion star would be ejected from the system with a high speed. The first direct observation confirming this idea was made in late 2004 when the *Hubble* telescope observed a rapidly moving star near the center of the supernova remnant observed by Tycho Brahe in 1572. **Figure 19-13a** is a composite of a ground-based optical image and the X-ray image from the *Chandra* X-ray telescope. The area enclosed in **Figure 19-13b** is outlined in part *a*. The object labeled "Tycho G" is observed to be moving at three times the speed of any of the neighboring objects and is taken to be the expected companion that fed the star until it exploded.

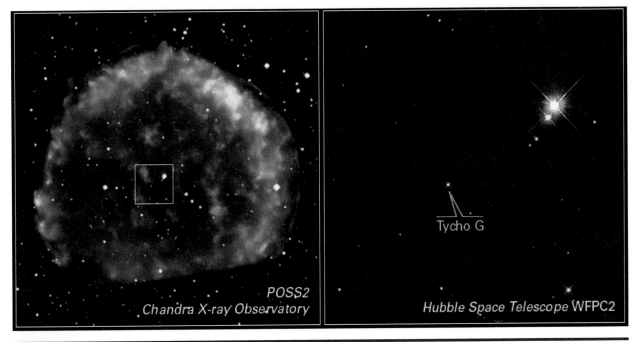

Figure 19-13 Tycho's supernova of 1572. *(a)* a combined optical and x-ray image. *(b)* The star thought to be the companion of the star that exploded in a Type Ia supernova.

Type II Supernovae and the Late Stages of Massive Stars' Evolution

On February 24, 1987, the 383-year dry spell without a nearby supernova ended when astronomers around the world were thrilled by a discovery made by a young Canadian observing assistant named Ian Shelton. Shelton was fortunate enough to be the first person to report the appearance of a supernova in the Large Magellanic Cloud, a nearby small, irregular galaxy in the southern hemisphere that can be seen with the naked eye (**Figure 19-14**). While not within our galaxy, SN 1987A (so named because it was the first supernova of 1987) is the next best thing because it is only some 170,000 light-years away. Furthermore, it was the first bright supernova since the invention of the telescope! Because the star was readily observable with the naked eye, there was no problem studying it with even small telescopes. Large telescopes allowed astronomers to study it in more detail than any other supernova. It was also the first supernova to be studied with a variety of spaceborne instruments: the *International Ultraviolet Explorer*, the Japanese *Ginga* X-ray satellite that had been launched shortly before the explosion, and rocket and balloon experiments.

Type II supernovae are defined as those that show prominent spectral lines of hydrogen. **Figure 19-15** shows the Type II spectrum of SN 1987A. Their light curves (**Figure 19-16**) are distinctly different from those of Type I supernovae rising more quickly to a maximum and declining more rapidly and irregularly. Sometimes their light curves have a hump on the way down. They are far less uniform in their properties than Type I objects. We infer from observations that they appear to be associated with massive stars, whereas the Type I supernovae are associated with low-mass white dwarfs. The properties of the light curve of SN 1987A are characteristic of a Type II supernova, although it is unusual because it did not become as bright as expected. SN1987A was also unusual because its progenitor star was a blue, not a red, giant.

Figure 19-14 SN 1987A both before and after the supernova explosion. Anglo-Australian Observatory/David Malin Images

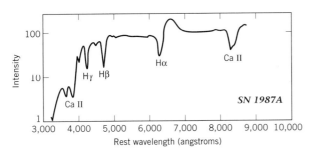

Figure 19-15 The spectrum of SN 1987A, a Type II supernova; note the presence of hydrogen lines.

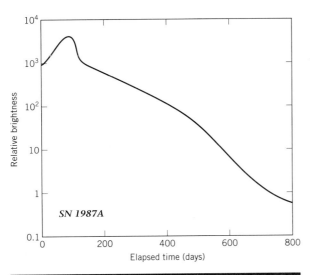

Figure 19-16 Type II supernova light curve.

Inquiry 19-8

Studies of chemical abundances in various filaments of gas throughout the Crab Nebula show them to have low hydrogen abundance and high helium abundance, compared to that of the Sun. How does such evidence help confirm the picture of stellar evolution that we have constructed in the last few chapters?

The observed low abundance of hydrogen and high abundance of helium in the various filaments of gas in the Crab Nebula is consistent with the idea that the star that exploded to produce it was an evolved star. During its lifetime, fusion depleted hydrogen and enhanced helium. The presence of other heavy elements in the nebulae surrounding other Type II supernovae shows that additional fusion processes must also have taken place.

Indeed, there are distinct differences between the spectra of various Type II supernovae, implying differences in the chemical abundances of the stars that exploded or the fusion processes during the explosion. There may be no single scenario for Type II supernova explosions as there seems to be for Type I supernovae. Rather, it may be that the details and end results depend on the exact mass of the star that explodes, the geometry of the explosion, and probably other factors as well.

In the previous chapter we found that giant stars are a later stage in the stellar aging process. Furthermore, we learned that in the cores of evolved giant stars temperatures are high enough (10^8 K) for the triple-alpha reaction to occur, in which three helium nuclei are fused into a carbon nucleus. But now, unlike lower mass stars, the fusion process for massive stars continues. Once a certain amount of carbon is present inside the star; other reactions can take place at nearly the same temperatures. In particular, the addition of an alpha particle (a helium nucleus) triggers the reactions

$$^{12}C + {}^4He \rightarrow {}^{16}O$$

$$^{16}O + {}^4He \rightarrow {}^{20}Ne$$

$$^{20}Ne + {}^4He \rightarrow {}^{24}Mg. \ldots$$

These so-called alpha-capture reactions (described in Chapter 18) can continue, giving ^{28}Si, ^{32}S, ^{36}Ar, ^{40}Ca, and even heavier elements.

Inquiry 19-9

The heavier an element is, the less of it is created because there are fewer building blocks from which to make it. How would you expect the cosmic abundance of elements to depend on their atomic weight? That is, would you expect to see a greater abundance of heavier elements or lighter elements in cosmic material?

These reactions, as well as others that take place at these and even higher temperatures, synthesize heavier elements inside massive stars until iron, ^{56}Fe, is created. The star's structure looks like an onion that contains numerous shells composed of different elements (**Figure 19-17**). At this point, the process of continuous element synthesis from one stage to another terminates. The reason is that in earlier evolutionary stages, whenever an inert core was produced, core collapse occurred and

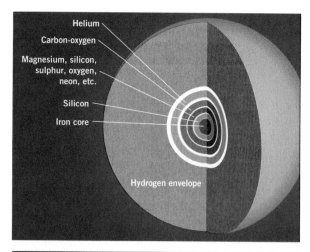

Figure 19-17 Massive stars may develop an iron core surrounded by a complex "onion skin" structure of different elements. The complex core structure is surrounded by an extended envelope of hydrogen. Theoretical calculations indicate that this structure is unstable.

Labels on figure: Helium, Carbon-oxygen, Magnesium, silicon, sulphur, oxygen, neon, etc., Silicon, Iron core, Hydrogen envelope

raised the temperature until the next bout of fusion began. The subsequent fusion reactions always produced more energy than was present in the motion of the particles.

With iron, however, the situation is different. To make heavier elements by the fusion of iron with other particles would require more energy than is given off by the fusion process. In other words, energy is absorbed by this process rather than released. For this reason, additional fusion reactions do not occur. Furthermore, under these conditions of high density and temperature, high-energy photons in the form of gamma rays are produced. The gamma ray triggers the iron nucleus to disintegrate (in a process called *photodisintegration*) into helium, neutrons, and neutrinos rather than form a still heavier nucleus:

$$\gamma + {}^{56}_{26}\text{Fe} \rightarrow 13{}^{4}_{2}\text{He} + 4\text{n} + \text{neutrinos}.$$

In addition, the gamma rays also cause the helium to photodisintegrate into protons and neutrons:

$$\gamma + {}^{4}_{2}\text{He} \rightarrow 2\text{p} + 2\text{n}.$$

The result is that when the inert iron core begins to collapse, contraction-halting fusion never occurs. The photodisintegration of iron consumes energy from the core, thus causing it to cool and collapse further. The further collapse tends to raise the temperature, which causes more photodisintegratation, which causes more collapse, and so on. Rapidly, in less than one second, the object implodes on itself.

Meanwhile, as the iron photodisintegrates, electrons within the increasingly dense core are pushed into protons, even those within the atomic nuclei, in the following reaction:

$$\text{electron} + \text{proton} \rightarrow \text{neutron} + \text{neutrino}$$

The neutrinos fly away from the core. Because neutrinos carry energy away from the core, they accelerate its energy loss and enhance its collapse. The result is a rapid implosion of the core. Eventually, the pressure of the imploding material at the center of the star becomes so great that the core "bounces." At the same time, the entire star above the shrinking core starts falling inward because it is no longer supported against gravity. This infalling matter strikes the rapidly outwardly expanding core, which transfers its outward momentum to the infalling envelope. This results in a massive explosion that produces an outward-moving shock wave. In addition, with the high density of matter inside the star, we expect neutrinos to encounter some resistance to their outward flight. A push from the large number of neutrinos, in combination with the shock wave from the core bounce, produces an incredibly strong explosion that throws a large fraction of the star's mass out into space. What is left depends mostly on the mass of the exploding star, and will be discussed later in this chapter.

Computer models suggest that stars with masses greater than approximately 8 solar masses should have an implosion followed by an explosion. Stars with less than 20 solar masses produce strong explosions, whereas those in the 20–45 solar mass-range are weaker because they have so much more material above the exploding core. Models of the supernova explosion indicate that stars more massive than 45 solar masses will not produce an observable supernova. The supernova formation rate also tends to agree with the star formation rate in this mass range. To put that another way, if only the most massive stars formed supernovae, we would not see as many as we do. If only low-mass stars formed supernovae, we should see many more than we do.

Inquiry 19-10

In what ways do the formation of Type I and Type II supernovae differ? How do these different formation mechanisms produce differences in the presence of hydrogen in the spectrum?

SYNTHESIS OF HEAVY ELEMENTS IN MASSIVE STARS

Although heavier elements cannot be produced by fusion reactions as they occur throughout a star's lifetime, nucleosynthesis of elements beyond ^{56}Fe can occur by a process that adds neutrons to elements. Because a neutron has no charge, it can enter a nucleus—for example, that of iron—at a low speed without disrupting it. Many heavy, stable elements (such as cobalt, barium, lead, and zirconium) are formed by the slow addition of neutrons to atoms, in a process called the **s-process** (s for slow). After a neutron is captured, the nucleus is transformed to a new and more stable element by emitting high-energy electrons. This process of neutron capture can synthesize elements that go as high in the periodic table as bismuth (^{209}Bi), beyond which elements become radioactively unstable. Technetium and barium, discussed in the previous chapter, are formed by the s-process. Technetium's presence in stars proves that the s-process of neutron addition operates, since no other nuclear process can produce it.

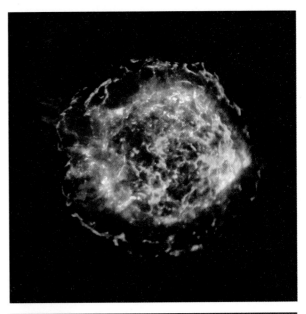

Figure 19-18 An x-ray image of the supernova remnant Cassiopeia A. The blue regions are the highest exergy X-rays while the red ones are the lowest. The red material on the left edge has more iron than normal, while the greenish-white region to the lower left contains extra silicon and sulfur. Compare with the radio image in Figure 19-10.

Inquiry 19-11

In general, we would expect that the addition of helium nuclei (^{4}He) to such elements as ^{12}C, ^{16}O, and so on, would produce nuclei whose atomic weight is a multiple of 4. Yet other nuclei, such as ^{17}O and ^{18}O, are found in nature. How might the existence of these nuclei be explained?

Inquiry 19-12

The most common form of oxygen is ^{16}O. Keeping this fact in mind, what conclusion can you draw about how most of the oxygen in our universe was created?

There is also a rapid process of adding neutrons to atoms. A supernova explosion involves extremely high temperatures and densities. We saw that the collapse of the iron core produces neutrons. In fact, a huge number of neutrons are produced quickly. These neutrons can be absorbed rapidly by the other elements, thus producing heavy elements that can be manufactured only under the conditions present during a supernova explosion. Production of these heavy elements takes place because of the rapid addition of neu-

trons in a nuclear process known as the **r-process** (r for rapid). One of the elements built up during these catastrophic moments of stellar evolution is ^{235}U, famous on Earth for its use in the first atomic bombs and in nuclear power plants. Both gold and silver can be produced by the r-process, too.

Observations of the supernova remnant Cas A (**Figure 19-18**), made by the *Chandra* X-ray observatory, have provided useful information for understanding supernovae and heavy element production. Astronomers have been able to observe the distribution of the elements within the Cas A remnant and work backward to deduce the structure of the star just prior to the explosion. For example, the brightest and most compact filaments of gas contain little or no iron, while these same regions have significant silicon and sulfur. These observations point to a formation in the interior where the temperature was some 3 billion degrees. Other filaments, however, contain iron, which tells us they formed yet deeper inside, where the temperature was 4–5 billion degrees. These iron-rich regions from the innermost part of the star are now at the outermost parts of the remnant, which means it has been moving at the greatest speeds. Similar types of observations of other supernovae remnants, which resulted from explosions of stars of other masses, are helping to provide a more complete story of pre-explosion stellar structure.

Heavy-element production during a supernova explosion, along with the elements' subsequent ejection into interstellar space, place supernovae in an important position in the ecology of the Galaxy. Eventually, the interstellar medium will incorporate these supernova-produced heavy elements into molecular clouds. When these clouds produce new stars, they will be formed from materials of an earlier generation of stars, in the ultimate recycling process.

CONFIRMATION OF THE TYPE II MODEL: SUPERNOVA 1987A

The general theoretical model of a Type II supernova explosion predicted that neutrinos escaping the explosion would carry 100 times as much energy as all other forms of energy produced in the explosion. The most exciting observation connected with Supernova 1987A was the simultaneous detection by Japanese and American neutrino telescopes of bursts of neutrinos—even before the supernova was detected optically. From the 21 neutrinos detected on Earth and the known distance to SN1987A, astronomers calculated that the total neutrino emission from the supernova indeed carried the predicted amount of energy. This may not seem like many neutrinos, but the fact that it is

so much higher than usual highlights how difficult it is to build a neutrino telescope. This, together with the observation that this supernova was of Type II, is a dramatic confirmation of the general theoretical model of the Type II supernova explosion that we have described.

The model was wrong, however, because it assumed that the star that exploded would be a red supergiant. The progenitor was SK 269°202, a blue supergiant star. This does, however, account for the lower than expected luminosity of the supernova, because a blue supergiant is less massive than a red one. SK 269°202 had a mass of about 20 solar masses and a radius of slightly more than 40 solar radii.

The debris from the supernova explosion has been traveling at speeds of nearly 2000 km/sec, which is about 10 million km per hour! That debris, which became visible only some eight years after the explosion, is now detectable inside the ring of the image in **Figure 19-19**. The elliptical ring is thought to have been formed more than 20,000 years ago, probably from prior less-explosive ejection events. Although the ring was previously invisible, radiation from the supernova explosion excited the gas to visibility. Early in 1998, a small part of the ring was observed to become substantially brighter as the blast wave from the supernova began to collide with the ring and excite it. As the blast wave has continued to

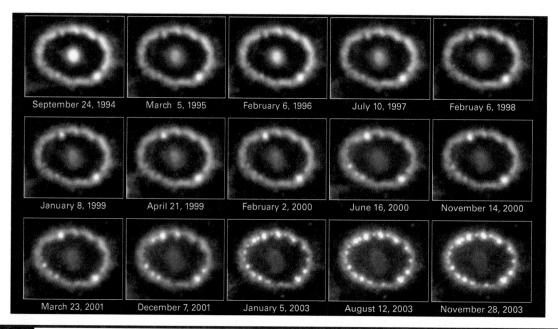

Figure 19-19 SN 1987A, as observed with the *Hubble* telescope from 1995 through 2003. The changes are caused by a collision of a shock wave from the star with a gaseous ring ejected some 20,000 years before the supernova explosion.

move outward over the intervening few years, more of the ring has been excited and brightenings have occurred in various wavelength regions. Through an analysis of the images in Figure 19-19, astronomers are able to understand, among other things, more about how supernovae and their blast waves interact with their surroundings.

The outer rings, which are in a different plane than the inner one, are thought to result when jets of high-speed particles from the rotating supernova collide with previously ejected material, which is then caused to glow. The observable effect is similar to what a camera would record in a time exposure photograph of a searchlight beam on high altitude clouds.

Light reaches Earth not only from the supernova itself, but also from photons scattered by dust in interstellar space. Because the path this reflected light travels is longer than the direct one from the supernova itself, it arrives later. Such "light echoes" from interstellar material are shown in **Figure 19-20**. Analysis of such data provides us with new information on the distribution of interstellar clouds in the general direction of the star.

Observations made at radio wavelengths three and one-half years after the explosion of Supernova 1987A revealed the beginning of for-mation of a supernova remnant. By observing this remnant over time, we will be better able to understand the development of such objects and to place older supernova remnants into their proper context. The fact that it took more than three years before such radio energy was seen is important, because it shows that gas in the vicinity of the star had been swept away by the star's strong stellar wind, just as expected. The radio emission was produced when the ejected material finally encountered dense clumps of gas in the interstellar medium.

19.3 Neutron Stars and Pulsars

The model of a Type II supernova explosion that we have described predicts different effects on different parts of the star. Although the outer envelope of the star is blown off, the central core undergoes a catastrophic collapse. In the last rush of nuclear reactions in the collapsing core, electrons and protons combine to make neutrons. This occurs even within the nuclei of the heavier elements. For example, a silicon nucleus in the central regions of the star might find its protons converted to neutrons. If this process continues long enough, the silicon nucleus will become unstable and decay into two smaller nuclei. Thus, the process—electron + proton → neutron + neutrino—not only destroys individual protons but also tends to break down heavier nuclei and convert them into neutrons. At some point various sources of outward pressure (high internal temperature and neutron degeneracy pressure) combine to stop the collapse and restore equilibrium.

These reactions had been known in theory for a number of years, and led astronomers Walter Baade and Fritz Zwicky in 1934 to hypothesize that after a supernova explosion had taken place, we might be left with a remnant star composed entirely of neutrons—a **neutron star**. The hypothesis suggested that such an object would have a mass less than about 3 solar masses, just as a white dwarf's mass must be less than 1.4 solar masses. Furthermore, the hypothesis said that the mass would be contained in a sphere only about 20 km in diameter. The density would therefore be enormous, equal to the density of a thimble that had been stuffed with a billion automobiles! This is 10^8 times the density of a white dwarf, which itself is 10^6 times as dense as water.

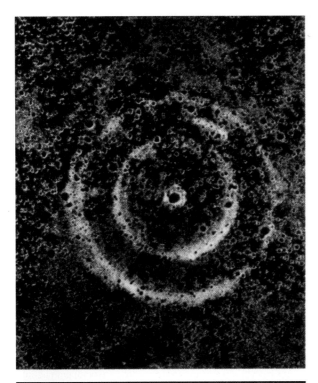

Figure 19-20 SN 1987A light echoes from interstellar matter.

The difficulty with this idea was to demonstrate that a neutron star was indeed left over after the explosion. Although such an object might have a high surface temperature, its extremely small surface area would make it a very faint object. It took about 30 years for astronomers to confirm this prediction observationally.

Astronomers had examined the various stars within the Crab Nebula and had not been able to find any clues that any of them might be a remnant star, although Zwicky did suggest that one of them was a likely candidate. Most of them were foreground or background stars not associated with the nebula at all. Furthermore, the Crab Nebula should have disappeared from visibility long ago, because the nebula was radiating energy so furiously that it should have cooled off quickly. Nevertheless, the nebular gases remained extremely turbulent, hot, and excited. This implied that, although no visible energy source was present, somewhere within the nebula there was a source of energy that continually excited the gas to glow.

Observations of the radio spectrum of the Crab Nebula show it to differ from that of a blackbody. Bodies that emit energy due to their temperature have a continuous spectrum whose intensity beyond a peak decreases with increasing wavelength (Chapter 13). For the Crab Nebula, however, the intensity is observed to *increase* with increasing wavelength (**Figure 19-21**). Furthermore, astronomers observe the radiation to be strongly polarized (see Figure 11-13 for images in polarized light). These observed characteristics are expected if the radiation from the Crab Nebula is due to the emission of energy by charged particles accelerating near the speed of light in a magnetic field. Such radiation is called **synchrotron radia-** **tion**; it an example of what is often called **non-thermal radiation** to distinguish it from that produced by a hot (thermal) body.

Synchrotron radiation from an accelerated electron is not radiated in all directions, but into a cone along the direction of the electron's motion (**Figure 19-22**). This directionality results in its characteristic spectrum and its polarization. Synchrotron radiation was first observed coming from synchrotrons, atom smashers built to study nuclear reactions by accelerating charged particles to near-light speeds and allowing them to collide with other particles at high energies. Synchrotrons use magnetic fields to accelerate charged particles to high velocities and to confine their motion. In the laboratory, physicists found that a charged particle accelerating at a high rate of speed through a magnetic field radiates away copious amounts of polarized energy whose intensity increases with increasing wavelength, exactly as now observed for the Crab Nebula. Thus, on the basis of the observations and theoretical ideas we conclude that the Crab Nebula environment contains charged particles accelerating at near-light speeds in an intense magnetic field.

PULSARS

The big breakthrough that tied many of the pieces in the supernova puzzle together took place in 1967, with the discovery of the objects that have come to be called **pulsars**. In that year a graduate student at Cambridge University, Jocelyn Bell, was working for astronomer Anthony Hewish and conducting a radio survey of the sky searching for fluctuations in the emission from radio sources (**Figure 19-23**). She noticed that a peculiar object (now known as CP 1919) appeared to be "pulsing"

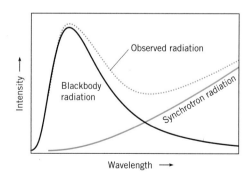

Figure 19-21 A schematic comparison of the spectrum of the Crab Nebula with that of a blackbody. The increase toward longer wavelengths is characteristic of synchrotron radiation.

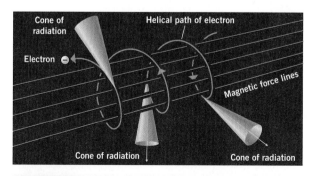

Figure 19-22 An electron follows a helical (spiral) path about a magnetic line of force, emitting synchrotron radiation into a narrow cone along the direction of motion.

at radio wavelengths in an extremely regular way. **Figure 19-24** shows one of the records of the object's radiation; it was distinctive in that the interval between the pulses of radiation (the period) was constant to an extremely high degree of precision. The period was determined to be an unchanging 1.33730113 seconds. Nothing of this regularity had ever been seen in nature before. In fact, the repeatability of the pulse period meant that the pulsars were providing more accurate time-keeping capabilities than any clocks on Earth. The question arose: was the source terrestrial or extraterrestrial, natural or artificial?

Inquiry 19-13

Because the radio telescope Jocelyn Bell used did not move, it detected objects as they passed over the telescope while the Earth rotated. The telescope observed the radio emission to occur nearly four minutes earlier on each successive night of observation. On the basis of this piece of information, what might you conclude about the location of the emitting source—terrestrial or extraterrestrial? Explain. (*Hint:* refer back to Section 4.4 if needed.)

The time intervals between pulses repeated so well that astronomers working on the data later confessed that they could not prevent themselves from considering the possibility that it might actually have been a coded message from an extraterrestrial civilization (the LGM theory, for Little Green Men). The obvious first step, however, was to look for other objects of this type; once astronomers knew what to look for, many others were found. Furthermore, pulsar periods were observed *not* to be constant but in many cases to increase continuously by tiny but measurable amounts. A natural cause now became a more plausible hypothesis. As a result of this and further work, Hewish was awarded the Nobel Prize in physics in 1974. (There is controversy to this day that Jocelyn Bell was not awarded the prize.)

How large are pulsars? We saw earlier in this chapter that an object's brightness theoretically cannot change faster than the time it takes light to travel across it. In the case of a pulsar, whose intensity changed over a time of 0.00007 seconds, the emitting region could not be larger than 0.00007 light-seconds across, which is about 20 km—in agreement with what Baade and Zwicky had predicted! Regardless of what was

Figure 19-23 Jocelyn Bell Burnell, the discoverer of pulsars.

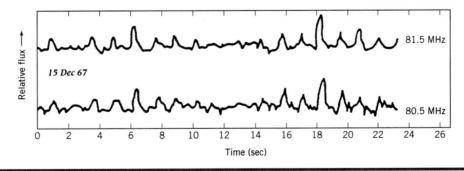

Figure 19-24 A graph of radiation intensity over time for CP 1919, which first drew astronomers' attention to the remarkable objects known as pulsars. Note the periodic nature of the intensity. Reprinted by permission from *Nature*, Vol 217, p. 709, 1968 Macmillan Journals Ltd.

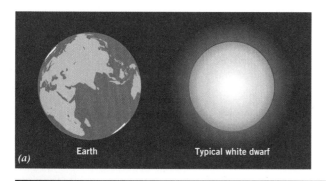
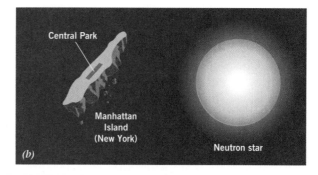

Figure 19-25 (a) The relative sizes of Earth and a white dwarf star. (b) The sizes of a neutron star and Manhattan Island, New York.

causing the variation in the light signal, the conclusion was still valid. Whatever a pulsar was, it was considerably smaller than a white dwarf star. Suddenly, theoretical speculation about the possibility of incredibly compact neutron stars was revived and made very real. **Figure 19-25** emphasizes the small sizes of white dwarfs and neutron stars by showing them relative to the Earth and Manhattan in New York city.

One early hypothesis about the nature of pulsars described them as pulsating white dwarfs or neutron stars. Models of such pulsating stars, however, showed that they would not be able to pulsate as rapidly as the 0.033-second rate observed for the pulsar found in the Crab Nebula. This is a perfect example of observations falsifying a hypothesis.

Another hypothesis was that a pulsar is a rotating white dwarf or neutron star. Because a white dwarf rotating with the period of the Crab pulsar would break apart, that hypothesis could not be valid.

The hypothesis that pulsars are rotating neutron stars makes several predictions. A rotating neutron star should have an intense magnetic field, as would be the case if the Sun were shrunk to a neutron star's size. In addition, just as magnetic braking was seen to slow the Sun's rotation, a similar mechanism would slow the neutron star's rotation. Both these predictions are verified by observations. In addition, the change in rotation period would mean that rotational energy is being lost. The law of conservation of energy would require that the lost energy reappear elsewhere. When astrophysicist Thomas Gold computed the amount of energy loss represented by the slowdown of the Crab Nebula pulsar, he discovered it to be almost exactly the amount required to explain the observed luminosity of the Crab Nebula. In other words, using the observed slowdown of the Crab pulsar and the assumption that the pulsar is a neutron star, the long-standing problem of the energy

source for the Crab Nebula was solved, thus affirming the hypothesis that pulsars are indeed rotating neutron stars. More than 1,500 pulsars are now known, and the hypothesis that they are rotating neutron stars has been well confirmed.

Do supernovae remnants contain pulsars? As we have seen, the Crab Nebula does. In the constellation of Vela, where a cloud of expanding gas had also been suggested as a supernova remnant, another pulsar was found near it's center. The Vela pulsar is 10 times closer to Earth than the Crab Nebula, and its explosion must have been an awesome sight. If it were an explosion of the same force as the Crab, the star would have appeared from Earth to be approximately 100 times brighter than the supernova of 1054 C.E., as seen from Earth.

Inquiry 19-14

What are the arguments that the Crab pulsar is a rotating neutron star?

Inquiry 19-15

The Vela pulsar is much slower than the Crab pulsar. What conclusion do you draw about the age of the Vela pulsar?

The discovery of the pulsar in the Crab Nebula by radio astronomers challenged optical astronomers to identify the remnant star and observe its variations in visible light. By synchronizing their observations with the observed radio pulse time of the object, astronomers were able to detect optical pulsing for one of the stars in the Crab (**Figure 19-26**). Finally, the star that exploded has been identified; it was the one suggested decades before by Zwicky.

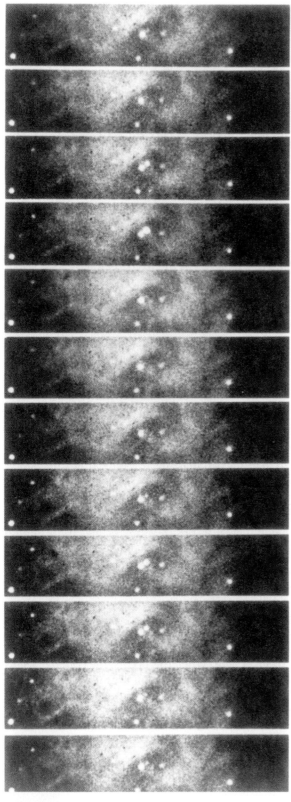

Figure 19-26 The pulses from the central star of the Crab Nebula, seen at optical wavelengths.

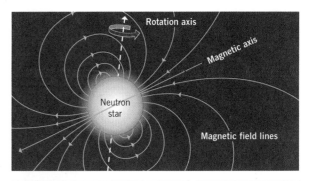

Figure 19-27 The central regions of a pulsar, showing the rotating neutron star and its magnetic field.

Many speculations concerning supernova explosions and neutron stars have now been tied together. However, not all supernovae remnants have observable pulsars, nor do all pulsars have observable remnants. For these reasons, astronomers wait anxiously for the discovery of a neutron star at the location of SN 1987A.

It is also possible that no condensed remnant will remain, analogous to a Type I supernova. One hypothesis for how this can happen is that the core implosion is not symmetrical. The search for possible mechanisms for complete dispersal of all the matter in a star back into the diffuse interstellar medium was begun when no condensed remnant was found associated with a number of supernova nebulae. This is a situation where the observations sparked the development of the theory to explain them.

WHY PULSARS "PULSE"

How does a rotating neutron star produce the observed brightness variations of a pulsar? The most successful models appear to be those in which a rotating neutron star is enclosed in a region of plasma (ionized gas), as in **Figure 19-27**. A rotating neutron star is shown together with the magnetic lines of force that are supposed to surround the star. If lines of magnetic force are threaded through a region of ionized gas (such as one that composes a star), then the gas and the magnetic field lines tend to move together. If the gas moves, the lines of magnetic force are dragged along with it.

This means that when the central core of a supernova collapses, the magnetic field will collapse with it. As the magnetic field is compressed together into the collapsing core, its strength increases, building up to an enormous value. It is not known exactly what form the magnetic field lines might take

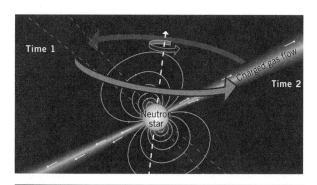

Figure 19-28 Viewed from a distance, the material that flows out along the magnetic axis of a neutron star is accelerated and generates a sweeping beam of radiation that gives the effect of a rotating searchlight. In the inset, the rotating beam is shown at two times.

around a collapsed star, but a simple pattern has been drawn in Figure 19-27. Notice that the rotational axis of the star and the axis of symmetry of the magnetic field do not coincide—neither do they for Earth, Jupiter, Uranus, and Neptune.

The outer surface of a neutron star is expected to consist of nondegenerate ionized gas. What would happen if some of these charged particles were to evaporate from the hot surface of the neutron star? As the charged particles flow away from the star, they are caught up by the magnetic field lines and forced toward the magnetic poles, where the field is stronger. The particles are accelerated by the rapidly rotating magnetic field, until at a certain distance from the star their speed approaches the speed of light. When this occurs, the gas begins to radiate synchrotron radiation near the poles, like two searchlights sweeping around through space. If one of these rotating beacons happens to sweep past Earth, we will see a pulsar, as in **Figure 19-28**.

Inquiry 19-16

What does the searchlight mechanism of pulsar emission imply about the completeness of our pulsar searches and the number of stars that might become pulsars?

Inquiry 19-17

How might you explain the fact that no pulsars are known whose periods are more than a few seconds long?

Even if a pulsar's radio beacon is aimed away from us, there are still observational techniques that can detect its presence. In particular, all strong pulsars apparently are immersed in a cloud of electrons that accelerate near the speed of light and that radiate X-rays. X-ray surveillance satellites, which can detect the X-rays from the electron clouds, provide indirect evidence of the presence of a neutron star. Even so, only about 25 percent of supernova remnants show evidence of neutron stars inside. At present, the significance of this fact is uncertain.

A MODEL OF A NEUTRON STAR

We have already talked about the unbelievably high density of neutron stars, but we wish to illustrate further what truly odd objects they are. Because we have at present no way of creating material of such tremendous densities in a laboratory, predictions based on model calculations are the only guide we have to understand these stars. A neutron star may have an atmosphere, but because of its strong surface gravity it would be only a few centimeters thick. Underneath it is the star's outer envelope, which would actually be *crystalline;* that is, a solid rather than a gas, with the particles composing it arranged in a rigid, regular structure.

Evidence for the crystalline outer layer of neutron stars is provided by a surprising event occasionally seen in observations of pulsars. Although in general the rotation of pulsars slows gradually and uniformly with the passage of time, in some cases they speed up suddenly, after which the pulsations gradually slow down again. These "glitches" have been interpreted as the result of adjustments of the crystalline shell of the star, accompanied by a slight collapse of the star. These episodes in the life of a neutron star have been called *starquakes,* by analogy to earthquakes.

Inquiry 19-18

What physical law causes a pulsar's rotation to speed up if, as a result of a starquake, the star collapses slightly?

BINARY AND MILLISECOND PULSARS

The first discovery of a **binary pulsar**, in which two neutron stars are bound together in a close binary system, was exciting for astronomers because they could now determine the masses of neutron stars from observations using a direct technique. In this binary system, known as PSR 1913116 and in which only one of the neutron stars is observed to be a pulsar, the sum of the masses is about 2.8 solar masses. We infer that both stars in this system are neutron stars of 1.4 solar masses each. A number of other binary systems are also known to include neutron stars.

An extremely important observation of the original binary pulsar system is that the orbital period of the two stars is observed to decrease in exactly the manner predicted by Einstein's theory of general relativity for an object emitting gravitational waves, which are extremely feeble waves predicted to be emitted whenever a mass is accelerated. Undoubtedly, similar observations will result for the other systems. Thus, the binary pulsars have turned out to provide an unexpected but important test of the theory of relativity.

Another unusual group of pulsars contains the "millisecond pulsars," one of which is PSR 19371214. With a period of 0.001558 seconds (1.558 milliseconds), this pulsar rotates on its axis at the phenomenal rate of 642 revolutions per second! If it were spinning only 10 percent faster, in fact, the forces generated would tear it apart. This pulsar appears to be very old—an age of 10^9 years has been suggested—but its magnetic field is so small that it loses energy slowly. Nevertheless, because the picture we gave earlier was one of pulsar periods lengthening over time, why is this old pulsar rotating so rapidly? The explanation is that millisecond pulsars are members of binary systems. The low-mass companion finally reached a stage where it could dump mass into an accretion disc which, when transferred to the neutron star surface, increased the rotation speed. (Think of the way you can increase the speed of a spinning basketball by hitting it periodically.) Many such systems are turning up in the centers of globular clusters, where the density of stars is high and where close binary star systems should be abundant.

Even better was the announcement in 2004 of a binary system in which both components are observed pulsars. One pulsar, known as A, rotates every 23 milliseconds, while the other one (B) rotates once every 2.8 seconds. A preliminary conjecture on the formation is that A accreted matter from a companion star and thus spun faster and faster. The B pulsar formed from the companion after it underwent a supernova explosion.

MAGNETARS

Up to now, neutron stars have provided astronomers with a way to understand a variety of observed phenomena, but not all of them. Various types of observations that have puzzled astronomers for a number of years can now all be understood and synthesized in terms of a new class of object confirmed to exist only as recently as 1998. These objects, called **magnetars**, are super-magnetized neutron stars, having magnetic fields 1,000 times greater than that of a normal neutron star. Such strong magnetic fields were theorized to exist in neutron stars that were rapidly rotating when formed. Thus, "normal" neutron stars form from objects rotating with periods of, say, 0.020 seconds (the period of the Crab pulsar when born), whereas a magnetar would form if the rotation period were less. In as little as 30,000 years, the magnetic field brakes the spin and the magnetar becomes a virtually undetectable neutron star. Although only 10 magnetars are currently known, they are all located in supernova remnants that have ages of less than 10,000 years. Astronomers hypothesize that hundreds—and perhaps hundreds of millions—of *dead* magnetars could exist. Furthermore, some astronomers conjecture that the supernova remnants lacking observed pulsars contain dead magnetars, which are simply normal, invisible neutron stars. Future observations will test these ideas.

An understanding of magnetars may help astronomers understand a number of other observed events. The extraordinary strength of magnetism they exhibit can hold the surface temperature high at 10 million degrees, which can explain the X-rays coming from them. Furthermore, the extreme magnetic fields can produce powerful stresses in the surface material. Such stresses can strain and crack the surface, which could produce starquakes similar to those mentioned previously. Some other classes of objects, such as the SGR (Soft Gamma Repeaters) and the AXP (Anomalous X-ray Pulsars), may actually be magnetars going through their aging process and slowing down. Thus, this class of object may bring together a number of apparently disparate objects.

PLANETS AROUND PULSARS

Astronomers have observed pulsars for which the pulses arrived some 0.003 seconds later than expected. Given the tiny uncertainties in measured pulsar periods, this short delay time is significant. These data show the presence of three small planets orbiting a neutron star. (These observations were mentioned in the concluding section of Chapter 6.) There are a number of ideas as to how such planets could have formed, and observations are now underway to narrow down the possibilities. One idea as to how planets might form in this unfriendly environment suggests that energetic particles from the neutron star erode the companion's gas, which eventually condenses into one or more planets. Another is that the planet formed with its star (the standard planetary formation model) and survived all of the evolutionary stages of that star, even its death.

19.4 Black Holes

The objects known as black holes have captured the imaginations of many people. Our investigation will first describe what they are and their properties. Then we will see how astronomers might go about detecting objects that are thought to emit no light. During this discussion, we hope to dispel some popular misconceptions about black holes.

THE THEORETICAL PREDICTION OF BLACK HOLES

We saw that theory had predicted the possibility of bizarre objects like neutron stars some 30 years before their existence was verified. Similar predictions and later discoveries had occurred earlier in physics with the neutrino and with antimatter. For these reasons, when the solution of an apparently valid equation suggests that a certain type of physical behavior *can* take place, scientists expect that somewhere in the vast universe it actually *may* take place. In a sense, the universe is so large and so old that if something can happen, it probably will.

Given this history, astronomers were intrigued with another idea that kept popping up in calculations from time to time—the possibility that under some circumstances a dense gaseous configuration could begin to collapse and there might be nothing to stop the collapse.

We have seen that in the center of a supernova explosion there is a dense imploding remnant, but high internal temperature and neutron degeneracy pressure combine to stop the collapse and restore equilibrium, thus producing a neutron star. Yet the possibility remains that there might be circumstances in which the core of an imploding star could either be pushed past this equilibrium point or else simply have so much mass that nothing could stop the tremendous inward crush caused by the gravitational forces. Remember that if a white dwarf gains mass beyond its maximum 1.4 solar masses, it collapses into a Type I supernova. What happens in the collapse of an object of mass greater than the neutron-star limit of three solar masses?

As a star's core collapses, its *surface* gravity will increase rapidly as its size decreases. Eventually, the speed required for anything to escape from the stellar core's surface (its escape velocity) will become greater than the speed of light. Because, according to the theory of relativity, nothing can travel faster than the speed of light, not even light would be able to escape. Astronomers call such a collapsed object a **black hole**.

Because no known force can stop the implosion, it continues until the mass becomes a point of zero radius and infinite density called a **singularity**. Surrounding the singularity is a region of empty space from which light cannot escape. The radius of this region, which is considered the radius of the black hole, is determined by the object's mass and is known as the **Schwarzschild radius**.[3] As an example, a one-solar-mass black hole will have a 3 km Schwarzschild radius, and the size increases directly with the mass. Light emitted inside this radius will be trapped. However, light emitted a few Schwarzschild radii away will be able to escape. Because communication from inside the black hole to the universe outside is impossible, the Schwarzschild radius is often referred to as the **event horizon**—outside observers are unable to see events occurring inside.

Thus, between the singularity and the event horizon is empty space. Except for the mass at the singularity, a black hole itself is, theoretically, just empty space! In fact, black holes are quite simple. For example, black holes can be fully described by three quantities: the mass, the amount of electrical charge present, and the amount of angular momentum. No information about what went into the black hole is maintained.

---

[3]The Schwarzschild radius is given by $2Gm/c^2$ where G is the gravitational constant, m is the object's mass, and c is the speed of light. Thus doubling the mass doubles the size of the Schwarzschild radius.

The preceding description is that of what we can refer to as a *classical* black hole. There are problems with this classical description, however. One is the concept of the singularity of zero size and infinite density. Another problem is with the idea that all information about what made the black hole is lost. These two problems make many astrophysicists uncomfortable with the classical description and, therefore, deeper understanding is being sought. That deeper understanding may come from something known as *string theory*, which is a way of unifying the ideas of relativity theory and quantum mechanics—something that has eluded scientists to this day. Under string theory, a black hole's mass exists in a small, but nonzero, volume—the density, then, is high but not infinite. Furthermore, information about the matter prior to its forming the black hole is not lost but can, in theory, be recoverable. Thus, our understanding of black holes is very much in its infancy.

To understand the unusual properties of a black hole, let us first see how mass affects the space near it. The modern theory of how mass affects space is Albert Einstein's theory of general relativity. In its most complete form it is expressed as a set of multidimensional equations, but some of its consequences can be discussed in a simple way.

According to this theory, the photon travels through space that has been changed, or "warped," by the presence of mass. Far away from a dense mass like a star, or near a less dense object like a planet, Newton's theory of gravity adequately describes the motion of objects and light. However, a very massive object warps space so much that the effect becomes measurable. The space closest to the mass is warped the most. Also, the more dense the mass is, the more the space is warped. All objects and light travel in a straight line, but the straight line, if it goes through curved space, is also curved. An analogy is helpful: the path of a billiard ball on a table made of a stretched sheet of rubber would be different from the path if the rubber top were warped by a heavy brick placed on it (**Figure 19-29**).

One of the first predictions Einstein made from his theory was that when radiation passes very close to the Sun, its path should be bent 1.75 seconds of arc. A total eclipse of the Sun in 1919 offered astronomers the first opportunity to verify the prediction. When the Moon covers the bright disk of the Sun, the sky near the Sun darkens and it is possible to see stars close to the Sun's edge.

Figure 19-29 *(a)* A ball traveling in "flat" space always travels in the same direction. *(b)* If the table top is made of rubber, a brick in the center "warps" the space through which the ball travels. In this way, the path changes as it passes through the warped area.

Figure 19-30 illustrates what should happen if the light rays from these stars appear to bend as they go through the curved space near the Sun during an eclipse. By comparing a photograph taken during the eclipse with one taken with the same telescope when the Sun has moved away from that part of the sky, the theory predicts that we should be able to measure a change in the apparent positions of the stars.

During the 1919 total solar eclipse, a research team led by English astronomer Sir Arthur Eddington (**Figure 19-31**) performed this extremely difficult experiment, and they observed exactly the predicted effect. This experiment has been refined and repeated successfully numerous times. More refined measurements have been made without requiring a total eclipse by using the light from the nuclei of distant galaxies (objects called *quasars* that are discussed in Chapter 22).

Through this observation and other experiments, scientists conclude that general relativity is a better approximation to nature than Newton's laws under most situations. (General relativity fails on scales the size of an atom.) We will expand upon Einstein's ideas of the geometry of space in Chapter 23 when we discuss cosmology.

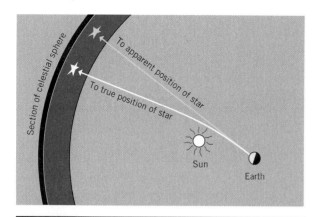

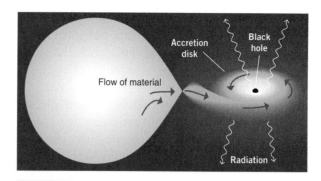

Figure 19-32 The most likely way to detect the presence of a black hole is by observing X-rays emitted by the hot gas that falls into such an object.

Figure 19-30 The bending of light rays from a distant star when the light passes near the Sun alters the apparent position of the star from its true position.

Figure 19-31 Sir Arthur Eddington, one of the more prominent astrophysicists of the 20th Century.

The popular idea of a black hole is that of a cosmic vacuum cleaner, something that sucks in everything near it. This concept is true only for matter within or near the event horizon. Transforming the Sun into a black hole would not cause the Earth to be sucked in because there would be no change in the gravitational attraction felt by the Earth at its distance of 1 AU.

Black holes can form in at least two ways, depending on the star's mass. Remember, stars in the range of 8 to 20 solar masses produce strong explosions and a neutron star. Stars in the range of

20 to 45 solar masses will result in a weak explosion in which the core initially becomes a neutron star. Since such a star has so much mass above the core, some of that material can fall back onto the neutron star and thus cause its mass to increase above the neutron star mass limit of about three solar masses. Once that happens, the neutron star collapses to a black hole. Black holes formed in this way are referred to as *fallback* black holes. For stars more massive than 45 solar masses, no explosion is observed, and the core collapses directly into a black hole.

It is also possible that no condensed remnant will survive the core implosion, as was mentioned earlier in this chapter.

Black holes of various masses can exist—from stellar masses up through gigantic supermassive black holes ranging from millions and up to billions of solar masses at the centers of galaxies. At the other end of the scale are "mini" black holes that were thought to form in the Big Bang. Here we look at observations of stellar black holes, whereas those in the cores of galaxies will be examined in Chapter 22.

OBSERVATIONS OF STELLAR BLACK HOLES

If an object cannot radiate energy, then we must be clever to observe it; for example, we may ask what effect a black hole might have on mass that is near it, but not yet inside the Schwarzschild radius.

Consider **Figure 19-32**, in which a black hole and a giant star orbit each other in a binary system. Such a configuration could be formed if a massive star in a fairly wide binary evolved into a black hole, leaving its less massive companion to live out its life normally. Much of the mass of the original star would be in the black hole singularity, so it still would be a gravitationally functioning member of the binary system, and the two objects would continue to orbit each other. (General relativistic effects might come into play if the two components of the binary are very close to each other.) However, if the companion were to become a giant, or for any reason spill mass onto the black hole, a situation similar to that of a nova would develop. Mass would be dumped onto the black hole (through an accretion disc) and, upon falling in, would disappear. However, before it disappeared from our sight, the accreting mass would be compressed and heated to extreme temperatures by friction with other particles.

Inquiry 19-19

If friction heated gases moving into a black hole to a temperature of 10^6 K, in what wavelength region should astronomers look for the radiation?

Since one signature of a black hole is the high temperature of the gas falling into it, strong X-ray sources found by satellite surveys have been extensively studied. One of the brighter X-ray objects known is Cygnus X-1 (the first X-ray source located in the constellation Cygnus). Its X-ray radiation fluctuates both randomly and also with a period of 5.6 days, indicating its membership in a binary system in which the X-rays are periodically eclipsed. Although it is often difficult to associate a visible object with an X-ray source, Cygnus X-1 turned out be at the same location as a known blue supergiant star called HDE 226868 that is a single-lined spectroscopic binary with a 5.6-day period. The fact that only one set of spectral lines is observed is important because it means the companion star cannot be seen in visible light.

If the hypothesis is that Cygnus X-1 consists of an invisible object and a blue supergiant, how is it possible to tell whether the invisible companion is a black hole? Perhaps it is merely a neutron star, for matter falling onto a neutron star will also generate a high enough temperature to give off X-rays. The fact that the blue supergiant is very massive and thus young is persuasive, but the blue supergiant could have accreted mass from its now-dead partner. The one way to tell is to find the mass of the invisible star. If it could be demonstrated that the companion star had a mass greater than about three solar masses, theory says it must be a black hole.

Inquiry 19-20

Why couldn't the invisible object with a mass greater than three solar masses be a white dwarf or a neutron star?

Solving for the mass of both stars in a binary system is not possible if spectral lines of only one of the stars can be seen. However, it is possible to set some limits on the mass of the invisible object. The blue supergiant star in Cygnus X-1 should have a mass of around 30 solar masses (assuming that it is not some kind of peculiar object). This mass, along with the observed Doppler shifts in its spectrum, implies that the mass of the invisible star in the Cygnus X-1 system is 16 solar masses. Neither a neutron star nor a white dwarf can be anywhere near this massive, so the only remaining possibility is that the collapsed object in Cygnus X-1 is a black hole.

There are other ways to tell when one component of a binary system may be a black hole. For example, the star V404 Cygnus is an X-ray source that emits more energy in X-rays than a million stars like the Sun do in all spectral regions. The observed orbital motion of 420 km/s means that the collapsed object must have a mass of about 12 solar masses. V404 Cygnus is a stronger black hole candidate than Cygnus X-1 because it requires us to make fewer assumptions about the mass of the observable component. All in all, there are about 20 binary systems thought to be strong stellar-remnant black hole candidates.

At present, we have a number of theories as to how a black hole might form, and it is possible that there is more than one path to a black hole. One possibility is that during a supernova explosion, a region from the central core of the star might be able to implode into a black hole in one quick transition. Another is that a neutron star near the upper limit of mass allowed for neutron stars could gravitationally accrete more material from its vicinity (perhaps from a companion star) and collapse into a black hole once the mass exceeds the neutron star limiting mass. If a star survived the red giant and planetary nebula mass-loss phases and still had more than three solar masses worth of material, it might be able to contract continuously into a black hole under the action of its own self-gravitational attraction.

Do black holes exist? Intriguing evidence points toward their existence. There are, however, a number of other ways to explain the observations, although many of these explanations are ad hoc and contrived (for example, that Cygnus X-1 is a multiple star system made up of the supergiant and two or three neutron stars; or that such objects are *Q stars,* objects composed of protons, electrons, and neutrons held together by the strong nuclear force rather than by gravity). The black hole explanation is the simplest and therefore, by Occam's razor, preferred. Thus, it is fair to say that the overwhelming majority of astronomers accept that black holes do exist and have been found.

19.5 Mystery Solved: Beasts in the Cosmic Zoo

In recent years the "zoo" of objects that appear to involve a compact object (white dwarf, neutron star, or black hole) in a binary system has continued to grow. In addition to the novae and Type I supernovae, the ranks have been augmented by objects known as X-ray bursters, gamma ray bursters, and two enigmatic objects known as SS433 and Geminga. This section provides a good example of how astronomy moves forward. The previous edition of this book, which came out in 2000, discussed these four "beasts," but explanations still eluded astronomers. Geminga has now been determined to be a fairly young (300,000 years old) and nearby (100 parsecs) neutron star, perhaps a pulsar whose radio beacon does not sweep over the Earth. In 2004, the astronomical community largely agreed on a unifying model for the other three of them: SS433, Cygnus X-3, and gamma ray bursters. We now know that both SS433 and Cygnus X-3 are examples of a class of objects called *microquasars.*

Microquasars

Quasars (quasi-stellar radio sources) have provided astronomers with decades worth of contemplation and frustration. These are objects that are now thought to be, as you'll read in more detail in Chapter 22, million to billion solar-mass black holes in the cores of distant galaxies. They have the mass of a galaxy packed into the size of the Solar System, jets of gas moving at near-light speeds and an accretion disc that feeds the black hole. Objects that appear to be miniature versions

of quasars have been discovered in our own galaxy and have been dubbed **microquasars** because of their high luminosity, rapid fluctuations in brightness, and the presence of thin jets of gas caused, in part, by the presence of an accretion disc. Some microquasars may contain a massive neutron star rather than a black hole. One difference between quasars and microquasars is that these are stellar objects in a binary system within our galaxy. Some 20 microquasars have now been identified in the Milky Way.

The strangeness of SS433 was first noticed in 1978, and it was recognized as the first known microquasar in 1992. SS433 has a variable energy output in the optical, radio, and X-ray regions. It also shows two sets of emission lines at displaced wavelengths that indicate the existence of clouds of material in the system streaming at speeds as much as 26 percent of the speed of light. Also, the velocities vary periodically over a 162-day period, from a blueshift of 8 percent the speed of light to a redshift of 16 percent the speed of light. The best model for this bizarre behavior is shown in **Figure 19-33**. A normal A-type supergiant star is in a binary pair with a compact object, whose mass has recently been found to be three solar masses—right on the borderline between a neutron star and a black hole. However, another recent study derives a mass of 8 solar masses, which says the compact star is a black hole! Future research will determine what it really is. The object is near the center of a supernova remnant, which supports the theory that it is a compact stellar remnant. Material from the normal star streams toward the compact object, feeding an accretion

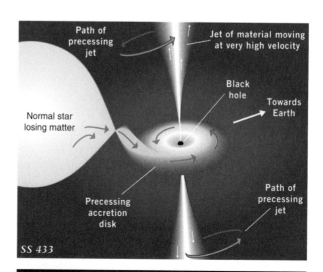

Figure 19-33 A model of SS433.

disk. Two jets of material are ejected perpendicular to the accretion disk at approximately 26 percent the speed of light. The accretion disk itself *precesses,* or changes its plane of rotation, under the gravitational influence of the normal star, much as Earth's axis of rotation precesses under the influence of the Moon. This causes the direction of the two jets of material to change, which, in turn, causes the velocities we observe to vary periodically.

The object Cygnus X-3, which is also now known to be a microquasar, is observed throughout the electromagnetic spectrum, from the longest radio wavelengths to the shortest gamma-rays. Its brightness varies considerably at all wavelengths. However, the object is shrouded behind a thick dust cloud that makes it invisible in optical wavelengths. With a period of 4.8 days, it was assumed to be a neutron star in a binary system, but we now think it contains a black hole. Observing at infrared wavelengths that penetrate the obscuring dust, astronomers have seen broad emission lines of both neutral and ionized helium, along with a deficiency of hydrogen. These observations are reminiscent of the type of star known as a Wolf-Rayet star, an extremely hot star with such a strong wind that the outer layers have been blown away, exposing the helium-rich core. It is this star that provides the infalling material to the black hole.

GAMMA RAY BURSTERS

Military satellites designed to detect the emission of gamma rays from nuclear explosions on Earth have been circling the globe since the early 1960s. Their observations of gamma rays from space that were made public in the 1970s provided evidence of a new class of astronomical objects called **gamma ray bursters**. As their name implies, the objects blast copious amounts of gamma rays over time intervals of seconds.

Additional sources were discovered when NASA launched the *Compton Gamma Ray Observatory* (*CGRO*) in the early 1990s. Instruments onboard *CGRO* discovered more than 2500 gamma ray bursters. Importantly, and frustratingly, the locations of the bursts never repeated. Furthermore, their distribution over the sky was uniform, and thus much different from that of other X-ray sources, which are mostly concentrated toward the plane of the Milky Way galaxy (**Figure 19-34**).

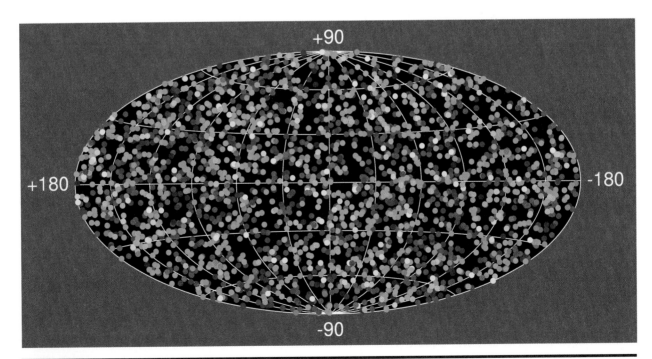

Figure 19-34 More than 2700 gamma ray sources are uniformly spread around the sky. The most energetic bursts are in red; weak ones are in purple.

Fundamental to learning the nature of the objects was a determination of their distances, from which the luminosity could be determined using the inverse square law of light. Although a uniform distribution in the sky generally means objects are located far beyond our galaxy, a more direct determination was needed. That possibility came about only in 1997, when the Italian-Dutch satellite *BeppoSAX* recorded a burst in both gamma rays and X-rays of an object known as GRB970508. Because X-ray detectors are able to provide more accurate positions on the sky than can gamma ray detectors, astronomers now knew where to point their telescopes to search for optical counterparts.

From optical spectra, astronomers were able to prove that at least this one gamma ray burster, and thus by inference all of them, were indeed far beyond the Milky Way—at 2 billion light-years! The luminosity of GRB970508 was thus found to be as much as 100 times that of a typical supernova. A 1998 observation of another gamma ray burster was also found to be even farther away (12 billion light-years) and more luminous yet. In late January 1999, some astronomers were able to observe a gamma ray burst in optical radiation while it was happening!

Such information constrains the models that theorists use to explain observations. One suggested explanation, the merger of two neutron stars or a neutron star and a black hole, has now been ruled out. In its place, theorists have suggested the existence of super supernovae, which they call **hypernovae**—the biggest bang since the Big Bang

itself. Hypernovae result from the collapse of stars at the high-mass end of the range of allowed masses—30 to 50 solar masses. The association of hypernovae and gamma ray bursters received observational confirmation when, on March 29, 2003, the gamma ray burster known as GRB 030329 burst forth from a distance of 2.6 billion light-years. Astronomers at the European Southern Observatory obtained optical spectra of it and found the gradual emergence of a hypernovalike spectrum. Furthermore, the observations indicate that the gamma ray burst and the hypernova explosion occurred within two days of one another, thus showing a direct connection between them. These observations are consistent with, and provide confirmation for, a model of black hole formation for the most massive systems known as the *collapsar* model, which involves the collapse of a rapidly rotating star of mass greater than 25 solar masses.

A General Model

These examples of "weird beast in the astronomical zoo" serve to emphasize two points. First, the universe is filled with many strange and exotic objects that provide astronomers with an understanding of concepts not otherwise fathomable. Second, many of these objects involve mass-transferring, interacting binary systems. Although details differ from one object to another, mass transfer between members of a binary system is the overriding concept that ties many of them together. An example of the process is shown in **Figure 19-35**, which presents a scenario for the evolution of the binary microquasar X-ray source

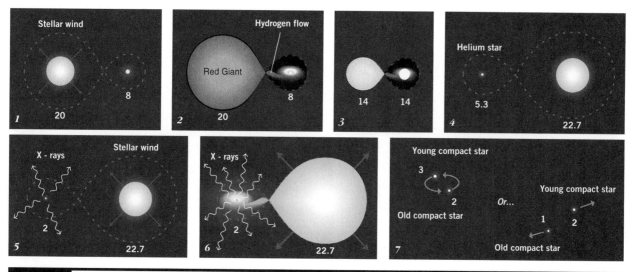

Figure 19-35 An evolutionary model for Centaurus X-3, showing how mass transfer changes a binary system and produces a variety of observable effects.

Centaurus X-3. It begins with two stars having 20 and 8 solar masses, respectively. The 20-solar-mass star evolves more quickly and begins to eject a stellar wind (1). Eventually it becomes a red giant, fills its Roche lobe (2), and begins to dump matter onto the companion until the masses are equalized (3) and a slower rate of mass transfer occurs. When enough mass has been transferred (4), a helium-rich core is exposed, and we have a Wolf-Rayet star that eventually becomes a supernova and leaves behind a neutron star. The evolu-

tion of the original secondary star now speeds up so that material from its stellar wind falls onto the neutron star, which appears as an X-ray source (5). The original secondary expands to fill its Roche lobe (6) and then rapidly transfers matter back to the original primary; this is Centaurus X-3 today. Mass may be lost from the entire system at this state. Finally, the system will evolve into a compact object (7) in which the stars may orbit each other, or perhaps break their gravitational bonds and separate from one another.

Chapter Summary

- **Figure 19-36** summarizes our entire discussion of stellar evolution from the main sequence onward. The figure shows the post-main-sequence evolution and final stages for stars of various masses. Keep in mind that the boundaries between mass ranges are not firmly known. Development of this summary consumed the careers of numerous astronomers over many decades; more complete understanding of these evolutionary phases is the job of future generations of astronomers.

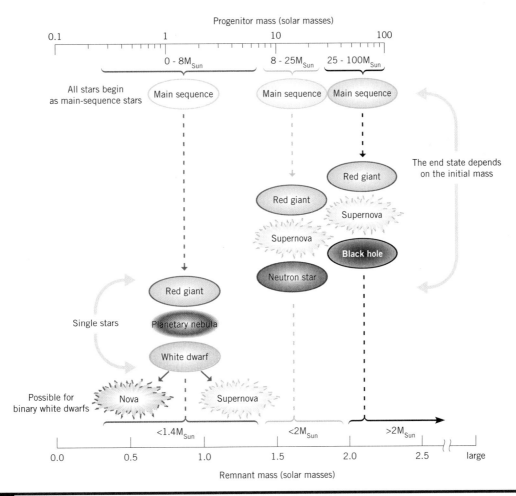

Figure 19-36 A summary of stellar evolution from the main sequence through stellar death.

Observations

- **Novae** are stars that increase in brightness as much as 10^5 times in a day or two. Their spectra show expansion velocities of hundreds up to a thousand km/s. All novae occur in binary star systems. While only some appear to be recurrent, astronomers think they may all be. Novae explosions are superficial and do not destroy the star.

- Supernovae are objects whose brightness increases by billions of times over the pre-outburst brightness. **Type I supernovae** have no hydrogen in their spectra, whereas **Type II supernovae** do show hydrogen. The spectra of gas in supernovae remnants show the presence of heavy elements, which were formed by nuclear burning during the explosion. Supernova explosions are nearly destructive of the entire star.

- **Pulsars** are objects for which astronomers observe precisely timed emissions of radio energy. The periods between pulses generally increase with time. Observations show some pulsars to have a sudden change in the pulse period.

- The observed increase of a pulsar's continuous spectrum with increasing wavelength, along with the radiation's polarization, is interpreted as **synchrotron radiation** from charged particles accelerating near the speed of light in a magnetic field.

- X-ray and gamma ray telescopes in orbit about the Earth have made observations of a variety of objects. Many of these objects change brightness rapidly, in a fraction of a second.

Theory

- Nova explosions are thought to occur when matter from one star in a binary system collects in an **accretion disc** surrounding a companion **white dwarf**, with the subsequent buildup of hydrogen on its surface. Once the temperature and pressure at the bottom of this accreted hydrogen layer reach high-enough values, a thermonuclear explosion occurs, causing the ejection of matter from the white dwarf.

- A *Type I supernova* results when a white dwarf accretes enough mass to reach the Chandrasekhar limiting mass of 1.4 solar masses. Because all Type I supernovae are formed from the same mechanism, they are all similar.

- A **Type II supernova** results from the collapse of a massive star containing 10 to 20 solar masses. Such an object has produced an inert iron core that disintegrates and emits neutrinos. The energy loss from the star's interior causes it to collapse catastrophically and then to explode. Stars with more mass result in **hypernovae.**

- For the collapsing remnant of a supernova explosion to produce a **neutron star**, the mass cannot exceed about three solar masses. During the collapse, its magnetic field intensifies.

- A massive star that undergoes a supernova explosion that leaves behind a core more massive than three solar masses is thought to produce a black hole. The object collapses into a **singularity**, and is surrounded by an **event horizon** whose radius (called the **Schwarzschild radius**) depends only on the remnant's mass.

- Matter falling into a **black hole** from a surrounding **accretion disc** will be heated to high temperatures and emit radiation in the X-ray and gamma ray spectral regions.

Conclusions

- Single stars get rid of mass by ejecting a planetary nebula, while binary stars eject matter by (recurrent) novae explosions.

- Both theory and observation point to stars as the location of element synthesis. Furthermore, only in supernovae and hypernovae explosions are conditions right for the formation of the heaviest elements.

- Because an object can be no larger than the light-travel time across it, we infer that pulsars are about 20 km in diameter.

- Pulsars are inferred to be rotating neutron stars and to have magnetic and rotation axes that do not coincide. The star's rotation causes a beam of radiation to point periodically toward the observer. From observations of binary pulsars, neutron star masses are computed and found to be less than the maximum allowed by theory. Observations of binary pulsars are consistent with Einstein's theory of relativity.

- Cygnus X-1 is an X-ray emitting object with a period of 5.6 days. Its optical counterpart is a single-lined spectroscopic binary with a 5.6-day period. On the basis of the observed X-ray and optical spectra, we infer that Cygnus X-1 has a mass of at least five solar masses and therefore is likely to be a black hole. Other black hole candidates also exist. The evidence for the existence of black holes is compelling but not yet definitive.

- Many of the peculiar objects astronomers observe can be understood in terms of mass transfer between one star and a compact object (white dwarf, neutron star, black hole).

Summary Questions

1. What do we mean by the term *nova?* What are two pieces of evidence that novae eject mass?

2. Why do astronomers conclude that the size of an object is about equal to the light-travel time across it?

3. What model for novae phenomena best allows astronomers to understand them?

4. What are the observed features of the Crab Nebula? What evidence exists that it is indeed the object observed about 1,000 years ago by Chinese astronomers?

5. What are the observed differences between ordinary novae and supernovae? Explain them in terms of the violence of the explosion, the increase in brightness of the object, and the amount of mass expelled.

6. In what ways do Type I and Type II supernovae differ? Your answer should be based on their observed properties and the models we have to explain them.

7. How do supernovae provide us with direct evidence for the synthesis of heavy elements within stars? Name and describe three basic processes by which heavy elements are synthesized.

8. Describe the interior structure of a neutron star.

9. What evidence is there that pulsars are actually rotating neutron stars? What observational evidence is there for the existence of pulsars?

10. How do light and matter behave near a black hole? Why can nothing escape from a black hole?

11. How might astronomers detect a black hole? What evidence is there that the Cygnus X-1 system actually consists of a giant star and a black hole, rather than some other collapsed object such as a white dwarf or a neutron star?

12. What is the basic model that explains most of the objects in the astronomical zoo?

Applying Your Knowledge

1. Outline the steps leading to the inference that Cygnus X-1 and V404 Cyg might or might not be black holes.

2. What evidence might Tycho Brahe have used when he said that the supernova of 1572 was farther away than the Moon? (Remember, this was prior to the telescope!)

3. Your body consists of numerous chemical elements, including carbon, oxygen, and iron. Discuss the origin of these elements in terms of their formation in stars.

▶ 4. Suppose the spectrum of a nova contains the Hβ spectral line observed at a wavelength of 4856 Å. Remembering that Hβ is normally at a wavelength of 4861 Å, calculate how rapidly the gaseous envelope expands.

▶ 5. Novae usually show broad emission lines caused by the expanding gas. Suppose such a spectral line is observed to be 10 Å wide. Using the observed width of the emission line, determine the speed of expansion of the gaseous envelope relative to the star itself.

▶ 6. What would be the maximum size of a pulsar whose brightness changes substantially in 0.00007 seconds?

▶ 7. What would be the mass of a black hole whose binary companion is a 20-solar-mass star separated by 0.175 AU and whose period is five days?

Answers to Inquiries

19-1. Presence of emission lines; absorption lines that are Doppler shifted relative to emission lines; presence of gaseous nebula visible on images taken around the star.

19-2. The star would be 100 times more luminous.

19-3. As the gas expands away from the star, it will get thinner, and we might expect the line in the spectrum due to the gas to get weaker. We might also expect the gravitational force of the star to slow down expansion of the gases. This would cause a change in the amount of the Doppler shift of the lines that are due to the gases.

19-4. Recurrent novae might be explained by a periodic process of mass dumping from the companion to the compact object, followed by an outburst, followed by a repeat of this process for an indefinite number of times. Eventually, either the companion loses enough mass to stop dumping it onto the white dwarf, the companion itself evolves beyond a red giant, or the white dwarf exceeds the Chandrasekhar mass.

19-5. If many novae are recurrent, then the number that we count overestimates the number of stars that "go nova" in our galaxy.

19-6. It is certainly true that the white dwarf star must have started out with the larger amount of mass. It probably did undergo a stage of mass transfer onto the other, initially smaller star. The evolution of that star then would have been sped up. In addition, the dumping of material consisting of heavier elements onto the surface of the initially smaller star might alter the chemical composition of its atmosphere.

19-7. Because supernovae are very unusual events, we might expect that if there is a remnant star in the Crab Nebula, it should appear to be peculiar or unusual compared to most stars. In fact, Fritz Zwicky, applying this principle, actually identified the remnant star of the Crab Nebula in the 1930s. Also, galactic objects close together in space usually have about the same radial velocity due to their motion around the center of the Milky Way.

19-8. Heavy elements are thought to have been produced in the interiors of massive stars. The Crab Nebula confirms this production, because we see these elements in the remnant, and because of the elements' dispersal into space.

19-9. We would expect heavier elements to be less abundant than lighter elements.

19-10. Type II supernovae result from the collapse of a single massive star as a natural part of its evolution. Theory tells us that abundant neutrinos will be produced during the core collapse, and that they will aid in blowing the star apart. Because the outer layers of the precursor were hydrogen, the spectrum will include hydrogen. Type I supernovae, by contrast, result when matter from one star is dumped onto a white dwarf, whose mass is increased above the Chandrasekhar limiting mass. Because the white dwarf contains little hydrogen, the spectrum will not contain hydrogen.

19-11. Some process other than the addition of helium nuclei must be at work in the production of ^{17}O and ^{13}O. The addition of neutrons may be one such process.

19-12. Probably most of the oxygen is produced by reactions between helium nuclei and the ^{12}C produced by the triple-alpha process.

19-13. The four-minute difference is the difference between the solar and sidereal day. It is reasonable to conclude that the emission is from an extraterrestrial source.

19-14. The Crab pulsar is observed to be slowing, and the Crab Nebula itself emits more energy than can be accounted for. Using the observed slowdown of the pulsar, and the expected size of a neutron star (say, 20 km), astronomers could compute the amount of energy lost as the pulsar rotation slowed. This energy loss was nearly identical to the luminosity of the Crab Nebula.

19-15. Presumably the Vela pulsar, being as much as 10,000 years old, is much older than the Crab pulsar.

19-16. If Earth is not in the region swept out by the beam from a pulsar, we may not see or notice the beam. This, in turn, would mean that we would *undercount* the number of pulsars. It might be possible to determine if this is the case by comparing the ages and number of known pulsars with the expected number of supernovae per century in our galaxy. Regrettably, this is difficult to do because the supernova rate itself is very uncertain.

19-17. As a rotating neutron star slows down, it loses rotational energy. Eventually, it will reach a point at which it has too little energy to accelerate the charged particles to near-light speeds. When this happens, the pulsar would no longer emit radio emission.

19-18. Conservation of angular momentum.

19-19. Wien's law tells us the X-ray region.

19-20. Theory indicates that the limiting mass for white dwarfs is about 1.4 solar masses. The neutron star limit is between two and three solar masses; it is more uncertain because it depends on assumptions that are less well understood.

PART FIVE

Discovering the Nature and Evolution of Galaxies and the Universe

The stars discussed in Part Four are grouped together into galaxies, the building blocks of the observable universe as a whole.

Our study of galaxies begins at home in our own galaxy, the Milky Way. In Chapter 20 we learn how astronomers have determined our location within our galaxy, and we examine the optical and radio evidence that our galaxy has a spiral shape. In discussing how our galaxy's mass is determined, we find that most of the material in the Milky Way is in some invisible form. Differences in the chemical content and orbital properties of stars are discussed and used to probe the formation and chemical evolution of the Milky Way. Because gas and dust between the stars play an important role in the Galaxy's structure, and because they profoundly affect observations, the interstellar medium is further discussed here.

The study of galaxies beyond our own is the subject of Chapter 21. After learning about the different shapes of galaxies, we investigate how their distances are found. This discussion ends with the all-important Hubble relation, which connects the observed motions of distant galaxies to their distances. Using the observed properties of different types of galaxies, we describe current ideas on how they form and evolve. We also discuss the observation that most galaxies are found in groups and clusters, and describe how such environments may affect the lives of galaxies.

Because not all galaxies are like those discussed in Chapter 21 (which are what we will call *normal* galaxies), Chapter 22 examines the others, which are usually called *peculiar* galaxies. Some of the peculiarities come about from galactic collisions while others emit tremendous amounts of energy that astronomers understand as resulting from the presence of supermassive black holes in their centers. Much of Chapter 22 concerns our attempts to understand the physical processes responsible for these vast amounts of energy.

In a sense, all discovery in astronomy is aimed at understanding the origin and evolution of the universe, the subject we discuss in Chapter 23. The purpose of this chapter is to present the generally accepted cosmological ideas and the evidence in favor of them. An important goal of this chapter is to provide you with an understanding of the evidence that the general idea of the big bang theory of cosmology is valid. The concepts of the geometry of curved space are also presented, along with current ideas concerning the fate of the universe.

The Milky Way: Our Galaxy

20

Be it ever so humble, there's no place like home.

John Howard Payne (from the opera Clari, The Maid of Milan, 1823)

Thomas Wright is generally credited with the first published statements, in 1750, describing our galaxy as a lens-shaped, or disklike, system of stars. He also suggested that the Sun and other stars travel in extended orbits about the *Universal Center of Gravitation,* which we would later call the *galactic center.* His thinking was both bold and correct, for although Galileo's telescopic observations over a century earlier had shown the Milky Way to be composed of many faint stars, further progress had been slow. By the 1780s, William Herschel had confirmed, using systematic star counts, that the Milky Way galaxy[1] was shaped like a lens, and he popularized the notion in his writings.

We have already examined various component objects within our own stellar system, which we call the **Milky Way,** because when viewed from a dark location the stars are so bright they look like milk spilled across the sky. We see these objects through a filter of dust that permeates the Milky Way. The growth of technology since World War II has given astronomers tools to penetrate the dust—detectors and telescopes sensitive to radiation at radio, infrared, and millimeter wavelengths. Technology has also made it possible to assemble bits and pieces of knowledge into a coherent overall picture of our galaxy.

But even today many questions about the Galaxy are difficult to answer because of our location *inside* it. One motivation for studying other galaxies (the topic of the next chapter) is to understand the Milky Way better.

Can we make an educated guess as to the shape of our galaxy? Looking at other galaxies, we find that they consist mostly of two types: spirals (**Figure 20-1a**) and ellipticals (**Figure 20-1b**). Spirals appear flattened, with a bulge in the center and extensive regions of gas and dust. Ellipticals appear to have a more uniform appearance and to lack extensive dusty regions. Because images of our night sky show it to be filled with dark, dusty regions (**Figure 20-2 and Figure 3-16**), we may hypothesize that our galaxy is a spiral. Investigating this hypothesis and understanding our location within the Galaxy are among our goals in this chapter.

20.1 Our Place in the Milky Way

Where are we located within our galaxy? How big is it? How rapidly are we moving? How long does it take for us to complete one orbit around the galactic center? Do other stars near us have similar motions? What is our galaxy's mass? The answers to these questions, which we now seek, provide fundamental information about our home galaxy.

Chapter opening photo: A nebula in the Milky Way.

[1]Whenever we talk about "the galaxy" in reference to the Milky Way, it is capitalized.

THE DISTRIBUTION OF GLOBULAR CLUSTERS

In Chapter 3 we described attempts by Galileo. Herschel and Kapteyn to determine the structure of the Milky Way by means of star counts, and we discussed how the dust in the Galaxy made those results thoroughly misleading. These studies indicated that the Galaxy was about 10,000 light-years in diameter, with the *Sun in the center.*

In the early 1920s, before astronomers had figured out a way to correct for the effects of dust, Harvard astronomer Harlow Shapley was already pursuing a different line of research. He was studying globular clusters, the giant spherical star clusters containing approximately 100,000 stars (and introduced in Chapter 18 in the context of stellar evolution). Figure 18-31 shows a photo-

(a)

(b)

Figure 20-1 (a) A spiral galaxy. (b) An elliptical galaxy.

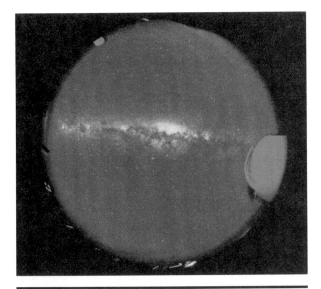

Figure 20-2 A fisheye view of the sky showing the Milky Way with its strong absorption by dust along the galactic plane.

graph of the globular cluster 47 Tucanae, which is easily observed in the southern sky. There are around 200 globular clusters in the Galaxy.

Decades before Shapley began his work, Herschel had noted that most globular clusters are located in one half of the sky, with many being found in the direction of Sagittarius. Shapley hypothesized that globular clusters are distributed evenly around the center of the Galaxy, so that determining the distribution of these clusters would reveal both the extent and the center of the Milky Way. In order to test this inspired guess, Shapley therefore needed to find distances to individual globular clusters. He exploited the fact that the Cepheid variables discussed in Chapter 18 have a relationship between their period and intrinsic luminosity and can thus be used for measuring distances. The details of the technique are discussed in the next chapter.

Shapley assumed that the short-period variables found in globular clusters were Cepheids and thereby determined the distances to relatively nearby globular clusters. Unfortunately, many globular clusters were too distant for these variable stars to be detected by Shapley. To find distances to these remote clusters, Shapley assumed that all globular clusters had similar physical sizes. We know, through the angular size formula of Chapter 3, that the observed angular diameter of a cluster depends on this physical size and its distance from us. By combining observed angular diameters and distances to nearby globular clusters, Shapley determined the average physical size of these clusters. By measuring the apparent angular diameter of remote clusters, the assumption that all globulars have the same physical size allowed Shapley to estimate their distances. This kind of indirect approach is common in distance determinations as we discuss in the next chapter. The method is summarized in **Figure 20-3**.

Inquiry 20-1

What assumptions are made in determining the distance to a cluster in this way?

Inquiry 20-2

What is the physical size of a globular cluster at a distance of 5,000 ly if its observed angular size is 0.057 degrees? (*Hint:* Use the angular size formula.)

(a) To find the *distance* to a cluster containing RR Lyrae stars:

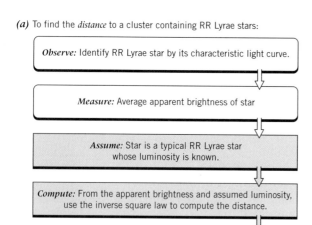

Observe: Identify RR Lyrae star by its characteristic light curve.

⬇

Measure: Average apparent brightness of star

⬇

Assume: Star is a typical RR Lyrae star whose luminosity is known.

⬇

Compute: From the apparent brightness and assumed luminosity, use the inverse square law to compute the distance.

⬇

Repeat: For all RR Lyrae stars in a given cluster: average the distance to find the distance to the cluster.

(b) To find the *linear size* of a cluster:

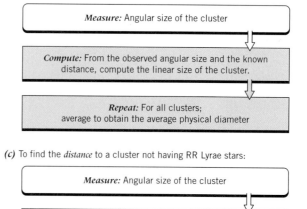

Measure: Angular size of the cluster

⬇

Compute: From the observed angular size and the known distance, compute the linear size of the cluster.

⬇

Repeat: For all clusters; average to obtain the average physical diameter

(c) To find the *distance* to a cluster not having RR Lyrae stars:

Measure: Angular size of the cluster

⬇

Assume: Cluster has the average physical diameter.

⬇

Compute: From the observed angular size and the assumed physical diameter, compute the distance to the cluster.

Figure 20-3 Summary of distance and size determination for globular clusters. *(a)* For clusters having RR Lyrae stars. *(b)* Finding a cluster's physical size. *(c)* Whenever no RR Lyrae stars are observed in a cluster.

Inquiry 20-3

What assumption did Shapley make when he estimated the distance to the farthest globular clusters from their apparent sizes?

Shapley found that some of the globular clusters were located at great distances from the Sun, and that the Galaxy must therefore be much larger than had been previously thought. More significantly, he found that these clusters fall into a roughly spherical distribution centered on a location far from the Sun (**Figure 20-4**). Consequently, the assumption that globular clusters are evenly distributed about the center of the Milky Way led Shapley to conclude that the Sun was not at the center of the Galaxy, contrary to the conclusion of Kapteyn and others. The excess of globular clusters noted by Herschel in Sagittarius marks the direction of the galactic center. Shapley is often regarded as a twentieth century Copernicus in the sense that he removed our Solar System from the privileged central position within the Galaxy.

In closing this discussion, it is worth noting that Shapley's numerous assumptions did lead to some systematic uncertainties. Most importantly, the variable stars that Shapley observed in nearby globular clusters were not Cepheids but RR Lyrae stars (Chapter 18) as indicated in Figure 20-3. Since RR Lyraes are less luminous than Cepheids

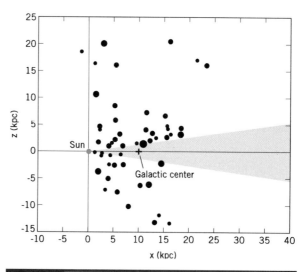

Figure 20-4 The distribution of globular clusters as determined by Harlow Shapley. Note that the center of the distribution is not at the Sun.

(see Figure 21-8), this led Shapley to overestimate the size of the Milky Way by about a factor of 2. Modern measurements place the Sun about 28,000 light-years from the galactic center.

Inquiry 20-4

What principle would have guided Shapley in placing the center of the Galaxy at the center of the observed distribution of globular clusters rather than at the center of the observed distribution of stars?

The Motion of the Sun Around the Galaxy

The Sun orbits around the center of the Galaxy. Its motion can be determined by observing the Sun's movement in relation to something stationary. The problem is, nothing in space is motionless! However, by observing the apparent motions of many globular clusters and distant galaxies distributed around the sky, we find that the Sun moves with a speed of about 220 km/s around the center of the Milky Way.

Knowing the distance of the Sun from the galactic center we can find the distance it travels in one rotation. Because we now know how fast the Sun moves, we can find how long it takes. The resulting galactic year is about 230 million years.

The Mass of Our Galaxy

We now have enough information to estimate the Milky Way's mass. We know that the period of the Sun around the galactic center is about 230 million years (2.3×10^8 years). We also know that the distance from the Sun to the center of the Galaxy is 28,000 light-years, or about 1.8×10^9 astronomical units (there are about 63,000 astronomical units in a light-year). With these data, we may solve for the mass of our Galaxy using Newton's version of Kepler's third law (Section 5.5) as follows

$$M_{\text{gal}} + M_{\text{sun}} = \frac{\text{Distance}^3}{\text{Period}^2}.$$

The Sun's mass is negligible compared to that of the Galaxy, so we may cancel it out of the equation and get

$$M_{\text{Gal}} = \frac{(1.8 \times 10^9)^3}{(2.3 \times 10^8)^2} = 1.0 \times 10^{11} \text{ solar masses.}$$

If we assume that an average star has one solar mass, then our result translates into approximately 100 billion stars. This is a rather crude assumption, but to within a factor of two or so our calculation is reliable.

The Motion of Other Stars

Because we know the size of the Milky Way and the speed of the Sun around its center, and we have an educated guess about its shape, we can now ask: How are the other stars near the Sun moving? The stars whose motions around the Galaxy are similar to that of the Sun are moving along with the Sun in the **galactic disk**, a flattened distribution about 1,000 ly thick, containing most of the observable matter in the Galaxy. The galactic disk therefore defines a plane in the same way that the ecliptic defines a plane in the Solar System.

The Doppler effect gives us a means to determine the traffic patterns obeyed by stars in different locations of the Galaxy. Because the motions of objects to some extent reflect their origins, studies of stellar motions using the Doppler effect provide information that helps us unravel the story of how the Galaxy evolved.

If the Galaxy rotated like a solid wheel, then no star would ever change its position relative to any other; like wooden horses on a carousel, they would remain at constant distances from each other. However, we observe Doppler shifts in the light of other stars. Remember, a Doppler shift provides information about relative motion *toward* or *away from* the observer. It follows that if the Sun and stars are moving relative to one another, their distances must also be changing. A natural question to ask is whether the stars move around the center of the Galaxy in a manner analogous to the motion of the planets around the Sun, with the stars farthest out moving the slowest.

Figure 20-5 illustrates how we might answer this question for stars in the Sun's neighborhood. For simplicity, the stars are shown as having circular orbits. We will hypothesize that stars closer to the center of the Galaxy move faster in their orbits, as do planets in the Solar System. A star at position A would then show a Doppler shift to longer (redder) wavelengths because the Sun would be leaving it behind. A star at F, on the other hand, would show a Doppler shift to shorter (bluer) wavelengths, indicating that the Sun and the star are approaching one another. This is so because the star at F moves more rapidly than the Sun and overtakes it. Stars in the direction of the galactic center, toward point G, show no radial velocity because they move parallel to the Sun, neither approaching nor receding from it. A star at E will also exhibit no radial velocity, because it is at the same distance as the Sun from the galactic center and therefore orbits the Galaxy with the same velocity.

> **Inquiry 20-5**
>
> What radial velocity, relative to the Sun, would a star have if it were located at point *B* in the diagram? Point *C*? Point *D*? Point *H*? Explain your reasoning.

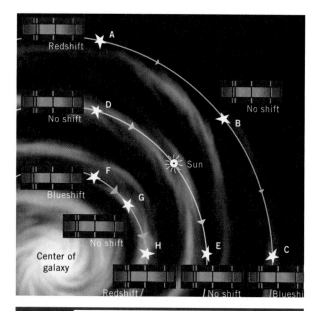

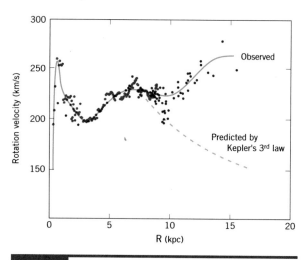

Figure 20-6 The rotation curve of the Milky Way.

Figure 20-5 The motions of stars near the Sun produce various Doppler shifts in stellar spectra, depending on the object's location relative to the Sun. Note the position of the absorption lines relative to the comparison lines.

> ## You should do Discovery 20-1, Galactic Rotation, at this time.

To compare observations with theoretically predicted motions, astronomers graph the rotation speeds of stars and gas clouds at various points throughout the Galaxy. The resulting graph, called a **rotation curve**, is shown for the Milky Way in **Figure 20-6**. The first feature we notice is that in the central regions, the velocity change with distance is approximately a straight line. This is what you would obtain if you drew a rotation curve for a rigid body such as a rotating plate: the farther from the center, the greater the speed. Thus the interior regions of the Milky Way rotate like a solid body. That does not mean the nuclear region is solid, of course, but it does tell us that material is less centrally concentrated than in the Solar System where most of the mass is in the Sun. Similar rotation curves are also observed in the central regions of other spiral galaxies.

Beyond the location of the Sun the orbital motions are observed to remain nearly constant, rather than decreasing with distance as expected from Kepler's laws. Whereas deviations from Kepler's laws in the nuclear region of the Galaxy

were expected and easily understood, the constant orbital motions beyond the Sun came as a shock to astronomers. This is because for objects orbiting at large distances from the galactic center, nearly all of the visible material in the Milky Way lies within their orbit. Consequently, like planets orbiting the Sun, we would expect to see a decrease in rotation velocity in accordance with Kepler's third law. We illustrate this point in Figure 20-6 where the predicted rotation curve (i.e., that produced by the mass of visible material) is compared to the observed one. Astronomers have concluded that the Milky Way galaxy is surrounded by an extensive dark **galactic halo**. Because the total amount of visible matter in the halo, such as globular clusters and stars, falls far short of the amount required to explain the observed rotation curve, astronomers hypothesize the existence of unseen material we call **dark matter**.

DARK MATTER

The distribution of dark matter in the Galaxy is difficult to determine simply because it appears to extend beyond the region where we have a sufficient number of visible objects to probe it. However, observations of the motion of distant globular clusters, gas clouds, high-velocity stars, and small satellite galaxies suggest that this dark halo is roughly 10 times more extended than the stellar disk, and that its distribution is roughly spherical. The total mass in this dark halo is estimated to be around 10^{12} solar masses, or about 10 times greater than all the visible material in the Milky Way. This general picture is supported by observations of the rotation curves of other spiral galaxies discussed in the next chapter.

The nature of this dark matter has proven even more difficult to pin down. Early candidates included very faint stars, brown dwarfs, and stellar remnants such as white dwarfs, neutron stars and black holes. Such objects are often referred to as *MACHOs* for MAssive Compact Halo Objects. Various arguments cast doubt on the viability of individual members of this list, such as the fact that black holes would accrete gas as they crossed the Milky Way disk and produce large amounts of X-ray emission. An ingenious experimental approach to address the entire class of MACHOs involves a prediction from Einstein's theory of general relativity called **microlensing**. When a MACHO passes in front of a more distant star, microlensing causes light from the background star to be magnified. Several different groups of astronomers have exploited this technique and many microlensing events have been observed. However, most astronomers have concluded that there are not enough such events for MACHOs to explain most of the Galactic dark matter.

The discipline of elementary particle physics may provide the most promising dark matter candidates. Known particles such as protons and electrons do not have the right properties; thus the halo dark matter is not hydrogen. However, physicists have suggested the existence of an entire zoo of particles that do have the right properties to make up the halo dark matter. Such particles must be massive (by elementary particle standards) and interact weakly with matter (otherwise we would have already detected them). Such particles are referred to as Weakly Interacting Massive Particles, or *WIMPs*. Our best evidence suggests that in the competition to be successful dark matter candidates, the WIMPs have beaten the MACHOs. We will return to the nature of dark matter in Chapter 23, since it is of great importance cosmologically.

HIGH-VELOCITY STARS

The majority of stars near the Sun move around the galactic center in nearly circular orbits at more or less the same speed of about 220 km/s. Relative to the Sun, however, the neighboring stars appear to be moving slowly. The situation is similar to cars speeding around a racetrack. Although cars at the Indianapolis 500 race are going more than 200 miles per hour relative to their pit crews, their speed relative to each other is small.

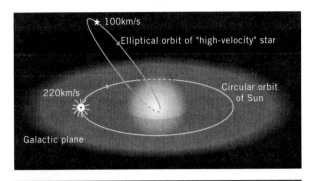

Figure 20-7 The orbit of a typical high-velocity star is highly elliptical and tends to lie outside the galactic plane.

Globular clusters, by contrast, move around the Galaxy in eccentric orbits inclined at random angles to the galactic disk. Such orbits cause most globular clusters to appear to be moving with high speeds relative to the Sun. For this reason, the globular clusters are called high-velocity objects. Similarly, stars that belong to the halo and whose tilted, eccentric orbits happen to carry them through the disk near the Sun will also have high speeds, similar to those of the clusters. Astronomers call these stars **high-velocity stars** because *relative to the Sun* they appear to move at high speeds.

How is it possible to determine whether a given star belongs to the halo or the disk? By observing its motion relative to the Sun. An object observed to be a high-velocity star belongs to the halo (**Figure 20-7**). We have been able to distinguish several thousand halo stars from disk stars on the basis of their apparent motions relative to the Sun. The ability to distinguish which part of the Galaxy a star belongs to is a prerequisite to understanding the formation and subsequent evolution of the Milky Way.

20.2 Interstellar Gas and Dust

The space between the stars, the **interstellar medium**, is not empty but contains atoms, molecules, dust grains, and cosmic rays. In Chapter 17 we saw that the interstellar medium provides the material from which stars form, and into which they lose matter by means of stellar winds, planetary nebulae ejection, and supernovae explosions.

We are now returning to a discussion of the interstellar medium because of a dual role it plays in studies of the Milky Way and other galaxies. First, the dust scattered throughout the Galaxy hinders the study of the *Milky Way's* structure through the extinction of star light passing through it. If we were to neglect the effects of the interstellar medium, our interpretations of observations would be subject to systematic uncertainties, as was the case for early star counts. Second, the gas and molecules scattered throughout the Galaxy provide methods that allow astronomers to study the Galaxy's structure. The known interstellar medium amounts to about 10 percent the stellar mass of the Milky Way. A study of the interstellar material itself tells us much about the conditions present in the Galaxy today and at times in the distant past. The following sections provide information useful in understanding the structure of the Milky Way, as well as information on the components of the interstellar medium itself.

INTERSTELLAR DUST: EXTINCTION OF STARLIGHT

Images of emission nebulae frequently reveal the existence of dust along with the gas. In **Figure 20-8** of the Trifid Nebula, the dark lanes across the middle as well as the mottled structure are caused by dust particles that block radiation coming from behind. A famous example of obscuration by dust is illustrated in **Figure 20-9**, a dense cloud of dust called the Horsehead Nebula.

One effect that dust has on passing radiation is illustrated in **Figure 20-10**. Some of the photons passing through dust clouds are scattered or reflected from the dust grains. These photons generally leave the dust in a direction different from the radiation's original direction, thus the star's intensity is diminished. Other photons are absorbed. Their energy goes into heating the dust grains slightly.

Figure 20-9 The Horsehead Nebula, a region of gas and dust in the constellation of Orion.

Figure 20-8 The Trifid Nebula in Sagittarius is a good example of an emission nebula—a region where stars, gas, and dust interact.

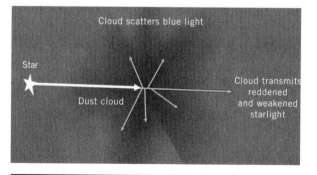

Figure 20-10 Scattering of light by dust particles both dims and reddens starlight as it passes through the interstellar medium.

Inquiry 20-6

What would be the effect on our estimate of the distance to a star if its light was affected by an unknown amount of intervening dust?

Inquiry 20-7

If our estimates of distances to stars were adversely affected by an unknown amount of dust, our knowledge of what stellar properties would be affected?

INTERSTELLAR DUST: REDDENING

The amount of light scattering depends on wavelength and the size of the dust grains. Although radiation of all wavelengths is scattered, shorter (bluer) wavelengths are more readily scattered than longer (redder) wavelengths (see Figure 20-10). You see the effect of the scattering of blue light by dust and molecules every clear day—scattering is what makes the sky blue. Removal of blue photons by the scattering process is enhanced at sunset because sunlight moves through a greater path length in the Earth's atmosphere, thus causing the Sun's reddish color.

Likewise, as starlight passes through interstellar dust, blue photons are scattered away, thus causing the light reaching an observer to be redder than it would have been without dust. The reddening of starlight gives us a clue to the size of the dust grains, because the particle size determines the extent to which light of different wavelengths is scattered. For example, a beam of light passed through an assembly of particles the size of bowling balls would have its intensity lessened, but the decrease would not depend on wavelength. However, when the size of the scattering particles is about the same size as the wavelength of the light, blue light is scattered more effectively than red light. For this reason, astronomers deduce that the dust particles in interstellar space must average about 3,000 Å (angstroms, or 3×10^{-7} m) in diameter.

Inquiry 20-8

What would be the effect on our estimate of a star's temperature if its light was reddened by dust, compared to a similar estimate for an unreddened star?

Early in the twentieth century, astronomers could not tell whether a star was red because it was cool or because its light had passed through interstellar dust. Unscrambling these two effects was not possible until a thorough understanding of stellar spectra was developed. Once spectra were understood, astronomers could deduce the temperatures of stars from the absorption lines in their spectra rather than from their color. Knowing the star's temperature, astronomers could then predict what its *true* color should be. If its *measured* color did not match its predicted true color, the difference in the colors could be attributed to reddening by interstellar dust.

From the amount by which the starlight was reddened, astronomers could estimate the total amount of absorption and scattering, and therefore the star's true brightness. The correct distance could then be determined. In this manner, astronomers began to account for the dust and to gain a truer picture of the distribution of stars in the Galaxy.

INTERSTELLAR DUST: REFLECTION NEBULAE

The bluish "haze" near the Trifid Nebula in Figure 20-8 results from starlight reflected from dust surrounding a hot star. Light reflected from dust produces a **reflection nebula**. Astronomers know the light is reflected from dust because the light's spectrum is that of the bright illuminating star and because the light is polarized.

Inquiry 20-9

Why is the nebulosity bluer than the star from which it comes?

An important piece of information about interstellar dust particles is available only from reflection nebulae. By studying the reflected light, we discover that the particles are highly reflective. In fact, a dust grain reflects approximately 90 percent of the light that falls on it. This demonstrates that the particles have smooth surfaces, and that whatever their chemical composition, they are probably covered with an outer coating of some type of ice (see **Figure 20-11**). Note that *ice* does not have to mean water ice. A number of simple compounds (methane, ammonia) could form ices at the temperatures we see in these nebulae.

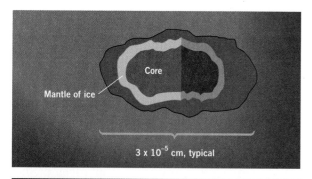

Figure 20-11 A suggested model for a typical interstellar dust grain.

Scattering by dust also polarizes light. By analyzing the polarization, we can determine more about the particles' sizes and chemical composition. The polarization results confirm the idea that the dust grains have an icy surface with, perhaps, graphite (a crystalline form of carbon) in the interior.

INTERSTELLAR DUST: POLARIZATION OF STARLIGHT

Astronomers obtain additional information about interstellar dust particles by studying the polarization exhibited by starlight passing directly through the dust. Starlight that has been reddened after passing through long paths of dust is also generally polarized to some degree, indicating that the polarization is indeed caused by the interstellar dust grains. To produce polarization, the dust particles must be somewhat elongated in shape and must also have their long axes aligned with one another. Only then will the light become polarized.

Inquiry 20-10

Why would this mechanism not work if the dust grains were spherical?

This whole picture requires something that "lines up" the dust grains over large regions of interstellar space. Such an agent can be found in the interstellar magnetic field, to be discussed more fully later in this chapter. Because our focus now is on the dust grains themselves, we can say that the existence of the polarization tells us something about the shape of the grains (they are elongated) and something about their chemical composition (at least some of the atoms in the grain must be iron or an element that can be magnetized). **Figure 20-12** summarizes the production of polarized light by magnetically aligned grains.

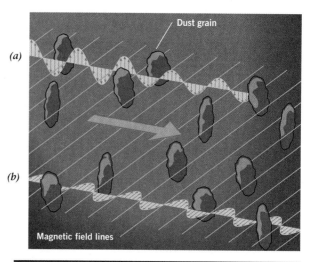

Figure 20-12 Interstellar grains tend to be aligned perpendicular to magnetic field lines. Radiation with its electric-field vibration lined up with the long axis of the grains *(a)* tends to be removed from the beam. Radiation not so aligned *(b)* tends to pass through undiminished. This absorption produces polarization.

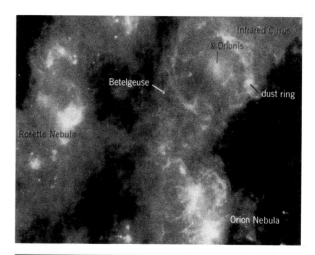

Figure 20-13 An infrared view from IRAS of the Orion region. Bright areas are where star formation occurs. Note, too, the wispy infrared cirrus.

INTERSTELLAR DUST: THERMAL RADIATION

Dust heated by the starlight it absorbs radiates energy in the infrared spectral region. By observing the distribution of infrared radiation throughout the Milky Way, astronomers are able to determine the distribution of dust in the Galaxy. In addition, information on the properties of dust itself is obtained. **Figure 20-13** shows a false-color image of the region containing the constellation

Orion as viewed by the *Infrared Astronomy Satellite* (*IRAS*). The brightest areas, corresponding to star formation regions, are bright because hidden protostars heat the surrounding dust, which in turn radiates energy in the infrared. Also shown are filaments of dust that are reminiscent of cirrus clouds; for this reason, these features are called infrared cirrus.

INTERSTELLAR DUST: CHEMICAL COMPOSITION AND FORMATION

What is interstellar dust made of? A dust grain is an aggregation of several hundred to perhaps several thousand atoms bound together into some sort of solid matrix. Such a particle does not produce simple spectral lines like an atom. Yet knowledge of the composition of the dust particles in the Galaxy is crucial to understanding how and where they are formed, and to determining the true chemical composition of the material in the interstellar medium.

In the early years of studies with orbiting ultraviolet telescopes, astronomers discovered a spectral feature at 2,200 Å that laboratory studies showed matched that of graphite, which is made of carbon. From these studies we assume that graphite was at least one of the components of interstellar dust. This result is consistent with the polarization data discussed previously.

Some stars are surrounded by dust shells that heat up and then reradiate at infrared wavelengths. While the radiation emitted generally has a continuous spectrum, on some occasions the spectrum shows a distinct hump or excess of radiation above the stellar spectrum at around 100,000 Å (see Figures 17-16 and 18-15 for examples). Comparison with laboratory spectra suggests that this is due to the presence of silicate minerals in the dust cloud. An example of a common silicate is SiO_2 (quartz).

How could grains containing carbon and silicon be formed? Graphite grains are thought to form in the carbon-rich atmospheres of certain giant pulsating stars (the long-period variable stars mentioned in Chapter 18). During part of their pulsation cycle, these stars expand and their outer atmospheres cool off, which allows carbon atoms to stick together and form graphite flakes. Later on, the star contracts and its outer regions heat up. Increased radiation pressure from the star exerts pressure on the carbon flakes and literally pushes them out of the stellar atmosphere and into interstellar space.

Silicate grains have been discovered in the atmospheres of long-period variables, thus demonstrating that interstellar dust probably has no single, uniform composition or origin. This is, in all probability, the reason why it has proved so difficult to make a definitive model of an interstellar dust particle.

Inquiry 20-11

The energy that holds one atom or molecule to another in a dust grain is not very large, and it may easily be exceeded by the random energy of collisions in a gas that is a thousand degrees or more in temperature. What does this imply about the fate of dust grains in a hot cloud?

INTERSTELLAR GAS

The interstellar medium also contains large amounts of gas. Most of the gas is hydrogen, but other elements are also present. For example, sometimes stellar spectra exhibit lines of sodium and calcium that are formed in interstellar gas clouds. The interstellar lines are distinguishable from those formed within the star itself because the interstellar lines are extremely narrow. In these **diffuse interstellar clouds** the hydrogen and other chemical elements are found as atoms rather than molecules. In other words, individual hydrogen atoms float around without bonding to other atoms.

The coldness of interstellar space means that much of the hydrogen is not atomic but molecular, *provided* that gas densities are sufficiently high. Higher temperatures would break the molecules apart through violent collisions, whereas lower densities make it less likely hydrogen atoms collide and form molecules. Low densities also leave any molecules that do form vulnerable to dissociation by the ultraviolet radiation that permeates the Galaxy. These molecules are not broadly scattered around the Galaxy but form the giant molecular clouds from which stars form (Chapter 17). Other molecules, too, are present in these clouds. A variety of different molecules are observed in the interstellar medium, including formaldehyde and ethyl alcohol. Table 17-1 lists others, which were discussed when we looked at star formation.

When interstellar hydrogen is close to a hot star, the hydrogen is ionized and forms an HII region.

Inquiry 20-12

Why would spectral lines formed in diffuse interstellar gas clouds be narrow? (*Hint:* From Chapter 14, why are lines of dwarf stars broader than those formed in supergiant stars?)

Inquiry 20-13

Why does hydrogen gas near a *hot* star produce an HII region?

Inquiry 20-14

We frequently find several interstellar absorption lines in a spectrum, all due to the same atomic transition, but at slightly different wavelengths. What conclusion do you draw from this observation?

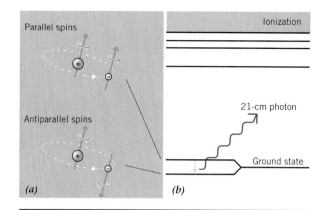

Figure 20-14 *(a)* The proton and electron can have their spins in the same (parallel) or opposite (antiparallel) direction. *(b)* Each situation has a different energy, so the energy level of the ground state is split in two. A transition between these states produces a photon corresponding to a wavelength of 21 cm.

Interstellar lines often show multiple components occurring at slightly different wavelengths. Each component is produced by a diffuse gas cloud at a different distance and moving with its own velocity. Radiation from each cloud is Doppler shifted by a different amount. Because of absorption by dust, we cannot satisfy our goal of determining the overall structure of the Milky Way when looking only at visible wavelengths. Fortunately, the vast amount of neutral hydrogen within the Galaxy has some properties we can exploit to study the Milky Way further.

21-CM HYDROGEN EMISSION

While pacing the beaches of occupied Holland during World War II, the young Dutch astronomer Hendrik Van de Hulst had ample time for reflection. He wondered whether there was any way to observe the cool neutral hydrogen gas in its atomic form that must be so abundant in the Galaxy. After analyzing the details of the hydrogen-energy-level diagram, he concluded that the ground state of neutral hydrogen was not a single level, as we drew it in Chapter 13, but was in fact split into two closely spaced energy states.

Researchers in the 1920s and 1930s had been able to show that both the proton and the electron behave as if they were spinning like miniature tops (**Figure 20-14***a*). As a result, each of these charged particles also has a tiny magnetic field. The magnetic fields of the proton and electron interact and affect the energy levels of the hydrogen atom. For

example, if the proton and the electron are both spinning in the same direction—a situation we describe by saying the spins are *parallel*—the atom has a slightly higher energy level than if they are spinning in opposite directions, which we describe as *antiparallel* (**Figure 20-14***b*). This gives the hydrogen atom another way to radiate, because if an atom in the antiparallel ground state collides with another atom, it can absorb a small amount of energy from the collision and end up in the parallel ground state. This is a metastable energy level, not unlike those discussed earlier (Chapter 13, Section 3), and the atom stays at the high energy state for a long time before ultimately making a transition downward.

Because the energy levels for the parallel and antiparallel states are close to one another, the emitted photon has a low energy. A low-energy photon has a low frequency or a long wavelength—in this case 21 cm in the radio part of the electromagnetic spectrum. The significance of van de Hulst's work was his prediction that all hydrogen atoms absorb and emit at the radio wavelength of 21 centimeters.

Observation of **21-cm radiation** revolutionized our knowledge of the Milky Way and other galaxies for several reasons. First, it is a spectral feature of the most abundant element in the universe. Second, it indicates the presence of *neutral atomic hydrogen gas*, which had previously been difficult to study. Third, this radiation is not absorbed by interstellar dust but passes easily through it.

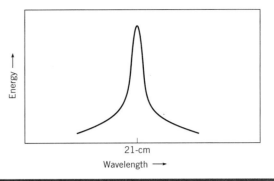

Inquiry 20-15

Why should 21 cm radiation pass so easily through interstellar dust, while visible light is so strongly absorbed and scattered? (*Hint:* Recall the discussion of diffraction in Chapter 11 and the sizes of dust particles earlier in this chapter.)

Figure 20-15 The narrow core of the line comes from clouds of cold gas, but the broad "wings" must come from hotter material.

The passage of 21-cm radiation through interstellar dust is enormously significant because it means that 21-cm radiation emitted anywhere in the Galaxy can reach Earth and be detected by radio telescopes. The discovery of 21-cm radiation made it possible for the first time to study the structure of nearly the entire Galaxy, unimpeded by the presence of absorbing interstellar dust particles. Later in this chapter we will see how it is used to study the overall structure of the Milky Way.

THE STRUCTURE OF THE INTERSTELLAR MEDIUM

To complete our interlude on the interstellar medium, we examine its structure. An analysis of interstellar absorption lines shows that diffuse interstellar gas clouds come in a range of sizes, with masses varying from just a few solar masses up to thousands of times the mass of the Sun. But for simplicity, it is frequently useful to speak of an average, or typical, cloud, which can be characterized as having a diameter on the order of 50 light-years and a total mass of perhaps 400 solar masses. The gas density averages about 10 atoms per cubic centimeter—a very good vacuum using Earth's standards, though more dense than the average of 1 atom per cubic centimeter for interstellar space as a whole. These clouds have temperatures of about 100 K. There are many such clouds; a volume in the Galaxy equivalent to a cube 1,000 light-years on each side contains approximately 1,000 such clouds, occupying approximately 2 percent of the volume of the cube.

Observations of 21-cm radiation tend to show emission profiles similar to that sketched in **Figure 20-15**. The sharp emission spike comes from a cool diffuse cloud. The broader shoulder of emission covering a greater range of wavelengths comes from atoms at temperatures of about 1,000 K. The hotter atoms move with higher speeds, which then produce larger Doppler shifts that broaden the spectral line. From these line profiles we therefore conclude that the space between the diffuse clouds is filled with a hotter, lower-density gas, which we call the **warm intercloud gas**.

Also embedded in the warm intercloud gas are the cold molecular clouds in which stars and star clusters form. Like the diffuse clouds, molecular clouds come in a range of sizes. Compared to their diffuse cousins, molecular clouds are larger, with typical diameters around 150 light-years. They are also about three orders of magnitude denser than diffuse clouds. The combination of a somewhat greater size and a much higher density implies that molecular clouds have a much greater mass than diffuse clouds, typically around 10^5 solar masses. Temperatures in these clouds are around 20 to 50 K.

Inquiry 20-16

The masses of typical star clusters in the Galaxy are several hundred to several thousand solar masses. These masses are considerably less than typical masses of giant molecular clouds, but star clusters are observed to form from such molecular clouds. What does this tell you about star formation in Milky Way molecular clouds?

The final component of the interstellar medium is so hot that it emits most of its radiation in the X-ray portion of the electromagnetic spectrum. Observations made with high-altitude rockets and satellites reveal this X-ray emission is produced throughout the Galactic disk from interstellar gas with temperatures in excess of a million Kelvin. This material is called *coronal gas* because its temperature is similar to that of the Sun's corona (a low-density gas surrounding the Sun).

Why does coronal gas have such a high temperature? Supernovae explosions, as we have seen, release tremendous amounts of energy in the form of photons and shock waves. With an average of one supernova exploding every 50 years or so, a continuous supply of energy goes into the interstellar medium. In addition, high-speed winds with velocities as great as 3,000 km/sec come from the hottest stars. All this energy churns the interstellar medium. Furthermore, these events produce what astronomers call interstellar bubbles and superbubbles, cavities filled with hot, rarefied gases scattered around the Galaxy. Thus, the interstellar medium is not a quiet place but one filled with high energy and activity.

Now that we have some understanding of the interstellar medium and its effect on observations, we can return to our study of the Milky Way's stellar components.

20.3 The Structure of the Milky Way System

Herschel and Kapteyn counted stars to determine the structure of the Milky Way. Their methods are basically valid even today, as long as we account for interstellar absorption.

> **You should do Discovery 20-2, The Distributions of Different Objects Around the Galaxy, at this time.**

From images of other galaxies, we see that the most luminous stars, as well as gas and dust, appear to be located in spiral-shaped arms. Given the evidence of extensive gas and dust within the Milky Way, we will hypothesize that ours is a spiral galaxy; that is, a galaxy with spiral arms, rather than a dustless elliptical galaxy. To detail the spiral structure of our galaxy, and to determine our location relative to the spirals, we must observe objects that we think lie within the spiral pattern. Furthermore, we require objects that may be seen at relatively large distances. Finding a variety of such objects is the subject of the next section.

Inquiry 20-17

Considering all the types of objects studied in the past few chapters, which types of objects might you expect to be (a) observable at large distances, and (b) located within the spiral arms?

OPTICAL EVIDENCE FOR THE SPIRAL STRUCTURE OF THE GALAXY

Objects that trace out the spiral structure of a galaxy are known as **spiral tracers**. Our immediate goal is to identify a number of spiral tracers that optical astronomers can use. The detailed study of the spiral nature of the Milky Way was pioneered by the Dutch astronomer Jan Oort, who was also famous for his studies of the origins of comets.

The first type of optical spiral tracer is the OB association, because O and B stars are young and expected to be in the thin galactic plane near the dust and gas from which they formed. Similarly, young, open clusters observed within both the Milky Way and other galaxies are located only in the thin galactic plane, where material for star formation is plentiful. (Because open clusters are confined to the plane of the Galaxy, they are often referred to as **galactic clusters**.) But because these stars are located near dust, their observed brightnesses must be corrected for the reddening and absorbing effects of dust. In this way, the spectroscopic parallax technique discussed in Chapter 15 provides distances that are free of systematic uncertainties caused by absorption of light by dust.

Because HII regions are associated with star-forming regions present in the galactic plane, they might be expected to be located in spiral arms. The locations of HII regions in the spiral galaxy M31 (**Figure 20-16**) agree with this view. We will therefore assume that HII regions in the Milky Way are tracers of spiral structure. Dark nebulae are also associated with star-forming regions and may be observed against bright stellar backgrounds at large distances. Therefore, they, too, may be useful tracers of spiral structure.

Finally, some of the luminous Cepheid variables are found within the spiral pattern. Their distances may be calculated using an important technique discussed in the next chapter. When a picture of

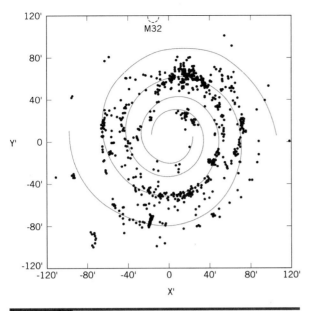

Figure 20-16 The distribution of HII regions in the nearby spiral galaxy in Andromeda, M31.

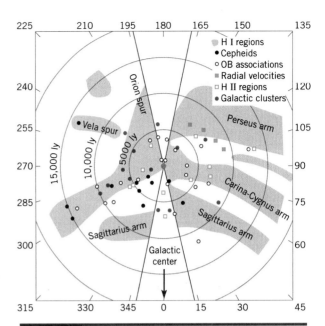

Figure 20-17 The distribution of various types of objects near the Sun, showing evidence for spiral structure (after Bok). The center of the Galaxy is located about 1.5 cm below the bottom of the figure.

the locations of all these types of objects was made, a striking result began to emerge: these objects appeared to be distributed along pieces of a spiral pattern.

Figure 20-17 is a diagram of the locations of the types of objects just described. The view is that of an observer above the Milky Way. Objects are not randomly placed but appear to fall onto pieces of spiral arms. However, these spiral fragments are far from being simple or well defined. Furthermore, because only objects close to the Sun can be observed at visual wavelengths because of absorption by dust, only that part of the spiral pattern located near the Sun can be discerned.

Inquiry 20-18

Astronomers have concluded that the primary location of star formation in our galaxy at the present time is in the spiral arms. What reasoning must have led them to this conclusion?

Inquiry 20-19

What type of astronomical object discussed in this chapter would you expect to see in the spiral arms as a result of the simultaneous presence of hot O and B stars along with the considerable amount of hydrogen gas?

RADIO EVIDENCE FOR THE SPIRAL STRUCTURE OF THE GALAXY

We have learned that neutral hydrogen emits and absorbs radiation at a radio wavelength of 21-cm, and that this radiation penetrates the light-absorbing dust present within the plane of the Milky Way. Observations at 21-cm are therefore useful for studying those parts of the Galaxy containing neutral hydrogen. Because 21-cm observations of other galaxies show the hydrogen to be located in a spiral pattern (**Figure 20-18** shows 21-cm contours for the galaxy M51), we can expect the hydrogen in the Milky Way likewise to lie in a spiral pattern. Mapping the distribution of 21-cm emission within the Milky Way is clearly the way to study the spiral arm structure.

Figure 20-19a indicates how the 21-cm-emission feature can be used to map out the spiral structure of the Galaxy. Two different clouds of neutral hydrogen, at positions A and B, will both emit 21-cm radiation. However, the spectral line emitted by the gas at A will be more intense because we are looking *along* the length of a spiral arm and therefore seeing through more gas than we do for cloud B. Cloud A will also show a larger Doppler shift to longer wavelengths, not only because A probably has a greater orbital velocity than B (being nearer the center of the Galaxy; recall Figure 20-6), but also because A's motion carries it directly away

Figure 20-18 Radio emission from hydrogen at 21-cm from M51 overlayed on an optical photograph.

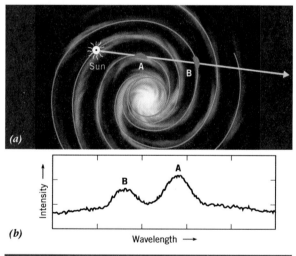

(a)

(b)

Figure 20-19 (a) Plotting the location of hydrogen gas in the Galaxy from 21-cm observations. The gas in different parts of the Galaxy will have different Doppler shifts and different intensities, enabling positions within the Galaxy to be estimated. (b) The composite emission line detected by a radio telescope.

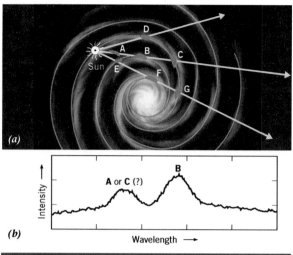

(a)

(b)

Figure 20-20 Resolving ambiguities concerning the distribution of gas requires an assumption that the gas is distributed in a regular way in the Galaxy. (a) Plotting the location of hydrogen gas. (b) The composite emission line.

from the Sun. The variation of intensity with wavelength of the spectral line that would be seen by the radio telescope is shown in **Figure 20-19b**.

There are some problems in applying these ideas to the real Milky Way. For example, the spectral lines formed by clouds *A, B,* and *C* in **Figure 20-20a** are shown in **Figure 20-20b**. The large bump in the spectrum at the longest wavelength is due to gas cloud *B*. However, it is not possible to tell just from the spectrum whether the shorter wavelength bump is due to gas at point *A* or at point *C* (although it would have to be at one of those two points because its velocity relative to the Sun is uniquely related to its distance from the center of the Galaxy). Gas at either point *A* or point *C* would show the *same* Doppler shift. In practice, ambiguities of this type must be resolved by assuming that the gas is distributed in a regular way in the Galaxy. For example, other observations in other directions (say in the direction of point *D*) might establish the existence of a continuous spiral arm going through points *D* and *C*.

Inquiry 20-20

Consider the line of sight *EFG* in Figure 20-20a. Notice that points *E* and *G* both lie in the arms and are the same distance from the galactic center. What problems might arise in interpreting the spectrum along this line of sight? What would the spectrum look like?

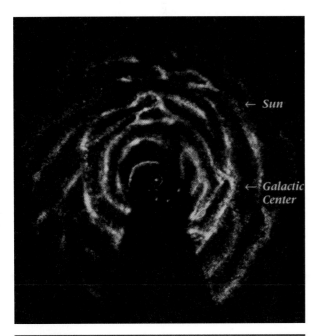

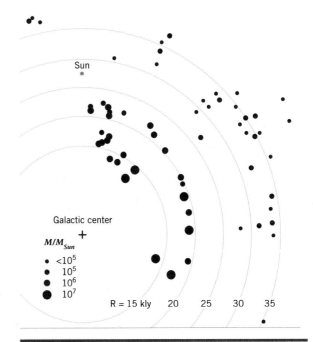

Figure 20-21 A map of the Galaxy based on 21-centimeter emissions. The blank sector is a region where accurate data cannot be obtained because the radial velocities are so close to zero.

Figure 20-22 The distribution of molecular clouds in the Milky Way.

Figure 20-21 illustrates the distribution of neutral hydrogen in our galaxy, based on 21-cm observations. The positions of the Sun and the galactic center are indicated. When drawn like this, the results are a little bit confusing. Rather than the obvious and straightforward spiral pattern that some external spiral galaxies exhibit, we see a much more complex maze of smaller arms. Is our galaxy really a spiral like the others we see, or is it fundamentally different? To what extent is this complex picture created by the ambiguities that were discussed earlier?

We have no way to resolve the ambiguities of our model near the galactic center, because all the spiral arms in that direction contain material that is moving tangentially to our line of sight and hence has no Doppler shift. There is no way we can tell how far away the material is or how many clouds there are along such a line of sight. We know there is hydrogen in this direction, but we cannot tell where it is located. If our Galaxy had shown a clearer pattern of spiral arms, we could have extrapolated that pattern into this unknown region; but given the observed distribution of gas there is no way to do this.

Further observational evidence of spiral structure comes from the distribution of molecular clouds. Because they emit at radio wavelengths,

their emissions also pass through the dust. The observed distribution of molecular clouds, shown in **Figure 20-22**, is also consistent with a spiral pattern and that of neutral hydrogen.

THE CAUSE OF SPIRAL STRUCTURE

We have ample evidence for spiral structure in our galaxy. The young, hot main-sequence stars and their HII regions arrange themselves in a spiral pattern; neutral hydrogen and molecular emissions both show evidence of spiral structure. The question we have not yet asked is: Why?

One early hypothesis was simply that the motion of the material in the Milky Way tended to create spiral structures. Consider the extreme case shown in **Figure 20-23**, in which a set of stars distributed in a straight line moves about the center of the Galaxy. Clearly the details of the evolution of this pattern depend on the nature of the rotation curve, but, with the exception of the extreme case of solid body rotation, stars close to the center make one complete orbit before stars further away have made the round trip. This is simply because more distant stars have further to travel. As a result, the initial straight line of stars tends to stretch out into "arms" that at first sight resemble spiral structure. However, on closer examination this simple model fails to explain the spiral structure that we currently observe.

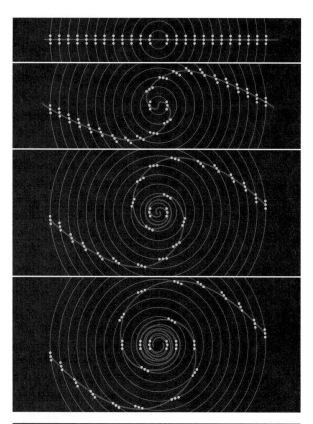

Figure 20-23 A long strip consisting of stars *(a)* would begin to wind up *(b)* and show spiral structure after some time if the inner part of the Galaxy rotates more rapidly than the outer part *(c, d)*.

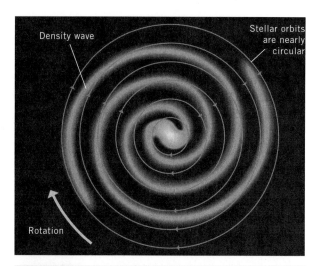

Figure 20-24 In the density-wave theory of star formation, interstellar material and gas and dust clouds overtake the wave from behind. The molecular clouds are compressed in the wave and some collapse into stars.

Why does this simple model fail? Because the older stars in our galaxy exceed 10 billion years in age, the galactic system itself is at least that old. We know the Sun is about 4.6 billion years old and orbits the center of the Galaxy in about 230 million years. This means that since it was formed, the Galaxy has rotated about 50 times and the Sun has revolved around it some 20 times. If the model were correct, then the arms of the Galaxy would be wound tightly around the galactic nucleus. In fact, however, the arms of most spiral galaxies are rather loosely wound, which contradicts the predictions of this hypothesis. What we have just described is often referred to as the *wind-up problem.*

THE DENSITY-WAVE THEORY

In the mid-1960s, astronomers C.C. Lin and Frank Shu, refined earlier ideas that the spiral pattern of stars in the Galaxy could be explained by the existence of a spiral-shaped compression wave moving through the material of the galactic disk. This compression wave is similar to the seismic P-waves described in Chapter 8. As the wave moves through

a certain part of the Galaxy, it compresses the gas it passes, just like a sound wave compresses the air it passes through. The increased density from the compression might cause interstellar clouds to begin to collapse and trigger star formation.

Figure 20-24 illustrates very schematically the simple form of such a wave. The material of the galactic disk is imagined to be rotating clockwise. The mathematics of the situation implies that the wave travels *more slowly* than the gas, so that a gas cloud approaches the wave *from behind,* enters into the wave, and is compressed. The simple model illustrated here is called the *two-armed spiral shock model,* because the passage of the wave causes a shock wave in the gas, much like the sonic boom produced by a jet plane.

Like any good theory, the density-wave theory makes a number of predictions. If a gas cloud (such as a giant molecular cloud) passes through the density wave and has star formation triggered, stars of all masses will form. However, remember that the luminosity of a main-sequence star depends approximately on the fourth power of its mass, and that the O and B stars, which are massive, will be by far the brightest objects that form. Thus, when we look at a galaxy, the light in the spiral arm regions will be dominated by the O and B stars. An easy observational confirmation of this idea is that the observed colors of the spiral arm regions of other spiral galaxies are indeed bluish (see Figure 20-18 and the book's frontispiece).

Another prediction results when you remember that the lifetimes of these O and B stars are short compared to the 300 million years or so it would take the wave to make one pass around the Galaxy. Thus, the O and B stars flare up, live their lives, and die on such a short time scale that the wave has hardly had time to move on at all. On a galactic time scale, the O and B stars are like flashbulbs that shine briefly and tell us where the wave *has just been*. The density-wave theory then predicts that the regions where we see bright O and B stars should be concentrated along the *inside* edges of spiral arms. Observations show this to be the case for many spiral galaxies.

Finally, the theory predicts that regions where the wave is currently compressing the gas and dust will be somewhat darker than other regions. Again, observations are in agreement with this prediction.

We should not think of the spiral arms as always being composed of the same stars. University of Texas astronomer Frank Bash has compared the locations through which a density wave passes to a hospital maternity ward; that is, a place where women and babies are always found and birth is constantly taking place. But the women and babies are not always the same individuals—some leave to be replaced by others. Alternately, think of people walking through a corridor and encountering stairs (**Figure 20-25**). They slow down and bunch up. Over time, different people pass through the same area, but the region of highest density remains stationary. Similarly, the O and B stars that light up the arms so dramatically have short lifetimes and, although the arms are always outlined by such stars, new stars are constantly being born to replace the O and B stars that have died out. Thus, a *spiral arm* is ephemeral—a region of a galaxy where interstellar gas is being illuminated by recently formed O and B stars that are strung out along the path through which the density wave has just passed, like a string of pearls, as astronomer Bart Bok has so aptly described it.

The origin of these density waves is still a matter of debate among astronomers. In some ways they are similar to sound waves passing through the air in a room. Like sound waves, some source must produce them. One possibility is that bars in the centers of spiral galaxies like the Milky Way (see Chapter 21) might "stir up" the surrounding gas, thereby causing the density wave to travel through the disk as described above.

Figure 20-25 A density wave composed of people.

Various researchers have noted that many spiral galaxies simply do not exhibit the clean, straightforward pattern of long, easy-to-trace spiral arms that the density-wave theory predicts. Indeed, Figure 20-21 shows that even our own galaxy does not. In the next chapter, we will discuss another possible mechanism of spiral arm formation based on supernova explosions.

THE GALACTIC HALO AND THE NUCLEAR BULGE

We previously defined the Galactic halo in terms of the roughly spherical distribution of globular clusters. We have seen that stars passing near the Sun and having high velocities relative to it are also part of the halo. Astronomers have found RR Lyrae stars that are not associated with any globular cluster to be halo members. Because only a small fraction of stars are RR Lyrae stars, this suggests that there are probably large numbers of faint stars in the halo. Surveys of faint objects, including observations made with the *Hubble Space Telescope*, indicate, however, that the total mass in halo stars is no more than about 10^9 solar masses, or about one hundred times more than combined mass of all the globular clusters. Comparing this number to the 10^{11} solar masses in stars in the galactic disk shows that the *density* of stars (i.e., the number of stars in a given volume) is much lower in the halo than in the disk, particularly when one realizes that the volume encompassed by the visible halo is much greater than that of the disk. Radio emission from the halo suggests that there is some gas there, although the exact amount is uncertain. It is clear, however, that there is essentially no ongoing star formation in the visible halo.

The extent of the galactic halo is also a subject of debate. The most distant globular clusters are found around 300,000 ly from the galactic center, which is *more* distant than our two neighboring satellite galaxies, the Large and Small Magellanic Clouds. It is therefore likely that these satellite galaxies are actually contained *within* the halo of the Galaxy, albeit as distinct entities. A visible halo radius around 300,000 ly is consistent with estimates of the extent of the dark halo discussed earlier, although, as noted previously, such estimates are quite uncertain at this time.

In 1989 the *COBE* satellite produced a wide-angle image of the Milky Way in infrared radiation (**Figure 20-26**). The image covers the entire sky; the view is similar to spreading the globe of the Earth onto a flat piece of paper. This view shows the concentration of old and cool stars in the galactic disk (since such stars radiate most of their energy in the infrared part of the spectrum) and the presence of a flattened **nuclear bulge** in the direction of the constellation Sagittarius.

The bulge is the least well understood component of the Milky Way, primarily because it is difficult to observe at optical wavelengths. In terms of its shape, Figure 20-26 shows that the bulge is a flattened spheroid. In other words, it is basically the shape you would produce if you took a rubber ball and squashed it in one direction. Notice that the bulge is elongated in the same plane as the Galactic disk. Unlike the halo, the bulge shows significant rotation, although overall it is not rotating as rapidly as the disk. The total mass in stars in the bulge is around 10^{10} solar masses. In addition to these stars, the bulge region contains more than

10^8 solar masses of molecular gas, hot X-ray-emitting gas, and the Galactic center itself, to which we now turn our attention.

THE CENTER OF THE GALAXY

As we have already noted, the central regions of our galaxy are hidden from the view of the optical astronomer by interstellar dust. However, the development of radio and infrared observing techniques since World War II has allowed us to penetrate the dust to the very center of the Milky Way. Furthermore, technological advances have greatly improved the ability of the instruments observing at these long wavelengths to resolve detail.

The Galactic center turns out to be a region of extreme complexity. **Figure 20-27** shows a large-scale, high-resolution radio map made with the 27 linked radio telescopes of the *VLA*. The most intense region is known as Sagittarius A (Sgr A). Other features include filamentary structures (such as the one dubbed the *Pelican*) that are either parallel or perpendicular to the galactic plane. Such alignments are probably caused by the filaments responding to strong magnetic fields near the center of the Galaxy. Supernovae remnants and HII regions are also observed.

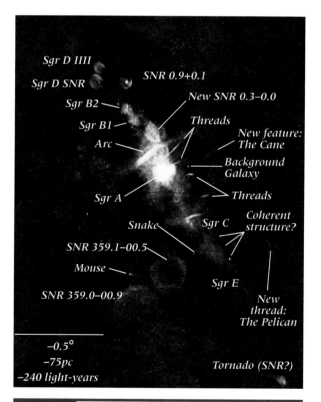

Figure 20-27 A radio image of the galactic center region.

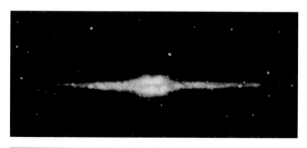

Figure 20-26 The Milky Way observed towards the galactic center as observed by the COBE satellite in the infrared.

Sgr A radiates strongly at both radio and infrared wavelengths and is one of the brightest radio and infrared sources in the sky, in spite of its great distance. Its brightness indicates that events involving tremendous amounts of energy are occurring in the Galactic center. We can recognize two sources of radiation: (1) ionized gas heated by hot stars and supernovae and (2) high-speed electrons accelerated by strong magnetic fields (synchrotron emission). When Sgr A was probed in more detail, the long-wavelength radio radiation that could penetrate through the dust showed us still more complexity on finer and finer scales (**Figure 20-28**). A more condensed region inside Sgr A, now named Sgr A*, is thought to be the dynamical center of the Galaxy. Only one second of arc from Sgr A* is an object referred to as IRS 16, a complex of at least 15 components, all of which exhibit strong stellar winds. The nature of these objects is as yet unknown.

Answers to the questions we have raised can come from an examination of the spectrum. Sgr A exhibits a continuous spectrum but not due to thermal energy. Its continuous spectrum is nonthermal radio emission, meaning that the brightness is *not* related to its temperature. The observed spectrum's increase in intensity as we go to longer radio wavelengths, along with its polarization, indicates that it is the same synchrotron radiation that we discussed in connection with the Crab Nebula. We conclude that the radiation is due to charged particles accelerating at near the speed of light in a magnetic field.

Researchers also detected SNR 0.3-0.0, a supernova remnant very close to the core (Figure 20-27). This and other detections show that the center of our galaxy is being churned by the detonation of massive stars. The high rate of supernovae going off should not come as any surprise—presumably the high density of material in the galactic nucleus leads to a high rate of star formation. Although the energy from Sgr A* is strongly concentrated in a small region of space, it still radiates an amount of energy tens of millions of times greater than that emitted by the Sun. Whatever the nature of Sgr A*, it must truly be exceptional.

Attempts to explain the high concentration of so much energy in such a small volume of space have produced a proliferation of competing hypotheses. For example, a dense group of hot stars surrounded by dust shells has been proposed. The most promising scenario, however, is that Sgr A* contains a supermassive black hole. There are two main reasons to suppose this is the case. First, the stars and gas in the vicinity of Sgr A* are moving rapidly. As we have seen in other situations, high speeds imply a large force of gravity and thus a large mass. From observations of the motion of gas and stars near Sgr A*, astronomers conclude that a mass in excess of 10^6 solar masses sits at the heart of this object. Such a large mass contained within a relatively small volume is immediately suggestive of a black hole. Perhaps more importantly, there is compelling evidence that many galaxies contain supermassive black holes at their centers (see Chapter 22). The presence of these objects in other galaxies boosts astronomers' confidence that the Milky Way also contains a supermassive black hole.

Figure 20-29 summarizes the Milky Way as we now think it looks. The principal visible components are the halo, the nuclear region or bulge, and the disk including the spiral arms. These visible components sit within the more extensive dark-matter halo.

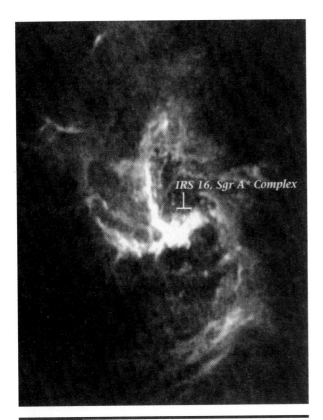

IRS 16, Sgr A* Complex

Figure 20-28 A radio image of the galactic center. Important features described in the text are indicated.

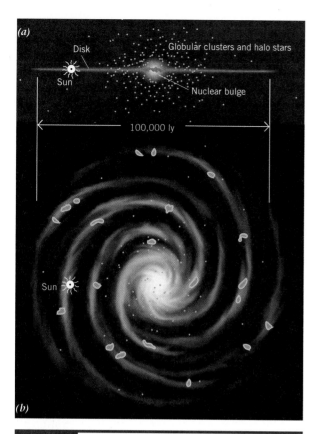

(a)

Disk

Globular clusters and halo stars

Sun

Nuclear bulge

100,000 ly

Sun

(b)

Figure 20-29 Artist's illustration of our galaxy, showing the locations of the globular clusters and halo, the nuclear bulge, the disk, HII regions, and the spiral arms. The extended halo, whose size is unknown, is not shown in this illustration.

STELLAR POPULATIONS

In studies culminating during World War II, Mount Wilson astronomer Walter Baade noticed an apparent division of stars into two strikingly different classes, which he called **Population I** and **Population II**. Population I objects are those confined most strongly to the plane of the Galaxy, and, in particular, those found in the spiral arm regions. Population II objects are those less strongly confined to the galactic plane, and, in particular, include the objects in the nuclear and halo regions of the Galaxy.

For example, the bright O and B stars that recently formed from molecular clouds in the spiral arm regions are Population I objects. Also strongly concentrated in the spiral arm regions and grouped together as Population I are the bright supergiant stars, young open clusters, some Cepheid variables, and the low-velocity stars. Examples of Population II objects not confined to the galactic plane but extending into the galactic halo and into the nuclear regions include a sub-

class of Cepheids called *W Virginis stars,* the globular clusters, RR Lyrae variables, and the high-velocity stars.

Significant differences in chemical compositions exist between the two stellar populations. We have stressed that most stars (and indeed virtually all objects that have been detected in the universe, with the exception of terrestrial-type planets) have a somewhat similar composition, with approximately 70 percent of the mass being hydrogen and approximately 28 percent helium. All the rest of the atoms in the periodic table contribute only about 2 percent of the mass of the universe.

When we examine the abundances of heavy elements in stars more closely, we find that, although the heavy elements are a minor component of all stars, there are significant differences between the two classes of stars. For example, Population II stars are extremely deficient in the heavy elements, with an abundance as much as 10,000 times smaller than in the Sun and other Population I stars. Since astronomers refer to all heavy elements as *metals,* Population II stars are said to have a low metallicity.

Inquiry 20-21

Where did the heavy elements found in Population I stars form?

Modern studies have shown that this two-population model is too simplistic and that there are other populations of stars intermediate in properties between these two extremes. For example, most astronomers think that the Milky Way disk is more accurately described as two disks: the thin disk that is essentially the component described earlier, and a thick disk with older, lower-metallicity stars that, as the name suggests, extends further from the galactic plane. More generally, as our understanding of the nature of stellar populations in the Milky Way and other galaxies has improved, the traditional classification of Population I and II stars has been found to be incomplete and inadequate. For example, in our galaxy young, high-metallicity, Population I stars are found almost exclusively in the galactic disk. However, there are also some disk stars that are old and have low metallicities—the characteristics of Population II stars.

A more useful approach to describing stellar populations is simply to specify the heavy-element

Table 20-1   Summary of Properties of the Visible Components of the Milky Way

Component	Ages of Stars	Heavy Element Abundance	Gas and Dust	Shape	Motion
DISK	Range from old to newly formed; light dominated by young stars	Mostly 1–2%. Oldest stars have lower values.	Plentiful	Flattened	Rapid rotation
HALO	Old with little variation	0.1% and below	Mostly absent	Roughly spherical Eccentric orbits	Little net rotation
BULGE	A range from old to newly formed	From less than 0.1% to 2% or more	Both present	Flattened spheroid	Rotating, but less rapidly than disk

content, age, and motion of stars rather than forcing them into Population I or Population II. This is the approach we will take in this and later chapters. In Table 20-1 we summarize the properties of the visible components of the Milky Way. The differences between these components help us unravel the formation history of our galaxy, as discussed in the next section.

20.4 The Formation and Evolution of the Galaxy

An understanding of the formation and evolution of the Milky Way requires us to explain why there is a disk, halo, and bulge, and why the stellar content and other properties of these components differ. In the following sections we present a general picture based on our current understanding.

THE CHEMICAL EVOLUTION OF THE GALAXY

As many stars end their lives, a portion of the heavy elements that they have manufactured by nuclear fusion in their cores is returned to the interstellar medium by means of a variety of mass-ejection mechanisms: stellar winds, planetary nebulae, and novae and supernovae explosions. Thus, the next generation of stars forms out of gas clouds that have been "enriched" in heavy elements. These, in turn, make more heavy elements, and so we are led to expect that the percentage of the Galaxy's mass that is bound up in heavy elements increases with time.

Inquiry 20-22

(a) Why would you not expect to find a B-type star in the Milky Way halo? (b) Would you be surprised to find an M-type star in the Milky Way disk?

Inquiry 20-23

Why is a high-metallicity star in the halo less likely than a low-metallicity star in the disk?

As we have already seen, this general picture has observational support. Young stars in the galactic disk do indeed have higher fractions of heavy elements than old stars in the visible halo. More detailed studies support the trend of an increase of metallicity with time, but also reveal some subtleties. For example, it appears that the metallicity of disk stars reached levels similar to that in the Sun fairly rapidly when the disk was still young, but that subsequent increases have been modest. Thus, a complete picture of chemical evolution needs to be more sophisticated and include other processes, such as the possibility of low-metallicity gas falling onto the disk after its formation.

THE FORMATION OF THE MILKY WAY

The properties of the various components of the Galaxy provide one starting point for understanding the formation of the Milky Way. However, as we have seen in many other contexts, one area of astronomy is not divorced from all the others, and so additional information is also important in building a model of how the Galaxy formed. Most relevant to the current discussion is the field of cosmology (see Chapter 23), which provides the broader framework within which galaxy formation occurs.

Let us first consider an oversimplified picture and see how far it takes us. Studies of other galaxies and cosmological considerations suggest that **protogalaxies**, which are galaxies-to-be in an early stage of their formation, consist of roughly spherical dark matter halos containing hydrogen and helium gas. Interactions with nearby proto-galaxies produce a small amount of rotation in the halo and the gas. On the one hand, the dark-matter halo is unable to get rid of energy and collapse. The gas, on the other hand, *can* collapse within the halo because it can radiate away its energy; that is, it can cool. As the gas collapses, it must conserve angular momentum, so, like a spinning ice skater pulling her arms close to her body and spinning faster, the gas will also "spin up." Like rapidly spun pizza dough, the gas will take on a flattened distribution. The physics of the situation is very similar to that in our discussion of the formation of the Solar System in Chapter 6. If this rapidly rotating, flattened disk of gas fragments and forms stars, we have a natural explanation of the formation of the dominant visible component of the Milky Way.

What about the visible halo? We have established that stars and globular clusters in the visible halo are old and contain few heavy elements. Further, their orbits imply that they formed out of material that had little ordered rotation, at least compared to the disk. The slowly rotating, roughly spherical gas cloud of the protogalactic stage of the Milky Way is likely to give rise to these general characteristics. In particular, we are assuming that no stars had formed at this time, so the cloud would be free of heavy elements. If a small fraction of this cloud formed stars and globular clusters *before* collapsing into a spinning disk, the observed properties of the visible halo would be explained fairly naturally.

The formation of the nuclear bulge is still a hotly debated topic among astronomers, and there is currently little consensus concerning the dominant mechanism. Some models propose that a collision with a smaller galaxy may be involved, or that the bulge is a kind of dumping ground for material falling into the Galaxy. Other scenarios propose that the location of the bulge at the center of the surrounding dark halo leads to rapid and violent star formation, or that processes unique to the center of the Galaxy funnel gas into the bulge where it forms stars. Finding the definitive answers is hard because of the difficulty in observing this component of the Galaxy, so that studying the bulges of other spirals may be an equally promising approach to understanding the formation of the galactic bulge. We return to this idea in the next chapter.

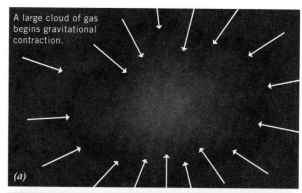

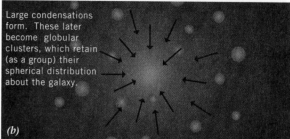

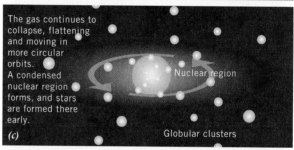

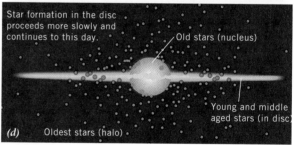

Figure 20-30 A hypothesized sequence of events in the formation of the Galaxy.

The picture of the formation of the Milky Way, is illustrated schematically in **Figure 20-30**. Although this model is simplified, it provides a useful starting point in understanding the formation of the Galaxy in particular and spiral galaxies in general. Indeed, many of its features may be fairly close to what actually happens. However, some of the detailed properties of the Milky Way are difficult to explain in this simple model. For example, if the spherical gas cloud out of which the visible halo formed was composed only of hydrogen and helium, why do we not see any stars with zero metallicity? One important refinement is to adopt a more sophisticated cosmological framework. As we will discuss further in Chapters 21 and 23, it is now generally agreed that bright spirals like the Milky Way do not form from a single protogalactic object, but are built up from many smaller fragments that may have already formed stars and globular clusters. Further, for some galaxies evolutionary changes occur between their formation and the present, so we have to include such effects when we use the observed properties of galaxies to probe the processes involved in their formation.

20.5 The Galactic Magnetic Field

We have mentioned the existence of magnetic fields in a variety of objects within the Milky Way. For example, the radio source Sgr A at the center of the Galaxy appears to be radiating by means of the nonthermal synchrotron mechanism in which energy is radiated by charged particles accelerated in magnetic fields to speeds near that of light. This emission mechanism was discussed in connection with the continuous emission from the filaments of the Crab Nebula, and as an important component of the emission from pulsars. In these cases, we were dealing with magnetic fields associated with individual radiation sources. However, magnetism is widespread throughout the structure of the Galaxy itself. This widespread magnetic field is important because it undoubtedly played a significant role in the Milky Way's formation.

There are two direct lines of evidence for a pervasive magnetic field in our galaxy. One is the Zeeman effect that was discussed for stars in Chapter 14. What would normally be observed in the spectrum as a single line at a discrete wavelength is replaced by several closely spaced lines, each representing a transition between the split energy levels. From these observations, astronomers conclude that the general interstellar magnetic field appears to have a strength that is about 10^{-5} times the average magnetic field at the surface of the Sun. This is large enough to have observable effects, but not so large as to enable the lines of magnetic force to affect the motions of gas clouds.

Another type of observation indicating the presence of important magnetic fields in the Galaxy is polarization, which was mentioned in this chapter's discussion of dust. We often observe starlight arriving at Earth to be polarized as a result of its passage through interstellar dust. The only satisfactory explanation of this polarization assumes that interstellar dust particles are elongated and of such a composition that they are aligned by magnetic fields. Organized magnetic fields in the Galaxy will align interstellar dust particles over wide reaches of space, resulting in polarization (Figure 20-12).

20.6 Cosmic Rays

Experiments in the early part of the twentieth century revealed that at high altitudes an *electroscope*— a simple device for measuring static electricity—lost its static charge more rapidly than at lower altitudes. The rate of loss was greater than could be explained by solar ultraviolet radiation, and at first it was supposed that high-energy photons from the cosmos were responsible. These photons were assumed to have wavelengths even shorter than gamma rays. But then physicists found that the loss of charge also depended on the magnetic latitude. Because we do not expect the Earth's magnetic field to affect uncharged particles such as photons, it follows that the phenomenon must be due to energetic charged particles impinging on the Earth. The paths of such charged particles would be curved by the Earth's magnetic field and would tend to be concentrated toward the Earth's magnetic poles. Scientists making these observations referred to the phenomenon they were observing as **cosmic rays**.

OBSERVATIONS OF COSMIC RAYS

Cosmic rays rain down on Earth from all directions. We now know that the name cosmic *ray* is a misnomer, dating from the early days of the study of radioactivity, well before photons and high-energy material particles had been distinguished clearly. We have also learned that these particles contain as much energy as all the radiation emitted by all the stars in the Milky Way!

Inquiry 20-24

Where on the Earth's surface would you expect a charged electroscope to lose its static charge most rapidly—near the equator or near the poles? Explain.

The discovery of these energetic particles is significant not only because it opened up the field of high-energy particle physics and advanced Einstein's theory of special relativity, but because it opened a window to the distant universe by allowing us to examine particles of matter from far-flung parts of the Galaxy.

Cosmic ray particles arrive at Earth traveling at speeds close to the speed of light. To study them before they interact with the Earth's atmosphere requires observing from altitudes greater than 100,000 feet, often using balloons. Studies show that the majority of cosmic ray particles are protons, the nuclei of hydrogen atoms. Approximately 15 percent are alpha particles—helium nuclei—whereas only 1 percent or so are nuclei of heavier elements. Overall, it appears that the abundances of various nuclei in cosmic rays are in agreement with the cosmic abundances that have been determined by other means.

Origin of Cosmic Rays

Although there are such things as bursts of cosmic rays, one of the most striking and significant properties of them is that the incoming particles are *isotropic;* meaning that equal numbers of particles fall on the Earth from all directions in space. Thus, the incoming material comes from no preferred direction in space.

> ## Inquiry 20-25
>
> Can the Sun be the source of most of these cosmic rays? Why or why not?

Cosmic rays appear to be a property of the Milky Way as a whole. We can rule out the Sun as the cause of these rays by direct observations of the types of particles that stream toward us from it.

We find, for example, that the average energy of particles from the Sun is much less than the average energy of a cosmic ray particle, and that under no circumstances does the Sun produce particles as energetic as some seen in cosmic rays. Also, the cosmic ray particles show high abundances of the light elements lithium, beryllium, and boron that bear no resemblance to the abundances of these elements in the Sun. We are led to feel that, although the Sun might be the source of a small percentage of the observed cosmic rays, it certainly cannot be an important contributor to it. A model that does account for the isotropy of the cosmic ray particles assumes that they are significantly affected by the presence of the interstellar magnetic field that is found throughout the Galaxy. Although this field has been accurately measured only in a few places, it is easy to imagine that magnetic forces pervade the entire galaxy, including the halo region. If this is true, energetic charged particles would be deflected by the magnetic lines of force, so that when they arrive at Earth the direction from which they come would be completely random and unrelated to where the cosmic rays were produced.

Yet this is not a complete explanation, because it does not account for the origin of these particles, nor how they acquired their great energies. An occasional cosmic ray particle enters the Earth's atmosphere moving at a speed of 99.999999999 percent the speed of light; its vast energy comes from its enormous velocity.

How can particles acquire such velocities? One suggestion is that they are created in the blast wave of a supernova explosion, which is one of the most energetic events that can happen in a galaxy. However, the rate of supernova explosions is such a puzzle to us that it is difficult to tell whether there are enough of them to have created all the cosmic rays we see.

Galactic Rotation

When you have completed this discovery you will be able to do the following:

- Describe the rotation curve for a rotating solid body and for a differentially rotating body.
- Describe the rotation curve of the Milky Way and explain why astronomers think the Galaxy contains dark matter.

To better understand the observed rotation of the Milky Way, you will now discover how bodies that follow Kepler's laws move, as well as the rules that govern the rotation of solid bodies.

Keplerian Motion

A. Using data on the planets from Appendix C, make a graph of orbital velocity on the vertical axis and distance (in AU) along the horizontal axis. Draw a smooth curve through the points.

B. Describe in detail the appearance of your graph. You have discovered how velocity changes with distance for bodies that move according to Kepler's laws. The shape of curve you found results whenever bodies orbit a more massive one, such as planets orbiting a star.

Solid-Body Rotation

Consider a rotating plate or frisbee (see **Discovery Figure 20-1-1**), and in particular the points *A, B, C,* and *D*. Because the object is a solid body, when point *A* reaches *A′*, *B* will reach *B′*, and so forth.

C. Draw a graph showing the speed of rotation as the distance from the center increases. (Your axes should be the same as in part A above.) Describe the graph.

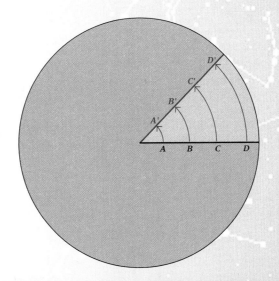

| **Discovery Figure 20-1-1** | The rotation of a solid object, such as a plate or frisbee. In the time point A rotates to point A', all the other points will rotate through the same angle. |

The Rotation of the Milky Way

A graph of rotation speed with distance from the galactic center is called a **rotation curve,** and is shown in Figure 20-6. You will now apply what you have discovered about rotation to the observed galactic rotation curve.

D. Describe the shape of the rotation curve of the Milky Way in the region from the center out to about 1 parsec, from a few parsecs out to the Sun, and beyond the Sun. Does the shape of the rotation curve in any of these parts follow either the rotation law for solid bodies or that for bodies following Kepler's laws?

E. How would the distribution of mass in a galaxy have to be modified in order to change a Keplerian rotation curve to one that looks like the Milky Way's observed rotation curve at distances greater than that of the Sun?

You have just discovered the reasons astronomers consider the Milky Way to contain a massive halo of matter surrounding the Galaxy.

The Distributions of Different Objects around the Galaxy

When you have completed this discovery you will be able to do the following:

- Describe the distribution of open clusters, globular clusters, dark nebulae, HII regions, and stars within the Milky Way.
- Describe the apparent distribution of galaxies on the celestial sphere.

Astronomers study the structure of the Milky Way by analyzing the way in which the various types of objects in our galaxy are distributed around it. For example, the location throughout the Galaxy of open clusters is shown in **Discovery Figure 20-2-1a.** Also shown in the figure are the celestial equator and the galactic equator, which lies along the plane of the Galaxy. We see that the open clusters are not distributed randomly but are confined along the galactic plane. (The few objects that seem to deviate from the trend appear to do so because they are relatively close to the Sun.)

- **Discovery Inquiry 20-2a** Describe the distribution of the *globular* clusters shown in **Discovery Figure 20-2-1b.** Compare and contrast the distribution with that of the open clusters.
- **Discovery Inquiry 20-2b** Describe the distribution of the planetary nebulae shown in **Discovery Figure 20-2-1c.** What conclusions do you draw relative to the open clusters and globular clusters.

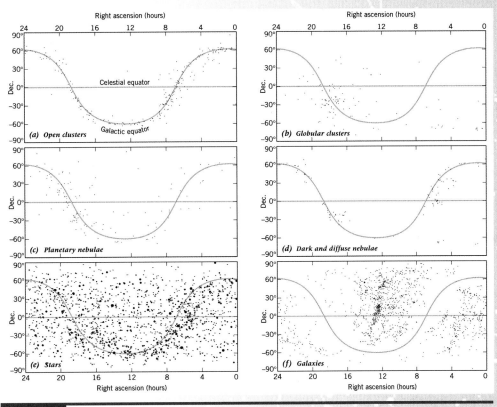

Right ascension (hours)

(a) Open clusters — Celestial equator, Galactic equator

(b) Globular clusters

(c) Planetary nebulae

(d) Dark and diffuse nebulae

(e) Stars

(f) Galaxies

Right ascension (hours)

Discovery Figure 20-2-1 The distribution on the sky of a variety of astronomical objects. The ecliptic and galactic planes are indicated.

- **Discovery Inquiry 20-2c** Describe the distribution of the dark and diffuse nebulae shown in **Discovery Figure 20-2-1d**. What conclusions do you draw relative to the open clusters and globular clusters?

- **Discovery Inquiry 20-2d** Describe the distribution of the stars shown in **Discovery Figure 20-2-1e**. What conclusions do you draw relative to the open clusters and globular clusters?

- **Discovery Inquiry 20-2e** What general conclusions can you draw about the structure of the Milky Way from the distribution of the objects studied earlier?

- **Discovery Inquiry 20-2f** The distribution of the galaxies differs from that of objects belonging to the Milky Way. How might you explain the distribution of galaxies shown in Discovery **Figure 20-2-1f?**

If you had further knowledge about the properties of these objects, you would have been able to draw more thorough conclusions. The rest of the chapter provides you with this additional information.

Chapter Summary

Observations

- From studies of **globular cluster** distances and their distribution in the sky, **Harlow Shapley** found the Sun to be located some 30,000 light-years from the center of the **Milky Way**. He was able to determine cluster distances by finding **RR Lyrae stars** in the clusters. This method worked because the variable stars he studied had a relation between period and luminosity, which allowed luminosity, and thus distance to be found.

- The motion of the Sun around the center of the Galaxy is found from studies of the motions of globular clusters and distant galaxies. From these studies we find the Sun moves with a speed of about 220 km/s.

- From the observed motions of stars astronomers determined the galactic **rotation curve**, a figure showing how an object's speed depends on distance from the galactic center.

- **Interstellar dust** affects starlight by extinction, **reddening**, and **polarization** of light waves. In addition, dust is observable by its scattering of light in **reflection nebulae**.

- **Interstellar gas** is observable in the form of **HII regions** and **molecular clouds**, and by means of the emission by hydrogen of **21-cm radiation**, which has the ability to pass through dust without being absorbed.

- The spiral structure of the Milky Way can be studied by means of optical spiral tracers (**OB associations, galactic clusters, HII regions, dark nebulae,** and **Cepheid variables**) and **radio spiral tracers** (21-cm radiation from neutral hydrogen, molecular clouds).

- The **galactic center** is a strong emitter of infrared and radio emission, in addition to gamma rays.

- **Cosmic rays** consist of high-energy particles, usually electrons, protons, and the nuclei of some elements. They are observed to come to the Earth from all directions in space. Their intensity is related to the latitude on Earth from which they are observed.

Theory

- Spiral structure is generally understood by means of **spiral density waves** that are hypoth-esized to move with respect to the gas and dust in the galactic plane.

- The hydrogen atom's ground state consists of two close energy levels, which occur because when an atom's proton and electron both spin in the same direction, the energy is different from when they spin in opposite directions.

Conclusions

- It takes 200 to 250 million years for the Sun to complete one revolution about our galaxy. From this period and the distance of the Sun from the galactic center, astronomers have determined a mass for the Galaxy of 10^{11} solar masses.

- The interstellar medium has a complex structure consisting of **diffuse clouds** with temperatures of about 100 K, surrounded by a **warm intercloud gas** at a temperature of 1,000 K, surrounded in turn by hot **coronal gas** at a million kelvins.

- From the observation that the **rotation curve** does not decrease with increasing distance, astronomers infer the presence of an extended **galactic halo** containing substantial amounts of **dark matter** beyond the Sun's orbit.

- Gamma rays from the galactic center appear to come from matter-antimatter annihilation.

- From the observation that radio emission from the galactic center increases in intensity with increasing wavelength and is polarized, astronomers infer that the radiation is produced by **synchrotron emission**.

- The presence of a magnetic field in the Galaxy is deduced from observations of the polarization of starlight, the 21-cm Zeeman effect, synchrotron radiation, and the distribution of cosmic rays in the Galaxy.

- The Galaxy is thought to have formed through the collapse of a rotating cloud of material. Because the globular clusters formed first, their orbits were determined by the material's motion at the time of formation; these orbits are therefore elliptical at great distances from the galactic center. After the Galaxy collapsed into a flat disk, newly formed stars moved into the galactic plane. Such a picture is supported by the observed differences between stellar populations.

Summary Questions

1. Why did early star counts lead to an erroneous picture of the size of the Galaxy and the location of the Sun within it? How was Harlow Shapley able to improve this picture?

2. What is the modern picture of our galaxy? Include the Galaxy's dimensions and the location of the Sun, halo, disk, and nucleus.

3. What are several kinds of evidence for the existence of interstellar dust? How can the size of individual interstellar dust grains be estimated? What are the various lines of evidence that tell us about the composition of interstellar dust grains?

4. How do interstellar absorption lines give us information about the interstellar medium? What conditions (temperature, mass, density, diameter) are present in interstellar clouds? Why are interstellar absorption lines generally difficult to observe from the ground?

5. What is the observational evidence for a multi-component interstellar medium?

6. What is the process by which neutral atomic hydrogen produces 21-cm emission? Why is 21-cm emission so important in astronomy?

7. What evidence is there for the existence of spiral structure in the Galaxy? Why do astronomers think star formation is taking place in the spiral arms?

8. How can study of the motions of stars using the Doppler effect be used to study the structure of the Milky Way?

9. How are observations of neutral hydrogen (H I) used to study the rotation of the Galaxy and the distribution of gas within it? What are the main results of these observations, and what are the difficulties in interpreting them?

10. What evidence do astronomers have that unusual events occur near the galactic center?

11. What are the old (incorrect) ideas about spiral structure and why don't they work? Describe the density-wave theory of spiral structure.

12. What are three kinds of evidence for the existence of an overall galactic magnetic field?

13. What is the significance of cosmic rays? Where are they thought to form?

14. What are the differences between stars in the galactic disk and those in the halo? You should distinguish them on the basis of age, chemical composition, association with gas and dust, motions, and orbital characteristics. How did these two populations come to be, and how are the characteristics of each related to the origin and evolution of our Galaxy?

Applying Your Knowledge

1. The star in the center of a reflection nebula is generally a B star, typically B3 or cooler in spectral type. Why is it unlikely that a reflection nebula would be caused by a hotter star?

2. Show your understanding of how we analyze the spiral structure of the Milky Way by explaining Figures 20-19 and 20-20.

3. Astronomers are confident there are still undiscovered globular clusters within the Milky Way. Where might such clusters exist? Explain. (*Hint:* Make a drawing of the Milky Way as seen edge-on, showing the Sun and the globular clusters.)

4. Explain how two stars can have identical spectra but different colors.

5. Examine Figure 20-5. For which positions (*A, B*, etc.) will stars appear to move through the greatest angular distance relative to distant background stars during one year? The least angular distance?

6. Explain why the caption in Figure 20-15 says that the wings must come from hotter material than the core of the line.

7. Explain how the methods used by Herschel and Kapteyn might be applied by a person lost in a forest to find the way out. What assumptions would the lost person be making?

▶ 8. What is the distance to a globular cluster inside of which an RR Lyrae star is observed that appears 3×10^{-17} as bright as the Sun? (*Hint:*

You may want to refer to Section 15.7, as well as the H-R diagram for a globular cluster.)

▶ 9. Using the distance of the Sun from the galactic center, and the speed with which it moves, what is the length of the galactic year?

▶ 10. Consider two stars having identical spectra but different colors although they are at the same distance from Earth. Suppose the redder star appears two times fainter. How much farther away will the redder star appear to be? Explain why the redder one appears to be more distant.

▶ 11. If a dust particle consists of gas atoms that stick together during collisions, and a typical atom is 1 to 2 Å in diameter, what is the number of atoms in an average-sized dust grain? (Make an order-of-magnitude estimate, not a precise calculation.)

▶ 12. Using the scale on Figure 20-27, determine the physical size of the smallest resolvable object in the galactic center region. Express your answer in light-years, astronomical units, and kilometers.

▶ 13. If distances within the Milky Way were in error by a factor of two (i.e., if the Galaxy were actually twice as large as we thought), what would be the effect on the computed mass? (Assume the observed periods of observed objects remain the same.)

▶ 14. If a star at a distance of 15 light-years has a radial velocity of 150 km/s, by what percentage does its distance from the Sun change in 100 years?

Answers to Inquiries

20-1. The principal assumption is that the stars identified as RR Lyrae variables in the cluster have the same properties as those known elsewhere, so we can assume that we know their luminosity correctly.

20-2. Angular size = 57.3° physical size/distance. Therefore, physical size = 0.057 × 5,000 ly/57.3° = 5 ly.

20-3. An important assumption is that all globular clusters have nearly the same size, and that the nearby clusters are similar to the distant ones. Because the first assumption, strictly speaking, is not true, Shapley's use of it is valid only in a statistical sense, much as one can talk about the height of an "average" adult human being. (We now know the second assumption is wrong!)

20-4. Shapley was guided by the Copernican principle that the Earth must not be considered to be located in a favored location in the Galaxy.

20-5. Point B: no Doppler shift or relative velocity. Point C: blueshift, as the Sun overtakes the star. Point D: no shift because it has the same orbital speed as the Sun. Point H: redshift because, it moves faster than the Sun and is pulling away from it.

20-6. Because the star would appear fainter than it should be, we would estimate its distance to be greater than it really was.

20-7. Uncertainties in distance would produce uncertainties in, for example, luminosity, size, and mass.

20-8. We would estimate a temperature that would be too low.

20-9. Because short wavelength radiation scatters more readily.

20-10. There would be no long axis to preferentially absorb the radiation. Polarization is an effect that depends on asymmetry.

20-11. High temperature will destroy grains over time.

20-12. They would be narrow because few if any collisions would occur to cause the lines to broaden.

20-13. Hot stars produce ultraviolet photons that are able to ionize the hydrogen and produce the HII region.

20-14. Several absorption features result when clouds of interstellar gas between the star and the observer move with slightly different velocities.

20-15. The wavelength of 21-cm radiation is much longer than the size of interstellar dust particles. Therefore, it is not effectively scattered or reflected by them.

20-16. This suggests that most of the gas in these clouds does not fragment into stars; that is, the star formation efficiency is low.

20-17. (a) Supergiants, OB associations, young open clusters, Cepheids, dark clouds, HII regions. (b) Same as part a, although the Cepheids might not be strictly in the plane.

20-18. The presence of very young stars, known to live only for short periods of time, near the spiral arms could not be explained otherwise. They cannot move more than a small fraction of the distance between spiral arms before they die.

20-19. Regions of ionized gas (HII regions)

20-20. Because E and G have the same distance from the Galaxy's center, the relative radial velocities of the two arms are the same. Therefore, the peaks in the spectrum would coincide. Only by observing in slightly different directions would the two peaks be seen to separate so that the ambiguity could be resolved. The spectrum would contain two peaks, one (from E and G) that was not shifted because the objects at E and G have the same orbital speed as the Sun, and a redshifted peak from F.

20-21. The heavy elements found in Population I stars (which are more recently formed) were produced in the interiors of previous generations of stars. The heavy elements thus formed were later ejected into the interstellar medium.

20-22. (a) B-type stars are massive and have short lifetimes. Since there is no current star formation in the Galaxy's halo, we would not expect to find a B-type star there. (b) No, most stars that form are M dwarfs, so they are plentiful in the disk.

20-23. Two reasons. Low-metallicity halo stars have eccentric orbits and thus may pass through the disk. Also, the oldest disk stars do have low metallicity.

20-24. Near the poles, because the magnetic field of Earth channels cosmic rays to move along the lines of force. This is also the reason why aurorae appear near the poles.

20-25. No. The paragraph following the inquiry provides an answer.

21

Galaxies

[The] analogy [of the nebulae] with the system of stars in which we find ourselves . . . is in perfect agreement with the concept that these elliptical objects are just [island] universes— in other words, Milky Ways. . . .

Immanuel Kant, Universal Natural History and Theory of the Heavens, 1755

21.1 The Historical Problem of the Nebulae

Eighteenth-century astronomers observed large numbers of fuzzy (nebulous) objects, which were known as *nebulae,* and asked: What is the nature of these objects? Do they consist of gas or are they stars? In the mid-eighteenth century, the philosopher Immanuel Kant, influenced by the astronomer Thomas Wright of Durham, England, proposed that many of the nebulae that had been observed were actually stellar systems similar to our own but external to it—in Kant's words, "island universes," at great distances from the Milky Way galaxy. It took almost 170 years of patient and careful research by both theoretical and observational astronomers to verify the validity of this hypothesis, and another 50 years before realistic models of galaxies could be constructed.

Probably the single most important factor in resolving the problem of the nature of the nebulae was the development of photography. By the 1880s, photographic emulsions were sensitive enough to permit time exposures of faint objects, allowing astronomers to penetrate more deeply into space than they could with the telescope-aided eye. **Figure 21-1a** is a drawing of the nebula known as the Whirlpool, M51, made by the Earl of Rosse, who observed the heavens with a giant 72-inch telescope. **Figure 21-1b** shows an image of the same object as does the book's frontispiece. The image shows much more detail than the drawing, and many of the details in the drawing are incorrect. Photography made it possible to produce a permanent record of the light from distant objects, a record that later could be analyzed at leisure and in more detail.

Even with photography, it was frequently impossible to distinguish between different types of objects

(a)

(b)

Figure 21-1 *(a)* The Whirlpool Nebula as drawn by the Earl of Rosse. *(b)* The same galaxy in a modern image.

solely on the basis of their appearance. Figure 18-31 shows the globular cluster 47 Tucanae; **Figure 21-2** is the giant elliptical galaxy M87 (also known as NGC 4486, meaning the 4,486th object in the *New General Catalogue*). 47 Tucanae consists of several hundred thousand stars and is about 20,000 light-years away, whereas M87 consists of more than 100 billion stars and is around 40 million light-years away. Telling them apart from their appearance alone is difficult.

As we move beyond the confines of the Milky Way we find that there are billions of galaxies scattered throughout the Universe. Some exhibit the flattened, spinning disk of gas and stars that we find in our galaxy, but others are almost completely spherical and hardly rotate at all. In this chapter we describe the great variety of galaxy shapes and sizes as well as how galaxies form and change over time.

Chapter opening photo: The Andromeda galaxy, M31, is the nearest large spiral galaxy to the Milky Way. Corel

Figure 21-2 | NGC 4486, an elliptical galaxy.

Inquiry 21-1

Is there any evidence in Figure 18-31 and Figure 21-2 that indicates to you that the first object is actually much closer? What is it?

We can see many small bright spots in the outer parts of the elliptical galaxy shown in Figure 21-2; most of these dots are individual globular clusters. This image dramatically shows the difference in size between a globular cluster and an elliptical galaxy.

When spectroscopic observations were begun at the end of the nineteenth century, astronomers found that objects such as the Orion Nebula and Ring Nebula had bright-line spectra of an excited, rarefied gas. But objects such as the great nebula in Andromeda, M31 (Figure 1-15), showed dark-line spectra similar to those of stars, indicating that they were, in fact, made up of stars. As we have already seen, the difficult questions were, "How many stars?" and "How far away?" Were those nebulae that were made of stars only modest stellar clusters in our own galaxy, or were they enormous and very distant stellar systems?

Ultimately, the only way of resolving this question was to measure distances to these systems. We will discuss how this is accomplished in detail in Section 21.3. In the early 1920s, however, there was still considerable controversy about the distance to and the nature of the stellar nebulae. The famous Shapley–Curtis debates centered on this issue, with Shapley arguing for the relative proximity of these objects and Heber D. Curtis of Lick Observatory advancing the view that the stellar nebulae were *island universes* (galaxies) beyond the Milky Way. When reliable distances were finally obtained, Curtis's view was confirmed, and the field of extragalactic (beyond our galaxy) astrophysics was born.

21.2 The Morphology of Galaxies

When confronted with a variety of newly discovered types of objects, astronomers first establish a system for classifying them. Biologists do the same with the bewildering array of plants and animals present on Earth; geologists, too, have established comprehensive systems for classifying rocks and minerals. Classification is a technique for trying to make sense out of the apparent chaos that nature presents to us. The most important classification scheme for galaxies is one based on the morphology (shape) of galaxies. Such a classification turns out to be useful because morphological characteristics of galaxies correlate with other properties such as the motion of the stars they contain and the rate of current star formation. The most obvious division of galaxy types in this scheme is between spirals and ellipticals, but within each category significant variations exist. These galaxies form what we will often refer to as *normal* galaxies, to distinguish them from the *peculiar* galaxies discussed in the following chapter.

SPIRAL GALAXIES

The classification of **spiral galaxies** has been subdivided primarily on the basis of the size of the nucleus relative to the entire galaxy, and the openness of the spiral arms. In **Figure 21-3** the main categories of spirals are shown. The designation S is employed for spirals; the letter following S designates the subclass. Sa galaxies are those with a large, dominant nuclear bulge and tightly wound spiral arms. Sc spirals have a small nucleus (relative to the entire galaxy) and loosely wound spiral arms. Between Sa and Sc spirals are the Sb spirals, similar in form to the Andromeda galaxy, with a less prominent nucleus and more-open spiral arms than that of the Sa spirals.

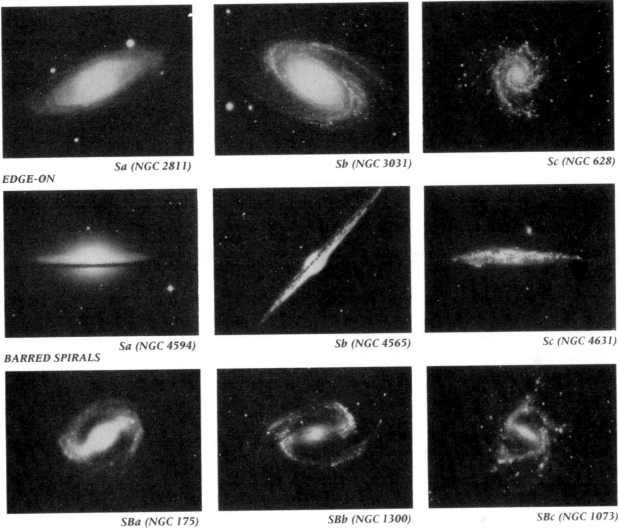

FACE-ON

Sa (NGC 2811)   Sb (NGC 3031)   Sc (NGC 628)

EDGE-ON

Sa (NGC 4594)   Sb (NGC 4565)   Sc (NGC 4631)

BARRED SPIRALS

SBa (NGC 175)   SBb (NGC 1300)   SBc (NGC 1073)

Figure 21-3   The main types of spiral galaxies, seen from various angles. The first two rows are "normal" spirals, while the bottom row shows barred spirals.

Other spirals, shown along the bottom row in Figure 21-3, have a large bar through their central regions. They are called **barred spirals** and are designated SBa, SBb, and SBc. Studies of the distribution and orbits of gas and stars in the central parts of the Milky Way have led astronomers to conclude that it is a barred spiral. The physical origin of bars is still a controversial issue, but it appears that barred spirals are common and may constitute half of all spiral galaxies.

ELLIPTICAL GALAXIES

The majority of nonspiral galaxies are classified as **elliptical galaxies** and are designated by the letter E, followed by a number that indicates how flat the galaxy appears. Type E0 galaxies are spherical in appearance, with galaxies getting increasingly flat-

tened as one goes from E1 to E7. The range of flattening of these systems is illustrated in **Figure 21-4**. A comparison of Figures 21-3 and 21-4 reveals that elliptical galaxies have a smoother, less blotchy appearance than do most spirals. As we will discuss shortly, this visual difference arises from the presence of gas, dust, and young stars in spirals, none of which are found in significant amounts in normal ellipticals.

Inquiry 21-2

Is there any way to tell whether an E0 galaxy is really spherical, or whether it is actually a football-shaped galaxy seen end-on? What other possibilities exist?

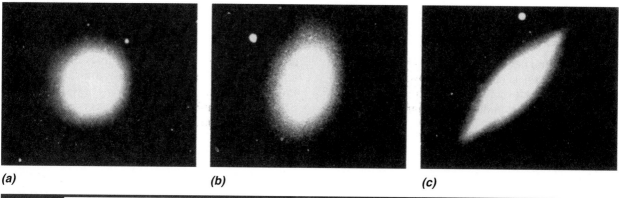

(a)   **(b)**   **(c)**

Figure 21-4 The main types of elliptical galaxies. *(a)* An E0 galaxy, NGC 4278. *(b)* An E3 galaxy, NGC 4406. *(c)* NGC 3115, a galaxy once classified as E7 now classified as S0.

Figure 21-5 An irregular galaxy, NGC 6822.

IRREGULAR GALAXIES

Irregular galaxies, as the name suggests, do not have a clear morphological signature like spirals and ellipticals, so to some extent this category includes all galaxies that do not fit into either of the two galaxy types just introduced. However, many irregulars, designated Irr, are similar to spirals in that they contain gas, dust, and young stars. Further, their gas and stellar distributions are, in many cases, somewhat flattened. What these galaxies lack is the *highly* flattened disk and spiral structure that defines spiral galaxies. **Figure 21-5** shows an example of an irregular galaxy.

S0 GALAXIES

There is also a kind of "transition" galaxy that morphologically can be thought of as halfway between spirals and ellipticals. Such galaxies are called **S0 galaxies** and have a flattened disk form like the spi-

rals, but no indication of spiral arms. They also have a large nucleus. NGC 3115 shown in Figure 21-4*c* was once classified as an E7 but is now regarded as an S0. Notice that the shape of NGC 3115 is quite similar to the Sa galaxy NGC 4594 shown in Figure 21-3, but NGC 3115 lacks the dark dust lane of NGC 4594, which is indicative of a spiral disk.

DWARF GALAXIES

As we will discuss further, the masses of galaxies cover a significant range. Astronomers find it convenient to distinguish between the bright or normal galaxies and lower luminosity galaxies usually referred to as dwarfs. Irregular galaxies tend to be of lower mass than spirals and ellipticals anyway, but the least massive examples of this class are referred to as *dwarf irregulars* (dIrrs). Like their larger counterparts, they have no apparent spiral structure, but do contain significant amounts of gas. *Dwarf ellipticals* (dEs) resemble ordinary ellipticals in morphology and overall appearance and are mostly free of gas. These objects are distinguished from *dwarf spheroidals* (dSphs), the latter having a much lower density of stars.

HUBBLE'S TUNING FORK DIAGRAM

The modern classification of galaxies was initiated by Mt. Wilson astronomer **Edwin Hubble** in the 1920s (**Figure 21-6**). As a way of displaying the types of galaxies classified, he arranged them in a diagram known as the **tuning fork diagram** (**Figure 21-7**). Barred and unbarred spirals make up the two tines of the fork, with S0 galaxies placed as a transition type between spirals and ellipticals. Modern modifications to this diagram place irregular galaxies on the end opposite the ellipticals.

Figure 21-6 Edwin Hubble.

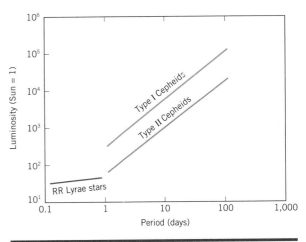

Figure 21-8 The period-luminosity relation. Notice that there are two different relations for the two types of Cepheids.

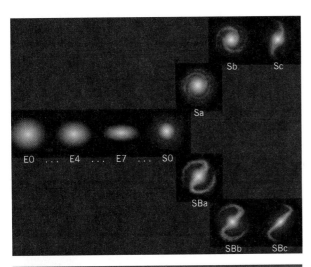

Figure 21-7 Hubble's tuning-fork diagram. Note the parallel sequences of spirals.

Hubble sought an evolutionary explanation for the various galaxy types. He suggested that elliptical galaxies are in the early stages of formation and that they later flatten out and become spirals; eventually, he thought, they break up and become irregulars. After studying other properties of galaxies—most importantly, variations in their spin and stellar content—we will see that this idea cannot be correct. In fact, the visual appearance of galaxies is not sufficient to answer many of the questions one can ask about them. For example, we cannot even determine their intrinsic size from their appearance, nor can we estimate their distance.

21.3 Distances of the Galaxies

Much of our knowledge of the universe depends on having a reasonable understanding of the distances to galaxies. In addition to being critical for probing the universe on the largest possible scales (see Chapter 23), such distances are required so that we can determine the *intrinsic* properties (luminosity, mass, and so on) of galaxies. We have already discussed this idea in the context of distances to stars that allow us to measure intrinsic luminosity from apparent brightness. Because galaxy distances are so important, we now spend some time describing the complex task of measuring them.

CEPHEID VARIABLES

During a study of Cepheid variable stars in the Small Magellanic Cloud in 1912, Henrietta Leavitt, an astronomer working for Shapley, made a discovery that would rank among the most important in the history of modern astronomy. She discovered that Cepheids obey what is called a **period-luminosity relation**, as illustrated in **Figure 21-8**. The Cepheids that vary in brightness over a time scale of just a few days have relatively low luminosities (on the order of 100 times the Sun's luminosity), whereas long-period Cepheids (periods around 100 days) have luminosities on the order of 10,000 times that of the Sun. This means that if we take enough images to be able to determine the period of the brightness variation of a

Cepheid in a galaxy, we can find out what its actual luminosity is by reading it from Figure 21-8. A calculation using the inverse square law of light compares the luminosity with the observed apparent brightness to find how far away the star is, and hence the distance to the galaxy it is in.

For example, suppose we observe a galaxy in which a Cepheid varies with a period of 10 days. Furthermore, suppose we observe the average brightness of the star to be 10^{-18} times that of the Sun. The first step in the solution is to assume (for this example only) that the period–luminosity relation for a Type I Cepheid applies to this star. We then use the period–luminosity relation to determine that its luminosity is on the order of 10,000 times that of the Sun. Using B, L, and d for the observed brightness, luminosity, and distance, respectively, from the inverse square law we find

$$\frac{B}{B_{Sun}} = \frac{L / L_{Sun}}{(d / d_{Sun})^2}$$
$$10^{-18} = \frac{10^4}{(d / d_{Sun})^2}$$
$$(d / d_{Sun})^2 = \frac{10^4}{10^{-18}} = 10^4 \times 10^{18} = 10^{22}$$

from which we determine that $d/d_{sun} = 10^{11}$; then, $d = 10^{11}$ AU $= 93 \times 10^{17}$ miles $= 1.6 \times 10^6$ ly.

When the period–luminosity relationship was first discovered, however, it was not realized that nature has provided us with two different types of Cepheids. Those seen in the Magellanic Clouds turned out to be the Type I objects that are significantly more luminous than the Type II objects they had been thought to be. Assuming that the Magellanic Cloud Cepheids were Type II objects made the calculated distance to the Magellanic Clouds much too small, giving the impression that they were part of our own galaxy. This is one reason why Shapley argued that the *nebulae* were not distant galaxies in his debates with Curtis (Section 21.1). When it was realized that they were actually the more luminous Type I objects, astronomers understood that the clouds were really farther away.

Inquiry 21-3

What kind of uncertainty had astronomers made in assuming the Cepheids were of Type II rather than Type I—random or systematic?

OTHER DISTANCE INDICATORS FOR GALAXIES

Cepheid variables can be used to determine the distance to any galaxy that is near enough for individual Cepheids to be detected and measured. With the *Hubble Space Telescope,* Cepheids can be used to determine galaxy distances to about 80 million light-years.

A Cepheid is an example of what astronomers call a **standard candle**—an object whose known intrinsic characteristics can be used along with observations to determine distances. We use the concept of a standard candle in our everyday lives: when we drive at night, we estimate, at least in part, the distances of other cars based on our knowledge of the (assumed) intrinsic brightness of the oncoming headlights. The fainter the apparent brightness of the headlights, the further away the vehicle is from us. The process of determining the *intrinsic* luminosity at a given period for Cepheids is quite complex, but is essentially based on the methods used for finding any stellar distance discussed in Chapter 15.

To study stellar systems more distant than those in which we can locate Cepheids, other brighter prospective standard candles must be found and their intrinsic characteristics determined. Then, when we see a similar object in a distant galaxy, we can assume its luminosity is known and calculate the distance. In practice, this known luminosity must be determined observationally. Ideally one would obtain a sample of galaxies that contain Cepheids, so their distances can be measured, determine the apparent luminosity of the potential standard candle within these galaxies, and then combine this apparent brightness with the galaxy distances to determine the intrinsic luminosity of the standard candle. Such a calibration of standard candles brighter than Cepheids underpins the extension of distance measurements beyond that achievable with Cepheids. A large number of such standard candles have been used with varying degrees of success over the years. In what follows, we focus on those that are the most effective and currently in use.

To expand upon this important point of the calibration of standard candles, consider the following example. You have decided to measure distances by observing the apparent brightness of a flashlight held by a willing friend. Having determined the intrinsic luminosity of the flashlight, a similar calculation to the one given for the Cepheid allows you to calculate your friend's distance. The flashlight in this example corresponds to a standard candle such as a Cepheid.

As your friend moves further away, it becomes increasingly difficult to measure the apparent brightness of the flashlight simply because it becomes too faint. Fortunately, at a distance at which you *can* still observe the flashlight, you notice that your friend is standing next to an automobile that has its headlights on. You determine that one headlight has an apparent brightness 100 times greater than the flashlight. Since both objects are at the same distance, you conclude that the intrinsic luminosity is one hundred times greater than the flashlight. You have thus calibrated the headlight (your new standard candle) using the known properties of the flashlight (the original standard candle).

If your friend now backs up the automobile, observing the apparent brightness of one headlight will again allow you to determine the distance to your friend. Most importantly, you will be able to measure much greater distances than if you only used the flashlight. This is exactly the same principle that astronomers use in calibrating increasingly luminous standard candles, thereby extending astronomical distance measurements to remote galaxies. Unfortunately, every calibration is inexact and introduces uncertainty.

Inquiry 21-4

There is an indication in the *HIPPARCOS* data that our older values for the distances of Cepheids (and their associated galaxies) were about 10 percent too small. If the Cepheids are indeed more distant than we thought, and hence intrinsically more luminous, what would this do to our estimates of their ages?

So far, the brightest individual objects discovered are supernovae. Type Ia supernovae are thought to form when a white dwarf in a binary system accretes enough mass from its companion that it exceeds the Chandrasekhar limit (Chapter 18). This leads to a collapse followed by a violent explosion. It has been determined that the maximum brightness attained by Type Ia supernovae is roughly the same for all such objects, so that they can be used as standard candles. The high intrinsic luminosity of supernovae means distances in remote galaxies can be determined. As we will see in Chapter 23, dis-

tances measured in this way have revolutionized our understanding of the universe as a whole.

The problem with supernovae is that they are relatively rare events. Astronomers cannot observe a given galaxy and simply wait for a supernova to go off. Consequently, other methods of distance determination must be used. One important technique relies on the rotation speed of spiral galaxies. In general, the brighter a spiral galaxy is, the more mass it has. We know from our discussion of Kepler's laws and gravitation that more mass requires that orbiting objects travel faster. Consequently, one expects there to be a relationship between the intrinsic luminosity of a spiral galaxy and its rotation velocity. Such a relationship was established by Brent Tully and Richard Fisher and is now known as the **Tully-Fisher relation**, after the astronomers who discovered it. It has been calibrated using Cepheid-based distances to relatively nearby spirals. By measuring the rotation velocity of a given spiral galaxy, the intrinsic luminosity of the galaxy can be calculated from the Tully-Fisher relation. As usual, the combination of intrinsic luminosity and apparent luminosity gives the distance to the galaxy. In practice, the width of the 21-cm line is observed to determine the rotation velocities of distant spirals. The idea is illustrated schematically in **Figure 21-9**.

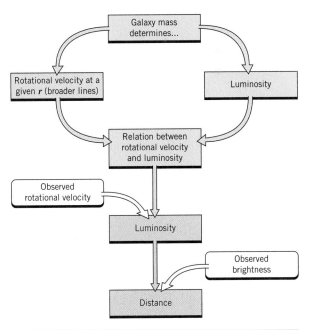

Figure 21-9 The flow of logic for finding galaxy distances using the Tully-Fisher method. The top half of the figure relies on theory, while the bottom half uses observations, which are placed in uncolored rectangles.

A similar technique can be used to measure the distance to ellipticals. Bright ellipticals do not rotate significantly, but the random velocities of stars within them probe the gravitational field (and thus mass) in the same way as the rotation speeds of stars and gas clouds in spirals. The relationship between stellar velocities and galaxy luminosity for ellipticals is known as the **Faber-Jackson relation**. Using it to measure distances is complicated by the fact that ellipticals do not contain Cepheids, but other methods have been used to calibrate the relation.

THE VELOCITY–DISTANCE RELATIONSHIP

One indirect method for determining galaxy distances is also largely responsible for our modern views on the universe as a whole (see Chapter 23). In the early part of the twentieth century, V. M. Slipher photographed the spectra of dark-line *nebulae* (whose nature was still unclear at that time) in order to measure their radial velocities. From these spectra, Slipher was able to measure the wavelengths of the absorption lines, which he found to be shifted from their usual positions. Assuming the Doppler effect produced these shifts, Slipher showed that some of the objects were moving at high rates of speed. The largest speeds he measured were in excess of 1,000 km/s, considerably greater than the highest speeds of any stars in our galaxy, and well in excess of the escape velocity from the Galaxy.

Hubble expanded upon these observations after Cepheid variable stars had established that the nebulae whose spectra contained absorption lines were external stellar systems we now call galaxies. With measured distances for some galaxies, Hubble and Milton Humason were able to demonstrate by the year 1929 that there was a correlation between the speed at which a galaxy was moving and its distance from us. They found that the more distant the galaxy, the more rapidly it was moving away from us. **Figure 21-10** illustrates this correlation for objects at different distances. Notice that the more distant galaxies appear smaller and fainter, on the average; also, in their spectrum, as judged by the Doppler shift of common spectral lines, progressively greater Doppler shifts are seen. A graph of this relationship is shown in **Figure 21-11**. This relationship, which is roughly linear, may be expressed mathematically as follows:

$$V = H_0 d$$

where V is the velocity in km/s determined from the observed Doppler shift, d is the distance in millions of light-years, and H_0 is called the **Hubble constant**. This important relationship is referred to as the **Hubble law**, or the **velocity-distance relation**. The numerical value of the Hubble constant is around 20 km/s/Mly (70 km/s/Mpc). This says that for every million light-years (Mly) farther we view, the speed will increase by 20 km/sec.

One consequence of this relationship is that it allows us to measure the distance to a galaxy simply by measuring the speed with which it is moving away from us. It should be stressed that such measurements naturally include some uncertainty, but for distant galaxies the recession speed, which is relatively easy to measure, is an effective way of determining galaxy distances.

Inquiry 21-5

How might the velocity–distance relation be used to estimate the distance to a galaxy?

Figure 21-10 shows that the galaxies are *receding* from us; that is, the spectra of most galaxies exhibit **redshifts**. We will see in Chapter 23 that this observation leads us directly to the concept of an expanding universe.

Figure 21-12 summarizes the determination of distances in the universe in what is often called the **distance pyramid** or the **cosmological distance ladder**. The standard candle at each level of the pyramid is on the left side; the distance to which it is valid is on the right. The pyramid structure emphasizes the firm foundation of distances to nearby objects, and the increasing uncertainty as distance increases. Uncertainty increases with distance not only because the observations themselves are more uncertain, but also because each level of calibration multiplies the uncertainty.

Beginning with our knowledge of Earth's diameter, radar is used to find most accurately the length of the astronomical unit. Once this is known, stellar parallaxes are observed to find distances to the nearest stars and to draw the H-R diagram. Once we have the H-R diagram, we can use it with the technique of spectroscopic parallaxes to determine distances to variable stars and star clusters. With these distances known, the properties of Cepheid variables stars allow us to find distances

CLUSTER NEBULA IN	DISTANCE IN LIGHT-YEARS	RED-SHIFTS

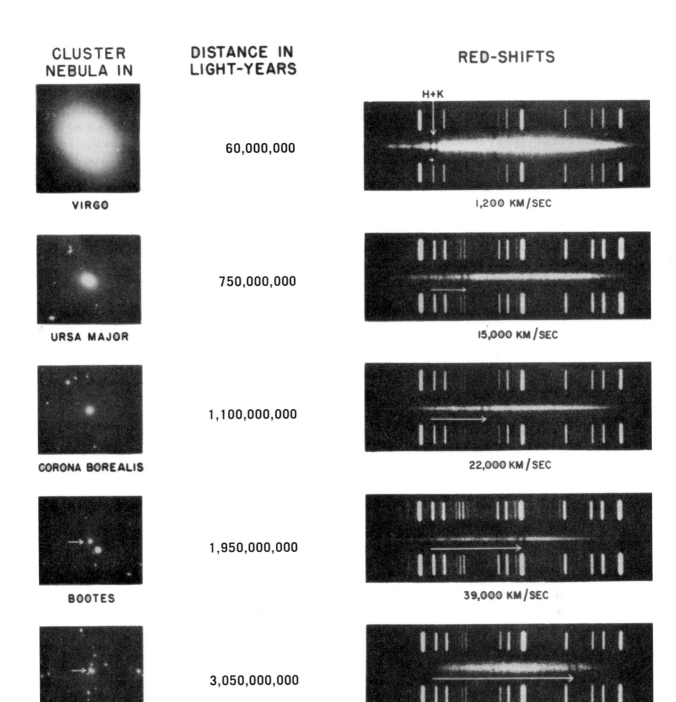

VIRGO — 60,000,000 — H+K — 1,200 KM/SEC

URSA MAJOR — 750,000,000 — 15,000 KM/SEC

CORONA BOREALIS — 1,100,000,000 — 22,000 KM/SEC

BOOTES — 1,950,000,000 — 39,000 KM/SEC

HYDRA — 3,050,000,000 — 61,000 KM/SEC

Figure 21-10 Photographs of various galaxies and their spectra, showing the increasing amount of redshift with increasing distance. The distance is inferred from the decreasing apparent size of the galaxies. The two close spectral lines indicated by "H + K" are caused by calcium in the stars composing the galaxy. Because the comparison spectrum is not calcium, the rest wavelengths of the galaxy lines do not coincide with lines in the comparison spectrum. The left end of the arrow indicates the approximate location of the rest wavelength of the shorter-wavelength calcium line.

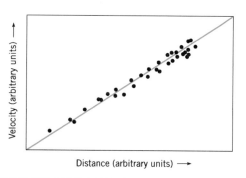

Figure 21-11 The Hubble velocity-distance relationship.

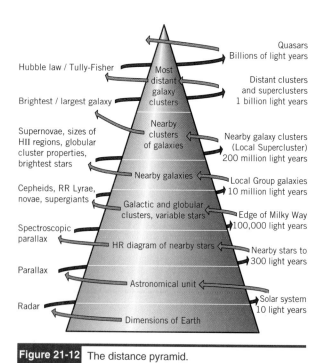

Figure 21-12 The distance pyramid.

to relatively nearby galaxies. From here, the properties of supernovae, the Tully-Fisher relation, and the Faber-Jackson relation are used to find distances to more distant galaxies. Note that not all the standard candles shown in Figure 21-12 are discussed here, either because they have become obsolete with the advent of more reliable methods, or because their physical basis is beyond the scope of this text.

To review a number of the steps in the distance pyramid, you should do Discovery 21-1, Distances Throughout the Universe, now.

21.4 General Galaxy Attributes

Our ability to measure distances to galaxies means that we can determine their key properties such as luminosity, radius and mass. Knowledge of these properties is critical in developing an understanding of how galaxies form and evolve, as well as providing insights into the origin of the *differences* between different types of galaxy.

LUMINOSITY

In principle, we can determine the total luminosity of a galaxy from its apparent brightness and its distance. In practice, this task is complicated by the fact that, unlike stars, galaxies do not have a well-defined edge. If our observations are carried out in the optical, we must also be concerned about absorption of radiation by dust within the Milky Way. Despite these caveats, observations of galaxies lead to some general conclusions about their luminosities. The observed range of galaxy luminosity is large and depends on the type of galaxy considered. Spirals span the range from about 100 times less luminous than the Milky Way to roughly twice as luminous. Disklike irregulars are typically less luminous than spirals. The brightest ellipticals are more luminous than any spiral with values up to 10 times that of the Milky Way. Dwarf ellipticals and dwarf spheroidals can be as faint as one-millionth the luminosity of the Milky Way.

DIMENSIONS

Galaxy dimensions are found from observations of the angular size along with the computed distance and the angular size formula. The lack of a well-defined edge to galaxies means that some *operational* definition must be introduced. Typically, this involves finding the distance from the center of a galaxy to where the galaxy's brightness drops to some specific low level. The issue is complicated by the fact that the extent of spirals is greater at radio wavelengths than at optical wavelengths, reflecting the fact that gas clouds extend further than the stars. However, the usual way of quoting the diameters of galaxies gives a size that includes the vast majority of the stars, With such a definition, we find that spiral galaxies have diameters that range from about ⅓ to 3 times that of the Milky Way.

Irregular galaxies tend to be smaller than spirals, with typical sizes from about 0.25 down to 0.05 times the diameter of the Milky Way.

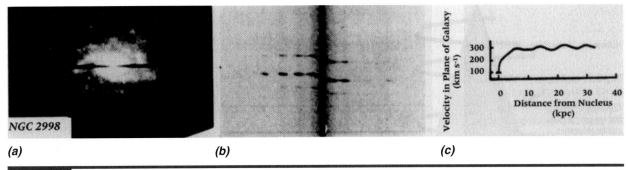

Figure 21-13 Galaxy rotation curve. *(a)* Photograph of a galaxy with the spectrograph slit across it. *(b)* The actual spectrum showing various Doppler shifts for different parts of the galaxy. *(c)* Rotation curve derived from the Doppler shifts.

Ellipticals have a large range in size, from about 0.02 for the dwarfs to four times the diameter of the Milky Way for the brightest ellipticals. Note that all these diameters include an uncertainty arising from the uncertainty in *distance* to a given galaxy. This highlights another reason why astronomers have spent so much effort obtaining reliable distance estimates.

MASS

The determination of galaxy mass is simple in principle but complicated in practice. If you were to count all the stars you could see and assume some average mass per star, you could determine a mass for all the *visible* material. What we need is a method that includes the nonluminous matter, too.

Because Kepler's third law as modified by Newton is applicable in any situation in which the gravitational field of one body controls the movement of another, we can apply it to stars moving around a galaxy, or to one galaxy revolving about another. Just as we saw that the mass of a planet having a moon can be found from Kepler's third law as modified by Newton, for a star traveling about a galaxy we have this equation:

$$(M_{\text{star}} + M_{\text{galaxy}})P^2 = A^3.$$

The masses (M_{star}, M_{galaxy}) are in units of the Sun, the period P is in years, and the distance from the galaxy's center A is in astronomical units. We will now apply this equation in a variety of ways to find the masses of galaxies.

We can find galactic masses using a method similar to that used to find the mass of the Milky Way. To find the mass of a single galaxy requires observing the motion (radial velocity) of individual stars orbiting the galaxy. Also, observing the angular distance of a given star from the galaxy's

center, and knowing the galaxy's distance, we can find the distance of the star from the galaxy's center. With this distance, and the star's observed radial velocity, we can determine the star's orbital period around the galaxy. Putting the star's period and distance from the galactic center into Kepler's third law allows us to compute the galaxy's mass interior to the orbit of the observed star.

Inquiry 21-6

How do you find the distance of a star from the center of a galaxy if you observe the star's angular distance from the center and know the galaxy's distance?

Inquiry 21-7

How do you find a star's period around a galaxy if you know the star's velocity and its distance from the galaxy's center?

This technique is crude and limited to only nearby galaxies. Uncertainties in the distance propagate into uncertainties in the mass.

Inquiry 21-8

Why is this technique limited to nearby galaxies?

A rotation curve, as we saw in Chapter 20, is a graph of the speed of galaxy rotation at each point within the galaxy. The rotation curve can be determined for spiral galaxies by making Doppler shift measurements of the galaxy spectra. **Figure 21-13** shows how the spectrograph slit is placed, and what the resulting spectrum and rotation curve look like for a spiral galaxy. The mass is then

determined through comparison of the observed rotation curve with galaxy models having various masses. This method provides an estimate of the total mass of both luminous and nonluminous material within the region where the rotation curve is measured. Since there may be more nonluminous material beyond the measured rotation curve, such mass estimates are often regarded as lower limits. Note that gas clouds in spiral disks usually extend beyond the region where rotation curves of stars can be measured, so these mass determinations often probe regions well into the dark halo.

As mentioned in the previous section, bright ellipticals have little rotation so one cannot determine their mass from rotation curves. However, the greater the mass of an elliptical, the more rapidly stars move (the basis of the Faber-Jackson relation discussed in Section 21.3), so measuring the velocities of stars in ellipticals provides some information about their mass—called a **dynamical mass**. The problem with this technique is that it can only be applied to the inner regions of ellipticals where there are plenty of stars. Further, ellipticals lack the remote cold gas clouds found in spirals that are used to determine spiral galaxy masses. One relatively recent approach to overcome these problems is to use the observed velocities of globular clusters around ellipticals to obtain mass estimates of these galaxies. Globular clusters are useful probes because they are relatively easy to identify and observe, they are plentiful around ellipticals, and some globulars are found at appreciable distances from the center of ellipticals.

The greatest uncertainty in quoting galaxy masses comes from our limited knowledge about how far their dark halos extend. With this in mind, the following numbers provide a rough benchmark for the range of galaxy masses. Typical spiral galaxies have masses ranging from double that of the Milky Way down to 1/100 of it. Irregular galaxies contain less mass, typically 0.5 to 0.0005 that of the Milky Way. The largest range is that of elliptical galaxies, which can go from a paltry 10^{-5} in the case of dEs up to 50 times the mass of the Milky Way for the brightest ellipticals.

Magnetic Fields

Because the light of stars within our galaxy is often polarized, we have inferred that the Milky Way has a magnetic field. From polarization studies, we know that a magnetic field must align dust grains that have elongated shapes. The magnetic fields of

Figure 21-14 An image of the spiral galaxy M83 indicating the flow of the galactic magnetic field as determined from radio polarization observations.

other galaxies manifest themselves by polarizing the light of their stars. In addition to using radio observations, astronomers have studied the magnetic field of M31 by analyzing the polarization of light from the galaxy's globular clusters as the radiation passes through the galaxy's magnetic field.

Figure 21-14 shows a polarization map determined from radio observations for the galaxy M83. The lines superimposed over the image indicate the direction of the magnetic field around the galaxy. The magnetic field is ordered over large distances and follows a spiral pattern. Similar data that examine the small-scale structure find the magnetic field to be chaotic in regions where star formation is rampant and supernovae occur. This is probably due to shock waves that produce bubbles in the interstellar medium, which then disturb the ordered magnetic field.

21.5 The Stellar Content of Galaxies

In Chapter 20 we found that the nature of stars in the Milky Way depends on where those stars are. Stars in the visible halo tend to be old and contain

few heavy elements, whereas relatively young, high-metallicity stars dominate the disk, at least in terms of the optical radiation it emits. The central bulge is less well understood but appears to contain stars with a range of ages and metallicities. We also saw in Chapter 20 that the observed properties of Milky Way stars provide critical input into models of how our galaxy formed. Studying the stellar populations of other galaxies helps us understand the origin of the differences between the various morphological types of galaxy and, more generally, provides insight into how galaxies of all types form and evolve.

SPIRAL GALAXIES

Spiral galaxies other than our own show many similarities to the Milky Way. In terms of their stellar populations, most spirals exhibit the three distinct components found in the Milky Way of disk, halo, and bulge. Like the Milky Way, the disk region is generally where current star formation is actively going on. As we have pointed out, stars tend to form in clusters. When a cloud collapses and fragments into a cluster of stars, objects of different masses will be formed. However, when we view this star cluster from a distance, the light will be dominated by the most massive stars, the O and B stars from the most luminous end of the main sequence. But these are also stars with high surface temperatures, so they are blue in color. The blue light from such stars dominates the light we receive from the spiral arm regions.

Inquiry 21-9

Where within the Milky Way do massive stars exist? Where are such stars absent? What do you conclude from this?

The bulge regions of spiral galaxies can be complex, but it is clear that a significant amount of star formation occurred long ago. The stars we see there now tend to older and redder than the young stars that dominate the light from the disk. The visible halo of all but the nearest spirals is difficult to observe because of the low density of stars, but generally observations indicate old, low-metallicity stars. Observations of globular clusters around spirals are currently limited but confirm this view. Thus, we can see a correlation between age, color, and location of objects in a spiral galaxy.

ELLIPTICAL GALAXIES

Elliptical galaxies resemble the nuclear regions of spirals in that they are reddish in color and are dominated by old stars. The cold molecular gas clouds in which stars form in spirals are not present in ellipticals, so there is no ongoing star formation in normal elliptical galaxies. Although the stars in most ellipticals are old, they appear to span a broad range in heavy element content, distinguishing them from the halos of spirals. Direct observations of stars in ellipticals are difficult simply because there are few such galaxies close to the Milky Way, but observations of globular clusters around ellipticals indicate some of these objects have low metallicities characteristic of the Milky Way halo, while others have metallicities in excess of the Milky Way disk. Recent observations also indicate that some ellipticals contain globular clusters that are only a couple of billion years old, suggesting that the traditional picture of ellipticals containing *only* old stars may not apply to all such galaxies. As usual, as observations improve, our ideas about astronomical objects have to be refined.

IRREGULAR GALAXIES

The majority of irregular galaxies are relatively low-luminosity, gas-rich, disklike systems. The presence of cold gas clouds ensures that star formation can occur, which means that young, massive, blue stars usually dominate the light from these galaxies, as is the case for spiral disks. However, old, low-metallicity stars and star clusters are also found in some irregulars. Some galaxies classified as irregulars are more helpfully regarded as *peculiar* and are discussed in Chapter 22.

Table 21-1 summarizes some of the most important features of galaxies of various types.

21.6 The Formation and Evolution of Galaxies

How can the different types of galaxies be understood? What causes the differences in form and stellar content that we have seen? When studies of the various types of galaxies were first carried out, the presence of young stars in spirals and the absence of such stars in ellipticals led some astronomers to suggest that the difference was due to evolution, with young galaxies looking like spirals and later evolving into ellipticals.

Table 21-1 Properties of Galaxies

PROPERTY	IRREGULAR	SPIRAL	ELLIPTICALS
Amount of gas	Substantial	Substantial	Little
Amount of dust	Substantial	Substantial	Little
Oldest stars	10^{10} years	10^{10} years	10^{10} years
Youngest stars	Newly formed	Newly formed	10^{10} years (some exceptions with $1-2 \times 10^9$ years)
Amount of rotation	Variable	Rapid in disks Moderate in bulges Little in halos	Little to none in high luminosity ellipticals Little to moderate in low luminosity ellipticals
Color	Bluish	Bluish disc Reddish halo and bulge	Reddish
Spectral type	A-F	A-K	G-K
Range in diameter[1]	0.05 to 0.5	0.3 to 3	0.03 to 10
Range in luminosity[1]	10^{-4} to 10^{-2}	10^{-2} to 2	5×10^{-5} to 10
Range in mass[1]	10^{-3} to 0.5	10^{-2} to 2	10^{-5} to 50

[1]Values are in units of the value for the Milky Way

However, it is apparent that the terms *older* and *younger* must be used with some care in describing galaxies. If we determine the ages of the oldest stars in both spiral and elliptical systems, we find that both types of galaxies contain equally old stars, around 10^{10} years old (Table 21-1). The stars in the halo region and some of the stars in the disk of our spiral galaxy are approximately the same age as the stars in an elliptical galaxy. Thus, all the galaxies that we see began to form at about the same epoch in the history of the universe, so in that sense all galaxies are equally old. However, the spiral and irregular galaxies have managed to continue star formation up until the present time.

In our discussion of the formation of the Milky Way we noted that the basic physics that controls the collapse of a protogalactic gas cloud naturally leads to the formation of a flattened, spinning disk. In other words, while the details may be complex, spiral galaxies seem to be the natural result of gas collapsing within a surrounding dark halo. Thus the question of how *all* galaxies form may be usefully recast to ask: Why are ellipticals so different from spirals? To some extent, this is a classic nature versus nurture issue. That is, were the differences between spirals and ellipticals set at birth, or did the differences evolve due to environmental effects? This topic is still hotly debated, but some clues to its resolution are described next.

OBSERVATIONAL INPUT

Before we address the theoretical considerations that are important in addressing this question, let us first summarize some of what we know observationally about the differences between ellipticals and spirals. Most obviously, spirals and ellipticals have different shapes. Additionally, spiral disks rotate rapidly, and include young stars, whereas bright ellipticals have little if any net rotation and contain no young stars. One further observation that we will return to later is that ellipticals tend to be found in regions where the density of galaxies is high, such as large clusters of galaxies, whereas spirals are more commonly found in regions of low galactic density such as the Local Group.

THE DENSITY-ANGULAR MOMENTUM PROBLEM

In the previous chapter we discussed the idea that protogalaxies rotate slowly, so that a rapidly rotating disk is the natural result of collapsing gas "spinning up." One early idea to explain the reason why ellipticals are roundish and have little rotation was that their gas did not collapse much from the protogalactic stage. The problem with this suggestion is that the density of stars within ellipticals is similar to the density of stars within spirals. If protospirals and protoellipticals were similar, this implies that the differences in rotation and shape cannot be due to the gas collapsing less in ellipticals. If this were the case, we would expect stellar densities in ellipticals to be significantly lower than densities in spirals.

An alternative suggestion is that ellipticals started off denser, perhaps because they formed earlier in the history of the universe. As the universe expands (See Chapter 23), its average density and the density of galaxies forming within it are expected to drop, so in principle this idea has some merit. If ellipticals are initially denser than spirals, they would need to collapse less than spirals in order to reach their current observed densities, and would thus not experience such a degree of spinning up.

To explain this point further, recall the ice skater analogy that we used in Chapter 20 to explain the formation and rotation of the Milky Way disk. A slowly spinning ice skater can spin faster by pulling her arms closer to her body. In the same way, slowly rotating gas spins faster when it collapses. The greater the amount of collapse, the greater is the increase in the rate of rotation. Thus, if ellipticals start off denser than spirals, they would need to collapse less to reach their current observed densities, and thus would have a slower rate of rotation.

There are at least two difficulties with this suggestion. First, our current understanding of cosmology is not consistent with massive elliptical galaxies forming so early on in the universe (see Chapter 23). Second, if protoellipticals were denser than protospirals, the dark-matter halos of ellipticals *now* would be expected to be denser than the halos of spirals. If this were the case, the dark-matter halos of ellipticals would have stronger effects than are observed on the motions of the stars within these galaxies.

If differences in initial density fail to explain the differences between spirals and ellipticals, an obvious alternative is that protoellipticals initially had less angular momentum (spin) than protospirals. This suggestion has been studied extensively with computer simulations, and it does not appear to work. These studies indicate that protogalaxies will acquire a range of initial angular momenta, but that the range is far too small to explain the dramatic differences in rotation properties between spirals and ellipticals.

The only plausible resolution of the density-angular momentum problem seems to be that, in ellipticals, angular momentum was somehow transferred from the visible regions to somewhere else. In other words, there does not seem to be a way of preventing protogalaxies from spinning up (at least if we are to simultaneously account for their stellar densities), so the only alternative is to somehow get rid of the angular momentum in the visible regions of ellipticals.

GALAXY INTERACTIONS AND MERGERS

One method of removing the angular momentum from the visible regions of galaxies has long been recognized and may hold the key to understanding the differences between spirals and ellipticals. Some of the peculiar galaxies discussed in the next chapter have odd shapes because they are in the process of merging. In some instances (such as the Antennae shown in **Figure 21-15**), it is clear that two spiral galaxies are coming together to form a single object. The signs for such a merger include "tidal tails" of stars and gas, other peculiar morphologies such as shells and rings of stars, and in the case of ongoing mergers of spiral galaxies, a huge increase in the amount of star formation. Such objects are usually referred to as **starburst galaxies** and have long been associated with spiral mergers.

Why do mergers of spirals produce starbursts? To address this question, recall that star formation occurs in giant molecular clouds. When the interstellar medium of the two galaxies collides, shock waves are sent throughout the interstellar gas that compresses these cold molecular clouds. The clouds then fragment and form stars. Thus, mergers of this kind act as the necessary trigger for molecular clouds to undergo star formation. However, unlike the Milky Way, where an individual cloud may be triggered by, for example, a nearby supernova or a passing density wave, in a major merger a large number of clouds are triggered at roughly the same time—hence the huge increase in the star formation rate.

Although many starburst galaxies are objects still undergoing a major merger, we observe other objects that show signs for a merger in the past that are now relatively subdued. Such "merger remnants" are interesting to the current discussion because they show some resemblance to elliptical galaxies. Specifically, they tend to have little rotation, to be much less flattened than spiral disks, and to have relatively small amounts of cold gas.

Figure 21-15 A *Hubble Space Telescope* image of the merging galaxies known as The Antennae.

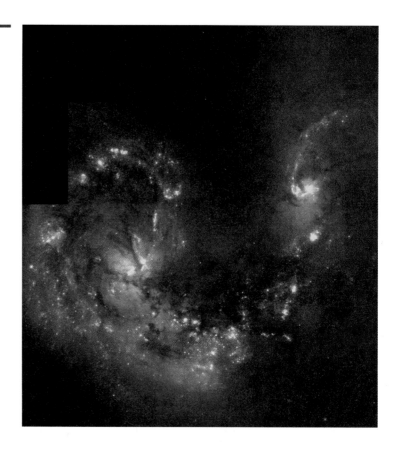

To understand how the merger of spirals might have produced such a merger remnant and eventually an elliptical galaxy, it is useful to return once more to our ice skater analogy. Imagine two rapidly spinning ice skaters having the misfortune to crash into one another. The rapid and ordered rotation of each is likely to be destroyed by such a collision. Computer simulations show that a similar process occurs when two spinning spiral galaxies collide. The resulting object has little rotation, at least in its inner regions, and the highly flattened morphology of the two disks gets transformed into a stellar distribution that is rounder. **Figure 21-16** shows snapshots from such a computer simulation, as well as a ground-based image of the Antennae, a pair of merging galaxies mentioned previously. This particular object is still in its starburst phase, but similar simulations can also reproduce more evolved merger remnants.

The big remaining question is whether the merger of two spirals eventually produces an elliptical galaxy. Again, much of the research in this area has been carried out using computer simulations, which indicate that ellipticals may indeed form in this way. Intuitively, it seems reasonable that such a process will produce something that is rounder and has less ordered rotation than spirals,

but how can such a scenario account for the fact that most normal ellipticals contain only old stars? Starbursts, after all, are so named because of their huge rates of star formation.

The answer to this question may lie in the violence of the star formation event that is produced by merging spirals. In starbursts, stars form more efficiently than they do in normal spirals like the Milky Way. This high efficiency has two consequences. First, a considerable amount of gas is used up as it gets turned into stars. Second, the strong stellar winds from young, massive stars, along with supernova explosions, may drive any remaining gas out of the starburst galaxy. To summarize, a starburst galaxy stops forming stars because it literally runs out of gas.

Inquiry 21-10

What are the two distinct processes that lead to a starburst galaxy running out of gas?

Inquiry 21-11

Why must starburst galaxies run out of gas if they are to evolve into normal elliptical galaxies?

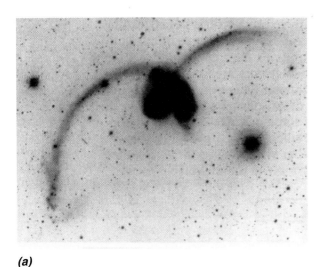

(a)

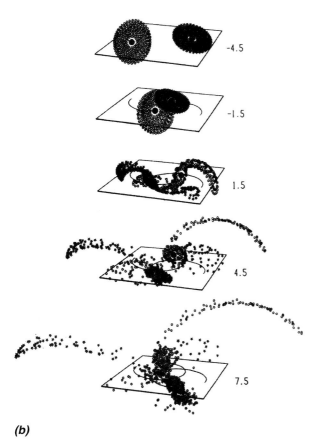

-4.5

-1.5

1.5

4.5

7.5

(b)

Figure 21-16 (a) A ground-based image of the Antennae. (b) A computer simulation of merging spirals, the end result of which closely resembles the Antennae.

One traditional objection to a spiral-spiral merger origin for elliptical galaxies comes from observations of globular cluster systems. It has been known for some time that elliptical galaxies have a greater number of globular clusters than do spiral galaxies. If two galaxies like the Milky Way collided, the resulting object would be expected to have around 400 globular clusters (since the Milky Way has about 200). What is observed, however, is that an elliptical with the mass of two Milky Ways has considerably more than 400 globular clusters. It is as if our two ice skaters were each initially wearing costumes with 200 sequins on each, but on examining the two skaters after their collision, we observed 800 or so sequins.

A possible solution to this problem is that globular clusters form when spiral galaxies merge. In the early 1990s, evidence that this occurs was uncovered by *Hubble Space Telescope* observations of nearby starburst galaxies. **Figure 21-17** shows a *Hubble* image of the central regions of the Antennae. Many of the bright blue dots in this image are young globular clusters. Consequently, it appears that elliptical galaxies *can* form from the merger of two (or possibly more) spiral galaxies. Of course, this does not guarantee that ellipticals *do* form in this way. Indeed, the formation mechanism responsible for ellipticals is still a controversial topic among astronomers.

Although there is still debate as to whether ellipticals are merged spirals, there is general agreement that galaxy mergers and interactions can change the appearance of galaxies, *and* that such events are probably quite common, particularly when the

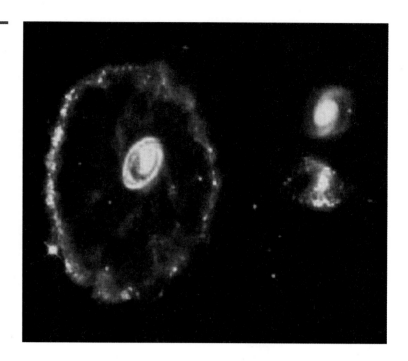

Figure 21-18 The Cartwheel galaxy. Such ring galaxies are produced by galaxy interactions.

universe was younger. Partly this is simply because the expansion of the universe (Chapter 23) implies that galaxies were closer together in the past than they are now, so that they were more likely to run into each other. Observational support for this idea is provided by observations of very distant galaxies. When we look at such galaxies we are seeing them as they were when they were much younger, simply because it takes light billions of years to reach us. One important finding in the current context is that the number of close pairs of galaxies was higher in the past than it is now. One way of reducing the number of galaxies is to suppose that some of these pairs have merged to form a single object.

Although galaxy interactions are relatively rare at the present epoch, objects like the Antennae indicate unambiguously that such events do occur at present. Another example of a recent interaction is shown in **Figure 21-18**, which is an optical image of the Cartwheel galaxy, so named because of the distinctive bright ring visible in the figure. There are other galaxies that exhibit this phenomena, and the most popular explanation is that a small galaxy has fallen into a spiral galaxy. The effect is somewhat similar to dropping a stone into a pond and producing circular ripples. Again, this reminds us that when using the properties of galaxies to investigate their formation, we must be aware of the possibility that subsequent events may have altered galaxies since the time of their birth.

Figure 21-19 The spiral galaxy M101.

DIFFERENCES BETWEEN SPIRALS: SUPERNOVA-INDUCED STAR FORMATION

So far, we have treated the formation of spiral galaxies as if all such objects were similar. For example, M51 in Figure 21-1 and NGC 628 in Figure 21-3 show galaxies with straightforward patterns of two well-defined spiral arms that can be followed out from the nucleus for a considerable distance. However, we find a much more complex spiral pattern in M101 (**Figure 21-19**). The arms seem to be made of bits and pieces, and it is not possible to trace one spiral arm for any significant distance. Thus, the different galaxies appear to be responding to different influences.

Although a spiral density wave may explain the star formation pattern in galaxies such as Figures 21-1 and 21-3, it fails seriously for galaxies such as Figure 21-19. The density-wave theory suggests that spirals should have two cleanly defined arms that can be traced to great distances, and it seems fairly clear that not all spirals fit this description. Numerous astronomers have proposed a different star formation mechanism in the more complex spirals, called *supernova-induced stochastic star formation.*

This theory begins by pointing out that a supernova is an incredibly energetic event. It releases so much energy in such a short period of time that it cannot be considered to be merely a disturbance in the structure of a galaxy; it must be a dominant force, at least for a certain period of time. Figure 19-9 is an image of the Veil Nebula, a remnant of an exploding star. The emission of light that we see is a consequence of the explosion's shock wave propagating through the interstellar gases and exciting them to radiate. This shock wave has a "snowplow effect", piling up and compressing matter ahead of itself. Star formation may be initiated by shock waves squeezing the gas clouds. If a supernova triggers more star formation, some of those new stars will be massive stars, and some fraction of those will evolve rapidly to a supernova event of their own. Thus, we have a scenario somewhat like a chain reaction, with an explosive event rapidly triggering other explosive events, which, in turn, do the same again.

In general, predicting *where* on the perimeter of the supernova shock wave the next supernova will form is difficult. Therefore, theories just imagine it to be something that can randomly take place anywhere around the supernova. This randomness gives the process a random or stochastic nature. Visualizing the outcome of this chain of events taking place in a galaxy's disk is difficult, but it is not difficult to program the scenario into a computer and display the pattern of star formation at certain time intervals (say, every few million years). The results of such calculations are shown in **Figure 21-20**, where the pattern of bright stars is shown at different times. Remember that stars and gas clouds close to the center of a spiral galaxy make a complete orbit before objects further out have completed one orbit. Thus, the regions of star formation are stretched out by the rotation, reproducing the pattern of bits and pieces of spiral arms. In a galaxy forming stars in this

Figure 21-20 A computer calculation that simulates the spiral structure of M101 caused by exploding supernovae that trigger new star formation.

way, computer simulations show that the spiral pattern tends to form and fade, being more defined at certain times and less defined at other times.

21.7 Clusters of Galaxies

Galaxies are not spread evenly through space. In some regions of the universe called *voids* (see Chapter 23), galaxies are almost completely absent, while in others tens of thousands of galaxies are bound together in clusters by their mutual gravitational attraction. The Milky Way belongs to a small system known as the **Local Group**, which we now examine.

OUR NEIGHBORHOOD: THE LOCAL GROUP

Once reliable methods for determining distances to nearby galaxies were developed, astronomers proceeded to map the universe, starting with the neighborhood of the Milky Way. It was during this mapping that we learned we belong to a group or small cluster. The diameter of the Local Group is approximately 3 million light-years. Its two largest members are familiar galaxies—our own and M31 in Andromeda. These large "sister spirals" are the dominant masses in the Local Group,

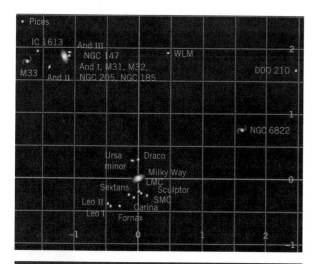

Figure 21-21 Artist's conception of the appearance of the Local Group, as seen from the outside.

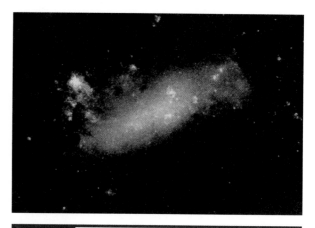

Figure 21-22 The Large Magellanic Cloud, a nearby irregular galaxy.

and associated with each are a number of smaller, less massive stellar systems (**Figure 21-21**). Because the Local Group is a gravitationally bound group of galaxies, the galaxies within the Local Group are not all moving away from one another as one might conclude from the Hubble law. The gravitational forces between the galaxies essentially overcome the expansion of the universe, much in the same way as the Earth's gravitational force will pull a baseball back to the ground, despite the fact that both the Earth and the baseball are orbiting the Sun. For this reason, some Local Group galaxies show blueshifts rather than redshifts. M31, for example, is approaching us at a speed of around 275 km/s. In a few billion years, M31 and the Milky Way will collide.

Because the Local Group galaxies are all relatively nearby, we are able to gain insight into some of the smaller, fainter galaxy types that are too dim to be seen at greater distances. For example, two external galaxies close to our own are the **Large Magellanic Cloud** and the **Small Magellanic Cloud**. The Large Cloud is shown in **Figure 21-22**. These galaxies are naked-eye objects in the Southern Hemisphere sky; because both are less than 200,000 light-years from the Sun, it is not difficult to study individual stars in them.

The Magellanic Clouds are irregular galaxies that appear chaotic, with no obvious symmetry (although the Large Cloud does show some suggestion of a central bar). In them, we find many star clusters, gaseous nebulae, and supergiant stars. These young objects indicate that star formation is still important in the Magellanic Clouds, and we

find that irregular galaxies generally have a large fraction of their mass in the form of interstellar gas. Active star formation in a few regions of dense clouds gives them their irregular appearance. The Magellanic Clouds also have globular clusters; however, there are two populations of them: one old like those in the Milky Way, and the other relatively young.

Inquiry 21-12

Do the globular clusters in the Milky Way all have nearly the same age, or is there a substantial range of ages? What do you conclude about their formation from your answer?

The nearby galaxies also provide information about interactions between galaxies. Radio astronomers have observed hydrogen gas flowing between the Magellanic Clouds and the Milky Way. This **Magellanic stream** shows that these galaxies have interacted with each other in the past. In some ways it may be more useful to think of the Magellanic Clouds as *within* the halo of the Milky Way. As we noted previously, estimates of the size of the Galaxy's dark halo suggest it extends beyond the Magellanic Clouds, and some distant Galactic globular clusters are also more remote than these satellite galaxies.

Also associated with the Milky Way are several dwarf elliptical and spheroidal galaxies such as Sculptor, Fornax, Leo I, and Leo II. The dwarf spheroidal galaxy in Sagittarius is currently being disrupted by the Milky Way, and in a few billion years will cease to be recognizable as an independent entity. It seems likely that such **galaxy cannibalism** may be a relatively common occurrence. This is important to keep in mind, because it indicates that galaxies are not isolated objects and that we cannot ignore interactions with other galaxies when we try to understand their formation and evolution. We will return to this point shortly.

The Andromeda galaxy also has several companions. NGC 205 and M32 (NGC 221), seen in Figure 1-15, are small elliptical galaxies, somewhat larger than typical dwarf ellipticals, and they are comparable in mass to the Magellanic Clouds. Therefore, it is apparent that the Local Group is basically a binary system dominated by two large spirals—ours and M31—along with 30 or so smaller galaxies. As we will see next, the Local Group is a rather pathetic cluster compared to some of the other gravitationally bound systems of galaxies that astronomers have identified.

LARGER CLUSTERS: THE CLUSTER ENVIRONMENT AND GALAXY EVOLUTION

In Section 21.6 we noted that interactions between galaxies, such as galaxy mergers, can have a profound influence on the evolution of galaxies. In large clusters of galaxies, tens of thousands of galaxies are found relatively close together, so that galaxy interactions are expected to be common. Additionally, large clusters contain significant amounts of hot gas. This gas may also lead to evolutionary changes in galaxies as they move through it. In order to understand galaxy evolution, we therefore need to understand the cluster environment in which many galaxies are found.

In the centers of massive clusters, the large amounts of mass in a relatively small volume mean that galaxies experience large gravitational forces. Consequently, galaxies in such environments move at high speeds—around 1,000 km/s. Two galaxies colliding at such high speeds are unlikely to merge into a single object. The large distances between individual stars means that such galaxies will pass through one another. (At lower speeds, the gravitational forces between two galaxies allow them to hang on to each other, eventually settling down into a single object.) If the galaxies are spirals, the gas within them *will* be affected by the collision since it is distributed throughout spiral disks. Such collisions may lead to the gas being left behind as the spirals move through one

another. Such gas removal will have an important impact on the subsequent evolution of galaxies affected in this manner.

Inquiry 21-13

What kind of changes will a spiral galaxy undergo if a significant amount of its gas is removed?

Another mechanism for removing gas from galaxies in clusters is provided by the low-density, high-temperature intergalactic medium (discussed later). As a gas-rich galaxy such as a spiral moves through this low-density medium, a process known as *ram-pressure stripping* can remove some of the interstellar medium from the galaxy. It may seem strange that a gas of low density can remove material of a higher density. The phenomenon is similar to a cyclist peddling furiously while wearing a hat. At a high enough speed, the surrounding air will blow off the cyclist's hat. Clearly, air is much less dense than even the flimsiest of hats, but it has no difficulty in rendering cyclists hatless.

Evidence that the cluster environment has some impact on galaxy evolution comes from the observation that dense clusters contain a much higher fraction of elliptical galaxies than is found in lower density environments such as the Local Group. The reasons for this are not entirely clear, particularly given that collision speeds are usually too high for spirals to merge and form ellipticals. One possibility, discussed further shortly, is that dense clusters we see today formed from the clustering of smaller groups of galaxies. The collision speeds in these smaller groups would be lower and may promote the formation of ellipticals through spiral-spiral merging. Simply removing gas from spirals through the processes already discussed is unlikely to produce an elliptical galaxy, but anemic spirals (ones deficient in gas) are found in clusters and seem to provide direct evidence that gas removal occurs. Other effects arising from close encounters between galaxies may include the removal of the outer stellar regions of galaxies. Stars and planetary nebulae have been detected apparently drifting between galaxies within clusters, suggesting they have been "tidally stripped" from their original parent galaxies through some kind of galaxy interaction.

MASSES OF GALAXY CLUSTERS

The most obvious way to determine the mass of a cluster of galaxies is to determine the mass in stars of each member galaxy and add them together. However, we have already seen that most of the mass in individual galaxies is dark matter. Even if we could add up the total masses of the individual gasses (using rotation curves and the other techniques discussed earlier), we cannot rule out the possibility that additional matter exists *between* galaxies.

A better technique is to extend the ideas used to obtain the total masses of individual galaxies by studying the motion of objects within a cluster. The obvious objects to use for such studies are the cluster galaxies themselves. Clusters of galaxies have little rotation, so the approach is most similar to that used for elliptical galaxies. Recall that for ellipticals, the speeds of objects such as globular clusters provide a method of determining the *dynamical* mass of these galaxies. The faster the globular clusters move, the greater the mass of the galaxy. Similarly, the faster galaxies move, the greater the mass of the galaxy cluster in which they reside.

The first astronomer to use this technique was Fritz Zwicky in the 1930s. He found that the *dynamical* mass estimates of clusters of galaxies obtained in this way far exceeded the mass observed in the visible components of galaxies. In fact, his work was the first indication that much of the mass in the universe was nonluminous. More

recent work has confirmed this original finding, even when intergalactic matter (discussed next) is included in the mass budget. Given the evidence that individual galaxies are surrounded by dark-matter halos, this finding is not really surprising.

Further evidence for the existence of dark matter in galaxy clusters comes from studies of **gravitational lensing**. In Chapter 19 we discussed how general relativity predicts that massive objects like a black hole or the Sun can bend light. When mass is distributed in an extended way, such as in a galaxy cluster, the distribution can act like a lens, both magnifying and distorting images of objects behind the lensing object (see **Figure 21-23**). In long-exposure images of dense galaxy clusters, gravitational lensing produces circular arcs as light from more distant galaxies is distorted by a closer cluster that acts as the lens. This is illustrated in **Figure 21-24**. It turns out that the details of the

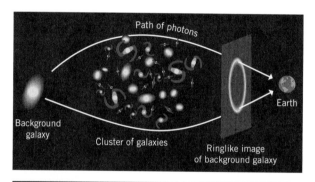

Figure 21-23 Gravitational lensing producing luminous arcs in clusters of galaxies.

Figure 21-24 Gravitational lensing in Abell 2218 showing arcs caused by foreground gravitational lenses.

distortion depend on how much mass the lensing cluster contains. Consequently, observations of the lensed galaxies provide a mass estimate for the closer cluster. Such estimates are consistent with those derived using the motions of galaxies within clusters already described, thereby providing independent evidence for the presence of dark matter in galaxy clusters.

INTERGALACTIC MATTER

In addition to a few stars and planetary nebulae between galaxies in clusters, astronomers have detected extremely hot, low-density gas referred to as intergalactic or intracluster gas. The high temperature of this gas—typically around 10 million degrees or so—means that it emits most of its energy in the X-ray part of the spectrum. **Figure 21-25** shows an optical image of the cluster Abell 1367 with contours superimposed outlining the region of X-ray emission. The total mass of the X-ray emitting gas in this cluster is comparable to the mass in stars in the individual galaxies within the cluster.

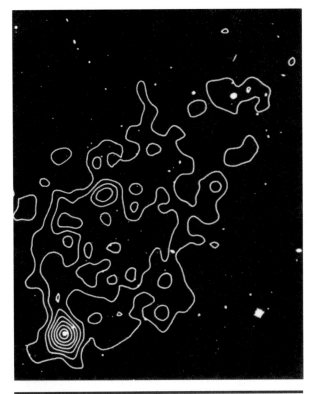

Figure 21-25 X-ray contours superimposed on an optical image of the galaxy cluster Abell 1367.

The origin of this gas, the total mass of which is comparable to all the stars within the cluster galaxies, is not completely understood. Some of it may be left over from the galaxy formation process, but the presence of heavy elements in the gas suggests that much of it must have been processed through stars in galaxies. Supernova explosions provide one method of spewing such enriched gas into the cluster environment. Gas removal by ram-pressure stripping and galaxy collisions also probably played a role in getting this enriched gas into the intergalactic medium.

Inquiry 21-14

Why does the presence of X-rays tell us that intergalactic gas has such a high temperature?

21.8 Clusters of Clusters: Superclusters

Some researchers have pointed out that within 70 million light-years of Earth there are a number of small- to medium-sized clusters of galaxies, but that there is a considerable gap of empty space between these clusters and the next significant collection of galaxies. The late University of Texas astronomer Gerard deVaucouleurs suggested that all these small clusters of galaxies, our own included, are actually part of a larger "cluster of clusters" we call the **Local Supercluster** of galaxies. With a diameter of about 100 million light-years, it contains a total mass perhaps 5,000 times that of the Milky Way.

The late Mt. Palomar astronomer George Abell, along with Harold Corwin and Ron Olowin, listed 4,076 rich clusters in both hemispheres, each of which may contain several thousand galaxies out to a distance of about 3 billion light-years. These galaxy counts do not, of course, include those rich clusters that may be hiding behind the dust plane of the Milky Way, nor those clusters that are poorer in member galaxies. The rich clusters of galaxies and their spatial relationship to each other gave us our first inkling of large-scale structure in the universe.

Figure 21-26 shows the results of a survey of relatively bright galaxies observable with moderate-sized telescopes from the Lick Observatory south of San Francisco. There are more than a million

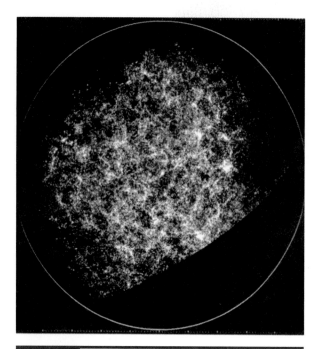

Figure 21-26 The distribution of a million galaxies.

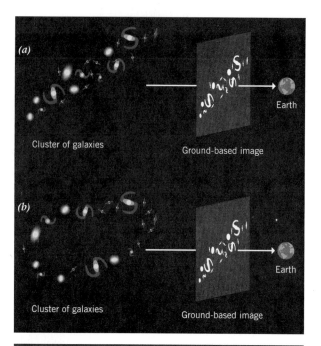

Figure 21-27 Different actual distributions of galaxies as shown in *(a)* and *(b)* can produce identical apparent galaxy distributions.

galaxies shown here. The galaxies are not distributed uniformly over the sky but fall into clumps and chains. On a small scale the appearance changes from place to place, but large areas in one location are basically similar to large areas in another. The importance of this statement will become clear in Chapter 23. The sharp curved edge is the limit of what could be observed from only the Northern Hemisphere.

Interpreting this figure is not straightforward, for reasons demonstrated in **Figure 21-27**. Suppose, for example, an observer examines a part of the sky in which galaxies are distributed in a line (part *a*). When this observer images the sky, the galaxies *appear* to be in a straight line. In part *b*, we see a case in which the galaxies are distributed in a plane around a circle. When imaged, they will *appear* to be in a straight line, too. We must always remember that we observe the positions of objects *projected onto the plane of the sky.*

Inquiry 21-15

What observations can astronomers make to determine the distribution of galaxies along the line of sight; that is, the distribution with distance?

We can calculate the true distribution of galaxies in space by obtaining spectra, measuring redshifts, and applying the Hubble law to determine distances. Our problem is to present the three-dimensional reality on a two-dimensional piece of paper. What we can do is observe galaxies contained in a narrow slice of the celestial sphere, as shown in **Figure 21-28**. Here we are looking at galaxies contained in a wide strip of right ascension but a narrow zone of declination (see Section 4.6). Once we have the spectrum of each object in the strip, we can graph its distance from us against its angular position on the sky (i.e., the right ascension).

Astronomers have obtained spectra for nearly 14,000 galaxies in two such slices in the northern and southern hemispheres (**Figure 21-29**). This remarkable illustration shows that galaxies are not uniformly distributed away from us, but that they fall along strings and circles, giving a "clumpy" or "bubbly" appearance. It is easy in such a diagram to see the clumping of galaxies into clusters, and of clusters into superclusters.

One large line of galaxies that covers the entire slice at a velocity of about 7,500 km/s in the northern part has been named the *Great Wall*, because it appears to divide the slice in two. This structure is at least 500 million light-years long, but only some 15 million light-years thick. A similar southern

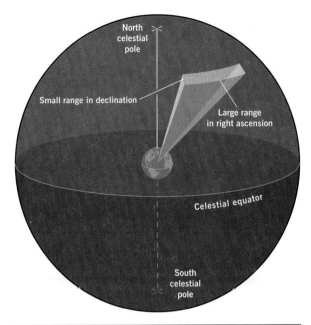

Figure 21-28 Defining a slice of the celestial sphere for use in finding the distribution of galaxies with distance.

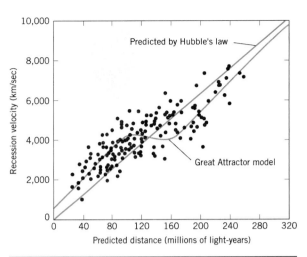

Figure 21-30 The velocities of galaxies in the direction of the Virgo supercluster differ from those expected from the Hubble law. Galaxies closer than about 150 million light-years are moving away faster than expected while those more distant than 150 million light-years are moving away slower than expected.

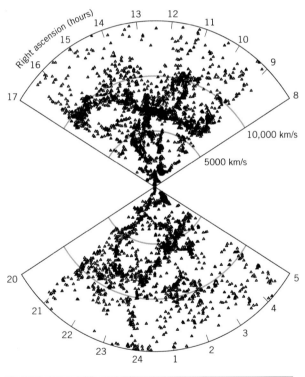

Figure 21-29 This figure shows two thin slices of the sky, as defined in the previous figure, that reveal the large-scale bubble structure of the universe. The declination is between 22° and 40° in both hemispheres.

wall extends several hundred million light-years. Furthermore, there are large regions that galaxies appear to shun. These **voids** are regions up to 500 million light-years in size containing few, if any, galaxies. A soapy froth of spherical voids with a characteristic diameter of about 150 million light-years appears to be a fundamental characteristic of our universe.

In studying the motions of the Local Group and other nearby galaxy clusters in the direction of (and closer than) the Virgo Cluster, in 1950 astronomer Vera Rubin first noticed that they all have speeds somewhat greater than expected from Hubble's velocity–distance relation. Furthermore, galaxies in the same direction but beyond the Virgo Cluster appeared to be moving somewhat more slowly than expected on the basis of their distance (**Figure 21-30**). It is as though the Local Group and the other nearby clusters are being accelerated by something, while the more distant ones are being decelerated. In other words, all these distant galaxies appear as though they are being pulled backward toward the Virgo Cluster. The interpretation of these observations has been that there is a large unseen mass, sometimes called the **Great Attractor**, which is drawing us toward it.

Figure 21-31 The Hubble Deep Field, obtained by the *Hubble Space Telescope.*

21.9 Cluster and Supercluster Formation

The formation of clusters and superclusters is intimately linked to the nature of the dark matter that pervades the universe. We will return to this issue in detail when we discuss the cosmological importance of dark matter in Chapter 23. For now, we note that the best evidence suggests that we live in a universe in which "structure," from dwarf galaxies to superclusters of galaxies, formed from the **bottom up**. This is simply a way of saying that small masses (such as dwarf galaxies) formed first, then merged or clustered together to form larger galaxies, clusters of galaxies, and eventually superclusters. There is even evidence that this process is still continuing. Some clusters of galaxies, for example, are found to be in the process of growing, either through galaxies falling in from the outside, or through two smaller clusters merging together to form a larger one.

Both computer simulations and theoretical studies indicate that the voids and walls we see on the largest scales can also be produced through this bottom-up formation. The entire process is being driven by gravitation, so that regions of the universe that are denser than average tend to add material and grow. This growth deprives lower-density regions of material, thereby eventually producing voids. Gravity ensures that, on large scales in the universe, the rich get richer.

21.10 The Hubble Deep Field and Ultra Deep Field

The variety and beauty of galaxies contained in the universe are visible in **Figure 21-31**. This image results from about 100 hours of data collection by the *Hubble Space Telescope* spread over 10 days in 1995. Known as the Hubble Deep Field, it shows galaxies fainter than anything that had ever been observed before. A companion image observed in the southern hemisphere is the chapter-opening photograph to Chapter 23.

The Hubble Deep Field (north) is a small region of the sky near the bowl of the Big Dipper. The region is so small that about 225 of them would fill the apparent area of the full moon. The area represented here is about one-forty-millionth of the entire sky. Thus, if this image is typical of the rest of the universe, the 1,600 galaxies in the image means there are around 100 billion galaxies in the universe.

More recently, the *Hubble Telescope* obtained the Ultra Deep Field (**Figure 21-32**), which combines data from both the visual and infrared spectral regions into an image showing galaxies never before observed. Collecting photons for a million seconds, the telescope makes an image that sees galaxies shortly after the Big Bang, when the first stars began to radiate. Thus, the 10,000 galaxies in the image will provide a treasure-trove of new data for astronomers to investigate.

Figure 21-32 The Hubble Ultra Deep Field, obtained by the *Hubble Space Telescope,* sees fainter galaxies than ever observed before.

Inquiry 21-16

Given that it took 100 hours of exposure time to obtain the Hubble Deep Field, how long would it take for the *Hubble Space Telescope* to observe the entire sky to the same extent?

Distances throughout the Universe

When you have completed this activity, you will be able to do the following:

- Discuss how astronomers find distances from the nearest stars to the most distant galaxies.

Throughout this book you have read about a number of techniques astronomers use to determine distances. In many cases, the same basic technique is modified to apply to different objects at greater distances. While doing the Discovery Inquiries that follow, you may want to look at previous chapters for help in remembering the techniques to be applied.

Nearest Stars

- **Discovery Inquiry 21-1a** If an astronomer measures the parallax to a star to be 0.025 second of arc, what is the star's distance?
- **Discovery Inquiry 21-1b** Suppose this star is observed to be 10^{-18} times as bright as the Sun. What is its luminosity relative to the Sun?

If we repeat what we have done in the previous two Discovery Inquiries for a large number of stars, and if we also know the spectral types for each of the stars, we can then draw an H-R diagram. Furthermore, we will assume that the H-R diagram for these stars is the same as the H-R diagram for any average sample of stars. In other words, we will assume that a large sample of distant stars is the same as a sample of nearby stars.

Fainter Stars within the Galaxy

We now observe a fainter star, whose apparent brightness is observed to be 10^{-20} times the brightness of the Sun. A spectrum of this star shows strong but extremely narrow spectral lines of hydrogen.

- **Discovery Inquiry 21-1c** What is the spectral type of the star?
- **Discovery Inquiry 21-1d** What is the luminosity class of the star?
- **Discovery Inquiry 21-1e** What is the star's luminosity relative to the Sun?
- **Discovery Inquiry 21-1f** What is the star's distance in parsecs?

The previous technique, the spectroscopic parallax, allows us to find the distance to any star for which we have a spectrum from which we can estimate the luminosity class.

Variable Stars

The spectroscopic parallax also allows us to find the distance to star clusters, some of which contain variable stars. For example, globular clusters contain RR Lyrae variable stars, which are about 30 times as luminous as the Sun. If an RR Lyrae star is observed in a cluster whose distance is unknown, we can determine its distance.

- **Discovery Inquiry 21-1g** A faint RR Lyrae star in a distant cluster is observed to be 10^{-19} as bright as the Sun. What is the distance to the cluster?

A Cepheid variable is one whose period of light variation is determined directly by its luminosity. Thus, by observing the apparent brightness and period of light variation for a Cepheid, and using the period–luminosity relation to find the luminosity, we can determine the distance.

- **Discovery Inquiry 21-1h** A Type I Cepheid variable having a period of 10 days is observed to be 10^{-20} as bright as the Sun. What is the distance to the star?

Star Clusters

Once we have the distance to a star cluster, we can find its intrinsic properties, such as linear diameter and luminosity. These properties will then be available for use further up the distance pyramid.

- **Discovery Inquiry 21-1i** If the angular diameter of the cluster in Discovery Inquiry 21-1g is 2 minutes of arc, what is the cluster's linear diameter?
- **Discovery Inquiry 21-1j** When looking at the light from the entire cluster, we find it to be 10^{14} times as bright as the Sun. What, then, is the luminosity of the cluster in terms of the Sun's luminosity?

Galaxies

The distance to a galaxy having Cepheid variables can be found using the same technique already described. Once the distance is found, the galaxy's linear size can be calculated. This linear size can be used to find distances to farther galaxies.

- **Discovery Inquiry 21-1k** What is the distance to a galaxy that appears to be identical to M31, but whose angular diameter is 100 times smaller?

Clusters of Galaxies

Similar techniques can be used to find distances to clusters of galaxies.

- **Discovery Inquiry 21-1l** If the largest spiral in the Leo Cluster of galaxies has an angular diameter three times smaller than the largest spiral in the Coma Cluster, whose distance is 322 Mly, what is the distance to Leo?
- **Discovery Inquiry 21-1m** If the brightest galaxy in the Leo Cluster is eight times fainter than the brightest galaxy in the Coma Cluster, what is the distance to Leo? What is the percentage difference from the previous determination?

Hubble Law

Once the distances to relatively nearby galaxies are determined, and their Doppler shifts are found from their spectra, we can graph the velocity–distance relation. We can then use this relation to find the distance to more distant galaxies.

- **Discovery Inquiry 21-1n** A galaxy is observed to be receding from Earth with a velocity of 55,000 km/s. Compute the galaxy's distance for a Hubble constant of 20 km/s/Mly.

Chapter Summary

Observations

- Galaxies were classified by **Edwin Hubble** into **spiral**, **elliptical**, and **irregular**. Furthermore, the spirals are divided into **normal** and **barred spirals**. The **S0 galaxies** are transition types between spiral and elliptical galaxies.

- The **Hubble law**, also known as the **velocity–distance relation**, relates the distances of galaxies and their velocities determined by the Doppler effect.

- The Milky Way is part of a small cluster of galaxies called the **Local Group**. It contains around three dozen known galaxies, most of which are dwarfs. The Large and Small Magellanic Clouds, which are observable with the naked eye, are also members.

- The properties of the various types of galaxies are summarized in Table 21-1.

- Intergalactic matter is observed by means of X-rays characteristic of a hot (million degree) gas between galaxies.

- Clusters of galaxies form into superclusters. We are in the **Local Supercluster**, which is some 10^8 ly across.

- Observations of the distribution of galaxies show a clumped or bubbly appearance. The **Great Wall** is such a clumped structure. Observations also show the presence of large regions of space in which few if any galaxies exist; these are called **voids**, and may be hundreds of millions of light-years in extent.

Theory

- The factors that determine the morphological type of a galaxy are still hotly debated. Galaxy mergers and interactions likely play some role, and may "convert" spirals into ellipticals.

- The **Tully-Fisher** and **Faber-Jackson** relation each provide a means of determining distance that is independent of the Hubble law.

- Not all spiral galaxies are alike; some have well-defined two-armed spirals, while others are more chaotic. The shapes of the two-armed spirals may be explained by the density-wave theory, and the more chaotic arms are better understood as having arisen from **supernova-induced star formation**.

- Clusters and superclusters appear to form from the clustering of smaller groups of galaxies.

Conclusions

- The understanding that the spiral nebulae are in fact **island universes** beyond the Milky Way galaxy came about when individual Cepheid variable stars were identified in the Andromeda galaxy. Using the known period–luminosity relation, a distance of 2 million light-years was computed.

- The methods used to determine the distances of galaxies are summarized in the distance pyramid in Figure 21-12.

- Galaxy luminosities, which are readily determined once the distance is known, range from nearly a million times less luminous than the Milky Way to some ten times greater. Their sizes are found from the observed angular diameter and the computed distance, and range from less than a tenth the size of the Milky Way to about four times greater. Masses, which are found from examination of the rotation curve of a spiral, the velocity spread in an elliptical, or the application of Kepler's third law, range from one-hundred-thousand times less massive than the Milky Way (dwarf elliptical galaxies) to some 50 times greater (giant elliptical galaxies).

- From the colors, shapes, and amounts of gas and dust present in galaxies, astronomers infer that elliptical galaxies had a high star formation rate early in their lives, while the rate within spirals was considerably lower.

- The amount of matter in a galaxy cluster may be found from the velocity spread of galaxies within the cluster. Such observations, which always indicate the presence of more mass than is luminous, provide evidence for **dark matter** in galaxy clusters. **Gravitational lensing** of distant galaxies by intervening dark matter causes luminous arcs around galaxy clusters.

Summary Questions

1. How was the extragalactic nature of the galaxies discovered? Why were astronomers led astray by some observations?

2. What galaxy characteristics are used for their classification?

3. What are the main differences between the types of stars found in spiral and elliptical galaxies?

4. What is the chain of observations necessary to estimate the distances to galaxies, from the nearest to the farthest? What is the impact of different types of errors on the distance pyramid?

5. What is the velocity–distance relation, and how is it used to find distance?

6. What specific types of galaxies inhabit the Local Group?

7. Explain why elliptical galaxy formation requires some method of removing angular momentum from the visible regions of these objects.

8. How do astronomers determine the fundamental properties of mass, size, and luminosity for galaxies and galaxy clusters?

9. How do astronomers study the distribution of galaxies, galaxy clusters, and superclusters in space? What are some current results of such studies?

10. What is the evidence that indicates the presence of nonluminous matter in galaxies and galaxy clusters?

11. How might the evolution of galaxies in large clusters differ from those that are relatively isolated?

12. What effects might supernovae have had on the formation of spiral arms?

Applying Your Knowledge

1. If the Cepheid in the example in the text were actually a Type II Cepheid rather than the assumed Type I Cepheid, would its distance be larger or smaller than the originally computed distance?

2. Look at the image of M31 in Figure 1-15. What assumption might you make to allow you to determine the inclination of the galaxy to our line of sight?

3. Is it possible that the drawing of the Whirlpool galaxy in Figure 21-1a was completely accurate when the Third Earl of Rosse made it in the middle of the nineteenth century, and that the galaxy changed to the appearance shown in Figure 21-1b in the intervening 150 years? Present explicit arguments based on material discussed in this book.

4. Classify everything you have in your own room into, at most, 10 categories. How might this classification scheme be useful to you?

5. Explain how our modern understanding of spectra allows us to say that there are fundamentally different classes of objects within the group of objects astronomers used to refer to as *nebulae*.

6. Explain the distance pyramid in your own words without a drawing.

7. How do you implicitly use the idea of a standard candle to judge the distance to an oncoming car at night?

▶ 8. Solve the example of the Cepheid distance given in the text, assuming the star is a Type II Cepheid rather than the assumed Type I. By what percentage does the distance change?

▶ 9. Suppose the brightest blue star in a galaxy is 20 times brighter than the brightest Cepheid. How much farther can astronomers see the blue star than a Cepheid?

▶ 10. What would be the angular size of a galaxy identical to the Milky Way if it were at a distance of 10 million light-years? If such a galaxy were observed to have an angular diameter 1/100 this size, what would its distance be? What assumptions did you make in finding this distance?

▶ 11. Look at the image of M31 in Figure 1-15. Compute the inclination of the galaxy to our line of sight. What assumption are you making in finding this angle? (*Note:* This problem requires basic knowledge of trigonometry.)

▶ 12. If a supernova in a galaxy is observed to be 10^{18} times fainter than the Sun, what is the galaxy's distance, assuming the supernova's luminosity is 10^9 times that of the Sun?

▶ 13. Suppose a galaxy is observed whose spectral lines of ionized calcium that normally would be at 3968 Å are actually at 4268 Å. What is the distance to this galaxy? What are you assuming in finding the distance?

Answers to Inquiries

21-1. You might have noted that individual stars can be seen in the globular cluster.

21-2. The answer is neither easy nor obvious. For example, there might be differences in the expected spread in stellar velocities for each case. Another possibility is that it could be flat like a plate, but seen face-on.

21-3. A systematic error.

21-4. If Cepheids are more luminous than we thought, they will burn their fuel faster and therefore be younger than we thought.

21-5. First measure the galaxy's velocity away from us. Then, assuming the velocity–distance relationship applies to the galaxy in question, find its velocity on the velocity–distance graph and read the corresponding distance from the graph.

21-6. As first described in Chapter 3, physical size is proportional to angular size times distance.

21-7. Just as you know a car traveling 60 mph will go 60 miles after one hour, the distance the galaxy will travel in one orbit will be the circumference of the circle it travels around the galaxy's center. The period will be the distance traveled (circumference) divided by the observed velocity.

21-8. Only for nearby galaxies can you resolve individual stars.

21-9. Massive (blue) stars are in the disk, in particular within the spiral arms, but are not found in the halo. Since massive stars have short lifetimes, one can conclude that the disk is where stars are currently forming.

21-10. Gas gets converted into stars (and ultimately stellar remnants). Remaining gas is blown out of the system by the winds from young stars and supernova explosions.

21-11. Normal ellipticals ceased forming stars around 10 billion years ago (and do not have cold gas). To prevent continuing star formation, gas must be removed so that starbursts can evolve into ellipticals.

21-12. As discussed in Chapter 20, the majority are around 12 billion years old. Thus, most were formed at about the same time, when the Milky Way was young.

21-13. The removal of gas from a spiral will reduce the rate at which it forms stars, so that it will look less blue and may lose its spiral pattern.

21-14. Wien's law relating temperature and the wavelength of maximum intensity.

21-15. Obtain a spectrum and measure the Doppler shift. The distribution along the line of sight can be determined by interpreting Doppler shift in terms of galaxy motion away from us and applying the velocity–distance relationship.

21-16. If the entire sky is 40 million times larger than the HDF area, it will take 100 hours $\times$ $40 \times 10^6 = 4 \times 10^9$ hours = 500,000 years of actual exposure time.

Peculiar and Unusual Extragalactic Objects

22

Eventually, we reach the dim boundary—the utmost limits of our telescopes. There, we measure shadows, and we search among ghostly errors of measurement for landmarks that are scarcely more substantial. The search will continue. Not until the empirical resources are exhausted, need we pass on to the dreamy realms of speculation.

Edwin P. Hubble, Realm of the Nebulae, *1936*

For many years, astronomical theories emphasized the majestic, unchanging calm of the universe as we came to appreciate the enormous time scales over which cosmic changes took place. Sophisticated technology, however, has revealed that dynamic activity and incredibly energetic phenomena are occurring in many extragalactic systems, leading to a new discussion of a "violent" universe. In this chapter, we describe the observational characteristics of these "active galaxies" and related objects. We also discuss theoretical ideas concerning how the vast amounts of energy associated with many of these systems are produced. In doing so, we will see how astronomers have attempted to produce a unified model for apparently diverse phenomena.

22.1 Peculiar-Looking Galaxies

Astronomers have observed many galaxies whose optical appearances are peculiar. Even among the irregular galaxies, these galaxies stand out as highly unusual. **Figure 22-1a** shows an old photograph of an object located in the constellation of Cygnus and known at one time as VV72. The object is so distant that images from this time show very little detail, thus understanding the origin of the peculiar morphology was initially difficult. A more recent image is shown in **Figure 22-1b**. Combined with an analysis of the spectrum from this object, most astronomers have concluded that a galaxy merger or interaction is probably responsible for the odd appearance of this galaxy. Specifically, it has been suggested that an ordinary elliptical swallowed a smaller, gas-rich galaxy.

A nearby example of a peculiar galaxy known as NGC 5128 is shown in **Figure 22-2**. Since this galaxy is so much closer than VV72, it is much easier to study it in detail and understand how it developed its current appearance. The underlying starlight of NGC 5128 is fairly typical of an ordinary elliptical galaxy, with the exception that some stars and a handful of globular clusters seem to be relatively young. The most obvious visual feature, however, is the broad band of dust that crosses the galaxy and blocks starlight. Recall from the previous chapter that dust is found in spiral and irregular galaxies. The presence of young stars and dust in NGC 5128 strongly suggests that a smaller gas-rich galaxy has fallen into a preexisting elliptical. If NGC 5128 were as distant as VV72, it would have a similar optical appearance. Indeed, this is one rea-

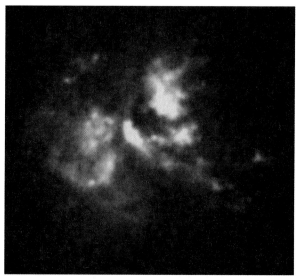

(a)

(b)

Figure 22-1 The galaxy VV72 may be the result of a collision between two galaxies. *(a)* An old image that illustrates the difficulty in understanding the origin of peculiarities in distant galaxies. *(b)* A more recent image.

Chapter opening photo: The peculiar galaxy Arp 220 is the result of two colliding spiral galaxies.

Figure 22-2 NGC 5128, a strong radio galaxy known as Centaurus A (Cen A).

of radio astronomy, an amateur astronomer named Grote Reber built a radio telescope and systematically mapped the sky in radio wavelengths.

Inquiry 22-1

If most radio sources are extragalactic, what does this imply about the amount of energy they emit?

Inquiry 22-2

Why would a galaxy composed of a mixture of ordinary stars like those studied in this book be expected to emit at least some energy in the radio region of the electromagnetic spectrum?

son astronomers think the peculiar morphologies of NGC 5128 and VV72 have the same origin.

We saw other examples of galaxy mergers and interactions producing unusual-looking objects in Chapter 21. However, we should not assume that collisions are responsible for all peculiar galaxies unless we have direct evidence that such processes may have been involved. Further, galaxies do not only emit radiation in the optical. Basing all our conclusions on the visual appearance of a galaxy alone ignores important information obtained at other wavelengths.

22.2 Radio Galaxies

The first galaxy discovered to emit radio waves was the Milky Way itself. In 1931, a Bell Telephone engineer named **Karl Jansky**, who was testing antennae for long-distance telephone communications, discovered that every day his equipment detected a strong source of radio static. Each day when the interruption came, it occurred four minutes earlier—exactly what one would expect of an astronomical source, which followed the sidereal rather than the solar clock. In actuality, Jansky was observing radio signals from the gas in the plane of the Milky Way. By accident, he had founded the science of radio astronomy.

At the end of World War II in 1945, surplus radar equipment became available and astronomers began to survey the sky for radio sources. As discussed in Chapter 12 when we introduced the ideas

THREE BRIGHT RADIO GALAXIES: CYGNUS A, CENTAURUS A, AND M87

Naturally, the brightest objects were found first, and the first *extragalactic* radio object discovered was in the constellation of Cygnus. It was designated Cygnus A (Cyg A). Astronomers wanted to see whether there was an optically observable object coincident with the location of Cyg A. Finding an optical counterpart to a radio source was not easy in the early days of radio astronomy, however, because the resolving power of radio telescopes was poor and astronomers could not obtain accurate positions. Once this was accomplished, however, astronomers found that Cyg A and the peculiar-looking VV72 just described were the same object.

Between the late 1960s and early 1970s, the resolving ability of radio telescopes had improved to the point where contour maps of the radio emission from Cygnus A (**Figure 22-2**) showed it to come from two lobes placed symmetrically on either side of the optical galaxy. These radio lobes appeared to be disconnected from the optical galaxy, and to extend many light-years from it.

The distance to the optically visible galaxy was found using the observed Doppler shift to compute the galaxy's velocity, and then applying the Hubble law. From its distance of 750 million light-years and the observed intensity of its radio radiation, astronomers used the inverse square law of light to find that Cyg A emits about as much radio energy as the total energy emitted by the Milky Way over *all* wavelengths. The large amount of energy emitted at radio wavelengths compared with the energy at visible wavelengths defines what we call a **radio galaxy**.

Another strong radio source, Centaurus A (Cen A), coincides with the elliptical galaxy NGC 5128 in Centaurus (Figure 22-2). We have already noted that this galaxy is classified as a peculiar elliptical, primarily because of the prominent dust lane, but its visual appearance gives no obvious indication of why it is such a strong radio emitter. It, too, has the symmetrical double-lobed structure of Cygnus A that is characteristic of the majority of strong extragalactic radio sources. Furthermore, the double-lobed structure of Cen A occurs on many size scales. As you view Cen A at successively higher resolution, the double-lobed structure continues into the central regions, as shown in the contours of **Figure 22-4**. Finally, in its innermost region, the galaxy shows a single, intense, elongated piece of material off to one side, referred to as a **jet**, that is also aligned with the lobes. While not all double-lobed radio sources have optically

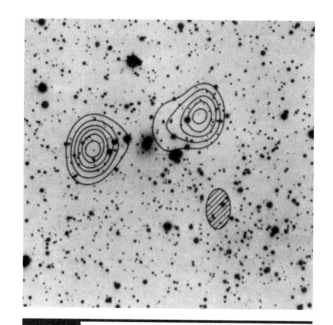

Figure 22-3 The typical double-lobed structure of a radio galaxy. In this photograph, the radio contours of the radio galaxy known as Cyg A are shown superimposed on a photograph of the optical galaxy between the radio lobes. Notice that the region of radio emission is much larger that the visible galaxy.

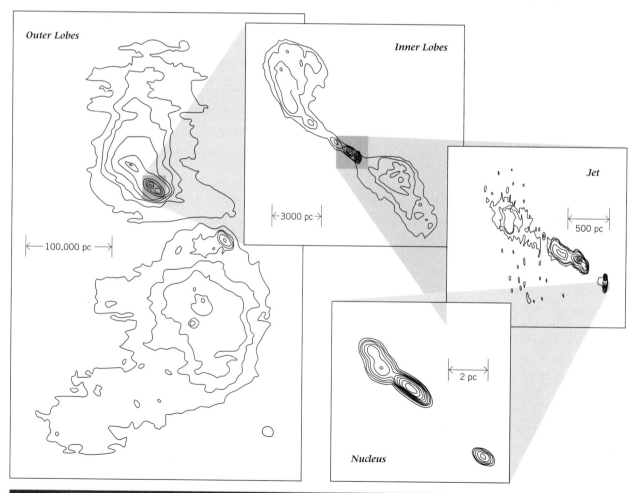

Figure 22-4 The double-lobed structure of Cen A is shown on a variety of scales, from the outermost lobes to the innermost jet.

identified galaxies associated with them, astronomers think that all such radio sources are associated with galaxies. Astronomers assume that the "missing" galaxies are simply too faint to be observed at present. This idea can be tested as advances in technology allow astronomers to detect fainter galaxies.

As already discussed, the prominent dust lane in Centaurus A and the presence of some young stars have led to the suggestion that a preexisting elliptical galaxy swallowed a smaller disk galaxy. Such an event would allow an old elliptical to undergo a small burst of star formation. Whether galaxy mergers have a connection to the double-lobed radio phenomena is uncertain, but such events do provide a mechanism for funneling gas into the central regions of objects like Centaurus A. We will discuss the relevance of this fact shortly.

At radio frequencies, the intensity of the radiation from these galaxies *increases* with increasing wavelength, opposite to the behavior of ordinary thermal emission for all but the coldest black bodies (see Figure 19-21). In addition, the radiation is observed to be polarized. This strongly suggests that the radiation is synchrotron emission similar to that observed in the Crab Nebula. Such radiation is produced by electrons accelerating rapidly through a magnetic field. This information is a useful clue in understanding the vast amounts of energy emitted by these objects, but is not sufficient to tell us the source of this energy. We will return to this topic in Section 22.4.

Inquiry 22-3

What conditions are required to produce synchrotron radiation?

Even if some of the strong radio sources we see are the result of recent collisions between galaxies, many are not. An example is the giant elliptical galaxy NGC 4486, more commonly known as M87, which dominates the Virgo Cluster of galaxies (see Figure 21-2). This giant elliptical galaxy is one of the most luminous known galaxies at optical wavelengths, giving off more energy than the brightest known spiral galaxy. It also has strong radio emission, which is concentrated into a jet that protrudes from the central regions of the object. A photograph of M87 using a short exposure to reveal just the bright core of the galaxy (**Figure 22-5a**) shows the jet more clearly. **Figure 22-5b** is a high-resolution image of the jet showing its many components.

Figure 22-5b shows another important feature found in many radio galaxies. Notice that the jet appears to emanate from a small, bright point at the center of NGC 4486. These compact central sources typically emit considerable amounts of energy over a large range of wavelengths and are sometimes highly variable. Such objects are referred to as **active galactic nuclei** or **AGN**, so the galaxies that contain such nuclei are often called **active galaxies**. Although all radio galaxies

(a)

(b)

Figure 22-5 *(a)* A short-exposure ground-based image of NGC 4486 showing the jet. *(b)* A high-resolution Hubble Space Telescope image of the jet.

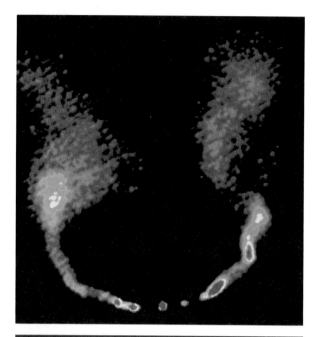

Figure 22-6 The head-tail radio galaxy NGC 1265.

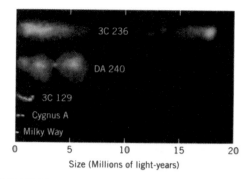

Figure 22-7 A schematic diagram showing the extent of some radio galaxies compared with the Milky Way.

Figure 22-8 A high-resolution image of Cyg A. Note the conduit carrying material from the optical galaxy to the radio lobes.

are thought to be associated with AGN, not all AGN are found in radio galaxies. We return to this point shortly.

Inquiry 22-4

What is one possible hypothesis for the formation of the jet in M87?

One additional class of radio galaxy is that shown in **Figure 22-6**. This galaxy is an example of what is called a **head-tail galaxy**. Its form appears to come from its rapid motion through the intergalactic medium, in which its gas and the associated magnetic field is swept back by the pressure of the intergalactic gas. The more rapid the motion, the more strongly the magnetic field is swept back, and the straighter the tail.

One of the surprising aspects of the peculiar radio galaxies is their sheer size. **Figure 22-7** shows the scale of some double-lobed radio sources. Note the comparison with the Milky Way.

HIGH-RESOLUTION OBSERVATIONS OF RADIO EMISSION FROM GALAXIES

Modern radio interferometers have provided far greater resolution than the observations of early radio astronomy. Astronomers now carry out observations with resolution often greatly exceeding that of optical observations. The detail apparent in recent images of radio galaxies is remarkable, as **Figure 22-8** of Cyg A shows. The greater resolution has occurred thanks to technology that has improved clocks and computers. Improved clocks allow us to time the arrival of energy at telescopes spread over the entire Earth in order to synchronize and combine the collected signals later. Improved computer technology allows us to handle the torrent of data that radio interferometers produce.

Astronomers have long been puzzled as to how material flowed from the central source to the outlying radio lobes. Clues leading to an answer come from high-resolution observations of Cyg A, which show conduits carrying material from the optical galaxy out to the distant lobes (Figure 22-8). We return to models of AGN and the cause of the ejection of material in Section 22.5.

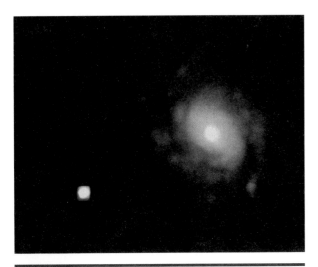

Figure 22-9 The Seyfert galaxy NGC 1068.

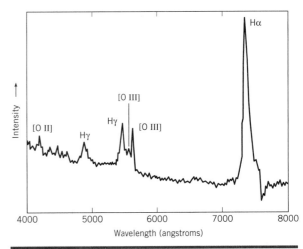

Figure 22-10 The spectrum of the nucleus of a Seyfert galaxy.

22.3 Seyfert Galaxies and Other Active Galaxies

During World War II, Carl Seyfert, an astronomer at Vanderbilt University, noted some galaxies that were distinctive enough in appearance and spectra to merit an entirely new category. Most of these **Seyfert galaxies** look like normal spirals, and some look like ellipticals. The Seyfert galaxy NGC 1068 is shown in **Figure 22-9**. Seyfert galaxies are characterized by two observational features. First, the nucleus of the galaxy is bright compared to the rest of the galaxy, and in a short-exposure image the nucleus even appears starlike. Second, the spectrum of the nucleus in a Seyfert galaxy shows both a continous spectrum and emission lines (**Figure 22-10**). Such emission lines are not found in the spectra of normal spirals. The emission lines show that there is a large amount of hot gas in the nuclear region; the widths of the lines indicate that the gases in the nucleus are in a state of violent agitation with speeds of thousands of km/s.

Inquiry 22-5

Why do broad lines indicate violent agitation in a Seyfert galaxy?

Distances to Seyfert galaxies can be found from the Doppler shift of the spectrum. From their distances and observed brightnesses we find that

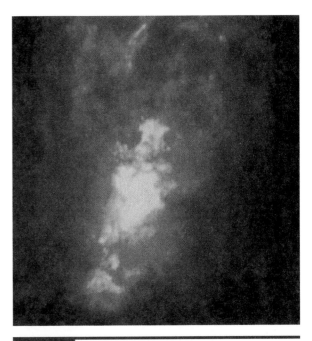

Figure 22-11 The center of NGC 1068 as observed by HST reveals previously unknown detail. Compare with the ground-based image in 9.

Seyfert galaxies are more luminous than normal spiral and elliptical galaxies. Furthermore, their radiation appears to come from an extremely small region, perhaps some 5 to 100 times the size of the Solar System. The generally accepted explanation is that there is a massive black hole in the nucleus; the radiation we observe comes from material falling toward the black hole and interacting with an accretion disk before crossing the event horizon. One piece of evidence for this model comes from HST images of the Seyfert galaxy NGC 1068 (**Figure 22-11**), which show clouds of ionized gas in the innermost regions. Other observations indicate that these clouds trace out a cone of radiation coming from an obscured area thought to contain the supermassive black hole. The model to produce this cone of clouds implies the presence of a disk. Ionizing radiation is emitted perpendicular to the disk, analogous to the way in which bipolar nebulae were produced in the star formation processes discussed in Chapter 17 (see also Section 22.5).

BL Lac objects also produce significant amounts of radiation from a small region, but lack the emission lines found in the spectra of Seyfert galaxies. The name comes from the object *BL Lacertae,* which was once thought to be an ordinary variable star. Its high redshift showed it to be an extragalactic, and thus highly luminous, object. Some BL Lac objects change in brightness by 100 percent in less than a month or even 20 percent in a day. Because the emitting region of an object cannot be larger than the time it takes light to travel across it (see Chapter 19), this means many BL Lac objects have sizes of no more than a light-day. If we block the light of the BL Lac object with a small obscuring spot at the telescope's focus, we are able to see faint radiation coming from the areas surrounding the obscuring spot, just as we are able to see the solar corona during an eclipse of the Sun. When we analyze this faint radiation, we find its spectrum to be similar to that of an elliptical galaxy. Astronomers have therefore concluded that BL Lac objects are AGN within elliptical galaxies.

22.4 Quasars

The first step in the development of any new field of science is to catalog the observations. During the 1950s, large numbers of new sources of radio radiation were discovered and cataloged. The next step was for astronomers to attempt to find an optical object corresponding to every radio source. We have already noted that the low resolution of early radio telescopes made the identification of many objects difficult or ambiguous. However, a high-resolution interferometer built at Cambridge University in England in the 1950s made it possible to identify many of the newly discovered radio sources with optical objects. Some turned out to be external galaxies, but one of them, 3C48 (the 48th object in the *Third Cambridge Catalog*), had no obvious optical counterpart other than what appeared to be a faint star.

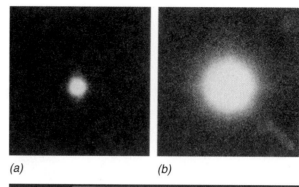

(a)   *(b)*

Figure 22-12 Two quasars. *(a)* 3C48. *(b)* 3C273, showing the jet.

The discovery of a strong radio source that looked like a star created a stir of interest in the astronomical community, because most previous radio sources had turned out to be galaxies. Normal stars were not expected to emit strongly at radio wavelengths. Other radio "stars" were soon found. The brightest known, 3C273, was found to be a point source (as opposed to an extended source like a galaxy) by monitoring its brightness very carefully as the sharp edge of the Moon passed in front of it. So starlike did these objects appear that the name **quasi-stellar radio source**, or **quasar**, was introduced. Although the first quasars were discovered at radio wavelengths, it has since been found that only about 10 percent of these objects are strong radio sources. The *radio-quiet* quasars are more accurately referred to as quasi-stellar objects, or QSOs, although the term *quasar* is sometimes used generically.

Because quasars are stellar in appearance, images of them rarely look particularly interesting. **Figure 22-12a** shows a photograph of 3C48. **Figure 22-12b** shows a photograph of the brightest quasar, 3C273, which does have a structure in the form of a small jet emerging from one side.

Inquiry 22-6

What does the jet in 3C273 remind you of?

THE SPECTRA, DISTANCES, AND LUMINOSITIES OF QUASARS

Once a radio object has been identified optically, the next step is to obtain its spectrum. These radio stars interested astronomer Maarten Schmidt of Mount Palomar Observatory, home of one of the

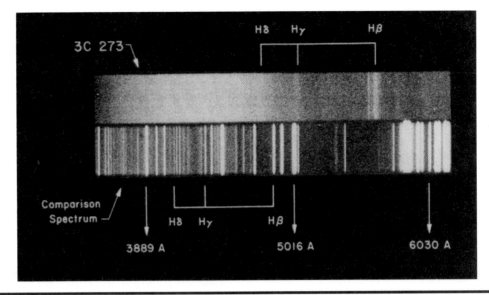

Figure 22-13 The spectrum of the quasar 3C273, which has a redshift 16% the speed of light.

few telescopes that was powerful enough to photograph their spectra. The spectrum of 3C273 (**Figure 22-13**) was found to consist of a continuous background and several bright emission lines, not unlike the spectrum of a Seyfert galaxy. However, the wavelengths of the emission lines did not seem to coincide with those of any known chemical elements.

Schmidt realized that there was a simple interpretation—the bright lines were not due to some mysterious element unknown on Earth but were simply considerably redshifted Balmer lines of hydrogen. Astronomers realized that these objects were not stars at all, but extragalactic objects emitting huge amounts of energy.

> **You should do Discovery 22-1,
> Measuring a Quasar's Redshift,
> at this time.**

Astronomers often express the redshift of a galaxy as $z = \Delta\lambda/\lambda$ the ratio of the shift of wavelength to the rest wavelength. For stars, $\Delta\lambda/\lambda$ is typically about 10^{-4}, and the shift is generally explained in terms of the Doppler effect. Then $\Delta\lambda/\lambda = V/c$, so that in this example, the velocity would be $10^{-4} c = 30$ km/s.

The redshift of the spectral lines in the quasar 3C273 was, however, found to be $z = \Delta\lambda/\lambda = 0.16$.

Interpreted as a Doppler shift, this means that the object is receding from us at a speed of 16 percent the speed of light. Assuming that the relationship between velocity and distance discovered by Hubble is applicable to 3C273, Schmidt was then able to determine its distance.

> **Inquiry 22-7**
>
> What is the distance to 3C273? Use all the information you have and a Hubble constant of 20 km/s/Mly.

If you attempted the above inquiry you should have discovered that this object is more than 2 *billion* light-years away. From its distance and its observed brightness of 10^{-16} that of the Sun, the object's luminosity can be calculated. The result is it radiates between 10 and 100 times more energy than the entire Milky Way galaxy. Schmidt had discovered a new class of superluminous extragalactic objects.

> **You should do Discovery 22-2,
> A Scale Model of Distances in the
> Universe, at this time.**

The quasar 3C48 was found to have a redshift of $\Delta\lambda/\lambda = 0.37$. If this is a Doppler shift, this quasar is moving away from us at a speed of 37% of the speed of light, placing it more than twice as far away as 3C273. Astronomers have observed numerous quasars having redshifts, $\Delta\lambda/\lambda$, greater than one. The largest known redshift (as of 2004) has $\Delta\lambda/\lambda = 6.4$. Does this mean that the object is receding from us at more than six times times the speed of light? No. The reason is that the Doppler shift formula we have used up to now, $\Delta\lambda/\lambda = v/c$, is valid only when velocity is far less than the speed of light. For larger values, the *relativistic* Doppler effect formula (given in Discovery 22-1) must be used. This formula properly includes effects arising from Einstein's theory of relativity.

It is also important to note that the shift in wavelength associated with quasars is not the same as the Doppler shift as we originally introduced it in Chapter 14. The Doppler shift that we measure for stars in the Milky Way, for example, arises from the motion of stars through the Milky Way relative to our telescope. The redshifts observed for quasars and other distant objects are the result of the Universe itself expanding. Such redshifts are referred to as **cosmological redshifts** and objects that are sufficiently remote to exhibit the effect are said to be at **cosmological distances**.

The spectrum of one of the most distant quasars detected to date is shown in **Figure 22-14**. The narrow peak marked "H" is normally at a wavelength of 1,216 Å in the ultraviolet. In the spectrum of this quasar, the line has a wavelength of about 7,200 Å, so the redshift is $\Delta\lambda/\lambda = (7{,}200 - 1{,}216)/1{,}216 = 4.9$. By employing the relativistic

Doppler effect formula you will find this corresponds to 94 percent the speed of light. Such a large redshift also implies that when the light we observe left this quasar, the universe was a fraction of its current age, the precise value depending on cosmological factors discussed in the next chapter. This reflects the amount of time it takes light to cross these great distances.

The Variability of Quasars

Because quasars are stellar in appearance, many of the brighter ones had gone unrecognized on old photographs of the sky, looking just like uninteresting faint stars. However, some major observatories have permanent photographic collections spanning long time intervals, and when the brightest quasar, 3C273, was found on old photographs, it soon became evident that it varied significantly in brightness in only a few months. Other quasars showed similar irregular light variations, sometimes changing brightness by 50 percent a day. Indeed, the *HEAO-2* satellite discovered a quasar that varies its X-ray brightness significantly in as little as 200 *seconds*.

Inquiry 22-8

What does the rapid variability of these objects indicate about their size? If an object varies its brightness significantly in only 3.7 days (a hundredth of a year), what is its maximum size? Express this size in terms of the size of our planetary system (40 AU) and also in terms of the distance to the closest star, α Centauri.

We can now see why quasars are perplexing to astronomers. The great distances to quasars, combined with their observed brightness, imply that they emit as much energy as 1,000 times the energy of the entire Milky Way, but from a region that is as small as one-millionth the diameter of the galaxy (less than 0.1 ly). As you can imagine, these discoveries left astronomers in something of a state of shock.

The Central Monster

Clearly, the greatest challenge in uncovering the nature of quasars is to understand how they generate such enormous amounts of energy from such a small volume. At the time quasars were discovered, the most efficient energy source familiar to

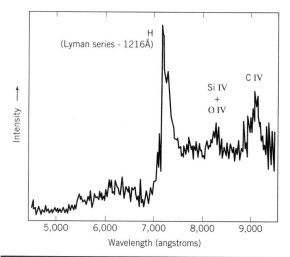

Figure 22-14 The spectrum of the high-redshift quasar PC 1247 + 3406, which has a redshift of more than 90% the speed of light.

astronomers was the fusion process that occurs in stars (Chapter 16). Recall that when four hydrogen nuclei fuse to become one helium nucleus the *lost mass* is converted into pure energy according to $E = Mc^2$. However, the mass involved in such a reaction is only a small fraction of the mass of the original four hydrogen nuclei. Detailed calculations led to the conclusion that nuclear fusion could *not* account for the energy output of quasars.

Gravity can provide a more efficient source of energy. When one drops an object close to the surface of the Earth, the object accelerates as it falls. This means that its energy of motion (kinetic energy) increases. The principle is used to great effect in hydroelectric power plants in which the kinetic energy of falling water is converted to electrical energy. In order to generate the amounts of energy emitted by quasars, the hypothesized falling material must reach great speeds. This, in turn, requires a strong gravitational field. In addition, the small physical extent of quasars means this strong gravitational field must be produced by a relatively small object. Putting all these factors together, astronomers came up with a possible candidate. The vast majority of astronomers are convinced that quasars are powered by a **supermassive black hole**. Specifically, to generate sufficient energy, the black hole that powers quasars must have a typical mass of 10^9 solar masses or more. These objects are sometimes referred to as *monsters,* partly because they are so much bigger than *stellar* black holes resulting from the death of massive stars.

One attractive aspect of the supermassive black hole model is that it leads to a number of testable predictions. For example, as material begins its descent into the black hole it is likely to have some initial angular momentum. As this material falls inward it will rotate more rapidly in order to conserve angular momentum, eventually settling into an **accretion disk (Figure 22-15)**. We have discussed the same basic idea in many previous parts of this book. When material falls inwards it tends to produce a rotating disk, whether we are discussing the formation of our Solar System or a star, mass transfer between binary stars, or the formation of the Milky Way. In the specific case of an accretion disk around a supermassive black hole, the rotation speed will be sufficiently large that friction will heat the gas to the point where the disk emits much of its radiation at X-ray wave-lengths. Thus, this model predicts that powerful quasars are also strong X-ray emitters. Observations confirm this prediction.

A critical feature of the black hole model is that it naturally associates quasars with galaxies. There are two related reasons for this. The first is that one plausible place where a supermassive black hole might form is at the center of a bright galaxy. Thus, quasars are viewed as a class of AGN. If this is indeed the case, it makes sense to suppose that other, less powerful, AGN also derive their energy from central black holes. We return to this idea of a "unified model" of AGN shortly.

An important question raised by the idea that quasars are AGN is: Where is the surrounding galaxy? After all, one defining feature of quasars is that they are pointlike. However, the high luminosities of quasars, which greatly exceed those of typical galaxies, allow for the possibility that the underlying galaxy has been overlooked. One of the most important programs carried out with the *Hubble Space Telescope* was to search for *quasar fuzz* around quasars. The name stems from the fact that the hypothesized galaxies surrounding quasars would, at these great distances, resemble little more than a fuzzy patch of light. *HST* and ground-based observations have detected the predicted fuzz around relatively nearby quasars and have further revealed that many of the host galaxies have peculiar morphologies consistent with galaxy interactions or mergers. An example of quasar fuzz is shown in **Figure 22-16**.

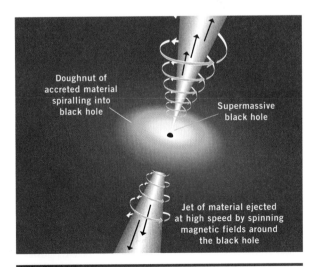

Figure 22-15 A black hole accretion disk of a quasar. Note the jet of gas squirted out at near the speed of light by the violent processes near the black hole.

Figure 22-16 An HST image, obtained with the Wide Field Planetary Camera, of the quasar 12291204. Faint light surrounding the quasar comes from the galaxy that surrounds the luminous quasar.

Further insight into the nature of quasars follows directly from the observation that most are at great distances. As already mentioned, this means the light we observe from quasars was emitted when the universe was much younger than it is now. More specifically, astronomers can use the *distribution* of quasar redshifts to infer that quasar activity was at a peak when the universe was a fraction of its current age (the exact time of this peak being dependent on cosmological parameters discussed in the next chapter). The important point is that quasars were more prevalent when the universe was younger, suggesting that conditions in the early universe were more favorable for producing quasars than they are now. This has a natural explanation in the model described above.

The period of peak quasar activity roughly coincides with the epoch at which we expect many young galaxies to populate the universe. It is probable that at this time galaxies contained more gas than their present-day counterparts, simply because there had been less time for gas to get locked up in stars. Further, the frequency of galaxy mergers is expected to be higher in the early universe than today (Chapter 21). Such mergers may

cause gas to flow towards the center of the resulting galaxy, thereby fueling a pre-existing black hole. Some support for this picture was provided by the *Infrared Astronomy Satellite* (*IRAS*), which detected quasarlike objects emitting primarily in the infrared. Spectra of these objects revealed the absorption lines expected from galaxies, whereas the high infrared luminosities implied the presence of dust and thus a high rate of star formation. Since galaxy mergers are known to trigger starbursts, mergers are good candidates for producing the properties of these objects.

Not all astronomers have always accepted this view of quasars. The enormity of the inferred energies associated with quasars may have been the main reason why some astronomers looked for alternative explanations. One suggestion was that quasars were not at cosmological distances at all, but were relatively nearby objects. This relieves the energy requirements, since if quasars are not cosmological their apparent brightness implies much lower intrinsic luminosities than the standard interpretation. However, removing the need for an immensely powerful energy source comes at the considerable price of rejecting the notion that quasar redshifts are cosmological. In other words, advocates of *local* quasars had to explain why their spectral lines exhibited massive redshifts. Although ingenious ideas that claimed to do exactly that were advanced by a handful of astronomers, the observations summarized above make the standard cosmological interpretation difficult to refute.

There is one additional piece of evidence supporting cosmological distances for quasars that effectively closes the case. Since the 1980s astronomers have realized that certain images that appeared to show two or more nearby quasars were actually multiple images of the *same* quasar. This is the *gravitational lensing* discussed in Section 21.7. This phenomenon is a prediction of Einstein's general theory of relativity and can occur when a massive object such as a galaxy sits between a quasar and the Earth. Consequently, a quasar lensed in this way must be more distant than the intervening galaxy. In cases in which the redshifts of the quasar and the lensing galaxy have been measured, the quasar has a higher redshift than the galaxy. This gives compelling support to cosmological redshifts for quasars.

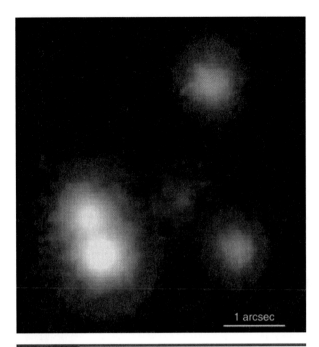

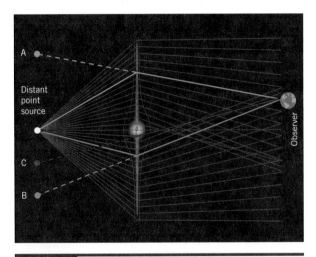

1 arcsec

Figure 22-17 The gravitational lensing system PG11151080, in which the galaxy in the center produces four images of a more distant quasar.

Figure 22-18 How a nearby galaxy produces three images of a quasar. The observer sees light from the quasar that appears to come from the three directions labeled A, B, and C.

Figure 22-17, which shows a system known as PG1115+1080, is a composite image in red and infrared light taken with the *Subaru* Telescope of the National Astronomical Observatory of Japan. The image appears to contain four quasars, but they are all *images* of the same quasar formed by the closer reddish galaxy in the center. Thus, the galaxy is acting like a lens and imaging a more distant quasar into the four images.

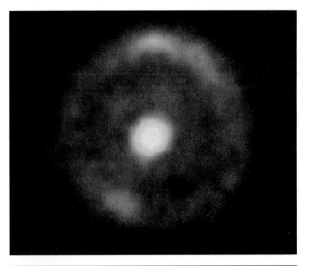

Figure 22-19 The gravitational lensing system B19381666, in which a foreground galaxy in perfect alignment with a distant one images it into a ring called an Einstein ring.

The idea behind gravitational lensing is shown in **Figure 22-18**, in which a disc-shaped galaxy produces three images of a more distant quasar as its light passes through the galaxy. Exactly what type of image is formed depends on the relative locations of the imaged object, the lens galaxy, and the observer. In some cases multiple images appear (as above), and in other cases the image will be smeared out; in yet others, a perfect ring, called an **Einstein ring**, can be formed (**Figure 22-19**). The observation of these lensing systems provides further evidence that Einstein's General Theory of Relativity is a valid description of nature.

Inquiry 22-9

Given that quasars are at great distances, are astronomers observing old or young objects?

22.5 A Unified Model of AGN

As mentioned briefly in the last section, one attraction of the supermassive black hole model of quasars is that it is readily extended to other types of AGN. Part of the motivation for a unified AGN model, other than its economy, is that the various objects described above have certain features in common. For example, all produce a large amount of radiation from a small region of space. Further, some quasars exhibit relativistic jets similar to those discussed in the context of radio galaxies.

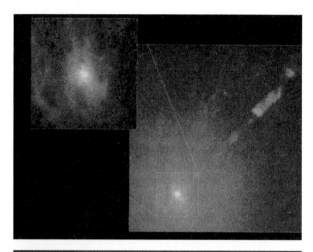

Figure 22-20   An HST image of the nucleus of M87 showing gas that appears to be spiraling about a massive central object, thought by astronomers to be a black hole.

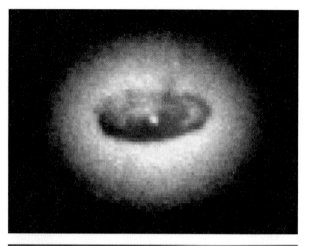

Figure 22-21   The core of the galaxy NGC 4261, as observed with the Hubble Space Telescope.

The giant elliptical galaxy M87 is a relatively nearby galaxy that helps illustrate some of these ideas. Data obtained by *HST* in 1994 of the nucleus of this galaxy (**Figure 22-20**; see also Figure 22-5) shows gas that appears to be spiraling around something. Spectra of the gas show it to have different velocities on opposite sides, indicating rotational motion of 450 km/s at 60 ly from the center. From these observations, astronomers infer that the central object must have a mass of 2.5 to 3.5 billion solar masses in a region the size of the Solar System. A massive black hole is the most reasonable conclusion.

M87 also shows a relativistic jet (Figure 22-20). Such jets fit naturally into the black hole model. As already discussed, material falling toward the central black hole will tend to form a flattened-out accretion disk. The situation is similar to disks around young stars discussed in Chapter 17, where we found that the disk will tend to confine out-flowing material into two narrow jets called a bipolar outflow. The energy produced close to a black hole in an AGN can also shoot out matter at high energy along the spin axes (Figure 22-15). This ejected matter becomes the narrow jet of energetic particles. An observation consistent with this model is the *HST* image of the core of the elliptical galaxy NGC 4261 (**Figure 22-21**). This image shows a 300 ly-diameter disc surrounding a bright spot that might harbor a supermassive black hole. In addition, observations show radio lobes

perpendicular to the disk and extending to a distance of nearly 200,000 ly from the visible galaxy.

Some quasars appear to be putting more energy into their jet than they exhibit in their main body. Where the energy appears may depend on the details of the accretion process; that is, how the material flows into the black hole. However, the existence of these jets may also soften the severity of the basic energy problem of the quasars. For example, if a quasar happens to be pointing an energetic jet our way, we might think that it is more luminous than it really is, because we would assume that it is radiating uniformly in all directions when it is not.

The idea that all AGN are powered by super-massive black holes has obvious attractions, but such a scenario must also account for the *differences* between AGN. A popular suggestion is that radio galaxies, BL Lac objects, and quasars are really the same type of object, but observed from different angles. **Figure 22-22** illustrates the model. Observations explained by the model include the fact that quasars are strong at both visible and radio wavelengths; that radio galaxies are strong in the radio but weak in visual wavelengths; and that some radio galaxies show two jets while quasars never show more than one. In the model, the central engine is surrounded by a thick ring of dust that causes jets to be emitted along the ring axis (again, this is similar to the bipolar nebulae observed in star-forming regions). Observers along

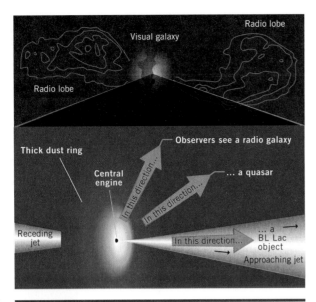

Figure 22-22 A model suggesting that radio galaxies, BL Lac objects, and quasars are the same type of object observed from different angles.

the jet axis see a BL Lac object. Observers more than 45° from the jet axis see a radio galaxy because the ring of dust hides the nucleus. Observers less than 45° from the jet axis see a quasar, because the central source is no longer hidden. This may also explain why quasars exhibit no

more than one jet when we would expect them to have two (as is the case for most radio galaxies). The angle at which we view quasars means that the second jet may be obscured by the central source and accretion disk. Such a unified model pulls together a great variety of observations and has great advantages over separate models.

One final question raised by this scenario is how many galaxies contain a supermassive black hole at their center. Observationally astronomers estimate that about 1 percent of galaxies fall into some category of AGN. However, we also have mentioned that quasars show that their activity peaked several billion years ago. In dead quasars, the central black hole is assumed to still sit inside a galaxy. Further, nonactive galaxies like the Milky Way also contain central, supermassive black holes, although typically these have masses two or three orders of magnitude less than the monsters in quasars. These pieces of evidence have led many astronomers to assume that the vast majority of bright galaxies contain supermassive black holes, and at some time in the past such galaxies went through an active phase. If this is correct, the observation that around 1% of galaxies contain AGN really means that galaxies go through an active phase for about 1% of their total lifetime.

Measuring a Quasar's Redshift

When you have completed this Discovery, you will be able to do the following:

- Better understand how astronomers obtain quasar redshifts by measuring one.

Once Maarten Schmidt realized that the spectral lines he was observing were red-shifted hydrogen lines, and once he assumed the redshift was due to a Doppler shift, he was able to determine the quasar's velocity. You will follow similar procedures using the spectrum of the brightest quasar, 3C273, given in Figure 22-13. (Although the procedure is not really correct, it is instructive.)

The first step in finding the velocity is to determine the scale of the spectrum using the comparison spectrum. The scale gives the number of angstrom units in each millimeter on the print. To find the scale, measure the number of millimeters between the comparison lines at 3,889 Å and 6,030 Å.

- **Discovery Inquiry 22-1a** What is the number of angstroms per millimeter on the photograph?

Next, choose one of the hydrogen lines (Hβ at a rest wavelength of 4,861 Å, Hγ at 4,340 Å, or Hγ at 4,101 Å). The rest positions of these lines are marked under the comparison spectrum. Measure the spectral line shift in millimeters from the rest position of one of the hydrogen lines to the emission line in the quasar spectrum. (Make your measurement parallel to the edge of the spectrum.) Then, use the scale previously determined to compute the shift in angstroms.

- **Discovery Inquiry 22-1b** What is the shift of the hydrogen line in millimeters? In angstroms?

The redshift is given by $\Delta \lambda / \lambda$, the change in the wavelength divided by the rest wavelength. Compute this quantity for the hydrogen line you measured.

- **Discovery Inquiry 22-1c** What is the measured redshift, $\Delta \lambda / \lambda$, for 3C273?
- **Discovery Inquiry 22-1d** Repeat this calculation for the other two hydrogen lines. What is the average redshift? Estimate the error in the measured redshift by computing

$$\frac{(\text{Maximum value} - \text{Minimum value})}{1.4}.$$

Let us now (incorrectly!) interpret this redshift as a Doppler shift caused by the relative motion of the quasar relative to the observer.

- **Discovery Inquiry 22-1e** Because quasar redshifts are large, the usual Doppler shift formula must be replaced by one that accounts for the theory of relativity. If $Z = \Delta \lambda / \lambda$, the expression to find the velocity becomes

$$\frac{v}{c} = \frac{(1+Z)^2 - 1}{(1+Z)^2 + 1}.$$

Use this formula to compute the quasar's velocity relative to the speed of light, v/c.

- **Discovery Inquiry 22-1f** To see how the measured uncertainty in the redshift affects the computed velocity, compute the maximum and minimum redshifts, $(Z +$ uncertainty) and $(Z -$ uncertainty), where uncertainty is the number found in Discovery Inquiry 22-1d. Use each of these numbers in the formula to find the maximum and minimum possible velocity consistent with your measured redshift.

We have said that interpreting the observed redshift as a Doppler shift is not completely correct. The reason is that quasars are thought to be at cosmological distances, such that the redshift is caused by the cosmological expansion of space. Nevertheless, from this activity you have correctly measured redshifts of the spectral lines. Also, some conclusions drawn from the derived velocities will be both correct and useful.

A Scale Model of Distances in the Universe

Now that we have studied the most distant objects yet observed, we can summarize distances with a scale model. If needed, refer to the distance pyramid in Chapter 21. This activity may be done with at least one other person.

Obtain a standard-size roll of toilet paper. The length of this roll will represent the distance from Earth to the most distant object known. The package label should tell you the number of tissues on the roll. Use this information to determine the distance (in light-years) represented by each sheet. Then, using a long hallway if available (if such a hall is not available, you can go back and forth within the longest area available), begin unrolling the paper, and mark the location of typical representatives of each of the following objects: a quasar, the edge of the Local Supercluster, a distant cluster of galaxies, M31, the center of the Milky Way, the nearest star, and the edge of the solar system.

- **Discovery Inquiry 22-2a** How many light-years does each sheet represent?
- **Discovery Inquiry 22-2b** On the scale of your model, how wide a line represents the size of the Milky Way? The size of the solar system?

Chapter Summary

Observations

- Peculiar-looking galaxies often turn out to be **radio galaxies**, objects that emit vastly more energy than would be expected from the stars composing them.

- Most radio galaxies have two radio-emitting lobes located symmetrically on either side of the optical galaxy. Head-tail galaxies appear the way they do because the galaxy's motion through the intergalactic medium and magnetic field causes material to stream behind, forming the tail.

- Some galaxies appear to be colliding.

- Radio images that show jets of material from the optical galaxy out to the radio-emitting lobes provide evidence of a physical connection between the galaxy and the distant lobes.

- **Seyfert galaxies** are most often spiral galaxies with bright nuclei and spectra that show gases in turbulent motion.

- **Quasars** appear to be starlike objects whose spectra show highly redshifted emission lines. Quasars often show multiple sets of absorption lines, vary over intervals of days to months, and are surrounded by a faint galaxy.

Theory

- The unified model of AGN suggests that all AGN are powered by central, supermassive black holes.

- Einstein's theory of general relativity predicts that a quasar's radiation, if passing by a massive galaxy, can be bent and formed into multiple images of the quasar. A number of quasars clearly show evidence of such gravitational lensing.

Conclusions

- If observed quasar redshifts are interpreted in terms of a redshift caused by the expansion of the universe, then the light we observe was emitted when the universe was much smaller and younger than it is today. The luminosity inferred shows the quasars to have luminosities 10 to 1,000 times that of the Milky Way. Furthermore, variations over times as short as a few hours show the emitting regions to be small, only a few light-hours in size.

- The most generally accepted model of a quasar is the nucleus of a young galaxy in which there is a massive black hole surrounded by an accretion disk.

- A unified model that explains the variety of radio-emitting objects is given in Figure 22-22, in which a central energy source is surrounded by a thick ring of dust. The type of radio object observed is then determined by the angle from which the observer sees the disk.

Summary Questions

1. What are some of the principal types of radio galaxies? Why is it sometimes difficult to associate a radio galaxy with an optical counterpart?

2. What are the observations that distinguish Seyfert galaxies from other galaxies?

3. What distinctive observations distinguish quasars from other objects? (You should include redshift, spectrum, brightness variations.)

4. What are the characteristics of quasars inferred from their observations (distance, luminosity, size, energy source, their true nature)?

5. What are several theories for quasars? Evaluate them on the basis of the observational evidence.

6. What is meant by a gravitational lens? What can astronomers learn from studying lensed quasars?

7. What is the single model that gives a unifying picture of the various radio-emitting galaxies?

Applying Your Knowledge

1. Hypothesis: Quasars are at cosmological distances. Question: What evidence have astronomers collected in favor of this hypothesis? Why were some astronomers initially reluctant to accept the idea that quasars are at great distances?

2. How do astronomers infer the size of a quasar's emitting region from observations of the variability of its brightness?

3. Summarize the evidence that active galaxies contain massive black holes in their nuclei.

4. What evidence is there that all bright galaxies may contain supermassive black holes?

5. Consider four classes of objects: normal galaxies like the Milky Way, Seyfert galaxies, radio galaxies, and quasars. Arrange these objects in order of increasing energy output. Sketch a diagram that shows the relationship.

▶ 6. The following are observations for a fictitious quasar: redshift is 0.6 the speed of light; observed brightness is 10^{-17} that of the Sun. Compute the distance to the quasar and its luminosity in terms both of the Sun and of the Milky Way galaxy.

▶ 7. Suppose astronomers observe hydrogen lines in the spectrum of a quasar to have three components at wavelengths of 5,200 Å, 5,201 Å, and 5,205 Å. The usual wavelength is at 4,861 Å. If the three components are due to absorption along the line of sight to the quasar, what is the distance between each of the absorbing clouds? What assumption are you implicitly making in obtaining your answer?

▶ 8. Consider a simplified galaxy having only 10 stars. If F_{12} is the force between stars numbered 1 and 2, write down all the other forces that a computer program must consider in analyzing the movements of stars in a collision.

Answers to Inquiries

22-1. The amount of radiation emitted from these extragalactic objects must be significantly greater than that emitted by sources within the Milky Way because of the great distances to extragalactic objects.

22-2. Because all hot bodies emit some radiation at all wavelengths, such a galaxy would emit in the radio, too. However, the amount would be small compared with that in the visual spectral regions.

22-3. As discussed in Chapter 19, charged particles must be accelerated in a magnetic field to speeds near that of light.

22-4. The most likely explanation is that an accretion disk confines the ejected material so that it shoots out of the center of M87 in a jet.

22-5. Broad lines are produced by rapidly moving atoms, some of which are moving toward the observer while others are moving away. The faster they move, the greater the Doppler shifts and the broader the lines.

22-6. It is reminiscent of the jet in NGC 4486 (M87).

22-7. The Hubble law is $V = HD$, so $D = V/H$. Using $V = 0.16c$, we have $D = 0.16(3 \times 10^5)/20$ km/s/Mly $= 2.4 \times 10^3$ Mly $= 2.4 \times 10^9$ ly.

22-8. They must be small. In the example given, the size would be only 1/100 of a light-year. Because α Cen is 4.3 light-years away, the size is about 0.04 light-year. Expressed in other time units, we have a size of about 365/100 = 3.7 light-days = 90 light-hours = 5,400 light-minutes. Because light takes 8 minutes to travel 1 AU, the size of the quasar's emitting region will be roughly 5,400/8 = 675 AU.

22-9. We observe the objects as they were when the light was emitted billions of years ago. Thus, we see the objects as they were when they were young.

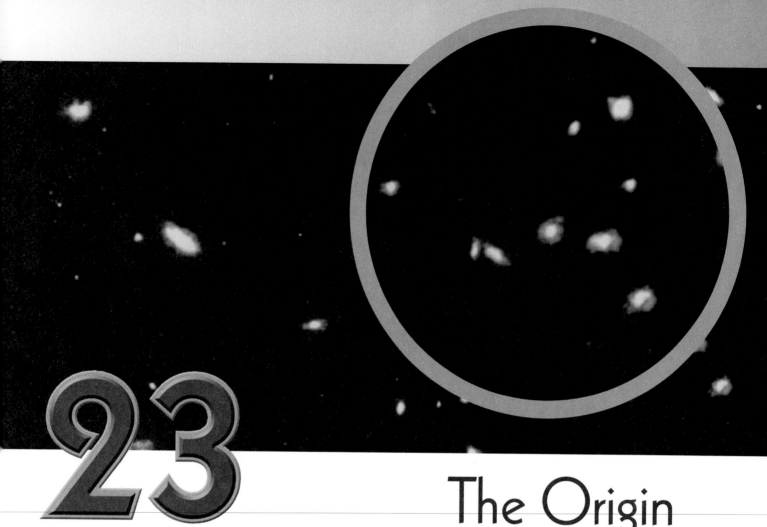

23

The Origin
and Evolution
of the Universe

The astronomers said:
"Give us matter, and a little motion,
and we will construct the universe."

Ralph Waldo Emerson

The universe is everything. As scientists, we cannot deal with everything, but only those things we can detect and measure. For this reason, in this chapter we will be concerned primarily with the observable universe, because scientific models must be based on observation.

However, when we expand our horizons to include the entire universe, physics and astronomy can become highly speculative. This is inevitable, partly because at the limits of our telescopic view objects are faint and data are scarce, so it is sometimes difficult to obtain observational constraints on the many theoretical possibilities. Furthermore, as we reach back in time toward the earliest stages of the evolution of the universe, we enter a realm in which the fundamental laws of physics are not fully understood.

One of the most remarkable aspects of the study of cosmology is the revolution that has occurred over the last decade or so. Thanks largely to technological advances, astronomers now think that we have a pretty good idea of the nature of the universe, how it is evolving, and how it will ultimately die.

23.1 Expansion of the Universe

Astronomers take the concept of an expanding universe for granted. In this section, we see what the evidence is and the consequences of it.

THE VELOCITY–DISTANCE RELATIONSHIP

In Chapter 21 we saw that Hubble found a linear (straight-line) relationship between the velocity at which a galaxy moves away from Earth and its distance. The **Hubble law**, illustrated in **Figure 23-1** can be expressed as follows:

$$V = H_0 d$$

where the constant of proportionality between the velocity V and the distance d is designated by the letter H_0, called the **Hubble constant**.

The determination of the value of H_0 has been one of the central goals of cosmologists ever since the expansion of the universe was first discovered. In principle, the procedure is straight forward. First we must measure the distances to a large number of galaxies using the methods discussed in Chapter 21 and summarized in Figure 21-12. The velocity of each galaxy is then measured using the Doppler shift of lines in the galaxy spectrum. A graph such as Figure 23-1 can then be constructed and the value of H_0 calculated by finding the slope of the line.

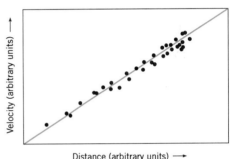

Figure 23-1 The Hubble velocity-distance relationship.

One of the primary missions of the *Hubble Space Telescope* has been to determine the value of H_0 by obtaining distances to galaxies using Cepheid variable stars. At the time the telescope was launched, estimates of H_0 varied by more than a factor of two between different observational studies. The results obtained using the *Hubble Space Telescope*, along with ground-based observations, have now pinned down the value of H_0 to an uncertainty of about 10 percent or so. This is a remarkable achievement in a relatively short span of time and represents the dedicated work of a large number of astronomers. Numerically, the value of H_0 *is* about 20 km/s/Mly—that is, for each million light-years of its distance from us, a galaxy recedes by an additional 20 km/s. For example, a galaxy 1 million light-years from Earth would, on average, recede at 20 km/s, while one at 2 million light-years would recede at 40 km/s, and so on.

Chapter opening photo: The Hubble Deep Field, which shows galaxies never before seen.

Why did astronomers invest so much time and effort in determining the value of the Hubble constant? First, the numerical value of H_0 is a measure of the *scale* of the universe. Imagine that we observe a galaxy with some cosmological recession velocity. Looking at Hubble's law we see that the larger the value of H_0 the smaller the distance to that particular galaxy. The value of H_0 also tells us something about *age* of the universe. The association of the Hubble constant with age is not difficult to see. We again write the Hubble law:

$$V = H_0 d.$$

Furthermore, we know that an object moving at a constant speed travels a distance $d = Vt$. Solving this for velocity, we get the following:

$$V = \frac{d}{t}.$$

If the expansion rate of the universe is constant, as we have assumed, the last two equations for velocity must give the same answer, implying $H_0 = 1/t$. We will have more to say about this assumption below. If H_0 is 20 km/s/Mly per million light-years, then the assumption of a constant expansion rate implies the universe is about 15 billion (1.5×10^{10}) years old. Larger values of H_0 correspond to a lower inferred age of the universe.

> You should do Discovery 23-1, The Age of the Universe, at this time.

Inquiry 23-1

The distance of the Hercules Cluster of galaxies is about 300 million (3×10^8) light-years. At what velocity is it traveling away from us?

Inquiry 23-2

The redshift of quasar 3C273 is about 16 percent the velocity of light. Approximately how far away is it?

Inquiry 23-3

If H_0 gets smaller as the universe becomes older, would the value of H_0 as measured from very distant galaxies be *larger* or *smaller* than that determined from nearby galaxies?

The galaxies in the Local Group move like bees in a swarm. Just as each bee has its own motion relative to every other one, even though the swarm itself moves as one object, each of the Local Group galaxies has its own motion relative to every other galaxy. For example, M31, at a distance of 2.5 million ly, would be expected from Hubble's law to be moving away from us at about 50 km/s. However, the galaxy's own motion within the Local Group, which is caused by the gravitational forces of the other galaxies in the Local Group acting on it, causes M31 to approach us at a speed of 275 km/s. Galaxies outside our own local gravitational cluster *are* all receding from us. An immediate consequence of this observation is that we live in an *expanding* universe. The only way to escape this conclusion would be to find an acceptable alternative interpretation of the observed redshift, and to date no one has succeeded in doing so.

> You should do Discovery 23-2, The Expansion of the Universe, at this time.

CONSEQUENCES OF THE HUBBLE LAW

Because we see galaxies receding from us in every direction, it is tempting to conclude that we are in the center of the universe. This view is incorrect, and it can be demonstrated that any observer located anywhere in an expanding universe would see the same thing. A simple analogy illustrates this point. Consider galaxies to be like raisins in a rising loaf of bread (**Figure 23-2a**). Assume that as the bread rises, it expands in all directions. The raisins are carried along by the expanding dough but are not moving through it. Thus, each raisin gets farther away from every other raisin in the loaf. After one hour, all dimensions have doubled, but the raisins themselves have remained the same size (**Figure 23-2b**). The distances through which the raisins have moved during the hour, and their speeds, are shown in **Figure 23-2c**. You can readily see that the greater the distance between *any* two raisins, the more rapidly they will move away from each other. No matter which raisin you examine, the same velocity–distance relation will be observed. No true center can be defined from the observed expansion. The same holds true for galaxies in an expanding universe; galaxies are not

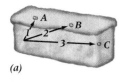

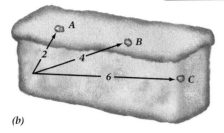

Raisin	Distance traveled	Velocity
A	1	1
B	2	2
C	3	3

(c)

Figure 23-2 An analogy to the expanding universe of galaxies: raisins in a loaf of bread. *(a)* Initial locations of raisins. *(b)* Positions after the loaf has risen for one hour. *(c)* Table of distances traveled and the velocities of raisins.

moving through space but are carried along by the expansion of space itself. Further, because they are systems held together by gravity, they do not expand.

You may still wonder why we cannot talk about a center of the expanding universe. After all, if we take our loaf of bread out of the oven it is not difficult to determine its center. This illustrates the limits of the analogy we have just used. The loaf of bread is expanding into something—specifically, the space inside the oven. In fact, all of our everyday experiences of expansion involve something getting bigger within the three-dimensional geometry of the world around us. The universe cannot be expanding into something because the universe is everything there is; it is simply getting progressively bigger. As it does so, cosmological objects such as distant galaxies become further apart. One additional issue to keep in mind is that when we observe galaxies in space, we do not see a snapshot of the universe as it actually is now. The light we observe from distant galaxies left those objects millions or billions of years ago, so we are seeing them where and as they were then.

Bear in mind that astronomers work under a fundamental limitation—we cannot see the whole universe. Because light itself travels at a finite speed, we can only observe those parts of the universe from which light has had time to reach us. In other words, the finite age of the universe means we can only probe those parts of it from which light has had time to reach us on Earth.

Think again about the expansion of space, but now attach a wave to space. As space itself expands, the wave will be stretched and thus redshifted. Therefore, the redshift of light from distant galaxies is caused by the expansion of space itself, not motion through space.

23.2 The Big Bang

If the distances between the galaxies increase with time, then in the past the galaxies must have been closer together. In other words, the density of galaxies was higher, and the universe was more compressed in the past. Carrying this further back in time, we are forced to conclude that approximately 14 billion years ago all the matter of the universe was packed together into a state of astonishingly high temperature, pressure, and density.

Such extreme conditions could never be created in a laboratory, yet theory has enabled us to reconstruct a picture of what these conditions were like. The temperature once must have exceeded trillions of degrees and the density hundreds of trillions of grams per cubic centimeter. This hot, dense state is often referred to as the **primeval fireball**. It contained radiation, along with matter in the form of ordinary elementary particles such as electrons and neutrinos, and a number of more exotic ones.

We hypothesize that the primeval fireball began its expansion in what we now call the **Big Bang**. The Big Bang was not an explosion in the usual sense. As already discussed, the universe is not expanding *into* something. The Big Bang did not happen somewhere; it happened everywhere. Another way of looking at this is to say that the Big Bang did not occur somewhere, it happened *somewhen*.

The sequence of events that took place in the early moments of the Big Bang followed each other with extreme rapidity. Nevertheless, we are able to detect the effects of some of its stages in what we observe around us today.

The Big Bang is a *theory,* a hypothesis that makes predictions and has strong observational support. Our current state of knowledge is imperfect, but while questions and problems abound, they do not make the overall structure of the theory invalid. The history of science tells us that eventually the Big Bang theory *as we know it today* will be refined and possibly replaced by something better. That does not mean that *all* our current ideas are wrong. Remember that while Newtonian physics was replaced by Einsteinian physics, Newton's laws still correctly predict the future behavior of physical systems when applied to the appropriate realm of nature.

The Three Cornerstones of the Big Bang

Before we describe our current ideas on how the universe evolved from its initial hot, dense state to its current condition, it is useful to describe the observational foundations upon which the Big Bang theory is based.

The Recession of the Galaxies

In some ways the discipline of modern cosmology began in the 1920s when it was determined that the universe was filled with galaxies external to the Milky Way. Right on the heels of this finding was the equally important discovery of the recession of galaxies and the Hubble law. As we have already noted, the fact that all galaxies are, on average, moving apart from each other is the first piece of evidence for the Big Bang model. This is because if we imagine "running the expansion backward" we find that at some point in the finite past the universe was in an incredibly hot, dense state.

This interpretation of the expansion of the universe was not always accepted by all astronomers. A rival theory to the Big Bang known as the **Steady State theory** also explained the expansion of the universe. In this picture, the universe has always existed and has always been expanding. If such a universe had a constant amount of mass in it, expansion over an infinite amount of time would spread this mass out so thinly that the density of the universe would be basically zero. Since this is clearly not the case, advocates of the steady state theory suggested that matter is created spontaneously throughout the universe in such a way that, at any given time, the universe looks the same everywhere and maintains a constant density of galaxies. Consequently, like the Ptolemaic and Copernican models before them, the Big Bang and Steady State theories could both account for a fundamental observation. In the cosmological case, this observation is simply that the universe is expanding. Fortunately, the two models made different predictions about other observable properties of the universe that allowed astronomers to rule out the Steady State theory. One of these observations involved radiation traveling across the universe for billions of years.

Figure 23-3 Penzias and Wilson, along with the radio receiver with which they discovered the 3-K background radiation.

The Cosmic Background Radiation

In 1964, Arno Penzias and Robert Wilson, working for Bell Telephone Laboratories, were using the 20-foot horn antenna shown in **Figure 23-3** to search for low-intensity radio signals from the sky. They wanted to push the sensitivity of the system as far as possible and look for extremely faint cosmic radio sources. To this end, they devoted a great deal of attention to reducing the receiver's internal noise (similar to the static heard in a cheap radio), which obscured faint signals.

By cooling the equipment, among other techniques, Penzias and Wilson succeeded in reducing the static noise of their instrument to low levels. Tuned to one specific wavelength, they scanned the sky with it. While not expecting to find anything other than a few localized sources, in fact they found a low-level signal coming from every direction in the sky. This signal had an interesting property—the intensity was **isotropic**; that is, it was exactly the same no matter what direction in space they looked.

Penzias and Wilson then made observations at other wavelengths to construct the spectrum of this radiation, which, they found, looked like the spectrum of a blackbody (Chapter 13). They identified the wavelength of the spectrum's peak intensity and used Wien's law to compute the temperature that a body emitting such radiation would have; the result turned out to be 3.5 K. It appeared that every point on the celestial sphere was emitting radiation as if it were a body at a temperature of

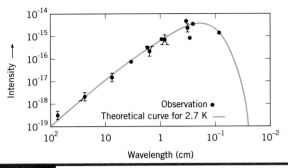

Figure 23-4 Spectrum of the 3-K background radiation.

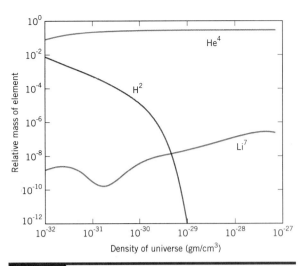

Figure 23-5 Dependence of the abundances of some light elements on the density of the universe as predicted by the Big Bang theory.

3.5 K. The radiation they detected is sometimes referred to as the "three-degree" or "cosmic background radiation." **Figure 23-4** shows the spectrum of the radiation, along with the uncertainties of the measurements.

Astronomers quickly noted an important consequence of the isotropic nature of the radiation. Because precisely the same signal is received from all directions in space, the radiation cannot be associated with any specific source in our galaxy. For any such signal to be so uniform over the sky it must originate at cosmological distances. In fact, the existence of such radiation had been predicted by theorists investigating the implications of the Big Bang. Penzias and Wilson had detected the leftover radiation from the Big Bang, redshifted to a lower temperature by the expansion of the universe. By doing so, they brought general acceptance to the validity of the Big Bang model and were awarded the Nobel Prize in Physics in 1978.

One additional interesting result comes from the observation of the background radiation. As discussed further below, the temperature of the universe when this radiation was emitted is calculated to be around a thousand times greater than its temperature now. This means that the universe has expanded by a factor of 1,000 as the radiation has been traveling to Earth, or that the redshift at which the radiation was emitted was around 1,000. That makes it by far the oldest radiation we have ever received, having been emitted long before any of the stars or galaxies were formed.

The Abundances of the Light Elements

The third observational cornerstone of the Big Bang is a bit more subtle than the other two, and for this reason we will have more to say about it later. However, the basic idea is the following. According to the Big Bang model, the universe started off in a hot, dense state, since which time it has been expanding and cooling. When the universe was around a minute old, most of the ordinary matter was in the form of neutrons and protons (hydrogen nuclei), moving rapidly because of the high temperatures. Under such conditions, the fusion of hydrogen into helium can take place, somewhat like the process that occurs inside stars.

The Big Bang makes specific predictions about the fraction of helium produced in the early universe. In fact, the amount of helium-3 (2 protons and 1 neutron), helium-4 (2 protons and 2 neutrons), as well as some other light elements such as lithium can be calculated. **Figure 23-5** shows the results of such calculations. When astronomers measured the amount of these light elements, they found excellent agreement with the predictions of the Big Bang model. For example, the *Hubble Space Telescope* observed interstellar deuterium (1 proton and 1 neutron) in the spectrum of the bright star Capella (**Figure 23-6**). From these data astronomers found there to be merely 15 deuterium atoms for every million hydrogen atoms, in agreement with predictions of the Big Bang theory. Along with the other observational foundations of the Big Bang model, the beautiful agreement between observed light element abundances and the predictions of this model have convinced astronomers that the Big Bang is an excellent description of the evolution of the universe.

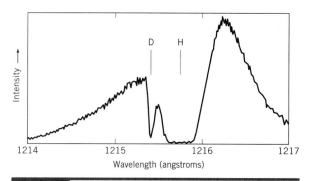

Figure 23-6 An absorption line of interstellar deuterium (D) observed by the Hubble Space Telescope in the spectrum of the bright star Capella. The broad line is hydrogen (H).

THE FOUR FORCES OF NATURE

To understand how the universe behaves during its earliest moments, we must first discuss the **four fundamental forces** in the universe. This is because all these forces play a part in the early history of the universe. We have already looked in some detail at two of them: the gravitational and the electromagnetic forces. These forces are long-range forces, with both following an inverse square law. Technically, this means that the gravitational or electromagnetic force between two particles, assuming they are charged, only falls to zero when they are an infinite distance apart. It turns out that the electromagnetic force is *much* stronger than the gravitational force. For example, the electromagnetic force between a proton and electron in a hydrogen atom is roughly *40 orders of magnitude* greater than the gravitational force between these particles. Why, then, does gravity control the behavior of the universe on large scales? To resolve this apparent conundrum, recall that gravity is purely attractive, whereas the electromagnetic force can be attractive or repulsive. When a lot of matter gets together in a star or a galaxy, each little piece exerts an attractive gravitational force. However, the positive and negative charges of all the particles making up this matter tend to cancel one another out, so that for large bodies gravity is much more important than electromagnetism. And nature will not allow a large number of, say, positively charged particles to get together and exert a powerful force on a similar agglomeration of negatively charged particles, simply because a large chunk of particles with the same charge will blow apart through the mutual repulsion of the particles.

In contrast to the long-range nature of gravity and electromagnetism, the two remaining known forces of nature are primarily of importance only on atomic and subatomic scales. The **strong nuclear force** is, as the rather unimaginative name suggests, a strongly attractive force some 100 times the strength of the electromagnetic force. It is the force that holds atomic nuclei together, overcoming the electromagnetic repulsion between protons that would otherwise prevent nuclei more massive than hydrogen from forming. The **weak nuclear force** comes into play over similar distances and is responsible for radioactivity and the production of neutrinos. It is some 10^7 times weaker than the electromagnetic force.

Developments in theoretical and experimental physics over the last few decades have led most physicists to conclude that the four forces of nature were not always distinct. By smashing particles together in huge accelerators, researchers can study how matter behaves at high energies. Along with a deeper theoretical understanding of fundamental particles, these experiments have produced many important new ideas that can be applied to the early universe. One critical breakthrough was the unification of the electromagnetic and weak forces into what is now known as the **electroweak force**. Earlier clues that electromagnetism and the weak nuclear force might be two aspects of a single force included the observation that the relative strengths of these forces became closer as particles were smashed together with higher and higher energies. At sufficiently high energies, the relative strengths become the same. Although details of this unification are complex, the idea is not as strange as it may at first appear. For example, before it was understood that electromagnetism is a single force, electricity and magnetism were viewed as distinct forces.

Particle accelerators can reach the energies of electroweak unification and have verified the predictions of the electroweak theory that describes this force. This success partly motivates the conviction of most physicists that the electroweak and strong forces will also be unified through a **Grand Unified Theory** or **GUT**. A second motivation is that, just as the relative strengths of the electromagnetic and weak forces converge to a common value as energies increase, the strengths of the electroweak and strong force also appear to be converging as even higher energies are achieved.

Perhaps the central goal of theoretical physics is to go even further than the unification of the strong and electroweak forces and develop a **Theory of Everything (ToE)** that unifies gravity with the other forces. In other words, physicists assume that the four forces of nature that are distinct in our current low-energy universe are in fact four aspects of a single *superforce*. Attempts to achieve this unification include **string theory** and the broader **M-theory** described in Brian Greene's popular book *The Elegant Universe*. Whether such theories are on the right track is still hotly debated, but all physicists agree that there is currently a serious gap in our understanding of the laws of nature. This is because to understand how the force of gravity works on the smallest scales, we need a quantum theory of gravity. Although string theory and other lines of investigation show some promise, no complete theory currently exists. As we now discuss, this sets a limit to how far back in the history of the universe we can currently probe.

THE EARLY MOMENTS OF THE UNIVERSE

As we apply the laws of physics to earlier moments in the universe, we must extrapolate those laws into realms where direct observational tests are difficult or impossible. Consequently, our levels of confidence in the details of our model drop as the extrapolation is extended. Despite this important caveat, the fact that we can say anything concrete about the universe when it was a tiny fraction of a second old is a stunning intellectual achievement.

At any stage in the expansion of the universe there is an important length scale that is given by the speed of light multiplied by the age of the universe. Since the speed of light sets a cosmic speed limit, the length represents the size of the universe over which signals can have traveled. When the universe was about 10^{-43} seconds old, this size was comparable to something physicists call the **Planck length**. On such tiny scales, we require the elusive quantum theory of gravity to understand how the universe behaves. In the absence of such a theory, 10^{-43} seconds (the **Planck time**) is therefore the earliest time at which we can currently discuss the universe. This is also the time at which gravity "freezes out" from the original hypothesized superforce. In other words, the universe is controlled by gravity and the unified strong and electroweak forces. The temperature of the universe at this time was in excess of 10^{32} K.

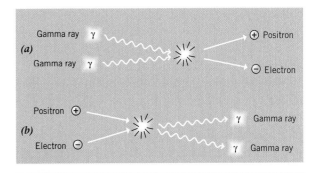

Figure 23-7 Positron and electron pairs. *(a)* Pair production; *(b)* Annihilation

Two important opposing processes, called **annihilation** and **pair production** (indicated schematically in **Figure 23-7**) dominate much of the physics of the early history of the universe after the Planck time. If a particle of matter collides with its antiparticle (which is antimatter), the particles annihilate each other, giving off a burst of radiation. The amount of energy produced is given by Einstein's equation $E = Mc^2$, where M represents the total mass of the two particles being converted into energy. The equation is equally valid for the reverse process of changing energy into matter via pair production. During pair production, photons of sufficiently high energy (gamma rays) interact with each other or a material particle and create new particles and their corresponding antiparticles.

The energy of photons in the universe is directly related to the temperature of the universe. Because the universe cools as it expands, the mass of particles that can be formed via pair production drops as the universal expansion progresses. At the earliest times, massive particles that cannot be produced in the most powerful particle accelerators on Earth would have been common, along with more familiar light particles such as electrons, and quarks that are the building blocks of protons and neutrons. As the temperatures fell, these very massive particles would have annihilated with their antimatter partners, but with lower photon energies no new particle-antiparticle pairs of the most massive particles could form.

At 10^{-35} second, when the temperature had cooled to about 10^{28} K, another momentous event took place. The strong nuclear force "froze out" from the GUT force, while the weak and electromagnetic forces remained united as the electroweak force. When these forces separated, it has been suggested that a large amount of energy was released. One popular idea is that this energy caused the universe to enter a short "inflationary" era during which it expanded rapidly; the inflationary era is further discussed in Section 23.5.

We have considerable experimental evidence about the physics that dominated this era. The most powerful particle accelerators now operating are capable of energies high enough to produce the particles of the electroweak force. These particles have been detected, and the details of their physical interactions with matter are being thoroughly investigated.

At 10^{-10} second, the temperature had dropped to 10^{15} K, and the final freezing-out took place when the electromagnetic and weak nuclear forces separated. At this time, photons were still so energetic that they created quarks and antiquarks (the antimatter partners of quarks) by pair production processes. Later, after the universe had cooled further and the photons were less energetic, only electrons and their corresponding antimatter particles, positrons, could be created. The quarks combined with each other to form protons and neutrons, and with antiquarks to form mesons, which are unstable particles formed by a quark and an antiquark.

During the entire process of cooling and expansion of the universe that we have discussed (which took 10^{-15} second), the creation of particles through pair production was balanced by the opposite process of annihilation, in which matter and antimatter particles combined to create radiant energy. The equilibrium between these two processes at each stage determined the balance between particles and radiation in the early moments of the universe. Before the universe was a tenth of a second old, the primeval mixture of matter and photons had cooled to a temperature of roughly 100,000 million degrees (10^{11} K) and a density of more than 100 million times that of water on Earth. Because matter and radiation were in equilibrium, the spectrum of the emitted radiation was that of a high-temperature blackbody.

The alert reader might have noticed an apparent problem with this history of the early universe. If matter and antimatter are produced by photons through pair production, and if matter particles annihilate with their antimatter partners to release energy, why is there any matter left in the universe? It turns out that the GUTs that are used to provide the theoretical framework for our study of the very early universe also predict a small imbalance between the amount of matter and antimatter in the early universe. The excess matter amounts to about one part per billion more matter than antimatter, but it is sufficient to form all the material objects discussed in this book. One way of testing the GUTs that lead to this idea is that they also predict that the proton is unstable. A given proton is expected to remain a proton for at least 10^{34} years, so proton decay is very rare. However, such a decay rate *is* within the reach of current experiments. The rarity of the event means that a huge number of protons need to be studied at once, as in the huge tank of very pure water used in the Super-Kamiokande experiment (mentioned in Chapter 16 and Figure 16–9).

THE EARLY UNIVERSE LEAVES RELICS

By the time it was one second old, the expanding universe had cooled to 10^{10} K, and at the end of three minutes, the temperature was approximately 10^9 K. As discussed briefly above, this was the period in the history of the universe when the light elements such as helium and lithium were formed as the entire universe acted like a fusion reactor. This process left us with roughly 25 percent of the normal matter in the universe as helium-4 and the rest mostly hydrogen. Why did this fusion process not make heavier elements such as carbon and oxygen? The problem is that there are no stable atomic nuclei with an atomic number (number of protons plus neutrons) of 5 or 8. If a proton, for example, fuses with a helium-4 nucleus, the resulting nucleus decays rapidly, making it unlikely for further fusion to make a bigger nucleus. The same is true if two helium-4 nuclei fuse. The net result is that atoms like carbon and oxygen that are the basis of life did not form in the early universe, but rather inside massive stars (as discussed in Chapters 16–19). The reason stars can make carbon and oxygen when the early universe cannot is that they have a much greater time to "cook" these heavier elements.

In the earliest stages of the universe, radiation was more important than matter. As the universe expanded, the density of both radiation and matter decreased. Because the redshift also caused radiation to lose energy as the expansion occurred (remember that longer wavelength radiation is less energetic than short wavelength radiation), after a few thousand years the density of radiation eventually became less than the density of matter. Thus, matter eventually came to dominate over radiation, as is the case today. **Figure 23-8** indicates the trend of the early universe with the progression of time and summarizes the discussion up to now.

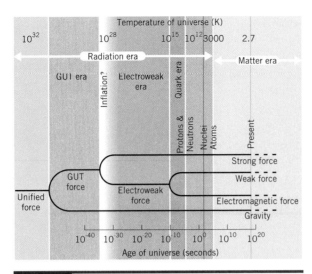

Temperature of universe (K)

Figure 23-8 The development of the early universe. At first, all forces of nature were unified. At 10^{-43} second, the gravitational and GUT forces split. At 10^{-35} second, the strong and electroweak forces split, and an inflationary era began. At 10^{-10} second the electromagnetic and weak forces split, leaving the four forces as we know them today. At 10^{-3} second, quarks combined to make protons and neutrons, and at 3 minutes, nuclei of deuterium and helium formed. At 300,000 years the nuclei were able to capture electrons to form atoms.

Inquiry 23-4

In stars, how is the chasm at atomic weight 5 overcome, and how are heavier nuclei created? (*Hint:* Review the triple-alpha process discussed in Chapter 18.)

During the first 300,000 years, the temperature was high, and a plasma of protons (hydrogen nuclei) and electrons were the main material constituents. Because electrons are highly effective scatterers of photons, radiation within the electron/proton soup was scattered randomly in all directions, just as light is scattered by water droplets in fog. Just as fog is opaque to light, the electrons formed an opaque curtain to the radiation. Because matter and radiation were coupled to each other and were in a state of equilibrium, the radiation emitted had a spectrum of a black body.

At about 300,000 years after the Big Bang, the temperature of the universe had dropped to about 3,000 K. At such temperatures, electrons and protons can combine to form hydrogen *atoms* without the constant risk of ionization that occurs through collisions at higher temperatures. Because hydrogen *atoms* do not scatter photons as effectively as electrons, the universe now became transparent (**Figure 23-9**). This "surface of last scattering" is somewhat analogous to the photosphere of the Sun and other stars. This is the point at which photons can escape the fog and stream freely through the universe. This also represents the point where matter and radiation *decoupled.* Consequently, radiation from this surface has a spectrum very close to that of a blackbody (see Chapter 13).

The cosmic background radiation's spectrum is as important as the very presence of this radiation in providing persuasive evidence in favor of the Big Bang model. At the decoupling of matter and radiation, the universe would be filled with radiation whose spectrum was that of a 3,000 K blackbody. As the universe expanded, the radiation's temperature would decrease. Since the universe today is 1,000 times larger than at the time of the final scattering, we can predict that today the entire universe should be filled with radiation having a temperature 1,000 times less than the temperature at which radiation formed, or around 3-K—just as observed.

Inquiry 23-5

At what wavelength should 3-K radiation from decoupling be observable? What part of the electromagnetic spectrum is this?

You should do Discovery 23-3, Observing the Cosmic Background Radiation, at this time.

THE EPOCH OF GALAXY FORMATION

We have seen that during its earliest phases the universe was largely radiation, but that after a few hundred thousand years matter became the most important component. As we discuss further in Section 23.4, the smoothness of the cosmic background radiation indicates that the distribution of matter was also smooth when matter and radiation decoupled. As we look at the universe around us, however, what really stands out is how lumpy it is,

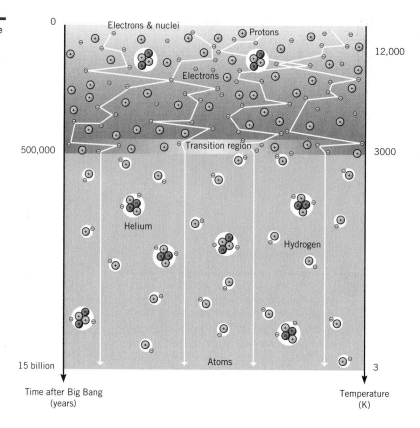

Electrons & nuclei

Protons

0

12,000

Electrons

Transition region

500,000

3000

Helium

Hydrogen

Atoms

15 billion

3

Time after Big Bang
(years)

Temperature
(K)

from the presence of cats, people, and planets to galaxies and superclusters. How did these inhomogeneities (lumps) form out of a universe that began its life much smoother?

There are two important points to consider. First, while it is the case that stars and galaxies are fairly notable lumps, if we look at the present universe on very large scales (exceeding the size of superclusters) the overall *distribution* of matter is quite smooth. A forest provides a useful analogy. Trees clearly represent local lumps since they are much denser than the surrounding air. Further, there are regions of a forest where trees are packed in more closely than in less-densely populated clearings. But when one looks at the forest from a sufficient distance, on average one part looks very much like another.

The second issue is that the smoothness of the early universe is expected to change as the universe expands—thanks to the relentless effects of gravity. If a region of the universe is slightly more dense than average, it will expand less rapidly than the average universal expansion. Gravity slows its expansion locally. Such regions tend to attract more matter towards them and so they grow more massive. Eventually, such regions may collapse to form entire galaxies.

In Chapter 21 we described some current ideas on the galaxy formation process. From a cosmological perspective, we can add that this process occurred sometime after the universe became transparent. The peak of quasar activity, the number of disturbed-looking galaxies in the Hubble Deep Field and Ultra Deep Field, and other observations of the amount of radiation in the universe, suggest that galaxy formation was probably at its most frenetic sometime between a redshift of around three and one. Although galaxy interactions are still having some impact on galaxy evolution at the current time, it seems most galaxy formation is over. Larger structures like superclusters, however, are probably still in the process of growing and becoming more settled.

In this section, we have discussed the *history* of the universe, from the earliest times we can hope to probe with the current laws of physics to the formation of galaxies and larger structures. Perhaps an even more fascinating question is: What is the future of the universe? Physics is primarily concerned with predicting the future, be it the ability of a building to withstand a hurricane or the position of the planets on December 13, 2063. Before we discuss the future of the universe, however, a brief aside concerning its geometry will be helpful.

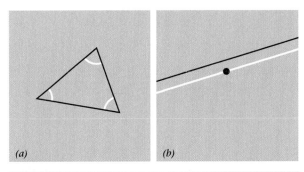

Figure 23-10 The geometric properties of flat space. *(a)* The sum of the angles is 180°. *(b)* Parallel lines never meet.

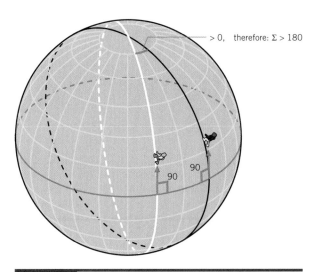

Figure 23-11 The geometric properties of positively curved space, showing that great circles intersect.

23.3 The Geometry of Space and the Shape of the Universe

Astronomers are often asked, "What is the universe expanding into?" After all, if the universe has an "end," there must be an "edge" and therefore something "outside" it. We will see that it is actually possible to have a universe that is finite in extent and yet still have nothing outside it. Certain pitfalls open up when we try to apply our three-dimensional powers of spatial visualization to the universe, and we will try to shed some light on how such concepts as curved or warped space can have meaning. In fact, we will discover that the concept of *outside* the universe is a meaningless one, no matter what model we accept.

WHAT IS MEANT BY CURVED SPACE?

Most of us have studied the rules of geometry on a flat surface, which we can call **flat space**. Flat space has two important characteristics. The first is that in a simple triangle the sum of the angles is exactly 180° (**Figure 23-10a**). Second, parallel lines never meet (**Figure 23-10b**).

To give meaning to the concept of curved space is not difficult. Let us consider the simple analogy of the appearance of Earth to an observer standing on it. Although a line on the ground might appear straight, we know that it is actually a section of a great circle[1] on Earth. **Figure 23-11** shows two great circles on the surface of Earth. Observers at the equator examining each of these lines would

have the impression that they were parallel, yet the curves eventually intersect at the North Pole. By examining only a local region of space you may not be able to see curvature over larger distances.

Let us pursue this analogy further. Suppose you were to start walking in any direction on Earth. It would appear to you that you were traveling on a flat surface, and yet you would eventually return to the point from which you departed, proving that the surface was curved. As you returned to your starting point, you would certainly observe that you had not reached any place that you could call the "end" of the Earth; yet Earth's surface is indisputably finite in area. In an analogous way, it is possible for the universe to be finite in volume—never having anything that could be called either an *end* or an *outside*.

The sphere is an example of a surface having **positive curvature**. Although movement along this surface can continue forever, the surface is finite in extent. The angles of a triangle drawn on a surface with positive curvature add up to more than 180°. Other surfaces can have negative curvature, an example of which is a saddle (**Figure 23-12**). On such a surface, the sum of the angles in a triangle is *less* than 180°.

The universe as a whole can be described in terms of these geometric concepts. On one hand, a universe with a positive curvature is one that we speak of as a **closed universe**, one that is finite. An **open universe**, on the other hand, has a negative curvature and is infinite in extent. For the intermediate case in which the curvature is zero we have a **flat universe**.

---

[1]A great circle on a sphere is a circle whose center coincides with the center of the sphere. An equator is an example. So, too, is a circle passing through both north and south poles.

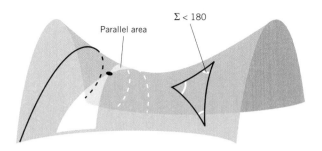

Σ < 180

Parallel area

Figure 23-12 The geometric properties of negatively curved space. Because none of the many lines passing through the point in the light-shaded area will ever intersect the black line, they are all "parallel" to it.

Visualizing curvature in space is difficult. One way to try it is to consider the trajectories of light rays as they move through space. In this sense we can describe a black hole as having a positive curvature that results from "warping" the space around it. Such warping bends the trajectories of light rays passing near it and coming from it. In the same way, the Sun warps the space around it.

23.4 The Future of the Universe

The science of cosmology has made significant advances in the last two decades. To understand those advances, we now look first at the traditional view before looking at the current revolution in cosmological thinking.

THE TRADITIONAL VIEW

Since the late 1990s, a revolution in our understanding of the large-scale dynamics of the universe has occurred. Since this change is so new, it is useful to explain why astronomers held what one may call a *traditional view* before elaborating on these exciting new developments. In addition to illustrating how significant the revolution in cosmology has been, it is also worth remembering that sometimes new ideas and findings turn out to be wrong.

The traditional view of the expansion of the universe was simply that the expansion rate must be slowing down. In other words, it was generally assumed that the universe was decelerating. To understand why, consider the simple exercise of throwing a ball upward from the surface of the Earth. The upward speed of the ball will decrease until it stops and falls back to Earth, thanks to the Earth's gravity. If one could throw the ball with a sufficiently high speed, it would escape the gravitational pull of the Earth altogether, but the speed

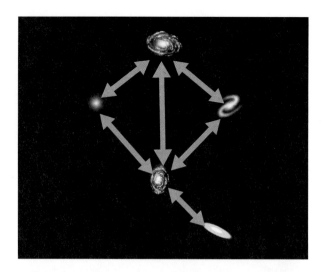

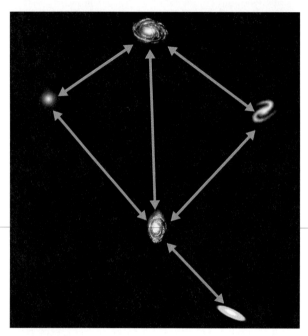

Figure 23-13 As the universe expands, the gravitational force between galaxies weakens (as shown by the arrows), yet it doesn't ever disappear entirely.

at which it departed would be continually slowed by the pull of the Earth. The same used to be assumed of the universe: Gravity, acting between all the matter in the universe, tends to cause a decrease in the expansion rate (see **Figure 23-13**).

According to this traditional view, deceleration can produce three possible outcomes. First, the deceleration could be so small that the universe will expand forever. Second, gravity could cause the expansion to slow and halt as the universe's age approaches infinity. Lastly, the deceleration could cause the expansion to halt and reverse direction, eventually leading to a *big crunch*. In this picture, the determining factor in which of

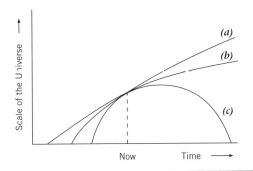

Figure 23-14 The expansion of the universe in the absence of a cosmological constant. *(a)* open (expands forever), *(b)* flat (critical density), *(c)* closed (collapses to big crunch).

these three options actually occurs depends on the density of matter in the universe. The more matter, the greater the gravitational attraction between different regions of the universe, and thus the greater the deceleration.

One method for determining the fate of the universe is simply to add up all the mass and determine whether there is enough to halt and reverse the universal expansion. Astronomers usually express the mass density of the universe as a fraction of the value that marks the *dividing line* between the ever-expanding and recollapsing possibilities just described. If the universe had exactly the mass density between these extremes, sometimes called the *critical density*, this would be written symbolically as $\Omega_m = 1$, where Ω is the Greek capital letter omega.

We have already discussed the difficulties in weighing galaxies and clusters of galaxies in Chapter 21. Even if we use dynamical methods to determine masses so that dark matter is included in the inventory, there is still the possibility that our dynamical tracers do not extend far enough for all the mass to be detected. However, many different techniques indicate that the density of matter in the universe is less than the critical density, so that in this traditional view we anticipate that the expansion of the universe is slowing, but that the universe will expand forever. See **Figure 23-14**.

Inquiry 23-6

What would be the effect on the expansion of the universe if the strength of the gravitational force decreased with time?

A COSMIC REVOLUTION

Although the observations that we will shortly discuss are recent, the origins of the cosmic revolution stem from 1917 when Einstein applied his new theory of gravity—general relativity—to the universe. Much to his irritation, he found that general relativity did not predict a static universe. Recall that the Hubble law was not established until the 1920s, so that in 1917 it was generally assumed (based on observations of that time) that the universe was unchanging. Einstein found that a static distribution of matter would not remain static. The gravitational attraction between parts of the universe would cause an initially static universe to collapse in on itself. Since there was no evidence that such a calamity was taking place, Einstein modified his equations slightly to include a *cosmological constant.*[2] This term, usually denoted Λ (Greek lambda), acts against gravity and allows a static universe. When Hubble and others found that the universe was expanding, Einstein described his introduction of the cosmological constant as the biggest mistake he had made in physics. With a zero value for the cosmological constant, he was concerned that the equations of general relativity lead to the *prediction* that the universe is either collapsing or expanding.

The traditional view of cosmological expansion assumes a zero cosmological constant, so that the expansion rate is slowed by the gravitational pull of all the matter within the universe. In the late 1990s, however, investigations of the Hubble law using extremely distant supernovae turned cosmology on its head. To understand why, we should first point out an unfortunate piece of cosmological terminology. We usually speak about the Hubble *constant* as a measure of the expansion rate of the universe. However, from the previous discussion it is apparent that the Hubble constant is unlikely to be constant. A change in the expansion rate of the universe means that the Hubble *parameter, H(t),* also changes with time. When we write the Hubble law as $V = H_0 d$, the subscript 0 indicates we are referring to the expansion rate of the universe at the present epoch. The expression $V = H(t)\, d$ indicates that over the history of the universe, $H(t)$ changes.

---

[2]For those readers who may find the modification of Einstein's equations to be something like cheating, the situation is a little more subtle. The cosmological constant term actually arises naturally in general relativity, but Einstein felt the equations were more elegant if this term had a value of zero.

When we measure the recession velocity of supernovae at great distances, the finite speed of light means that we are also looking back in time. One consequence is that by measuring the Hubble parameter over a range of distances we can, in principle, look for any changes in $H(t)$ and thus directly observe changes in the expansion rate of the universe. Probing these great distances has only become possible in the last few years. However, as more such studies have been undertaken, most astronomers are now convinced by the evidence that the expansion rate of the universe is currently *increasing*. This is in direct contrast to the traditional view in which the presence of matter slows the universal expansion. It appears that something is *pushing* the universe so that it expands at an ever-increasing rate.

What could be causing the universe to speed up in this manner? The simple answer is Einstein's cosmological constant, but that does not tell us what physical process is responsible for the acceleration. Two common terms for the cause of the increasing expansion rate are *dark energy* and *quintessence,* but there is currently little consensus amongst astronomers about what this mysterious energy is. However, if these results are correct and the universe is expanding at an ever-increasing rate, it will mean that Einstein's self-described biggest mistake will be a fundamental component in our description of the expansion of the universe.

It is important to emphasize that one reason the distant supernovae results have been accepted by most astronomers so readily is that there are independent studies, based on completely different physical principles, that confirm the idea of an accelerating universe. The most compelling of these involves the properties of the cosmic background radiation. When Penzias and Wilson first detected this radiation, its cosmological origin was apparent from its *isotropy;* that is, the same amount of radiation was coming from all directions. Subsequent observations confirmed this view showing that the background radiation is very *smooth.* In other words, the temperature of the background radiation varies hardly at all from one part of the sky to another. This means that the distribution of matter in the universe at 300,000 years after the Big Bang was also very smooth.

The ***Cosmic Background Explorer*** (***COBE***) satellite, launched in 1989, studied many aspects of the background radiation, including its tempera-

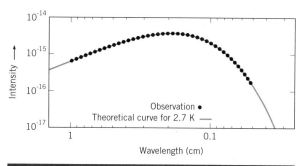

Figure 23-15 The spectrum of the 3-K background radiation as determined by COBE. The circles are the observed data points, while the curve is the best fitting blackbody curve.

ture. **Figure 23-15** shows the resulting observed data as filled circles. From these data, we may compute the temperature of the blackbody that agrees best with the data, and thus derive the temperature of the cosmic background radiation. The resulting temperature is $2.74 \pm .06$ K; the agreement at all frequencies is exceptional.

When the *COBE* data are examined in different directions in space, the data show small, but real, deviations from perfect isotropy. **Figure 23-16a** shows *COBE's* direct measurements placed on a map of the Milky Way with the galactic equator along the center line. Color-coded pink regions have temperatures up to 0.003 K lower than average, while blue areas are as much as 0.003 K above average. These variations are not intrinsic to the background but are due to Earth's motion of some 360 km/s relative to the radiation. The motion is toward the blue region in the south galactic hemisphere in Figure 23-16a. Subtraction of this **dipole anisotropy** caused by Earth's motion produces Figure 23-16b, which shows the concentration of microwave emission *from the Milky Way* having temperature variations as small as 0.0003 K. Subtracting this local Galactic emission results in Figure 23-16c, which shows temperature variations of 1 part in 100,000 for the universe as a whole.

More recent studies of the background radiation, such as that carried out with the ***Wilkinson Microwave Anisotropy Probe*** (***WMAP***) have confirmed and extended the *COBE* results. The importance of such studies is that they give us a map of the universe when it was a few hundred thousand years old. We also have a map of the universe as it is now through the observed distribution of galaxies and larger structures in the universe. It is pri-

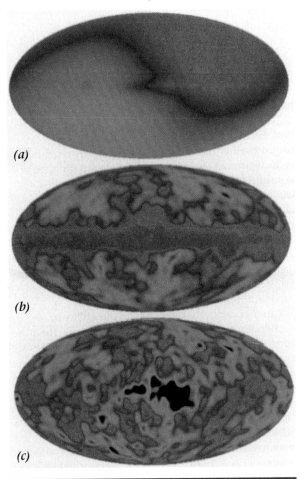

Figure 23-16 The variation in the strength of the 3-K radiation over the sky, shown in the coordinate system of the Milky Way. *(a)* The observed data in which blue represents the location to which the Sun is moving relative to the radiation; red is the direction opposite to the motion. *(b)* What remains after subtracting part a; the microwave emission from the Milky Way itself. *(c)* What remains after subtracting part b; the microwave background from the universe itself.

marily the action of gravity that takes us from the decoupling of matter and radiation to our universe today, along with dark energy. The increase in the lumpiness of the universe depends on the amount of dark matter and dark energy. Thus, our knowledge of the evolution of the universe from 300,000 years to now, combined with the fact that we have maps of the lumpiness at both ends of this evolutionary journey, allows us to determine how much matter and dark energy is in the universe. We have already mentioned that dynamical measurements of the mass density of the universe give values around $\Omega_m = 0.3$. This value is consistent with

results stemming from properties of the cosmic background radiation. We can express the amount of dark energy in the same units. The result turns out to be $\Omega_\Lambda = 0.7$. Notice that the sum of these quantities is 1.0. It turns out that this total density has the same geometric effects as a $\Lambda = 0$ universe with a critical mass density. In other words, although the universe will expand forever, its geometry is flat. This turns out to be extremely important in testing one of the oddest ideas associated with the early universe, discussed in the next section.

In Section 23.1 we mentioned that, for the case of a universe expanding at a constant rate, the age of the universe is simply $1/H_0$. In the traditional view of cosmic expansion in which gravity slows this expansion, the actual age of the universe is less than $1/H_0$.

To elaborate on this point, consider the following analogy. You invite three friends to dinner, each of whom lives four miles away. Looking out the window a few minutes before the roast potatoes have reached a perfect golden brown, you see your three friends walking briskly toward your front door at four miles per hour. Knowing that all three are keen on exercise and that it is a lovely summer evening, you assume that all three walked to your house at four miles per hour. Consequently, you conclude that each began their four-mile journey one hour ago. However, it transpires during dinner conversation that Adam left his house late and drove to dinner. His journey took only seven minutes. Daphne spent the early part of her journey inspecting wild flowers and left her house two hours before her arrival, speeding up during the last half of the journey in order to get to dinner on time. Only Karon walked at a steady four miles an hour all the way and thus left her house an hour before arriving at yours.

Karon's trip corresponds to the case in which the expansion of the universe is constant. The speed at which you saw Karon arrive at your house is the same at which she walked during the entire route, and so you correctly calculated the time of her trip. If we lived in such a universe, $1/H_0$ would be the age of the universe. Adam's journey is an extreme case of the situation in which the universe slows down as it expands—that is, the traditional view in which matter in the universe slows down the universal expansion. His speed as he arrives for dinner might lead us to assume his journey took an

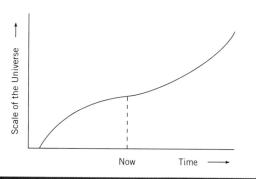

Figure 23-17 The expansion history of the universe.

hour, but once we know the first part of his journey was covered at a higher speed, we can correctly conclude his actual journey took less than this. Daphne's walk to dinner is comparable to the universe in which we live. Because the first part of her walk was slower, it took her longer than one hour to complete the journey. Similarly, the age of the universe is greater than $1/H_0$.

Detailed studies of the expansion history of the universe indicate that the actual situation is somewhat more complex. Combining results from distant supernovae and observations of the microwave background radiation, astronomers conclude that for the first couple of billion years, the expansion rate slowed, and that it is only more recently that the expansion has been speeding up. This is because initially, when the universe was more compressed, the higher density of matter meant gravity could slow the expansion, whereas at later times when matter was more spread out, the pushing effect of *dark energy* became dominant. This is illustrated in **Figure 23-17**. Putting together all the pieces of this puzzle, including the observed values for H_0 along with Ω_m and Ω_Λ, astronomers can calculate the age of the universe. The answer comes out to be around 14 billion years.

Why is this important? Apart from being an interesting result in its own right, the modern view that the expansion of the universe is speeding up solves an old problem. With the traditional view of the universe, estimates of its age derived from the Hubble constant were often less than 10 billion years. This was somewhat embarrassing, because the age of globular clusters in the Milky Way (see Chapter 18) has been determined to be around 12 billion years. In fact, the problem was not with the Big Bang model, but with the assumption that the expansion rate of the universe was slowing down.

23.5 The Early Universe Revisited: Inflation

The Big Bang model has been the paradigm of cosmology for many years, primarily because of the impressive agreement between the three observational cornerstones described earlier and the predictions of the model. There are a couple of problems, however, which may require a refinement to the model.

THE FLATNESS PROBLEM

The universe appears to be flat—not just nearly flat, but extremely flat. Why should that be? Is it possible that the universe was made to have exactly the density of matter and energy required for it to be flat?

Let us ignore the cosmological constant for a moment and imagine that when the universe formed it had a density just slightly greater than the critical density. That slight deviation at the beginning would have grown, and, as time went on, the universe today would appear to be closed. In fact, such a universe may have collapsed to a big crunch before human beings had time to evolve and ask all these questions about the universe. However, if the density had been slightly less than critical, that deviation would have grown until the universe today would look open. Thus, any deviation from critical density at the beginning would have rapidly grown away from flatness. The situation is no better if we include the cosmological constant. The sum of Ω_m and Ω_Λ is observed to be very close to 1, so that the universe is very close to flat, but such closeness now requires exquisite fine-tuning so that the universe was extremely close (or identically) flat during the early stages of the Big Bang.

Astronomers do not like this kind of fine-tuning. There is something suspicious about the universe being so close to this "special" case of flatness. If, when you looked out your window waiting for your friends to arrive for dinner, you observed that every car that went by was dayglo pink, you would likely find that suspicious, too. You would probably ask yourself *why* this was the case. The observation that our universe is very close to being flat is known to cosmologists as the **flatness problem**.

THE HORIZON PROBLEM

Whether you look at the celestial sphere in one direction or the opposite direction, the cosmic microwave background radiation is highly uniform. Why? How does one part of the universe "know" that it needs to have the same temperature (to one part in 100,000) as a point in the universe a large distance away? For example, consider a cool room with a heater at one end. The other side of the room "knows" to become warmer only when enough time has passed for the information carried by the heat to reach the other side. Before enough time has gone by for the information to get there, the opposite wall will remain cool, because it is beyond what we call the heater's *horizon*.

When the microwave background radiation was scattered for the last time, the universe was about 300,000 years old. Regions separated by, say, 10 billion light-years today were then separated by 10 billion divided by 1,000, or 10 million light-years. But, at this time, the universe was only 300,000 years old! Thus, information could not have "told" another part of the universe 10 million light-years away what temperature it should be. In other words, when we observe the microwave background radiation from widely separated parts of the sky, we are looking at parts of the universe that had never communicated with each other when they sent background radiation on its long journey across space. This is known as the **horizon problem**.

Since this is a difficult point to understand, let us again consider your dinner party. When your three friends arrived simultaneously for dinner, you may not have been surprised since you told them to arrive promptly at seven and you have a reputation for making a big deal about such things. However, if all three of your friends arrived at your front door wearing matching dayglo pink jumpsuits, you would conclude that they had been in communication with each other before they had left their respective houses. And you would probably ask them if they knew anything about the dayglo pink cars.

INFLATION

These problems with the original Big Bang theory can be solved by a refinement usually referred to simply as **inflation**. In the inflationary model, the universe undergoes a brief but extremely rapid expansion phase in which the distance between different regions increases 10^{50} times in 10^{-32} second. One can think of this phase as one in which the universe has a temporary, but extremely large, cosmological constant distinct from the one just discussed. It has been suggested that energy released when the strong and electroweak forces separate from one another may power this expansion. What this means is that our entire observable universe (that from which light signals can reach us within the age of the universe) was initially a tiny region in which all parts had time to communicate with each other before inflation occurred. Consequently, there is no horizon problem.

Inflation also solves the flatness problem. Recall the issue here is that the universe is remarkably flat, despite the fact that any departures from flatness in the early universe tend to get greatly magnified as the universe evolves. To understand how inflation helps, consider an ant on a small balloon. This ant is smarter than the average ant and has deduced, by drawing triangles and measuring the sum of their interior angles, that the surface of the balloon is curved. Imagine that this balloon now inflates rapidly, so that it is the size of a star. The ant, of course, is still a tiny creature on this huge sphere. If the ant again draws triangles and measures the interior angles as before, it will conclude the surface on which it lives is flat. In the same way, any initial curvature of the universe gets "stretched out" and thus flattened by inflation. To put it another way, inflation ensures that whatever the initial curvature of the universe, the current curvature will be indistinguishable from flat. There is, therefore, no fine-tuning problem.

23.6 The Halley-Olbers Paradox

After delving into the details of how the geometry of space might be determined, it is refreshing to find that sometimes simple lines of reasoning lead to conclusions of great significance. For example, consider a question asked first by Edmund Halley and then later in 1826 by the German astronomer Heinrich Olbers: Why is the sky dark at night? If space is imagined to be without curvature and infinite in extent, this question is by no means trivial. **Figure 23-18** illustrates concentric circles around an observer in flat space such that the distance

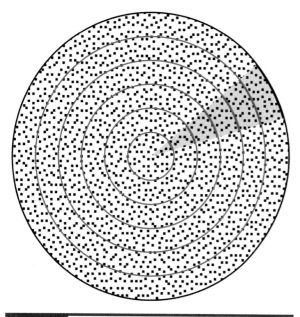

Figure 23-18 Concentric zones of equal thickness around the Earth have uniformly distributed stars whose number increases with the square of the distance.

between each circle is the same. We will assume the stars are uniformly distributed. Because the volume of each region increases with the distance squared, the number of stars in each region also increases with the square of the distance. But the inverse square law of light tells us that the brightness of stars decreases with the distance squared. Because the increasing number of stars is exactly cancelled by their decreasing brightness, each zone, no matter how far away, adds the same amount as any other zone to the total light observed on Earth. The end result is that the night sky should look like wall-to-wall Sun, and we should be roasting in a 6,000 K oven! The paradox, then, is that the night sky is dark when it should be bright.

How can we resolve the paradox that the night sky is dark when it might be expected to be bright? We cannot invoke the absorbing properties of interstellar dust to escape the apparent paradox, because within a finite amount of time the dust would come into equilibrium with the radiation and be heated to the temperature of the Sun's surface.

Might the expansion of the universe resolve the paradox? Photons from a receding galaxy will, of course, be redshifted out of the visual region of the spectrum. If you look farther, therefore, objects will appear fainter. Furthermore, the expansion reduces the density of photons in space the farther away you look. However, it can be calculated that

the actual reduction of the energy flow is not large enough to relieve the paradox.

The accepted solution to the paradox is that the universe is still young. Because light travels with finite speed, there is a physical limit (a horizon) to the *observable* universe—the distance light can travel since the universe formed. Therefore, the continuous zones in Figure 23-18 do *not* continue forever and the night sky is not bright.

23.7 The Multiverse

To conclude this chapter, we describe briefly some recent speculations that suggest that our universe may be one of many universes. Not only are these ideas speculative, they are also rooted in some of the most complex areas of theoretical physics, so our description is necessarily a simplification.

In Section 23.5 we discussed the horizon and flatness problems and explained that these constitute fine-tuning problems with the simple Big Bang model. In essence, scientists do not like to find that physical systems have special values or properties unless they can explain why this is so. Although the inflationary scenario overcomes the flatness and horizon problems, other fine-tuning issues still bother astrophysicists and have led, at least in part, to ideas that are even more exotic than inflation.

We described how inflation can smooth out the universe so that, whatever the initial geometry, the universe today is predicted to be spatially flat. This assumes, of course, that the universe survives long enough for inflation to occur at all. At first sight, this may not seem like much of a problem. After all, inflation is predicted to occur when the universe was a tiny fraction of a second old. However, when physicists look at all the possible universes that are consistent with the laws of physics as we know them, they find that many universes will have expanded and then collapsed to a big crunch before the epoch at which inflation could have occurred. Further, a lot of possible universes would go through inflation, but have so much dark energy that galaxies could not have formed in the rapid expansion.

The "short-lived" universes and the "rapidly expanding" ones have one important feature in common; neither type is likely to have led to the emergence of life. The final chapter of this book is devoted to a discussion of the conditions under which life might arise, but for now we simply note that life *has* arisen at least once in the universe, and that some physicists think that this fact cannot be ignored when we discuss the universe. This is because we once again come up against a fine-tuning problem. For life to have had a chance to emerge, the universe must have existed longer than the short-lived universes. Further, the universe cannot have too much dark energy, because without the formation of galaxies, and stars within those galaxies, it is hard to see how life could have evolved (see Chapter 24).

It seems, therefore, that out of all the possible universes, ours is one of a class of "special" ones in which conditions are suitable for the emergence of life. In fact, if one delves further into this issue, it turns out there are many aspects of our universe that seem "fine-tuned" for the emergence of life. For example, if the strong nuclear force had been slightly stronger, protons would have combined with one another to form "diprotons" so that there would be little hydrogen in the universe. This is problematic for life, both because stars would have much shorter main sequence lifetimes, and because without hydrogen there would be no water.

Inquiry 23-7

Why would the absence of hydrogen lead to short main sequence lifetimes for stars? Why is this problematic for the emergence of life? (*Hint:* see Chapter 24 and the discussion of the timeline for life).

Some physicists feel that once we have a Theory of Everything, we will understand why the strengths of the fundamental forces have their observed values, and why there is the observed amount of dark energy in the universe. However, even if this turns out to be the case, it still troubles many scientists that out of all the conceivable values for these fundamental quantities, they have somehow combined to give a life-sustaining universe.

One solution to this problem that has been gaining popularity is the idea of the multiverse. In this picture, our universe is one of many. By this, we mean not just one of many *imaginable* universes, but one of many physical universes that actually exist. This is a radical idea, not least because traditionally the *universe* has meant everything there is, but it overcomes the fine-tuning problems described. Specifically, if there are many (possibly an infinite number of) universes, then *some* are bound to have the right conditions for the emer-

gence of life, and we live in one of those universes. If this is the case, it is the multiverse that is "everything there is," and not the universe as we suggested at the beginning of this chapter.

Simply overcoming an annoying problem is obviously not a sufficient reason for a theory—particularly one as radical as the multiverse—to be taken seriously. It turns out that the same theoretical ideas that give rise to inflation, and which may provide us with a theory of quantum gravity, are consistent with the idea of multiple universes springing up out of nothingness.

You may have been told in high school science class, "Nature abhors a vacuum." This is a rather pompous way of saying that if you pump all the air out of a glass jar, air will force its way back in again through any hole, no matter how small. Current speculations arising from quantum physics suggest that Nature abhors "nothing" even more. Just as air will eventually leak back into our evacuated jar, these ideas may imply that something will eventually replace nothing. In other words, the universe is an inevitability. And the formation of a universe may be something that happens not just once, but an uncountable number of times, thereby producing the multiverse.

Throughout this book we have cautioned the reader that science must deal with that which can be observed. In other words, scientific theories are grounded in observation. In this discussion we have clearly entered a speculative realm where direct observation of other universes may forever be impossible. We make no apology for this. Indeed, we would be remiss if we did not present the reader with these fascinating areas of modern physics. However, one may wonder whether this discussion might fit more naturally in a book devoted to questions involving trees falling in forests and whether or not a sound is produced.

As science has developed, it has spread into realms that were once reserved for philosophers. It may be that we are jumping the gun and that the theory of the multiverse has not yet progressed to the point where it has left the bookshelf devoted to philosophy. Sir Martin Rees, British Astronomer Royal and noted cosmologist, recently gave the opinion that discussions of the multiverse were "speculative science, not metaphysics." Even if the claims that the multiverse is a scientific theory are premature, the fact that our species can engage in such speculations enhances our appreciation of the universe and, perhaps, of ourselves.

The Age of the Universe

This Discovery will help you gain a further appreciation for the age of the universe, and the role played by humanity during that time. This activity may be done with at least one other person.

Obtain a standard-size roll of toilet paper. The length of this roll will represent the age of the universe. The package should specify the number of tissues on the roll. Use this information to determine the number of years represented by each sheet. Then, using a long hallway if available (if such a hall is not available, you can go back and forth within the longest area available), begin unrolling the paper, and mark each of the following events in the history of the universe: the formation of the 3-K background radiation, the formation of galaxies, the formation of the Milky Way's globular clusters, the formation of the Milky Way, the formation of stars, the formation of the Solar System. As an added section, use the library if necessary to obtain the information to mark when life on Earth first appeared, when dinosaurs roamed the Earth, the appearance of humanoids, the time when the Neanderthals lived, the high point of the Greek civilization, and the time of your birth.

- **Discovery Inquiry 23-1a** On the scale of your model, how wide a line represents all of human history? The last 100 years?

DISCOVERY 23-2

The Expansion of the Universe

When you have completed this Discovery you should be able to do the following:

- Explain, using words and a graph, the reasoning astronomers use to conclude that the universe is expanding.

We can use Figure 21-10 to demonstrate the expansion of the universe. The five elliptical galaxies shown were selected from five different populous clusters of galaxies. Each is one of the largest and most luminous of the elliptical galaxies in its cluster. If we assume that these five objects are comparable in actual size and luminosity, then their *apparent* size is going to depend on their distance. The smaller the galaxy on the photograph, the farther away it is.

Measure the size of each galaxy with a millimeter scale, trying to read the size to a tenth of a millimeter. Calculate the quantity 1 divided by the size of the galaxy and make a table of your results. This quantity will get numerically larger as the galaxy's distance increases.

The spectrum of each galaxy is also shown in Figure 21-10, with the position of two absorption lines of ionized calcium (Ca II) in the spectrum indicated by an arrow. Measure the redshift shown by each galaxy by measuring the length of the arrow above each spectrum (in millimeters). Include these values in your table.

Plot your values of 1/size on the horizontal axis of a piece of graph paper and the corresponding values of the redshift (as measured by the length of the arrows in millimeters) on the vertical axis. Label the scales on your graph paper so that your plot extends over most of the page.

Clearly, the more distant objects show a larger redshift. If the redshift is interpreted as speed away from Earth, this means that the more distant the galaxy, the faster it is receding. In other words, the clusters of galaxies are all moving away from each other; the universe is expanding. The only way to escape the logic of this argument is to find some other physical effect that will cause the redshift seen in the spectrum, and no one has come up with a successful alternate interpretation.

The photos that you are examining in this activity are comparable in quality to the data that were originally used to deduce the expansion of the universe. As a consequence, you can see that the actual experimental proof of this enormously significant idea is not difficult at all. The genius and the progress come from having the *idea* in the first place.

- **Discovery Inquiry 23-2a** Explain why your graph demonstrates the expansion of the universe.

Observing the Cosmic Background Radiation

We can detect the remnants of the Big Bang with an ordinary television receiver. Turn a TV set with a good antenna (not cable) to an unused channel. Set the contrast to its maximum, and then turn the brightness down to the point where the snow on the screen appears as white flecks on a black background. About one out of every hundred of the flecks of snow actually represents the detection of an individual photon from the remnants of the Big Bang. The other 99 percent represent primarily radio noise (static) generated in the receiver itself, and other environmentally produced interference. It was these interfering effects that Penzias and Wilson had to eliminate almost completely from their equipment before they could unambiguously detect the background radiation. Indeed, before recognizing the true source of the noise they detected with their equipment, they had to eliminate all other possibilities. At one time, they even thought that the droppings from a flock of pigeons that had nested in their antenna might have been responsible!

Chapter Summary

In forming a comprehensive summary of cosmology, some applicable information is carried over from the two previous chapters.

Observations

- The **Hubble law** is a graph showing the relationship between the velocity with which a galaxy is moving away from the Milky Way versus its distance. The Hubble law is a relationship showing that the greater a galaxy's distance, the greater its velocity. The value of the Hubble constant is uncertain and generally thought to be about 20 km/s/Mly.

- All quasars are generally considered to be at large distances with none nearby. Galaxy counts show there to be more galaxies at large distances than nearby.

- Helium accounts for some 25 to 30 percent of the mass of the universe, with hydrogen accounting for most of the rest.
- **Penzias and Wilson** observed a general background radiation in the universe at radio wavelengths. According to the *COBE* satellite, this radiation is almost perfectly isotropic and has a spectrum of a blackbody of temperature 2.74 + 0.06 K. The *COBE* satellite has found small temperature fluctuations in the microwave background radiation.
- Viewed on a large scale, the universe appears to be **homogeneous** and **isotropic**.
- The density of matter in the universe can be found by studying the motion of objects within galaxies, and the motion of galaxies within **clusters of galaxies**. It can also be inferred from observations of the expansion of the universe and the **microwave background radiation**.
- Galaxies in clusters have velocities larger than expected on the basis of the amount of observable matter within the clusters.
- Galaxies appear to be accelerating rather than decelerating as the universe ages.

Theory

- A summary of the history of the universe is in **Table 23-1**.
- The universe was once in an extremely hot, dense state, since which time it has been expanding.
- The four forces of nature were probably unified until a short time after the Big Bang.
- The **Steady-State theory**, in which the universe remains the same all over, requires the continuous creation of matter to hold the density of the expanding universe constant.
- The observed acceleration of the universe can be explained by Einstein's **cosmological constant**.
- An **open universe** is one with a **negative curvature** and is infinite; a **closed universe** has a **positive curvature** and is finite. In a **flat universe**, the laws of Euclidean geometry hold. The critical density is the density of a flat universe with zero cosmological constant.

Table 23-1   Summary of the History of the Universe

Cosmic Time	Years Ago	Redshift ($\Delta\lambda/\lambda$)	Event
0	14×10^9	infinite	Big Bang
10^{-43} s	14×10^9	10^{32}	Planck epoch
$10^{-43} - 10^{-35}$ s	14×10^9		GUT era
10^{-35} s	14×10^9		Inflation occurs?
10^{-6} s	14×10^9	10^{13}	Proton-antiproton pair annihilation
1 s	14×10^9	10^{10}	Electron-positron pair annihilation
1 minute	14×10^9	10^9	Helium and deuterium formed
10,000 years	14×10^9	10^4	Universe becomes matter-dominated
300,000 years	14×10^9	10^3	Decoupling of matter and radiation; 3-K radiation comes from this time
$1-2 \times 10^9$ years	$11-12 \times 10^9$	20	First stars form
$2-6 \times 10^9$	$8-12 \times 10^9$	$1-3$	Peak of galaxy formation
2×10^9 years	12×10^9		Milky Way protogalaxy begins collapse
2×10^9 years	12×10^9	3	Halo globular clusters form in Milky Way
4×10^9 years	10×10^9	1	Formation of Milky Way disk
9×10^9 years	4.7×10^9		Protosolar nebula collapses
9×10^9 years	4.6×10^9		Planets form
9×10^9 years	4.3×10^9		Intense cratering on Moon and planets
10×10^9 years	4.1×10^9		Oldest terrestrial rocks form
11×10^9 years	3.5×10^9		Microscopic life forms
12×10^9 years	2.3×10^9		Oxygen-rich atmosphere develops
13×10^9 years	2.1×10^9		Macroscopic life forms
13×10^9 years	2×10^9		Oldest fossil record
14×10^9 years	200×10^6		First mammals
14×10^9 years	60×10^6		First primates
14×10^9 years	200,000		Homo sapiens

- Observations of galaxies at large distances are capable of distinguishing between positive, negative, and flat universes.

Conclusions

- While modifications to the Big Bang theory are required to bring some of its details into agreement with observations, the general outline of the theory is in excellent agreement with three independent lines of evidence: (1) the abundances of the elements in the oldest stars; (2) the observed Hubble velocity–distance relation; and (3) the agreement of the observed 3-K microwave background radiation with theoretical predictions.

- From *COBE* and *WMAP* satellite observations we conclude that sufficient fluctuations were present within the Big Bang to allow galaxies to form.
- An extensive amount of dark matter in galaxy clusters is inferred from the motions of galaxies within them.
- Observations strongly suggest that the universe is accelerating.
- The **inflationary** scenario accounts for its apparent flatness and its high degree of isotropy.
- The universe's expansion is accelerating.

Summary Questions

1. What is the crucial role of measurement and observation in the definition of the universe?

2. What is the interpretation of Hubble's redshift observations?

3. What are the various observable differences to be expected between a Steady State universe, a Big Bang universe that expands forever, and a Big Bang universe that eventually collapses on itself? Your discussion should include observations of changes in density and expansion velocity, and the 3-K background radiation. What are the difficulties in making observations of each?

4. What are the main events during the expansion of the universe? What were the conditions prevailing when each event occurred? What evidence of the events can be observed today?

5. Why must elements heavier than helium have been created in the stars rather than during the Big Bang?

6. What is the probable future of the universe? How did astronomers reach their conclusions on this question?

7. How is the shape of the universe related to its history and the amount of mass it contains? What are some geometrical analogies to various types of universes?

8. What is the 3-K background radiation? Describe it both from the observational and theoretical viewpoints. What was the significance of the observations made by *COBE* and *WMAP?*

9. What are the observations that astronomers have made that lend credence to the Big Bang theory?

10. What is the Halley-Olbers paradox? What are various ideas considered to resolve it successfully?

Applying Your Knowledge

1. What influence would the aging of galaxies have on observations of galaxy counts at greater and greater distances? Would such aging tend to make the universe seem to decelerate at a greater or lesser rate than it actually does?

2. Summarize the predictions the Big Bang theory makes. What do observations have to say about each of these predictions?

3. The energy of a photon is inversely proportional to the wavelength of light. Let's speculate that as light travels through the universe it becomes *tired*—that is, it loses energy. Predict the consequences of tired light on our observations of both distant and nearby galaxies. (*Note:* there are observational and theoretical reasons that strongly limit any such variation.)

4. The wavelength of a photon emitted in a transition is inversely proportional to the mass of an electron. Let's speculate that the mass of an electron increases with time. Predict the consequences of increasing electron mass on our observations of the spectra of both distant and nearby galaxies.

5. Suppose that rather than beginning from a strongly condensed condition, the universe began as a ball of gas having the size it has today. Suppose, further, that matter was spread out uniformly (except for a few local irregularities) and with zero motion (that is, it was static). Describe the subsequent history of this universe and the reasons for your description.

6. Suppose the velocity–distance relation for the universe were as given in **Figure 23-19**. Explain the consequences of this velocity–distance relation.

7. Some Creationists argue that because astronomers observe the mass within galaxy clusters to be insufficient to hold the clusters together, such clusters can exist only if the universe is considerably younger than astronomers say. How might you respond to such an argument?

8. In Chapter 21 we saw that galaxy masses could be found from studies of rotation curves for spiral galaxies and velocity dispersion for elliptical

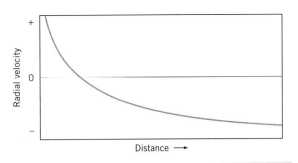

Figure 23-19 A hypothetical velocity-distance relation.

galaxies. If galaxies containing massive black holes were studied with these techniques, would the computed masses include the masses of the black holes or would their masses be neglected? Explain. Would the presence of black holes at the centers of galaxies bring the universe's overall density above the critical density?

9. Explain the Hubble law using points along a line, in a manner similar to that used in the raisin bread analogy.

10. Compute how far a photon could travel over the age of the universe.

11. Estimate the total number of atoms in the universe! (*Hint:* To solve this problem, you need to make a number of educated guesses.)

12. Compute the maximum age of the universe (in years) for values of H_0 = 15 km/s/Mly and 25 km/s/Mly.

Answers to Inquiries

23-1. Each million light-years corresponds to 20 km/s; therefore, 300 million light-years corresponds to a velocity of 300×20 = 6,000 km/s.

23-2. The velocity is $0.16c = 0.16 \times 300,000$ = 45,000 km/sec. From the Hubble law, $d = V/H_0 = 45,000/20 = 2,250$ Mly = 3×10^9 ly.

23-3. Larger, because those galaxies are observed as they were when the universe was younger and expanding with a higher velocity.

23-4. The difference is that the density in the cores of stars is higher, allowing more than two particles to collide at once. The triple-alpha reaction (Chapter 18), for example, forms carbon from short-lived beryllium. Further, stars have longer to make heavy elements than the element creation epoch in the early universe.

23-5. From $\lambda_{max} T = 3 \times 10^7$ ÅK, we have λ_{max} = 3×10^7 Å K 3 K = 10^7 Å = 0.1 cm, in the short end of the radio (microwave) region of the spectrum.

23-6. It would make it more difficult for gravity to halt the expansion of the universe and would increase the likelihood that it would expand forever.

23-7. The lifetime for helium burning in stars is much shorter than that for hydrogen burning. Thus, if stars are initially composed mostly of helium, they would have short lifetimes. Our one example of intelligent life required several billion years of evolution. With short-lived stars, this would not be possible.

PART SIX

Discovering if There Is Life Elsewhere in the Universe

We end at one of the frontiers of human discovery, the search for life on other planets. Since the only life we know—so far!—is that on our own planet, we start here first. Then we review the prospects for life in the rest of our Solar System, and how people's assessment of this has changed over the last hundred years or so.

The last step is the big one—the leap to the stars and considering which ones will be most likely to be the cradle for life that we could recognize. Then, last but not least, the search for extraterrestrial *intelligent* life is discussed. Included here are interesting questions such as whether and how to communicate with other civilizations, and why many scientists think this is a worthwhile scientific quest, yet do not think aliens have already visited Earth.

The Search for **24**
Extraterrestrial Life

You are a child of the universe,
no less than the trees and the stars;
you have a right to be here.
And whether or not it is clear to you,
no doubt the universe is unfolding as it should.

Unknown, Desiderata, *circa 1600*
Translated by Max Ehrmann, 1927

24.1 What Is Life?

"What is life?" is a question that seems easy at first glance, but when you start to think about it, it is not easy at all. However, to find life on other worlds, we must first carefully define what it is we are looking for. The easiest place to start is with a description of the only life we know, life on Earth. The description must be broad enough to include every example of life we have discovered so far. We do not want to overlook any of our neighbors just because they do not look like us!

Inquiry 24-1

What experiment would you design, if you were taking the photograph in **Figure 24-1**, to determine if there is an animal in this picture?

When astrobiologists speak of life, they mean not only animals, but also plants, fungi, and single-celled creatures such as bacteria. However, the only example we are sure about is life that flourishes in the unique location of the Earth. The deductive scientific process requires us to test a hypothesis about life by testing the validity of its predictions. However, tests of all hypotheses about life other than that on Earth will have to wait until we find another example of such life.

All life on Earth is broadly called the **biosphere**, which can be defined both in terms of the

Chapter opening image: Steven Vincent Johnson's *Evening Visitation,* an artist's impression of extraterrestrial life on the moon of a gas giant in another solar system.

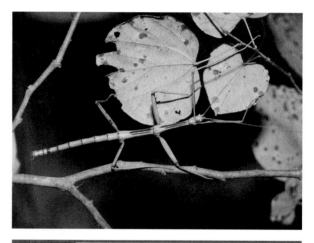

Figure 24-1 A brownish "walking stick" sitting on a green twig.

places where life can be sustained and the organisms themselves.

In the same way that the constellations have two names as in Appendix B, Earth has an alternative Latin name, Terra. Thus, **extraterrestrial life** refers to life that is neither on nor from the Earth. Although it is difficult to apply the scientific process to questions concerning extraterrestrial life (because we do not yet have an example of it), nonetheless we can start by thoroughly understanding life on Earth and attempting to extrapolate the rules to extraterrestrial environments.

LIFE ON EARTH

Biologists usually define life on Earth as something that has at least one cell, is internally organized, reproduces using inherited information encoded in the molecule color and old deoxyribonucleic acid (DNA), grows and develops, uses energy, interacts with the environment, responds to evolutionary

(a)

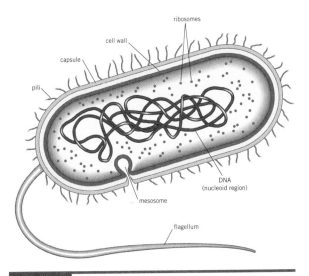

Figure 24-3 This is a bacterium, an example of a prokaryote.

(b)

Figure 24-2 Examples of life on Earth. *(a)* This one-celled diatom is a type of alga so tiny it cannot be seen with a microscope using light; this photo was taken with a scanning electron micrograph. *(b)* This giant sequoia in Tuolumme Grove towers over other terrestrial species.

forces, and is born and dies. This is true of all life, from tiny one-celled diatoms to humans to the great redwoods (**Figure 24-2**). When you thought about how to figure out if the walking stick in Figure 24-1 was alive, you probably thought about experimenting to see if it responded to you moving the twig—that is, you were testing to see if it interacted with its environment.

Life on Earth can be classified depending on the number and type of cells it has. Cells without a nucleus are called **prokaryotes**, which means "before a nucleus," and can be either **bacteria** or **archaea** (pronounced ar-KEY-a). One cell makes up the entire creature. A sketch showing typical components of a bacterium is at **Figure 24-3**.

Every other form of life, including humans, is called a **eukaryote** (you-CARry-oat) and has one or more cells, with a distinct nucleus as in **Figure 24-4** and is encased in a membrane rather than a rigid cell wall. A prokaryote cell is usually about one-tenth the size of one eukaryote cell. Prokaryotes can use a variety of energy sources, from light to the chemical bonds of metals, but eukaryotes use oxygen (O_2) in the process of respiration to release energy. Prokaryotes also outnumber eukaryotes; if we were to judge the dominance of life on Earth by how many individuals are living, Earth would be described as always being in the "Age of Bacteria."

The three-branched diagram in **Figure 24-5** summarizes the relationships among organisms of these three domains. Biologist Carl Woese discovered and added archaea, constructing this version of the universal "Tree of Life" in 1990 from the decoding of mitochondrial RNA. It largely confirms the fossil record for tracing the evolution of eukaryotes on Earth. The distance between the tips of each branch represents the distance in time between the emergence of each of the three domains of life. Animals, plants, and fungi

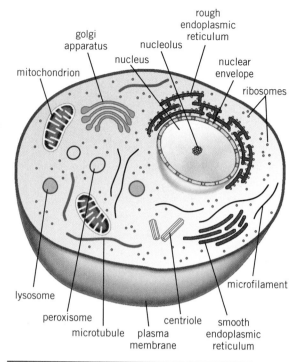

Figure 24-4 This is a typical eukaryotic cell with each function usually separated from the other by an internal membrane.

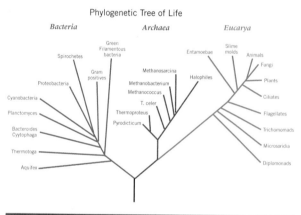

Phylogenetic Tree of Life

Bacteria   *Archaea*   *Eucarya*

Figure 24-5 The phylogenetic tree of life.

appeared 600 million years ago in an event called the Cambrian Explosion; all other forms evolved in the three billion years before that.

Current results show that using a tree as in Figure 24-5 to represent genetic evolution is an oversimplification. A network with multiple interconnections seems to be more realistic, as the evidence shows older life forms can genetically incorporate newer ones. Since this analysis can only include living life forms, it cannot identify ancestors that are extinct.

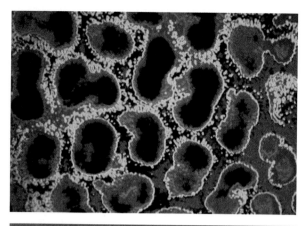

Figure 24-6 Influenza viruses.

A virus is an example of an organism that may or may not be alive (**Figure 24-6**). It meets most of the definition of life, except that it is not a cell and its genetic material can be DNA or a relative, ribonucleic acid (RNA). Although this electron micrograph makes them look flat, they are actually spherical and about one-tenth the size of the prokaryote in Figure 24-3, and a hundred times smaller than the eukaryotic cell in Figure 24-4. It is amazing how something so small can make us so sick! Since viruses can only reproduce inside cells, it is possible that they evolved after bacteria, or perhaps their ancestors were the first, simple RNA-based life.

24.2 The Ingredients for Life

We have explored what life on Earth is, and how it is broadly categorized. We now look at life on Earth to discover what this one biosphere tells us about how to discover possible life beyond the Earth. In return, this exploration will also shed light on the conditions under which life arose on Earth. We start by exploring the range of environments in which life on Earth now thrives, and use that information to identify three factors that are essential for life—energy, liquid water, and carbon.

THE EXTREME RANGE OF LIFE ON EARTH

The range of environments in which life on Earth can survive is the minimum range of environments for which we should reasonably search for extraterrestrial life. More importantly, studying these examples lets us identify a short and general list of absolute requirements for life.

In 1998, the NASA Astrobiology Institute was founded. One of its missions is to further the study of the distribution of terrestrial life. In the last few years, life has been found in amazing places—it survives and even thrives everywhere from Antarctic rocks, Yellowstone's boiling hot springs, and lakes meters below the surface of the land. Life at the extreme ends of the range is called an *extremophile*. Can this life exist outside of the temperature range of liquid water (0–100° C) on the surface of the Earth? In what range of temperatures does this life exist? In what environments is it possible for life to survive? What are the minimum requirements for these extremophiles?

The record for the maximum temperature in which life can thrive, 121° C, is currently held by the archaeon aptly named Strain 121. (See Appendix A3 for the relationships between temperature scales.) The single-celled Strain 121, pictured in **Figure 24-7**, was found near a deep-sea vent, a thin spot in the deep ocean floor where

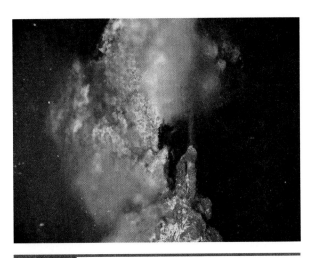

Figure 24-8 This 'smoker' projects sulfide and other minerals as hot fluid into the frigid, dark deep-sea waters.

water heated by the underlying magma of the Earth's asthenosphere flows up into the oceanic crust (Figure 8-7). Here, the pressure is so high that, just as in a pressure cooker, the boiling point of water is raised above 100° C. This hot water is rich in iron and sulfur, which would make it look black—if there were any natural light by which to see it. These hydrothermal vents are called *black smokers,* not because they are burning, but because the heated, mineral-rich water is less dense than the cold seawater and so it rises in a plume (**Figure 24-8**). Eukaryotes live in this environment as well, notably the vestimentiferan worms seen in **Figure 24-9**.

Figure 24-7 The archaeon Strain 121, the balloon-like microbe at the upper left, with multiple whip-like flagella. The main body of the archaeon is about 1 micrometer (10^{-6} m) across. Reprinted with permission from Kazem Kashefi and Derek Lovley, *Science,* vol. 301, p. 934

Inquiry 24-2

Why is Strain 121 aptly named?

Figure 24-9 Vestimentiferan worms. The light is from the submarine that photographed them.

Figure 24-10 Aerial photo taken at the eastern end of Lake Vida in Antarctica, looking north towards McMurdo Sound (the lake is in the foreground).

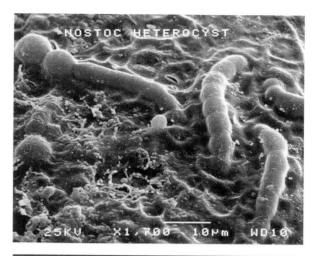

Figure 24-11 Filament-like lake-ice microbes found in ice from the region where Lake Vida is located.

We commonly think about life as being on the surface of the land and in the sea, wherever light can reach. However, the thriving ecosystems around black smokers are not the only ecosystems that thrive without light. You yourself host such an ecosystem. Bacterial life exists within our bodies, especially in our digestive tracts. Without the many beneficial types of bacteria, we cannot digest certain foods. It is even becoming evident that we have adapted to what used to be thought of as harmful bacteria, and will become sick if they are removed.

Life also exists deep underground. In Lidy Hot Springs, Idaho, archaea live 220 m underground and get energy from geothermal hydrogen. Scientists exploring Lake Vida, a saltwater lake under 19 m of permanent ice in Antarctica pictured in **Figure 24-10**, retrieved 2,800-year-old microbes frozen into the ice. The microbes (**Figure 24-11**) revived when exposed to liquid water. The lake water is seven times saltier than seawater, so salty that it stays liquid down to a chilly −10° C.

Even the fractures within rock itself can be infested with bacteria. In 1996 a drilling project in a South African gold mine found a different type of gold: bacteria growing so deep in the Earth that the only energy sources are hydrogen released by water reacting with the minerals, and carbon dioxide. Three kilometers down, this environment is so poor in nutrients that these bacteria may reproduce once every thousand years. The pressure is 300 times that on the Earth's surface, and the tempera-ture is a hellish 60° C (**Figure 24-12**). Neverthe-less, life is there.

All of these extremophiles, microscopic or not, have three things in common: an energy source, liquid water, and carbon. We will now look more closely at each of these factors.

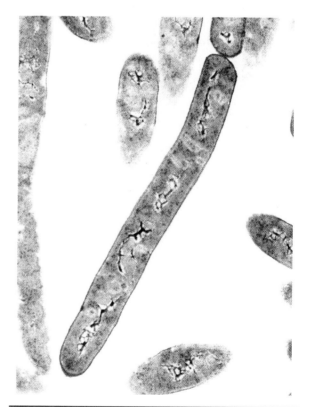

Figure 24-12 Bacillus infernus, informally translated as "the bug from hell."

A SOURCE OF ENERGY

The energy for life on Earth comes from three sources: radiation from the Sun, the decay of radioactive elements, and energy left over from the process of forming the Earth (Section 8.2). Organisms tap these energy reserves in a variety of ways.

Since we live on the surface of the Earth, we are most familiar with the strategy of using sunlight as an energy source. Plants and many bacteria use the chemical process of photosynthesis to produce sugar and starches from carbon dioxide (CO_2) and water. Sugar and starches are then stored and used over time; oxygen gas (O_2) is released into the atmosphere. Animals and fungi consume the plants—or other animals that ate plants—to get these compounds and the energy stored in them. Some bacteria and archea use light directly. They get carbon from organic compounds, which are any molecules that incorporate carbon. Frequently, the organic compounds are debris from other living things, rather than directly from carbon dioxide in the air.

Sunlight is also used indirectly. Since 39 percent of the light from the Sun is reflected and never reaches the surface of the Earth (Appendix C), this means that 61 percent of the light is absorbed not only by plants and animals directly but also by the land, atmosphere, and oceans. This energy is exploited by organisms for purposes as simple as direct warming (sunbathing), or more indirectly by benefiting from the climates to which they have adapted. The absorbed sunlight drives climates that provide hospitable environments in otherwise too-cold or too-hot regions. This is why global warming, an increase in retained energy in the Earth's overall energy budget, is predicted to cause massive climatic change and extinctions of species.

Several other prokaryotes that do not use sunlight have already been mentioned. They can use inorganic chemical reactions involving iron, nitrogen, sulfur, or ammonia (NH_3) for energy. They indirectly use the energy trapped under the mantle of the Earth (e.g., the black smokers in the deep-sea vents are driven by energy from the Earth's hot upper mantle—the asthenosphere; Figure 8-3), although black smokers also release ancient organic materials made using sunlight, then buried by the activity of tectonic plates.

To date, no life on Earth has been discovered that directly uses radioactivity as an energy source.

LIQUID WATER

All terrestrial life is organized into cells, and all active cells use liquid water. Water transports useful molecules around the cell, facilitates the intake of desired substances and the expulsion of waste products in and out of the organism, regulates the internal temperature of the organism, and provides support. No known life on Earth incorporates other phases of water—ice or steam—into its internal structure.

Water is often called the *universal solvent,* and this is one of the few times when a common description is apt. What makes water special? A water molecule consists of two hydrogen atoms and one oxygen atom (H_2O) bonded tightly together. However, the side of the molecule with the two hydrogens has a slight positive charge and the oxygen side is slightly negative. These slightly positive and negative sides of the water molecule help water dissolve compounds like table salt (NaCl) and molecules like glucose (the sugar in blood), amino acids, and proteins.

This polarity also makes the water molecules electrically stick to each other, as shown in **Figure 24-13**. The extra energy needed to pull these hydrogen bonds apart raises the boiling temperature of water to $100°$ C, increasing the range of environments in which water can perform its life-giving functions. However, the upper limit for life is different for eukaryotes, since the upper limit for

Figure 24-13 Models of water molecules (H_2O), with hydrogen shown in gray and oxygen in red.

them is determined not by the boiling point of water, but by the temperature where their membranes break down, about 50°C. This is why the vestimentiferan worms (Figure 24-9) live in cool water near black smokers, but not in the very hot water that the tougher cell wall of Strain 121 can withstand.

On the other end of the temperature scale, water is also unusual in that the solid (ice) it forms is less dense than the liquid. Usually, a solid takes up less space than a liquid of the same substance, but the unusual nature of the water molecule causes it to form large, four-sided crystals. If water inside a cell freezes and expands, it can burst the cell the same way a frozen bottle of water bursts its container.

Pure water on Earth under the pressure of the Earth's atmosphere at sea level is liquid in the range of temperatures 0°C to 100°C. It is tempting to conclude that we therefore should look for extraterrestrial life only where the temperature falls in this range at least some of the time. However, the range of temperature that water can be a liquid is affected by dissolving other materials in it or by changing the external pressure, as we saw in the black smoker ecosystem. For example, a green stink bug found in Alaska manufactures a protein that lowers the freezing point of its blood so it can remain active when temperatures drop to −29°C, and survive at lower temperatures. If the external pressure decreases, like the situation on Mars where the atmospheric pressure is one-hundredth the atmospheric pressure of Earth, water will boil into a gas. If the pressure increases, as it does deep underwater or underground, water can remain liquid well above 100°C.

Inquiry 24-3

Why do you think salt is spread over ice on sidewalks in the winter? Would this work in Antarctica during the winter?

Inquiry 24-4

Why does food cook faster in a pressure cooker than in a regular pot?

CARBON

All life on Earth must acquire carbon to live, either directly from carbon dioxide or indirectly by con-

suming other life. The first scientist NASA paid to design experiments to detect life, Wolf Vladimir Vishniac in 1959, narrowed down the search to carbon-based life because of carbon's unique chemical characteristics. Carbon has four electrons available to bond with itself and other elements to form large and stable molecules in two basic shapes, rings or chains. Ring-shaped carbons are the basis for the four amino acids in DNA (**Figure 24-14**), RNA, and sugar. Carbon in the form of a chain is the backbone for fat molecules. However, these molecules are not so stable that they cannot release their information or energy. If the information contained in DNA and the energy stored in fat is locked up so tightly that it cannot be accessed, the organism dies.

The rules of chemistry would lead us to expect that the four other elements in the same periodic-table group as carbon—silicon, germanium, tin, and lead—would also be candidates for the basis for life. However, even the most similar element, silicon, forms compounds slightly less stable than carbon. Silicon is found incorporated into life forms on Earth such as the shell of the diatom in Figure 24-2(a), but silicon is not in the molecules

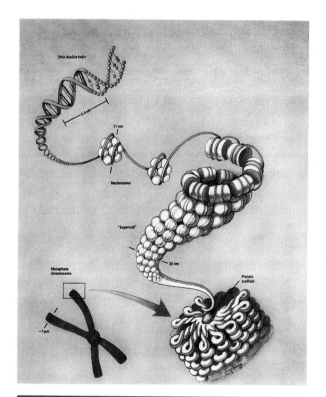

Figure 24-14 A model of deoxyribonucleic acid (DNA) as it coils up into a chromosome. The scale size changes from top to bottom.

that provide energy or the genetic code. It is also one-tenth as abundant as carbon (Figure 14-10). Silicon may be the basis for alien life—and this was the premise for a *Star Trek* 1967 television episode "The Devil in the Dark"—but this speculation is not based on what has been discovered on Earth. Silicon, like carbon, easily combines with oxygen. However, silicon forms SiO_2, the solid mineral quartz. When carbon combines with oxygen, it forms the gas CO_2, the basis for plant respiration.

Germanium is six orders of magnitude less abundant in the universe than carbon and, like silicon, forms a crystalline structure when an oxide, GeO_2. It has no known biological role. Tin and lead are metals, and do not readily form the basis for complex molecules. These characteristics are not useful for storing energy or information. In addition, the cosmic abundance of tin and lead in the universe is low, and so we predict that they will not be found to be the basis for other life forms.

Other common elements found in the basic molecules of life include hydrogen, oxygen, nitrogen, phosphorus, and sulfur, in that order. Not surprisingly, these are among the most abundant elements in the universe (Figure 14-10).

To summarize, life on Earth comes in a stunning variety of shapes and sizes, and seems to occupy most possible niches—not to mention some environments that seem to be unlikely cradles of life. Life on Earth is so pervasive that it seems as if finding life on other planets is not a matter of if, but when.

> All readers are encouraged
> to do Discovery 24-1 at the end
> of the chapter *before* reading
> the rest of this chapter.

24.3 The Search for Extraterrestrial Life Inside Our Solar System

Having identified three factors necessary to life—energy, liquid water, and carbon—we then can start searching locations in the universe where life may be likely. We begin by exploring our own solar system.

Chemist Stanley L. Miller, in 1959, summarized three possibilities for extraterrestrial life inside our Solar System:

1. It could be exactly like life on Earth. In this case we suspect life arose once (where it initially arose would still be controversial) and then spread amongst the planets. The question as to where life originally arose is left still unanswered.

2. Extraterrestrial life could be completely different, implying that life arose at least twice. This would be encouraging news for the possibility of there being even more extraterrestrial life.

3. Alien life could be similar to Earth, but with significant differences. This would also imply that life arose twice, but from similar conditions. Those conditions might then in general be conducive to life and provide excellent clues for where to look for yet more extraterrestrial life.

One final aspect of life as we know it must be emphasized: the only life forms we know include complex carbon molecules. A reasonable assumption in designing a life-detection experiment is that life would be expected to have a signature of carbon-based (organic) materials.

We start by asking the question: Where in our solar system are the conditions we have discussed for life present? Keeping in mind the extremophiles found on Earth, there could be a number of tantalizing possibilities. Let us start by examining the evidence for life gathered so far from observations and experiments of the Moon and its rocks, Mercury, Venus, and Mars.

THE MOON AND MERCURY

The rocks brought back from the Moon by the Apollo missions not only show no signs of organic materials, but they are exceptionally dry compared to rocks on the Earth (Section 8.8). This would seem to rule out life on the Moon.

On a hopeful note, data from the Clementine mission (**Figure 24-15**) have been interpreted as water ice at the South Pole of the Moon. The area (shown in blue) is inside an exceptionally deep crater the size of Puerto Rico (**Figure 24-16**). The possibility of life underneath a frozen lake able to

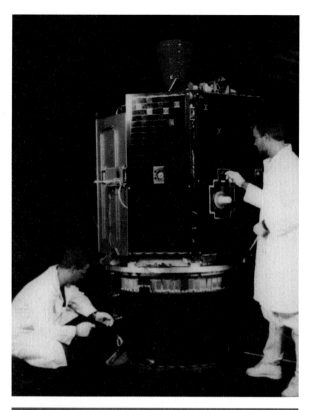

Figure 24-15 The Department of Defense's Moon orbiter *Clementine*, being prepared for launch in 1994.

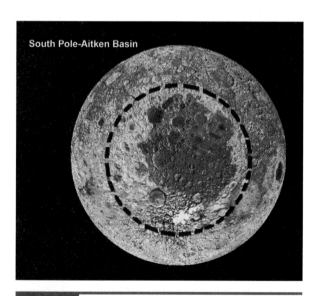

South Pole-Aitken Basin

Figure 24-16 An unusual view of the South Pole of the Moon. The dashed line outlines Aitken Basin.

hold as much as 120,000 cubic meters is reminiscent of the life found under Lake Vida in Antarctica on Earth, shown in Figure 24-10. Even if this ice is not the site of lunar life, this much water could sustain a lunar base for humans for some time. However, follow-up observations using the Arecibo radio telescope from Earth did not confirm the presence of any concentrated ice deposits.

This is an interesting but not uncommon situation in astronomy, where intriguing data are obtained from instruments not specifically designed for that purpose, in this case *Clementine* and Arecibo. The data suggest or imply a particular result—ice on the poles of the Moon or lack of it—but there are other interpretations of the data. For example, a telescope located at Arecibo cannot look directly down on the poles of the Moon, and might not have seen ice deep in a crater. *Clementine* was not designed to search for water deposits. The conservative nature of astronomy requires direct confirmation of surprising results, as you will read about on the next pages in the search for water on Mars. The question of water at the poles of the Moon is unresolved.

Mercury provided another surprise. In 1991, radar observations of its north pole also showed bright reflections, best explained as water ice (Section 9.4). These seem to be clearly associated with impact craters seen in visible light, and are consistent with the idea that, because Mercury spins perpendicular to the plane of its orbit around the Sun, the bottoms of craters at its poles are never in sunlight. The next proposed mission to Mercury, MErcury Surface, Space ENvironment, GEochemistry, and Ranging (*MESSENGER*), would study the composition of these deposits.

VENUS

Venus is the biggest disappointment to those of us hoping for company. Venus's overall characteristics—size, mass, cloudy atmosphere, and location in the Solar System—make it the twin of Earth. Surely there would be great swamps teaming with tropical life!

The Soviet *Venera* series of probes between 1961 and 1983, however, revealed a surface temperature of 457° C, as hot as Mercury and hot enough for molten rock to flow for thousands of kilometers without solidifying (Figures 9-14 and 9-25). The atmosphere is carbon dioxide at a pressure similar to that deep under the oceans of Earth, and those white clouds turned out to be sulfuric acid (Figure 9-12). There is not even a chance for polar ice, since the thick atmosphere and slow rotation results in a similar climate over the entire surface.

The likelihood of discovering if the ancient Venus had conditions more conducive to life

"I'm going to Venus. He's going to Mars."

Figure 24-17 Neither Venus nor Mars would make relaxing vacation sites—and neither one requires "interstellar travel." It is far more likely that humans will someday walk on Mars than on Venus. www.cartoonstock

Figure 24-18 The view of Utopia Planitia from *Viking Lander 2*. (The lander is in the foreground.)

seems slim, since the entire surface appears to have been resurfaced with molten rock over the last half billion years. Unlike Mars, Venus's shield volcanoes may still be active, adding more gas to its already thick atmosphere. For these reasons, the search for life on nearby terrestrial worlds is concentrated on Mars.

MARS

The possibilities for life on Mars, especially in its ancient and wetter past, are more encouraging, although the tourism shown in **Figure 24-17** is not likely soon! In 1976, two *Viking* landers were sent to Mars to perform experiments that considered the three requirements for life separately: an energy source, the presence of carbon (organic compounds), and liquid water.

One energy source for life is respiration, the process by which energy is released from energy-storing molecules. For humans, the process uses oxygen (O_2) and releases carbon dioxide (CO_2). The Viking missions' two identical landers (**Figure 24-18**) each performed the same three life-detection experiments. How to interpret the results of these experiments, consistent for both landers, remains controversial to this day.

The first of the Viking life-detection experiments, the Gas-Exchange Experiment, added nutrients to samples of soil from Mars and Earth and then compared the changes in the amounts of oxygen and carbon dioxide gas in the chamber. Then the chamber was sterilized by heating to kill any life, and the experiment repeated. The Earth soil first showed O_2 being exchanged into CO_2—respiration—then no change after sterilization. The Martian soil had an increase in both gases before and after sterilization, consistent with a chemical change.

The Labeled Release Experiment was similar, but radioactive carbon (C) was included in the nutrients. Radioactive CO_2 was measured in similar amounts for both Earth and Martian soil, with no radioactive CO_2 being detected after sterilization.

The third experiment, Pyrolitic Release, replaced the atmosphere above a Martian soil sample with identical gases, except that the CO_2 and carbon monoxide (CO) had radioactive carbon. The Martian soil still reacted after heat was applied—*pyrolitic* means the process of causing chemical change by adding heat—but at a much lower rate than the Earth sample.

The 1976 *Viking* landers also included an experiment that attempted to detect six organic compounds in the soil using a gas chromatograph/mass spectrometer. The results were (for once) clear; none of the six compounds were present in the soil sampled by either of the *Viking* landers.

To summarize: The Gas-Exchange and the Pyrolitic Release experiments showed that Martian soil was different than terrestrial soil, but the Labeled Release Experiment showed no difference. One experiment seems to indicate life in the Martian soil, but this contradicts the result of the other two experiments where there was a reaction. Six organic compounds common on Earth are not on Mars.

Inquiry 24-5

List some assumptions made by the designer of these experiments.

Inquiry 24-6

Which of the assumptions from Inquiry 24-5 may not be correct?

Inquiry 24-7

Based on the results of these experiments, what do you conclude about the presence of life on Mars?

Inquiry 24-8

What experiment would you design to answer the question, "Is there life now on Mars"?

The next piece of evidence about life on Mars was, amazingly enough, delivered to Earth about 13,000 years ago. In 1984, a team of meteorite hunters collected a 1.9 kg stone in the Allen Hills region of Antarctica, shown in **Figure 24-19**. In 1996 a team of NASA geologists announced that this stone, ALH84001, not only was from Mars (as are 12 other collected meteorites), but contained evidence of biological activity. Extraordinary claims require extraordinary evidence, as astronomer Carl Sagan liked to say. What is the extraordinary evidence supporting this extraordinary claim?

The comparison of isotope ratios of gas trapped in ALH84001 was identical to that of Mars's atmosphere measured by the *Viking* landers, and quite different from Earth or any other planet. Using rates of weathering on Earth, it was calculated that it had been in Antarctica for 13,000 years. The effect of cosmic rays on the exterior of such a stone on a journey between Mars and Earth inferred a travel time of 16 million years. Radioactive dating (Section 7.5) indicated that this stone solidified 4.5 billion years ago, with signs of fracturing and heating since that time. To this day this is the oldest rock known, since no rocks this age on Earth have been found and we assume that they were melted down by Earth's tectonic plate activity.

The story pieced together from this evidence is that ALH84001 solidified on Mars, was blasted out of Mars's gravity by an impact, went into a solar orbit in-between Mars and Earth, and eventually was trapped in the Earth's gravity and landed on Antarctica. But the most exciting part of ALH84001's history is its unusual mineral composition.

ALH84001 has carbonate globules in the rock, and its fractures are younger than the rock itself. This same type of carbonate on Earth is usually created by biological processes. Further, the globules are layered with iron compounds that cannot form under the same nonbiological conditions. The interior fractured surfaces of the stone are also rich in carbon-ring molecules, which are not found in other meteorites from the Allen Hills. The most visually striking evidence is the presence of 100-nm-long strings, looking like tiny versions of fossilized terrestrial bacteria (**Figure 24-20**). Similar evidence was subsequently found in two other, much younger Martian meteorites. The team of NASA geologists concluded that these meteorites held the long-awaited evidence of Martian life.

Skepticism was immediate and authoritative for a number of reasons. The miniature size of the structures seemed too small to contain the minimum cellular features required for terrestrial life. There is a possibility that the 'microfossils' are artifacts of the imaging process required to see such small features. Amino acids found deep

Figure 24-19 The meteorite ALH84001.

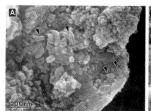

Figure 24-20 Microscopic image of the interior of ALH84001. Reprinted with permission from "Search for Past Life on Mars: Possible Relic Biogenic Activity in Martian Meteorite ALH84001," McKay D.S. et al. (1996) *Science 273*, 924 (1996).

Figure 24-21 A picture of a Pathfinder *Sojourner* rover inserted into the foreground of a photograph of a Lunar Roving Vehicle used in the Apollo 15, 16 and 17 missions. A non-manned mission can be a lot smaller and cheaper!

Figure 24-22 "El Capitan", the 10-centimeter tall rock where the rover *Opportunity* found jarosite.

inside ALH84001 matched those of the Antarctic ice, implying terrestrial contamination. Resolution of the controversy awaits the return of rocks from Mars. If life on Mars is found, then meteorites might be one means to transfer it between the planets, as the first of the possibilities put forth by Stanley L. Miller.

In 1997, NASA had a successful mission to Mars with the *Pathfinder* and its microrover, *Sojourner,* which is added into **Figure 24-21** to compare its size with that of a lunar rover. *Sojourner* examined the surface and rocks, and the data indicated that it was examining water-worn rock conglomerates and sand created by liquid water.

Mars has two polar ice caps (Figure 9-1c), and these are logical places to search for life. Indeed, such a mission to the edge of Mars's South Pole was planned and launched. However, the *Mars Polar Lander* and the two probes it presumably launched into the atmosphere on its way down to the surface were all lost and are assumed to have crashed into, rather than landed on, Mars in 1999. This is not unusual, as about one-third of the missions to Mars have failed, including recent missions by Japan and the European Space Agency. Although disappointing, no lives were lost, as these were robotic missions.

The success of *Sojourner* was followed in 2004 by Mars Exploration Rovers *Opportunity* and *Spirit*. Like *Viking*, these were two separate launches, and both rovers landed near the equator but on opposite sides of the planet. The mineral hematite was found in abundance, but although hematite probably formed in water, there are some

other plausible ways for it to form on the surface of Mars without flowing or standing water. When the mineral **jarosite**, which forms on the Earth in dilute sulfuric acid in ground water, was found in Meridiani Planum, however, the presence of water on the surface of Mars millions of years ago was confirmed (**Figure 24-22**).

Although Earth is the only terrestrial planet with liquid water throughout most of its history, Mars is now confirmed as the only other terrestrial planet known to have had abundant water in its distant past.

Inquiry 24-9

Why do you think the *Opportunity* and *Spirit* robots did not have respiration experiments like the *Viking* landers did?

THE JOVIAN MOONS

The terrestrial planets are not the only places in the Solar System where the basic conditions for life are met. Extremophiles on Earth might be able to exist on the larger moons of the outer planets, particularly Jupiter's Europa (an ice-covered part of Earth is at **Figure 24-23** for comparison) and Saturn's Titan.

You may wonder why the Jovian planets themselves are not included as possible locations for life, especially since Figure 10-8 shows that Saturn has water clouds—drops of H_2O—at a reasonable 27° C. Unfortunately, the strong storms on the Jovian planets (Figures 10-4, 10-6, and 10-7) constantly sweep material up and down through these regions: there does not seem to be enough stability to support life.

The Jovian moons with water do have stable conditions as well as a strong possibility of liquid

Figure 24-23 Compare this close up of pack ice on Earth to the surfaces of Europa in Figure 10-17. Differences are the lack of an atmosphere on Europa in which snow would form, and therefore the lack of clouds and helicopters on Europa.

water. The complex but uncratered surface of Europa (Figure 10-16) can be explained by a deep liquid ocean of water overlaid by a crust of water ice. A definite, variable magnetic field in both Europa and Ganymede also indicates an ice-encrusted saltwater sea on top of a hot rocky core. Tidal heating from interactions with Jupiter could release energy into this ocean from underneath and thus keep it liquid.

The only other possible life-bearing moon is Saturn's Titan (Figure 10-21f). Its nitrogen atmosphere extends out 200 km from its surface, and is 1.5 times as dense as Earth's nitrogen-dominated atmosphere. There the similarity ends, with the frigid −180° C of Titan resulting in an opaque smog of methane (CH_6) and ethane (C_2H_6). It is immediately clear that carbon is also here. However, the possibility of liquid water is highly unlikely. Life on Titan, if any, would not be anything like life on Earth. The joint *Cassini-Huygens* mission arrived at Saturn in July 2004, and the data from it are now being studied to better understand Titan's chemistry.

24.4 The Search for Life Outside Our Solar System

The basic requirements for life—energy, liquid water, and carbon—guide the requirements for a suitable place to host life. Although life does exist

on Earth without being directly dependent on sunlight, all searches for extraterrestrial life assume that it will be on another planet (or its moon) orbiting a star. This is so because the average temperature in the clouds between the stars is extremely cold, from −263° C to −173° C (Section 17.1). Complex molecules can form (Table 17-1), but this is far below the minimum of about −13° C that terrestrial life can survive. Which stars moving in this frigid medium are capable of cradling life?

Since our star has a planet with life, we can infer that all stars of similar mass, age, and chemical composition could also be capable of having a planetary system with life. We will go beyond this to use the requirements for life on Earth, the results so far of our search for life in our Solar System, and our understanding of stars to assess if and where each particular type of star could have a planet with life. Then, using our understanding of galaxies, we will discuss which types of galaxies and where in them to search for stars with life-bearing planetary systems.

STELLAR HABITABLE ZONES

If planets can form, the size of the planets formed depends on the chemical composition of the initial material, which is almost the same as the chemical composition of the star around which the planet is forming. In our galaxy, stars can have 0.001 to 2 percent of their mass in elements heavier than hydrogen and helium. For example, the Sun has about 1 percent of its mass in heavier elements. If there are not enough of the heavy elements like carbon, planets do not appear to be able to form. If a star has too high a percentage of heavy elements, then giant planets many times the mass of Jupiter form. Not surprisingly, it seems that the Sun has just the right percentage of heavy elements to form some terrestrial planets and some giant planets. In addition, observations show that the stars now observed to have planets have abundances of heavy elements similar to that of the Sun. Assuming that a planetary system similar to our own is optimized for life, our search can be limited to stars with no less than 1 percent and no more than 2 percent of elements other than hydrogen and helium.

Assuming a star with a chemical composition similar to our own, the requirement for liquid water is the limiting factor. The stellar habitable zone is the volume around a star where liquid water can exist. Closer to the star, water exists as a gas. Farther away, only ice exists. For our solar system, the stellar habitable zone starts outside the orbit of Venus and goes out to Europa and Ganymede. Planets at the location of Venus might inevitably suffer from a greenhouse effect, and we may soon discover that Jovian moons cannot support life, but for now we are choosing to be generous about defining the just-right zone between hot and cold, sometimes called the *Goldilocks Zone.*

The stellar habitable zone for the solar system is represented by the light blue disk in **Figure 24-24**. Based on our review of likely locations for life in the previous section, it starts at about 0.8 AU from the Sun and goes out to about 5 AU. The rate at which the Sun emits energy per second is called its *luminosity,* and this determines the location of its stellar habitable zone. Because each star's habitable zone is determined primarily by its luminosity, and most stars are less luminous than the Sun, most stars have smaller habitable zones.

All stars go through a birth–life–death cycle that can take from 2 million to 1,000 billion years or more (Table 16-1). This is determined primarily by a star's initial mass, with the more massive stars having the internal pressure and temperature to burn their fuel faster. Stars that burn their fuel faster are more luminous. This additional energy flow results in a higher surface temperature. The relationship of luminosity and size to temperature is presented graphically on the Hertzsprung-Russell diagram (Figure 15-18). The well-populated center strip is called the *main sequence,* with the short-lived massive stars at the top left, and the least massive stars at the bottom right. Although the details of each star's life cycle are different, all stars spend most of their lives fusing hydrogen to helium in their core. This is why, assuming the stellar birthrate is constant, most stars are on the main sequence at any given time. For example, the Sun is about halfway through its 10-billion-year life on the main sequence. Since planets take time to form, and dying stars are not hospitable to life, we will discuss where life is possible around main-sequence stars.

The relationship between mass, luminosity, and surface temperature is plotted in Figure 15-19, and shows that stellar luminosities for stars on the main sequence range from 10^{-5} all the way up to 10^4 times the luminosity of the Sun. Clearly, the stellar habitable zone also has a large range, with low-mass stars having tiny, close-in stellar habitable zones, and massive stars having larger habitable zones located far away from the central star.

It would seem that we should first search for life around stars having the largest habitable zones, but this is not the best strategy for two reasons. First, the most massive stars have short lifetimes, perhaps too short a time for planets to form a habitable crust. The second reason is that they are rare: only about 3 percent of stars in our galaxy on the main sequence are more massive than the Sun.

Stars having about half the mass of the Sun or less make up about 85 percent of the stars in our galaxy (Section 15.5). These stars have extremely long lifetimes, but their habitable zone is so small that the probability for a habitable planet forming precisely the right distance from these stars is expected to be low.

This narrows down the search for life around other stars to the 12 percent of stars that are slightly more or less massive than the Sun. There is still another consideration; where a star is located inside a galaxy.

GALACTIC HABITABLE ZONES

Galaxies at this time in the universe come in three basic types, spiral (or disk), elliptical, and irregular (Chapter 21). Let us examine which type of galaxy and where inside a galaxy the basic requirements for life, energy, liquid water, and carbon are most likely to be found. These, plus some other factors, define the galactic habitable zone. There are four requirements for a galactic habitable zone. The

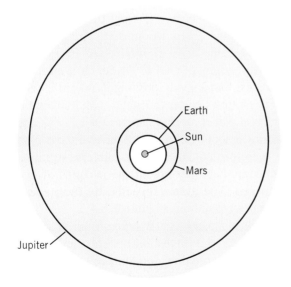

Figure 24-24 The light blue disk halfway between Venus and Earth to just outside Jupiter illustrates our solar system's current habitable zone.

Labels in figure: Earth, Sun, Mars, Jupiter

first is stars of solar-like chemical abundance, the second is sufficient space between stars, the third is the absence of intense star formation, and the fourth is the absence of significant amounts of damaging radiation.

Stars with solar chemical abundances are generally found in the disk of spiral galaxies (where our star Sol is located) and in irregular galaxies. Elliptical galaxies and the spheroidal component of disk galaxies have a high proportion of stars with high heavy-element abundances (Section 21-5). Intense star formation includes the formation of many massive stars that end their short lives in supernovae explosions. These explosions can release as much energy over a month as an entire galaxy, and would sterilize any life on the surface of planets too close to them. Avoiding supernovae rules out merger and starburst galaxies.

The centers of galaxies produce intense amounts of X-ray and gamma-ray radiation that are dangerous to life and its formation. Thus, the galactic habitable zone must be sufficiently far from these radiation sources to provide life a safe haven. This limits our search to the disk component of spiral galaxies. This brings us back again to the guidance of the one example of life we yet know; our Sun is located in the disk of a spiral galaxy, a place with low stellar density, a low star formation rate, and away from the radiation-producing galactic nucleus. The galactic habitable zone is the volume of a disk galaxy where there is a steady low rate of star formation for energy and sufficient heavy elements including oxygen for water (H_2O) and carbon for the molecules of life.

THE SEARCH FOR LIFE ON WORLDS OUTSIDE OUR SOLAR SYSTEM

The planets inside our solar system are not the only planets that are known. As is explained in Section 6.6, we now have evidence for more than a hundred extraterrestrial planets. Can any of these possibly support life?

The primary method of detection, the Doppler shift in spectral lines of the star due to the planet gravitational pull on it, is currently sensitive to planets down to the mass of Neptune and Uranus. (This method is explained in detail in Section 14.3.) Surprisingly, all but two of the systems have their giant planet(s) extremely close to its star, often closer than Mercury is to our Sun; such planets are referred to as *hot Jupiters*. Even though it is easier to detect such a planet when it is close to its

star, the searches are now sensitive enough to find giant planets at the distance of Saturn; they are rare. It appears that hot Jupiters are the norm in our neighborhood of the Milky Way.

Inquiry 24-10

Why is it easier to detect a planet when it is closer to its star?

Another method of detecting extrasolar planets within our technical ability right now is the transit of a planet across the face of its own star. This is the same idea as when the spectrum of the star changes when an eclipsing binary star moves between us and its partner (Figures 15-5 and 15-6), but for a planet the effects are harder to detect.

As the starlight passes through the edge of the cool atmosphere of the planet, molecules in the atmosphere absorb unique wavelengths (frequencies) of light (**Figure 24-25**). For example, our atmosphere, as seen by distant aliens, would show absorption lines in the spectrum indicating the possibility of life primarily from oxygen (O_2), carbon dioxide (CO_2), and water (H_2O). Our star is too hot to have molecules, so such missing light must be the signature of a planet's atmosphere.

It is interesting to look at the first three such planets to have been detected. One is a planet 0.7 times the mass and 1.3 times the size of Jupiter orbiting the star HD 209458 with a period of 3.5 days. This means it is eight times closer to its star than Mercury is to our star. The second has about the same mass of Jupiter and is part of the planetary system of OGLE-TR-56b. It is also one of the fastest found so far, orbiting its star in 29 hours, and is 14 times closer to its star than Mercury. The third is a planet around the star OGLE-TR-3, also orbiting at about 29 hours. All three of these planets are therefore considered hot Jupiters, so close to their star that they are in the process of being destroyed. The atoms revealed by their spectral signatures are sodium and iron.

These planetary systems are so unlike our own that they call into question our theories of planetary system formation. Jupiters are supposed to be able to form only far from their star. Is our theory of planetary system formation wrong? Scientists, being conservative, would rather consider a correction to the accepted theory. One of the currently

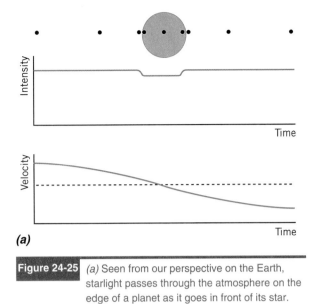

(a)

Figure 24-25 *(a)* Seen from our perspective on the Earth, starlight passes through the atmosphere on the edge of a planet as it goes in front of its star.

popular variations is that these gas giants did form about 10 AUs away from their star, but migrated inward over time. The effect of such a migration on any terrestrial planets would be to eject them from the solar system, so these hot Jupiter systems can probably be ruled out as the home of Earthlike planets.

However, the transiting planet method is considered the one most likely to identify Earth-mass planets within the next 10 years.

Inquiry 24-11

Why is it not surprising that the first transiting planets discovered would be orbiting very close to their respective stars?

24.5 The Search for Extraterrestrial Intelligence

The most exciting discovery ever would be that of other intelligent life somewhere else in the universe. How do we approach making such a discovery? How would we recognize such life? Should we purposely broadcast our own presence to the rest of the Milky Way? These are not only scientific questions, but questions that belong in the realm of politics, sociology, philosophy and, yes, even religion. Scientifically, this is a situation

where theory directs observations, not the other way around, because we are short on observations: We know of only one planet where what we like to call intelligent life has arisen.

THE HISTORICAL SEARCH FOR EXTRATERRESTRIAL INTELLIGENCE

The modern Western scientific search for other civilizations started with Caroline and William Herschel, a prominent British sister-and-brother team of observational astronomers. Using their state-of-the-art telescope, they observed Mars's polar caps and atmosphere and by 1784 had concluded that it was inhabited—a reasonable assumption, given that the apparently similar Earth is inhabited.

Inspired by the Italian Giovanni Schiaparelli's announcement in 1877 that there were *canali* on Mars, the wealthy amateur astronomer Percival Lowell initiated an observing program to study them. Lowell was not a professional astronomer, but he had studied mathematics at Harvard and was passionately knowledgeable about astronomy and its methods. In 1894, Lowell built an observatory in then-remote Flagstaff, Arizona, a location so high and dry that even today it is used for certain projects requiring its superior seeing. This was a bold step, as the best translation of *canali* is "channels," and even Schiaparelli did not imply that they were artificial or even carrying water. The human brain tends to connect faint, disconnected features into patterns, and Schiaparelli and Lowell were both working at the faintest limit of their telescopes.

Using Flagstaff's superior observing conditions, Lowell soon published detailed maps of Martian canals, as shown in **Figure 24-26**. Notice how the canals are all straight and therefore thought to be artificial. They were numbered and named by Lowell, with the dark areas presumed to be vegetation and the light areas desert. Although in 2004 NASA's rover *Opportunity* found Martian rocks that could only form in water, that evidence is from Mars's ancient past. Lowell believed to the end of his life that Mars had artificial canals carrying water from the polar caps to a thirsty, dying Martian civilization. He even published books such as *Mars as the Abode of Life* in 1908. However, other astronomers and better instrumentation (Figure 9-5d) soon showed that although Mars has seasons and polar caps, perfectly straight artificial canals did not exist.

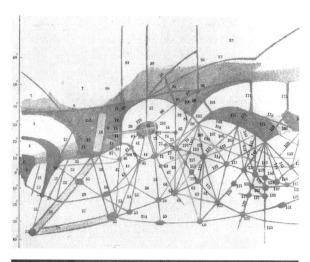

Figure 24-26 A map of Mars from Percival Lowell's 1895 book *Mars*.

False alarms by otherwise reputable scientists continued into the next century. Nikola Tesla, the inventor of the fluorescent lamp and pioneer of the use of alternating current that is now the international standard for electricity, claimed in 1901 that he had heard interplanetary signals in 1899. However, the radio wavelength he was using was too long to penetrate the Earth's atmosphere (*long-wave radio*, shown in Figure 11-8).

In 1959, physicists Giuseppe Cocconi and Philip Morrison calculated that interstellar communication using radio waves would be practical, since such waves could get through the interstellar clouds of gas and dust.

The first publicized search capable of detecting extraterrestrial signals was the next year, when astronomer Frank Drake searched two Sunlike stars for a week, looking for artificial signals at the "21-cm line" radio wavelength (1,420 MHz) of atomic hydrogen (discussed in Chapter 20). He called it *Project Ozma* after Princess Ozma, a fictional character in the *Wizard of Oz* books who does not know she is a princess. Although he failed to find any artificial signals, the **search for extraterrestrial intelligence (SETI)** had begun.

The 1960s marked the height of the USA–USSR space race, and SETI was another arena where this competition played out. In 1964, astrophysicist Nikolai Kardashev suggested the existence of supercivilizations, classifying them into three categories by how much energy they could make available to announce their existence. In 1965, the Soviet news agency announced the discovery by astronomer Gennady Sholomitskii of a distant object that he identified as the beacon of such a supercivilization. As you can imagine, many other astronomers immediately started examining this object. Called CTA-102, it was soon determined to be a natural phenomenon called a quasar (Chapter 22). Nevertheless, a strong USSR SETI program was established in 1968 when astronomers embarked on an ambitious program to look at 12 stars in the radio region of the spectrum and expanded to an all-sky search in 1970. No further discoveries resulted from this study.

Fourteen other observing programs by a small number of interested observers in the United States and Europe continued through the 1970s. In 1984, the SETI Institute, an independent private organization, was established with the mission "to explore, understand and explain the origin, nature and prevalence of life in the universe."

Is Anybody Out There?

Before we set out on a full-scale search for other intelligent life, it makes sense to consider the likelihood that anybody is "out there" in our galaxy. You might think that we do not know enough to make this calculation, but just trying to do it allows us to identify the important pieces of the puzzle that must be considered.

Let us compare the search for extraterrestrial life to another search many of us make—the search for the perfect life partner from among the 6 billion people alive right now. This is the total possible pool, but of course this can be narrowed down. Most of us would like our life partner to be about our own age. If you are in the 20- to 24-year age bracket, you make up about 0.08 of the total population and so we multiply by 0.08. Assuming for our argument a preference for only one gender, if you are looking for a male partner, about 51 percent of the population will meet this criterion, so you now multiply by 0.51. (This is age-dependent: at the 20–24 age bracket, about 49 percent are women, but for 80-somethings, it is 65 percent. The odds for a heterosexual male or a lesbian improve with age.) Thus, we can narrow down the search to the quarter of a billion (2.5×10^8) men in the world in your age range.

Surely one of these will meet the other personal factors that we take into account when making this choice! The goal of this process is not to come up with a precise number, but to think about what factors are important when beginning the search for what sometimes seems to be the proverbial needle in a haystack. Scientists used this process to think about the search for extraterrestrial intelligence.

In 1961, astronomer Frank Drake wrote an equation containing various factors that would involve determining the number of intelligent civilizations that might currently inhabit the Galaxy and be willing and able to communicate with us (**Figure 24-27**). We will present a modernized version of his formula with the following quantities:

N = number of detectable civilizations in the Milky Way

N^* = number of stars in our galactic habitable zone

f_P = fraction of these stars with planets

n_e = number of planets in the ecosphere, i.e. the stellar habitable zone

f_l = fraction of planets with life

f_i = fraction of planets with intelligent life

f_c = fraction of planets bearing intelligent life which communicates

f_n = fraction of intelligent, communicative civilizations that exist now.

All of these factors are multiplied together to estimate the number of civilizations in our one galaxy that we expect to be able to communicate with now:

$$N = N^* f_p\, n_e\, f_l\, f_i\, f_c\, f_n.$$

Now estimates of the various quantities are needed. Going from left to right, we move from the best known to the least known variables. Also, we can group the quantities. For example, N^*, f_p, and n_e are all observable by astronomers. The next two, f_l and f_i, are biological factors. The last two factors are more complex, and probably involve cultural, sociological, technological and possibly religious factors.

Figure 24-27 Frank Drake with his equation to estimate the number of intelligent civilizations in the Milky Way with whom we could communicate.

Table 24-1 has two examples of calculations giving the range of values within which the real value might lie. Notice that not knowing even one number can significantly change the results. Many scientists consider the last factor, f_n, to be the most difficult one to determine.

The estimate produced by this equation is not as interesting as the fact that it points out where our knowledge is poor and more research is needed. Knowing, say, precisely the number of stars in our galactic habitable zone does not make the final number accurate if we have no idea about the fraction of intelligent, communicative civilizations that exist now. Frank Drake calls this problem—you do not know how hard it will be to find another intelligent civilization until you succeed—the *Paradox of SETI*.

MODERN SEARCHES FOR EXTRATERRESTRIAL INTELLIGENCE

Radio wavelengths continue to be the most popular choice for these programs, with impressive new technology expanding the speed at which millions

Table 24-1 Number of Detectable Civilizations in the Galaxy

N^*	f_p	n_e	f_l	f_i	f_c	f_n	$= N$
10×10^{10}	0.50	3	1	0.001	0.01	0.01	1.5×10^4 (Highest)
2×10^{10}	0.05	0.1	0.0001	0.0001	0.001	0.001	1×10^{-6} (Lowest)

of frequencies can be searched simultaneously for the beacons of distant civilizations. This is a far cry from Project Ozma's single-frequency capability, and is necessary due to the Doppler effect. Searching at one wavelength means any signal would have to be intended only for Earth and be continually adjusted for changes in relative velocity between the sender and the Earth. Recent major radio searches are described below. Optical searches, looking for pulses of laser light aimed at us rather than generalized beacons, are rapidly becoming practical as well.

With such a broad frequency range to choose from, how do we choose which frequencies to use? As long as we are looking from the surface of the Earth, the answer is limited by our atmosphere. Figure 11-8 shows that there are two regimes in the electromagnetic spectrum where our atmosphere is transparent: visible plus some infrared wavelengths and the microwave to television regions. These two windows, or *holes,* in our atmosphere let in natural as well as artificial signals.

The best place to detect artificial signals would be where the background signal from natural sources is the lowest. For the radio hole, this is around 10^9 Hz (**Figure 24-28**). It turns out that two natural lines do occur in that hole, the 18-cm emission from the hydroxyl molecule (OH) and the 21-cm emission of neutral hydrogen. A reasonable assumption is that artificial signals would preferentially be sent in that low background H to OH range, which leads to a SETI witticism. If we put H and OH together, we get H_2O, or water. If you were searching for animals in a jungle, you might look near a water hole. For this reason this hole centered at 10^9 Hz is usually called the *water hole.*

The SETI Institute's recently completed Project Phoenix was a targeted search. It examined about 1,000 Sunlike stars within 200 light-years of the Sun, scanning across 28 *million* frequencies from 1 to 3 MHz. It began in 1995 with the southern stars at Australia's Parkes 210-foot radio telescope, the largest radio telescope in the Southern Hemisphere (**Figure 24-29**). Searches of northern stars were started in 1996 to 1998, and the entire complement of stars was completed in March 2004. During these few years, no clearly extraterrestrial transmissions were found.

An alternative to a targeted search is to examine the data from other research programs for the presence of artificial signals. Over time, this piggybacking results in scanning a large portion of the sky, especially when the telescope is doing an all-sky survey. This approach is being used by University of California, Berkeley's Search for Extraterrestrial Radio Emissions from Nearby Developed Intelligent Populations (SERENDIP) project at the Arecibo 1,000-foot and the SETI Australia Centre's Southern SERENDIP program

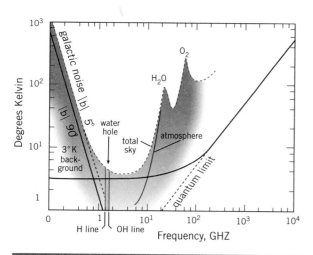

Figure 24-28 The "water hole" at around 10^9 Hz. Outside of this hole, the noise of our own galaxy and the absorption of our atmosphere interfere with getting signals.

Figure 24-29 The 64-meter radio telescope at Parkes, New South Wales, Australia.

at Parkes Radio Telescope. Again, signals clearly from extraterrestrials have not been found.

Inquiry 24-12

Define and distinguish between targeted searches and piggy-back searches.

Of course, it would be much more efficient to do SETI using a telescope designed and dedicated to such an observing program. The SETI Institute's Allen Telescope Array (**Figure 24-30**) currently has three prototype dishes at Hat Creek, California, and is eventually planned to total 350 20-foot dishes. This telescope is named after Paul Allen, a cofounder of Microsoft and financial rescuer of SETI when NASA stopped funding it in 1993.

The most common current optical SETI strategy is different than for radio SETI. Optical searchers look for short—billionth of a second—powerful bursts of laser light aimed directly at our solar system. This is similar to the approach suggested by physicists Robert Schwartz and Charles Townes in 1961, and the burst can feasibly outshine the ET's star for that billionth of a second. It also assumes that the extraterrestrial civilization considers our solar system to be a likely candidate for life. This approach is attractive because it seems possible that these bursts could be detected with 1- to 2-meter-class optical telescopes and currently available detectors.

Figure 24-30 The first dishes of the Allen Telescope Array in Hat Creek, California.

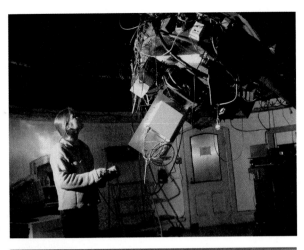

Figure 24-31 Shelley Wright, an undergraduate physics student at UC Santa Cruz, is moving Lick Observatory's 1-m telescope.

One difficulty with this approach is that many short bursts of light in the optical have been detected, but so far they appear to be false alarms due to, for example, cosmic rays. In 2000–2001, California's Lick Observatory, the SETI Institute University of California at Santa Cruz, and University of California at Berkeley sponsored a search using this approach. To avoid false alarms, the instrument at the back of the telescope shown in **Figure 24-31** is a triple-photomultiplier photometer; a flash of light must be caught in all three detectors to count as a positive signal. A number of other optical SETI projects, notably four funded by the Planetary Society including a dedicated 72-inch optical telescope, are underway.

Why have we not found anything so far? It may be that we have not been searching at exactly the right wavelengths, because the transmitting civilizations do not know the transparency of our atmosphere. Perhaps we have not been looking *when* they have been transmitting, since no SETI program has been continuously sustained for even a few years. Or maybe the transmitting civilization has been too far away for us to detect. **Figure 24-32** shows that so far we have only explored a tiny portion of our galaxy. The second two factors will be overcome with the Allen Telescope Array in the radio and the Planetary Society's Harvard 72-inch telescope, or any of the many piggy-backed programs that are proliferating. Of course, it could be that one of the factors in the Drake equation makes finding another contemporaneous civilization highly unlikely. Or, perhaps, we are alone.

Project Phoenix

Allen Telescope Array

Figure 24-32 The pink sphere represents how far out from our own Solar System Project Phoenix could detect another civilization: the blue sphere is how far out the Allen Telescope Array, when completed, could detect that same civilization. The center of the Milky Way can be seen in the background of this artist's drawing.

SHOULD WE GO THERE?

If we do decide that humans should travel to another planetary system, how hard would that be? This book's journey of discovery started with "a grand tour of the universe," with the scale of astronomical sizes in Figure 1-20. That figure shows that a *close* star, Alpha Centauri, is about 10^{15} km or 4.2 light-years away from us. But if we do go to the stars, we would be more interested in visiting a system more like our own, such as Gliese 777A, the G8 member of a wide binary that has one known planet with about Jupiter's mass located in about Jupiter's location, 5 AUs from its star. This star is about 52 light-years away, so a one-way trip for light takes 52 years. How long would such a journey take for humans?

NASA successfully tested an ion engine in 1998, a design suitable for interstellar travel at 1 percent the speed of light (**Figure 24-33**). With time added to gently accelerate and decelerate—human beings are fragile—it would take about

Figure 24-33 Artist's rendering of *Deep Space 1*, the NASA spacecraft that successfully tested the first ion engine in 1998. The image in the background is Comet Borrelly, which *Deep Space 1* visited in 2001.

10,000 years to reach Gliese 777A. Obviously this is not only a one-way trip, but also a trip where the astronauts would not live to see their goal! Solutions such as hibernation, deep-freeze, multigenerational ships, and so on, have been proposed, but none of them is yet technically feasible. Boosting to speeds close to the speed of light would shorten the time of the trip in the time frame of the astronauts, but would require more energy than we now know how to produce at a reasonable cost. The vast distances between the stars is not the last frontier, but a chasm separating us from any neighboring civilizations. This is one of the reasons why astronomers have difficulty accepting that aliens have already visited Earth.

Science fiction gets around this barrier to visiting other worlds by inventing technology like warp drives, or using shortcuts like wormholes that could conceivably connect distant regions of space. Many mathematical descriptions of the fabric of space–time involve additional dimensions, but we may not exist in them. Alternative ways to move through space may one day prove feasible, but to date they have technical or physical inconsistencies that do not yet have answers. For example, black holes certainly warp space, but humans could not survive in rapidly changing warped space any better than we can survive being crushed by a steamroller.

What Do We Do When We Get There?

The movies *E.T. the Extra Terrestrial* and *Men in Black* were certainly big hits, and the UFO Museum in Roswell, New Mexico, is a popular tourist destination. However, if the point of visiting alien worlds is to get information, this could be done more efficiently and cheaply by just exchanging messages, without ever going there. For example, think about the information the Apollo astronauts on the Moon brought back to Earth (Section 8.8). They brought back rocks. Even the scientific usefulness of this task was limited, because the areas from which these rocks were gathered were severely limited due to (appropriate) concerns about the astronauts' safety.

One of humanity's strengths is our mental flexibility, but when every gram of mass launched into space is agonized over, it is difficult to modify experiments on the fly, or perform different ones than were planned. Robots can do precise, preplanned tasks more cheaply and efficiently even in deadly environments like Venus and Titan. If a mission fails, the cost is not paid in human lives. The other consideration, particularly important when there is alien life, is that we not contaminate or accidentally destroy that other life. Life from Earth unintentionally riding on a robotic spacecraft is currently assumed to be destroyed by the harsh environment of space, yet we paradoxically think that life can survive a ride inside a meteorite.

Announcing Our Existence to the Universe

You might be wondering why we are spending so much time listening and no time transmitting. In fact, we have been transmitting signals since television was invented in the 1920s. Thus, we are continually sending message bottles into the vast ocean of space.

Social scientists also have laws; one of them is called the *law of unintended consequences,* which happens when changing one thing unexpectedly changes something else as well. Our television transmissions follow that law. In 1926, we began broadcasting television in the 100–1,000 MHz range of the electromagnetic spectrum (Figure 11-8). At these frequencies, the waves do not bounce off the atmosphere but rather travel in straight lines, which is why we can communicate with the astronauts on the Moon, or to our robots on Mars, but cannot easily beam television signals to a receiver only hundreds of kilometers around the curved surface of the Earth (**Figure 24-34**). Such television signals are not confined to Earth but have been traveling throughout space at the speed of light since their original transmission. Although it is unlikely that another life form could make sense of the signals, it would probably be clear that they are artificial and thus indicate our presence, even though we had not intended to do so.

We have also intentionally sent signals. In 1974 a three-minute burst of data was sent toward the globular cluster M13 (NGC6205; Figure 1-14) using the Arecibo radio telescope. The transmission was powerful enough to be detected by a similar telescope in M13, 25,000 light-years away. The signal's message, shown in **Figure 24-35**, was an encoded graphic including an outline of the radio dish, a human figure, our solar system, and DNA. Although M13 does contain hundreds of thousands of stars, it is unlikely to be the cradle of intelligent life because globular cluster stars are low in their abundance of carbon and other elements heavier than hydrogen and helium.

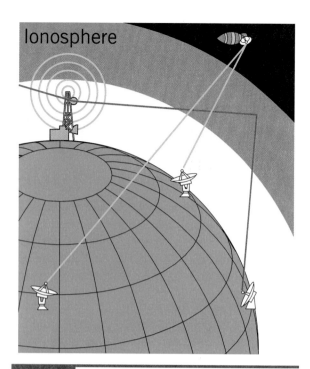

Ionosphere

Figure 24-34 An illustration of the path of television and radio light waves from broadcast towers. Some wavelengths reflect down from the atmosphere, while others travel straight into space.

Inquiry 24-13

Other than a lack of carbon, why would a globular cluster like M13 be an unlikely place to find an extraterrestrial civilization?

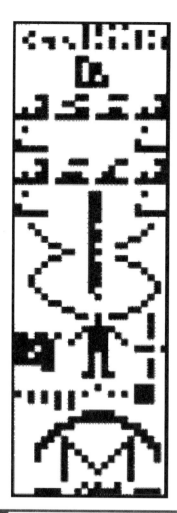

Figure 24-35 The re-constructed signal beamed at M13, a globular star cluster, using the Arecibo radio telescope in 1974. It shows information about humans and the telescope.

Figure 24-36 Carl Sagan, shown here holding the *Pioneer 10* and *11* plaque suggested by Eric Burgess and designed by Sagan and Frank Drake.

Since the 1970s the United States has launched four space probes carrying messages from the peoples of Earth to other intelligent beings who might happen to intercept them. The first, *Pioneer 10* in 1972, was the first probe to travel through the asteroid belt and take close-up images of Jupiter. *Pioneer 10* is also our first ambassador to other civilizations, carrying a plaque shown in **Figure 24-36**. The top of the *Pioneer* plaque sets the distance scale with the transitions between two hydrogen atoms, then below it the location of the Sun among 14 pulsars, an image of the spacecraft with the scale of two humans, then at the bottom the location of the planet of the spacecraft's origin and the spacecraft's trajectory. Its twin, *Pioneer 11,* left Earth a year later, toured Jupiter and Saturn, then also headed outward through the disk of the solar system.

Voyager 1 and *2,* two identical spacecraft, were launched in 1977. Like *Pioneer 10* and *11,* the

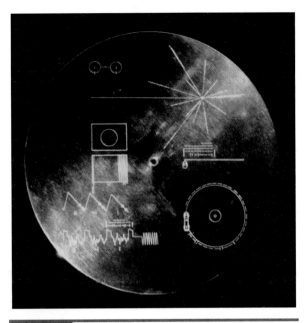

Figure 24-37 The album cover of the phonograph record attached to *Voyager 1* and *2*.

Voyagers also shot by the outer planets, recording images of them and sending them back to Earth. The *Voyagers* had a phonograph record and instructions for a playback device attached to the side (**Figure 24-37**). The record included 115 images of life and culture on Earth, natural sounds, musical selections, and spoken and printed greetings.

Have these probes made it to other star systems yet? **Figure 24-38** shows their path and January 2005 location. *Pioneer 10* and *11*, although they started before the *Voyagers*, were not accelerated as quickly and are now not as far away from Earth. After *Voyager 1* visited Jupiter and Saturn, it then continued mostly north out of the solar system in the direction of the constellation Ophiuchus. It will pass the 100 AU mark in summer of 2006. *Voyager 2* visited Jupiter, Saturn, Uranus, and Neptune, and then continued south out of the solar system

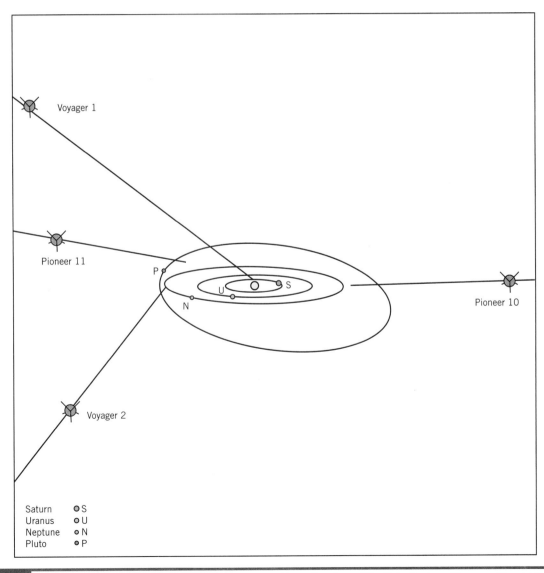

Saturn	⊙ S
Uranus	⊙ U
Neptune	⊙ N
Pluto	⊙ P

Figure 24-38 The paths and location of the first four human spaceships to head out of the Solar System, *Pioneers 10* and *11* and *Voyagers 1* and *2*. This scale is too large to show the orbits of Jupiter or the Earth.

toward the constellation Telescopium; it will get to 100 AU in 2012. For perspective, Pluto is at about 40 AU, a newly discovered member of the solar system, Sedna, is about 120 AU away from the Sun, and Gliese 777A is about 330,000 AU away.

Inquiry 24-14

Why might broadcasting our location be unwise?

What is the advantage to us in finding other civilizations? We could learn new technologies, explore alternative solutions to cultural, environmental, and other problems, and be exposed to new viewpoints about life. Less technically advanced civilizations on Earth have benefited from exposure to more advanced cultures; difficulties arose for both sides when the more advanced cultures attempted to exploit the resources of the less-advanced ones. Exchange of information alone, however, may not bring with it those often devastating results. It is also unlikely that a more advanced civilization would find us threatening—or even interesting!—especially as long as we are trapped in our own solar system. We may find other civilizations, but it is certain that we are unique—even if not alone.

DISCOVERY 24-1

Exploring the Conditions for Life

After completing this activity, you will be able to do the following:
- Describe how adding different common materials to water affects its freezing point.
- Determine a number of factors that will destroy simple terrestrial life.
- **Discovery Inquiry 24-1a** Carefully measure out five identical and freezable containers holding the same small amount of water (perhaps half-fill plastic drink bottles). Measure and add salt to the second container and stir it thoroughly. Then double the amount of salt for the third container, and so on. Place the containers in a freezer close together but not touching, check them every hour, and record when each container freezes. Prepare a graph showing how adding salt affects the time it takes to freeze water. What do you conclude about the ability of salt to change the freezing point of water?
- **Discovery Inquiry 24-1b** You will need packets of baking yeast, sugar, and water. (Baking yeast is one of a number of digestible types of fungi used in the preparation of many foods from bread to beer. It is sold in its dormant state.) Dissolve sugar and yeast into a container of warm water and cover it tightly with plastic wrap. Wait an hour, then observe and record changes in the container. These are the signs of life you will be looking for in the next series of experiments. Devise a series of experiments simulating more alien environments, varying the addition of water, temperature, amount of sugar, addition of other foods, and so on. Based on your experiments, summarize the minimum requirements for yeast to be activated from its dormant state.

Chapter Summary

Observations

- The only example of life that we know about is terrestrial life.
- The minimal requirements for terrestrial life are a source of energy, liquid water, and carbon.
- Life exists on Earth in a range of temperatures somewhat wider than that for liquid water at sea-level atmospheric pressure.
- There is currently little or no liquid water on most of the Solar System's planets or moons, with the exception of Earth and possibly Mars, Europa, and Ganymede.
- Earth and Mars are the only planets that are known to have had liquid water on their surfaces.
- Extrasolar planetary systems exist.
- As of spring 2005, no signs of currently living extraterrestrial life has been found, intelligent or otherwise.
- As of spring 2005, no signs of past extraterrestrial intelligent civilizations have been found.
- We have attempted, intentionally and unintentionally, to contact other civilizations.
- Life on Earth currently has three branches: **bacteria**, **archaea** (both are single-celled prokaryotes), and **eukaryotes**, which can be single- or multicelled organisms.

Theory

- **Extraterrestrial life** will be chemically similar to terrestrial life.
- Life may currently exist on Mars, Europa, and Ganymede.
- The **stellar habitable zone** is a volume of space surrounding most stars where the conditions for life are met.
- The most likely location in a galaxy for a planetary system harboring life is that of our solar system, the middle portion of a galactic disk, well away from the galactic nucleus and areas of active star formation.
- It is possible to estimate the number of galactic civilizations that can and would communicate with us.

Conclusions

- It is worthwhile searching for life on other planets, if the minimum requirements for life exist there.
- Extraterrestrial life is most likely to be single-celled organisms.

Summary Questions

1. What is the fundamental building block of life? What are the three basic types? How are they different from each other?
2. What are the three minimal requirements for life?
3. Explain how life can survive without sunlight.
4. How can water be liquid at $121°C$, $21°$ above what we usually call the boiling point of water?
5. What gives water its unusual chemical properties?
6. What makes carbon the elemental building block for complex biological molecules?
7. Which of the Solar System's planets and moons have humans visited? Which ones have been searched for life? Why or why not?
8. What characteristics of a star determine if it has a habitable zone, and how large its habitable zone is?
9. Which extraterrestrial planet historically and currently dominates the search for extraterrestrial life? Why is that so?
10. Why do most SETI programs use the radio part of the electromagnetic spectrum?
11. How many other civilizations have been found? Comment on why this is so.
12. Why is there so much more effort expended on looking for intelligent life rather than in sending out our own signals?

Applying Your Knowledge

1. If single-celled life similar to Earth's is found to have existed on Mars, what would be the increased significance of meteorites like ALH84001?

2. If single-celled life exactly like Earth's is currently found on Mars, what might you conclude?

3. What would be the significance of finding life on Titan?

4. Using your understanding of the Sun's life cycle, how would the stellar habitable zone in Figure 24-24 change over time?

5. Using your understanding of the galaxy's evolution, how would the galactic habitable zone in our galaxy change over time?

6. As the Nobel Prize-winning physicist Enrico Fermi asked about extraterrestrial civilizations: "So where is everybody?"

7. The Drake Equation has traditionally been applied only to our galaxy due to the vast distances to other galaxies. However, if you applied the Drake Equation to the entire universe, how many civilizations would there be?

8. If an extraterrestrial civilization were to be found, how would this change your life?

Answers to Inquiries

24-1. You would probably try to see if it responded to its environment, perhaps by moving the twig it is sitting on.

24-2. Strain 121 is named after the maximum temperature it can tolerate on the Centigrade temperature scale.

24-3. Salt is spread over ice because salt water has a lower freezing point than water. This will not work in Antarctica because it is colder than the freezing point of salt water.

24-4. The additional external pressure on the surface of the water forces it to stay liquid at a higher temperature than $100°C$. Food cooks faster at this higher temperature.

24-5. Assumptions include (1) Respiration is the energy process for Martian life. (2) Life is on the dry, cold surface of Mars rather than where liquid water may exist. (3) A definitive answer to the existence of Martian life can be found by two or more samples.

24-6. Martian life may rely on other energy sources, and may be in only a few locations. If Martian life arose when Mars was warmer and wetter, it probably would not be on the surface of Mars now.

24-7. A clear answer cannot be drawn from these experiments, which is part of the reason why they were not repeated for later experiments. *Beagle 2,* a British mission designed to search for life on Mars, proposed to land near the northern ice cap, where ancient oceans may once have existed.

24-8. Sending equipment that can visually look for single-cell life would make sense, but the current generation of such equipment is too fragile to withstand the stresses of space flight. This is why missions to bring back samples are desirable.

24-9. None of the affordable experiments would give definite proof that life was not currently on Mars, since a thorough search of the entire surface and subsurface will have to be done. Rather, it makes sense to investigate the history of conditions of life on Mars, particularly liquid water, to guide the decision to continue the search for Martian life or not.

24-10. Newton's law of gravity, $F = GM_1M_2/d^2$, describes gravity as being stronger when the two objects, in this case the planet and its star, are closer together (Section 5.5). Therefore, the planet and the star move faster around their mutual center of gravity and the Doppler effect is more pronounced and evident over a shorter observing time (Section 14.3).

24-11. The closer a planet is to its star, the shorter its orbital period, and therefore the more frequently it passes in front of its star from the perspective of any observer (Figure 15-6 applies for a planet and star system as well as a binary star system). Also, the closer a planet is to its star, the smaller its maximum angular separation and the more time it spends closer to the star (Figure 4-29).

24-12. Targeted searches identify specific locations, usually stars, that will be studied. Piggy-back searches simply look over the shoulder of other observing programs. The latter are usually surveys covering large fractions of the sky.

24-13. The central stellar density of many globular clusters is so high that planetary systems are unlikely to be stable over long periods of time.

24-14. It is likely that any other civilization responding to our broadcasts is hundreds of thousands of years ahead of us in their technology. It has been our experience on Earth that the less-technically advanced of two newly interacting civilizations undergoes rapid change at a pace determined by the more technically advanced civilization.

Appendix A

1. The Powers of Ten Shorthand for Numbers

Astronomers will inevitably be involved with numbers that are both large and small, because they are concerned with sizes ranging from the smallest to the largest conceivable. A compact notation has been developed to represent this wide range of numbers conveniently and to facilitate arithmetic using them.

To start off with a familiar example, consider the number "10 squared." This of course means $10 \times 10 = 100$. In our compact notation, the number is expressed as 10^2, where the superscript after the ten (called the exponent) indicates how many times the number 10 is multiplied by itself. Thus, 10^2 can be read as "ten squared," or "ten raised to the second power," or "ten to the two."

Similarly, "10 cubed" is $10 \times 10 \times 10 = 1{,}000 = 10^3$ (ten raised to the third power), and 10^6 (ten to the sixth power) is six 10s multiplied together, or one million.

Notice in these examples that the exponent gives the number of zeroes that follow the 1 when the number is written out in long form. Thus,

$$10^9 = \text{nine 10s multiplied together}$$
$$= 1{,}000{,}000{,}000$$
$$= 1 \text{ followed by } \textit{nine} \text{ zeroes.}$$

Powers of ten notation is a compact way to express large numbers. Thus, we may write the number $1{,}723{,}000{,}000{,}000$ as $1{,}723 \times 10^9$, or by moving the decimal point around we could also express the same number in the following equivalent ways:

$$1{,}723 \times 10^9$$
$$172.3 \times 10^{10}$$
$$17.23 \times 10^{11}$$
$$1.723 \times 10^{12}.$$

The last form is the preferred form of expressing the number, although all the others are also correct.

A further advantage of the powers of ten shorthand is that it makes arithmetic easy. Consider, for example, taking the product of 100 times 1,000:

$$100 \times 1{,}000 = 10^2 \times 10^3$$
$$= 10^{2+3}$$
$$= 10^5$$
$$= 100{,}000.$$

Thus, to multiply two numbers that are expressed in powers of ten notation, simply add the exponents together.

As a second example, consider $15{,}000 \times 3{,}000{,}000{,}000$. Expressing both numbers in powers of ten, we have

$$(1.5 \times 10^4) \times (3 \times 10^9)$$
$$= (1.5 \times 3) \times 10^{4+9}$$
$$= 4.5 \times 10^{13}.$$

The powers of ten notation also handles small numbers by using negative exponents. By definition, when we write 10^{-2}, we mean the reciprocal of 10^2. That is,

$$10^{-2} = 1/10^2 = 0.01.$$

Similarly, $10^{-7} = 0.0000001$. Thus, the exponent is the number of places the decimal point moves to the left of the 1.

The rule for adding exponents applies even if the exponent is negative. Thus, we can multiply large and small numbers together with equal ease. Consider, for example, the product of 0.00000013 times $30{,}000{,}000{,}000{,}000{,}000{,}000 = 1.3 \times 10^{-7} \times 3 \times 10^{19}$. This becomes

$$(1.3 \times 3) \times 10^{-7+19} = 3.9 \times 10^{12}.$$

Sometimes is it is necessary to take a large number to a power, such as $(3 \times 10^3)^4$ This will be $(3 \times 10^3)(3 \times 10^3)(3 \times 10^3)(3 \times 10^3) = (3 \times 3 \times 3 \times 3) \times (10^3 \times 10^3 \times 10^3 \times 10^3) = (3)^4 \times (10^3)^4 = 3^4 \times 10^{12}$. That is, the exponent of the power of ten is multiplied by the exponent, while the number before the exponent is raised to that power.

2. Some Useful Numbers

Physical Constants

The speed of light	$c = 3 \times 10^8$ m/s
	$= 3 \times 10^5$ km/s
The gravitational constant	$G = 6.7 \times 10^{-11}$ m^3 kg^{-1} s^{-2}
The mass of the hydrogen atom	$m_H = 1.67 \times 10^{-27}$ kg
The mass of the electron	$m_e = 9.1 \times 10^{-31}$ kg
The angstrom	$1 \text{ Å} = 10^{-8}$ cm
Planck's constant	$h = 6.62 \times 10^{-34}$ J s
Boltzman's constant	$k = 1.38 \times 10^{-23}$ J K^{-1}
Constant in Wien's law	$\lambda_{max} T = 2.9 \times 10^7$ Å K 2.9×10^{-3} m K

Astronomical Constants

The astronomical unit	$1 \text{ AU} = 1.5 \times 10^8$ km
The light-year	$1 \text{ ly} = 9.5 \times 10^{12}$ km
The parsec	$1 \text{ pc} = 3.26$ ly
Mass of Earth	$M_\oplus = 5.974 \times 10^{24}$ kg
Mass of Sun	$M_\odot = 1.989 \times 10^{30}$ kg
Radius of Earth (equatorial)	$R_E = 6,378$ km
Radius of Sun	$R_\odot = 6.96 \times 10^5$ km
Luminosity of Sun	$L_\odot = 3.827 \times 10^{26}$ watts

3. Relations between Units

Angles

A circle contains 360° (degrees)

$1° = 60'$ (minutes of arc)

$1' = 60''$ (seconds of arc)

1 radian = 57.3°

Length

1 kilometer (km) = 1,000 meters (m) = 0.62 mile (mi)

1 meter = 1.09 yards = 39.37 inches

1 centimeter (cm) = 0.01 meter ≈ 0.4 inch

1 millimeter (mm) = 0.1 cm = 0.001 m

1 micron (mm) = 10^{-6} m

1 mile = 1.6 km

1 inch = 2.54 cm

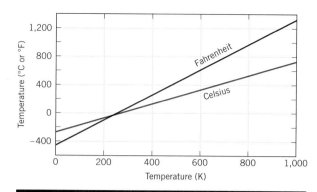

Figure A-1 Graph to convert between temperature scales for low temperatures.

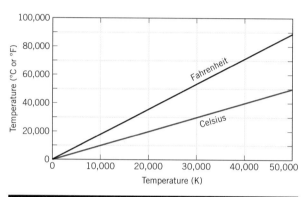

Figure A-2 Graph to convert between temperature scales for high temperatures.

Mass/Weight

1 kilogram (kg) = 1,000 grams (g)

A mass of 1 kg on Earth has a weight of 2.2 pounds.

Temperature

$$°C = \frac{(°F - 32)}{1.8}$$
$$°F = (1.8 \times °C) + 32$$
$$K = C + 273$$

(° is dropped when Kelvin scale is written)

The following graphs allow approximate conversion from K to °C or °F. **Figure A-1** is useful for Kelvin temperatures between 0 and 1,000, while **Figure A-2** is for higher temperatures. For example, to convert 600 K to the other scales, find 600 on the horizontal scale; reading up to the Celsius line gives a temperature of about 340° C. Reading to the Fahrenheit line gives slightly over 600° F. To convert 30,000° F to K, find 30,000 on the vertical axis; moving horizontally to the Fahrenheit curve gives an equivalent temperature of about 17,000 K.

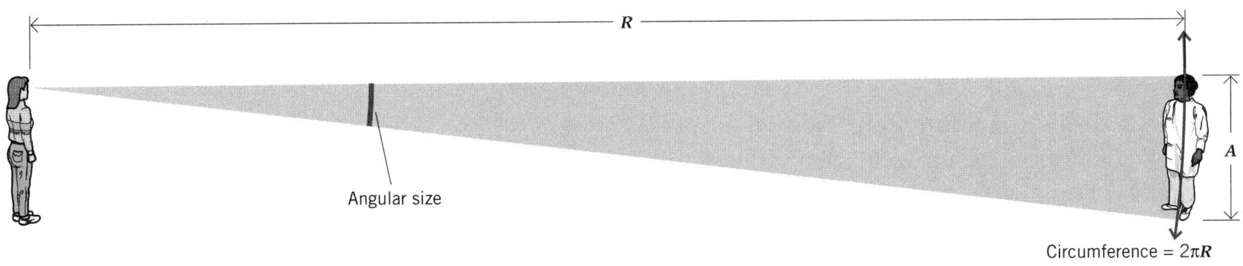

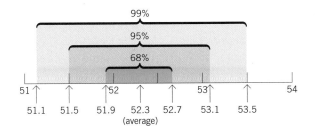

Figure A-3 Figure for the derivation of the angular size formula.

4. Conversion between Units

Conversion between units is easier if you write all units involved and separate each term with parentheses. As an example, express 25 light-years in kilometers:

$$(25 \text{ ly}) \times (9.5 \times 10^{12} \text{ km/ly}) = 2.6 \times 10^{13} \text{ km}.$$

Note that the ly in the numerator cancels the ly in the denominator. Further convert to astronomical units:

$$(2.6 \times 10^{13} \text{ km}) \times (1 \text{ AU}/1.5 \times 10^{8} \text{ km})$$
$$\approx 1.7 \times 10^{18} \text{ AU}.$$

5. Derivation of Angular Size Formula

To find how physical size, distance, and angular size are related, refer to **Figure A-3**, in which we draw about the observer a circle of radius R equal to the object's distance from the observer. From this drawing you can write the following simple proportion: The angular size of the person (in degrees) is to the physical height of the person *(A)* as the complete angle around the circle (360°) is to the circumference of the circle ($2\pi R$). Symbolically,

$$\frac{\text{Angular size}}{A} = \frac{360}{2\pi R}.$$

Solving for the angular size, we find

$$\text{angular size} = \frac{360 A}{2\pi R}$$
$$= 57.3 \, \frac{A}{R}.$$

In other words,

$$\text{angular size} = 57.3 \times \frac{\text{physical size}}{\text{Distance}}.$$

Figure A-4 The meaning of an error.

6. Computing Random Uncertainties

An average is not an exact number. How, then, can we express the uncertainty in a number? If the average of a set of measurements is, for example, 52.3, we express this as 52.3 plus or minus the random uncertainty. If such an uncertainty were 0.4, we would write it as

$$52.3 \pm 0.4.$$

The meaning of this expression is seen in **Figure A-4**, which shows that there is a 68 percent probability that the true value is between 52.3 − 0.4 = 51.9, and 52.3 + 0.4 = 52.7. From probability theory, if we consider twice this range from the average, there is a 95 percent chance that the true value is between 51.5 and 53.1. Although this is pretty good, it also means there is a 5 percent chance that the true value is *outside* these limits.

In a series of measurements, there will always be a largest and a smallest value, and the difference between these two numbers is called the *range* of the measurements. The range indicates the amount of variation about the average value that you are likely to encounter in a series of measurements, and it is therefore an indication of the amount of confidence that can be placed in any single measurement. Scientists characterize the variation about an average by a precisely defined quantity called the **standard deviation**, but for convenience we will adopt an approximate rule called **Snedecor's rough check**. This rule states that if you have five independent measurements of a quantity, the standard deviation of one measurement is *approximately* equal to the range of the measurements divided by two:

$$\frac{(x_{max} - x_{min})}{2}$$

where x_{max} and x_{min} are the values of the maximum and minimum measurements. If there are ten independent measurements, the Snedecor rough check is approximately equal to the range divided by three. In general, dividing the range by the square root of $(N - 1)$, where N is the number of observations, will give you a number that approximates the standard deviation if N is small. The computed value is the number that follows the plus-and-minus symbol in presenting a result.

7. Why Bodies of Different Masses Fall at the Same Rate

You can see the reason bodies of different masses fall at the same rate by combining Newton's second law with his law of gravity. The force of gravity as given by Newton's universal law of gravity must equal the force determined by the second law, $F = ma$. Setting the expressions equal to each other, we find the following:

$$ma = G\frac{mM}{d^2}.$$

Dividing each side by m, we find the acceleration:

$$a = G\frac{M}{d^2}.$$

The acceleration is thus independent of the mass of the falling body (m) and depends only on the size and mass of the gravitating body (M). In other words, bodies of different masses fall at the same rate!

8. Weight as the Force of Gravity

Because weight is the force that gravity exerts on your body, you can express your weight using Newton's equation. If W is your weight, m_{you} is your mass, M_{Earth} is the mass of the Earth, and R_{Earth} is your distance from the center of the Earth, then

$$W = G\frac{m_{you}M_{Earth}}{R^2_{Earth}}.$$

In a similar manner, you could write your weight on the Moon:

$$W_{moon} = G\frac{m_{you}M_{moon}}{R^2_{moon}}.$$

9. A Further Example of the Doppler Effect

Suppose a star is approaching Earth at a speed of 30 km/sec. Let's turn the problem around and determine by how much a spectral line will shift, and the wavelength at which the spectral line would be observed. Because the star is approaching Earth, we know that its observed wavelengths will be shorter than the rest wavelength. The ratio of speeds in the formula is

$$\frac{\text{Relative speed between the star and Earth}}{\text{Speed of light}}$$

$$\frac{-30 \text{ km/s}}{300,000 \text{ km/s}} = 10^{-4} = \frac{\text{Change in wavelength}}{\text{Rest wavelength}}.$$

The change in the wavelength will be -10^{-4} of the rest wavelength. Suppose the rest wavelength is 6,562.79 Å. The change in wavelength will be $6,562.79 \times (-0.0001) = -0.66$ Å, so the observed wavelength would be $6,562.79 - 0.66 = 6,562.13$ Å. (Here it is not correct to round off to the nearest whole angstrom!) This amount of shift in the position of a spectral line is easily detectable in a star's spectrum.

10. Spectroscopic Parallaxes

The examples of spectroscopic parallax in the text are for two simple possibilities. Readers desiring a more general and explicit mathematical treatment will find the following discussion useful. From the inverse square law we know that a star of luminosity L at a distance d will have an apparent (observed) brightness B given by this equation:

$$B \propto \frac{L}{d^2}.$$

If we have two stars, then we have

$$B_1 \propto \frac{L_1}{d_1^2}$$

$$B_2 \propto \frac{L_2}{d_2^2}.$$

Taking ratios of these two expressions we find the following:

$$\frac{B_2}{B_1} = \frac{L_2 / d_2^2}{L_1 / d_1^2}.$$

(In taking the ratio, the proportionality becomes an equality because all constants cancel out.) Using the example in which α Centauri is star 2 and the Sun is star 1 we have the following:

$$\frac{B_{\alpha\,Cen}}{B_{Sun}} = \frac{L_{\alpha\,Cen} / d_{\alpha\,Cen}^2}{L_{Sun} / d_{Sun}^2}$$

$$\frac{1}{9 \times 10^{10}} = \frac{1 / d_{\alpha\,Cen}^2}{1 / d_{Sun}^2} = \frac{d_{Sun}^2}{d_{\alpha\,Cen}^2}.$$

Solving for $d_{\alpha\,Cen}/d_{Sun}$ we find $d_{\alpha\,Cen}/d_{Sun} = 3 \times 10^5$ as described previously.

Similarly, for the second example in which Canopus and α Cen appear equally bright, we have the following:

$$\frac{B_{Canopus}}{B_{\alpha\,Cen}} = \frac{L_{Canopus} / d_{Canopus}^2}{L_{\alpha\,Cen} / d_{\alpha\,Cen}^2}$$

or

$$\frac{B_{Canopus}}{B_{\alpha\,Cen}} = \frac{L_{Canopus} / d_{\alpha\,Cen}^2}{L_{\alpha\,Cen} / d_{Canopus}^2}$$

$$1 = 1,200 \times \frac{d_{\alpha\,Cen}^2}{d_{Canopus}^2}$$

$$d_{Canopus} = \sqrt{1,200} \times d_{\alpha\,Cen}$$

$$d_{Canopus} = 35 \times d_{\alpha\,Cen}.$$

11. The Relation between Sidereal and Synodic Periods

The sidereal period, which describes the period of one object around another (e.g., the Moon around the Earth or a planet around the Sun) in relation to the background stars, is a fundamental quantity in describing orbital motion. However, because the object is observed from a moving platform (Earth), it is difficult to observe the sidereal period directly. However, the synodic period, which is the time interval between successive recurrences of a phase (e.g., full moon to full moon) or recurrences of a planetary configuration (opposition to opposition), is simple to observe. From the synodic period, then, the sidereal period can be found as follows:

Let S be a planet's synodic period; P is the sidereal period, and E is the Earth's sidereal period (365.25 days). Then we have the following:

$\frac{1}{S} = \frac{1}{P} - \frac{1}{E}$ for inferior planets (those closer to the Sun than Earth), and

$\frac{1}{S} = \frac{1}{E} - \frac{1}{P}$ for superior planets (those farther from the Sun than Earth).

As an example, suppose we observe that the time between oppositions for some object farther from the Sun is 400 days. We would then derive its sidereal period about the Sun to be:

$$\frac{1}{400} = \frac{1}{365.25} - \frac{1}{P}$$

$$\frac{1}{P} = \frac{1}{365.25} - \frac{1}{400}$$

$$= 0.0027379 - 0.0025000 = 0.00023785$$

$$P = 4,204 \text{ days.}$$

We can use the same idea if the periods are expressed in years. Thus, if the planet's synodic period was 1.1 years (and with $E = 1$ year) we would have this result:

$$1/1.1 = 1 - 1/P$$

$$1/P \; 1 - 1/1.1 = 1 - 0.909 = 0.0909$$

$$P = 11 \text{ years.}$$

Appendix B

Constellations

Constellation (Latin name)	Abbreviation	English Name or Description
Andromeda	And	Princess of Ethiopia
Antlia	Ant	Air Pump
Apus	Aps	Bird of Paradise
Aquarius	Aqr	Water Bearer
Aquila	Aql	Eagle
Ara	Ara	Altar
Aries	Ari	Ram
Auriga	Aur	Charioteer
Boötes	Boo	Herdsman
Caelum	Cae	Graving Tool
Camelopardalis	Cam	Giraffe
Cancer	Can	Crab
Canes Venatici	CVn	Hunting Dogs
Canis Major	CMa	Big Dog
Canis Minor	CMi	Little Dog
Capricornus	Cap	Sea Goat
Carina	Car	Keel of the Argonaut's ship
Cassiopeia	Cas	Queen of Ethiopia
Centaurus	Cen	Centaur
Cepheus	Cep	King of Ethiopia
Cetus	Cet	Sea Monster (whale)
Chamaeleon	Cha	Chameleon
Circinus	Cir	A Pair of Compasses
Columba	Col	Dove
Coma Berenices	Com	Berenice's Hair
Corona Australis	CrA	Southern Crown
Corona Borealis	CrB	Northern Crown
Corvus	Crv	Crow
Crater	Crt	Cup
Crux	Cru	Cross (Southern)
Cygnus	Cyg	Swan
Delphinus	Del	Porpoise
Dorado	Dor	Swordfish
Draco	Dra	Dragon
Equuleus	Equ	Little horse
Eridanus	Eri	River
Fornax	For	Furnace
Gemini	Gem	Twins
Grus	Gru	Crane
Hercules	Her	Hercules, son of Zeus
Horologium	Hor	Clock
Hydra	Hya	Water snake (female)
Hydrus	Hyi	Water snake (male)

Constellation (Latin name)	Abbreviation	English Name or Description
Indus	Ind	Indian
Lacerta	Lac	Lizard
Leo	Leo	Lion
Leo Minor	LMi	Little lion
Lepus	Lep	Hare
Libra	Lib	Scales
Lupus	Lup	Wolf
Lynx	Lyn	Lynx
Lyra	Lyr	Lyre or harp
Mensa	Men	Table mountain
Microscopium	Mic	Microscope
Monoceros	Mon	Unicorn
Musca	Mus	Fly
Norma	Nor	Carpenter's level
Octans	Oct	Octant
Ophiuchus	Oph	Holder of serpent
Orion	Ori	Orion, the hunter
Pavo	Pav	Peacock
Pegasus	Peg	The winged horse
Perseus	Per	The rescuer of Andromeda
Phoenix	Phe	Phoenix
Pictor	Pic	Easel
Pisces	Psc	Fishes
Piscis Austrinus	PsA	Southern fish
Puppis	Pup	Stern of the Argonaut's ship
Pyxis	Pyx	Compass of the Argonaut's ship
Reticulum	Ret	Net
Sagitta	Sge	Arrow
Sagittarius	Sgr	Archer
Scorpius	Sco	Scorpion
Sculptor	Scl	Sculptor's tools
Scutum	Sct	Shield
Serpens	Ser	Serpent
Sextans	Sex	Sextant
Taurus	Tau	Bull
Telescopium	Tel	Telescope
Triangulum	Tri	Triangle
Triangulum Australe	TrA	Southern triangle
Tucana	Tuc	Toucan
Ursa Major	UMa	Big bear
Ursa Minor	UMi	Little bear
Vela	Vel	Sail of the Argonaut's ship
Virgo	Vir	Virgin
Volans	Vol	Flying fish
Vulpecula	Vul	Fox

Appendix C

Orbital Data of the Planets

Planet	Semi-Major Axis (AU)	Semi-Major Axis (× 10⁶ km)	Sidereal Orbital Period (Tropical Years)	Sidereal Orbital Period (Days)	Synodic Orbital Period (Days)	Eccentricity	Inclination of Orbital Plane	Average Orbital Velocity (km/sec)	Inclination to Orbital Plane
Mercury	0.387	57.9	0.241	87.96	115.9	0.206	7°0'19"	47.89	0
Venus	0.723	108.2	0.615	224.7	583.9	0.007	3°23'41"	35.03	177°18'
Earth	1	149.6	1	365.26	—	0.017	0°0'0"	29.79	23° 27'
Mars	1.524	228	1.881	686.98	779.9	0.093	1°51'1"	24.13	25° 12'
Jupiter	5.203	778.3	11.86	4333	398.9	0.048	1°18'17"	13.06	3° 07'
Saturn	9.539	1427	29.46	10759	378.1	0.056	2°29'9"	9.64	26° 44'
Uranus	19.18	2884	84.01	30685	369.7	0.047	0°46'23"	6.81	97° 52'
Neptune	30.06	4521	164.8	60188	367.5	0.009	1°46'15"	5.43	29° 34'
Pluto	39.53	5959	248.6	90700	366.7	0.248	17°7'30"	4.74	122° 30'

Physical Data of the Planets

Planet	Radius (km)	Radius (Earth Radii)	Mass (Earth Masses)	Mass (kg)	Density (g/cm³)	Gravity (Earth=1)	Reflectivity (%)	Temperature (k)	Escape Speed (km/s)
Mercury	2439	0.38	0.0562	3.30×10^{23}	5.43	0.38	6	100–700	4.3
Venus	6052	0.95	0.815	4.87×10^{24}	5.24	0.91	76	726	10.4
Earth	6378	1	1	5.974×10^{24}	5.52	1	39	210–300	11.2
Mars	3374	0.53	0.1074	6.419×10^{23}	3.94	0.39	16	190–310	5.0
Jupiter	71492	11.19	317.82	1.899×10^{27}	1.33	2.54	51	110–150	59.5
Saturn	60268	9.46	95.2	5.69×10^{26}	0.70	1.07	50	95	35.5
Uranus	25559	4.01	14.5	8.66×10^{25}	1.30	0.9	66	58	2.1
Neptune	24764	3.88	17.1	1.03×10^{26}	1.76	1.14	62	56	2.3
Pluto	1151	0.18	0.0022	1.5×10^{22}	2.10	0.06	50	40	1.1

Major Planetary Satellites

Planet	Satellite	Distance (10³ km)	Distance (Planetary Radii)	Orbital Period (days)	Radius (km)	Mass (Moon = 1)	Bulk Density (g/cm³)
Earth	Moon	384.4	60.2	27.32	1737.4	1.00	3.34
Mars	Phobos	9.38	2.76	0.319	$13 \times 11 \times 9$	1.3×10^{-7}	1.9
	Deimos	23.46	6.93	1.262	$8 \times 6 \times 5$	2.7×10^{-8}	2.1
Jupiter	Metis	128	1.79	0.29	30	1.3×10^{-6}	—
	Adrastea	129	1.8	0.30	$13 \times 10 \times 8$	2.6×10^{-7}	—
	Amalthea	181	2.53	0.50	$13 \times 73 \times 67$	5.2×10^{-5}	3.00
	Thebe	222	3.11	0.67	55×45	1.0×10^{-5}	—
	Io	422	5.90	1.77	$1830 \times 1819 \times 1815$	1.21	3.53
	Europa	671	9.39	3.55	1565	0.67	3.03
	Ganymede	1070	14.97	7.15	2634	2.02	1.93
	Callisto	1883	26.34	16.69	2403	1.47	1.79
	Leda	11.094	155	239	5	7.8×10^{-8}	—
	Himalia	11.480	161	251	85	1.3×10^{-4}	~1
	Lysithea	11.720	164	259	12	1.0×10^{-6}	—
	Elara	11.737	164	260	40	1.0×10^{-5}	—
	Ananke	21.200	297	631 (R)	10	5.2×10^{-7}	—
	Carme	22.600	316	692 (R)	15	1.3×10^{-6}	—
	Pasiphae	23.500	329	735 (R)	18	2.6×10^{-6}	—
	Sinope	23.700	332	758 (R)	14	1.0×10^{-6}	—
Saturn	Pan	133.58	2.22	0.58	10	—	—
	Atlas	137.67	2.28	0.602	$19 \times 17 \times 14$	—	—
	Prometheus	139.35	2.31	0.613	$74 \times 50 \times 34$	—	—
	Pandora	141.70	2.35	0.629	$55 \times 44 \times 31$	—	—
	Epimetheus	151.42	2.51	0.694	$69 \times 55 \times 55$	—	—
	Janus	151.47	2.51	0.695	$97 \times 95 \times 77$	—	—
	Mimas	185.52	3.08	0.942	$209 \times 196 \times 191$	6.2×10^{-4}	1.2
	Enceladus	238.02	3.95	1.37	$256 \times 247 \times 245$	1.0×10^{-3}	1.2
	Tethys	294.66	4.88	1.888	$536 \times 528 \times 526$	0.01	1.2
	Telesto	294.66	4.88	1.888	$15 \times 13 \times 8$	—	—
	Calypso	294.66	4.88	1.888	$15 \times 8 \times 8$	—	—
	Dione	377.40	6.26	2.737	560	0.01	1.4
	Helene	377.40	6.26	2.737	$18 \times 16 \times 15$	—	—
	Rhea	527.04	8.74	4.518	764	0.03	1.3
	Titan	1221.83	20.25	15.95	2575	1.82	1.88
	Hyperion	1481.1	24.55	21.28	$180 \times 140 \times 113$	2.3×10^{-4}	—
	Iapetus	3561.3	59.02	79.33	718	0.03	1.2
	Phoebe	12952	214.7	550.5 (R)	110	5.4×10^{-6}	—
Uranus	Cordelia	49.75	1.95	0.34	13	—	—
	Ophelia	53.76	2.10	0.38	15	—	—
	Bianca	59.17	2.32	0.43	21	—	—
	Cressida	61.77	2.42	0.46	33	—	—
	Desdemona	62.66	2.45	0.47	27	—	—
	Juliet	64.36	2.52	0.49	42	—	—
	Portia	66.10	2.59	0.51	54	—	—
	Rosalind	69.93	2.74	0.56	27	—	—
	Belinda	75.26	2.94	0.62	33	—	—
	S/1986U10	76.42	2.99	0.64	20	—	—
	Puck	86.00	3.30	0.76	77	—	—
	Miranda	129.90	5.08	1.41	240×233	2.4×10^{-4}	—
	Ariel	190.9	7.48	2.52	581×578	0.02	—

Planet	Satellite	Distance (10³ km)	Distance (Planetary Radii)	Orbital Period (days)	Radius (km)	Mass (Moon = 1)	Bulk Density (g/cm³)
	Umbriel	266.00	10.41	4.14	585	0.01	—
	Titania	436.3	17.07	8.71	789	0.08	~1.5
	Oberon	583.60	22.79	13.46	761	0.08	~1.5
	Caliban	7200	282	580 (R)	~40	—	—
	Sycorax	12000	470	1288 (R)	~80	—	—
Neptune	Naiad	48.23	1.95	0.29	29	—	—
	Thalassa	50.07	2.02	0.311	40	—	—
	Despina	52.53	2.12	0.335	74	—	—
	Galatea	61.95	2.50	0.429	79	—	—
	Larissa	73.55	2.97	0.555	96	—	—
	Proteus	117.65	4.75	1.12	208	—	—
	Triton	354.8	14.33	5.88 (R)	1352	1.82	2.05
	Nereid	5513	222.67	360.1	170	2.8×10^{-4}	—
Pluto	Charon	19.6	33	6.39	593	0.026	2.1

[1](R) = Retrograde motion

Appendix D

The 40 Brightest Stars

Star	Name	Apparent Magnitude	Spectral Type	Luminosity (Sun = 1)	Distance pc (ly)
α CMa A	Sirius A	−1.46	A1 V	21	2.6 (8.6)
α Car	Canopus	−0.72	F0 II	1300	96 (312)
α Boo	Arcturus	−0.04	K1.5 III	100	11.3 (36.7)
α Cen A	Rigil Kentaurus	−0.1	G2 V	1.4	1.3 (4.4)
α Lyr	Vega	0.03	A0 V	49	7.8 (25.3)
α Aur	Capella	0.08	G8 III	134	12.9 (42.2)
β Ori A	Rigel	0.12	B8 Iab	53,000	237 (773)
α CMi A	Procyon	0.34	F5 IV-V	6.4	3.5 (11.4)
α Ori	Betelgeuse	0.58	M2 Iab	13,000	131 (427)
α Eri	Achernar	0.50	B3 V pec	640	44.1 (144)
β Cen AB	Hadar	0.60	B1III	9300	161 (525)
α Aql	Altair	0.77	A7 V	10	5.1 (16.8)
α Tau A	Aldebaran	0.86	K5 III	150	20.0 (65.1)
α Vir	Spica	1.04	B1 V	1600	80.4 (262)
α Sco A	Antares	1.09	M1 Ib	8500	185 (603)
α PsA	Fomalhaut	1.15	A3 V	12	7.7 (25.1)
β Gem	Pollux	1.16	K0 III	31	10.3 (33.7)
α Cyg	Deneb	1.26	A2 Ia	53,000	990 (3226)
β Cru	Mimosa	1.28	B0.5 III	1.1	108 (352)
α Leo A	Regulus	1.36	B7 V	150	23.8 (77.5)
α Cru A	Acrux	1.39	B0.5 IV	3700	125 (408)
α Cen B		1.40	K4 V	0.4	1.3 (4.4)
ε CMa	Adara	1.51	B2 II	6400	132 (431)
λ Sco	Shaula	1.6	B1 V	1900	216 (703)
γ Ori	Bellatrix	1.64	B2 III	2800	74.5 (243)
β Tau	Elnath	1.65	B7 III	300	40.2 (131)
β Car	Miaplacidus	1.70	A1 III	64	34.1 (111)
γ Cru		1.63	M4 III	230	27.0 (87.9)
ε Ori	Alnilam	1.70	B0 Ia	49,000	412 (1342)
α Gru	Al Na'ir	1.76	B7 IV	210	31.1 (101)
ζ Ori	Alnitak	1.79	O9.5 Ib	23,000	251 (817)
ε UMa	Alioth	1.79	A0p	59	24.8 (80.9)
α Per	Mirfak	1.82	F5 Ib	8400	182 (592)
α UMa	Dubhe	1.81	K0 III	160	37.9 (124)
ε Sgr	Kaus Australis	1.81	B9.5 III	77	44.3 (145)
γ Vel	Suhail al Muhlif	1.83	WC8	37,000	258 (840)
δ CMa		1.85	F8 Ia	120,000	550 (1791)
β Aur	Menkalinan	1.86	A2 V	41	25.2 (82.1)
α Cru B	Acrux	1.90	B1 V	1500	125 (408)
θ Sco	Sargas	1.86	F0 Ib	700	83.4 (272)

Appendix E

Star	Parallax (arcseconds)	Distance pc (ly)	Spectral Type	Apparent Magnitude	Luminosity (Sun = 1)
Sun	—	—	G2 V	−26.7	1
α Cen C	0.772	1.3 (4.2)	M5.5Ve	11.00	5.5×10^5
α Cen B	0.747	1.3 (4.4)	K0 V	1.30	0.45
α Cen A	0.747	1.3 (4.4)	G2 V	−0.1	1.6
Wolf 359	0.549	1.8 (5.9)	M5.5e	13.50	1.9×10^{-5}
Barnard's star	0.547	1.8 (6.0)	M4 Ve	9.50	4.5×10^{-4}
Lalande 21185	0.393	2.5 (8.3)	M2 V	7.50	5.5×10^{-3}
Sirius B	0.379	2.6 (8.6)	wd	8.70	2.9×10^{-3}
Sirius A	0.379	2.6 (8.6)	A1 V	−1.5	2.4×10^{-1}
UV Cet	0.374	2.7 (8.7)	M6e	13.00	4.2×10^{-5}
BL Cet	0.374	2.7 (8.7)	M6e	12.50	5.7×10^{-5}
Ross 154	0.336	3.0 (9.7)	M3.5e	10.95	4.8×10^{-4}
Ross 248	0.316	3.2 (10.3)	M5.5 V	12.30	1.1×10^{-4}
ε Eri	0.310	3.2 (10.5)	K2 V	3.70	3.0×10^{-1}
Lacaille 9352	0.304	3.3 (10.7)	M2 V	7.40	1.3×10^{-2}
Ross 128	0.299	3.3 (10.9)	M4	11.10	3.6×10^{-4}
EZ Aqr A	0.290	3.5 (11.3)	M5.0V	13.33	
Luyten 789–6	0.290	3.5 (11.3)	M5e	12.20	1.3×10^{-4}
Procyon B	0.286	3.5 (11.4)	wd	10.70	5.5×10^{-4}
Procyon A	0.286	3.5 (11.4)	F5 IV	0.40	$7.7 \times 10^{+0}$
61 Cyg B	0.286	3.5 (11.4)	K7 V	6.00	3.9×10^{-2}
61 Cyg A	0.286	3.5 (11.4)	K5 V	5.20	8.2×10^{-2}
Σ 2398 B	0.283	3.5 (11.5)	M4 V	9.70	1.5×10^{-3}
Σ 2398 A	0.283	3.5 (11.5)	M3.5 V	8.90	3.0×10^{-3}
GQ And	0.281	3.6 (11.6)	M3.5 V	11.00	4.2×10^{-4}
GX And	0.280	3.6 (11.6)	M1.5 V	8.10	6.6×10^{-3}
ε Ind C	0.276	3.6 (11.8)	T6.0	—	—
ε Ind B	0.276	3.6 (11.8)	T1.0	—	—
DX Cnc	0.276	3.6 (11.8)	M6.5 V	14.78	—
ε Ind	0.276	3.6 (11.8)	K4.5 V	4.69	1.4×10^{-1}
τ Ceti	0.274	3.6 (11.9)	G8 V	3.50	4.5×10^{-1}
Recons 1	0.271	3.7 (12.0)	M5.5 V	13.03	—
YZ Cet	0.269	3.7 (12.1)	M4.5 V	12.02	3.2×10^{-4}
BD +5 1668	0.263	3.8 (12.4)	M3.5 V	9.90	1.4×10^{-3}
Kapteyn's star	0.255	3.9 (12.8)	M1	8.90	3.9×10^{-3}
AX Mic	0.253	3.9 (12.9)	M1	6.70	2.6×10^{-2}
Kruger 60B	0.248	4.0 (13.1)	M5e	11.30	4.2×10^{-4}
Kruger 60A	0.248	4.0 (13.1)	M4	9.80	1.6×10^{-3}

Appendix F

Some Local Group Galaxies

	Type	Luminosity (Sun = 1)
M31 (Andromeda)	Sb	2.1×10^{10}
Milky Way	Sb/Sc	1.3×10^{10}
M33 = NGC 598	Sc	2.8×10^{9}
Large Magellanic Cloud	Ir	1.3×10^{9}
M32 = NGC 221	E2	2.8×10^{8}
NGC 6822	Ir	2.8×10^{8}
NGC 205	Spheroidal	2.6×10^{8}
Small Magellanic Cloud	Ir	2.3×10^{8}
NGC 185	Dwarf spheroidal	1.0×10^{8}
NGC 147	Dwarf spheroidal	8.5×10^{7}
IC 1613	Ir	7.0×10^{7}
WLM	Ir	3.4×10^{7}
Fornax	Dwarf spheroidal	2.3×10^{7}
And I	Dwarf spheroidal	4.1×10^{6}
And II	Dwarf spheroidal	4.1×10^{6}
Leo I	Dwarf spheroidal	3.7×10^{6}
DDO 210	Ir	3.1×10^{6}
Sculptor	Dwarf spheroidal	1.5×10^{6}
And III	Dwarf spheroidal	1.0×10^{6}
Pisces	Ir	9.3×10^{5}
Sextans	Dwarf spheroidal	7.7×10^{5}
Phoenix	Dwarf Ir / Dwarf spheroidal	7.0×10^{5}
Tucana	Dwarf spheroidal	4.9×10^{5}
Leo II	Dwarf spheroidal	4.4×10^{5}
Ursa Minor	Dwarf spheroidal	2.8×10^{5}
Draco	Dwarf spheroidal	2.1×10^{5}
Carina	Dwarf spheroidal	8.5×10^{4}
EGB 0427+63	Dwarf Ir	—
Sagittarius	Dwarf spheroidal	—

Periodic Table of the Elements

The periodic table of elements, color-coded to show the time (e.g. Big Bang), place (small mass stars, large mass stars, or supernovae) or mechanism (cosmic rays) for element formation. The numbers are the atomic number, which is the number of protons in the nucleus.

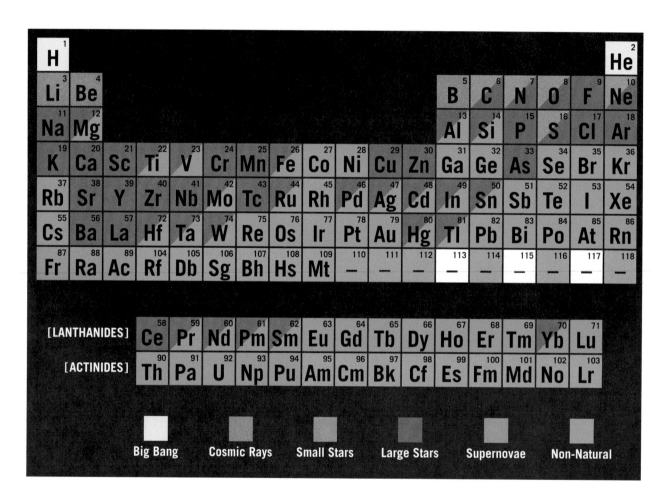

Star Maps for the Northern Hemisphere

Index to Star Maps   Choose the date and time; the appropriate map is then specified.

DATE	8 P.M.	9 P.M.	10 P.M.	11 P.M.	12 A.M.	1 A.M.
Jan 5		1		2		3
Jan 20	1		2		3	
Feb 5		2		3		4
Feb 20	2		3		4	
Mar 5		3		4		5
Mar 20	3		4		5	
Apr 5		4		5		6
Apr 20	4		5		6	
May 5		5		6		7
May 20	5		6		7	
Jun 5		6		7		8
Jun 20	6		7		8	
Jul 5		7		8		9
Jul 20	7		8		9	
Aug 5		8		9		10
Aug 20	8		9		10	
Sep 5		9		10		11
Sep 20	9		10		11	
Oct 5		10		11		12
Oct 20	10		11		12	
Nov 5		11		12		1
Nov 20	11		12		1	
Dec 5		12		1		2
Dec 20	12		1		2	

Note: The stars shown on the star maps are all north of declination –40°. Next to each map number is the range of right ascension shown on the map. Thus, the notation ($6^h – 14^h$) means that stars with right ascension between 6^h, 7^h, 8^h, ... 14^h are shown. The notation ($22^h – 10^h$) means that stars of right ascension 22^h, 23^h, 0^h, 1^h, 2^h, ... 10^h are shown. The first number is at the western horizon, while the second one is on the eastern horizon.

Bright Objects on the Star Charts

Sorted by Right Ascension

Star	Right Ascension	Declination	Chart Closest to the Meridian
M31	1	41	11
Polaris	3	89	All
Pleiades	4	24	1
Aldebaran	5	16	1
Capella	5	46	1, 2
Rigel	5	−8	1, 2
Betelgeuse	6	7	2
Sirius	7	−17	2, 3
Castor	8	32	3
Pollux	8	28	3
Procyon	8	5	3
Regulus	10	12	4
Arcturus	14	19	6
Vega	19	39	8
Altair	20	9	9
Summer Triangle	20	28	9
Deneb	21	45	9
Fomalhaut	23	−30	10, 11

Sorted by Star Name

Star	Right Ascension	Declination	Chart Closest to the Meridian
Aldebaran	5	16	1
Altair	20	9	9
Arcturus	14	19	6
Betelgeuse	6	7	2
Capella	5	46	1, 2
Castor	8	32	3
Deneb	21	45	9
Fomalhaut	23	−30	10, 11
M31	1	41	11
Pleiades	4	24	1
Polaris	3	89	All
Pollux	8	28	3
Procyon	8	5	3
Regulus	10	12	4
Rigel	5	−8	1, 2
Sirius	7	−17	2, 3
Summer Triangle	20	28	9
Vega	19	39	8

Directions:

Hold map so that the direction you are facing is at the bottom.

January   20   8 P.M.
January   5   9 P.M.
December 20   10 P.M.
December   5   11 P.M.
November 20   12 P.M.
November   5   1 A.M.

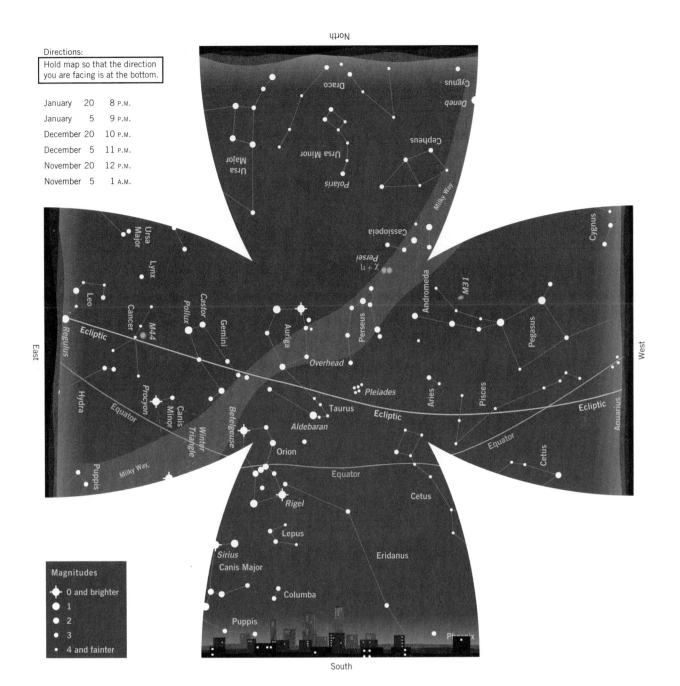

Map 1 [22ʰ–10ʰ]

Directions:

Hold map so that the direction you are facing is at the bottom.

February 20	8 P.M.
February 5	9 P.M.
January 20	10 P.M.
January 5	11 P.M.
December 20	12 P.M.
December 5	1 A.M.

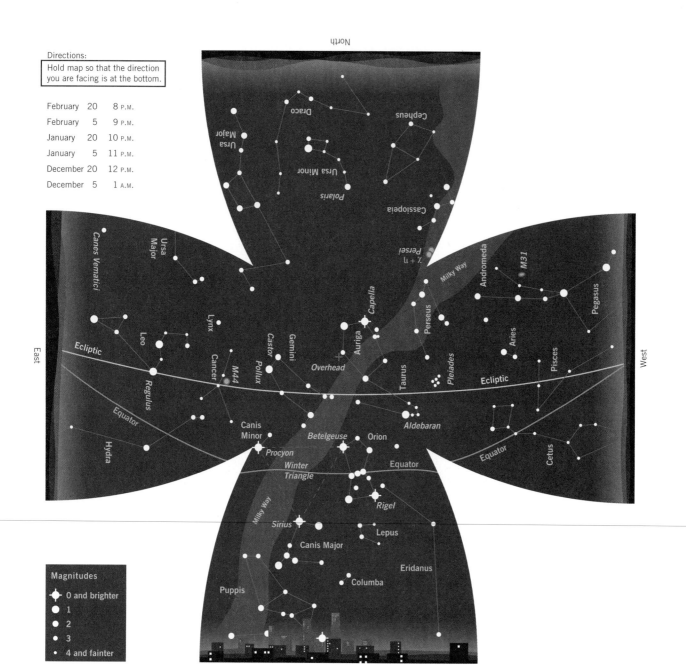

Magnitudes

- ✦ 0 and brighter
- ⬤ 1
- ● 2
- • 3
- · 4 and fainter

Map 2 [0ʰ–12ʰ]

North

Directions:

Hold map so that the direction
you are facing is at the bottom.

March	20	8 P.M.
March	5	9 P.M.
February	20	10 P.M.
February	5	11 P.M.
January	20	12 P.M.
January	5	1 A.M.

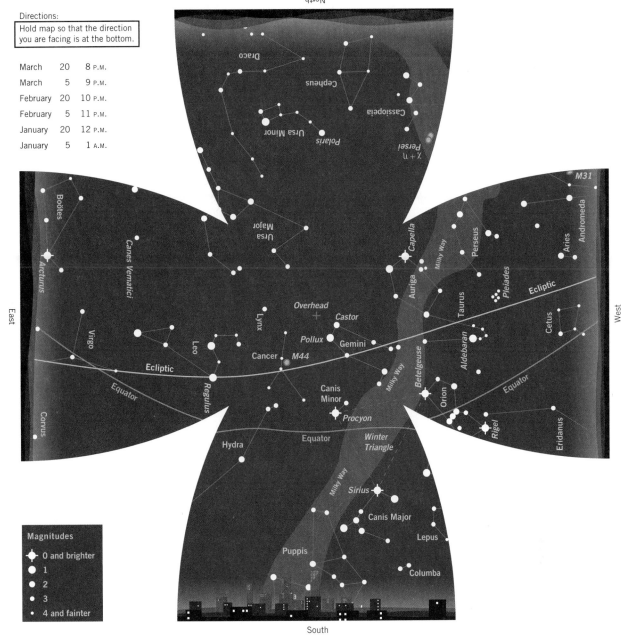

Magnitudes

- ✦ 0 and brighter
- ● 1
- ● 2
- • 3
- · 4 and fainter

South

Map 3 [2ʰ–14ʰ]

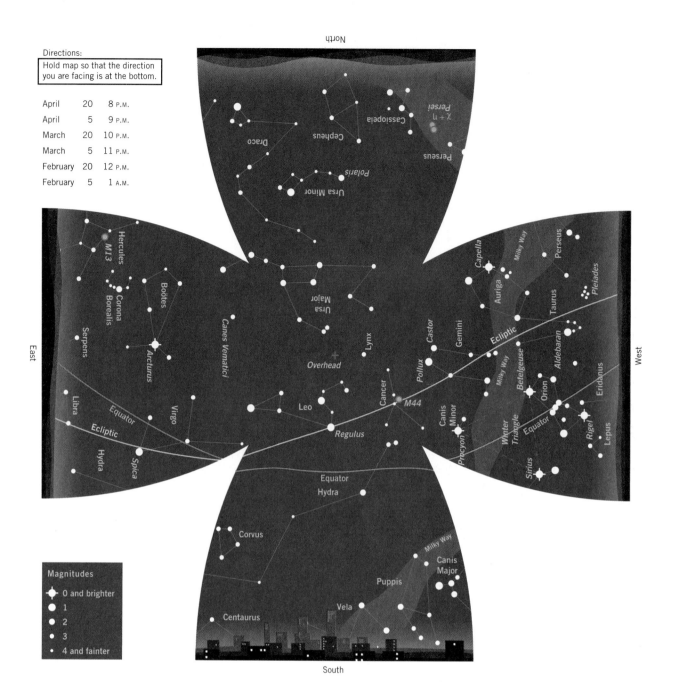

Directions:

Hold map so that the direction you are facing is at the bottom.

April	20	8 P.M.
April	5	9 P.M.
March	20	10 P.M.
March	5	11 P.M.
February	20	12 P.M.
February	5	1 A.M.

Perseus

χ + η Persei

Cassiopeia

Cepheus

Draco

Polaris

Ursa Minor

Hercules

M13

Corona Borealis

Boötes

Serpens

Canes Vematici

Ursa Major

Capella

Milky Way

Perseus

Auriga

Taurus

Pleiades

Castor

Gemini

Ecliptic

Aldebaran

Lynx

Pollux

Betelgeuse

Milky Way

Orion

Eridanus

Overhead

Arcturus

East

Virgo

Leo

Cancer

M44

Canis Minor

Rigel

Lepus

West

Libra

Equator

Ecliptic

Regulus

Procyon

Winter Triangle

Equator

Sirius

Hydra

Spica

Equator

Hydra

Corvus

Milky Way

Canis Major

Puppis

Vela

Centaurus

Magnitudes

- ✦ 0 and brighter
- ● 1
- ● 2
- ● 3
- · 4 and fainter

Map 4 [4ʰ–16ʰ]

Directions:

Hold map so that the direction
you are facing is at the bottom.

May   20   8 P.M.
May   5   9 P.M.
April   20   10 P.M.
April   5   11 P.M.
March   20   12 P.M.
March   5   1 A.M.

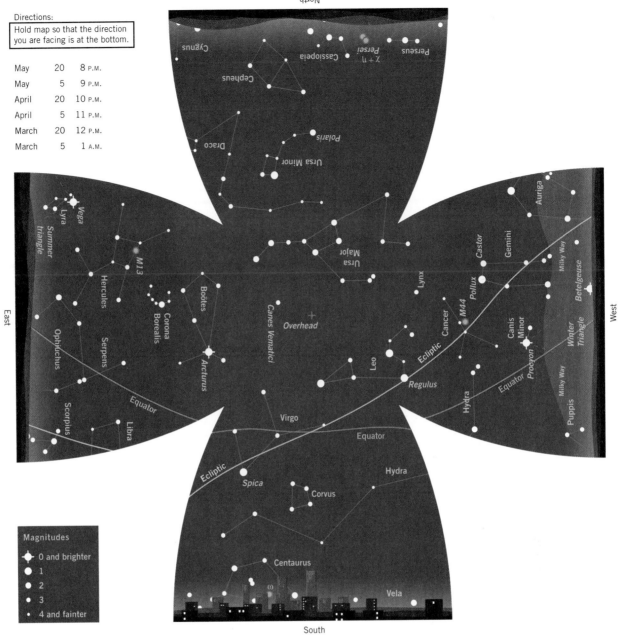

Magnitudes

○◆ 0 and brighter
○   1
•   2
·   3
·   4 and fainter

Map 5 [6ʰ–18ʰ]

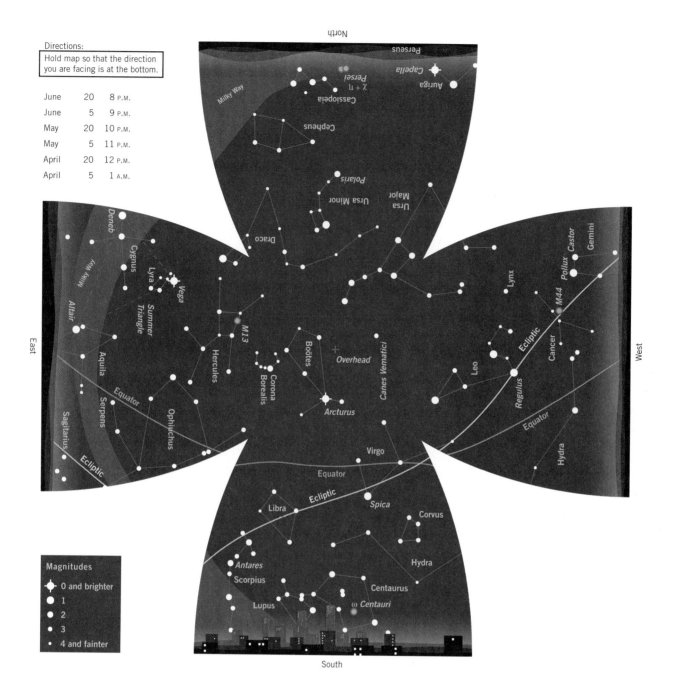

Directions:
Hold map so that the direction you are facing is at the bottom.

June	20	8 P.M.
June	5	9 P.M.
May	20	10 P.M.
May	5	11 P.M.
April	20	12 P.M.
April	5	1 A.M.

North

Perseus

Capella
Auriga
χ + η
Persei
Cassiopeia

Cepheus

Polaris

Ursa Minor
Ursa Major

Draco

Milky Way

East

Deneb
Cygnus
Lyra
Vega
Summer Triangle
Altair
Aquila
Equator
Serpens
Sagittarius
Ecliptic
M13
Hercules
Corona Borealis
Boötes

Ophiuchus

Overhead

Arcturus

Canes Vematici

Lynx

Gemini
Castor
Pollux
M44
Cancer
Ecliptic
Leo
Regulus
Equator

West

Hydra

Virgo

Equator

Libra
Ecliptic
Spica
Corvus

Hydra

Antares
Scorpius
Lupus
Centaurus
ω Centauri

South

Magnitudes
✦ 0 and brighter
● 1
● 2
• 3
· 4 and fainter

Map 6 [8ʰ–20ʰ]

Directions:
Hold map so that the direction you are facing is at the bottom.

July	20	8 P.M.
July	5	9 P.M.
June	20	10 P.M.
June	5	11 P.M.
May	20	12 P.M.
May	5	1 A.M.

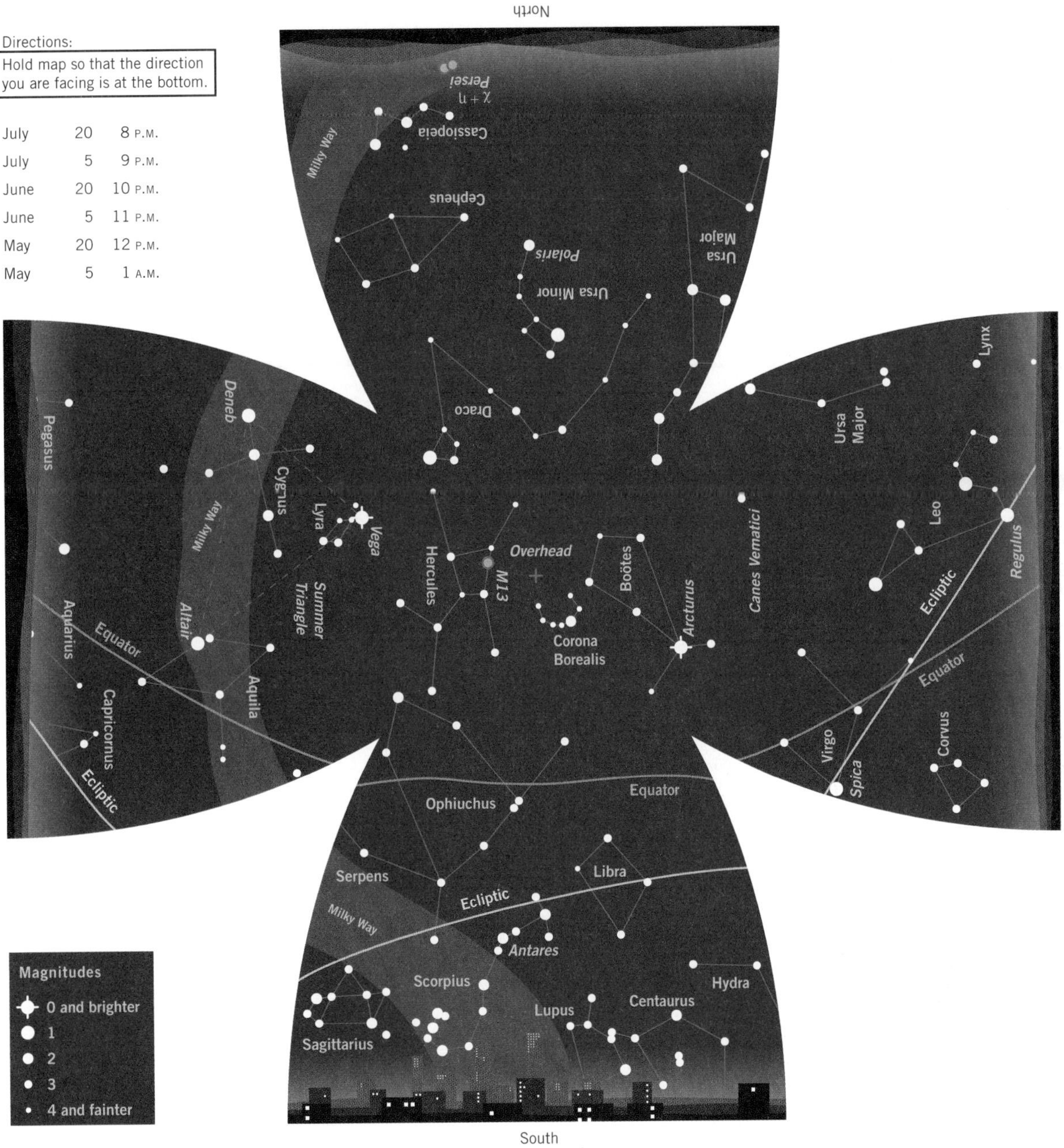

North

χ + η
Persei
Cassiopeia
Milky Way
Cepheus
Polaris
Ursa Minor
Ursa Major
Draco

East

Pegasus
Deneb
Cygnus
Milky Way
Lyra
Vega
Summer Triangle
Altair
Aquila
Aquarius
Equator
Capricornus
Ecliptic

Hercules
M13
Overhead
Corona Borealis
Boötes
Arcturus
Canes Venatici

Lynx
Ursa Major
Leo
Regulus
Ecliptic
Equator
Virgo
Spica
Corvus
Hydra

West

Ophiuchus
Equator
Serpens
Libra
Milky Way
Ecliptic
Antares
Scorpius
Lupus
Centaurus
Sagittarius

South

Magnitudes
- 0 and brighter
- 1
- 2
- 3
- 4 and fainter

Map 7 [10ʰ–22ʰ]

Directions:

Hold map so that the direction you are facing is at the bottom.

August	20	8 P.M.
August	5	9 P.M.
July	20	10 P.M.
July	5	11 P.M.
June	20	12 P.M.
June	5	1 A.M.

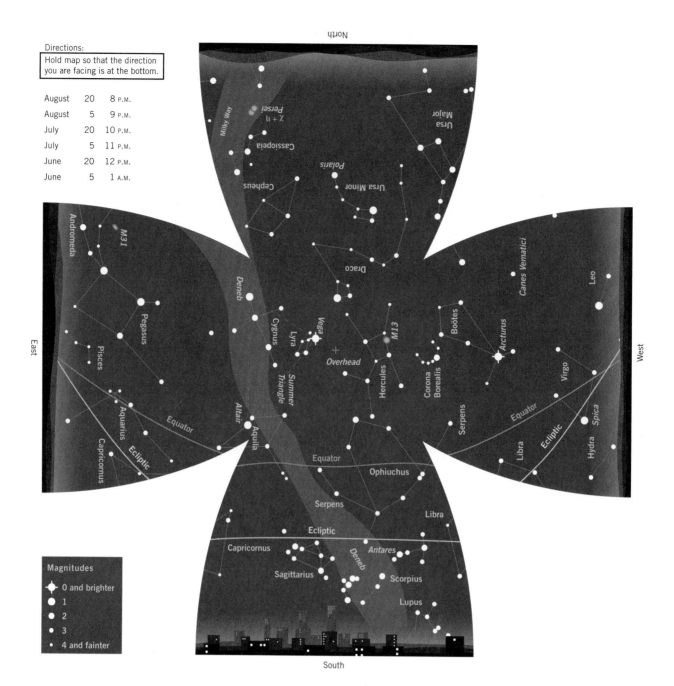

Magnitudes
- 0 and brighter
- 1
- 2
- 3
- 4 and fainter

Map 8 [12ʰ–0ʰ]

Directions:

Hold map so that the direction you are facing is at the bottom.

September	20	8 P.M.
September	5	9 P.M.
August	20	10 P.M.
August	5	11 P.M.
July	20	12 P.M.
July	5	1 A.M.

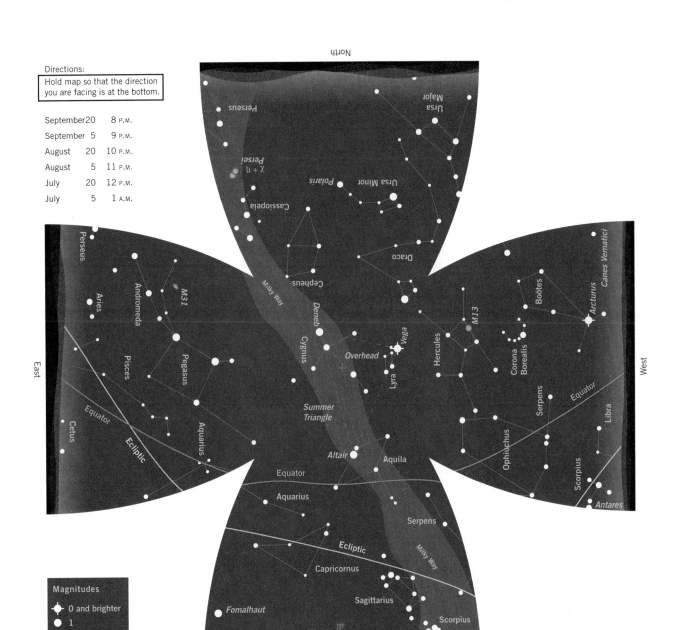

Magnitudes
- 0 and brighter
- 1
- 2
- 3
- 4 and fainter

Map 9 [14ʰ–2ʰ]

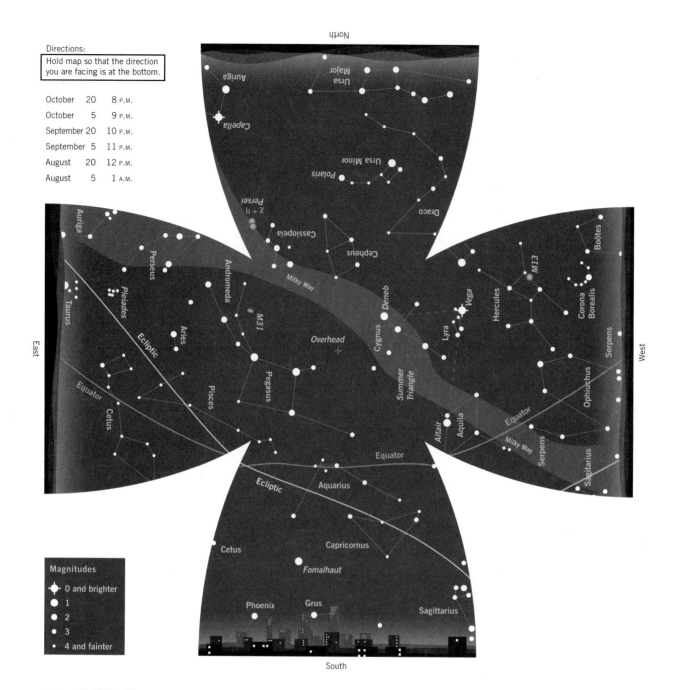

North

Directions:
Hold map so that the direction you are facing is at the bottom.

October   20   8 P.M.
October   5   9 P.M.
September 20   10 P.M.
September   5   11 P.M.
August   20   12 P.M.
August   5   1 A.M.

Ursa Major

Auriga

Capella

Ursa Minor

Polaris

Draco

χ + η Persei

Cassiopeia

Cepheus

Boötes

Auriga

Perseus

Andromeda

Milky Way

Deneb

M13

Corona Borealis

Pleiades

Taurus

M31

Cygnus

Vega

Lyra

Hercules

Serpens

Ecliptic

Aries

Overhead

Summer Triangle

Ophiuchus

East

Pisces

Pegasus

West

Equator

Cetus

Altair

Aquila

Equator

Serpens

Sagittarius

Equator

Aquarius

Milky Way

Ecliptic

Capricornus

Cetus

Fomalhaut

Magnitudes

◆ 0 and brighter
● 1
● 2
• 3
· 4 and fainter

Phoenix   Grus

Sagittarius

South

Map 10 [16ʰ–4ʰ]

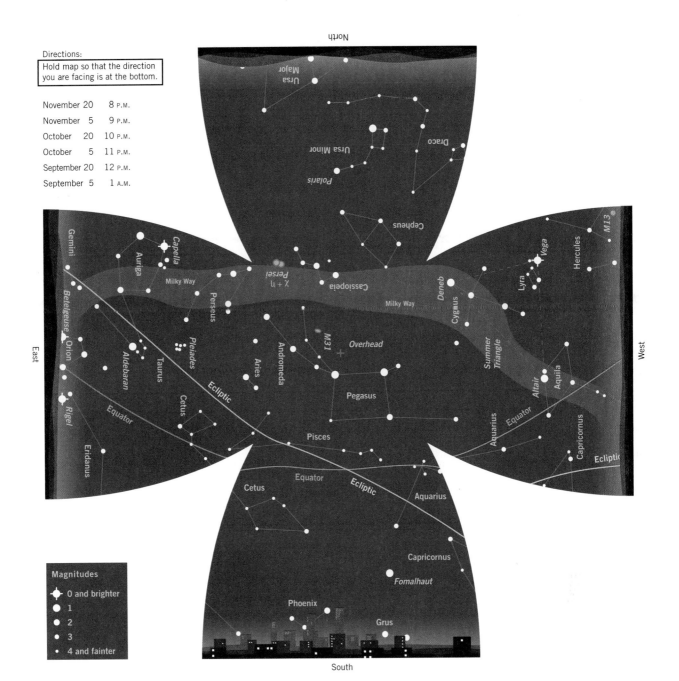

Directions:

Hold map so that the direction you are facing is at the bottom.

November 20   8 P.M.
November   5   9 P.M.
October   20 10 P.M.
October   5 11 P.M.
September 20 12 P.M.
September   5   1 A.M.

Ursa Major

Draco

Ursa Minor

Polaris

Cepheus

M13

Vega

Hercules

Lyra

Gemini

Capella

Auriga

Milky Way

Perseus

χ + h

Persei

Cassiopeia

Milky Way

Deneb

Cygnus

Summer Triangle

East

Betelgeuse

Orion

Aldebaran

Taurus

Pleiades

Aries

Andromeda

M31

Overhead

Pegasus

Altair

Aquila

West

Rigel

Eridanus

Cetus

Ecliptic

Equator

Pisces

Aquarius

Equator

Capricornus

Ecliptic

Equator

Ecliptic

Cetus

Aquarius

Capricornus

Fomalhaut

Magnitudes
- ✦ 0 and brighter
- ● 1
- ● 2
- • 3
- · 4 and fainter

Phoenix

Grus

South

Map 11 [18ʰ–6ʰ]

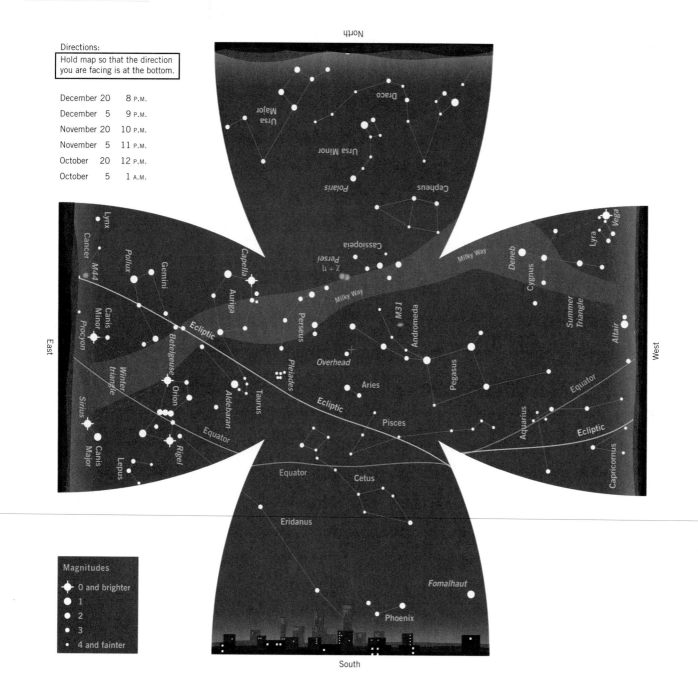

Directions:

Hold map so that the direction you are facing is at the bottom.

December 20	8 P.M.
December 5	9 P.M.
November 20	10 P.M.
November 5	11 P.M.
October 20	12 P.M.
October 5	1 A.M.

Ursa Major

Draco

Ursa Minor

Polaris

Cepheus

Cassiopeia

χ + h Persei

Milky Way

Vega

Lyra

Deneb

Cygnus

Summer Triangle

Altair

Lynx

Cancer

M44

Pollux

Gemini

Capella

Auriga

Ecliptic

Perseus

Milky Way

M31

Andromeda

Overhead

Pegasus

Canis Minor

Procyon

Betelgeuse

Winter Triangle

Pleiades

Aries

Equator

East

Orion

Aldebaran

Taurus

Ecliptic

Pisces

Aquarius

Ecliptic

West

Sirius

Canis Major

Rigel

Lepus

Equator

Equator

Cetus

Capricornus

Eridanus

Fomalhaut

Phoenix

Magnitudes

- ✦ 0 and brighter
- ● 1
- ● 2
- • 3
- · 4 and fainter

South

Map 12 [20ʰ–8ʰ]

Glossary

aberration of light: A small shift in the apparent direction to an object caused by the motion of Earth and the finite speed of light. This effect was used by Bradley in 1727 to make the first historical proof that Earth moved through space.

absorption-line spectrum: The spectrum formed by narrow regions in a spectrum when a continuous spectrum passes through a cooler gas.

acceleration: The rate of change of velocity with time (the change in velocity divided by the elapsed time).

accretion: An aggregation of material that is increasing its mass by gathering and incorporating smaller pieces of material—either by gravity or by collisions.

accretion disk: Material accreted by a larger mass and caused to spiral into a flattened-disk shape by the conservation of angular momentum.

Active Galactic Nuclei (AGN): One of a class of galaxies with very energetic behavior observed in the nuclear region, including high levels of nonthermal continuous emission and unusual radio jets.

active optics: Telescopic systems that can adjust the shape and focus of their optical components rapidly to compensate for distortion in the incoming signal due to turbulence and other motions in the Earth's atmosphere. Also called *adaptive optics*.

Algol: An Arabic name meaning *Demon Star*. Algol is an eclipsing binary system whose variations in brightness can be seen by the naked eye.

Allende meteorite: A meteorite with unusual isotopic ratios suggesting that a supernova may have played a part in forming the early Solar System.

Almagest: The Arabic title of Ptolemy's greatest book.

almanac: A table giving the position at various times of a celestial body. Also called an *ephemeris*.

Alpha Centauri: A bright triple-star system close to the Sun.

alpha particle: The nucleus of a helium atom, composed of two protons and two neutrons.

alpha-capture reactions: Nuclear reactions in which alpha particles are added to a heavier nucleus.

altitude: The angular distance of an object above the horizon.

amino acids: Complex molecules that form the basic building blocks of protein molecules. Conditions inside the interstellar clouds are favorable for the formation of these molecules.

Anaximander: A member of the Pythagorean school, credited with early theories on the composition and motion of the universe.

Andromeda Galaxy: (also *Andromeda Nebula* and *M31*). This diffuse, naked-eye object is revealed by telescopic observations to be a giant, spiral galaxy nearly 3 million light-years away.

angle of incidence: The angle between an incident ray of light and a perpendicular to the surface at the point of incidence.

angle of reflection: The angle between a reflected ray of light and the perpendicular to the surface at the point of reflection.

angstrom: (abbreviated Å) A unit for measuring the wavelengths of light. It is 10^{-8} cm. On this scale, blue light is around 4500 Å and red light is around 6500 Å.

angular momentum: A measure of the amount of spin or orbital motion possessed by a body due to its inertia. Physicists describe it as a *conserved* property. See *conservation*.

angular size: The angle at the eye subtended by rays drawn from the outside of an object. Same as *apparent size*.

annihilation: The total conversion of mass to energy brought about by the collision of matter with antimatter.

annular eclipse: An eclipse of the Sun in which the angular size of the Moon's disk is not quite large enough to block the solar photosphere. In an annular eclipse, the Sun is not dimmed enough to allow the corona to be seen.

antielectron: The antiparticle of an electron.

antimatter: Material with charge and various other properties reversed when compared with normal matter. When matter and antimatter collide, they annihilate each other.

Ap (A-peculiar stars): A-type stars that have some anomalies in their chemical composition when compared with normal A-stars.

aphelion: The greatest distance from the Sun reached by an object orbiting the Sun. See also *perihelion*.

apogee: The most distant point in the orbit of an Earth-orbiting body. See also *perigee*.

Apollo asteroids: A group of asteroids with orbits that come close to the Earth.

Apollonius of Perga: Early Greek credited with the first suggestion of epicycles and deferents as a device to explain the motions of celestial bodies.

apparent magnitude: The brightnesses of celestial bodies measured on a scale first suggested by Hipparchus. In this system (which turns out to be logarithmic), fainter stars are characterized by larger numbers. Two stars that differ by one magnitude differ by a factor of 2.512 in their brightness ratio. A difference of five magnitudes corresponds to a factor of 100 in brightness.

apparent size: See *angular size.*

apparent solar time: Also called *sundial time.* The hours of the day as measured by the apparent progress of the Sun across the sky.

archaea: A type of single-celled life without a nucleus; most extremophiles are archaea. The other type of prokaryote is bacteria.

Ariel: One of the larger moons of Uranus.

Aristarchus: An early Greek astronomer who carried out estimates of the relative sizes and distances of the Earth, Moon, and Sun and espoused a sun-centered solar system model.

Aristotle: The most influential of Greek philosophers. He produced enduring studies in biology, geometry, and logic. His work in physical science and astronomy, however, was less successful.

asteroid: A body whose diameter ranges from that of a grain of sand up to that of Texas. Asteroids populate the region between Mars and Jupiter. Also called *minor planets.*

asteroid belt: A region of asteroids located between the orbits of Mars and Jupiter.

asthenosphere: A region of soft material in the Earth's upper mantle.

Astronomical Unit (AU): The average distance from Earth to the Sun—about 150 million km (or 93 million miles).

asymptotic giant branch (AGB): A region in the H-R diagram of a globular cluster. It is a post–main-sequence evolutionary track followed by stars having both an H- and an He-burning shell.

atmosphere: The gaseous outer regions of a planet or a star.

atmospheric extinction: A reduction in the apparent brightness of the light from a celestial object due to the absorption and scattering of light out of the incoming beam of energy by the Earth's atmosphere.

atmospheric windows: Spectral regions of the atmosphere that are largely or partly transparent to radiation.

atom: The smallest subdivision of matter (the smallest piece) that still retains the chemical properties of the element.

atomic number: The number of protons in the nucleus of an element.

atomic weight: A scale of the relative weights of the atoms of different chemical elements. For an individual atom, the atomic weight would be the atomic mass number—the number of protons plus neutrons in the nucleus. For example, the most frequently occurring form of carbon has a nucleus with 6 protons and 6 neutrons and an atomic weight of 12. There is a less frequently occurring form of carbon with 6 protons and 7 neutrons and an atomic weight of 13. The "average" atomic weight for carbon depends on the frequency of occurrence in nature of the different types of carbon.

AU: Abbreviation for Astronomical Unit.

aurora: High-energy charged particles from space impact upon and excite atmospheric molecules, causing them to radiate. This radiation is more commonly seen near the magnetic poles, since that is where the magnetic field lines tend to direct the incoming particles. The aurora borealis is seen around the north magnetic pole, and the aurora australis is seen in south magnetic regions.

auroral ring: Auroral radiation occurring in a ring-shaped region around the magnetic pole.

autumnal equinox: The moment when the Sun crosses the celestial equator from north to south, marking the beginning of fall. See also *equinox.*

average: From a series of measurements, the value most likely to be closest to the true value. It helps to negate the effects of high and low values.

axis: A line about which a body rotates.

azimuth: Angular distance along the horizon of a celestial object, measured clockwise (eastward) from north.

B.C.E.: Before the Common Era, which coincides with the time before Christian era. The meaning is the same as B.C.

background 3-K radiation: Faint radio static with a highly isotropic distribution, thought to be remnant radiation from the early stages after the big bang.

bacteria: A type of single-celled life without a nucleus, probably the oldest type of life still alive on the Earth. The other type of prokaryote is archaea.

Balmer lines: A set of spectral lines due to transitions of the hydrogen atom between level 2 and higher energy levels.

barium stars: Red giant stars showing abnormally large abundances of barium and other chemical elements not normally found in most stars.

Barnard's star: A star close to the Sun with wiggles in its motion, suggesting the possibility of other masses (in particular, planets) orbiting the star.

barred spirals: Galaxies with a bar of stars running through the nucleus. The spiral arms appear to start from the bar.

Barringer Crater: A one-mile-diameter crater in Arizona created by meteoric impact.

basalt: Igneous rock (produced at high temperatures) that is common in the outer parts of the Earth and the Moon.

basin: Large, shallow depression seen on terrestrial planets, created by impact or plate tectonics.

Bayeux Tapestry: A tapestry woven at the time of the Battle of Hastings, showing a representation of the 1066 C.E. appearance of Halley's comet.

Becklin-Neugebauer source (B-N source): A bright point source of infrared emission seen in the Orion Nebula. It is probably either a new star or a new cluster of stars in the process of formation.

Beta Lyrae: A star whose rapid rotation changes the shape of its spectral line profiles.

Beta Persei: See *Algol.*

Beta Pictoris: A star circled by a disk of dusty material that may form planets.

Big Bang: A model that postulates the origin of the universe in a state of enormous temperature and density (see *primeval fireball*) that produced the expanding universe we observe today.

big crunch: What results when an expanding universe reverses direction, so that density and temperature again increase.

Bighorn medicine wheel: A construction of rocks high in the Wyoming Big Horn Mountains that may have been constructed by Plains Indians for astronomical, calendrical, and ritualistic purposes.

binary accretion model: Postulates that the formation of the Earth–Moon system occurred by accretion of both bodies from the same dust and gas cloud.

binary pulsar: A highly evolved binary pair of stars in which one of the two has become a pulsar.

binary star: Two stars in mutual orbit under the attraction of gravitational forces.

binary X-ray source: A highly evolved binary pair of stars in which one star has become an X-ray-emitting, collapsed object such as a white dwarf, neutron star, or black hole. The X-rays are thought to be emitted by a hot accretion disk in the system.

biosphere: The volume below and above the surface of a planet where life can be sustained. This can also refer to all of the organisms themselves.

bipolar flow: Gas flowing outward in opposite directions seen in objects that may be Young Stellar Objects (YSOs).

BL Lac objects: Active galaxies that show rapidly varying, highly polarized nonthermal emission from their nuclei.

black holes: Collapsed objects whose gravitational field is so intense that not even light can escape from them. They cannot be observed directly, but their effect on their environment can be studied.

black smokers: Plumes of heated water, dark with dissolved minerals, venting from a fracture in the ocean floor.

blackbody: A conceptualized *perfect radiator,* which is found useful in theoretical studies of radiation. A blackbody is one that absorbs 100 percent of the radiation it encounters. It re-emits that energy in a spectral pattern that depends only on temperature.

blackbody spectrum: A continuous distribution of energy given off by a blackbody. The amount of energy emitted at each wavelength by a blackbody is given by Planck's Law.

blink microscope (blink comparator): An instrument that projects two different photos of the same area of the sky in rapid succession. Stars with brightness variations will be seen to "blink" through the viewer.

Bohr model of the atom: An early 'solar system' model for the atom. Negatively charged electrons were envisioned as orbiting the massive and positively charged nucleus of the atom. Only certain orbits were allowed.

Bok globules: Small dark regions seen in interstellar clouds, many of which are observed to surround a young star.

bolide: A meteor that explodes in the Earth's atmosphere.

bottom-up hypothesis: A model of galaxy cluster formation in which galaxies form before clustering itself occurs.

bow shock: The region surrounding a planet's magnetosphere where the high speed (supersonic) particles of the solar wind encounter the planet's magnetic field, which slows the particles to subsonic speeds.

Brackett series: A set of spectral lines due to transitions of the hydrogen atom between level 3 and higher energy levels.

Brahe, Tycho: Sixteenth-century astronomer famous for his accurate naked-eye observations. Considered to be the father of modern observational astronomy. Collaborated with Johannes Kepler to complete the Copernican Revolution.

braided F-ring: One of Saturn's rings that shows time-variable shape and behavior.

breccia: A rock composed of different types of mineral fragments pressed together.

bright-line spectrum: Same as *emission-line spectrum*.

brown dwarf: A low-mass object that is larger than a typical planet but that cannot have hydrogen fusion in its core. As a consequence, it is an almost-star with low temperature and luminosity.

burster: Sporadic outbursts of high-intensity X-ray emission. The sources are still a matter for argument.

butterfly diagram: The pattern created when the latitude at which sunspots form are graphed as a function of time.

C.E.: The Common Era, which coincides with the Christian era. The meaning is the same as A.D.

Callisto: The outermost Galilean satellite.

Caloris Basin: A large multi-ring basin on Mercury.

canali (canals): Long, dark features seen on Mars by some astronomers. Schiaparelli called them *canali*, which is Italian for channels. Other people made the transition to *canals* in their hypothesizing and went on to postulate life and modern civilizations on Mars.

capture model: Imagines the Moon to have been captured by the Earth's gravity.

carbonaceous chondrites: A group of meteorites with a relatively high carbon abundance and chondrules. See *chondrules*.

Cas A: A strong radio source that is the remnant of a supernova explosion that took place in 1667.

Cassegrain: A telescope arrangement that uses a secondary mirror and a small hole in the primary mirror to focus light behind the primary mirror.

Cassini division: A gap in Saturn's rings named after its discoverer—Giovanni Cassini.

cataclysmic variables: Stars that become unstable and violent in their final evolutionary stages.

CCD (charge coupled device): A two-dimensional semiconductor chip that is used as an extremely sensitive photon detector.

celestial equator: This great circle on the celestial sphere is the projection of the Earth's equator onto the plane of the sky. It is halfway between the celestial poles.

celestial pole: The projection onto the celestial sphere of the spin axis of the Earth. The North Celestial Pole (NCP) is currently located near the star Polaris, while the SCP is not near a bright star. The long, slow motion of the Earth, called *precession*, will cause these two celestial poles to slowly move with respect to the stars over a 26,000-year cycle.

celestial sphere: The Bowl of the Sky. To the ancients, this is the apparent sphere of sky encircling the Earth on which the stars appear to be fixed.

Cen A: A strong radio source resembling an elliptical galaxy, but with large regions of dust obscuration.

Cen X-3: A powerful X-ray source in Centaurus composed of a giant star orbited by an X-ray-emitting neutron star.

center of mass: The imaginary point on the sky about which two gravitationally bound bodies will orbit.

centripetal force or acceleration: A force or acceleration directed toward the center of a circular path.

Cepheid instability strip: The region of the H-R diagram where Cepheid variable stars and Cepheid-like variables are located.

Cepheid variables: Luminous giant stars whose energy output varies in a regular and recognizable manner. They are important for determining the distances to nearby galaxies.

Ceres: The first discovered and largest asteroid.

Chandrasekhar limit: The largest mass that theory will permit a white dwarf to acquire, roughly 1.4 solar masses.

Charon: Pluto's moon.

Chiron: An object discovered in 1977 in an orbit that lies mostly between Saturn and Uranus. Initially classified as an asteroid but later exhibited behavior more like the nucleus of a comet.

chondrites: See *carbonaceous chondrites*.

chondrules: Silicate nodules found in a particular class of meteorites that may be a good sample of primeval solar system composition.

chromatic aberration: A defect of telescopes made with lenses. Different colors of light are refracted (bent) by different amounts, so that they are brought to different focal points.

chromosphere: The layer of the Sun's atmosphere just above the photosphere where most of the absorption lines are formed.

circumpolar stars: Those stars that never go below the horizon in their 24-hour daily motion because of their proximity to the celestial pole.

Circus Maximus: A surface feature on Uranus's satellite, Miranda.

closed universe: A universe that possesses enough mass that the curvature of space would be positive, and the universe would be finite.

CNO cycle: A cycle of nuclear reactions in which H fusion produces He through reactions that use C nuclei as a catalyst.

COBE: *COsmic Background Explorer satellite.*

collisional dissociation: The breakup of a molecule through a collision.

collisional ionization: An atom or an ion loses an electron due to an energetic collision.

coma: The bright diffuse halo of a comet. Also called the *head.*

comets: Small (a few miles in size) solar system bodies composed of dust and frozen gases that evaporate and glow as they near and go around the Sun. Some comets have regular and predictable reappearances, whereas others enter the solar system without warning and from unpredictable directions.

comparative planetology: An approach to the study of planetary bodies, having the goal of understanding their similarities and differences.

conduction: The transfer of heat by particle (often electron) collisions.

conjunction: Two bodies apparently as close to each other in the sky as possible.

conservation: A conserved quantity is one whose numerical value stays constant in a system with no external forces present. Examples are linear and angular momentum, and energy. The momentum can be passed between bodies, but it does not disappear.

constellations: Arbitrary groupings of stars that have passed down to us from ancient times. Different cultures on Earth historically evolved different constellations and myths related to them. For convenience, astronomers have adopted the ancient Greek constellations and given them well-defined boundaries.

constructive interference: When two similar waves encounter each other and interact in such a way as to combine their strength. See also *destructive interference.*

continental drift: A slow movement of Earth's great land masses that has produced the present continents we now see from the breakup and drifting apart of one giant landmass.

continental plates: The large-scale, moving pieces of surface crust, which are mostly above sea level.

continuous creation of matter: A hypothesis that tries to explain the expansion of the universe as being due to the pressure caused by new matter that is hypothesized to be constantly and spontaneously appearing. See also *Steady State universe.*

continuous spectrum: A spectrum in which energy is emitted at all wavelengths. With the naked eye, we see a rainbow of color. It is distinct from the absorption- and emission-line spectrum.

convection: The transfer of heat energy by the movements of hot material.

co-orbital satellites: Satellites inhabiting the same orbits. Examples occur with Saturn's satellites Dione and Tethys.

Copernican hypothesis: The hypothesis put forth by Copernicus in 1543 that the planets all orbit around the Sun.

Copernicus **satellite:** An orbiting telescope for observing stars in the ultraviolet spectral region.

Cordillera Mountains: Lunar mountains forming the multiringed Orientale Basin.

coriolis effects: Observed effects due to living on a rotating frame of reference. Two examples would be (1) the deflections exhibited by artillery shells, and (2) the circulations of winds on the Earth.

corona: The Sun's tenuous outer atmosphere, which is visible only at the time of a total solar eclipse. It has a temperature near 2 million degrees.

coronagraph: A telescopic device to block most of the light from the bright surface of the Sun and reveal its fainter, outer layers without an eclipse.

coronal holes: Regions of the Sun's corona where strong loops of magnetic field produce a lower than normal density in the hot coronal plasma.

cosmic abundance: Most of the universe exhibits a chemical abundance that is (by number of atoms) roughly 90 percent hydrogen, almost 10 percent helium, and less than 1 or 2 percent of all the other elements. The planets form an exception to this generalization.

cosmic background radiation: See *background 3-K radiation.*

cosmic rays: Particles from space moving at close to the speed of light, impinging on the Earth with tremendous energies from all directions in space.

cosmological constant: A term added by Einstein to his equations of general relativity to force the mathematics to provide a static solution rather than a collapsing one.

cosmological distance: If an object's distance is so great that the effects of general relativity are important, the object is said to be at a cosmological distance.

cosmological redshift: A redshift caused by the expansion of space rather than by relative motion.

cosmology: The study of the structure and evolution of the universe.

coudé: A focal arrangement for a telescope that sends the light down the polar axis into an observing room. This is the longest possible focal length for a telescope and produces the largest possible image.

Crab Nebula: An energetic gas cloud with a pulsar in the center, which has been found to be the remains left over from the spectacular 1054 C.E. supernova explosion in Taurus.

crater: A depression in a planetary or satellite surface, created by the impact of a fast-moving body or by volcanism.

crater chains: A line of craters. In many cases, they provide evidence for some sort of volcanic activity. Some are caused by secondary impacts, or by an impactor that breaks up before hitting the surface.

critical density: The particular value for the density of matter in the universe that would just bring the observed expansion of the universe to a halt.

critical mass: The mass of material required to reach the critical density.

critical observation: An observation that, by itself, is sufficient to favor one model over another.

curved space: According to Einstein's general theory of relativity, the distribution of mass in the universe determines the curvature of space. As bodies move, they must follow the local curvature of space, which will not necessarily be the familiar three dimensions of Euclidian geometry.

Cyg A: A powerful source of synchrotron radiation in Cygnus that may be two galaxies colliding with each other.

Cygnus X-1, X-3: X-ray sources in Cygnus that probably consist of a collapsed object emitting X-rays and orbiting a giant star.

dark cloud: An interstellar cloud of dust and gas that is dense enough to be opaque.

dark matter: Not yet detected or explained, the existence of this material is revealed by the rapid accelerations of galaxies seen in clusters of galaxies. The amount of this material will be critical in determining the future fate of the universe.

dark-line spectrum: See *absorption-line spectrum*.

Death Star: See *Nemesis*.

declination: The angular distance of a celestial object north or south of the celestial equator.

decoupling (of radiation and matter): The epoch early in the life of the universe when the matter becomes largely transparent to the radiation in the universe.

deductive reasoning (deduction): Inference from a general statement to a specific prediction.

deferent: In the ancient geocentric theory of the universe, this is a large circle approximately centered on the Earth, along which a smaller circle, the epicycle, would move.

degenerate electron gas: Electrons compressed into a state of density so high that the ordinary gas laws break down. In the degenerate state, a change of temperature does not result in a change of pressure.

Deimos: One of the two small moons of Mars (the other is Phobos).

density: The amount of mass in a unit volume; mass divided by the volume (mass per unit volume).

density wave: A giant (galaxy-sized), spiral-shaped compression wave that may create spiral arm structures by compressing giant gas clouds and causing the onset of star formation.

destructive interference: If two similar interacting waves are out of phase with each other, the result will be a mutual cancellation.

deuterium: An isotope of hydrogen having one neutron. It is useful in age-dating celestial objects.

diffraction: The bending or spreading-out of waves when they pass through apertures or move close to edges of objects and obstacles.

diffraction grating: A surface that is marked with thousands of fine lines per inch. These lines diffract light as if they were many single slits working in unison, causing a spectrum to form.

diffuse nebula: A gas cloud excited to radiate by absorbing the ultraviolet emission from a hot star in or near the gas cloud. The gas is ionized, and as it recombines, it emits a bright-line spectrum.

dipole anisotropy: An apparent anisotropy in the 3-K microwave background radiation caused by the motion of Earth.

dirty snowball; dirty iceberg: A model for the nucleus of a comet that envisions its structure to be a frozen ball of gas, dust, and rocks that tends to break down when the comet nears the Sun. The evaporated material forms the tail of the comet.

discrete emission: Energy given off only at certain specific wavelengths.

dispersion: The separation of light into its component colors by the action of a prism or a grating.

dissociation: Breaking a molecule apart either by the absorption of a photon or a collision with something.

distance pyramid (distance ladder): A diagram showing how the methods for determining distances to distant objects depend on the distances to nearby ones.

Doppler effect: A shift in the observed wavelength of radiation caused by radial motion of the source of light with respect to the observer.

double quasar: An optical illusion produced by a strong gravitational field in space. The light from a quasar is bent by the field, producing an extra, spurious image.

double shell source: When fusion occurs in two shells surrounding a star's nucleus.

double-lined spectroscopic binary: A spectroscopic binary in which absorption lines from both stars are visible. Such systems provide vital information on the stellar masses.

DQ Herculus: The first stellar nova to be shown to be a binary star system.

Drake equation: A type of equation first written by astronomer Frank Drake containing the various factors that would involve determining the number of intelligent civilizations that might currently inhabit the Galaxy and be willing and able to communicate with us.

dredge-up: Products of fusion reactions deep inside a star, sometimes carried up closer to the surface of the star by convective currents, changing the chemical composition of the material and affecting the reactions that are possible.

dual nature of light: See *wave-particle duality*

dwarf: A small, low-luminosity star. Sometimes used to describe main-sequence stars of luminosity class V.

dwarf elliptical: Small, elliptical galaxies containing in some cases as few as a million stars.

dynamo model: A model that postulates the generation of magnetic fields in planets and other objects by the circulation of conducting fluids inside the object.

eccentricity: The deviation from circular shape (or flattening) of an ellipse.

eclipse: When one celestial body passes in front of another, blocking its light as seen from the Earth.

eclipsing binary: A binary system where the plane of the orbit is nearly in our line of sight, causing each star to periodically pass in front of the other, blocking its light. The light curve of the system will exhibit dips at regular intervals.

ecliptic: The apparent path of the Sun on the celestial sphere with respect to the fixed stars.

Einstein observatory: An X-ray telescope that orbited outside the atmosphere from 1978 to 1981.

ejecta blanket: Material excavated by an impact explosion and deposited over the surface.

electromagnetic radiation: Energy in the form of a wave, propagating through space at the speed of light.

electromagnetic waves: See *electromagnetic radiation.*

electron: A small, negatively charged particle. In the Bohr model of the atom, the electrons of an element orbit the positively charged, massive nucleus of the atom.

electroweak force: At the high energies permeating the universe in the initial moments after the Big Bang, the electromagnetic force and the weak nuclear force were initially combined into one force, the electroweak force.

ellipse: A geometric curve in the shape of a squashed circle. The shape is determined by the eccentricity. See *eccentricity.*

elliptical orbit: Bodies that are gravitationally bound to other bodies and moving in closed orbits will exhibit an orbital path that is an ellipse.

elongation: The angular distance between a planet and the Sun as viewed from the Earth.

emission nebula: A rarified gas cloud excited to radiate by absorbing radiaton from a nearby hot star.

emission-line spectrum: Energy emitted at specific, discrete wavelengths by transparent gases. Each chemical element has its own pattern of emission lines.

Enceladus: A satellite of Saturn that may be significantly affected by tidal friction.

encounter theory: Proposed explanations of the origin of the solar system in which a collision or near approach of another star to the Sun gravitationally pulls off some material from the Sun. This material is hypothesized to condense and form planets.

energy level diagram: A graphical display of the energy levels of an atom.

energy levels: A specific amount of energy associated with each possible electron orbit. Each chemical element has its own set of stable energy levels.

epicenter: A point on the Earth's surface from which seismic waves appear to radiate, located directly above the true center of the earthquake disturbance.

epicycle: In the ancient Ptolemaic (geocentric) theory of the universe, the epicycle is a small circle that moves around the deferent, while the planet moves around the epicycle. This arrangement can produce the retrograde motion shown by planets.

equatorial bulge: The equatorial diameter of the Earth is slightly larger than the polar diameter.

equatorial coordinate system: A coordinate grid on the sky analogous to latitude and longitude on the Earth. The north–south position of an object is called its declination, while the east–west location is called its right ascension.

equatorial mounting: A mount for a telescope that enables it to follow the motion of a celestial object across the sky. Most mounts will feature a polar axis that points toward the north celestial pole.

equinox: Two points on the celestial sphere where the ecliptic plane and the celestial equator intersect. The spring (vernal) equinox occurs near March 21 when the Sun is crossing the celestial equator going from south to north. The autumnal equinox, at about September 21, occurs when the Sun crosses the celestial equator going from north to south. At these times, day and night are of equal length everywhere on the Earth (hence the name, meaning *equal night*).

Eratosthenes: An ancient Greek astronomer who produced an accurate estimate of the size of the Earth by comparing shadows at two locations on Earth.

erg: A tiny unit of energy.

Eros: An Apollo asteroid that passes close to the Earth. Measurements of its parallax gave us our first accurate determinations of the scale of the orbits in the solar system.

escape velocity: If an orbiting body possesses a speed greater than this velocity, then it will escape from the body it is orbiting. Such orbiting bodies will move in parabolic or hyperbolic paths.

eukaryote: An organism, either single- or multicelled, with a distinct nucleus. The cells are encased in a membrane rather than a rigid cell wall.

Europa: The smallest of the Galilean satellites of Jupiter.

event horizon (Schwarzschild radius): A critical distance characterizing a gravitating mass collapsing into a black hole. When the object shrinks beyond this critical distance, it is no longer visible to us, and light and other information is no longer able to escape from within the object itself.

evolutionary track: The path traced out with time on an H-R diagram by an evolving star, which is changing its temperature and luminosity.

excitation: The process of exciting an atom to a higher energy level, either by absorbing a photon (photo-excitation) or by collisions (collisional excitation).

excited state: An energy level (state) other than the ground state.

expansion of the universe: Everything in the universe is observed to be moving away from everything else.

extinction: The dimming of light observed when it passes through a medium. The interstellar medium causes dimming and reddening of starlight passing through it, as does the atmosphere of the Earth.

extrasolar planet: A planet around a star other than the Sun.

extraterrestrial life: Life that is neither on nor from the Earth.

extremophiles: Life at the extreme ends of the temperature or other range of habitable environments. These are often archaea.

faint object camera; faint object spectrograph: Two instrument systems of the Hubble Space Telescope.

fireball: An unusually bright meteor.

fission: The nucleus of a heavy atom splitting apart and forming other, lighter nuclei (and giving off energy in the process).

fission model: A theory of the origin of the Moon, which proposes that the material for the Moon was thrown outward by a rapidly spinning Earth.

five-minute oscillation: An oscillation (expansion and contraction) of the outer layers of the Sun.

flare: Unpredictable energetic outbursts seen in some stars and on the surface of the Sun.

flash spectrum: The instant before a total solar eclipse when the dark-line spectrum is replaced by a spectrum of curved emission lines from the hot gases above the observable surface.

flat space: Ordinary three-dimensional Euclidean space.

flatness problem: The cosmological problem of explaining why the universe appears to be so flat, that is, why the density of the universe is so close to the critical density.

fluorescence: A radiation process whereby an atom or molecule absorbs a photon and then re-emits the energy at longer wavelengths.

focal length: The distance from the primary mirror or lens at which the light from a distant source is brought to a focus.

focus: The point in space where the light gathered by a telescope is formed into a clear image.

forbidden lines: Certain emission lines that are only seen in very low-density gases such as astronomical nebulae. These spectral features are not normally seen on Earth because of the higher densities (hence the misnomer 'forbidden').

frequency: The number of waves passing the observer per second. The frequency of electromagnetic waves is measured in Hertz, which is the number of cycles per second.

FU Orionis star: A small group of stars observed to increase brightness strongly in a short time. They are objects associated with star formation.

full Moon: When the Moon is geometrically opposite to the Sun in the sky, Earth observers can see the disk of the Moon fully illuminated.

fusion: Light nuclei merging together to form a heavier nucleus.

galactic cannibalism: Smaller galaxies consumed by colliding with or being drawn into larger galaxies. This process may cause changes in the shape and form of the larger galaxy and change its evolutionary pattern.

galactic cluster: Small (a few hundred stars at most) groups of associated stars found in the galactic plane. They have formed from gas clouds in the plane and exhibit a variety of ages, from billions of years to recently formed.

galactic corona (galactic halo): State-of-the-art imaging of many galaxies reveals the existence of extensive, spherical regions of emission surrounding the bright, obvious portions of galaxies. The Milky Way appears to have such a region. Depending on physical conditions, these regions could contain large amounts of mass.

galactic equator: The great circle on the sky that traces out the fundamental plane of the Milky Way.

galactic plane (galactic disk): The flattened, pancake-shaped distribution of gas, dust, and young stars in our galaxy.

galactic rotation curve: A graph of the speed of the stars and other material versus their distance from the center of the galaxy.

galaxy: Giant aggregates of stars, gas, and dust in a variety of forms. They range in size from objects that contain perhaps a million solar masses worth of material up to perhaps a trillion stars.

galaxy cluster: Clusters of galaxies, ranging from small groups of only a few up to thousands of galaxies.

Galilean satellites: The four brightest, largest satellites of Jupiter.

gamma ray: Extremely high-energy photons with wavelengths on the order of the size of an atom (approximately 1 Å).

gamma-ray burster: An object that suddenly emits a huge amount of gamma-ray energy in a short burst.

Ganymede: The largest of the Galilean satellites.

gas (ion) tail: A comet tail consisting of ions. Its structure and shape are determined by the material's interaction with the solar wind.

gegenschein: A small region of diffuse light, located 180 degrees away from the Sun in the sky, caused by sunlight reflected from dust particles in the ecliptic plane.

Geminga: A powerful source of high-energy gamma rays and pulsed X-rays.

general theory of relativity: Einstein's theory of gravity. General relativity is used to describe the motion of bodies in the presence of mass, which warps the surrounding space and time, thus giving space a curvature.

geocentric theory: Early theories that imagined the Earth to be the unmoving center of the universe, and that the stars and planets revolved around the Earth.

geodesic: The shortest distance between two points. A generalization of the concept of a straight line.

geomagnetic axis: The line connecting the two magnetic poles on Earth.

giant elliptical galaxies: Huge extragalactic aggregates of stars that present an elliptical distribution of light. The largest of these may contain many trillions of stars.

giant impact theory: A theory for the formation of the Moon that postulates a collision between the young Earth and a Mars-sized object. Material from the Earth and the passing object combine to form a disk, which then condenses into the Moon.

giant molecular cloud: Aggregations of molecules into clouds of more than 100,000 solar masses of material, these objects are the largest discrete entities in the Galaxy. They are thought to be the site of star formation.

giant star: After leaving the main sequence, normal stars undergo changes in their physical state that cause them to swell in size, to sizes on the order of 10 solar radii. See also *supergiant stars*.

gibbous: The phase between the quarter phases and full in which the Moon's surface as seen from Earth is more than half illuminated.

Global Oscillation Network Group (GONG): A worldwide network of solar telescopes for continuous monitoring of the Sun.

globular cluster: A star cluster, often containing more than 100,000 stars. Within the Milky Way, they orbit in elliptical orbits and have ages in excess of 10 billion years.

globules: See *Bok globules*.

gnomon: A stick or rod held perpendicular to Earth's surface for the purpose of casting shadows and determining time.

Grand Unified Theory: A theory that attempts to unify the four forces of nature as different manifestations of a single force.

granulation: The cellular appearance of the solar photosphere, brought about by a boiling motion in which bright blobs of hot material are rising and cooler, dark material is sinking back into the Sun.

gravitation: The mutual attraction between any two bodies possessing mass. In nonrelativistic environments, the size of the attraction and the accelerations of the bodies involved can be calculated from Newton's law of gravity.

gravitational instability: A physical condition where material is on the verge of gravitational collapse.

gravitational lens: An object whose mass is sufficient to bend radiation from background objects and form images of the background object.

gravitational potential energy: The energy residing in gravitating masses that can be called upon to do work.

gravitational radiation (gravitational waves): Waves predicted by the general theory of relativity that should be emitted when large masses accelerate in strong gravitational fields.

gravitational redshift: A redshift caused by a photon losing energy when escaping from a large gravitational field.

great circle: The largest circle one can trace on a sphere. Its center is at the center of the sphere.

Great Dark Spot: An atmospheric phenomenon observed in the atmosphere of Neptune.

Great Red Spot: A cyclonic storm in the upper atmosphere of Jupiter that has been observed for several hundred years.

Great Wall: In a figure of the distribution of galaxies with distance, an extensive alignment of galaxies appearing as a wall separating the nearby universe from more distant regions.

greatest (eastern or western) elongation: When the angular distance of Mercury or Venus from the Sun (as seen from Earth) achieves its maximum value.

greenhouse effect: The trapping of radiation near the surface of a planet due to the absorbing properties of the overlying atmosphere.

ground state: The lowest energy level of an atom.

GUT: See *Grand Unified Theory*.

H alpha (Hα): The red line of the hydrogen Balmer series, located at 6563Å.

H I region: Regions and gas clouds primarily composed of neutral atomic hydrogen.

H II region: Regions and gas clouds primarily composed of ionized hydrogen, symbolized by H II. The clouds of ionized hydrogen are produced by nearby, high temperature O and B stars.

H₂ (hydrogen): Molecular hydrogen—two H atoms bound together.

habitable zone, galactic: The location in a galaxy where life is most likely to be found because there is energy, liquid water, and carbon.

habitable zone, stellar: The volume around a star where liquid water and therefore life can exist.

half-life: The time it takes for one-half of a quantity of radioactive material to decay.

Halley's comet: A comet that approaches the Sun every 75 to 76 years. It has been observed repeatedly since ancient Chinese astronomers recorded it.

halo: A spherical region around the galaxy containing globular clusters and some stars. There may be other constituents of which we are unaware.

head-tail galaxy: A radio galaxy moving through intergalactic space will have its outer portions (where the radio emission comes from) swept backward into the shape of a tail.

heat: The thermal energy of an object—the sum total energy of the random motions of all the object's particles.

Hebrew calendar: A lunar calendar that requires the insertion of additional months from time to time to keep pace with the solar calendar.

heliocentric theory: A model of the Solar System with the Sun at the center and the planets (including the Earth) going around it. In ancient Greece, this hypothesis was put forward by Aristarchus.

helioseismology: The study of the interior of the Sun as revealed by vibrations (oscillations) seen at its surface.

helium burning: The fusion of helium into carbon via the triple-alpha process.

helium flash: An explosive episode of helium burning taking place in the degenerate core of a low-mass star.

Herbig-Haro objects: Emission regions observed to move away from young stellar objects in a bipolar flow.

hertz: A unit of frequency for electromagnetic waves, equal to one cycle per second.

Hertzsprung-Russell diagram: See *H-R diagram*.

Hidalgo: An asteroid with an elongated orbit that is inclined 40° to the ecliptic plane.

highlands: On the Moon, the regions of higher elevation that were not covered by the lava flows that created the maria.

high-velocity stars: Stars that show unusually large velocities relative to the Sun. They are part of the galactic halo and not moving in the plane of the Galaxy as is the Sun.

HIPPARCHUS: A Greek astronomer with many achievements: the first star catalog, the magnitude scale, discovery of precession.

Hipparcos: A satellite that measured stellar distances.

Hirayama asteroid families: Groups of asteroids with highly similar orbits.

horizon: The great circle that represents the intersection of the Earth and the sky.

horizon problem: The cosmological problem of explaining why the microwave background is so uniform, given that the size of the universe at the time the background radiation was formed was larger than the light travel time across it. In other words, why is the background radiation so uniform, given that two separate regions emitting it are too far apart for information to be transferred between them?

horizontal branch: A region of the H-R diagram of globular clusters. It consists of evolved low-mass stars burning helium in the core and hydrogen in a shell.

H-R diagram: A graph of the luminosity of stars plotted against their surface temperatures (or spectral types, or color). The many forms of the diagram are often called collectively by the initials of its co-discoverers, Hertzsprung and Russell.

Hubble constant: The constant H in the mathematical expression of Hubble's law, $V = Hd$. See also *Hubble law*.

Hubble Deep Field: Two areas of the sky (one north, the other south) that the Hubble Space Telescope observed for about 100 hours to observe fainter than previously done.

Hubble law: Describes the expansion of the universe in which we see that the speed of recession of a galaxy is proportional to its distance from us.

Hubble Space Telescope: The first complete observatory in space, equipped with all the instrumentation needed to observe celestial objects in many different ways.

Hyades: A nearby, open cluster of stars in the constellation Taurus.

hydrogen burning: Nuclear fusion in which four protons (hydrogen nuclei) are converted to helium with the emission of energy and neutrinos.

hydrostatic equilibrium: A stable condition where the inward force of gravity at every point in a star is balanced by the outward forces of gas and radiation pressure.

hyperbolic orbit: The mathematical curve that will be followed by an object that is orbiting at speeds greater than the velocity of escape.

Hyperion: A satellite of Saturn.

Iapetus: A satellite of Saturn, having its forward-facing hemisphere dark-colored and the other one light-colored.

ICE: *International Comet Explorer satellite.*

ideal gas: Under normal conditions, the pressure of a gas is directly dependent on its temperature and density. Also known as a *perfect gas*.

igneous rock: Rock formed by the cooling of molten material.

inductive reasoning (induction): Reasoning from particular observations to general conclusions or principles.

inertia: The resistance to any change in an object's motion. Inertia is a property of mass.

inferior conjunction: When Mercury and Venus come into conjunction with the Sun and are at their closest to Earth.

inferior planets: Planets closer to the Sun than Earth is—i.e., Mercury and Venus.

inflationary universe: The hypothesis that a period of extremely rapid expansion occurred early in the history of the universe (followed by the slower expansion we see today).

instability strip: The region of the H-R diagram where Cepheid variables and Cepheid-like variables are found.

intensity: the number of photons striking a detector each second. Intensity should not be confused with the energy of a photon.

interference: The interaction of two similar waves. See *constructive* and *destructive interference*.

interferometer: An instrument that uses the properties of interference to increase resolution. Radio interferometers consist of two or more telescopes examining the same source at the same time. The interference between the different dishes enables astronomers to examine the target with very high resolution.

intergalactic matter: Material in space between clusters of galaxies.

interstellar absorption lines: Sharp absorption lines seen in the spectra of some stars because of absorption of energy by interstellar gas.

interstellar extinction: Radiation removed from a beam of light due to its passage through the dust component of the interstellar material.

interstellar medium: The atoms, ions, and molecules in space between the stars.

inverse square law of light: The brightness of a radiating body varies inversely with the square of its distance from us.

Io: The innermost Galilean satellite of Jupiter, heated by gravitational and magnetic forces into an active volcanic state.

ion: An atom with a net charge, due to the loss or gain of one or more electrons.

ionization: The process in which an electron in an atom absorbs sufficient energy for the electron to escape.

ionization energy: The energy necessary to remove an electron from an atom or ion.

ionosphere: The layer of the Earth's atmosphere (at and beyond 100 km) consisting of charged particles.

IRAS: *The InfraRed Astronomical Satellite,* which mapped most of the sky at long wavelengths.

iridium: A chemical element found in greater abundance in meteorites than on the Earth. It may allow us to locate ancient impacts more successfully and draw more conclusions about their consequences for life on Earth.

irregular galaxies: Galaxies with odd shapes that do not fit into the conventional schemes of classification by appearance.

Ishtar Terra: A large plateau in the northern hemisphere of Venus.

isotope: Two atoms having the same number of protons but different numbers of neutrons are said to be isotopes.

isotropy: Having equal value in all directions. Isotropic radiation would be radiation impinging on the Earth equally from all directions.

IUE: *The International Ultraviolet Explorer satellite.*

jarosite: A mineral that forms on the Earth in dilute sulfuric acid in groundwater. It has also been found on Mars.

Jean's length: Condensations and perturbations in a gas that are larger than the Jean's length will be inclined toward gravitational collapse.

Jean's mass: The amount of material in a length equal to the Jean's length.

jet: A narrow, high-energy beam of particles and radiation.

jet stream: High-speed flows of material in the upper atmosphere.

Jovian planets: Solar system planets that are larger, rotate more rapidly, and are less dense than the terrestrial planets such as Earth. They have dense atmospheres and chemical compositions similar to that of the Sun.

Juno: The third asteroid discovered.

Kapteyn Universe: A model of the galaxy derived from star counts. The model, which suffered from systematic errors, showed the Sun to be at the center. Named after the twentieth-century Dutch astronomer J. C. Kapteyn.

Keck telescope: As of 1994, the world's largest telescope with a mirror 10 m in diameter.

kelvin: A temperature scale with no negative numbers. The temperature in Kelvins is the temperature in Celsius plus 273.

Kelvin-Helmholtz contraction time: The time required for a body to collapse to half its size under its own self-gravity.

Kepler: Johannes Kepler was the seventeenth-century astronomer who used Tycho Brahe's observations to discover the laws governing the motions of the planets in the Solar System.

Kepler's laws: Three empirically determined laws, based on observations of Tycho Brahe, that describe the motions of planets around the Sun.

Keplerian (orbital) motion: Motion that follows Kepler's laws.

kinetic energy: Energy possessed by a body by virtue of its motion.

Kirchhoff's laws: Extensive laboratory work by Kirchhoff discovered several modes of emitting radiation (continuous, dark-line, and bright-line) and the conditions under which each would be produced.

Kirkwood gaps: Regions in the asteroid belt that contain no asteroids.

KL (Kleinman-Low) Nebula: A source of strong infrared radiation near the Orion Nebula.

KREEP: Lunar material with a composition high in potassium (K), rare earth elements (REE), and phosphorus (P).

Kuiper Airborne Observatory: A flying infrared telescope system.

Kuiper belt: A disk-shaped region surrounding the Solar System thought to contain short-period comets.

Lagrangian points: Five points in space where the combined gravitational attraction of two bodies will cancel each other out.

Large Magellanic Cloud (LMC): See *Magellanic Clouds.*

laser: Stands for Light Amplification by Stimulated Emission of Radiation. It is a device that produces light in a very narrow range of wavelengths.

laser-ranging: An experiment where laser light is bounced off a distant object (such as the Moon) to determine precisely how far away that object is.

latitude: Angular distance north or south of the equator.

law of parsimony: Of two equally successful hypotheses, the simplest one is preferred.

law of superposition: When interpreting the surface of the Moon or other firm surfaces, the assumption that the topmost features are the most recently formed.

law of unintended consequences: According to social science, changing one thing can unexpectedly change something else as well.

L-dwarf stars: The spectral classification for certain brown dwarf stars.

Leonids: An August meteor shower famous for its occasional, spectacular displays.

light-gathering power: A measure of the ability of a lens or telescope to gather light. It depends upon its area, which for a circular mirror is πR^2, where R is the radius of the lens or telescope.

light-year: The distance light travels in one year (9.53×10^{12} km).

limb: The edge of an object.

limb darkening: The edges of the Sun are less bright because the radiation we receive from there comes from outer, cooler layers. The center of the disk of the Sun is brighter because the radiation comes from deeper, hotter layers.

line profile: A detailed display with wavelength of the amount of energy absorbed at each and every point within an absorption line. Such details can reveal atmospheric density, rotation, magnetic fields, and other effects.

lithosphere: The solid, rocky layers of Earth just above the asthenosphere.

Local Group: Our galaxy is a member of a small group of about two dozen galaxies. This group also contains the Magellanic Clouds and the Andromeda Galaxy.

Local Supercluster: Our Local Group is also a member of a cluster of galaxy clusters dominated by the extensive Virgo Cluster of galaxies.

longitude: East–west location on Earth along the equator, measuring from the circle that runs through Greenwich, England.

long-period variable star: A cool, giant star that may take more than a year to go through its cycle of brightness variation.

luminosity class: Stars of the same surface temperature show considerable variation in total energy output. The luminosity classification was invented to be able to label those differences. Class I contains superluminous supergiants, while class V designates feeble stars from the bottom of the main sequence.

luminosity: The total energy emitted per second by a celestial object.

lunar eclipse: The Moon orbits through Earth's shadow, and we see a circular line of shadow cross the Moon.

lunar occultation: The passing of the Moon in front of a star.

Lunar Orbiter: A series of satellites that provided an extensive survey of most parts of the lunar surface in the 1970s.

Lunar Ranger: The earliest explorations of the Moon featured rockets with TV cameras in the nose that would crash into the lunar surface, transmitting pictures as long as they were able.

Lyman series: Transitions of the hydrogen atom involving the ground state. The resulting spectral lines are in the ultraviolet spectral region.

Maat Mons: A volcanic mountain on Venus.

Magellanic Clouds (Large and Small): Two small irregular galaxies that are probably bound gravitationally to the Milky Way. The SMC actually consists of two galaxies. They are naked-eye objects in the southern hemisphere.

Magellanic Stream: A ribbon of hydrogen gas running from the Milky Way to the Magellanic Clouds.

magma: Molten rock.

magnetar: Neutron star with a super-high magnetic field 100 to 1,000 times stronger than that of a normal neutron star.

magnetic braking: A process in which particles ejected by a star interact with its magnetic field to slow the star's rotation.

magnetic field: A force field that can affect the motion of charged particles and magnetic materials.

magnetic lines of force: Lines of equal magnetic field intensity.

magnetosphere: A zone surrounding a planetary body in which solar wind particles are trapped by the planet's magnetic field.

magnification: An increase in the apparent size of an object. In a telescope system, the magnification is the focal length of the objective lens divided by the focal length of the eyepiece.

magnitude: A scale of apparent brightness used by astronomers since the time of Hipparchus. If two objects differ by one magnitude in apparent brightness, they differ in brightnesses by a factor of 2.512. (*Note:* This is not the same as *order of magnitude.*)

main sequence: The region of the H-R diagram consisting of stable stars of different masses. A normal star spends most of its life on the main sequence.

major axis: The longest dimension of an ellipse, measured through the foci.

mantle: The region of Earth between the crust and the core.

maria: *Singular:* mare. Latin for *seas,* these are dark-colored lowlands on the Moon's surface that have been covered by lava flows.

Mariner: Spaceprobes to Venus and Mars in the late 1960s.

mascons: Concentrations of mass just under the Moon's surface discovered by the anomalous motions of *Lunar Orbiters.*

mass: A measure of the amount of material in a body. It is a measure of the concept of inertia.

mass–luminosity relationship: The luminosity of a star is approximately proportional to the fourth power of its mass; that is, $L \propto M^4$.

mass–radius relation: For white dwarfs. The more massive the white dwarf, the smaller its radius.

Maunder minimum: A decline in sunspot and other solar activity from 1645 to 1715, which had consequences for Earth weather.

meridian: The great circle of the sky through the two celestial poles and the zenith. It divides the sky into eastern and western halves.

metal-poor stars: Stars whose chemical composition is low in metal abundance relative to the Sun.

metal-rich stars: Stars with above-average abundances of metals.

metamorphic rock: Rock that originally formed by igneous or sedimentary processes and changed in form due to the presence of high pressures or temperatures.

metastable states: Low-lying energy states where the atom will remain for an unusually long time before emitting radiation and undergoing transitions to the ground state.

meteor: A streak of light left by a meteoroid passing through Earth's atmosphere.

meteor shower: An encounter between Earth and a stream of particles that have evaporated from a comet but continued to orbit the Sun. This encounter produces a large number of meteors that appear to be radiating from one particular direction called the radiant.

meteorite fall: A meteorite that is observed to hit the surface of Earth and then is recovered.

meteorite find: A meteorite that attracts our attention by its unusual appearance and hence is discovered serendipitously.

meteorite: A meteoroid that survives the passage through Earth's atmosphere and reaches the surface.

meteoroid: A small chunk of material (rocky or metallic) that travels through interplanetary space.

micrometeorites: Minute particles of meteoritic material that are strongly decelerated by the atmosphere. They are able to radiate away their heat before striking the surface.

Milky Way: Originally, the prominent band of diffuse light that runs across the sky. More recently, the spiral system of perhaps 1 trillion stars in which our Solar System is located.

millisecond pulsars: Pulsars whose signals vary in periods of milliseconds.

Mimas: A moon of Saturn showing a giant impact crater.

minor planet: An asteroid.

Mira variables: A long-period variable star, of which Mira is the prototype.

Miranda: A small Uranian moon with a highly distressed surface.

model: It is often a metaphor that provides a mental picture to explain phenomena. A model, as is the case of a theory or working hypothesis, makes predictions that can be tested. Examples of models include the wave and particle models of light.

molecular bands: Groups of closely spaced absorption lines due to molecules that are seen in some spectra.

molecular clouds: Giant interstellar clouds with a large fraction of their matter in the form of molecules.

molecules: Two or more atoms bound together chemically.

momentum: For an object moving in a straight line, the *linear* momentum is mass times velocity: For an object moving in a circular orbit about the Sun, the *angular* momentum is mass times speed times distance from the Sun.

month, sidereal: The length of time for the Moon to complete one orbit around the Earth in relation to the stars.

month, synodic: the length of time that elapses between a lunar phase and the next occurrence of the same phase. For example, the time between two successive new moons.

multiple mirror telescope: A telescope that acquires a large light-gathering power by collecting photons in many small mirrors that are oriented to bring all of the different beams of light to a common focal point.

nadir: The opposite direction from the zenith (straight down). The direction that is indicated by a weight hanging freely in Earth's gravitational field.

naked-eye: Observations made from without any technical aids such as binoculars or a telescope.

nebula: A diffuse, nonstellar object.

nebular hypothesis: The theory suggesting that the Sun and planets were formed as condensations out of a cloud of dust and gas that was flattened into a disk of material by the rotation of the cloud.

negative curvature: A geometry in which two parallel beams of light would diverge from each other with distance and in which the sum of the angles in a triangle are less than 180 degrees.

Nemesis: A star hypothesized to orbit the Sun and to perturb the orbits of comets to produce a suggested 65-million-year period of mass extinctions on Earth.

Nereid: A satellite of Neptune.

neutrino: An elusive elementary particle with no electric charge and little if any mass. Neutrinos are produced in certain nuclear reactions and carry energy away from the star.

neutrino oscillation: The ability of a neutrino to change from one type to another.

neutron: A nuclear particle with no electric charge and approximately the same mass as a proton.

neutron star: A star that derives most of its pressure support from degenerate neutrons. It is a collapsed object whose material approaches the density of the nucleus of an atom.

new Moon: The phase where the Moon and the Sun are in conjunction but it is the back side of the Moon that is illuminated.

Newton's first law: A body will maintain its current state of motion until it is disturbed by an external force. Also called *law of inertia*.

Newton's second law: The famous $F = ma$ describes how forces cause accelerations. It shows that the acceleration a body experiences, $a = F/m$, is directly proportional to the force and inversely proportional to its mass.

Newton's third law: If object 1 exerts a force on object 2, then object 2 exerts an equal and opposite force on object 1—*action = reaction*.

Newtonian reflector: A telescope that employs a small diagonal mirror to produce a focus outside the side of the telescope tube.

Nix Olympica: A giant but now inactive volcano on Mars. Also known as Olympus Mons.

nonthermal radiation: Radiation produced by interactions between charged particles moving near the speed of light in a magnetic field. The intensity of the radiation does not follow the blackbody curves exhibited by thermal radiation and is polarized.

noon: Apparent noon is when the Sun reaches its maximum altitude and shows the shortest shadow for the day. Noon on the civil clock incorporates effects such as location in a time zone and Daylight Savings Time.

north celestial pole: See *celestial pole*.

nova: A star that increases its brightness dramatically in a short time, and then slowly declines in light output. Many novae are recurrent. Current models of novae postulate that they are binary systems where one star dumps mass onto a nearby collapsed object.

nuclear bulge: The mass concentration at the center of a galaxy.

nuclear reaction: An interaction between a nucleus and another particle that results in energy emission or absorption and a change in the type of particles present. See *fusion* and *fission*.

nucleosynthesis: The manufacture (synthesis) of elements through a variety of nuclear reactions inside stars.

nucleus of a comet: The small core of frozen gases and dust that evaporates when the comet is near the Sun to produce the bright halo and tail features.

nucleus of a galaxy: The central regions where one finds a concentration of mass and often violent activity.

nucleus of an atom: The central concentration of mass in an atom, composed of protons and neutrons and having a net positive charge. It is surrounded by a distribution of electrons that are bound to it by means of a negative electrical charge.

OB associations: Recently formed loose associations of O and B stars. The associations do not have enough mass to prevent them from drifting apart in a relatively short time.

Oberon: A satellite of Uranus.

objective: The large lens or mirror that gathers light in a telescope system.

objective prism spectrograph: A telescope-prism combination in which a large prism is placed in front of the telescope objective to produce a photograph of the spectra of numerous stars rather than one of the stars themselves.

Occam's Razor: See *law of parsimony*.

occultation: When one celestial body passes in front of another. See, for example, *lunar occultation*.

oceanic plates: The large-scale, moving pieces of surface crust underneath the oceans. See *continental plates*.

Olbers' paradox: In a uniform distribution of stars throughout an infinite universe, all lines of sight from the Earth would terminate on the surface of a star, thus causing the sky to be everywhere as bright as the surface of the Sun. This simple conjecture leads quickly to sophisticated cosmological conjectures.

Olympus Mons: A large, extinct volcano on Mars.

Oort Cloud: A hypothesized cloud of primeval material surrounding the Solar System at a considerable distance from the Sun and thought to be a source of comet nuclei, which are perturbed by stars to pass near the Sun.

open cluster: Clusters of stars that form in the galactic plane. Also called *galactic clusters*.

open universe: A universe that possesses insufficient mass to halt the observed expansion. The curvature of space would be negative, and the universe would be infinite.

opposition: When two objects are separated in the sky by 180° as seen from Earth.

optical doubles: Two stars close together in the sky, which give the impression of being affiliated but are really at very different distances from the observer.

orbital angular momentum: See *momentum*.

orbital inclination: The angle between an orbital plane and some reference plane. For example, in the solar system the reference plane would be the ecliptic plane.

Orbiting Astronomical Observatory: One of the earlier telescope systems sent into high Earth orbit to escape the detrimental effects of the Earth's atmosphere.

order of magnitude: A concept that allows making approximate comparisons of numbers.

organic: Any compound or molecule that incorporates carbon.

Orientale Basin (Mare Orientale): A giant, multi-ringed basin on the Moon.

Orion Nebula: A giant cloud of gas and dust illuminated by a cluster of hot stars in the cloud. The hot stars are young, and there is evidence that new stars are continuing to form inside the nebula. The Orion Nebula is visible to the naked eye as the middle star in the Sword of Orion.

oscillating universe: A hypothesis of the evolution of the universe in which it expands and contracts endlessly, undergoing repeated fireballs.

outgasing: An expulsion of material from the inner, hotter regions of a planet.

ozone layer: A layer of Earth's atmosphere in which sunlight ionizes O_2 molecules and creates O_3 (ozone).

pair production: The interaction of two high-energy gamma rays to produce a pair of particles—an atomic particle and its antiparticle.

Pallas: The second largest asteroid, discovered in 1802.

Pangaea: The name given to the hypothetical supercontinent of crustal material on Earth that eventually separated and moved about to form the continents we see today.

parabola: Mathematically, a curve on which every point is equidistant from some point and some line. Dynamically, it is the orbit that would be followed by a body that just exactly achieved escape velocity.

paradigm: The state of a science at any particular time, comprising its current theories and observations and the conclusions drawn from them. A scientific revolution occurs when a field of science goes through a major change of paradigm, as for example when the Ptolemaic model was displaced by the Copernican model.

parallax: A shift in the apparent direction of an object due to a change in the position of the observer. Most commonly, the movement of nearby stars on the background of unmoving distant stars due to the motion of the Earth around the Sun. The term is often loosely used synonymously with distance.

parent molecule: A molecule inferred to exist from observations of the *daughter* molecules that we are able to see. The daughter is produced by a modification of the original parent molecules.

parsec: 3.26 ly, the distance at which the parallax of a star will equal one second of arc.

partial eclipse: When the eclipsed body is not totally covered by the eclipsing body.

Paschen series: The transitions of the hydrogen atom involving the third energy level and appearing in the near infrared.

penumbra: The region of partial shadow in an eclipse. Also, the lighter dark region surrounding a sunspot.

perfect gas: See *ideal gas*.

perigee: The nearest approach to the Earth by an Earth-orbiting body. See also *apogee*.

perihelion: The nearest approach to the Sun by a solar system object. See also *aphelion*.

period: The time interval required for a certain cyclic behavior to repeat itself. For example, the synodic period of Mars (the time interval between two successive oppositions) is 780 days.

period–luminosity relation: The correlation between the period of light variation and the luminosity of a Cepheid variable star. This property enables astronomers to judge the distances of nearby galaxies that have Cepheids in them.

Perseids: A meteor shower at its strongest in early to mid-August.

Pfund series: The series of transitions of the hydrogen atom that involve the fourth energy level. The lines occur in the infrared.

phases: The variation in the appearance of a solar system object seen by an observer as the object takes up different positions with respect to the observer and the Sun.

Phobos: One of the two small moons of Mars.

Phoebe: A moon of Saturn, consisting of a dark surface.

photodissociation: The separation of a molecule into two or more atoms or molecules caused by the absorption of a photon.

photoelectric effect: The process in which certain photosensitive materials will eject electrons when illuminated by blue light but not when illuminated by red light (no matter how strong the irradiation is). This behavior was explained by Einstein as implying that radiation comes in discrete particles, or chunks, known as photons.

photoionization: The absorption of a photon leads to the removal of an electron from an atom or ion.

Photon energy: the energy of an individual photon, given by $E = hf = hc/\lambda$. See *Planck's constant*.

photons: Discrete particles (bundles, chunks) of electromagnetic energy.

photosphere: The visible surface of the Sun. The region of the Sun's atmosphere that contributes most of the visible light that we see when we look at the Sun.

Pioneer 10, 11: 1970s spacecraft probes to Jupiter.

pixel: A picture element of a detector. It is the smallest detecting element on a photographic film or a CCD chip.

Planck curve: A mathematical expression describing the emission given off by radiating bodies as a function of temperature. Describes the spectrum of a blackbody.

Planck time: The first 10^{-43} second after the Big Bang, when the laws of quantum mechanics and relativity both determined what occurred.

Planck's constant: The energy of a photon is $E = hf$, where f is the frequency of the radiation and h is Planck's constant.

planet: The name means *wanderer* in Greek, since the ancients observed these bodies to move around with respect to the sphere of what appeared to be fixed stars.

planetary nebula: An expanding shell of gas ejected from the outer atmosphere of red giant stars in a late stage of stellar evolution.

planetesimals: Aggregates of mass (about asteroid-size) that are thought to combine to form protoplanets in the early history of the solar system.

Plaskett's star: A star of high mass (approximately 70 times the mass of the Sun). It is perhaps the most massive star known.

plasma: A gas in which all or many of the components are charged particles (electrons and ions).

plate tectonics: Many changes in the surface of the Earth can be explained in terms of the slow movements of the Earth's crustal plates.

Pleiades: A small, open cluster of stars in Taurus that are easily visible to the naked eye and are known as the Seven Sisters. This star cluster has been singled out as worthy of observation by virtually every skywatching culture over the ages. Its appearances just before dawn in the east and its disappearances just after dusk in the west just happen to coincide with important phases of the agricultural cycles in the northern temperate zone.

polar axis: A line pointing toward the north celestial pole about which telescopes can rotate to follow the daily motion of a star.

polar caps: Regions of ice and snow at the poles of a planet.

Polaris: The end star in the handle of the Little Dipper, Polaris is a second-magnitude star that coincidentally happens to be located within 1° of the north celestial pole at the current time.

polarization: The alignment of the planes of vibration of electromagnetic waves.

polarization of starlight: Starlight that has been polarized by scattering from dust particles in space.

pole star: See *Polaris*.

Population I stars: Stars associated with the galactic plane that are (on the average) younger in age and have a higher percentage of heavy elements in their chemical compositions.

Population II stars: Older stars with a lower percentage of heavy elements in their chemical compositions, preferentially located in the galactic halo.

positive curvature: Spacetime that closes on itself, forming a finite configuration. If two parallel beams of light set out in positively curved space, the two beams will slowly approach each other, like lines drawn on the surface of the Earth.

positron: An antielectron, a unit of antimatter with the same mass as the electron but with opposite charge.

potential energy: Energy that is stored in some way and available to do work—e.g., a ball held in the air has potential energy; when it is released, that potential energy is converted to kinetic energy of motion. See *gravitational potential energy*.

precession of the equinoxes: The 26,000-year revolution of the Earth's rotation axis causes the equinox points to move slowly around the celestial sphere with the same period. Precession of the equinoxes produces a change in the coordinates of stars with time.

pressure: The force per unit area exerted by a gas or a liquid.

pressure wave (P-wave): Waves that are compressional in nature—in particular, a seismic wave in the Earth or a sound wave.

prime focus: The first focus that is produced by a lens or a mirror, unmodified by other extra reflecting surfaces. The prime focus allows short photographic exposure times and results in small images.

primeval fireball: The state of incredibly high temperature and pressure out of which our expanding universe began.

prokaryotes: One-celled organisms whose DNA is not enclosed within a nucleus. This includes bacteria and archaea.

prominences: Solar material elevated above the surface by the magnetic forces found in active regions on the solar surface. Most easily seen at the limb.

proton: A positively charged nuclear particle of approximately the same mass as a neutron.

proton-proton (p-p) chain: A series of nuclear fusion reactions in low-mass stars. The chain produces energy in the process of creating a helium nucleus from four protons.

protoplanets: An early stage in the process of making a planet where the body is still accreting mass.

protostar: A cloud of gas and dust that is collapsing under its own gravity and destined to become a star when its central temperature becomes high enough to start nuclear reactions in its central regions.

protostellar disk: A disk of gas and dust surrounding a protostar.

protosun: The Sun when it was a protostar.

Proxima Centauri: The nearest star to the Earth.

Ptolemaic system: The model of the cosmos, passed on from ancient Greece, in which the Earth is the unmoving center of the universe and the planets and stars revolve around it.

pulsars: Objects that emit energetic and rapid pulses of radio radiation. They are thought to be rapidly rotating neutron stars surrounded by plasma and a strong magnetic field.

pyrolitic: A chemical process caused by adding heat.

Pythagoras: An early Greek philosopher-scientist who formed a school arguing that numbers are the basis of all understanding.

Q stars: Hypothetical objects to explain certain observations in a way other than calling on the concept of the black hole. They consist of protons, electrons, and neutrons held together by the strong nuclear force rather than gravity.

quadrature: The configuration of a superior planet in which the Sun and the planet are 90° apart in the sky, when viewed from the Earth.

quantum mechanics (quantum theory): A theory of physics that says that, on the subatomic level, the properties of the universe are not continuous but are instead found in discrete units called *quanta*. A photon is an example of quantized energy.

quarks: A fundamental particle in subnuclear physics. It is proposed that other particles are composed of the union of different types of quarks.

quasar: A starlike-appearing object observed to have a large redshift in a spectrum containing emission and absorption lines. The word is a contraction for quasi-stellar radio source but is applied even to objects that do not emit radio radiation. They are now known to be the nuclei of young galaxies.

radar observations: By bouncing radar waves off the surfaces of nearby celestial objects and determining the time it takes to return, details about the structure and elevation of the surface can be learned.

radial velocity: The relative line-of-sight velocity between a celestial object and the observer.

radian: A unit of angular measurement equal to 57.3°.

radiant: A point in the sky from which objects in a meteor shower appear to diverge. It is not a physical point but an effect of perspective.

radiation: Historically, this term was applied to both particles (e.g., beta radiation, which turned out to be emitted electrons) and energy propagating through space in the form of electromagnetic waves.

radiation belts: Equatorial regions of energetic charged particles trapped in the magnetic field of a planet. The strongest radiation belts in the solar system belong to Jupiter. For the Earth's belts, see *Van Allen belts*.

radiation pressure: The push exerted by the impact of photons.

radiation, cosmic 3-K background: See *background 3-K radiation*.

radiation, hydrogen 21 cm: Low-energy photons emitted at a radio wavelength of 21 cm by hydrogen atoms undergoing a *spin-flip* transition within the lowest energy level of the hydrogen atom.

radiative transport: The movement of energy from one place to another by the emission and absorption of radiation.

radio galaxies: Galaxies that give off anomalously large quantities of radio emission.

radio interferometer: Two radio telescopes linked together electronically. By comparing the interference patterns between the two signals, fine details in the pattern of the object's radio image can be discerned.

radio jets: Material that appears to be ejected at high speeds from active regions in radio galaxies.

radio telescope: A telescope constructed to detect the emission of electromagnetic energy of wavelengths ranging from millimeters to meters.

radioactive dating: A method of dating rocks by examining the quantity of radioactive decay byproducts in relation to the decaying material.

radioactive decay: A fission process in which a heavy element spontaneously splits into simpler components. The speed of the decay is governed by the half-life of the element. Every half-life, half of the remaining radioactive atoms will break down.

radioactive element: An element whose nucleus is unstable and spontaneously decays (changes) to a lighter element.

radioactivity: The spontaneous radiation of emission from an unstable element. Particles emitted in a decay process may include electrons, nuclei, and gamma rays.

random error: An error in a set of measurements caused by random variations in the way a set of measurements is obtained. All measurements have random errors, which are measured by the *standard deviation*.

ray: A thin pencil of light. Also, a plume of material ejected radially from the point of a meteoric impact.

real image: An image formed at a definite point in space by the convergence of light rays. Energy is concentrated at this point and can be examined by eye or captured on a photographic plate or other detector.

recombination: An atomic process in which an electron and an ion reunite, giving off energy.

recurrent novae: Stars that undergo the nova phenomenon repeatedly.

red giant: Low surface temperature, highly evolved stars of great size and luminosity.

reddening: In the scattering process that removes photons from a beam of radiation passing through a medium, the short wavelength radiation undergoes more scattering, leaving the color of the object to appear redder than it really is.

redshift: The Doppler shifting of spectral features to longer wavelengths due to the relative motion of the source of light away from the Earth.

redshift, cosmological: See *cosmological redshift*.

reflecting telescope: A telescope that uses a mirror to gather radiation.

reflection nebula: A dust cloud that reflects some starlight in our direction.

reflectivity: The fraction of sunlight falling on an astronomical body reflected back into space.

refracting telescope (refractor): A telescope that uses a lens to gather radiation.

refraction: The bending of light that passes from one medium into a different medium.

regolith: The crumbling and shattered top layer of lunar dirt created by constant bombardment of particles from space.

relativistic jets: Jets moving outward with speeds near that of light from active regions in galaxies.

resolving power (resolution): The ability of an optical system to observe fine detail, such as two stars close together.

resonance: A situation in which the orbital and rotational periods of an object are either equal or related by a simple ratio such as 2:1, 3:2, 3:1, etc. Examples include the Moon's rotation and revolution periods (1:1) and Mercury's rotation and revolution periods (3:2). A resonance also occurs when the orbital period of an object is a simple fraction of the orbital period of another body.

retrograde motion: The occasional westward movement in the motion of planets that generally move eastward relative to the background stars. The westward motions are caused by the Earth's movement in its orbit, when it laps or is lapped by another planet.

retroreflector: A reflector that will return a beam of light in exactly the same direction from which it came.

revolution: Orbital motion around a body or a center of mass.

Rhea: A cratered, icy moon of Saturn.

right ascension: The east–west coordinate in the equatorial coordinate system. It is the distance along the celestial equator (in hours of time running from 0 to 24) eastward from the vernal equinox.

rille: See *sinuous rille*.

ring galaxy: A galaxy containing a nucleus surrounded by a ring of material. Ring galaxies appear to form from galaxy collisions.

Ring Nebula: A well-known planetary nebula in the constellation Lyra.

ring system: A system of icy rocks and particles orbiting each of the giant planets.

Roche limit: A critical distance from a planet, inside of which tidal forces prevent material from accreting to form larger bodies. Inside the Roche limit, tidal forces are larger than the mutual forces between nearby gravitating bodies; outside the limit, mutual gravity allows bodies to accumulate more material and grow.

Roche lobe: An imaginary surface formed by the overlapping gravitational fields of the two stars in a binary system. The lobes look like the parts of a figure eight. Material inside each lobe is considered to belong to the nearest star; material can readily flow from one star to the other through the point of intersection.

ROSAT: An X-ray satellite named for Wilhelm Röntgen, who discovered X-rays. It produced the first all-sky X-ray survey.

rotation: The movement of a body about an axis.

rotation curve: A graph of the orbital speeds of objects in a spiral galaxy as a function of distance from the center of the galaxy.

r-process (rapid process): A nuclear reaction that builds up heavy elements by the rapid accumulation of neutrons.

RR Lyrae stars: Variable stars whose luminosity varies in regular cycles of less than 24 hours.

Russell-Vogt theorem: The structure of a normal star in equilibrium depends only on the mass, chemical composition, and age of the star.

SO galaxy: A galaxy with a large nucleus and a flattened disk but no indication of spiral arms in the disk.

SAGE: Soviet American Gallium Experiment, an experiment searching for solar neutrinos.

Sagittarius A (Sgr A): A group of several extremely energetic radio and infrared sources found at or near the center of our Milky Way galaxy.

scattering: The change in direction suffered by photons passing through gas atoms or dust particles.

Schmidt telescope: A telescope employing a spherical mirror and an aspherical correcting lens. Its advantages are short photographic exposure times and a wide field of view.

Schroeter's Valley: A lunar valley running from a crater to a mare. It is thought to be a channel for the flow of lava.

Schwarzschild radius: See *event horizon.*

Sculptor: A dwarf spheroidal galaxy in the Local Group.

secondary craters: Craters formed by the impact of material ejected from the impact that formed the primary crater.

sedimentary rock: A rock formed when rock debris, transported to a new site by water, wind, ice, or gravity, settles and becomes cemented.

seeing: The amount of distortion, blurring, and artificial enlargement of stellar images caused by the turbulent motions of gases in the Earth's atmosphere.

seismic event: A sudden, violent movement of dense material that moves through the surrounding material in a wave, such as an earthquake or the vibrations of the Sun.

seismic waves: Waves caused by earthquakes that propagate through and around the Earth. See also *pressure wave* and *shear wave.*

seismometer: An instrument to detect seismic events.

selection effect: A bias in data collection that preferentially exhibits certain evidence and withholds other evidence from observers. A simple example: the assumption that most meteoroids are composed largely of iron, because iron meteorites are easier to recognize and discover when they are lying on the ground.

self-gravity: The gravitational attraction each part of an object has on all the other parts.

semi-major axis: Half the longest axis of an ellipse, which is also equal to the average distance of the orbiting object from the center of mass.

sexagesimal system: A number system devised in ancient Mesopotamia using a base of 60 rather than the base of 10 that is used in the decimal system. It is particularly applied to situations involving angles and time.

Sextans: A dwarf spheroidal galaxy in the Local Group.

Seyfert galaxies: Spiral galaxies with brilliant, starlike nuclei exhibiting violent activity.

shear wave (s-wave): A type of seismic wave where the motion of the material is perpendicular to the direction of the movement of the wave.

shepherd satellites: Small moons that may be instrumental in shaping and maintaining the rings around giant planets.

shock wave: A wave produced in a medium (gas, liquid, solid, or plasma) that results from a sudden violent event. The event must take place in a time shorter than the time required for sound waves to traverse the region. In other words, the speed of the event must be greater than the speed of sound in the medium. Sonic booms are a terrestrial example of a shock wave. See also *bow shock.*

short-period comets: Comets having periods of less than 200 years. They may have come from the Kuiper belt.

sidereal day: The time interval between two successive meridian passages of a star (23 hours and 56 minutes).

sidereal period: The period of an object's movement with respect to the stellar background.

silicates: Rocks and compounds formed with silicon (e.g., sand).

single-line spectroscopic binary: A binary star in which the spectral lines of only one star are observed to shift back and forth due to the Doppler effect.

singularity: A point in space where extreme conditions are found and normal physical laws break down.

sinuous rille: A winding channel on a planetary or satellite surface, generally thought to be caused by past lava flows.

Sirius (Sirius A): The brightest star in the sky.

Sirius B: A white dwarf star in orbit with Sirius A.

Small Magellanic Cloud: See *Magellanic Clouds.*

Snedecor rough check: An approximation to the standard deviation. Given by the range in the observations divided by the square root of the number of observations minus 1.

solar day: The time interval between two successive meridian passages of the Sun (24 hours).

solar neutrino problem: Theory predicts a larger number of neutrinos from the Sun than is measured by current detectors. This has spawned a variety of possible explanations, including the possibility that neutrinos actually have a small mass.

solar wind: An outward flow of high-speed particles streaming from the Sun.

solstice, summer and winter: The moment of the year when the Sun is as far north of the celestial equator (summer in the northern hemisphere) as it can be and as far south (winter) as it can be. The northern hemisphere summer solstice occurs about June 21 (the longest day of the year), and the winter solstice occurs around December 21.

space-time: In Einstein's general theory of relativity, time is no longer an absolute quantity. It becomes another coordinate like the spatial coordinates. Instead of discussing space and time, we must now speak of space-time.

Spacewatch Camera: A telescope system, at the University of Arizona, designed to search for asteroids that come near the Earth.

spectral class (spectral type): A classification scheme for spectra that uses letters and numbers to express the surface temperatures of stars.

spectral line: A feature seen at a discrete wavelength in a spectrum.

spectral sequence: OBAFGKM is a sequence of letters that order the stars by decreasing temperature in the astronomical classification scheme of spectral types.

spectrograph: An instrument that disperses the radiation from celestial objects into a spectrum, which is then recorded on film or electronically.

spectroheliograph: An instrument that produces an image of the Sun in one single wavelength, using a sophisticated filtering technique.

spectroscopic binary: An apparently single star that is discovered to be a multiple system by the detection of a variable Doppler shift in the spectrum.

spectroscopic parallax: A method of estimating the distances of stars from spectroscopic information, which provides a star's temperature and the luminosity class.

spectroscopy: The specialized study of spectra.

spectrum: *Plural:* spectra. The energy of different wavelengths radiated by celestial objects. See *bright line spectrum, continuous spectrum,* and *dark line spectrum.*

spherical aberration: Distortions in images formed by spherical mirrors and lenses.

spicules: Jets of bright gas dancing in the chromosphere of the Sun, looking like a waving field of wheat.

spin angular momentum: Angular momentum of a body due to its rotational motion about a spin axis.

spiral arm: A spiral-shaped region in a spiral galaxy, where bright gas clouds and recently formed, young stars are found.

spiral galaxies: Galaxies that show spiral patterns of brightness.

spiral tracers: Young objects that trace out the locations of the arms of a spiral galaxy.

s-process: A nuclear reaction involving the slow accumulation of neutrons.

SS 433: A complex star system exhibiting both redshifts and blueshifts in its spectrum. It may be a normal star losing matter to an accretion disk around a collapsed object, with rapidly moving jets of material in the system.

standard candle: An object whose luminosity is well-known, so that it can be used as a stable reference to estimate the luminosities, and thus distances, of distant objects.

standard deviation: A statistical measurement of the precision of a set of measurements. A measure of the random errors of a measurement.

starburst galaxy: A class of galaxy characterized by periodic outbursts of star formation over widespread regions in the galaxy.

starspots: Dark, magnetically active regions seen on stars that may be similar to sunspots in their origin.

Steady State universe: A hypothesis that proposes that the expansion of the universe is caused by the pressure of new matter that is continually being spontaneously produced. In this model, the universe would look the same at all times past, present, and future. It would be infinitely old, without a beginning or an end.

Stefan-Boltzmann law: The total energy per unit area emitted each second by a radiating object varies as the fourth power of its temperature ($E \propto T^4$).

stellar model: A theoretical model of the interior properties of a star. The model is expressed as a table or graph of important stellar properties (temperature, pressure, density, and so on) as a function of distance from the center.

stellar wind: An outflow of charged particles from a star; similar to the solar wind.

stochastic star formation: See *supernova-induced star formation.*

Stonehenge: An ancient and huge stone structure in England. Apparent astronomical alignments have been found in the structure, and many hypotheses have been put forward about the astronomical sophistication of the builders. Many of the hypotheses remain controversial.

stony meteorite: A rocky meteorite that on casual examination looks like an Earth rock.

stony-iron meteorite: A meteorite with some iron and nickel mixed into its rocky composition.

Straight Wall: A long cliff on the Moon.

stratosphere: A region of the Earth's atmosphere, between about 20 to 50 km, in which temperature increases with height.

strong nuclear force: The force responsible for binding particles together in the nucleus.

subduction: The process by which cold lithospheric material sinks into the asthenosphere.

sunspot cycle: A regular variation in the number of sunspots, which repeats itself every 11 years or so. If variations in magnetic orientation are included, the sunspot cycle is 22 years.

sunspots: Dark regions on the photosphere of the Sun that are cooler than their surroundings and appear dark by contrast. Sunspots also possess strong magnetic fields.

supercluster: A cluster of galaxy clusters.

supergiant star: Highly evolved star with low surface temperature, high luminosity, and very large size.

superior conjunction: When a planet is in conjunction with the Sun, but on the other side of the Sun, as viewed from the Earth.

superior planets: Planets with orbits larger than Earth's.

supermassive black hole: A black hole containing millions, perhaps billions of solar masses. The *Hubble Space Telescope* has provided overwhelming evidence in 1994 of the existence of a supermassive black hole in the center of M87, the giant elliptical galaxy in the Virgo cluster of galaxies.

supernova: A star that undergoes a spectacular increase in its energy output due to a catastrophic explosion of the core.

supernova 1987A: A relatively nearby supernova in the Large Magellanic Cloud.

supernova remnant: The energetic and rapidly expanding gases produced by a supernova explosion. The appearance and the amount of radio emission as compared to visible wavelength emission depend upon the time since the explosion.

supernova-induced collapse: The compressional wave from a supernova explosion can squeeze gas clouds that it encounters and induce them to begin a gravitational collapse.

supernova-induced star formation: A statistical mode of star formation in which the supernova explosion of a massive star triggers the formation of other massive stars, which evolve rapidly to become supernovae in turn, causing the process to continue.

synchronous rotation: A satellite whose rotation period is exactly equal to its period of orbital revolution.

synchrotron radiation: Radiation emitted by charged particles moving in a strong magnetic field near the speed of light.

synodic period: For the Moon, the time required for the Moon to go from one phase back to the same phase (29.5 days). For a planet, the time required for it to go from a particular configuration with respect to the Sun back to that same configuration—e.g., from superior conjunction to superior conjunction.

systematic error: An error that causes a measurement to be always too large or too small.

T Tauri star: A young star that is still collapsing onto the main sequence.

T-dwarf stars: Brown dwarf stars having methane in their spectrum.

technetium: A radioactive element with a cosmically short half-life that is seen in stellar atmospheres. Its very presence indicates that nuclear reactions are taking place in a star.

temperature: A measure of the speeds of the particles in a gas.

terminator: The line dividing light and dark on an illuminated body.

terrestrial planets: Small, high-density planets that are rocky in composition and have relatively thin atmospheres (Mercury, Venus, Earth, Mars).

Tethys: A satellite of Saturn with a crater 40 percent its size.

thermal equilibrium: An equilibrium in which the energy radiated by a star equals the energy produced in the interior.

thermal radiation: Energy emitted by a body as a consequence of its temperature; the same as blackbody radiation.

thermal runaway: An unstable situation where an uncontrolled energy buildup occurs.

3-K background radiation: See *background 3-K radiation*.

tidal force: A force resulting from unequal gravitational forces (called differential gravitational forces) on opposite sides of a body.

tidal friction: Water slowed by the tidal forces on it from the Moon will cause a drag force on the Earth by rubbing across the continental shelves.

Titan: Saturn's largest moon—large enough to retain an atmosphere of its own.

Titania: One of the five large moons of Uranus.

top-down hypothesis: A model of galaxy cluster formation in which clustering occurs prior to the formation of the galaxies making up the cluster.

total eclipse: The complete blocking of one body by another when spatial alignments are just right, as seen from the Earth.

transition region: The region between the chromosphere and the corona in which temperature rises dramatically.

Trapezium: A cluster of four young, luminous stars, ionizing and illuminating the central regions of the Orion Nebula.

trigonometric parallax: A method of determining the distance to an object by noting the angle of displacement it shows with respect to distant background objects when it is viewed from two different locations of the observer.

triple-alpha process: A fusion reaction that converts three helium nuclei into a carbon nucleus.

Triton: Neptune's largest moon.

Trojan group: A group of asteroids that orbit the Sun in Jupiter's orbit such that they maintain a 60-degree angle ahead of and behind Jupiter. The asteroids are located at two Lagrangian points.

troposphere: The lowest level of the Earth's atmosphere, where weather occurs.

Tully-Fisher relation: The larger the luminosity of a galaxy, the larger the width of its 21-cm emission feature.

Tunguska: A region in Siberia that suffered a massive impact of some sort in 1908.

tuning fork diagram: A spatial display of the various shapes and forms exhibited by galaxies.

turbulence: A chaotic, convective motion of material.

turnoff point: When the stars in a cluster are plotted in an H-R diagram, they will generally exhibit a lower portion of a main sequence, terminated at the top end in a point where the evolution of more massive stars has caused them to move away from the main sequence towards the red giant region. This turnoff point can be used to determine the age of the cluster.

two-armed spiral shock model: Applying the density wave model to explain the simplest form of spiral structure that we observe—a plain, two-armed spiral system.

Tycho Brahe: The sixteenth-century astronomer who came to be called the "father of modern observational astronomy" by diligently obtaining observations that were as good as the naked eye can resolve. Brahe provided the observations that allowed Kepler to deduce for the first time the correct laws of motion for the planets.

Type I supernova: A supernova in which hydrogen lines are not seen in the spectrum. They show a sharp peak in brightness followed by a slow decline.

Type II supernova: A supernova in which strong hydrogen lines are seen in the spectrum. The brightness shows a broader peak up until a sharp decline in emission sets in.

ultraviolet radiation: Electromagnetic waves not visible to the eye, with wavelengths that are shorter than blue visible light.

Ulysses: The first spacecraft to orbit the Sun in a polar, rather than equatorial, orbit.

umbra: The region of total shadow in an eclipse. Also, the darkest part of a sunspot.

universe: The totality of everything that can conceivably be observed.

Valhalla: A multi-ringed basin on Jupiter's satellite, Callisto.

Valles Marineris: The extensive canyon system on Mars.

Van Allen belts: Regions of charged particles circling the Earth above the equator. The particles are captured and held by the Earth's magnetic field.

variable star: A star that varies in brightness.

velocity dispersion: The range of velocities exhibited by some group of objects being studied—e.g., the stars in a cluster or the galaxies in a galaxy cluster.

velocity: Speed is distance traveled per unit time, but velocity is a vector quantity that includes both speed and direction of motion.

velocity-distance relation: See *Hubble law*.

Venera: A series of Russian spacecrafts sent to study Venus in the 1970s.

vernal equinox: See *equinox*.

Very Large Array (VLA): A giant radio interferometer, consisting of 27 antennas mounted on railroad tracks upon which they move.

Very Long Baseline Interferometry (VLBI): The process of combining the signals from radio telescopes on different continents (and possibly even telescopes in Earth orbit) to achieve high resolution radio views of celestial objects.

Viking: American spacecraft that landed on the surface of Mars in 1976.

Virgo Cluster: A very large cluster of galaxies relatively near the Local Group.

virtual image: An image with no definite location in space. No energy passes through the apparent image.

viscosity: The property of a fluid that has a high resistance to flowing.

visual binaries: Double stars that can be seen as separate stars in a telescope.

voids: Large empty regions of intergalactic space.

volatiles: Material that condenses at relatively low temperatures.

volcanic domes: Raised regions on the Moon and planets that resembles the domes left on the Earth by volcanic activity.

Voyager: A pair of sophisticated interstellar probes that visited the Jovian planets in the 1980s.

W Virginus stars: Cepheid variables that belong to Population II.

wavelength: The distance between two crests of a wave (or any two similar points on the waveform). It is usually represented in astronomy by the Greek letter lambda (λ).

wave-particle duality: A conclusion of quantum mechanics—that all objects have an admixture of both wave and particle properties. This duality is most easily seen in light, where it shows a wave nature (from diffraction and interference) and a particle nature (from the photoelectric effect).

weak nuclear force: The force responsible for holding neutrons together.

weight: The force exerted on an object by gravity.

weightlessness: Not the lack of gravity, but the lack of acceleration relative to a local reference frame. For example, astronauts whose acceleration around the Earth is the same as that of their spacecraft will be weightless.

white dwarfs: A class of hot stars that are near the end of their evolution. They have lost their sources of nuclear energy and as a consequence have collapsed into extremely small configurations—the size of the Earth or smaller.

white light: Radiation containing all visible wavelength.

Widmanstätten patterns: Crystalline patterns seen in iron meteorites that indicate a past history of heating and slow cooling.

Wien's law: The wavelength at which the maximum amount of radiation is emitted is inversely proportional to the temperature of the radiating object.

winter solstice: See *solstice.*

Wolf-Rayet stars: Hot, large, and luminous stars with broad emission lines seen in their spectra caused by high-velocity stellar winds.

X-ray burster: Brief episodes of powerful X-ray emission from a celestial source.

X-rays: Highly energetic photons and electromagnetic waves of short wavelength (1 to 100 angstroms).

young stellar object (YSO): Point sources of infrared radiation associated with dark clouds and other phenomena connected with star formation.

Zeeman effect: The splitting of atomic energy levels caused by a magnetic field. The result is that numerous spectral lines are formed rather than a single line; when seen in low resolution, a line is broadened.

zenith: The point in the sky directly overhead, 90° from the horizon.

zero-age main sequence (ZAMS): The location of stars of different mass in the H-R diagram as they leave the contracting phase and first settle into an equilibrium state powered by nuclear fusion.

zodiac: Twelve constellations along the ecliptic that have been passed down from ancient cultures. These constellations divide the Sun's motion approximately into 12 months.

zodiacal light: A region of diffuse illumination stretching along the ecliptic and centered on the Sun— also known as the *false dawn.* It is caused by scattering of sunlight off dust particles in the ecliptic plane.

zone of avoidance: A region near the galactic plane of the Milky Way where few galaxies can be observed, due to extinction caused by the dense dust clouds in the plane of the galaxy.

Credits

FRONTMATTER

i: Hubble Heritage Team (AURA/STScI/NASA).
vii: Courtesy of Stephen Shawl, University of Kansas.

CHAPTER 1

1: Photograph taken by W. Liller at the Eastern Island Station of NASA's International Halley Watch program. 3: Photograph taken by W. Liller at the Eastern Island Station of NASA's International Halley Watch program. 4: NASA. 7: Solar & Heliospheric Observatory (SOHO). SOHO is a project of international cooperation between ESA and NASA. 8 left: NASA. 8 right: NASA. 9 left: NASA/JPL/Space Science Institute. 9 top right: Courtesy USGS Astrogeology Research Program. 9 bottom right: NASA. 10 left: NASA/JPL/Space Science Institute. 10 right: Photograph taken by W. Liller at the Eastern Island Station of NASA's International Halley Watch program. 11: Courtesy of European Southern Observatory. 12 left: Hubble/Space Telescope Science Institute/NASA. 12 right: Hubble Heritage Team (AURA/STScI/NASA). 13 left: Jay Gallagher (U. Wisconsin)/National Optical Astronomy Observatory/Association of Universities for Research in Astronomy/ National Science Foundation. Copyright WIYN Consortium, Inc., all rights reserved. 13 right: S. Kafka and K. Honeycutt, Indiana University/NOAO/NSF. Copyright WIYN Consortium, Inc., all rights reserved. 14 top left: Bill Schoening, Vanessa Harvey/REU program/NOAO/ AURA/NSF. 14 top right: Copyright © UKAT, Royal Observatory, Edinburgh. 14 bottom left: N.A. Sharp/ NOAO/AURA/NSF. 15: Omar Lopez-Cruz & Ian Shelton/NOAO/AURA/NSF. 16: Hubble/Space Telescope Science Institute/NASA.

CHAPTER 2

20: Photofest. 22: Courtesy of Stephen J. Shawl, University of Kansas. 24: NOAO/AURA/NSF. 28: Courtesy of Stephen J. Shawl, University of Kansas. 29: From Scientific Literacy and the Myth of the Scientific Method. Copyright © 1992 by the Board of Trustees of the University of Illinois. Used with permission of the University of Illinois Press. 30: NASA. 31: From The Gemini Syndrome by Culver and Ianna. © 1979 by the Pachart Foundation dba Pachart Publishing House. Reprinted by permission.

CHAPTER 3

41: The Granger Collection, New York. 42: Picture Collection, The Branch Libraries, The New York Public Library, Astor, Lenox and Tilden Foundations. 46 top left: Roger Ressmeyer/CORBIS. 46 top right: SPL/Photo Researchers, Inc. 50 top: Copyright Lund Observatory, Sweden. 50 bottom: NASA. 51: NASA.

CHAPTER 4

55: SPL/Photo Researchers, Inc. 56: George Holton/Photo Researchers, Inc. 57 left: SPL/Photo Researchers, Inc. 57 right: Airphoto-Jim Wark, all rights reserved. 58 left: Collection of the Frederick R. Weisman Art Museum at the University of Minnesota, Minneapolis. Transfer, Anthropology department. 58 right: Courtesy of William Jefferys. 60: SPL/Photo Researchers, Inc. 63: Courtesy of Joe Orman. 68: Adapted from Voyager from Carina Software. 70: Courtesy of the Whitin Observatory, Wellesley College. 72 top left: Corel. 72 top right: Corel. 73 left: Adapted from Voyager from Carina Software. 73 bottom: Adapted from Voyager from Carina Software.

CHAPTER 5

83: The Granger Collection, New York. 84: © The Trustees of The British Museum. 90: Photo by William Jefferys. Astrolabe courtesy of Mahboubian Gallery. 91: Bettmann/CORBIS. 93 left: The Granger Collection, New York. 93 right: The Granger Collection, New York. 94: CORBIS. 99: CORBIS. 107 left: Photograph courtesy of Stephen J. Shawl and the Reuben H. Fleet Science Center.

CHAPTER 6

115: NASA/JPL-Caltech. 116: Illustration by Medialab, ESA, 2001. 118: Starhome Observatory—AstroStock. 119: Adapted from Table 3.2, p. 45, of The Solar System by T. Encrenaz, et al. Springer, 1990. Originally published by CNRS Editions (France). 124: Starhome Observatory—AstroStock. 133: California & Carnegie Planet Search Team. 134 left: California & Carnegie Planet Search Team. 134 right: California & Carnegie Planet Search Team.

CHAPTER 7

140: Comet P/Halley as taken March 8, 1986, by W. Liller, Eastern Island, part of the International Halley Watch (IHW) Large Scale Phenomena Network. 141: Bridgeman-Giraudon/Art Resource, NY. 142 top: Courtesy of the Institute for Astronomy, University of Hawaii; photos by Dale Cruikshank (now at NASA/ Ames), Alex Storrs (now at Towson University), and Marc Buie (now at Lowell Observatory). 142 right: NASA. 144: NASA. and the Space Telescope Science Institute. 147: NASA. 148: Courtesy of Dr. H. U. Keller, copyright Max-Planck-Institute für Sonnensystemforschung, Lindau/Harz, Germany. Photo taken by the Halley Multicolour Camera on board ESA's Giotto spacecraft © 1986. 149: NASA/JPL. 150: NASA/JPL/Johns Hopkins University Applied Physics Laboratory. 151: NASA/JPL. 152: Adapted from Brouwer, The Astronomical Journal. 153: NASA. 155: Michael Collier/Stock Boston. 156: Jason Barnes. 157: NASA. 158: Photo by James Baker, Lillian, Alabama. 160 left: NASA. 160 center: NASA/Johnson Space Center. 160 right: Science Graphics, Bend Oregan. 161 left: Science Graphics, Bend Oregan. 161 bottom: Copyright © 1975 by Jerome Kuhl. Reprinted by permission.

CHAPTER 8

169: NASA/JPL. 172: Adapted from The Dynamic Earth: An Introduction to Physical Geology, B.J. Skinner and S.C. Porter (John Wiley Publishing, 1992). This material is used by permission of John Wiley & Sons, Inc. 174: Adapted from The Dynamic Earth: An Introduction to Physical Geology, B.J. Skinner and S.C. Porter (John Wiley Publishing, 1992). This material is used by permission of John Wiley & Sons, Inc. 175 top: Adapted from The Dynamic Earth: An Introduction to Physical Geology, B.J. Skinner and S.C. Porter (John Wiley Publishing, 1992). This material is used by permission of John Wiley & Sons, Inc. 175 right: Adapted from The Dynamic Earth: An Introduction to Physical Geology, B.J. Skinner and S.C. Porter (John Wiley Publishing, 1992). This material is used by permission of John Wiley & Sons, Inc. 177: Photo by J.D. Griggs, U.S. Geological Survey, Hawaiian Volcano Observatory. 178: Topinka, USGS/CVO, 1997, Modified from: Tilling, Heliker, and Wright, 1987, and Hamilton, 1976. 179: NASA/JPL. 182: Adapted from The Dynamic Earth: An Introduction to Physical Geology, B.J. Skinner and S.C. Porter (John Wiley Publishing, 1992). This material is used by permission of John Wiley & Sons, Inc. 183: Runk/Schoenberger-Grant Heilman Photography, Inc. 184: Dr. L.A. Frank University of Iowa. 186: Photo © UC Regents/Lick Observatory. 187 left: Photo © UC Regents/Lick Observatory. 187 right: NASA. 188 left: NASA. 188 right: NASA. 189 left: NASA. 189 right: Courtesy of the Astronomical Society of the Pacific. 190 top left: U.S. Naval Research Laboratory. 190 top right: NASA. 190 center right: NASA. 191: U.S. Department of Interior, U.S. Geological Survey. 192 left: NASA. 192 lower left: NASA. 192 lower right: NASA/USGS. 193 center: NASA. 193 bottom: Los Alamos National Laboratory. 194 top left: Brian J. Skinner. 194 top right: NASA. 194 bottom: U.S. Naval Research Laboratory. 196: U.S. Naval Research Laboratory. 199: Willey Benz. 200: Courtesy of the Lunar and Planetary Institute. From John Wood, in Origin of the Moon, edited by W.K. Hartman, R.J. Phillips, and G.J. Taylor: Houston. Updated grades in the final column were supplied by John Wood. 201: Copyright © 1999 The University of Texas at Austin, McDonald Observatory.

CHAPTER 9

207: NASA/JPL 208 left: New Mexico State University Observatory. 208 center: Starhome Observatory—AstroStock. 208 right: The Observatories of the Carnegie Institution of Washington. 209 top: Lowell Observatory. 209 bottom: Lowell Observatory. 212 top left: NASA/JPL. 212 top right: NASA. 212 lower left: NASA/Johnson Space Center. 212 lower right: NASA/JPL/Malin Space Science Systems. 220: NASA/JPL. 221 top left: NASA/JPL/Malin Space Science Systems. 221 top center: NASA/JPL/Malin Space Science Systems. 221 top right: Courtesy of the Astronomical Society of the Pacific. 222 left: NASA/JPL. 222 right: NASA/JPL. 224 top: NASA/JPL/Malin Space Science Systems. 224 left: NASA. 224 right: NASA. 225 left: NASA. 225 center: NASA. 225 right: NASA. 226 top left: NASA/JPL. 226 bottom left: NASA/JPL/Cornell. 226 top right: NASA/JPL/Malin Space Science Systems. 226 bottom right: NASA/JPL/Malin Space Science Systems. 227 top: NASA/JPL/Cornell/USGS. 227 bottom: Courtesy of the Astronomical Society of the Pacific. 228 left: NASA/JPL. 228 right: Davies, M.E., S.E. Dwornik, D.E. Gault, and R.G. Strom, Atlas of Mercury, NASA.SP-423 (1978). 229 left: Adapted from The Dynamic Earth: An Introduction to Physical Geology, B.J. Skinner and S.C. Porter (John Wiley Publishing, 1992). This material is used by permission of John Wiley & Sons, Inc. 229 top right: NASA. 229 center right: NASA. 229 bottom right: NASA. 230 top left: NASA/JPL. 230 center left: NASA/JPL. 230 bottom left: NASA. 231 left: Meszaros S.P. (1985) Photographic Catalog of Selected Planetary Size Comparisons. NASA. TM-86207. 231 right: NASA/JPL/Northwestern

University. **232 top left:** NASA/JPL. **232 left:** Astronomical Society of the Pacific. **232 right:** Astronomical Society of the Pacific. **233 left:** NASA/JPL. **233 right:** NASA/JPL. **234 top:** NASA. **234 bottom:** NASA. **235 left:** U.S. Department of Interior, U.S. Geological Survey. **235 right:** Reprinted with permission from Science Magazine, Vol. 279, No. 5357, March 13, 1998, p. 1681. Copyright © 1998 American Association for the Advancement of Science. **237 left:** Astronomical Society of the Pacific. **237 right:** Astronomical Society of the Pacific. **238:** Adapted from Cyril Ponnamperuma, Comparative Planetology (Academic Press, 1978).

Chapter 10

244: NASA/JPL/Caltech. **246 top left:** The Observatories of the Carnegie Institution of Washington. **246 top right:** Roger Ressmeyer/ CORBIS. **246 lower left:** NOAO/AURA/NSF. **246 lower right:** UCO/Lick Observatory image. **250 top:** Courtesy of the Astronomical Society of the Pacific. **250 right:** NASA/JPL. **251 left:** NASA/STScI. **251 right:** NASA/JPL. **252:** NASA/JPL. **254:** NASA/ JPL/Johns Hopkins University Applied Physics Laboratory. **255 left:** NASA/STScI. **255 right:** J.T. Trauger (JPL) and NASA. **257 top:** USGS. **257 bottom:** Courtesy of the Astronomical Society of the Pacific. **258 top:** NASA/JPL. **258 bottom:** NASA/ JPL/PIRL. **259:** University of Arizona. **260 top:** NASA/ JPL/PIRL University of Arizona. **260 left:** NASA/JPL. **260 right:** NASA/JPL. **260 bottom right:** NASA/JPL. **261:** NASA/JPL/Institute für Planetenforschung. **262:** NASA. **263 top left:** NASA/JPL/Space Science Institute. **263 top center:** NASA/JPL/Space Science Institute. **263 top right:** NASA/JPL/Space Science Institute. **263 center left:** NASA/JPL/Space Science Institute. **263 center:** NASA/JPL/Space Science Institute. **263 center right:** NASA/JPL/Space Science Institute. **263 bottom left:** NASA/JPL/Space Science Institute. **263 bottom center:** NASA/JPL/Space Science Institute. **263 bottom right:** NASA/JPL/Space Science Institute. **264:** NASA and Finley Holiday Films. **264:** NASA/JPL/Space Science Institute. **265 left:** ESA/NASA/University of Arizona. **265 right:** ESA/NASA/University of Arizona. **266 top left:** NASA/JPL/USGS. **266 top center:** Calvin J. Hamilton. **266 top right:** NASA/JPL. **266 center left:** NASA/JPL. **266 center:** NASA/JPL. **267:** NASA/JPL. **268 top:** NASA/JPL. **268 center:** NASA/JPL. **269 top:** NASA/JPL. **269 left:** NASA/JPL. **269 center:** NASA. **269 right:** NASA/JPL/Space Science Institute. **271 top left:** NASA/JPL/Space Science Institute. **271 top right:** NASA/JPL/Voyager. **271 lower left:** NASA/JPL/ University of Colorado at Boulder. **271 lower center:** NASA/JPL. **271 lower right:** NASA/JPL. **272 left:** NASA/JPL/Galileo/Cornell University. **272 right:** NASA. **273:** Courtesy of Stephen J. Shawl, University of Kansas. **274 left:** Lowell Observatory. **274 right:** Lowell Observatory. **275 left:** NASA/STScI. **275 right:** Dr. R. Albrecht, ESA/ESO Space Telescope European Coordinating Facility; NASA.

Chapter 11

283: Tony Stone Images/Getty Images. **285:** Tony Stone Images/Getty Images. **287:** Runk/ Schoenberger-Grant Heilman Photography. **288 left:** From Atlas of Optical Phenomena by Cagnet-Francon-Thrierr. Copyright © by Springer-Verlag New York, Inc. Reprinted by permission. **288 right:** From Atlas of Optical Phenomena by Cagnet-Francon-Thrierr. Copyright © by Springer-Verlag New York, Inc. Reprinted by permission. **289:** SPL/ Photo Researchers, Inc. **293:** Courtesy of Stephen J. Shawl, University of Kansas. **294 top left:** NOAO/ AURA/NSF. **294 top right:** NOAO/AURA/NSF. **294 bottom right:** NOAO/AURA/NSF.

Chapter 12

303: Courtesy of the European Southern Observatory. **307:** Roger Ressmeyer/CORBIS. **308:** NASA. **310:** Courtesy of the McDonald Observatory, University of Texas at Austin. **311:** Courtesy of the Astronomical Society of the Pacific. **312:** **314:** Courtesy of the European Southern Observatory. **316 top left:** Courtesy of Stephen J. Shawl, University of Kansas. **316 lower left:** Courtesy of Stephen J. Shawl, University of Kansas. **316 right:** NASA. **317:** Courtesy of the Astronomical Society of the Pacific. **318:** Reprinted, with permission, from the Annual Review of Astronomy and Astrophysics, Volume 5, © 1967, by Annual Reviews, www.annualreviews.org. **319:** Courtesy of the Astronomical Society of the Pacific. **320:** Courtesy of the Astronomical Society of the Pacific. **321 top:** NOAO/AURA/NSF. **321 bottom:** Courtesy of USGS and EROS Data Center. **323:** NASA. **324:** NASA. **325 left:** NASA/CXC/SAO. **325 right:** NASA.

Chapter 13

329: NGS Image Collection. **330 top:** Courtesy Bausch & Lomb. **330 bottom:** Courtesy Bausch & Lomb. **331 top:** Helmut Abt/NOAO/AURA/NSF. **331 bottom:** Courtesy of the Yerkes Observatory. **331:** Courtesy of the Yerkes Observatory. **337:** NOAO/AURA/NSF. **346:** Courtesy of the Observatories of the Carnegie Institution of Washington. **347:** NOAO/AURA/NSF.

CHAPTER 14

353: Courtesy of NOAO/AURA/NSF/the National Solar Observatory. 357: Courtesy of NOAO/AURA/NSF/the National Solar Observatory. 358: Courtesy of the Astronomical Society of the Pacific and Annie Jump Cannon. 359: Courtesy of Gerard de Vaucoulers. 365: Helmut Abt/NOAO/AURA/NSF. 367: Courtesy of the Astronomical Society of the Pacific. 371: Photo © UC Regents/Lick Observatory. 373 top: NOAO/AURA/NSF. 373 bottom: Courtesy of Stephen J. Shawl, University of Kansas. 375: NASA. 376: Courtesy of Gerard de Vaucoulers.

CHAPTER 15

381: Space Telescope Science Institute. 385 left: Courtesy of the Yerkes Observatory. 385 center: Courtesy of the Yerkes Observatory. 385 right: Courtesy of the Yerkes Observatory. 386: The Observatories of the Carnegie Institution of Washington. 389: Adapted from Nather, R.E., in The Astrophysical Journal (University of Chicago Press, 1970). 389: Courtesy of NASA, STScI, ESA, and Harvard-Smithsonian CfA.

CHAPTER 16

407: NASA. 412: Bettmann/CORBIS. 415: Courtesy of the Astronomical Society of the Pacific. 416: The Institute for Cosmic Ray Research of the University of Tokyo. 417: Courtesy of the Flying Wallendas. 418: Courtesy of Stephen J. Shawl, University of Kansas. 422: Peter Lloyd (Artist)/NGS Image Collection. 423: Courtesy of the Astronomical Society of the Pacific. 424 left: Courtesy of Project Stratoscope, copyright 2005, The Trustees of Princeton University, supported by NSF, NASA, and ONR. 424 right: Courtesy of Project Stratoscope, copyright 2005, The Trustees of Princeton University, supported by NSF, NASA, and ONR. 425 top: T. Rimmele (NSO), M. Hanna (NOAO)/AURA/NSF. 425 bottom: Courtesy of David H. Hathaway, NASA, Marshal Space Flight Center. 426 top: Courtesy of David H. Hathaway, NASA/ Marshal Space Flight Center. 426 left: Courtesy of David H. Hathaway, NASA/ Marshal Space Flight Center. 426 right: Courtesy of David H. Hathaway, NASA/Marshal Space Flight Center. 428 top: Courtesy of Peax Observatory, U.S. Air Force Cambridge Research Laboratories. 428 center: Courtesy, High Altitude Observatory, National Center for Atmospheric Research (NCAR), Boulder, Colorado, USA. NCAR is sponsored by the National Science Foundation. 428 bottom: Courtesy, High Altitude Observatory, National Center for Atmospheric Research (NCAR), Boulder, Colorado, USA. NCAR is sponsored by the National Science Foundation. 429: Corel. 430: NASA/Goddard Space Flight Center. 431: Solar & Heliospheric Observatory (SOHO). SOHO is a project of international cooperation between ESA and NASA. 432 left: Science Source/Photo Researchers, Inc. 432 right: NOAO/AURA/NSF. 432 bottom: Created with images from the Naval Research Laboratory, American Science and Engineering High Altitude Observatory and NORO. 433: NOAO.

CHAPTER 17

439: NASA/JPL/Hubble. 440: Anglo-Australian Observatory/David Malin Images. 441: Goddard Space Flight Center—NASA. 442: D. E. Brownlee, University of Washington. 443: T.A. Rector (NOAO/AURA/NSF) and Hubble Heritage Team (STScI/AURA/NASA). 444: With permission, from the Annual Review of Astronomy and Astrophysics, Volume 13, © 1975, by Annual Reviews. 446: UK Astronomy Technology Centre, Royal Observatory, Edinburgh. 450 top: Ronald Snell and F. Peter Schloerb. 450 bottom: Ronald Snell and F. Peter Schloerb. 451: Anglo-Australian Observatory/David Malin Images. 452: NASA/JPL-Caltech/B. Brandl (Cornell & University of Leiden). 453 left: Adapted from Stahler, Steven, The Astrophysical Journal, (University of Chicago Press, 1970). 453 right: Adapted, with permission, from Wilking and Shu in the Annual Review of Astronomy and Astrophysics, © 1987, by Annual Reviews. 454 left: Courtesy of Steve Heathcote, Cerro Tololo InterAmerican Observatory. 454 right: Courtesy of Steve Heathcote, Cerro Tololo InterAmerican Observatory. 457 top left: Space Telescope Science Institute. 457 top right: Space Telescope Science Institute. 457 center: NOAO/AURA/NSF. 459 left: Anglo-Australian Observatory/David Malin Images. 459 right: Adapted from Walker, M., in The Astrophysical Journal (University of Chicago Press, 1956). 459 bottom: T. Nakajima and S. Kulkarni (Caltech), S. Durrance and D. Golimowski (JHU), and NASA, STScI. 460: Atlas Image mosaic obtained as part of the Two Micron All Sky Survey (2MASS), a joint project of the University of Massachusetts and the Infrared Processing and Analysis Center /California Institute of Technology, funded by the National Aeronautics and Sp. 461: [left] C. Robert O'Dell, Shui Kwan (Rice Univ.), and NASA. [right] Robert Thompson, Marcia Rieke, Glenn Schneider, Susan Stolovy (Univ. Arizona); Edwin Erickson (SETI Institute /Ames Research Center); David Axon (STScI), and NASA. 462 top: Lynne Hillenbrand. 462 bottom: Space Telescope Science Institute; C.R. O'Dell/Rice University; NASA.

CHAPTER 18

466: NASA. and The Hubble Heritage Team (AURA/STScI). 474: Space Telescope Science Institute; J. Hester/ Arizona State University, NASA. 476 left: ESA/ISO, PHT, H. Izumiura. 476 bottom left: M. Karovska (Center for Astrophysics/NASA/ Space Telescope Science Institute. 477: Dana Berry (STScI). 478 top: NASA, NOAO, ESA, the Hubble Helix Nebula Team, M. Meixner (STScI), and T.A. Rector (NRAO). 478 center: The Hubble Heritage Team (STScI/AURA/ NASA). 478 bottom: NASA, ESA and The Hubble Heritage Team (STScI/AURA). 480 top: NASA. and The Hubble Heritage Team (STScI/AURA). 480 center: Space Telescope Science Institute and NASA. 480 bottom: NASA. and The Hubble Heritage Team (STScI/AURA). 481: Reprinted by permission of Icko Iben, Jr. 482: Photo © UC Regents/Lick Observatory. 484: American Institute of Physics, Emilio Segre, Visual Archives, Physics Today Collection. 486: Adapted from Allan Sandage, The Astrophysical Journal (University of Chicago Press, 1957). 486: Anglo-Australian Observatory/David Malin Images. 487 left: Dr. Bill Harris. 487 right: Adapted from the publications of the Astronomical Society of the Pacific. 488: Photo by Irene Little-Marenin.

CHAPTER 19

493: NASA. and The Hubble Heritage Team (STScI/AURA). 495 top: Adapted from C. Payne-Gaposchkin and S. Gaposchkin (Harvard College Observatory). 495 right: Palomar Observatory/ California Institute of Technology. 498: Alex Filippenko and Weidong Li, University of California at Berkley, Astronomy Department. 499 left: Jay Gallagher (U. Wisconsin)/National Optical Astronomy Observatory/Association of Universities for Research in Astronomy/ National Science Foundation. Copyright WIYN Consortium, Inc., all rights reserved. 499 right: N.A. Sharp, REU program/NOAO/AURA/ NSF. 500: Image courtesy of NRAO/AUI. 501: NASA, ESA, CXO and P. Ruiz-Lapuente (University of Barcelona). 502 left: Anglo-Australian Observatory/David Malin Images. 502 right: Anglo-Australian Observatory/David Malin Images. 505: NASA/CXC/GSFC/U. Hwang, et al. 506: NASA/R. Kirshner/Harvard-Smithsonian Center for Astrophysics. 507: Anglo-Australian Observatory/ David Malin Images. 509 top: Copyright © UKAT, Royal Observatory, Edinburgh. 509 bottom: From Nature, Vol. 217, p. 709 (Macmillan Journals Limited, 1968). 511: NOAO. 516: Bettmann/ CORBIS. 519: The BATSE science team at the NASA Marshall Space Flight Center. 520: From Sky and Telescope, Oct. 1984, p. 311. Copyright © by Sky

Publishing, Corp. Reprinted by permission. 521: From Understanding the Universe, 1st edition, by Flowers. Copyright © 1990, Reprinted with permission of Brooks/Cole, a division of Thomason Learning.

CHAPTER 20

527: Anglo-Australian Observatory/David Malin Images. 529: Anglo-Australian Observatory/David Malin Images. 531 top left: Anglo-Australian Observatory/David Malin Images. 531 center: Anglo-Australian Observatory/David Malin Images. 531 bottom: Roger Ressmeyer/CORBIS. 534: Adapted from a figure supplied by Harvey S. Liszt, and published data from Leo Blitz. 536 left: Anglo-Australian Observatory/David Malin Images. 536 center right: Anglo-Australian Observatory/David Malin Images. 538: IPAC/Caltech/JPL, courtesy of Jay M. Pasachoff. 543 left: Courtesy of the Carnegie Institution of Washington Observatory. 543 right: Bart Bok, reprinted from The Observatory. 544: Adapted from Astronomy and Astrophysics. Courtesy of D.S. Mathewson. 545: Gart Westerhout. 548 left: NASA. 548 right: U.S. Navy. 549: Farhad Yusef-Zadeh. 557: Produced by Voyager from Carina Software.

CHAPTER 21

562: Palomar Observatory/California Institute of Technology. 563 top: Palomar Observatory/California Institute of Technology. 563 bottom: NASA. 564: Anglo-Australian Observatory. Photograph by David Malin. 565: Palomar Observatory/California Institute of Technology. 566 top: Palomar Observatory/California Institute of Technology. 566 center: Hubble Heritage Team (AURA/STScI), and NASA. Material created with support to AURA/STScI from NASA contract NAS5-26555 is reproduced here with permission. 567: The Huntington Library/ Index Stock Imagery. 571: Palomar Observatory, California Institute of Technology. 573 left: Courtesy of The Observatories of the Carnegie Institution of Washington. 573 center: Courtesy of The Observatories of the Carnegie Institution of Washington. 573 right: Courtesy of The Observatories of the Carnegie Institution of Washington. 574: Anglo-Australian Observatory. Photograph by David Malin. 578: Brad Whitmore (STScI)/NASA. 579 top left: Courtesy of The Observatories of the Carnegie Institution of Washington. 579 center left: Dr. Alar Toomre. 579 top right: Brad Whitmore/NASA/STScI. 580 top: Kirk Borne (STScI), and NASA. 580 center right: NASA/JPL/Caltech. 581: From Fundamentals of Cosmic Physics, Physics 7, p. 241, by P.E. Seiden and H. Gerola. Copyright © 1982. Reprinted by permission. **582:** Anglo-Australian Observatory. Photograph by David Malin. **584:** W. Couch (University of New South Wales), R. Ellis (Cambridge University), and NASA. **585:** Harvard-Smithsonian Center for Astrophysics. **586:** P.J.E. Peebles. **587 left:** Margaret J. Geller, Smithsonian Astrophysical Observatory. **587 right:** Sky and Telescope, May 1990, p. 475, © by Sky Publishers. Reprinted by permission. **588:** R. Williams (STScI), the Hubble Deep Field Team, and NASA.

CHAPTER 22

595: Space Telescope Science Institute. **596 top:** Palomar Observatory, California Institute of Technology. **596 bottom:** R. Fosbury/European Southern Observatory. **597:** Anglo-Australian Observatory. Photograph by David Malin. **598:** Reprinted with permission, from the Annual Review of Astronomy and Astrophysics, Volume 5, © 1967, by Annual Reviews. **599 left:** Palomar Observatory, California Institute of Technology. **599 right:** Palomar Observatory, California Institute of Technology. **600 top left:** National Radio Astronomy Observatory/AUI and C. O'Dea and F. Owen. **600 right:** National Radio Astronomical Observatory/AUI. **601 top left:** Reprinted by permission of the Harvard-Smithsonian Center for Astrophysics. **601 top right:** Kitt Peak National Observatory/NOAO. **601 bottom:** NASA/STScI. **602 left:** Palomar Observatory, California Institute of Technology. **602 right:** Palomar Observatory, California Institute of Technology. **603:** Palomar Observatory, California Institute of Technology. **606:** Hubble/NASA/STScI. **607 top left:** Copyright Subaru Telescope, National Astronomical Observatory of Japan. **607 top right:** Hubble/NASA/STScI. **608 left:** Hubble/NASA/STScI. **608 top right:** Hubble/NASA/STScI.

CHAPTER 23

614: NASA. **618:** Reprinted with permission of Lucent Technologies Inc./ Bell Labs. **619:** Adapted by Smithsonian Astrophysical Observatory in Frontiers of Astrophysics, 1976, from P.J.E. Peebles, Physical Cosmology (Princeton University Press, 1971). **628:** Adapted by Smithsonian Astrophysical Observatory in Frontiers of Astrophysics, 1976, from P.J.E. Peebles, Physical Cosmology (Princeton University Press, 1971). **629:** NASA.

CHAPTER 24

643: Steven Vincent Johnson/OrionWorks. **644:** Buddy Mays/CORBIS. **645 top left:** Jim Zuckerman/CORBIS. **645 center left:** Galen Rowell/CORBIS. **646:** Howard Sochurek, Inc./ CORBIS. **647 left:** The Archaeon Strain 121, reprinted with permission from Kazem Kashefi and Derek Lovley, Science, vol. 301, p. 934. Copyright 2003AAAS. **647 center:** NOAA. **647 right:** © Woods Hole Oceanographic Institution. **648 left:** Peter Doran. **648 right:** Courtesy of the Priscu Research Group, Montana State University. **648 bottom:** Henry C. Aldrich. **649:** Illustration by Hirohito Ogasawara/Stanford Synchrotron Radiation Laboratory. **650:** Sofia Lambropoulou, National Technical University, Athens, Greece. **652 left:** Department of Defense. **652 bottom:** The Lunar and Planetary Institute. **653 left:** www.cartoonstock. **653 right:** NASA/JPL/GSFC. **654 left:** NASA-Johnson Space Center. **654 right:** Reprinted with permission from "Search for Past Life on Mars: Possible Relic Biogenic Activity in Martian Meteorite ALH84001," McKay D.S. et al. (1996) Science 273, 924 (1996). **655 left:** Courtesy of NASA/JPL-CalTech. **655 right:** NASA/JPL/Cornell. **656:** Dr. Collin Roesler, Bigelow Laboratory for Ocean Sciences. **659:** European Southern Observatory. **660:** Courtesy Eric Hutton, from Mars by Percival Lowell (1895). **661:** Seth Shostak, SETI Institute. **662:** CSIRO **663 bottom left:** Seth Shostak, SETI Institute. **663 top right:** Seth Shostak, SETI Institute. **664 top:** Seth Shostak, SETI Institute. **664 right bottom:** Don Foley. **666 top right:** Seth Shostak, SETI Institute. **666 bottom:** Jeff Albertson/CORBIS. **667 top left:** Courtesy of NASA/JPL-CalTech. **667 center left:** Courtesy of NASA/JPL-CalTech. **667 center right:** Courtesy of Heavens-Above.com. **667 bottom left:** Courtesy of Heavens-Above.com.

Index

h

i

j

Pre-main-sequence evolution, 447–448
   time scales, 450
Precession, 68
Precision, 47
   See also Accuracy
Prediction, 25
Presession, rotating Earth and, 108
Pressure/energy equilibrium, 417–419
Pressure wave (P-wave), 172
Prestellar core, 447–448
Primary waves, 172
Primeval fireball, 617
Primitive atmospheres, 129
Project Ozma (Drake), 660, 662
Project Phoenix, SETI Institute, 662, 664
Prokaryotes, 645, 649
Prominences, 431–432
Protogalaxies, 551–552
Proton-proton chain, 410–415
Protons, 338, 411–413
Protoplanetary disk, 125
Protostars, 123, 447–450
   See also Star formation
ProtoSun, 125
Pseudoscience, 21–22
   science and, 31–35
Ptolemaic system, 89
Ptolemy, Claudius, 89–90
   Almagest, 90
Ptolmaic model, 618
Pulsars, 317, 508–512
   millisecond, 513
   planets around, 514
Pulsating stars, 474–475
Pyramids, 57
Pyrolitic Release Experiment, 653
Pythagoras, 88

q

Quadrature, 74
Quantum mechanical tunneling, 414
Quasar activity, 624
Quasars, 317, 515, 602–607, 609, 660
   central monster (supermassive black
      hole), 604–607
   measuring redshift, 610–611
   microquasars, 518–519
   spectra, distances, and luminosities of,
      602–604
   variability of, 604
Quintessence, 628

r

r-process, 505
Radial velocity, 368
Radial velocity curve, 387
Radiant, 157–158
Radiation
   21 cm, 540–541
   blackbodies, 333–335
   in centers of galaxies, 658

in the early universe, 622, 623
electromagnetic, 289–291
microwave background radiation
      production, 623, 624
nonthermal, 508
point source of, 305–306
synchrotron, 508
thermal, 538–539
Radiation pressure, 143, 417–418, 446
Radiative transport, 419
Radio and optical observatory sites,
      320–322
Radio astronomy, 316–320
Radio dish, 317
Radio galaxies, 14, 597–600, 609
   high-resolution observations of radio
      emissions from, 600
Radio searches, 662
Radio wavelengths, 662
Radioactive dating, 654
Radioactivity as an energy source, 649
Random uncertainties, 47
Rapidly-expanding universe, 632
Rays, 286
   See also Light
Reber, Grote, 317
Recombination, 342
Red dwarfs, 11, 398, 460
Red giants
   becoming, 469
   evolution of, 469–471
Reddening, 537
Redshifts, 370–371, 571, 622
   quasars, 603–604, 610–611
Rees, Sir Martin, 633
Reflected light, 335
Reflecting telescopes
   compared to refracting, 310–311
   types of, 311–313
Reflection, 286
Reflection nebulae, 537–538
Refracting telescopes, compared to
      reflecting, 310–311
Refraction, 286
Regolith (rocky layer of Moon), 192
Relics of the early universe, 622–623
Renaissance, 75
Research, 34–35
Resolution (resolving power), 308–309
Resolving power (resolution), 308–309
Resonance, 213
Retroflector, 201
Retrograde motion, 73, 88–89, 91, 213
Rhea (Saturn moon), 264
Ribonucleic acid (RNA), 645, 646, 650
Right ascension, 67
Rilles (tiny water courses), 191–192
Ring arcs, 272
Ring Nebula, 12, 478, 564
Ring of fire, 177–178
Rings, of Jovian planets, 268–273
RNA. *See* Ribonucleic acid

Robots in space, 665
Roche limit, 273
Roche lobe, 476
Rock cycle, 176
Rockets, image sizes of, 305–306
Rocks
   age of, 177
   on Moon, 192–195
   rock cycle, 176
   types of, 176–177
Roemer, Ole, 291
Roentgen Satellite, 322–323
Roman Catholic Church, 90
Rosetta mission, 149
Rosette Nebula, 124
Rotation, 65
   galactic, 534, 555–556, 573
   terrestrial planets, 212–214
Rotation curve, 534
RR Lyrae stars, 475, 532
Runaway greenhouse effect, 219
Russell, Henry Norris, 360
Russell-Vogt theorem, 420–421
Rutherford, Ernest, 338
Rutherford's model, 338–339

s

s-process, 505
Sagan, Carl, 654, 666
SAGE. *See* Soviet-American Gallium
      Experiment
Sagittarius, 531–532, 536, 548
Sagittarius A, 548–549
Saturn, 9–10, 116–120, 280, 655, 656,
      666, 667
   magnetic comparison, 255
   moons of, 262–265
   rings of, 269–271
   See also Jovian planets and Pluto
Schiaparelli, Giovanni, 210, 659
Schmidt, Maarten, 602–603
Schmidt telescope, 312
Schroeter's Valley (Moon), 192
Schwartz, Robert, 663
Schwarzschild radius, 514
Science
   in the 20th century, 27–29
   astronomy as an observational
      science, 23–27
   characteristics of good science, 31
   pseudoscience and, 31–35
Scientific models, 615
Search for extraterrestrial intelligence
      (SETI), 659–668
Search for Extraterrestrial Radio Emissions
      from Nearby Developed
      Intelligent Populations
      (SERENDIP), 662–663
Seasons, 66–67
   of terrestrial planets, 214–215
Second law of motion (Newton), 100